Graduate Texts in Mathematics

Graduate Texts in Mathematics bridge the gap between passive study and creative understanding, offering graduate-level introductions to advanced topics in mathematics. The volumes are carefully written as teaching aids and highlight characteristic features of the theory. Although these books are frequently used as textbooks in graduate courses, they are also suitable for individual study.

For further volumes:
http://www.springer.com/series/136

Brian C. Hall

Quantum Theory for Mathematicians

 Springer

Brian C. Hall
Department of Mathematics
University of Notre Dame
Notre Dame, IN, USA

ISSN 0072-5285
ISBN 978-1-4899-9362-5   ISBN 978-1-4614-7116-5 (eBook)
DOI 10.1007/978-1-4614-7116-5
Springer New York Heidelberg Dordrecht London

Mathematics Subject Classification: 81-01, 81S05, 81R05, 46N50, 81Q20, 81Q10, 81S40, 53D50

Printed on acid-free paper

Springer is part of Springer Science+Business Media (www.springer.com)

For as the heavens are higher than the earth, so are my ways higher than your ways, and my thoughts than your thoughts, says the Lord.

Isaiah 55:9

Preface

Ideas from quantum physics play important roles in many parts of modern mathematics. Many parts of representation theory, for example, are motivated by quantum mechanics, including the Wigner–Mackey theory of induced representations, the Kirillov–Kostant orbit method, and, of course, quantum groups. The Jones polynomial in knot theory, the Gromov–Witten invariants in topology, and mirror symmetry in algebraic topology are other notable examples. The awarding of the 1990 Fields Medal to Ed Witten, a physicist, gives an idea of the scope of the influence of quantum theory in mathematics.

Despite the importance of quantum mechanics to mathematics, there is no easy way for mathematicians to learn the subject. Quantum mechanics books in the physics literature are generally not easily understood by most mathematicians. There is, of course, a lower level of mathematical precision in such books than mathematicians are accustomed to. In addition, physics books on quantum mechanics assume knowledge of classical mechanics that mathematicians often do not have. And, finally, there is a subtle difference in "culture"—differences in terminology and notation—that can make reading the physics literature like reading a foreign language for the mathematician. There are few books that attempt to translate quantum theory into terms that mathematicians can understand.

This book is intended as an introduction to quantum mechanics for mathematicians with little prior exposure to physics. The twin goals of the book are (1) to explain the physical ideas of quantum mechanics in language mathematicians will be comfortable with, and (2) to develop the necessary mathematical tools to treat those ideas in a rigorous fashion. I have

vii

attempted to give a reasonably comprehensive treatment of nonrelativistic quantum mechanics, including topics found in typical physics texts (e.g., the harmonic oscillator, the hydrogen atom, and the WKB approximation) as well as more mathematical topics (e.g., quantization schemes, the Stone–von Neumann theorem, and geometric quantization). I have also attempted to minimize the mathematical prerequisites. I do not assume, for example, any prior knowledge of spectral theory or unbounded operators, but provide a full treatment of those topics in Chaps. 6 through 10 of the text. Similarly, I do not assume familiarity with the theory of Lie groups and Lie algebras, but provide a detailed account of those topics in Chap. 16. Whenever possible, I provide full proofs of the stated results.

Most of the text will be accessible to graduate students in mathematics who have had a first course in real analysis, covering the basics of L^2 spaces and Hilbert spaces. Appendix A reviews some of the results that are used in the main body of the text. In Chaps. 21 and 23, however, I assume knowledge of the theory of manifolds. I have attempted to provide motivation for many of the definitions and proofs in the text, with the result that there is a fair amount of discussion interspersed with the standard definition-theorem-proof style of mathematical exposition. There are exercises at the end of each chapter, making the book suitable for graduate courses as well as for independent study.

In comparison to the present work, classics such as Reed and Simon [34] and Glimm and Jaffe [14], along with the recent book of Schmüdgen [35], are more focused on the mathematical underpinnings of the theory than on the physical ideas. Hannabuss's text [22] is fairly accessible to mathematicians, but—despite the word "graduate" in the title of the series—uses an undergraduate level of mathematics. The recent book of Takhtajan [39], meanwhile, has an expository bent to it, but provides less physical motivation and is less self-contained than the present book. Whereas, for example, Takhtajan begins with Lagrangian and Hamiltonian mechanics on manifolds, I begin with "low-tech" classical mechanics on the real line. Similarly, Takhtajan assumes knowledge of unbounded operators and Lie groups, while I provide substantial expositions of both of those subjects. Finally, there is the work of Folland [13], which I highly recommend, but which deals with quantum field theory, whereas the present book treats only nonrelativistic quantum mechanics, except for a very brief discussion of quantum field theory in Sect. 20.6.

The book begins with a quick introduction to the main ideas of classical and quantum mechanics. After a brief account in Chap. 1 of the historical origins of quantum theory, I turn in Chap. 2 to a discussion of the necessary background from classical mechanics. This includes Newton's equation in varying degrees of generality, along with a discussion of important physical quantities such as energy, momentum, and angular momentum, and conditions under which these quantities are "conserved" (i.e., constant along each solution of Newton's equation). I give a short treatment here of Poisson brackets and Hamilton's form of Newton's equation, deferring a full discussion of "fancy" classical mechanics to Chap. 21.

In Chap. 3, I attempt to motivate the structures of quantum mechanics in the simplest setting. Although I discuss the "axioms" (in standard physics terminology) of quantum mechanics, I resolutely avoid a strictly axiomatic approach to the subject (using, say, C^*-algebras). Rather, I try to provide some motivation for the position and momentum operators and the Hilbert space approach to quantum theory, as they connect to the probabilistic aspect of the theory. I do not attempt to *explain* the strange probabilistic nature of quantum theory, if, indeed, there is any explanation of it. Rather, I try to elucidate how the wave function, along with the position and momentum operators, *encodes* the relevant probabilities.

In Chaps. 4 and 5, we look into two illustrative cases of the Schrödinger equation in one space dimension: a free particle and a particle in a square well. In these chapters, we encounter such important concepts as the distinction between phase velocity and group velocity and the distinction between a discrete and a continuous spectrum.

In Chaps. 6 through 10, we look into some of the technical mathematical issues that are swept under the carpet in earlier chapters. I have tried to design this section of the book in such a way that a reader can take in as much or as little of the mathematical details as desired. For a reader who simply wants the big picture, I outline the main ideas and results of spectral theory in Chap. 6, including a discussion of the prototypical example of an operator with a continuous spectrum: the momentum operator. For a reader who wants more information, I provide statements of the spectral theorem (in two different forms) for bounded self-adjoint operators in Chap. 7, and an introduction to the notion of unbounded self-adjoint operators in Chap. 9. Finally, for the reader who wants all the details, I give proofs of the spectral theorem for bounded and unbounded self-adjoint operators, in Chaps. 8 and 10, respectively.

In Chaps. 11 through 14, we turn to the vitally important canonical commutation relations. These are used in Chap. 11 to derive algebraically the spectrum of the quantum harmonic oscillator. In Chap. 12, we discuss the uncertainty principle, both in its general form (for arbitrary pairs of noncommuting operators) and in its specific form (for the position and momentum operators). We pay careful attention to subtle domain issues that are usually glossed over in the physics literature. In Chap. 13, we look at different "quantization schemes" (i.e., different ways of ordering products of the noncommuting position and momentum operators). In Chap. 14, we turn to the celebrated Stone–von Neumann theorem, which provides a uniqueness result for representations of the canonical commutation relations. As in the case of the uncertainty principle, there are some subtle domain issues here that require attention.

In Chaps. 15 through 18, we examine some less elementary issues in quantum theory. Chapter 15 addresses the WKB (Wentzel–Kramers–Brillouin) approximation, which gives simple but approximate formulas for the eigenvectors and eigenvalues for the Hamiltonian operator in one dimension. After this, we introduce (Chap. 16) the notion of Lie groups, Lie algebras, and their representations, all of which play an important role in many parts of quantum mechanics. In Chap. 17, we consider the example of angular momentum and spin, which can be understood in terms of the representations of the rotation group SO(3). Here a more mathematical approach—especially the relationship between Lie group representations and Lie algebra representations—can substantially clarify a topic that is rather mysterious in the physics literature. In particular, the concept of "fractional spin" can be understood as describing a representation of the *Lie algebra* of the rotation group for which there is no associated representation of the rotation group itself. In Chap. 18, we illustrate these ideas by describing the energy levels of the hydrogen atom, including a discussion of the hidden symmetries of hydrogen, which account for the "accidental degeneracy" in the levels. In Chap. 19, we look more closely at the concept of the "state" of a system in quantum mechanics. We look at the notion of subsystems of a quantum system in terms of tensor products of Hilbert spaces, and we see in this setting that the notion of "pure state" (a unit vector in the relevant Hilbert space) is not adequate. We are led, then, to the notion of a mixed state (or density matrix). We also examine the idea that, in quantum mechanics, "identical particles are indistinguishable."

Finally, in Chaps. 21 through 23, we examine some advanced topics in classical and quantum mechanics. We begin, in Chap. 20, by considering the path integral formulation of quantum mechanics, both from the heuristic perspective of the Feynman path integral, and from the rigorous perspective of the Feynman–Kac formula. Then, in Chap. 21, we give a brief treatment of Hamiltonian mechanics on manifolds. Finally, we consider the machinery of geometric quantization, beginning with the Euclidean case in Chap. 22 and continuing with the general case in Chap. 23.

I am grateful to all who have offered suggestions or made corrections to the manuscript, including Renato Bettiol, Edward Burkard, Matt Cecil, Tiancong Chen, Bo Jacoby, Will Kirwin, Nicole Kroeger, Wicharn Lewkeeratiyutkul, Jeff Mitchell, Eleanor Pettus, Ambar Sengupta, and Augusto Stoffel. I am particularly grateful to Michel Talagrand who read almost the entire manuscript and made numerous corrections and suggestions. Finally, I offer a special word of thanks to my advisor and friend, Leonard Gross, who started me on the path toward understanding the mathematical foundations of quantum mechanics. Readers are encouraged to send me comments or corrections at bhall@nd.edu.

Notre Dame, IN, USA   Brian C. Hall

Contents

1

The Experimental Origins of Quantum Mechanics

Quantum mechanics, with its controversial probabilistic nature and curious blending of waves and particles, is a very strange theory. It was not invented because anyone thought this is the way the world *should* behave, but because various experiments showed that this is the way the world *does* behave, like it or not. Craig Hogan, director of the Fermilab Particle Astrophysics Center, put it this way:

> No theorist in his right mind would have invented quantum mechanics unless forced to by data.[1]

Although the first hint of quantum mechanics came in 1900 with Planck's solution to the problem of blackbody radiation, the full theory did not emerge until 1925–1926, with Heisenberg's matrix model, Schrödinger's wave model, and Born's statistical interpretation of the wave model.

1.1 Is Light a Wave or a Particle?

1.1.1 Newton Versus Huygens

Beginning in the late seventeenth century and continuing into the early eighteenth century, there was a vigorous debate in the scientific community

---

[1]Quoted in "Is Space Digital?" by Michael Moyer, *Scientific American*, February 2012, pp. 30–36.

B.C. Hall, *Quantum Theory for Mathematicians*, Graduate Texts in Mathematics 267, DOI 10.1007/978-1-4614-7116-5_1,
© Springer Science+Business Media New York 2013

over the nature of light. One camp, following the views of Isaac Newton, claimed that light consisted of a group of particles or "corpuscles." The other camp, led by the Dutch physicist Christiaan Huygens, claimed that light was a wave. Newton argued that only a corpuscular theory could account for the observed tendency of light to travel in straight lines. Huygens and others, on the other hand, argued that a wave theory could explain numerous observed aspects of light, including the bending or "refraction" of light as it passes from one medium to another, as from air into water. Newton's reputation was such that his "corpuscular" theory remained the dominant one until the early nineteenth century.

1.1.2 The Ascendance of the Wave Theory of Light

In 1804, Thomas Young published two papers describing and explaining his double-slit experiment. In this experiment, sunlight passes through a small hole in a piece of cardboard and strikes another piece of cardboard containing two small holes. The light then strikes a third piece of cardboard, where the pattern of light may be observed. Young observed "fringes" or alternating regions of high and low intensity for the light. Young believed that light was a wave and he postulated that these fringes were the result of *interference* between the waves emanating from the two holes. Young drew an analogy between light and water, where in the case of water, interference is readily observed. If two circular waves of water cross each other, there will be some points where a peak of one wave matches up with a trough of another wave, resulting in *destructive interference*, that is, a partial cancellation between the two waves, resulting in a small amplitude of the combined wave at that point. At other points, on the other hand, a peak in one wave will line up with a peak in the other, or a trough with a trough. At such points, there is *constructive interference*, with the result that the amplitude of the combined wave is large at that point. The pattern of constructive and destructive interference will produce something like a checkerboard pattern of alternating regions of large and small amplitudes in the combined wave. The dimensions of each region will be roughly on the order of the wavelength of the individual waves.

Based on this analogy with water waves, Young was able to explain the interference fringes that he observed and to predict the wavelength that light must have in order for the specific patterns he observed to occur. Based on his observations, Young claimed that the wavelength of visible light ranged from about $1/36,000$ in. (about 700 nm) at the red end of the spectrum to about $1/60,000$ in. (about 425 nm) at the violet end of the spectrum, results that agree with modern measurements.

Figure 1.1 shows how circular waves emitted from two different points form an interference pattern. One should think of Young's second piece of cardboard as being at the top of the figure, with holes near the top left and

FIGURE 1.1. Interference of waves emitted from two slits.

top right of the figure. Figure 1.2 then plots the intensity (i.e., the square of the displacement) as a function of x, with y having the value corresponding to the bottom of Fig. 1.1.

Despite the convincing nature of Young's experiment, many proponents of the corpuscular theory of light remained unconvinced. In 1818, the French Academy of Sciences set up a competition for papers explaining the observed properties of light. One of the submissions was a paper by Augustin-Jean Fresnel in which he elaborated on Huygens's wave model of refraction. A supporter of the corpuscular theory of light, Siméon-Denis Poisson read Fresnel's submission and ridiculed it by pointing out that if that theory were true, light passing by an opaque disk would diffract around the edges of the disk to produce a bright spot in the center of the shadow of the disk, a prediction that Poisson considered absurd. Nevertheless, the head of the judging committee for the competition, François Arago, decided to put the issue to an experimental test and found that such a spot does in fact occur. Although this spot is often called "Arago's spot," or even, ironically, "Poisson's spot," Arago eventually realized that the spot had been observed 100 years earlier in separate experiments by Delisle and Maraldi.

Arago's observation of Poisson's spot led to widespread acceptance of the wave theory of light. This theory gained even greater acceptance in 1865, when James Clerk Maxwell put together what are today known as Maxwell's equations. Maxwell showed that his equations predicted that electromagnetic waves would propagate at a certain speed, which agreed with the observed speed of light. Maxwell thus concluded that light *is* simply an electromagnetic wave. From 1865 until the end of the nineteenth

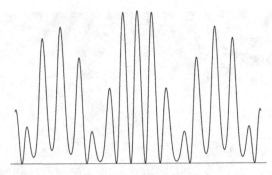

FIGURE 1.2. Intensity plot for a horizontal line across the bottom of Fig. 1.1

century, the debate over the wave-versus-particle nature of light was considered to have been conclusively settled in favor of the wave theory.

1.1.3 Blackbody Radiation

In the early twentieth century, the wave theory of light began to experience new challenges. The first challenge came from the theory of *blackbody radiation*. In physics, a blackbody is an idealized object that perfectly absorbs all electromagnetic radiation that hits it. A blackbody can be approximated in the real world by an object with a highly absorbent surface such as "lamp black." The problem of blackbody radiation concerns the distribution of electromagnetic radiation in a cavity within a blackbody. Although the walls of the blackbody absorb the radiation that hits it, thermal vibrations of the atoms making up the walls cause the blackbody to emit electromagnetic radiation. (At normal temperatures, most of the radiation emitted would be in the infrared range.)

In the cavity, then, electromagnetic radiation is constantly absorbed and re-emitted until thermal equilibrium is reached, at which point the absorption and emission of radiation are perfectly balanced at each frequency. According to the "equipartition theorem" of (classical) statistical mechanics, the energy in any given mode of electromagnetic radiation should be exponentially distributed, with an average value equal to $k_B T$, where T is the temperature and k_B is Boltzmann's constant. (The temperature should be measured on a scale where absolute zero corresponds to $T = 0$.) The difficulty with this prediction is that the average amount of energy is *the same for every mode* (hence the term "equipartition"). Thus, once one adds up over all modes—of which there are infinitely many—the predicted amount of energy in the cavity is infinite. This strange prediction is referred to as the *ultraviolet catastrophe*, since the infinitude of the energy comes from the ultraviolet (high-frequency) end of the spectrum. This ultraviolet catastrophe does not seem to make physical sense and certainly does not match up with the observed energy spectrum within real-world blackbodies.

An alternative prediction of the blackbody energy spectrum was offered by Max Planck in a paper published in 1900. Planck postulated that the energy in the electromagnetic field at a given frequency ω should be "quantized," meaning that this energy should come only in integer multiples of a certain basic unit equal to $\hbar\omega$, where $\hbar$ is a constant, which we now call Planck's constant. Planck postulated that the energy would again be exponentially distributed, but only over integer multiples of $\hbar\omega$. At low frequencies, Planck's theory predicts essentially the same energy as in classical statistical mechanics. At high frequencies, namely at frequencies where $\hbar\omega$ is large compared to $k_B T$, Planck's theory predicts a rapid fall-off of the average energy (see Exercise 2 for details). Indeed, if we measure mass, distance, and time in units of grams, centimeters, and seconds, respectively, and we assign $\hbar$ the numerical value

$$\hbar = 1.054 \times 10^{-27},$$

then Planck's predictions match the experimentally observed blackbody spectrum.

Planck pictured the walls of the blackbody as being made up of independent oscillators of different frequencies, each of which is restricted to have energies of $\hbar\omega$. Although this picture was clearly not intended as a realistic physical explanation of the quantization of electromagnetic energy in blackbodies, it does suggest that Planck thought that energy quantization arose from properties of the walls of the cavity, rather than in intrinsic properties of the electromagnetic radiation. Einstein, on the other hand, in assessing Planck's model, argued that energy quantization was inherent in the radiation itself. In Einstein's picture, then, electromagnetic energy at a given frequency—whether in a blackbody cavity or not—comes in packets or *quanta* having energy proportional to the frequency. Each quantum of electromagnetic energy constitutes what we now call a *photon*, which we may think of as a particle of light. Thus, Planck's model of blackbody radiation began a rebirth of the particle theory of light.

It is worth mentioning, in passing, that in 1900, the same year in which Planck's paper on blackbody radiation appeared, Lord Kelvin gave a lecture that drew attention to another difficulty with the classical theory of statistical mechanics. Kelvin described two "clouds" over nineteenth-century physics at the dawn of the twentieth century. The first of these clouds concerned aether—a hypothetical medium through which electromagnetic radiation propagates—and the failure of Michelson and Morley to observe the motion of earth relative to the aether. Under this cloud lurked the theory of special relativity. The second of Kelvin's clouds concerned heat capacities in gases. The equipartition theorem of classical statistical mechanics made predictions for the ratio of heat capacity at constant pressure (c_p) and the heat capacity at constant volume (c_v). These predictions deviated substantially from the experimentally measured ratios. Under the second cloud lurked the theory of quantum mechanics, because the resolution of this discrepancy is similar to Planck's resolution of the blackbody problem. As in the case of blackbody radiation, quantum mechanics gives rise to a correction to the equipartition theorem, thus resulting in different predictions for the ratio of c_p to c_v, predictions that can be reconciled with the observed ratios.

1.1.4   The Photoelectric Effect

The year 1905 was Einstein's *annus mirabilis* (miraculous year), in which Einstein published four ground-breaking papers, two on the special theory of relativity and one each on Brownian motion and the photoelectric effect. It was for the photoelectric effect that Einstein won the Nobel Prize in physics in 1921. In the photoelectric effect, electromagnetic radiation striking a metal causes electrons to be emitted from the metal. Einstein found that as one increases the *intensity* of the incident light, the *number* of emitted electrons increases, but the *energy* of each electron does not change. This result is difficult to explain from the perspective of the wave theory of light. After all, if light is simply an electromagnetic wave, then increasing the intensity of the light amounts to increasing the strength of the electric and magnetic fields involved. Increasing the strength of the fields, in turn, ought to increase the amount of energy transferred to the electrons.

Einstein's results, on the other hand, are readily explained from a particle theory of light. Suppose light is actually a stream of particles (photons) with the energy of each particle determined by its frequency. Then increasing the intensity of light at a given frequency simply increases the number of photons and does not affect the energy of each photon. If each photon has a certain likelihood of hitting an electron and causing it to escape from the metal, then the energy of the escaping electron will be determined by the *frequency* of the incident light and not by the *intensity* of that light. The photoelectric effect, then, provided another compelling reason for believing that light can behave in a particlelike manner.

1.1.5   The Double-Slit Experiment, Revisited

Although the work of Planck and Einstein suggests that there is a particlelike aspect to light, there is certainly also a wavelike aspect to light, as shown by Young, Arago, and Maxwell, among others. Thus, somehow, light must in some situations behave like a wave and in some situations like a particle, a phenomenon known as "wave–particle duality." William Lawrence Bragg described the situation thus:

> God runs electromagnetics on Monday, Wednesday, and Friday
> by the wave theory, and the devil runs them by quantum theory
> on Tuesday, Thursday, and Saturday.

(Apparently Sunday, being a day of rest, did not need to be accounted for.)

In particular, we have already seen that Young's double-slit experiment in the early nineteenth century was one important piece of evidence in favor of the wave theory of light. If light is really made up of particles, as blackbody radiation and the photoelectric effect suggest, one must give a particle-based explanation of the double-slit experiment. J.J. Thomson suggested in 1907 that the patterns of light seen in the double-slit experiment could be the result of different photons somehow interfering with one another. Thomson thus suggested that if the intensity of light were sufficiently reduced, the photons in the light would become widely separated and the interference pattern might disappear. In 1909, Geoffrey Ingram Taylor set out to test this suggestion and found that even when the intensity of light was drastically reduced (to the point that it took three *months* for one of the images to form), the interference pattern remained the same.

Since Taylor's results suggest that interference remains even when the photons are widely separated, the photons are not interfering with one another. Rather, as Paul Dirac put it in Chap. 1 of [6], "Each photon then interferes only with itself." To state this in a different way, since there is no interference when there is only one slit, Taylor's results suggest that each individual photon passes through *both* slits. By the early 1960s, it became possible to perform double-slit experiments with electrons instead of photons, yielding even more dramatic confirmations of the strange behavior of matter in the quantum realm. (See Sect. 1.2.4.)

1.2 Is an Electron a Wave or a Particle?

In the early part of the twentieth century, the atomic theory of matter became firmly established. (Einstein's 1905 paper on Brownian motion was an important confirmation of the theory and provided the first calculation of atomic masses in everyday units.) Experiments performed in 1909 by Hans Geiger and Ernest Marsden, under the direction of Ernest Rutherford, led Rutherford to put forward in 1911 a picture of atoms in which a small nucleus contains most of the mass of the atom. In Rutherford's model, each atom has a positively charged nucleus with charge nq, where n is a positive integer (the *atomic number*) and q is the basic unit of charge first observed in Millikan's famous oil-drop experiment. Surrounding the nucleus is a cloud of n electrons, each having negative charge $-q$. When atoms bind into molecules, some of the electrons of one atom may be shared with another atom to form a bond between the atoms. This picture of atoms and their binding led to the modern theory of chemistry.

Basic to the atomic theory is that electrons are particles; indeed, the number of electrons per atom is supposed to be the atomic number. Nevertheless, it did not take long after the atomic theory of matter was confirmed before wavelike properties of electrons began to be observed. The situation, then, is the reverse of that with light. While light was long thought to be a wave (at least from the publication of Maxwell's equations in 1865 until Planck's work in 1900) and was only later seen to have particlelike behavior, electrons were initially thought to be particles and were only later seen to have wavelike properties. In the end, however, both light and electrons have both wavelike and particlelike properties.

1.2.1   The Spectrum of Hydrogen

If electricity is passed through a tube containing hydrogen gas, the gas will emit light. If that light is separated into different frequencies by means of a prism, bands will become apparent, indicating that the light is not a continuous mix of many different frequencies, but rather consists only of a discrete family of frequencies. In view of the photonic theory of light, the energy in each photon is proportional to its frequency. Thus, each observed frequency corresponds to a certain amount of energy being transferred from a hydrogen atom to the electromagnetic field.

Now, a hydrogen atom consists of a single proton surrounded by a single electron. Since the proton is much more massive than the electron, one can picture the proton as being stationary, with the electron orbiting it. The idea, then, is that the current being passed through the gas causes some of the electrons to move to a higher-energy state. Eventually, that electron will return to a lower-energy state, emitting a photon in the process. In this way, by observing the energies (or, equivalently, the frequencies) of the emitted photons, one can work backwards to the change in energy of the electron.

The curious thing about the state of affairs in the preceding paragraph is that the energies of the emitted photons—and hence, also, the energies of the electron—come only in a discrete family of possible values. Based on the observed frequencies, Johannes Rydberg concluded in 1888 that the possible energies of the electron were of the form

$$E_n = -\frac{R}{n^2}. \tag{1.1}$$

Here, R is the "Rydberg constant," given (in "Gaussian units") by

$$R = \frac{m_e Q^4}{2\hbar^2},$$

where Q is the charge of the electron and m_e is the mass of the electron. (Technically, m_e should be replaced by the reduced mass μ of the proton–electron system; that is, $\mu = m_e m_p/(m_e + m_p)$, where m_p is the mass of the proton. However, since the proton mass is much greater than the electron mass, μ is almost the same as m_e and we will neglect the difference between the two.) The energies in (1.1) agree with experiment, in that all the observed frequencies in hydrogen are (at least to the precision available at the time of Rydberg) of the form

$$\omega = \frac{1}{\hbar}\left(E_n - E_m\right), \tag{1.2}$$

for some $n > m$. It should be noted that Johann Balmer had already observed in 1885 frequencies of the same form, but only in the case $m = 2$, and that Balmer's work influenced Rydberg.

The frequencies in (1.2) are known as the *spectrum* of hydrogen. Balmer and Rydberg were merely attempting to find a simple formula that would match the observed frequencies in hydrogen. Neither of them had a theoretical explanation for *why* only these particular frequencies occur. Such an explanation would have to wait until the beginnings of quantum theory in the twentieth century.

1.2.2 The Bohr–de Broglie Model of the Hydrogen Atom

In 1913, Niels Bohr introduced a model of the hydrogen atom that attempted to explain the observed spectrum of hydrogen. Bohr pictured the hydrogen atom as consisting of an electron orbiting a positively charged nucleus, in much the same way that a planet orbits the sun. Classically, the force exerted on the electron by the proton follows the *inverse square law* of the form

$$F = \frac{Q^2}{r^2}, \tag{1.3}$$

where Q is the charge of the electron, in appropriate units.

If the electron is in a circular orbit, its trajectory in the plane of the orbit will take the form

$$(x(t), y(t)) = (r\cos(\omega t), r\sin(\omega t)).$$

If we take the second derivative with respect to time to obtain the acceleration vector **a**, we obtain

$$\mathbf{a}(t) = (-\omega^2 r\cos(\omega t), -\omega^2 r\sin(\omega t)),$$

so that the magnitude of the acceleration vector is $\omega^2 r$. Newton's second law, $F = ma$, then requires that

$$m_e\omega^2 r = \frac{e^2}{r^2},$$

so that

$$\omega = \sqrt{\frac{Q^2}{m_e r^3}}.$$

From the formula for the frequency, we can calculate that the momentum (mass times velocity) has magnitude

$$p = \sqrt{\frac{m_e Q^2}{r}}. \tag{1.4}$$

We can also calculate the angular momentum J, which for a circular orbit is just the momentum times the distance from the nucleus, as

$$J = \sqrt{m_e Q^2 r}.$$

Bohr postulated that the electron obeys classical mechanics, *except* that its angular momentum is "quantized." Specifically, in Bohr's model, the angular momentum is required to be an integer multiple of $\hbar$ (Planck's constant). Setting J equal to $n\hbar$ yields

$$r_n = \frac{n^2 \hbar^2}{m_e Q^2}. \tag{1.5}$$

If one calculates the energy of an orbit with radius r_n, one finds (Exercise 3) that it agrees precisely with the Rydberg energies in (1.1). Bohr further postulated that an electron could move from one allowed state to another, emitting a packet of light in the process with frequency given by (1.2).

Bohr did not explain *why* the angular momentum of an electron is quantized, nor how it moved from one allowed orbit to another. As such, his theory of atomic behavior was clearly not complete; it belongs to the "old quantum mechanics" that was superseded by the matrix model of Heisenberg and the wave model of Schrödinger. Nevertheless, Bohr's model was an important step in the process of understanding the behavior of atoms, and Bohr was awarded the 1922 Nobel Prize in physics for his work. Some remnant of Bohr's approach survives in modern quantum theory, in the WKB approximation (Chap. 15), where the Bohr–Sommerfeld condition gives an approximation to the energy levels of a one-dimensional quantum system.

In 1924, Louis de Broglie reinterpreted Bohr's condition on the angular momentum as a wave condition. The *de Broglie hypothesis* is that an electron can be described by a wave, where the spatial frequency k of the wave is related to the momentum of the electron by the relation

$$p = \hbar k. \tag{1.6}$$

Here, "frequency" is defined so that the frequency of the function $\cos(kx)$ is k. This is "angular" frequency, which differs by a factor of 2π from the cycles-per-unit-distance frequency. Thus, the period associated with a given frequency k is $2\pi/k$.

In de Broglie's approach, we are supposed to imagine a wave superimposed on the classical trajectory of the electron, with the quantization

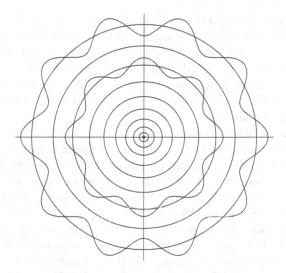

FIGURE 1.3. The Bohr radii for $n = 1$ to $n = 10$, with de Broglie waves superimposed for $n = 8$ and $n = 10$.

condition now being that the wave should match up with itself when going all the way around the orbit. This condition means that the orbit should consist of an integer number of periods of the wave:

$$2\pi r = n\frac{2\pi}{k}.$$

Using (1.6) along with the expression (1.4) for p, we obtain

$$2\pi r = n2\pi\frac{\hbar}{p} = 2\pi n\hbar\sqrt{\frac{r}{m_e Q^2}}.$$

Solving this equation for r gives precisely the Bohr radii in (1.5).

Thus, de Broglie's wave hypothesis gives an alternative to Bohr's quantization of angular momentum as an explanation of the allowed energies of hydrogen. Of course, if one accepts de Broglie's wave hypothesis for electrons, one would expect to see wavelike behavior of electrons not just in the hydrogen atom, but in other situations as well, an expectation that would soon be fulfilled. Figure 1.3 shows the first 10 Bohr radii. For the 8th and 10th radii, the de Broglie wave is shown superimposed onto the orbit.

1.2.3 Electron Diffraction

In 1925, Clinton Davisson and Lester Germer were studying properties of nickel by bombarding a thin film of nickel with low-energy electrons. As a result of a problem with their equipment, the nickel was accidentally heated to a very high temperature. When the nickel cooled, it formed into large crystalline pieces, rather than the small crystals in the original sample. After this recrystallization, Davisson and Germer observed peaks in the pattern of electrons reflecting off of the nickel sample that had not been present when using the original sample. They were at a loss to explain this pattern until, in 1926, Davisson learned of the *de Broglie hypothesis* and suspected that they were observing the wavelike behavior of electrons that de Broglie had predicted.

After this realization, Davisson and Germer began to look systematically for wavelike peaks in their experiments. Specifically, they attempted to show that the pattern of angles at which the electrons reflected matched the patterns one sees in x-ray diffraction. After numerous additional measurements, they were able to show a very close correspondence between the pattern of electrons and the patterns seen in x-ray diffraction. Since x-rays were by this time known to be waves of electromagnetic radiation, the Davisson–Germer experiment was a strong confirmation of de Broglie's wave picture of electrons. Davisson and Germer published their results in two papers in 1927, and Davisson shared the 1937 Nobel Prize in physics with George Paget, who had observed electron diffraction shortly after Davisson and Germer.

1.2.4   The Double-Slit Experiment with Electrons

Although quantum theory clearly predicts that electrons passing through a double slit will experience interference similar to that observed in light, it was not until Clauss Jönsson's work in 1961 that this prediction was confirmed experimentally. The main difficulty is the much smaller wavelength for electrons of reasonable energy than for visible light. Jönsson's electrons, for example, had a de Broglie wavelength of 5 nm, as compared to a wavelength of roughly 500 nm for visible light (depending on the color).

In results published in 1989, a team led by Akira Tonomura at Hitachi performed a double-slit experiment in which they were able to record the results *one electron at a time*. (Similar but less definitive experiments were carried out by Pier Giorgio Merli, GianFranco Missiroli and Giulio Pozzi in Bologna in 1974 and published in the *American Journal of Physics* in 1976.) In the Hitachi experiment, each electron passes through the slits and then strikes a screen, causing a small spot of light to appear. The location of this spot is then recorded for each electron, one at a time. The key point is that each individual electron strikes the screen at a *single point*. That is to say, *individual* electrons are not smeared out across the screen in a wavelike pattern, but rather behave like point particles, in that the observed location of the electron is indeed a point. Each electron, however, strikes the screen at a different point, and once a large number of the electrons have struck and their locations have been recorded, an interference pattern emerges.

It is not the variability of the locations of the electrons that is surprising, since this could be accounted for by small variations in the way the electrons

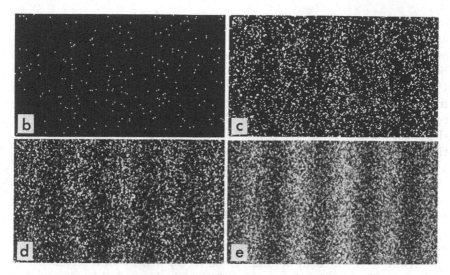

FIGURE 1.4. Four images from the 1989 experiment at Hitachi showing the impact of individual electrons gradually building up to form an interference pattern. Image by Akira Tonomura and Wikimedia Commons user Belsazar. File is licensed under the Creative Commons Attribution-Share Alike 3.0 Unported license.

are shot toward the slits. Rather, it is the distinctive *interference* pattern that is surprising, with rapid variations in the pattern of electron strikes over short distances, including regions where almost no electron strikes occur. (Compare Fig. 1.4 to Fig. 1.2.) Note also that in the experiment, the electrons are widely separated, so that there is never more than one electron in the apparatus at any one time. Thus, the electrons cannot interfere with one another; rather, *each electron interferes with itself.* Figure 1.4 shows results from the Hitachi experiment, with the number of observed electrons increasing from about 150 in the first image to 160,000 in the last image.

1.3   Schrödinger and Heisenberg

In 1925, Werner Heisenberg proposed a model of quantum mechanics based on treating the position and momentum of the particle as, essentially, matrices of size $\infty \times \infty$. Actually, Heisenberg himself was not familiar with the theory of matrices, which was not a standard part of the mathematical education of physicists at the time. Nevertheless, he had quantities of the form x_{jk} and p_{jk} (where j and k each vary over all integers), which we can recognize as matrices, as well as expressions such as $\sum_l x_{jl} p_{lk}$, which we can recognize as a matrix product. After Heisenberg explained his theory to Max Born, Born recognized the connection of Heisenberg's formulas to matrix theory and made the matrix point of view explicit, in a paper coauthored by Born and his assistant, Pascual Jordan. Born, Heisenberg, and Jordan then all published a paper together elaborating upon their theory. The papers of Heisenberg, of Born and Jordan, and of Born, Heisenberg, and Jordan all appeared in 1925. Heisenberg received the 1932 Nobel Prize in physics (actually awarded in 1933) for his work. Born's exclusion from this prize was controversial, and may have been influenced by Jordan's connections with the Nazi party in Germany. (Heisenberg's own work for the Nazis during World War II was also a source of much controversy after the war.) In any case, Born was awarded the Nobel Prize in physics in 1954 for his work on the statistical interpretation of quantum mechanics (Sect. 1.4).

Meanwhile, in 1926, Erwin Schrödinger published four remarkable papers in which he proposed a wave theory of quantum mechanics, along the lines of the *de Broglie hypothesis*. In these papers, Schrödinger described how the waves evolve over time and showed that the energy levels of, for example, the hydrogen atom could be understood as *eigenvalues* of a certain operator. (See Chap. 18 for the computation for hydrogen.) Schrödinger also showed that the Heisenberg–Born–Jordan matrix model could be incorporated into the wave theory, thus showing that the matrix theory and the wave theory were equivalent (see Sect. 3.8). This book describes the mathematical structure of quantum mechanics in essentially the form proposed by Schrödinger in 1926. Schrödinger shared the 1933 Nobel Prize in physics with Paul Dirac.

1.4   A Matter of Interpretation

Although Schrödinger's 1926 papers gave the correct mathematical description of quantum mechanics (as it is generally accepted today), he did not provide a widely accepted *interpretation* of the theory. That task fell to Born, who in a 1926 paper proposed that the "wave function" (as the wave appearing in the Schrödinger equation is generally called) should be interpreted statistically, that is, as determining the *probabilities* for observations of the system. Over time, Born's statistical approach developed into the *Copenhagen interpretation* of quantum mechanics. Under this interpretation, the wave function ψ of the system is not directly observable. Rather, ψ merely determines the probability of observing a particular result.

In particular, if ψ is properly normalized, then the quantity $|\psi(x)|^2$ is the probability distribution for the position of the particle. Even if ψ itself is spread out over a large region in space, any measurement of the position of the particle will show that the particle is located at a *single point*, just as we see for the electrons in the two-slit experiment in Fig. 1.4. Thus, a measurement of a particle's position does not show the particle "smeared out" over a large region of space, even if the wave function ψ *is* smeared out over a large region.

Consider, for example, how Born's interpretation of the Schrödinger equation would play out in the context of the Hitachi double-slit experiment depicted in Fig. 1.4. Born would say that each electron has a wave function that evolves in time according to the Schrödinger equation (an equation of wave type). Each particle's wave function, then, will propagate through the slits in a manner similar to that pictured in Fig. 1.1. If there is a screen at the bottom of Fig. 1.1, then the electron will hit the screen at a single point, even though the wave function is very spread out. The wave function does not determine where the particle hits the screen; it merely determines the probabilities for where the particle hits the screen. If a whole sequence of electrons passes through the slits, one after the other, over time a probability distribution will emerge, determined by the square of the magnitude of the wave function, which is shown in Fig. 1.2. Thus, the probability distribution of electrons, as seen from a large number of electrons as in Fig. 1.4, shows wavelike interference patterns, even though each individual electron strikes the screen at a single point.

It is essential to the theory that the wave function $\psi(x)$ itself is not the probability density for the location of the particle. Rather, the probability density is $|\psi(x)|^2$. The difference is crucial, because probability densities are intrinsically positive and thus do not exhibit destructive interference. The wave function itself, however, is complex-valued, and the real and imaginary parts of the wave function take on both positive and negative values, which can interfere constructively or destructively. The part of the wave function passing through the first slit, for example, can interfere with the part of the wave function passing through the second slit. Only *after* this interference has taken place do we take the magnitude squared of the wave function to obtain the probability distribution, which will, therefore, show the sorts of peaks and valleys we see in Fig. 1.2.

Born's introduction of a probabilistic element into the interpretation of quantum mechanics was—and to some extent still is—controversial. Einstein, for example, is often quoted as saying something along the lines of, "God does not play at dice with the universe." Einstein expressed the same sentiment in various ways over the years. His earliest known statement to this effect was in a letter to Born in December 1926, in which he said,

> Quantum mechanics is certainly imposing. But an inner voice tells me that it is not yet the real thing. The theory says a lot, but does not really bring us any closer to the secret of the "old one." I, at any rate, am convinced that *He* does not throw dice.

Many other physicists and philosophers have questioned the probabilistic interpretation of quantum mechanics, and have sought alternatives, such as "hidden variable" theories. Nevertheless, the Copenhagen interpretation of quantum mechanics, essentially as proposed by Born in 1926, remains the standard one. This book resolutely avoids all controversies surrounding the interpretation of quantum mechanics. Chapter 3, for example, presents the standard statistical interpretation of the theory without question. The book may nevertheless be of use to the more philosophically minded reader, in that one must learn something of quantum mechanics before delving into the (often highly technical) discussions about its interpretation.

1.5 Exercises

1. Beginning with the formula for the sum of a geometric series, use differentiation to obtain the identity

$$\sum_{n=0}^{\infty} ne^{-An} = \frac{e^{-A}}{(1 - e^{-A})^2}.$$

2. In Planck's model of blackbody radiation, the energy in a given frequency ω of electromagnetic radiation is distributed randomly over all numbers of the form $n\hbar\omega$, where $n = 0, 1, 2, \ldots$. Specifically, the likelihood of finding energy $n\hbar\omega$ is postulated to be

$$p(E = n\hbar\omega) = \frac{1}{Z}e^{-\beta n\hbar\omega},$$

$$Z = \frac{1}{1 - e^{-\beta\hbar\omega}},$$

where Z is a normalization constant, which is chosen so that the sum over n of the probabilities is 1. Here $\beta = 1/(k_B T)$, where T is the temperature and k_B is Boltzmann's constant. The *expected value* of the energy, denoted $\langle E \rangle$, is defined to be

$$\langle E \rangle = \frac{1}{Z}\sum_{n=0}^{\infty}(n\hbar\omega)e^{-\beta n\hbar\omega}.$$

(a) Using Exercise 1, show that

$$\langle E \rangle = \frac{\hbar\omega}{e^{\beta\hbar\omega} - 1}.$$

(b) Show that $\langle E \rangle$ behaves like $1/\beta = k_B T$ for small ω, but that $\langle E \rangle$ decays exponentially as ω tends to infinity.

Note: In applying the above calculation to blackbody radiation, one must also take into account the number of modes having frequency in a given range, say between ω_0 and $\omega_0 + \varepsilon$. The exact number of such frequencies depends on the shape of the cavity, but according to Weyl's law, this number will be approximately proportional to $\varepsilon\omega_0^2$ for large values of ω_0. Thus, the amount of energy per unit of frequency is

$$C\frac{\hbar\omega^3}{e^{\beta\hbar\omega} - 1}, \tag{1.7}$$

where C is a constant involving the volume of the cavity and the speed of light. The relation (1.7) is known as Planck's law.

3. In classical mechanics, the kinetic energy of an electron is $m_e v^2/2$, where v is the magnitude of the velocity. Meanwhile, the potential energy associated with the force law (1.3) is $V(r) = -Q^2/r$, since $dV/dr = F$. Show that if the particle is moving in a circular orbit with radius r_n given by (1.5), then the total energy (kinetic plus potential) of the particle is E_n, as given in (1.1).

2
A First Approach to Classical Mechanics

2.1 Motion in $\mathbb{R}^1$

2.1.1 Newton's law

We begin by considering the motion of a single particle in $\mathbb{R}^1$, which may be thought of as a particle sliding along a wire, or a particle with motion that just happens to lie in a line. We let $x(t)$ denote the particle's position as a function of time. The particle's velocity is then

$$v(t) := \dot{x}(t),$$

where we use a dot over a symbol to denote the derivative of that quantity with respect to the time t.

The particle's acceleration is then

$$a(t) = \dot{v}(t) = \ddot{x}(t),$$

where $\ddot{x}$ denotes the second derivative of x with respect to t. We assume that there is a force acting on the particle and we assume at first that the force F is a function of the particle's position only. (Later, we will look at the case of forces that depend also on velocity.)

Under these assumptions, Newton's second law ($F = ma$) takes the form

$$F(x(t)) = ma = m\ddot{x}(t), \tag{2.1}$$

where m is the mass of the particle, which is assumed to be positive. We will henceforth abbreviate Newton's second law as simply "Newton's law," since

B.C. Hall, *Quantum Theory for Mathematicians*, Graduate Texts in Mathematics 267, DOI 10.1007/978-1-4614-7116-5_2,
© Springer Science+Business Media New York 2013

we will use the second law much more frequently than the others. Since (2.1) is of second order, the appropriate initial conditions (needed to get a unique solution) are the position and velocity at some initial time t_0. So we look for solutions of (2.1) subject to

$$x(t_0) = x_0$$
$$\dot{x}(t_0) = v_0.$$

Assuming that F is a smooth function, standard results from the elementary theory of differential equations tell us that there exists a unique *local* solution to (2.1) for each pair of initial conditions. (A local solution is one defined for t in a neighborhood of the initial time t_0.) Since (2.1) is in general a nonlinear equation, one cannot expect that, for a general force function F, the solutions will exist for all t. If, for example, $F(x) = x^2$, then any solution with positive initial position and positive initial velocity will escape to infinity in finite time. (Apply Exercise 4 with $V(x) = -x^3/3$.) For a proof existence and uniqueness, see Example 8.2 and Theorem 8.13 in [28].

Definition 2.1 *A solution $x(t)$ to Newton's law is called a **trajectory**.*

Example 2.2 (Harmonic Oscillator) *If the force is given by Hooke's law, $F(x) = -kx$, where k is a positive constant, then Newton's law can be written as $m\ddot{x} + kx = 0$. The general solution of this equation is*

$$x(t) = a\cos(\omega t) + b\sin(\omega t),$$

where $\omega := \sqrt{k/m}$ is the frequency of oscillation.

The system in Example 2.2 is referred to as a (classical) *harmonic oscillator*. This system can describe a mass on a spring, where the force is proportional to the distance x that the spring is stretched from its equilibrium position. The minus sign in $-kx$ indicates that the force pulls the oscillator back toward equilibrium. Here and elsewhere in the book, we use the "angular" notion of frequency, which is the rate of change of the argument of a sine or cosine function. If ω is the angular frequency, then the "ordinary" frequency—i.e., the number of cycles per unit of time—is $\omega/2\pi$. Saying that x has (angular) frequency ω means that x is periodic with period $2\pi/\omega$.

2.1.2 Conservation of Energy

We return now to the case of a general force function $F(x)$. We define the *kinetic energy* of the system to be $\frac{1}{2}mv^2$. We also define the *potential energy* of the system as the function

$$V(x) = -\int F(x)\,dx, \tag{2.2}$$

so that $F(x) = -dV/dx$. (The potential energy is defined only up to adding a constant.) The *total energy* E of the system is then

$$E(x, v) = \frac{1}{2}mv^2 + V(x). \tag{2.3}$$

The chief significance of the energy function is that it is *conserved*, meaning that its value along any trajectory is constant.

Theorem 2.3 *Suppose a particle satisfies Newton's law in the form $m\ddot{x} = F(x)$. Let V and E be as in (2.2) and (2.3). Then the energy E is conserved, meaning that for each solution $x(t)$ of Newton's law, $E(x(t), \dot{x}(t))$ is independent of t.*

Proof. We verify this by differentiation, using the chain rule:

$$\frac{d}{dt}E(x(t), \dot{x}(t)) = \frac{d}{dt}\left(\frac{1}{2}m(\dot{x}(t))^2 + V(x(t))\right)$$

$$= m\dot{x}(t)\ddot{x}(t) + \frac{dV}{dx}\dot{x}(t)$$

$$= \dot{x}(t)[m\ddot{x}(t) - F(x(t))].$$

This last expression is zero by Newton's law. Thus, the time-derivative of the energy along any trajectory is zero, so $E(x(t), \dot{x}(t))$ is independent of t, as claimed. ∎

We may call the energy a *conserved quantity* (or *constant of motion*), since the particle neither gains nor loses energy as the particle moves according to Newton's law.

Let us see how conservation of energy helps us understand the solution to Newton's law. We may reduce the second-order equation $m\ddot{x} = F(x)$ to a pair of first-order equations, simply by introducing the velocity v as a new variable. That is, we look for pairs of functions $(x(t), v(t))$ that satisfy the following system of equations

$$\frac{dx}{dt} = v(t)$$

$$\frac{dv}{dt} = \frac{1}{m}F(x(t)). \tag{2.4}$$

If $(x(t), v(t))$ is a solution to this system, then we can immediately see that $x(t)$ satisfies Newton's law, just by substituting dx/dt for v in the second equation. We refer to the set of possible pairs of the form (x, v) (i.e., $\mathbb{R}^2$) as the *phase space* of the particle in $\mathbb{R}^1$. The appropriate initial conditions for this first-order system are $x(0) = x_0$ and $v(0) = v_0$.

Once we are working in phase space, we can use the conservation of energy to help us. Conservation of energy means that each solution to the system (2.4) must lie entirely on a single "level curve" of the energy function, that is, the set

$$\left\{ (x, v) \in \mathbb{R}^2 \,\middle|\, E(x, v) = E(x_0, v_0) \right\}. \tag{2.5}$$

If F—and therefore also V—is smooth, then E is a smooth function of x and v. Then as long as (2.5) contains no critical points of E, this set will be a smooth curve in $\mathbb{R}^2$, by the implicit function theorem. If the level set (2.5) is also a simple closed curve, then the solutions of (2.5) will simply wind around and around this curve. Thus, the set that the solutions to (2.5) trace out in phase space can be determined simply from the conservation of energy. The only thing not apparent at the moment is how this curve is parameterized as a function of time.

In mechanics, a conserved quantity—such as the energy in the one-dimensional version of Newton's law—is often referred to as an "integral of motion." The reason for this is that although Newton's second law is a second-order equation in x, the energy depends only on x and $\dot{x}$ and not on $\ddot{x}$. Thus, the equation

$$\frac{m}{2} (\dot{x}(t))^2 + V(x(t)) = E_0,$$

where E_0 is the value of the energy at time t_0, is actually a *first-order* differential equation. We can solve for $\dot{x}$ to put this equation into a more standard form:

$$\dot{x}(t) = \pm \sqrt{\frac{2(E_0 - V(x(t)))}{m}}. \tag{2.6}$$

What this means is that by using conservation of energy we have turned the original second-order equation into a first-order equation. We have therefore "integrated" the original equation once, that is, changed an equation of the form $\ddot{x}(t) = \cdots$ into an equation of the form $\dot{x}(t) = \cdots$. The first-order equation (2.6) is separable and can be solved more-or-less explicitly (Exercise 1).

2.1.3 *Systems with Damping*

Up to now, we have considered forces that depend only on position. It is common, however, to consider forces that depend on the velocity as well as the position. In the case of a damped harmonic oscillator, for example, one typically assumes that there is, in addition to the force of the spring, a damping force (friction, say) that is proportional to the velocity. Thus, $F = -kx - \gamma \dot{x}$, where k is, as before, the spring constant and where $\gamma > 0$ is the damping constant. The minus sign in front of $\gamma \dot{x}$ reflects that the damping force operates in the opposite direction to the velocity, causing the particle to slow down. The equation of motion for such a system is then

$$m\ddot{x} + \gamma \dot{x} + kx = 0.$$

If γ is small, the solutions to this equation display decaying oscillation, meaning sines and cosines multiplied by a decaying exponential; if γ is large, the solutions are pure decaying exponentials (Exercise 5).

In the case of the damped harmonic oscillator, there is no longer a conserved energy. Specifically, there is no nonconstant continuous function E on $\mathbb{R}^2$ such that $E(x(t), \dot{x}(t))$ is independent of t for all solutions of Newton's law. To see this, we simply observe that for $\gamma > 0$, all solutions $x(t)$ have the property that $(x(t), \dot{x}(t))$ tends to the origin in the plane as t tends to infinity. Thus, if E is continuous and constant along each trajectory, the value of E at the starting point has to be the same as the value at the origin.

We now consider a general system with damping.

Proposition 2.4 *Suppose a particle moves in the presence of a force law given by $F(x, \dot{x}) = F_1(x) - \gamma\dot{x}$, with $\gamma > 0$. Define the energy E of the system by*

$$E(x, \dot{x}) = \frac{1}{2}m\dot{x}^2 + V(x),$$

where $dV/dx = -F_1(x)$. Then along any trajectory $x(t)$, we have

$$\frac{d}{dt}E(x(t), \dot{x}(t)) = -\gamma\dot{x}(t)^2 \le 0.$$

Thus, although the energy is not conserved, it is decreasing with time, which gives us some information about the behavior of the system.

Proof. We differentiate as in the proof of Theorem 2.3, except that now $dV/dx = -F_1(x)$:

$$\frac{d}{dt}E(x(t), \dot{x}(t)) = \dot{x}(t)[m\ddot{x}(t) - F_1(x(t))].$$

Since F_1 is not the full force function, the quantity in square brackets equals not zero but $-\gamma\dot{x}$. Thus, $dE/dt = -\gamma\dot{x}^2$. ∎

We can interpret Proposition 2.4 as saying that in the presence of friction, the system we are studying gives up some of its energy to heat energy in the environment, so that the energy of our system decreases with time. We will see that in higher dimensions, it is possible to have conservation of energy in the presence of velocity-dependent forces, provided that these forces act perpendicularly to the velocity.

2.2   Motion in $\mathbb{R}^n$

We now consider a particle moving in $\mathbb{R}^n$. The position $\mathbf{x} = (x_1, \ldots, x_n)$ of a particle is now a vector in $\mathbb{R}^n$, as is the velocity $\mathbf{v}$ and acceleration $\mathbf{a}$. We let

$$\dot{\mathbf{x}} = (\dot{x}_1, \ldots, \dot{x}_n)$$

denote the derivative of $\mathbf{x}$ with respect to t and we let $\ddot{\mathbf{x}}$ denote the second derivative of $\mathbf{x}$ with respect to t. Newton's law now takes the form

$$m\ddot{\mathbf{x}}(t) = \mathbf{F}(\mathbf{x}(t), \dot{\mathbf{x}}(t)),\qquad(2.7)$$

where $\mathbf{F} : \mathbb{R}^n \times \mathbb{R}^n \to \mathbb{R}^n$ is some force law, which in general may depend on both the position and velocity of the particle.

We begin by considering forces that are independent of velocity, and we look for a conserved energy function in this setting.

Proposition 2.5 *Consider Newton's law (2.7) in the case of a velocity-independent force: $m\ddot{\mathbf{x}}(t) = \mathbf{F}(\mathbf{x}(t))$. Then an energy function of the form*

$$E(\mathbf{x}, \dot{\mathbf{x}}) = \frac{1}{2}m \left|\dot{\mathbf{x}}\right|^2 + V(\mathbf{x})$$

is conserved if and only if V satisfies

$$-\nabla V = \mathbf{F},$$

where ∇V is the gradient of V.

Saying that E is "conserved" means that $E(\mathbf{x}(t), \dot{\mathbf{x}}(t))$ is independent of t for each solution $\mathbf{x}(t)$ of Newton's law. The function V is the *potential energy* of the system.

Proof. Differentiating gives

$$\frac{d}{dt}\left(\frac{1}{2}m \left|\dot{\mathbf{x}}(t)\right|^2 + V(\mathbf{x}(t))\right) = m\sum_{j=i}^{n} \dot{x}_j(t)\ddot{x}_j(t) + \sum_{j=1}^{n} \frac{\partial V}{\partial x_j}\dot{x}_j(t)$$

$$= \dot{\mathbf{x}}(t) \cdot [m\ddot{\mathbf{x}}(t) + \nabla V]$$

$$= \dot{\mathbf{x}}(t) \cdot [\mathbf{F}(\mathbf{x}) + \nabla V(\mathbf{x})]$$

Thus, dE/dt will always be equal to zero if and only if we have

$$-\nabla V(\mathbf{x}) = \mathbf{F}(\mathbf{x})$$

for all $\mathbf{x}$. ∎

We now encounter something that did not occur in the one-dimensional case. In $\mathbb{R}^1$, any smooth function can be expressed as the derivative of some other function. In $\mathbb{R}^n$, however, not every vector-valued function $\mathbf{F}(\mathbf{x})$ can be expressed as the (negative of) the gradient of some scalar-valued function V.

Definition 2.6 *Suppose $\mathbf{F}$ is a smooth, $\mathbb{R}^n$-valued function on a domain $U \subset \mathbb{R}^n$. Then $\mathbf{F}$ is called **conservative** if there exists a smooth, real-valued function V on U such that $\mathbf{F} = -\nabla V$.*

If the domain U is simply connected, then there is a simple local condition that characterizes conservative functions.

Proposition 2.7 *Suppose U is a simply connected domain in $\mathbb{R}^n$ and $\mathbf{F}$ is a smooth, $\mathbb{R}^n$-valued function on U. Then $\mathbf{F}$ is conservative if and only if $\mathbf{F}$ satisfies*

$$\frac{\partial F_j}{\partial x_k} - \frac{\partial F_k}{\partial x_j} = 0 \qquad (2.8)$$

at each point in U.

When $n = 3$, it is easy to check that the condition (2.8) is equivalent to the *curl* $\nabla \times \mathbf{F}$ of $\mathbf{F}$ being zero on U. The hypothesis that U be simply connected cannot be omitted; see Exercise 7.

Proof. If $\mathbf{F}$ is conservative, then

$$\frac{\partial F_j}{\partial x_k} = -\frac{\partial^2 V}{\partial x_k \partial x_j} = -\frac{\partial^2 V}{\partial x_j \partial x_k} = \frac{\partial F_k}{\partial x_j}$$

at every point in U. In the other direction, if $\mathbf{F}$ satisfies (2.8), V can be obtained by integrating $\mathbf{F}$ along paths and using the Stokes theorem to establish independence of choice of path. See, for example, Theorem 4.3 on p. 549 of [44] for a proof in the $n = 3$ case. The proof in higher dimensions is the same, provided one knows the general version of the Stokes theorem. ∎

We may also consider velocity-dependent forces. If, for example, $\mathbf{F}(\mathbf{x}, \mathbf{v})$ $= -\gamma \mathbf{v} + \mathbf{F}_1(\mathbf{x})$, where γ is a positive constant, then we will again have energy that is decreasing with time. There is another new phenomenon, however, in dimension greater than 1, namely the possibility of having a conserved energy even when the force depends on velocity.

Proposition 2.8 *Suppose a particle in $\mathbb{R}^n$ moves in the presence of a force $\mathbf{F}$ of the form*

$$\mathbf{F}(\mathbf{x}, \mathbf{v}) = -\nabla V(\mathbf{x}) + \mathbf{F}_2(\mathbf{x}, \mathbf{v}),$$

where V is a smooth function and where $\mathbf{F}_2$ satisfies

$$\mathbf{v} \cdot \mathbf{F}_2(\mathbf{x}, \mathbf{v}) = 0 \qquad (2.9)$$

for all $\mathbf{x}$ and $\mathbf{v}$ in $\mathbb{R}^n$. Then the energy function $E(\mathbf{x}, \mathbf{v}) = \frac{1}{2} m |\mathbf{v}|^2 + V(\mathbf{x})$ is constant along each trajectory.

If, for example, $\mathbf{F}_2$ is the force exerted on a charged particle in $\mathbb{R}^3$ by a magnetic field $\mathbf{B}(\mathbf{x})$, then

$$\mathbf{F}_2(\mathbf{x}, \mathbf{v}) = q\mathbf{v} \times \mathbf{B}(\mathbf{x}),$$

where q is the charge of the particle, which clearly satisfies (2.9).

Proof. See Exercise 8. ∎

2.3   Systems of Particles

If we have a system if N particles, each moving in $\mathbb{R}^n$, then we denote the position of the jth particle by

$$\mathbf{x}^j = (x_1^j, \ldots, x_n^j).$$

Thus, in the expression x_k^j, the superscript j indicates the jth particle, while the subscript k indicates the kth component. Newton's law then takes the form

$$m_j \ddot{\mathbf{x}}^j = \mathbf{F}^j(\mathbf{x}^1, \ldots, \mathbf{x}^N, \dot{\mathbf{x}}^1, \ldots, \dot{\mathbf{x}}^N), \quad j = 1, 2, \ldots, N,$$

where m_j is the mass of the jth particle. Here, $\mathbf{F}^j$ is the force on the jth particle, which in general will depend on the position and velocity not only of that particle, but also on the position and velocity of the other particles.

2.3.1   Conservation of Energy

In a system of particles, we cannot expect that the energy of each individual particle will be conserved, because as the particles interact, they can exchange energy. Rather, we should expect that, under suitable assumptions on the forces $\mathbf{F}^j$, we can define a conserved energy function for the whole system (the *total* energy of the system).

Let us consider forces depending only on the position of the particles, and let us assume that the energy function will be of the form

$$E(\mathbf{x}^1, \ldots, \mathbf{x}^N, \mathbf{v}^1, \ldots, \mathbf{v}^N) = \sum_{j=1}^{N} \frac{1}{2} m_j \left| \mathbf{v}^j \right|^2 + V(\mathbf{x}^1, \ldots, \mathbf{x}^N). \quad (2.10)$$

We will now try to see what form for V (if any) will allow E to be constant along each trajectory.

Proposition 2.9 *An energy function of the form (2.10) is constant along each trajectory if*

$$\nabla^j V = -\mathbf{F}^j \quad (2.11)$$

for each j, where ∇^j is the gradient with respect to the variable $\mathbf{x}^j$.

Proof. We compute that

$$\frac{dE}{dt} = \sum_{j=1}^{N} \left[m_j \dot{\mathbf{x}}^j \cdot \ddot{\mathbf{x}}^j + \nabla^j V \cdot \dot{\mathbf{x}}^j \right]$$

$$= \sum_{j=1}^{N} \dot{\mathbf{x}}^j \cdot \left[m_j \ddot{\mathbf{x}}^j + \nabla^j V \right]$$

$$= \sum_{j=1}^{N} \dot{\mathbf{x}}^j \cdot \left[\mathbf{F}^j + \nabla^j V \right].$$

If $\nabla^j V = -\mathbf{F}^j$, then E will be conserved. ■

As in the one-particle case, there is a simple condition for the existence of a potential function V satisfying (2.11).

Proposition 2.10 *Suppose a force function* $\mathbf{F} = (\mathbf{F}^1, \ldots, \mathbf{F}^N)$ *is defined on a simply connected domain* U *in* $\mathbb{R}^{nN}$. *Then there exists a smooth function* V *on* U *satisfying*

$$\nabla^j V = -\mathbf{F}^j$$

for all j *if and only if we have*

$$\frac{\partial F_k^j}{\partial x_m^l} = \frac{\partial F_m^l}{\partial x_k^j} \tag{2.12}$$

for all j, k, l, *and* m.

Proof. Apply Proposition 2.7 with n replaced by nN and with j and k replaced by the pairs (j, k) and (l, m). ■

2.3.2   Conservation of Momentum

We now introduce the notion of the momentum of a particle.

Definition 2.11 *In an* N-*particle system, the* **momentum** *of the* jth *particle, denoted* $\mathbf{p}^j$, *is the product of the mass and the velocity of that particle:*

$$\mathbf{p}^j = m_j \dot{\mathbf{x}}^j.$$

The **total momentum** *of the system, denoted* $\mathbf{p}$, *is defined as*

$$\mathbf{p} = \sum_{j=1}^N \mathbf{p}^j.$$

Observe that

$$\frac{d\mathbf{p}^j}{dt} = m_j \ddot{\mathbf{x}}^j = \mathbf{F}^j.$$

Thus, Newton's law may be reformulated as saying, "The force is the rate of change of the momentum." This is how Newton originally formulated his second law.

Newton's third law says, "For every action, there is an equal and opposite reaction." This law will apply if all forces are of the "two-particle" variety and satisfy a natural symmetry property. Having two-particle forces means that the force $\mathbf{F}^j$ on the jth particle is a sum of terms $\mathbf{F}^{j,k}$, $j \neq k$, where $\mathbf{F}^{j,k}$ depends only $\mathbf{x}^j$ and $\mathbf{x}^k$. The relevant symmetry property is that $\mathbf{F}^{j,k}(\mathbf{x}^j, \mathbf{x}^k) = -\mathbf{F}^{k,j}(\mathbf{x}^k, \mathbf{x}^j)$; that is, the force exerted by the jth particle on the kth particle is the negative (i.e., "equal and opposite") of the force exerted by the kth particle on the jth particle. If the forces are assumed also to be conservative, then the potential energy of the system will be of the form

$$V(\mathbf{x}^1, \mathbf{x}^2, \ldots, \mathbf{x}^N) = \sum_{j<k} V^{j,k}(\mathbf{x}^j - \mathbf{x}^k). \tag{2.13}$$

One important consequence of Newton's third law is conservation of the total momentum of the system.

Proposition 2.12 *Suppose that for each j, the force on the jth particle is of the form*

$$\mathbf{F}^j(\mathbf{x}^1, \mathbf{x}^2, \ldots, \mathbf{x}^N) = \sum_{k \neq j} \mathbf{F}^{j,k}(\mathbf{x}^j, \mathbf{x}^k),$$

for certain functions $\mathbf{F}^{j,k}$. Suppose also that we have the "equal and opposite" condition

$$\mathbf{F}^{j,k}(\mathbf{x}^j, \mathbf{x}^k) = -\mathbf{F}^{k,j}(\mathbf{x}^j, \mathbf{x}^k).$$

Then the total momentum of the system is conserved.

Note that since the rate of change of $\mathbf{p}^j$ is $\mathbf{F}^j$, the force on the jth particle, the momentum of each individual particle is not constant in time, except in the trivial case of a noninteracting system (one in which all forces are zero).

Proof. Differentiating gives

$$\frac{d\mathbf{p}}{dt} = \sum_{j=1}^{N} \frac{d\mathbf{p}^j}{dt} = \sum_{j=1}^{N} \mathbf{F}^j = \sum_{j} \sum_{k \neq j} \mathbf{F}^{j,k}(\mathbf{x}^j, \mathbf{x}^k).$$

By the equal and opposite condition, $\mathbf{F}^{j,k}(\mathbf{x}^j, \mathbf{x}^k)$ cancels with $\mathbf{F}^{k,j}(\mathbf{x}^j, \mathbf{x}^k)$, so $d\mathbf{p}/dt = 0$. ∎

Let us consider, now, a more general situation in which we have conservative forces, but not necessarily of the "two-particle" form. It is still possible to have conservation of momentum, as the following result shows.

Proposition 2.13 *If a multiparticle system has a force law coming from a potential V, then the total momentum of the system is conserved if and only if*

$$V(\mathbf{x}^1 + \mathbf{a}, \mathbf{x}^2 + \mathbf{a}, \ldots, \mathbf{x}^N + \mathbf{a}) = V(\mathbf{x}^1, \mathbf{x}^2, \ldots, \mathbf{x}^N) \tag{2.14}$$

for all $\mathbf{a} \in \mathbb{R}^n$.

Proof. Apply (2.14) with $\mathbf{a} = t\mathbf{e}_k$, where $\mathbf{e}_k$ is the vector with a 1 in the kth spot and zeros elsewhere. Differentiating with respect to t at $t = 0$ gives

$$0 = \sum_{j=1}^{N} \frac{\partial V}{\partial x_k^j} = -\sum_{j=1}^{N} F_k^j = -\sum_{j=1}^{N} \frac{dp_k^j}{dt} = -\frac{dp_k}{dt},$$

where p_k is the kth component of the total momentum $\mathbf{p}$. Thus, if (2.14) holds, $\mathbf{p}$ is constant in time.

Conversely, if the momentum is conserved, then the sum of the forces is zero at every point, and so

$$\frac{d}{dt}V(\mathbf{x}^1 + t\mathbf{a}, \mathbf{x}^2 + t\mathbf{a}, \ldots, \mathbf{x}^N + t\mathbf{a})$$

$$= \sum_{j=1}^{N} \nabla^j V(\mathbf{x}^1 + t\mathbf{a}, \mathbf{x}^2 + t\mathbf{a}, \ldots, \mathbf{x}^N + t\mathbf{a}) \cdot \mathbf{a}$$

$$= -\left(\sum_{j=1}^{N} \mathbf{F}^j(\mathbf{x}^1 + t\mathbf{a}, \mathbf{x}^2 + t\mathbf{a}, \ldots, \mathbf{x}^N + t\mathbf{a}) \right) \cdot \mathbf{a}$$

$$= 0$$

for all t. Thus, the value of the quantity being differentiated is the same at $t = 0$ as at $t = 1$, which establishes (2.14). ∎

The moral of the story is that conservation of momentum is a consequence of translation-invariance of the system, where "translation invariance " means invariance under simultaneous translations of *every* particle by the *same* amount. (See Exercise 11 for a more general version of this result.) If the potential is of the "two-particle" form (2.13), then it is evident that the condition (2.14) is satisfied.

2.3.3 Center of Mass

We now consider an important application of momentum conservation.

Definition 2.14 *For a system of N particles moving in $\mathbb{R}^n$, the* **center of mass** *of the system at a fixed time is the vector $\mathbf{c} \in \mathbb{R}^n$ given by*

$$\mathbf{c} = \sum_{j=1}^{N} \frac{m_j}{M} \mathbf{x}^j,$$

where $M = \sum_{j=1}^{N} m_j$ is the total mass of the system.

The center of mass is a weighted average of the positions of the various particles. Differentiating $\mathbf{c}(t)$ with respect to t gives

$$\frac{d\mathbf{c}}{dt} = \frac{1}{M} \sum_{j=1}^{N} m_j \dot{\mathbf{x}}^j = \frac{\mathbf{p}}{M}, \tag{2.15}$$

where $\mathbf{p}$ is the total momentum.

Proposition 2.15 *Suppose the total momentum* **p** *of a system is conserved. Then the center of mass moves in a straight line at constant speed. Specifically,*

$$\mathbf{c}(t) = \mathbf{c}(t_0) + (t - t_0)\frac{\mathbf{p}}{M},$$

where $\mathbf{c}(t_0)$ *is the center of mass at some initial time* t_0.

Proof. The result follows easily from (2.15). ∎

The notion of center of mass is particularly useful in a system of two particles in which momentum is conserved. For a system of two particles, if the potential energy $V(\mathbf{x}^1, \mathbf{x}^2)$ is invariant under simultaneous translations of $\mathbf{x}^1$ and $\mathbf{x}^2$, then it is of the form

$$V(\mathbf{x}^1, \mathbf{x}^2) = \tilde{V}(\mathbf{x}^1 - \mathbf{x}^2),$$

where $\tilde{V}(\mathbf{a}) = V(\mathbf{a}, 0)$.

Now, the positions $\mathbf{x}^1, \mathbf{x}^2$ of the particles can be recovered from knowledge of the center of mass and the *relative position*

$$\mathbf{y} := \mathbf{x}^1 - \mathbf{x}^2$$

as follows:

$$\mathbf{x}^1 = \frac{\mathbf{c} + m_2\mathbf{y}}{m_1 + m_2}$$

$$\mathbf{x}^2 = \frac{\mathbf{c} - m_1\mathbf{y}}{m_1 + m_2}.$$

Meanwhile, we may compute that

$$\ddot{\mathbf{y}}(t) = \ddot{\mathbf{x}}^1 - \ddot{\mathbf{x}}^2 = -\frac{1}{m_1}\nabla\tilde{V}(\mathbf{x}^1 - \mathbf{x}^2) - \frac{1}{m_2}\nabla\tilde{V}(\mathbf{x}^1 - \mathbf{x}^2).$$

This calculation gives the following result.

Proposition 2.16 *For a two-particle system with potential energy of the form* $V(\mathbf{x}^1, \mathbf{x}^2) = \tilde{V}(\mathbf{x}^1 - \mathbf{x}^2)$, *the relative position* $\mathbf{y} := \mathbf{x}^1 - \mathbf{x}^2$ *satisfies the differential equation*

$$\mu\ddot{\mathbf{y}} = -\nabla\tilde{V}(\mathbf{y}),$$

where μ *is the* **reduced mass** *given by*

$$\mu = \frac{1}{\frac{1}{m_1} + \frac{1}{m_2}} = \frac{m_1 m_2}{m_1 + m_2}.$$

Thus, when the total momentum of a two-particle system is conserved, the relative position evolves as a one-particle system with "effective" mass μ, while the center of mass moves "trivially," as described in Proposition 2.15.

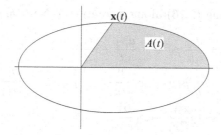

FIGURE 2.1. $A(t)$ is the area of the shaded region.

2.4 Angular Momentum

We start by considering angular momentum in the simplest nontrivial case, motion in $\mathbb{R}^2$.

Definition 2.17 *Consider a particle moving in $\mathbb{R}^2$, having position $\mathbf{x}$, velocity $\mathbf{v}$, and momentum $\mathbf{p} = m\mathbf{v}$. Then the **angular momentum** of the particle, denoted J, is given by*

$$J = x_1 p_2 - x_2 p_1. \tag{2.16}$$

In more geometric terms, $J = |\mathbf{x}|\,|\mathbf{p}|\sin\phi$, where ϕ is the angle (measured counterclockwise) between $\mathbf{x}$ and $\mathbf{p}$. We can look at J in yet another way as follows. If θ is the usual angle in polar coordinates on $\mathbb{R}^2$, then an elementary calculation (Exercise 9) shows that

$$J = mr^2 \frac{d\theta}{dt}. \tag{2.17}$$

It then follows that

$$J = 2m \frac{dA}{dt}, \tag{2.18}$$

where $A = (1/2)\int r^2\,d\theta$ is the area being swept out by the curve $\mathbf{x}(t)$. See Fig. 2.1.

One significant property of the angular momentum is that it (like the energy) is *conserved* in certain situations.

Proposition 2.18 *Suppose a particle of mass m is moving in $\mathbb{R}^2$ under the influence of a conservative force with the potential function $V(\mathbf{x})$. If V is invariant under rotations in $\mathbb{R}^2$, then the angular momentum $J = x_1 p_2 - x_2 p_1$ is independent of time along any solution of Newton's equation. Conversely, if J is independent of time along every solution of Newton's equation, then V is invariant under rotations.*

Proof. Differentiating (2.16) along a solution of Newton's law gives

$$
\begin{aligned}
\frac{dJ}{dt} &= \frac{dx_1}{dt}p_2 + x_1\frac{dp_2}{dt} - \frac{dx_2}{dt}p_1 - x_2\frac{dp_1}{dt} \\
&= \frac{1}{m}p_1p_2 - x_1\frac{\partial V}{\partial x_2} - \frac{1}{m}p_2p_1 + x_2\frac{\partial V}{\partial x_1} \\
&= x_2\frac{\partial V}{\partial x_1} - x_1\frac{\partial V}{\partial x_2}.
\end{aligned}
$$

On the other hand, consider rotations R_θ in $\mathbb{R}^2$ given by

$$
R_\theta = \begin{pmatrix} \cos\theta & -\sin\theta \\ \sin\theta & \cos\theta \end{pmatrix}.
$$

If we differentiate V along this family of rotations, we obtain

$$
\left.\frac{d}{d\theta}V(R_\theta \mathbf{x})\right|_{\theta=0} = \frac{\partial V}{\partial x}\frac{dx}{d\theta} + \frac{\partial V}{\partial y}\frac{dy}{d\theta} = -x_2\frac{\partial V}{\partial x_1} + x_1\frac{\partial V}{\partial x_2} = -\frac{dJ}{dt}(\mathbf{x}).
$$

Thus, the angular derivative of V is zero if and only if J is constant. ∎

Conservation of J [together with the relation (2.18)] gives the following result.

Corollary 2.19 (Kepler's Second Law) *Suppose a particle is moving in $\mathbb{R}^2$ in the presence of a force associated with a rotationally invariant potential. If $\mathbf{x}(t)$ is the trajectory of the particle, then the area swept out by $\mathbf{x}(t)$ between times $t = a$ and $t = b$ is $(b-a)J/(2m)$, where J is the constant value of the angular momentum along the trajectory. Since the area swept out depends only on $b - a$, we may say that "equal areas are swept out in equal times."*

Kepler, of course, was interested in the motion of planets in $\mathbb{R}^3$, not in $\mathbb{R}^2$. The motion of a planet moving in the "inverse square" force of a sun will, however, always lie in a plane. (This claim follows from the three-dimensional version of conservation of angular momentum, as explained in Sect. 2.6.1.)

In $\mathbb{R}^3$, the angular momentum of the particle is a vector, given by

$$
\mathbf{J} = \mathbf{x} \times \mathbf{p}, \tag{2.19}
$$

where $\times$ denotes the cross product (or vector product). Thus, for example,

$$
J_3 = x_1 p_2 - x_2 p_1. \tag{2.20}
$$

If, then, we have a particle in $\mathbb{R}^3$ that just happens to be moving in $\mathbb{R}^2$ (i.e., $x_3 = 0$ and $p_3 = 0$), then the angular momentum will be in the z-direction with z-component given by the quantity J defined in Definition 2.17.

The representation of the angular momentum of a particle in $\mathbb{R}^3$ as a vector is a low-dimensional peculiarity. For a particle in $\mathbb{R}^n$, the angular momentum is a skew-symmetric matrix given by

$$J_{jk} = x_j p_k - x_k p_j. \tag{2.21}$$

In the $\mathbb{R}^3$ case, the entries of the 3×3 angular momentum *matrix* are made up by the three components of the angular momentum *vector* together with their negatives, with zeros along the diagonal. [Compare, e.g., (2.20) and (2.21).]

Definition 2.20 *For a system of N particles moving in $\mathbb{R}^n$, the **total angular momentum** of the system is the skew-symmetric matrix $\mathbf{J}$ given by*

$$J_{jk} = \sum_{l=1}^{N} \left(x_j^l p_k^l - x_k^l p_j^l \right). \tag{2.22}$$

Theorem 2.21 *Suppose a system of N particles in $\mathbb{R}^n$ is moving under the influence of conservative forces with potential function V. If V satisfies*

$$V(R\mathbf{x}^1, R\mathbf{x}^2, \ldots, R\mathbf{x}^N) = V(\mathbf{x}^1, \mathbf{x}^2, \ldots, \mathbf{x}^N) \tag{2.23}$$

for every rotation matrix R, then the total angular momentum of the system is conserved (constant along each trajectory). Conversely, if the total angular momentum is constant along each trajectory, then V satisfies (2.23).

The proof of this result is similar to that of Proposition 2.18 and is left as an exercise (Exercise 10). We will re-examine the concept of angular momentum in the next section using the language of Poisson brackets and Hamiltonian flows.

2.5   Poisson Brackets and Hamiltonian Mechanics

We consider now the Hamiltonian approach to classical mechanics. (There is also the Lagrangian approach, but that approach is not as relevant for our purposes.) The Hamiltonian approach, and in particular the Poisson bracket, will help us to understand the general phenomenon of conserved quantities. The Poisson bracket is also an important source of motivation for the use of commutators in quantum mechanics.

In the Hamiltonian approach to mechanics, we think of the energy function as a function of position and momentum, rather than position and velocity, and we refer to it as the "Hamiltonian." If a particle in $\mathbb{R}^n$ has the usual sort of energy function (kinetic energy plus potential energy), we have

$$H(\mathbf{x}, \mathbf{p}) = \frac{1}{2m} \sum_{j=1}^{n} p_j^2 + V(\mathbf{x}). \tag{2.24}$$

Here, as usual, $p_j = m_j \dot{x}_j$. We now observe that Newton's law can be expressed in the following form:

$$\frac{dx_j}{dt} = \frac{\partial H}{\partial p_j}$$

$$\frac{dp_j}{dt} = -\frac{\partial H}{\partial x_j}. \tag{2.25}$$

After all, with H of the indicated form, these equations read $dx_j/dt = p_j/m$, which is just the definition of p_j, and $dp_j/dt = -\partial V/\partial x_j = F_j$, which is just Newton's law, in the form originally given by Newton. We refer to Newton's law, in the form (2.25) as *Hamilton's equations*.

Although it is not obvious at the moment that we have gained anything by writing Newton's law in the form (2.25), let us proceed on a bit further and see. Our next step is to introduce the *Poisson bracket*.

Definition 2.22 *Let f and g be two smooth functions on $\mathbb{R}^{2n}$, where an element of $\mathbb{R}^{2n}$ is thought of as a pair $(\mathbf{x}, \mathbf{p})$, with $\mathbf{x} \in \mathbb{R}^n$ representing the position of a particle and $\mathbf{p} \in \mathbb{R}^n$ representing the momentum of a particle. Then the **Poisson bracket** of f and g, denoted $\{f, g\}$, is the function on $\mathbb{R}^{2n}$ given by*

$$\{f, g\}(\mathbf{x}, \mathbf{p}) = \sum_{j=1}^{n} \left(\frac{\partial f}{\partial x_j} \frac{\partial g}{\partial p_j} - \frac{\partial f}{\partial p_j} \frac{\partial g}{\partial x_j} \right).$$

The Poisson bracket has the following properties.

Proposition 2.23 *For all smooth functions f, g, and h on $\mathbb{R}^{2n}$ we have the following:*

1. $\{f, g + ch\} = \{f, g\} + c\{f, h\}$ *for all $c \in \mathbb{R}$*

2. $\{g, f\} = -\{f, g\}$

3. $\{f, gh\} = \{f, g\}h + g\{f, h\}$

4. $\{f, \{g, h\}\} = \{\{f, g\}, h\} + \{g, \{f, h\}\}$

Properties 1 and 2 of Proposition 2.23 say that the Poisson bracket is bilinear and skew-symmetric. Property 3 says that the operation of "bracket with f" satisfies the derivation property (similar to the product rule for derivatives) with respect to pointwise multiplication of functions, while Property 4 says that "bracket with f" satisfies the derivation property with respect to the Poisson bracket itself. Property 4 is equivalent to the *Jacobi identity*:

$$\{f, \{g, h\}\} + \{h, \{f, g\}\} + \{g, \{h, f\}\} = 0, \tag{2.26}$$

as may easily be seen using the skew-symmetry of the Poisson bracket. The Jacobi identity, along with bilinearity and skew-symmetry, means that the space of C^∞ functions on $\mathbb{R}^{2n}$ forms a *Lie algebra* under the operation of a Poisson bracket. (See Chap. 16.)

Proof. The first two properties of the Poisson bracket are obvious and the third is an easy consequence of the product rule. Let us think about what goes into proving Property 4 by direct computation. (An alternative proof is given in Exercise 15.) We compute that

$$\{f, \{g, h\}\} = \sum_{j=1}^{n} \frac{\partial f}{\partial x_j} \frac{\partial}{\partial p_j} \left(\frac{\partial g}{\partial x_j} \frac{\partial h}{\partial p_j} - \frac{\partial g}{\partial p_j} \frac{\partial h}{\partial x_j} \right)$$
$$- \sum_{j=1}^{n} \frac{\partial f}{\partial p_j} \frac{\partial}{\partial x_j} \left(\frac{\partial g}{\partial x_j} \frac{\partial h}{\partial p_j} - \frac{\partial g}{\partial p_j} \frac{\partial h}{\partial x_j} \right).$$

Just the first term in the expression for $\{f, \{g, h\}\}$ generates the following four terms (all summed over j) after we use the product rule:

$$\frac{\partial f}{\partial x_j} \frac{\partial^2 g}{\partial x_j \partial p_j} \frac{\partial h}{\partial p_j} + \frac{\partial f}{\partial x_j} \frac{\partial g}{\partial x_j} \frac{\partial^2 h}{\partial p_j^2} - \frac{\partial f}{\partial x_j} \frac{\partial^2 g}{\partial p_j^2} \frac{\partial h}{\partial x_j} - \frac{\partial f}{\partial x_j} \frac{\partial g}{\partial p_j} \frac{\partial^2 h}{\partial x_j \partial p_j}. \quad (2.27)$$

We see, then, that the left-hand side of (2.26) will have a total of 24 terms, each summed over j. Each term will have a single derivative on two of the three functions, and two derivatives on the third function. There are three possibilities for which function gets two derivatives. Once that function is chosen, there are four possibilities for which derivatives go on the other two functions, with the function that gets two derivatives getting whatever derivatives remain (for a total of two x-derivatives and two p-derivatives). That makes 12 possible terms. It is a tedious but straightforward exercise to check that each of these 12 possible terms occurs twice in the left-hand side of (2.26), with opposite signs. To check just one case explicitly, in computing $\{h, \{f, g\}\}$, we will get a term like the second term in (2.27), but with (f, g, h) replaced by (h, f, g):

$$\frac{\partial h}{\partial x_j} \frac{\partial f}{\partial x_j} \frac{\partial^2 g}{\partial p_j^2}.$$

This term (in the computation of $\{h, \{f, g\}\}$) cancels with the third term in (2.27) (in the computation of $\{f, \{g, h\}\}$). ∎

The following elementary result will provide a helpful analogy to the "canonical commutation relations" in quantum mechanics.

Proposition 2.24 *The position and momentum functions satisfy the following Poisson bracket relations:*

$$\{x_j, x_k\} = 0$$
$$\{p_j, p_k\} = 0$$
$$\{x_j, p_k\} = \delta_{jk}.$$

Proof. Direct calculation. ∎

One of the main reasons for considering the Poisson bracket is the following simple result.

Proposition 2.25 *If* $(\mathbf{x}(t), \mathbf{p}(t))$ *is a solution to Hamilton's equation (2.25), then for any smooth function f on $\mathbb{R}^{2n}$ we have*

$$\frac{d}{dt} f(\mathbf{x}(t), \mathbf{p}(t)) = \{f, H\} (\mathbf{x}(t), \mathbf{p}(t)).$$

We generally write Proposition 2.25 in a more concise form as

$$\frac{df}{dt} = \{f, H\},$$

where the time derivative is understood as being along some trajectory.

Proof. Using the chain rule and Hamilton's equations, we have

$$\frac{df}{dt} = \sum_{j=1}^{n} \left(\frac{\partial f}{\partial x_j} \frac{dx_j}{dt} + \frac{\partial f}{\partial p_j} \frac{dp_j}{dt} \right)$$

$$= \sum_{j=1}^{n} \left(\frac{\partial f}{\partial x_j} \frac{\partial H}{\partial p_j} + \frac{\partial f}{\partial p_j} \left(-\frac{\partial H}{\partial x_j} \right) \right)$$

$$= \{f, H\},$$

as claimed. ∎

Observe that Proposition 2.25 includes Hamilton's equations themselves as special cases, by taking $f(x,p) = x_j$ and by taking $f(x,p) = p_j$. Thus, this proposition gives a more coordinate-independent way of expressing the time-evolution.

Corollary 2.26 *Call a smooth function f on $\mathbb{R}^{2n}$ a conserved quantity if $f(\mathbf{x}(t), \mathbf{p}(t))$ is independent of t for each solution $(\mathbf{x}(t), \mathbf{p}(t))$ of Hamilton's equations. Then f is a conserved quantity if and only if*

$$\{f, H\} = 0.$$

In particular, the Hamiltonian H is a conserved quantity.

Conserved quantities are also called *constants of motion*. See Conclusion 2.31 for another perspective on this result. Conserved quantities (when one can find them) are useful in that we know that trajectories must lie in the level surfaces of any conserved quantity. Suppose, for example, that we have a particle moving in $\mathbb{R}^2$ and that the Hamiltonian H and one other independent function f (such as, say, the angular momentum) are conserved quantities. Then, rather than looking for trajectories in the four-dimensional phase space, we look for them inside the joint level sets of H

and f (sets of the form $H(x, p) = a$, $f(x, p) = b$, for some constants a and b). These joint level sets are (generically) two-dimensional instead of four-dimensional, so using the constants of motion greatly simplifies the problem—from an equation in four variables to one in only two variables.

Solving Hamilton's equations on $\mathbb{R}^{2n}$ gives rise to a *flow* on $\mathbb{R}^{2n}$, that is, a family Φ_t of diffeomorphisms of $\mathbb{R}^{2n}$, where $\Phi_t(\mathbf{x}, \mathbf{p})$ is equal to the solution at time t of Hamilton's equations with initial condition $(\mathbf{x}, \mathbf{p})$. Since it is possible (depending on the choice of potential function V) that a particle can escape to infinity in finite time, the maps Φ_t are not necessarily defined on all of $\mathbb{R}^{2n}$, but only on some open subset thereof. If Φ_t *does* happen to be defined on all of $\mathbb{R}^{2n}$ (for all t), then we say that the flow is *complete*.

Theorem 2.27 (Liouville's Theorem) *The flow associated with Hamilton's equations, for an arbitrary Hamiltonian function H, preserves the $(2n)$-dimensional volume measure*

$$dx_1 dx_2 \cdots dx_n dp_1 dp_2 \cdots dp_n.$$

What this means, more precisely, is that if a measurable set E is contained in the domain of Φ_t for some $t \in \mathbb{R}$, then the volume of $\Phi_t(E)$ is equal to the volume of E.

Proof. Hamilton's equations may be written as

$$\frac{d}{dt} \begin{bmatrix} x_1 \\ \vdots \\ x_n \\ p_1 \\ \vdots \\ p_n \end{bmatrix} = \begin{bmatrix} \frac{\partial H}{\partial p_1} \\ \vdots \\ \frac{\partial H}{\partial p_n} \\ -\frac{\partial H}{\partial x_1} \\ \vdots \\ -\frac{\partial H}{\partial x_n} \end{bmatrix}. \tag{2.28}$$

This means that Hamilton's Equations describe the flow along the vector field on $\mathbb{R}^{2n}$ appearing on the right-hand side of (2.28). By a standard result from vector calculus (see, e.g., Proposition 16.33 in [29]), this flow will be volume-preserving if and only if the *divergence* of the vector field is zero. We compute this divergence as

$$\frac{\partial}{\partial x_1} \frac{\partial H}{\partial p_1} + \cdots + \frac{\partial}{\partial x_n} \frac{\partial H}{\partial p_n} - \frac{\partial}{\partial p_1} \frac{\partial H}{\partial x_1} - \cdots - \frac{\partial}{\partial p_n} \frac{\partial H}{\partial x_n}. \tag{2.29}$$

Since

$$\frac{\partial^2 H}{\partial x_j \partial p_j} = \frac{\partial^2 H}{\partial p_j \partial x_j},$$

the divergence is zero. ∎

The existence of an invariant volume has important consequences for the dynamics of a system. For example, for "confined" systems, an invariant volume implies that the system exhibits "recurrence," which means (roughly) that for most initial conditions, the particle will eventually come back arbitrarily close to its initial state in phase space. We will not, however, delve into this aspect of the theory.

Note that the divergence of X_H, computed in (2.29), vanishes in a very particular way, namely the sum of the jth and $(n + j)$th terms vanishes for all $1 \leq j \leq n$. This stronger condition turns out to be equivalent to the condition that the Hamiltonian flow Φ_t associated with an arbitrary smooth function on $\mathbb{R}^{2n}$ preserves the symplectic form ω, defined by

$$\omega((\mathbf{x}, \mathbf{p}), (\mathbf{x}', \mathbf{p}')) = \mathbf{x} \cdot \mathbf{p}' - \mathbf{p} \cdot \mathbf{x}'.$$

What this means, more precisely, is that for any $t \in \mathbb{R}$ and any $(\mathbf{x}, \mathbf{p}) \in \mathbb{R}^{2n}$, the matrix of partial derivatives of Φ_t at the point $(\mathbf{x}, \mathbf{p})$—thought of as a linear map of $\mathbb{R}^{2n}$ to $\mathbb{R}^{2n}$—preserves ω. This property of Φ_t, as it turns out, is equivalent to the property that Φ_t preserves Poisson brackets, meaning that

$$\{f \circ \Phi_t, g \circ \Phi_t\} = \{f, g\} \circ \Phi_t$$

for all $f, g \in C^\infty(\mathbb{R}^n)$. A map $\Psi : \mathbb{R}^{2n} \to \mathbb{R}^{2n}$ that preserves ω is called a *symplectomorphism* (in mathematics notation) or a *canonical transformation* (in physics notation). We defer the proofs of these claims until Chap. 21, where we can consider them in a more general setting.

Definition 2.28 *For any smooth function f on $\mathbb{R}^{2n}$, the **Hamiltonian flow** generated by f is the flow obtained by solving Hamilton's equation (2.25) with the Hamiltonian H replaced by f. The function f is called the **Hamiltonian generator** of the associated flow.*

Although *any* smooth function on $\mathbb{R}^{2n}$ can be inserted into Hamilton's equations to produce a flow, physically one should think that there is a distinguished function, the Hamiltonian H of the system, such that the flow generated by H is the time-evolution of the system. For any other function f, the Hamiltonian flow generated by f should not be thought of as time-evolution, but as some other flow, which might, for example, represent some family of symmetries of our system.

Proposition 2.29 *The Hamiltonian flow generated by the function*

$$f_{\mathbf{a}}(\mathbf{x}, \mathbf{p}) := \mathbf{a} \cdot \mathbf{p} \tag{2.30}$$

is given by

$$\mathbf{x}(t) = \mathbf{x}_0 + t\mathbf{a}$$
$$\mathbf{p}(t) = \mathbf{p}_0, \tag{2.31}$$

and the Hamiltonian flow generated by the function

$$g_{\mathbf{b}}(\mathbf{x}, \mathbf{p}) := \mathbf{b} \cdot \mathbf{x} \tag{2.32}$$

is given by

$$\mathbf{x}(t) = \mathbf{x}_0$$
$$\mathbf{p}(t) = \mathbf{p}_0 - t\mathbf{b}.$$

Proof. Direct calculation. ∎

What this means is that the Hamiltonian flow generated by a linear combination of the *momentum* functions consists of translations in *position* of the particle. That is to say, in the flow (2.31) generated by the function $f_{\mathbf{a}}$ in (2.30), the particle's initial position $\mathbf{x}_0$ is translated by $t\mathbf{a}$ while the particle's momentum is independent of t. Similarly, the Hamiltonian flow generated by a linear combination of the *position* functions [the function $g_{\mathbf{b}}$ in (2.32)] consists of translations in the particle's *momentum*.

Proposition 2.30 *For a particle moving in $\mathbb{R}^2$, the Hamiltonian flow generated by the angular momentum function*

$$J(\mathbf{x}, \mathbf{p}) = x_1 p_2 - x_2 p_1$$

consists of simultaneous rotations of $\mathbf{x}$ and $\mathbf{p}$. That is to say,

$$
\left[\begin{array}{c} x_1(t) \\ x_2(t) \end{array} \right] = \left[\begin{array}{cc} \cos t & -\sin t \\ \sin t & \cos t \end{array} \right] \left[\begin{array}{c} x_1(0) \\ x_2(0) \end{array} \right]
$$
$$
\left[\begin{array}{c} p_1(t) \\ p_2(t) \end{array} \right] = \left[\begin{array}{cc} \cos t & -\sin t \\ \sin t & \cos t \end{array} \right] \left[\begin{array}{c} p_1(0) \\ p_2(0) \end{array} \right]. \tag{2.33}
$$

Proof. If we plug the angular momentum function J into Hamilton's equations in place of H, we obtain

$$\frac{dx_1}{dt} = \frac{\partial J}{\partial p_1} = -x_2; \qquad \frac{dp_1}{dt} = -\frac{\partial J}{\partial x_1} = -p_2$$

$$\frac{dx_2}{dt} = \frac{\partial J}{\partial p_2} = x_1; \qquad \frac{dp_2}{dt} = -\frac{\partial J}{\partial x_2} = p_1$$

The solution to this system is given by the expression in the proposition, as is easily verified by differentiation of (2.33). ∎

Note that since the Hamiltonian flow generated by J does not have the interpretation of the time-evolution of the particle, the parameter t in (2.33) should not be interpreted as the physical time; it is just the parameter in a one-parameter group of diffeomorphisms. In this case, t is the angle of rotation. Thus, one answer to the question, "What *is* the angular momentum?" is that J is the *Hamiltonian generator of rotations*.

If f is any smooth function, then by the proof of Proposition 2.25, the time derivative of any other function g along the Hamiltonian flow generated by f is given by $dg/dt = \{g, f\}$. In particular, the derivative of the Hamiltonian H along the flow generated by f is $\{H, f\}$. Thus, f is constant along the flow generated by H if and only if $\{f, H\} = 0$, which holds if and only if $\{f, H\} = 0$, which holds if and only if H is constant along the flow generated by f. This line of reasoning leads to the following result.

Conclusion 2.31 *A function f is a conserved quantity for solutions of Hamilton's equation (2.25) if and only if H is invariant under the Hamiltonian flow generated by f. In particular, the angular momentum J is conserved if and only if H is invariant under simultaneous rotations of $\mathbf{x}$ and $\mathbf{p}$.*

We will return to this way of thinking about conserved quantities in Chap. 21. Compare Exercise 12.

The Hamiltonian framework can be extended in a straightforward way to systems of particles.

Proposition 2.32 *Consider the phase space for a system of N particles moving in $\mathbb{R}^n$, namely $\mathbb{R}^{2nN}$, thought of as the set of $(2N)$-tuples of the form*

$$(\mathbf{x}^1, \ldots, \mathbf{x}^N, \mathbf{p}^1, \ldots, \mathbf{p}^N)$$

with $\mathbf{x}^j$ and $\mathbf{p}^j$ belonging to $\mathbb{R}^n$. Define the Poisson bracket of two smooth functions f and g on the phase space by

$$\{f, g\} = \sum_{j=1}^N \sum_{k=1}^n \left(\frac{\partial f}{\partial x_k^j} \frac{\partial g}{\partial p_k^j} - \frac{\partial f}{\partial p_k^j} \frac{\partial g}{\partial x_k^j} \right)$$

and consider a Hamiltonian function of the form

$$H(\mathbf{x}^1, \ldots, \mathbf{x}^N, \mathbf{p}^1, \ldots, \mathbf{p}^N) = \sum_{j=1}^N \frac{1}{2m_j} \left| \mathbf{p}^j \right|^2 + V(\mathbf{x}^1, \ldots, \mathbf{x}^N).$$

Then Newton's law in the form $m_j \ddot{\mathbf{x}}^j = -\nabla^j V$ is equivalent to Hamilton's equations in the form

$$\frac{dx_k^j}{dt} = \frac{\partial H}{\partial p_k^j}$$

$$\frac{dp_k^j}{dt} = -\frac{\partial H}{\partial x_k^j}. \tag{2.34}$$

For any smooth function f, the derivative of f along a solution of Hamilton's equations is given by

$$\frac{df}{dt} = \{f, H\}.$$

The proof of these results is entirely similar to the one-particle case and is omitted.

2.6 The Kepler Problem and the Runge–Lenz Vector

2.6.1 The Kepler Problem

We consider now the classical Kepler problem, that of finding the trajectories of a planet orbiting the sun. Since the sun is very much more massive than any of the planets, we may consider the position of the sun to be fixed at the origin of our coordinate system. The sun exerts a force on a planet given by

$$\mathbf{F} = -k \frac{\mathbf{x}}{|\mathbf{x}|^3}. \tag{2.35}$$

Here $k = GmM$, where m is the mass of the planet, M is the mass of the sun, and G is the universal gravitational constant. Note that the magnitude of $\mathbf{F}$ is proportional to the reciprocal of the square of the distance from the origin; thus, the force follows an *inverse square law*. Since k contains a factor of the mass m of the planet, this quantity drops out of the equation of motion, $m\ddot{\mathbf{x}} = \mathbf{F}$. The potential associated with the force (2.35) is easily seen to be

$$V(\mathbf{x}) = -\frac{k}{|\mathbf{x}|}. \tag{2.36}$$

Since our potential V is invariant under rotations, the angular momentum vector $\mathbf{J} = \mathbf{x} \times \mathbf{p}$ is a conserved quantity (Theorem 2.21 with $N = 1$ and $n = 3$). If $\mathbf{J} = 0$, the particle is moving along a ray through the origin. In that case, either the particle will pass through the origin at some point in the future (if the initial momentum points toward the origin), or else the particle must have passed through the origin at some point in the past (if the initial momentum points away from the origin). Trajectories of this sort are called *collision trajectories*, and we will regard such trajectories as pathological.

We will, from now on, consider only trajectories along which the angular momentum vector is nonzero. Fixing the energy and angular momentum of the particle guarantees that the particle stays a certain minimum distance from the origin (Exercise 20). Meanwhile, since $\mathbf{J} = \mathbf{x} \times \mathbf{p}$, the position $\mathbf{x}(t)$ of the particle will always be perpendicular to the constant value of $\mathbf{J}$. We will therefore refer to the plane (through the origin) perpendicular to $\mathbf{J}$ as the "plane of motion."

2.6.2 Conservation of the Runge–Lenz Vector

We are going to obtain a description of the classical trajectories in an indirect way, using something called the Runge–Lenz vector.

Definition 2.33 *The **Runge–Lenz vector** is the vector-valued function on $\mathbb{R}^3 \backslash \{0\} \times \mathbb{R}^3$ given by*

$$\mathbf{A}(\mathbf{x}, \mathbf{p}) = \frac{1}{mk}\mathbf{p} \times \mathbf{J} - \frac{\mathbf{x}}{|\mathbf{x}|}.$$

Here $\mathbf{x}$ represents the position of a classical particle and $\mathbf{p}$ its momentum.

The significance of this vector is that it is a *conserved quantity* for the Kepler problem. Of course, whenever the potential energy is radial (a function of the distance from the origin), the angular momentum vector is a conserved quantity. What is special about the $1/r$ potential of the Kepler problem is that there is another conserved vector-valued quantity.

Proposition 2.34 *The Runge–Lenz vector is conserved quantity for Newton's law with force given by (2.35).*

Proof. Since $\mathbf{J}$ is conserved, we compute that

$$\dot{\mathbf{A}}(t) = \frac{1}{mk}\mathbf{F} \times \mathbf{J} - \frac{1}{|\mathbf{x}|}\frac{\mathbf{p}}{m} + \frac{\mathbf{x}}{|\mathbf{x}|^2}\sum_{j=1}^{3}\frac{\partial |\mathbf{x}|}{\partial x_j}\frac{dx_j}{dt}$$

$$= -\frac{1}{m}\frac{1}{|\mathbf{x}|^3}\mathbf{x} \times (\mathbf{x} \times \mathbf{p}) - \frac{1}{|\mathbf{x}|}\frac{\mathbf{p}}{m} + \frac{\mathbf{x}}{|\mathbf{x}|^2}\sum_{j=1}^{3}\frac{x_j}{|\mathbf{x}|}\frac{p_j}{m}$$

$$= \frac{1}{m}\left(-\frac{1}{|\mathbf{x}|^3}\mathbf{x}(\mathbf{x} \cdot \mathbf{p}) + \frac{1}{|\mathbf{x}|^3}\mathbf{p}(\mathbf{x} \cdot \mathbf{x}) - \frac{\mathbf{p}}{|\mathbf{x}|} + \frac{\mathbf{x}(\mathbf{x} \cdot \mathbf{p})}{|\mathbf{x}|^3}\right)$$

$$= 0.$$

Here we have used the identity $\mathbf{b} \times (\mathbf{c} \times \mathbf{d}) = \mathbf{c}(\mathbf{b} \cdot \mathbf{d}) - \mathbf{d}(\mathbf{b} \cdot \mathbf{c})$, which holds for all vectors $\mathbf{b}, \mathbf{c}, \mathbf{d} \in \mathbb{R}^3$. ∎

2.6.3 Ellipses, Hyperbolas, and Parabolas

We now use the Runge–Lenz vector to determine the trajectories for the Kepler problem.

Proposition 2.35 *The magnitude of the Runge–Lenz vector $\mathbf{A}$ satisfies*

$$|\mathbf{A}|^2 = 1 + \frac{2|\mathbf{J}|^2}{mk^2}E,$$

where $E = |\mathbf{p}|^2/(2m) - k/|\mathbf{x}|$ is the energy of the particle. Furthermore, if $\hat{\mathbf{x}} := \mathbf{x}/|\mathbf{x}|$ is the unit vector in the $\mathbf{x}$-direction, we have

$$\mathbf{A} \cdot \hat{\mathbf{x}} = \frac{|\mathbf{J}|^2}{mk|\mathbf{x}|} - 1 \tag{2.37}$$

for all nonzero $\mathbf{x}$. It follows from (2.37) that

$$|\mathbf{x}| = \frac{|\mathbf{J}|^2}{mk(1 + \mathbf{A} \cdot \hat{\mathbf{x}})}.$$

Note that from (2.37), $\mathbf{A} \cdot \hat{\mathbf{x}} > -1$ for all points $(\mathbf{x}, \mathbf{p})$ with $\mathbf{x} \neq 0$.
Proof. Using the identity $\mathbf{b} \cdot (\mathbf{c} \times \mathbf{d}) = \mathbf{d} \cdot (\mathbf{b} \times \mathbf{c})$, we see that

$$\hat{\mathbf{x}} \cdot (\mathbf{p} \times \mathbf{J}) = \mathbf{J} \cdot (\hat{\mathbf{x}} \times \mathbf{p}) = |\mathbf{J}|^2 / |\mathbf{x}|.$$

Since $\mathbf{J}$ and $\mathbf{p}$ are orthogonal, we get

$$|\mathbf{A}|^2 = \frac{1}{m^2 k^2} |\mathbf{p}|^2 |\mathbf{J}|^2 + 1 - \frac{2}{mk} \hat{\mathbf{x}} \cdot (\mathbf{p} \times \mathbf{J})$$

$$= 1 + \frac{2 |\mathbf{J}|^2}{mk^2} \left(\frac{|\mathbf{p}|^2}{2m} - \frac{k}{|\mathbf{x}|} \right)$$

$$= 1 + \frac{2 |\mathbf{J}|^2}{mk^2} E.$$

Using again the identity for $\mathbf{b} \cdot (\mathbf{c} \times \mathbf{d})$, we next compute that

$$\mathbf{A} \cdot \mathbf{x} = \frac{1}{mk} \mathbf{J} \cdot (\mathbf{x} \times \mathbf{p}) - \frac{\mathbf{x} \cdot \mathbf{x}}{|\mathbf{x}|}$$

$$= \frac{|\mathbf{J}|^2}{mk} - |\mathbf{x}|.$$

We may now divide by $|\mathbf{x}|$ to obtain the desired expression for $\mathbf{A} \cdot \hat{\mathbf{x}}$. It is then straightforward to solve for $|\mathbf{x}|$. ∎

Corollary 2.36 *Choose orthonormal coordinates in the plane of motion so that $\mathbf{A}$ lies along the positive x_1-axis. If r and θ are the polar coordinates associated with this coordinate system, then along each trajectory $(r(t), \theta(t))$, we have*

$$r(t) = \frac{|\mathbf{J}|^2}{mk} \frac{1}{1 + A \cos \theta(t)}, \tag{2.38}$$

where $A = |\mathbf{A}|$.

If $\mathbf{A} = 0$, any orthonormal coordinates can be used.

Proposition 2.37 *If $A := |\mathbf{A}| < 1$, (2.38) is the equation of an ellipse with eccentricity A and with the origin being one focus of the ellipse. If $A > 1$, (2.38) is the equation of a hyperbola, and if $A = 1$, (2.38) is the equation of a parabola.*

The orbit of the particle in the plane of motion is an ellipse if the energy of the particle is negative, a hyperbola if the energy is positive, and a parabola if the energy is zero.

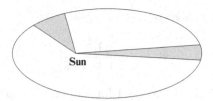

FIGURE 2.2. Elliptical orbit for the Kepler problem, with two equal areas shaded.

Kepler's first law is the assertion that planets move in elliptical trajectories with the sun at one focus, as shown in Fig. 2.2. The shaded regions indicate two equal areas that are swept out in equal times, in accordance with Kepler's second law (Corollary 2.19).

Recall that the eccentricity of an ellipse is $\sqrt{1-(b/a)^2}$, where a is half the length of the major axis and b is half the length of the minor axis. Thus, when $A = 0$, we have $b = a$, meaning that the ellipse is a circle.

Proof. We continue to work in a coordinate system in which $\mathbf{A}$ is along the positive x_1-axis. Then (2.38) becomes

$$\sqrt{x^2 + y^2} = \alpha \frac{1}{1 + A \frac{x}{\sqrt{x^2+y^2}}},$$

where $\alpha = |\mathbf{J}|^2 /(mk)$. From this we obtain

$$1 = \frac{1}{\alpha}\left(\sqrt{x^2 + y^2} + Ax\right).$$

Now we can solve for $\sqrt{x^2 + y^2}$, square both sides of the equation, and simplify. Assuming $A^2 \neq 1$, we obtain

$$\alpha^2\left(\frac{1}{1 - A^2}\right) = (1 - A^2)\left(x + \frac{A\alpha}{1 - A^2}\right)^2 + y^2. \tag{2.39}$$

This is the equation of an ellipse (if $A^2 < 1$) or a hyperbola (if $A^2 > 1$), where the center of the ellipse or hyperbola is the point $(-\alpha/(1 - A^2), 0)$. In light of the formula for $A := |\mathbf{A}|$ in Proposition 2.35, we obtain an ellipse if the energy of the particle is negative and a hyperbola if the energy is positive.

In the case $A^2 < 1$, we may readily compute the half-lengths a and b of the major and minor axes as

$$a = \frac{\alpha}{1 - A^2}; \quad b = \frac{\alpha}{\sqrt{1 - A^2}}.$$

From this, we readily calculate that the eccentricity is A. Now, the distance between the foci of an ellipse is the length of the major axis times the eccentricity, in our case, $2A\alpha/(1 - A^2)$. Since the center of the ellipse in (2.39) is at the point $(A\alpha/(1 - A^2), 0)$, the origin is one focus of the ellipse.

If $A^2 = 1$, then when we perform the same analysis, x^2 drops out of the equation and we obtain

$$x = \frac{1}{2A\alpha}\left(-y^2 + \alpha^2\right)$$

which is the equation of a parabola opening along the x-axis. This case corresponds to energy zero. ∎

Note that Proposition 2.37 does not tell us how the particle moves along the ellipse, hyperbola, or parabola as a function of time. We can, however, determine this, at least in principle, by making use of the angular momentum. After all, applying (2.17) in the plane of motion gives

$$\frac{d\theta}{dt} = \frac{1}{mr^2}\,|\mathbf{J}|, \tag{2.40}$$

where θ is the polar angle variable in the plane of motion. Since we have computed r as a function of θ in Corollary 2.36, (2.40) gives us a (first-order, separable) differential equation, from which we can attempt to solve to obtain θ—and thus also r—as a function of t.

2.6.4 Special Properties of the Kepler Problem

As we have said, the existence of another conserved vector-valued function—in addition to the conserved energy and angular momentum—is special to a potential of the form $-k/|\mathbf{x}|$. For a general radial potential, the energy and the angular momentum will be the only conserved quantities. Assuming $\mathbf{J} \neq 0$, the motion of a particle in any radial potential will always lie in the plane perpendicular to $\mathbf{J}$. Taking this into account, we think of our particle

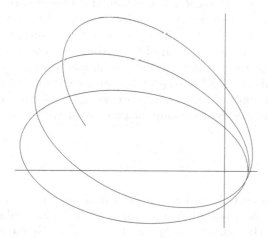

FIGURE 2.3. Trajectory in the plane of motion for a typical radial potential.

as moving in $\mathbb{R}^2$ rather than $\mathbb{R}^3$, and accordingly think of our phase space as being four-dimensional rather than six-dimensional. From this point of view, there are two remaining conserved quantities, the energy E and the scalar angular momentum J in the plane, as given by Definition 2.17. Thus, each trajectory will lie in a set of the form

$$\left\{ (\mathbf{x}, \mathbf{p}) \in \mathbb{R}^2 \times \mathbb{R}^2 \,\middle|\, E(\mathbf{x}, \mathbf{p}) = a, \; J(\mathbf{x}, \mathbf{p}) = b \right\}.$$

We refer to such a set as a *joint level set* of E and J. These sets are two-dimensional surfaces inside our four-dimensional phase space.

For a general radial potential, a trajectory $(\mathbf{x}(t), \mathbf{p}(t))$ in phase space may not be a closed curve, but may fill up a dense subset of the joint level surface on which it lives. In particular, the trajectory $\mathbf{x}(t)$ in position space will typically not be a closed curve. For example, $\mathbf{x}(t)$ may trace out a roughly elliptical region in the plane, but where the axes of the ellipse "precess," that is, vary with time. Such a trajectory is shown in Fig. 2.3, which should be contrasted with Fig. 2.2.

In the Kepler problem, even after restricting attention to the plane of motion, we still have one conserved quantity in addition to E and J, namely the direction of $\mathbf{A}$, which can be expressed in terms of the angle ϕ between $\mathbf{A}$ and the x_1-axis in the plane of motion. (Note that both terms in the definition of $\mathbf{A}$ lie in the plane of motion. Note also that the magnitude of $\mathbf{A}$ is, by Proposition 2.35, computable in terms of E and J.) The trajectories of the Kepler problem, then, lie in the joint level sets of E and J and ϕ, which are one-dimensional. When $E < 0$, the joint level sets of E and J are compact, in which case the joint level sets of E and J and ϕ are compact and one-dimensional, that is, simple closed curves.

Another special property of the Kepler problem is that the period of the closed trajectories (the trajectories with negative energy) is the same for all trajectories with the same energy (Exercise 21). This apparent coincidence can be explained by showing that the Hamiltonian flows (Definition 2.28) generated by $\mathbf{J}$ and $\mathbf{A}$ act transitively on the energy surfaces. These flows commute with the time evolution of the system, because they are all conserved quantities (Conclusion 2.31). Thus, any two points with the same energy are "equivalent" with respect to time evolution. Although we will not go into the details of this analysis, we will gain a better understanding of the flows generated by the components of $\mathbf{A}$ in Sect. 18.4.

2.7 Exercises

1. Consider a particle moving in the real line in the presence of a force coming from a potential function V. Given some value E_0 for the energy of the particle, suppose that $V(x) < E_0$ for all x in some closed interval $[x_0, x_1]$. Then a particle with initial position x_0 and

positive initial velocity will continue to move to the right until it reaches x_1. Using (2.6), show that the time needed to travel from x_0 to x_1 is given by

$$t = \int_{x_0}^{x_1} \sqrt{\frac{m}{2(E_0 - V(y))}} \, dy.$$

Note: This shows that we can solve Newton's equation in $\mathbb{R}^1$ more or less explicitly for time as a function of position, which in principle determines the position as a function of time.

2. In the notation of the previous problem, suppose now that $V(x) < E_0$ for $x_0 \leq x < x_1$, but that $V(x_1) = E_0$.

   (a) Show that if $V'(x_1) \neq 0$, then the particle reaches x_1 in a finite time.

   (b) Show that if $V'(x_1) = 0$, then the time it takes the particle to reach x_1 is infinite; that is, the particle approaches but never actual reaches x_1.

   Note: In Part (b), the point x_1 is an unstable equilibrium for the system, that is, a critical point for V that is not a local minimum.

3. Consider the equation of motion of a pendulum of length L,

$$\frac{d^2\theta}{dt^2} + \frac{g}{L} \sin \theta = 0,$$

where g is the acceleration of gravity. Here θ is the angle between the pendulum and the negative y-axis in the plane. This system has a stable equilibrium at $\theta = 0$ and an unstable equilibrium at $\theta = \pi$.

Consider initial conditions of the form $\theta(0) = \pi - \delta$, $\dot{\theta}(0) = 0$, for $0 < \delta < \pi/4$. Fix some angle θ_0 and let $T(\delta)$ denote the time it takes for the pendulum with the given initial conditions to reach the angle θ_0. (Here θ_0 represents an arbitrarily chosen cutoff point at which the pendulum is no longer "close" to $\theta = \pi$.) Show that $T(\delta)$ grows only logarithmically as δ tends to zero.

Note: Logarithmic growth of T as a function of δ corresponds to exponential decay of δ as a function of T. Thus, if we want T to be large, we must choose δ to be *very* small.

4. Consider a particle moving in the real line in the presence of a "repelling potential," such that there is an A with $V'(x) < 0$ for all $x > A$. Then a particle with initial position $x_0 > A$ and positive initial velocity will have positive velocity for all positive times. Suppose now that $V(x) = -x^a$ for all $x > 1$, for some positive constant

a. Suppose also that the particle is given initial position $x_0 > 1$ and positive initial velocity. Show that for $a > 2$, the particle escapes to infinity in finite time, but that for $a \leq 2$, the position of the particle remains finite for all finite times.

Hint: Use Problem 1.

5. Consider the equation $m\ddot{x} + \gamma\dot{x} + kx = 0$, where γ and k are positive constants (the damping constant and spring constant, respectively). Find the critical value γ_c of γ (for a fixed m and k) such that for $\gamma < \gamma_c$, we get solutions that are sines and cosines times a decaying exponential and for $\gamma > \gamma_c$, we get pure decaying exponentials.

6. Continue with the notation of Exercise 5. Given particular choices for m, γ, and k, let r be the rate of exponential decay of a "generic" solution to the equation of motion. Here, if the solution is of the form $ae^{-rt}\cos(\omega t) + be^{-rt}\sin(\omega t)$, the rate of exponential decay is r. If the solution is of the form $ae^{-r_1 t} + be^{-r_2 t}$, then $r = \min(r_1, r_2)$, since the slower-decaying term will dominate as long as a and b are both nonzero.

   For a fixed value of m and k, show that the maximum value for r is achieved by taking $\gamma = \gamma_c$. (This accounts for the terminology "critical damping" for the case in which $\gamma = \gamma_c$.)

7. Consider the $\mathbb{R}^2$-valued function $\mathbf{F}$ on $\mathbb{R}^2 \setminus \{0\}$ given by

$$\mathbf{F}(x_1, x_2) = \left(-\frac{x_2}{x_1^2 + x_2^2}, \frac{x_1}{x_1^2 + x_2^2} \right).$$

   Show that $\partial F_1 / \partial x_2 - \partial F_2 / \partial x_1 = 0$ but that there does not exist any smooth function V on $\mathbb{R}^2 \setminus \{0\}$ with $\mathbf{F} = -\nabla V$.

   Hint: If $\mathbf{F}$ were of the form $-\nabla V$, we would have

$$V(\mathbf{x}(b)) - V(\mathbf{x}(a)) = -\int_a^b \mathbf{F}(\mathbf{x}(t)) \cdot \frac{d\mathbf{x}}{dt}\, dt$$

   for every smooth path $\mathbf{x}(\cdot) : [a, b] \to \mathbb{R}^2 \setminus \{0\}$, by the fundamental theorem of calculus and the chain rule.

8. Consider a particle moving in $\mathbb{R}^n$ with a velocity-dependent force law given by
$$\mathbf{F}(\mathbf{x}, \mathbf{v}) = -\nabla V(\mathbf{x}) + \mathbf{F}_2(\mathbf{x}, \mathbf{v}),$$

   where the velocity-dependent term $\mathbf{F}_2$ acts perpendicularly to the velocity of the particle. (That is, we assume that $\mathbf{v} \cdot \mathbf{F}_2(\mathbf{x}, \mathbf{v}) = 0$ for all $\mathbf{x}$ and $\mathbf{v}$.) Let E denote the usual energy function $E(\mathbf{x}, \mathbf{v}) = \frac{1}{2}m|\mathbf{v}|^2 + V(\mathbf{x})$, unmodified by the presence of the velocity-dependent term in the force. Show that E is conserved.

9. (a) If r and θ are the usual polar coordinates on $\mathbb{R}^2$, compute $\partial\theta/\partial x_1$ and $\partial\theta/\partial x_2$.

    (b) If $\mathbf{x}(\cdot)$ denotes the trajectory of a particle of mass m moving in $\mathbb{R}^2$, show that

$$\frac{d}{dt}\theta(\mathbf{x}(t)) = \frac{1}{mr^2}J(\mathbf{x}(t), \mathbf{p}(t)).$$

10. Prove Theorem 2.21, by imitating the proof of Proposition 2.18. You may assume that every rotation can be built up as a product of repeated rotations in the various coordinate planes (i.e., rotations in the (x_j, x_k) plane, for various pairs (j, k), where the same plane may be used more than once).

11. Consider Hamilton's equations for N particles moving in $\mathbb{R}^n$, as in Proposition 2.32. Show that the total momentum $\mathbf{p} = \sum_{j=1}^N \mathbf{p}^j$ of the system is a conserved quantity if and only if the quantity

$$H(\mathbf{x}^1 + \mathbf{a}, \ldots, \mathbf{x}^N + \mathbf{a}, \mathbf{p}^1 + \mathbf{a}, \ldots, \mathbf{p}^N + \mathbf{a}), \quad \mathbf{a} \in \mathbb{R}^n,$$

    is independent of $\mathbf{a}$ for all $\mathbf{x}^1, \ldots, \mathbf{x}^N$ and $\mathbf{p}^1, \ldots, \mathbf{p}^N$ in $\mathbb{R}^n$.

    Hint: Use (the N-particle version of) Conclusion 2.31.

12. Let J denote the angular momentum of a particle moving in $\mathbb{R}^2$. Let R_θ denote a counterclockwise rotation by angle θ in $\mathbb{R}^2$.

    (a) If f is any smooth function on $\mathbb{R}^4$, show that

$$\{f, J\}(\mathbf{x}, \mathbf{p}) = \frac{d}{d\theta}f(R_\theta\mathbf{x}, R_\theta\mathbf{p})\Big|_{\theta=0}.$$

    (b) Let H be any smooth function on $\mathbb{R}^4$ and consider Hamilton's equations with this function playing the role of the Hamiltonian. Show that J is conserved (i.e., constant in time along any solution of Hamilton's equations) if and only if

$$H(R_\theta\mathbf{x}, R_\theta\mathbf{p}) = H(\mathbf{x}, \mathbf{p})$$

    for all θ in $\mathbb{R}$ and all $\mathbf{x}$ and $\mathbf{p}$ in $\mathbb{R}^2$. (This argument is a more explicit way to obtain Conclusion 2.31.)

13. Suppose that f and g are smooth functions on $\mathbb{R}^{2n}$ and that at least one of the two functions has compact support. Show that

$$\int_{\mathbb{R}^n}\int_{\mathbb{R}^n} \{f, g\}(\mathbf{x}, \mathbf{p})\, d^n\mathbf{x}\, d^n\mathbf{p} = 0.$$

Hint: Use integration by parts or Liouville's theorem.

14. Let X and Y be "vector fields" on $\mathbb{R}^n$, viewed as first-order differential operators. This means that X and Y are of the form

$$X = \sum_{j=1}^{n} a_j(\mathbf{x}) \frac{\partial}{\partial x_j}; \quad Y = \sum_{j=1}^{n} b_j(\mathbf{x}) \frac{\partial}{\partial x_j}.$$

[If $\tilde{X}(\mathbf{x}) = (a_1(\mathbf{x}), \dots, a_n(\mathbf{x}))$, then the operator X is the directional derivative in the direction of $\tilde{X}$. It is common to identify the vector-valued function $\tilde{X}$ with the associated first-order differential operator X.]

Show that the commutator $[X, Y]$ of X and Y, defined by

$$[X, Y] = XY - YX$$

is again a vector field (i.e., a *first*-order differential operator).

15. Given a smooth function f on $\mathbb{R}^{2n}$, define an operator X_f, acting on $C^\infty(\mathbb{R}^{2n})$, by the formula

$$X_f(g) = \{f, g\}.$$

That is to say,

$$X_f = \sum_{j=1}^{n} \left(\frac{\partial f}{\partial x_j} \frac{\partial}{\partial p_j} - \frac{\partial f}{\partial p_j} \frac{\partial}{\partial x_j} \right).$$

The operator X_f is called the *Hamiltonian vector field* associated with the function f. (Here, as in Exercise 14, we identify vector fields with first-order differential operators.)

(a) Show that for all $f, g \in C^\infty(\mathbb{R}^{2n})$, we have

$$X_{\{f,g\}} = [X_f, X_g],$$

where $[X_f, X_g] = X_f X_g - X_g X_f$.

Hint: By Exercise 14, all terms in the computation of $[X_f, X_g](h)$ involving second derivatives of h can be neglected, since they will always cancel out to zero.

(b) Use Part (a) to compute $\{\{f, g\}, h\} = X_{\{f,g\}}(h)$ and thereby obtain another proof of the Jacobi identity for the Poisson bracket.

16. Recall the definition of a Hamiltonian vector field X_f in Exercise 15.

(a) Consider a smooth vector field X on $\mathbb{R}^2$ (viewed as a first-order differential operator as in Exercise 14) of the form

$$X(\mathbf{x}) = g_1(x, p) \frac{\partial}{\partial x} + g_2(x, p) \frac{\partial}{\partial p}.$$

Show that X can be expressed as $X = X_f$, for some $f \in C^\infty(\mathbb{R}^2)$, if and only X is *divergence free*, that is, if and only if

$$\nabla \cdot X := \frac{\partial g_1}{\partial x} + \frac{\partial g_2}{\partial p} = 0.$$

Hint: As in Proposition 2.7, given a pair of functions h_1 and h_2 on $\mathbb{R}^2$, there exists a function f with $\partial f/\partial x = h_1$ and $\partial f/\partial p = h_2$ if and only if we have $\partial h_1/\partial p = \partial h_2/\partial x$.

(b) Show that there exists a smooth vector field X on $\mathbb{R}^4$ of the form

$$X = \sum_{j=1}^{2} \left(g_j(\mathbf{x})\frac{\partial}{\partial x_j} + g_{j+2}(\mathbf{x})\frac{\partial}{\partial p_j} \right)$$

such that

$$\nabla \cdot X := \sum_{j=1}^{2} \left(\frac{\partial g_j}{\partial x_j} + \frac{\partial g_{j+2}}{\partial p_j} \right) = 0$$

but such that there does not exist $f \in C^\infty(\mathbb{R}^4)$ with $X = X_f$.
Hint: You should be able to find a counterexample in which the coefficient functions g_j are linear.

17. Show that the space of homogeneous polynomials of degree 2 on $\mathbb{R}^{2n}$ is closed under the Poisson bracket.

18. Determine the Hamiltonian flow on $\mathbb{R}^2$ generated by the function $f(x, p) = xp$.

19. Let $\mathbf{J}$ denote the angular momentum vector for a particle moving in $\mathbb{R}^3$, namely $\mathbf{J} = \mathbf{x} \times \mathbf{p}$. Show that the components J_1, J_2, and J_3 of $\mathbf{J}$ satisfy the following Poisson bracket relations:

$$\{J_1, J_2\} - J_3; \quad \{J_2, J_3\} - J_1; \quad \{J_3, J_1\} = J_2.$$

20. In the Kepler problem, show that for each real number E and positive number J, there exists $\varepsilon > 0$ such that for all $(\mathbf{x}, \mathbf{p})$ with $E(\mathbf{x}, \mathbf{p}) = E$ and $|\mathbf{J}(\mathbf{x}, \mathbf{p})| = J$, we have $|\mathbf{x}| \geq \varepsilon$.

Hint: Suppose that $(\mathbf{x}_n, \mathbf{p}_n)$ is a sequence with $|\mathbf{J}(\mathbf{x}_n, \mathbf{p}_n)| = J$ and $|\mathbf{x}_n|$ tending to zero. Show that $E(\mathbf{x}_n, \mathbf{p}_n)$ tends to $+\infty$.

21. (a) Determine the area of the ellipse in the plane of motion in Proposition 2.37, in the case $A < 1$.

    (b) Show that the time T it takes the particle to travel once around the ellipse is given by

$$\frac{\pi}{\sqrt{2}}GM(-\tilde{E})^{-3/2},$$

where $\tilde{E}$ is the "massless energy" of the particle, given by

$$\tilde{E} = \frac{E}{m} = \frac{1}{2}|\dot{\mathbf{x}}| - \frac{GM}{|\mathbf{x}|}.$$

Note in the case where the trajectory in the plane of motion is elliptical, the energy of the particle is negative.

Note: The result of Part (b) is closely related to Kepler's third law.

3

A First Approach to Quantum Mechanics

In this chapter, we try to understand the main ideas of quantum mechanics. In quantum mechanics, the outcome of a measurement cannot—even in principle—be predicted beforehand; only the *probabilities* for the outcome of the measurement can be predicted. These probabilities are encoded in a *wave function*, which is a function of a *position* variable $\mathbf{x} \in \mathbb{R}^n$. The square of the absolute value of the wave function encodes the probabilities for the position of the particle. Meanwhile, the probabilities for the momentum of the particle are encoded in the *frequency of oscillation* of the wave function. The probabilities can be described using the *position operator* and the *momentum operator*. The time-evolution of the wave function is described by the *Hamiltonian operator*, which is analogous to the Hamiltonian (or energy) function in Hamilton's equations.

3.1 Waves, Particles, and Probabilities

There are two key ingredients to quantum theory, both of which arose from experiments. The first ingredient is *wave–particle duality,* in which objects are observed to have both wavelike and particlelike behavior. Light, for example, was thought to be a wave throughout much of the nineteenth century, but was observed in the early twentieth century to have particle behavior as well. Electrons, meanwhile, were originally thought to be particles, but were then observed to have wave behavior.

B.C. Hall, *Quantum Theory for Mathematicians*, Graduate Texts in Mathematics 267, DOI 10.1007/978-1-4614-7116-5_3, © Springer Science+Business Media New York 2013

The second ingredient of quantum theory is its probabilistic behavior. In the two-slit experiment, for example, electrons that are "identically prepared" do not all hit the screen at the same point. Quantum theory postulates that this randomness is fundamental to the way nature behaves. According to quantum mechanics, it is impossible (theoretically, not just in practice) to predict ahead of time what the outcome of an experiment will be. The best that can be done is to predict the *probabilities* for the outcome of an experiment.

These two aspects of quantum theory come together in the wave function. The wave function is a function of a variable $\mathbf{x} \in \mathbb{R}^n$, which we interpret as describing the possible values of the position of a particle, and it evolves in time according to a wavelike equation (the Schrödinger equation). The wave function and its time-evolution account for the wave aspect of quantum theory. The particle aspect of the theory comes from the *interpretation* of the wave function. Although it is tempting to interpret the wave function as a sort of cloud, where we have, say, a little bit of electron-cloud over here, and little bit of electron-cloud over there, this interpretation is not consistent with experiment. Whenever we attempt to measure the position of a *single* electron, we always find the electron at a single point. A single electron in the two-slit experiment is observed at a single point on the screen, not spread out over the screen the way the wave function is. The wave function does not describe something that is directly observable for a single particle; rather, the wave function determines the statistical behavior of a whole sequence of identically prepared particles. See Fig. 1.4 for a dramatic experimental demonstration of this effect.

In the two-slit experiment, for example, it is possible to determine how the wave function behaves as a function of time by solving the (deterministic) Schrödinger equation. Knowledge of the wave function of an individual electron, however, does not determine where that electron will hit the screen. The wave function merely tells us the probability distribution for where the electron might hit the screen, something that is only observable by shooting a whole sequence of electrons at the screen.

It is an oversimplification, but a useful one, to describe the wave–particle aspect of quantum theory in this way: a single electron (or photon, or whatever) acts like a particle, but a large collection of electrons behaves like a wave. A single measurement of a single electron always gives its position as a point, just as we would expect for a particle. This point, however, varies from one electron to the next, even if we shoot each electron toward the screen in precisely the same way. Repeated measurements of identically prepared electrons give a distribution that can, for example, exhibit interference patterns, just as we would expect for a wave. See, again, Fig. 1.4, which should be compared to Figs. 1.1 and 1.2.

It is interesting to note that at the macroscopic scale, where quantum effects are not apparent, light appears to be a wave, whereas electrons appear to be particles. This is the case even though both light and electrons are really wave–particle hybrids, described in probabilistic terms by a wave function. The difference between the two situations is that photons (the particles of light) have mass zero, whereas electrons have positive mass. This means that photons, unlike electrons, can easily be created and destroyed even at low energies. Thus, the discrete aspect of light—namely, that the energy in light comes only in discrete "quanta," namely the photons—is less evident than the corresponding discrete aspect of electrons.

3.2 A Few Words About Operators and Their Adjoints

In quantum mechanics, physical quantities—such as position, momentum, and energy—are represented by operators on a certain Hilbert space $\mathbf{H}$. These operators are unbounded operators, reflecting that in classical mechanics, these quantities are unbounded functions on the classical phase space. In this section, we look briefly at some technical issues related to unbounded operators and their adjoints. We will delay a full discussion of these technicalities (Chap. 9) until after we have understood the basic ideas of quantum mechanics.

Here and throughout the book, $\mathbf{H}$ will represent a Hilbert space over $\mathbb{C}$, always assumed to be separable. We follow the convention in the physics literature that the inner product be linear in the *second* factor:

$$\langle \phi, \lambda\psi \rangle = \lambda \langle \phi, \psi \rangle ; \quad \langle \lambda\phi, \psi \rangle = \bar{\lambda} \langle \phi, \psi \rangle$$

for all $\phi, \psi \in \mathbf{H}$ and all $\lambda \in \mathbb{C}$.

Recall (Appendix A.3.4) that a linear operator $A : \mathbf{H} \to \mathbf{H}$ is *bounded* if there is a constant C such that $\|A\psi\| \leq C \|\psi\|$ for all $\psi \in \mathbf{H}$. For any bounded operator A, there is a unique bounded operator A^*, called the *adjoint* of A, such that

$$\langle \phi, A\psi \rangle = \langle A^*\phi, \psi \rangle$$

for all $\phi, \psi \in \mathbf{H}$. The existence of A^* follows from the Riesz theorem (Appendix A.4.3), by observing that for each fixed ϕ, the map $\psi \mapsto \langle \phi, A\psi \rangle$ is a bounded linear functional on $\mathbf{H}$. A bounded operator is said to be *self-adjoint* if $A^* = A$.

For various reasons, both physical and mathematical, we want the operators of quantum mechanics operators to be self-adjoint. Once one sees the formulas for these operators, however, one is confronted with a serious technical difficulty: the operators are not bounded.

If A is a linear operator defined on all of $\mathbf{H}$ and having the property that $\langle \phi, A\psi \rangle = \langle A\phi, \psi \rangle$ for all $\phi, \psi \in \mathbf{H}$, then A is automatically bounded. (See Corollary 9.9.) To put this fact the other way around, *an unbounded self-adjoint operator cannot be defined on the entire Hilbert space.* Thus, to deal with the unbounded operators of quantum mechanics, we must deal with operators that are defined only on a subspace of the relevant Hilbert space, called the **domain** of the operator.

Definition 3.1 *An **unbounded operator** A on $\mathbf{H}$ is a linear map from a dense subspace $\mathrm{Dom}(A) \subset \mathbf{H}$ into $\mathbf{H}$.*

More precisely, the operator A is "not necessarily bounded," since nothing in the definition prevents us from having $\mathrm{Dom}(A) = \mathbf{H}$ and having A be bounded.

In defining the adjoint of an unbounded operator, we immediately encounter a difficulty: for a given $\phi \in \mathbf{H}$, the linear functional $\langle \phi, A \cdot \rangle$ may not be bounded, in which case we cannot use the Riesz theorem to define $A^* \phi$. What this means is that the adjoint of A, like A itself, will be defined not on all of $\mathbf{H}$ but only on some subspace thereof.

Definition 3.2 *For an unbounded operator A on $\mathbf{H}$, the **adjoint** A^* of A is defined as follows. A vector $\phi \in \mathbf{H}$ belongs to the domain $\mathrm{Dom}(A^*)$ of A^* if the linear functional*

$$\langle \phi, A \cdot \rangle ,$$

defined on $\mathrm{Dom}(A)$, is bounded. For $\phi \in \mathrm{Dom}(A^*)$, let $A^* \phi$ be the unique vector χ such that

$$\langle \chi, \psi \rangle = \langle \phi, A\psi \rangle$$

for all $\psi \in \mathrm{Dom}(A)$.

Saying that the linear functional $\langle \phi, A \cdot \rangle$ is bounded means that there is a constant C such that $|\langle \phi, A\psi \rangle| \leq C \, \|\psi\|$ for all $\psi \in \mathrm{Dom}(A)$. If $\langle \phi, A \cdot \rangle$ is bounded, then since $\mathrm{Dom}(A)$ is dense, the BLT theorem (Theorem A.36) tells us that $\langle \phi, A \cdot \rangle$ has a unique bounded extension to all of $\mathbf{H}$. The Riesz theorem then guarantees the existence and uniqueness of χ. The adjoint of an unbounded linear operator is a linear operator on its domain.

We are now ready to define self-adjointness (and some related notions) for unbounded operators.

Definition 3.3 *An unbounded operator A on $\mathbf{H}$ is **symmetric** if*

$$\langle \phi, A\psi \rangle = \langle A\phi, \psi \rangle$$

*for all $\phi, \psi \in \mathrm{Dom}(A)$. The operator A is **self-adjoint** if $\mathrm{Dom}(A^*) = \mathrm{Dom}(A)$ and $A^* \phi = A\phi$ for all $\phi \in \mathrm{Dom}(A)$. Finally, A is **essentially self-adjoint** if the closure in $\mathbf{H} \times \mathbf{H}$ of the graph of A is the graph of a self-adjoint operator.*

That is to say, A is self-adjoint if A^* and A are the *same operator* with the *same domain*. Every self-adjoint or essentially self-adjoint operator is symmetric, but not every symmetric operator is essentially self-adjoint. For any symmetric operator, $\mathrm{Dom}(A^*) \supset \mathrm{Dom}(A)$ and A^* agrees with A on $\mathrm{Dom}(A)$. The reason a symmetric operator may fail to be self-adjoint is that $\mathrm{Dom}(A^*)$ may be strictly larger than $\mathrm{Dom}(A)$.

Although the condition of being symmetric is certainly easier to understand (and to verify) than the condition of being self-adjoint, self-adjointness is the "right" condition. In particular, the spectral theorem, which is essential to much of quantum mechanics, applies only to operators that are self-adjoint and not to operators that are merely symmetric. If A is essentially self-adjoint, then we can obtain a self-adjoint operator from A simply by taking the closure of the graph of A, and we can then apply the spectral theorem to this self-adjoint operator. Thus, for may purposes, it is enough to have our operators be essentially self-adjoint rather than self-adjoint.

It is generally easy to verify that the operators of quantum mechanics (those representing position, momentum, and so forth) are symmetric on some suitably chosen domain. Proving that these operators are essentially self-adjoint, however, is substantially more difficult. Although establishing essential self-adjointness is a crucial technical issue, it is best not to worry too much about it on a first encounter with quantum mechanics. In this chapter, we will not concern ourselves overly with technical details concerning essential self-adjointness and the precise choice of domain for our operators, depending on Chap. 9 to take care of such matters. For now, we content ourselves with deriving some very elementary properties of symmetric (and thus also self-adjoint) operators.

Proposition 3.4 *Suppose A is a symmetric operator on* **H**.

1. *For all $\psi \in \mathrm{Dom}(A)$, the quantity $\langle \psi, A\psi \rangle$ is real. More generally, if $\psi, A\psi, \ldots, A^{m-1}\psi$ all belong to $\mathrm{Dom}(A)$, then $\langle \psi, A^m\psi \rangle$ is real.*

2. *Suppose λ is an eigenvector for A, meaning that $A\psi = \lambda\psi$ for some nonzero $\psi \in \mathrm{Dom}(A)$. Then $\lambda \in \mathbb{R}$.*

Proof. Since A is symmetric, we have

$$\langle \psi, A\psi \rangle = \langle A\psi, \psi \rangle = \overline{\langle \psi, A\psi \rangle}$$

for all $\psi \in \mathrm{Dom}(A)$. If $\psi, A\psi, \ldots, A^{m-1}\psi$ all belong to the domain of A, we can use the symmetry of A repeatedly to show that

$$\langle \psi, A^m\psi \rangle = \langle A^m\psi, \psi \rangle = \overline{\langle \psi, A^m\psi \rangle}.$$

Meanwhile, if ψ is an eigenvector for A with eigenvalue λ, then

$$\lambda \langle \psi, \psi \rangle = \langle \psi, A\psi \rangle = \langle A\psi, \psi \rangle = \bar{\lambda} \langle \psi, \psi \rangle.$$

Since ψ is assumed to be nonzero, this implies that $\lambda = \bar{\lambda}$. ∎

Physically, $\langle \psi, A\psi \rangle$ represents—as we will see later in this chapter—the expectation value for measurements of A in the state ψ, whereas the eigenvalue λ represents one of the possible values for this measurement. On physical grounds, we want both of these numbers to be real. If A is self-adjoint, and not just symmetric, then the spectral theorem will give a canonical way of associating to each $\psi \in \mathbf{H}$ a probability measure on the real line that encodes the probabilities for measurements of A in the state ψ.

3.3   Position and the Position Operator

Let us consider at first a single particle moving on the real line. The wave function for such a particle is a map $\psi : \mathbb{R}^1 \to \mathbb{C}$. Although this map will evolve in time, let us think for now that the time is fixed. The function $|\psi(x)|^2$ is supposed to be the probability density for the position of the particle. This means that the probability that the position of the particle belongs to some set $E \subset \mathbb{R}^1$ is

$$\int_E |\psi(x)|^2 \, dx.$$

For this prescription to make sense, ψ should be normalized so that

$$\int_{\mathbb{R}} |\psi(x)|^2 \, dx = 1. \tag{3.1}$$

That is, ψ should be a unit vector in the Hilbert space $L^2(\mathbb{R})$.

Now, if the function $|\psi(x)|^2$ is the probability density for the position of a particle, then according to the standard definitions of probability theory, the *expectation value* of the position will be

$$E(x) = \int_{\mathbb{R}} x \, |\psi(x)|^2 \, dx, \tag{3.2}$$

provided that the integral is absolutely convergent. More generally, we can compute any *moment* of the position (i.e., the expectation value of some power of the position) as

$$E(x^m) = \int_{\mathbb{R}} x^m \, |\psi(x)|^2 \, dx, \tag{3.3}$$

assuming, again, the convergence of the integral.

A key idea in quantum theory is to express expectation values of various quantities (position, momentum, energy, etc.) in terms of *operators* and the *inner product* on the relevant Hilbert space, in this case, $L^2(\mathbb{R})$. In the case of position, we may introduce the *position operator* X defined by

$$(X\psi)(x) = x\psi(x).$$

That is, X is the "multiplication by x" operator. The point of introducing this operator is that the expectation value of the position [defined in (3.2)] may now be expressed as

$$E(x) = \langle \psi, X\psi \rangle,$$

where the inner product is the usual one on $L^2(\mathbb{R})$:

$$\langle \phi, \psi \rangle = \int \overline{\phi(x)}\psi(x) \, dx.$$

(Recall that we are following the physics convention of putting the conjugate on the *first* factor in the inner product.)

We use the following notation for the expectation value of the operator X in the state ψ:

$$\langle X \rangle_\psi := \langle \psi, X\psi \rangle.$$

The higher moments of the position, as defined in (3.3), are also computable in terms of the position operator:

$$E(x^m) = \langle \psi, X^m\psi \rangle.$$

At this point, it is not clear that we have gained anything by writing our moments in terms of an operator and the inner product instead of in terms of the integral (3.3). The operator description will, however, motivate a parallel description of moments for the momentum, energy, or angular momentum of a particle in terms of corresponding operators.

It should be noted that, for a given $\psi \in L^2(\mathbb{R})$, $X\psi$ might fail to be in $L^2(\mathbb{R})$. This failure of X to be defined on all of our Hilbert space reflects that X is an unbounded operator, something that we discussed briefly in Sect. 3.2. Even if $X\psi$ is in $L^2(\mathbb{R})$, $X^m\psi$ might fail to be in $L^2(\mathbb{R})$ for some m. Nevertheless, for any unit vector ψ in $L^2(\mathbb{R})$, we have a well-defined probability density on $\mathbb{R}$, given by $|\psi(x)|^2$.

3.4   Momentum and the Momentum Operator

At any fixed time, the wave function $\psi(x)$ of a particle (according to the wave theory postulated by Schrödinger) is a function of a "position" variable x only. Although the wave function ψ directly encodes the probabilities for the position of the particle, through $|\psi(x)|^2$, it is not as clear how information about the particle's momentum is encoded. As it turns out, the momentum is encoded in the *oscillations* of the wave function. A crucial idea in quantum mechanics is the *de Broglie hypothesis*, which we introduced in Sect. 1.2.2 as a way of understanding the allowed energies in the Bohr model of the hydrogen atom. The de Broglie hypothesis proposes a particular relationship between the frequency of oscillation of the wave function—as a function of position at a fixed time—and its momentum.

Proposition 3.5 (de Broglie hypothesis) *If the wave function of a particle has spatial frequency k, then the momentum p of the particle is*

$$p = \hbar k, \tag{3.4}$$

where $\hbar$ is Planck's constant.

The Davisson–Germer electron-diffraction experiments, described in Sect. 1.2.3, strongly support not only the idea that electrons have wavelike behavior, but also the specific relationship (3.4) between the momentum of an electron and the spatial frequency of the associated wave. Of course, Proposition 3.5 is rather vague. To be a bit more precise, Proposition 3.5 is supposed to mean that a wave function of the form $\psi(x) = e^{ikx}$ represents a particle with momentum $p = \hbar k$. [Here, as in Chap. 2, "frequency" is in the angular sense. The cycles-per-unit-distance frequency is $\nu = k/(2\pi)$.]

Now, the function e^{ikx} is obviously not square integrable, so it is not strictly possible for the wave function [which is supposed to satisfy (3.1)] to be e^{ikx}. Let us therefore briefly switch to thinking of a particle on a circle, so that we can avoid certain technicalities. We think of the wave function ψ for a particle on a circle as a 2π-periodic function on $\mathbb{R}$, satisfying the normalization condition

$$\int_0^{2\pi} |\psi(x)|^2 \, dx = 1.$$

For any integer k, it makes sense to say that the normalized wave function $\psi(x) = e^{ikx}/\sqrt{2\pi}$ represents a particle with momentum $p = \hbar k$. In this case, we are supposed to think that the momentum of the particle is definite, that is, nonrandom. If the particle's wave function is $e^{ikx}/\sqrt{2\pi}$, then a measurement of the particle's momentum should (with probability 1) give the value $\hbar k$.

Now, the functions $e^{ikx}/\sqrt{2\pi}$, $k \in \mathbb{Z}$, form an orthonormal basis for the Hilbert space of 2π-periodic, square-integrable functions, which may be identified with $L^2([0, 2\pi])$. Thus, the typical wave function for a particle on a circle is

$$\psi(x) = \sum_{k=-\infty}^{\infty} a_k \frac{e^{ikx}}{\sqrt{2\pi}}, \tag{3.5}$$

where the sum is convergent in $L^2([0, 2\pi])$. If ψ is normalized to be a unit vector, then we have

$$\sum_{k=-\infty}^{\infty} |a_k|^2 = \|\psi\|_{L^2([0,2\pi])}^2 = 1. \tag{3.6}$$

For a particle with wave function given by (3.5), the momentum of the particle is no longer definite. Rather, we are supposed to think that a measurement of the particle's momentum will yield one of the values $\hbar k$, $k \in \mathbb{Z}$, with the probability of getting a particular value $\hbar k$ being $|a_k|^2$. Following elementary probability theory, then, the expectation values for the momentum should be

$$E(p) = \sum_{k=-\infty}^{\infty} \hbar k \, |a_k|^2 , \qquad (3.7)$$

and higher moments for the momentum should be

$$E(p^m) = \sum_{k=-\infty}^{\infty} (\hbar k)^m \, |a_k|^2 , \qquad (3.8)$$

assuming absolute convergence of the sum.

We would like to encode the moment conditions (3.7) and (3.8) in a *momentum operator* P, which should be defined in such a way that if the particle's wave function ψ is given by (3.5), then $E(p^m) = \langle \psi, P^m \psi \rangle$. We can achieve this relation if P satisfies

$$P e^{ikx} = \hbar k e^{ikx}, \qquad (3.9)$$

since then,

$$\langle \psi, P^m \psi \rangle = \sum_{k=-\infty}^{\infty} (\hbar k)^m \, |a_k|^2 = E(p^m). \qquad (3.10)$$

The (presumably unique) choice for P satisfying (3.9) is

$$P = -i\hbar \frac{d}{dx}.$$

Returning now to the setting of the real line, it is natural to postulate that the momentum operator P on the line should also be given by $P = -i\hbar \, d/dx$. This operator satisfies the relation

$$P e^{ikx} = (\hbar k) e^{ikx},$$

which is supposed to capture the idea that the wave function e^{ikx} has momentum $\hbar k$. Although the function e^{ikx} is not square-integrable with respect to x, the Fourier transform allows us to build up any square-integrable function as a "superposition" of functions of the form e^{ikx}. (*Superposition* is the term physicists use for a linear combination or the continuous analog thereof, namely an integral.) This means that [by analogy to (3.5)] we have

$$\psi(x) = \frac{1}{\sqrt{2\pi}} \int_{-\infty}^{\infty} e^{ikx} \hat{\psi}(k) \, dk, \qquad (3.11)$$

where $\hat{\psi}(k)$ is the Fourier transform of ψ, defined by

$$\hat{\psi}(k) = \frac{1}{\sqrt{2\pi}} \int_{-\infty}^{\infty} e^{-ikx} \psi(x) \, dx. \qquad (3.12)$$

(See Appendix A.3.2 for information about the Fourier transform.)

The Plancherel theorem (Theorem A.19) then tells us that the Fourier transform is a unitary map of $L^2(\mathbb{R})$ onto $L^2(\mathbb{R})$. Thus, for any unit vector $\psi \in L^2(\mathbb{R})$,

$$\int_{-\infty}^{\infty} |\psi(x)|^2 \; dx = \int_{-\infty}^{\infty} \left|\hat{\psi}(k)\right|^2 \; dk = 1.$$

In light of what we have in the circle case, it is natural to think that $|\hat{\psi}(k)|^2$ is essentially the probability density for the momentum of the particle. (To be precise, $|\hat{\psi}(k)|^2$ is the probability density for $p/\hbar$.)

We can now express the properties of the momentum operator entirely within the Hilbert space $L^2(\mathbb{R})$, without making explicit mention of the non–square-integrable functions e^{ikx}.

Proposition 3.6 *Define the momentum operator P by*

$$P = -i\hbar \frac{d}{dx}.$$

Then for all sufficiently nice unit vectors ψ in $L^2(\mathbb{R})$, we have

$$\langle \psi, P^m \psi \rangle = \int_{-\infty}^{\infty} (\hbar k)^m \left|\hat{\psi}(k)\right|^2 \; dk \qquad (3.13)$$

for all positive integers m. The quantity in (3.13) is interpreted as the expectation value of the mth power of the momentum, $E(p^m)$.

Equation (3.13) should be compared to (3.10) in the case of the circle. **Proof.** If ψ is in, say, the Schwartz space (Definition A.15), then, by applying Proposition A.17 m times, we see that the Fourier transform of the nth derivative of ψ is $(ik)^m \hat{\psi}(k)$, and so the Fourier transform of $P^m \psi$ is $(\hbar k)^m \hat{\psi}(k)$. Meanwhile, since the Fourier transform is unitary, we have

$$\langle \psi, P^m \psi \rangle = \int_{-\infty}^{\infty} \overline{\hat{\psi}(k)} (\hbar k)^m \hat{\psi}(k) \; dk,$$

which gives (3.13). (The assumption that ψ be in the Schwartz space is stronger than necessary. The reader is invited to use integration by parts and the definition of the Fourier transform to find weaker assumptions that allow the same conclusion.) ∎

3.5   The Position and Momentum Operators

In the following definition, we summarize what we have learned, in the two previous sections, about the position and momentum operators.

Definition 3.7 *For a particle moving in $\mathbb{R}^1$, let the quantum Hilbert space be $L^2(\mathbb{R})$ and define the* **position** *and* **momentum operators** X *and* P *by*

$$X\psi(x) = x\psi(x)$$

$$P\psi(x) = -i\hbar\frac{d\psi}{dx}.$$

Neither the position nor the momentum operator is defined as mapping the entire Hilbert space $L^2(\mathbb{R})$ into itself. After all, for $\psi \in L^2(\mathbb{R})$, the function $x\psi(x)$ may fail to be in $L^2(\mathbb{R})$. Similarly, a function ψ in $L^2(\mathbb{R})$ may fail to be differentiable, and even if it is differentiable, the derivative may fail to be in $L^2(\mathbb{R})$. What this means is that X and P are unbounded operators, of the sort discussed briefly in Sect. 3.2. They are defined on suitable dense subspaces $\mathrm{Dom}(X)$ and $\mathrm{Dom}(P)$ of $L^2(\mathbb{R})$. We defer a detailed examination of the domains of these operators until Chap. 9.

A vitally important property of this pair of operators is that they *do not commute*.

Proposition 3.8 *The position and momentum operators X and P do not commute, but satisfy the relation*

$$XP - PX = i\hbar I, \tag{3.14}$$

This relation is known as the *canonical commutation relation*.
Proof. Using the product rule we calculate that

$$PX\psi = -i\hbar\frac{d}{dx}\left(x\psi(x)\right)$$

$$= -i\hbar\psi(x) - i\hbar x\frac{d\psi}{dx}$$

$$= -i\hbar\psi(x) + XP\psi,$$

from which (3.14) follows. ∎

There are many important consequences of the relation (3.14), which we will examine at length in Chaps. 11–14 of the book. For now, we simply note a parallel between (3.14) and the Poisson bracket relationship in classical mechanics: $\{x, p\} = 1$, as follows directly from the definition of the Poisson bracket. This hints at an analogy, which we will explore further in Sect. 3.7, between the *commutator* of two operators A and B on the quantum side (namely, the operator $AB - BA$) and the Poisson bracket of two functions f and g on the classical side.

Proposition 3.9 *For all sufficiently nice functions ϕ and ψ in $L^2(\mathbb{R})$, we have*

$$\langle \phi, X\psi \rangle = \langle X\phi, \psi \rangle$$

and

$$\langle \phi, P\psi \rangle = \langle P\phi, \psi \rangle.$$

Proof. Suppose that ϕ and ψ belong to $L^2(\mathbb{R})$ and that the functions $x\phi(x)$ and $x\psi(x)$ also belong to $L^2(\mathbb{R})$. Then since x is real, we have

$$\int_{-\infty}^{\infty} \overline{\phi(x)}x\psi(x)\ dx = \int_{-\infty}^{\infty} \overline{x\phi(x)}\psi(x)\ dx,$$

where both integrals are convergent because they are both integrals of the product of two L^2 functions.

Meanwhile, for the second claim, let us assume that ϕ and ψ are continuously differentiable and that $\phi(x)$ and $\psi(x)$ tend to zero as x tends to $\pm\infty$. Let us also assume that ϕ, ψ, $d\phi/dx$ and $d\psi/dx$ belong to $L^2(\mathbb{R})$. We note that $\overline{d\phi/dx}$ is the same as $d\overline{\phi}/dx$. Thus, using integration by parts, we obtain

$$-i\hbar \int_{-A}^{A} \overline{\phi(x)}\frac{d\psi}{dx}\ dx = -i\hbar\ \overline{\phi(x)}\psi(x)\Big|_{-A}^{A} + i\hbar \int_{-A}^{A} \frac{\overline{d\phi}}{dx}\psi(x)\ dx.$$

Under our assumptions on ϕ and ψ, as A tends to infinity, the boundary terms will vanish and the remaining integrals will tend (by dominated convergence) to integrals over the whole real line. Thus,

$$\int_{-\infty}^{\infty} \overline{\phi(x)}\left(-i\hbar\frac{d\psi}{dx}\right)\ dx = i\hbar \int_{-\infty}^{\infty} \frac{\overline{d\phi}}{dx}\psi(x)\ dx$$

$$= \int_{-\infty}^{\infty} \overline{\left(-i\hbar\frac{d\phi}{dx}\right)}\psi(x)\ dx,$$

which is the second claim in the proposition. ∎

In the language of Definition 3.3, Proposition 3.9 means that X and P are *symmetric* operators on certain dense subspaces of $L^2(\mathbb{R})$ (the space of functions for which the proposition is proved). It is actually true that X and P are essentially self-adjoint on these domains. The proof of essential self-adjointness, however, will have to wait until Chap. 9.

3.6 Axioms of Quantum Mechanics: Operators and Measurements

In this section we consider the general "axioms" of quantum mechanics. These axioms are not to be understood in the mathematical sense as rules from which all other results are derived in a strictly deductive fashion. Rather, the axioms are the main principles of how quantum mechanics works. Here we look at the "kinematic" axioms, those that apply at one fixed time. There is one additional axiom, governing the time-evolution of the system, which we consider in the next section.

Axiom 1 *The state of the system is represented by a unit vector ψ in an appropriate Hilbert space* **H**. *If ψ_1 and ψ_2 are two unit vectors in* **H** *with $\psi_2 = c\psi_1$ for some constant $c \in \mathbb{C}$, then ψ_1 and ψ_2 represent the same physical state.*

The Hilbert space **H** is frequently called the "quantum Hilbert space." This does not, however, mean that **H** is some variant of the notion of a Hilbert space, the way a quantum group is a variant of the notion of a group. Rather, "quantum Hilbert space" means simply, "the Hilbert space associated with a given quantum system."

In Axiom 1, it should be noted that unit vectors in **H** actually represent only the "pure states" of the theory. There is a more general notion of a "mixed state" (described by a "density matrix") that we will consider in Chap. 19. We will follow the custom in most physics texts of considering at first only pure states.

Axiom 2 *To each real-valued function f on the classical phase space there is associated a self-adjoint operator $\hat{f}$ on the quantum Hilbert space.*

In almost all cases, the operator $\hat{f}$ is unbounded. This unboundedness is unsurprising when we realize that physically relevant functions f on the classical phase space (e.g., position and momentum) are unbounded functions. In the unbounded case, the notion of self-adjointness is rather technical; see Definition 3.3 in Sect. 3.2. In most applications, it is not really necessary to define $\hat{f}$ for *all* functions on the classical phase space, but only for certain basic functions, such as position, momentum, energy, and angular momentum. We will describe the quantizations of these basic functions in this chapter. If one really needs to define $\hat{f}$ for an arbitrary function f (satisfying some regularity assumptions), the standard approach is to use the Weyl quantization scheme, described in Chap. 13.

For a particle moving in $\mathbb{R}^1$, the classical phase space is $\mathbb{R}^2$, which we think of as pairs (x, p) with x being the particle's position and p being its momentum. The quantum Hilbert space in this case is usually taken to be $L^2(\mathbb{R})$ [*not* $L^2(\mathbb{R}^2)$]. In that case, if the function f in Axiom 2 is the position function, $f(x, p) = x$, then the associated operator $\hat{f}$ is the position operator X, given by multiplication by x. If f is the momentum function, $f(x, p) = p$, then $\hat{f}$ is the momentum operator $P = -i\hbar \, d/dx$.

In the physics literature, a function f on the classical phase space is called a *classical observable*, meaning that it is some physical quantity that could be observed by taking a measurement of the system. The corresponding operator $\hat{f}$ is then called a *quantum observable*.

Axiom 3 *If a quantum system is in a state described by a unit vector $\psi \in$* **H**, *the probability distribution for the measurement of some observable f satisfies*

$$E(f^m) = \left\langle \psi, (\hat{f})^m \psi \right\rangle. \tag{3.15}$$

In particular, the expectation value for a measurement of f is given by

$$\left\langle \psi, \hat{f}\psi \right\rangle. \tag{3.16}$$

Note that we have adopted the point of view that even in a quantum mechanical system, what one is measuring is the *classical* observable f. In the quantum case, however, f no longer has a definite value, but only probabilities, which are encoded by the quantum observable $\hat{f}$ and the vector $\psi \in \mathbf{H}$.

If ψ is a nonzero vector in $\mathbf{H}$ but not a unit vector, then (3.16) should be replaced by

$$\frac{\left\langle \psi, \hat{f}\psi \right\rangle}{\langle \psi, \psi \rangle} = \left\langle \tilde{\psi}, \hat{f}\tilde{\psi} \right\rangle,$$

where $\tilde{\psi} := \psi/\|\psi\|$ is the unit vector associated with ψ. It is convenient to assume that our vectors have been normalized to be unit vectors, simply to avoid having to divide by $\langle \psi, \psi \rangle$ in our expectation values.

Since $\hat{f}$ is assumed to be self-adjoint and every self-adjoint operator is symmetric, Proposition 3.4 tells us that the moments $E(f^m)$, and in particular the expectation value $E(f)$, are real numbers. Since $\hat{f}$ is assumed to be self-adjoint and not just symmetric, the spectral theorem (Chaps. 7 and 10) will give a canonical way of constructing a probability measure $\mu_{A,\psi}$ on $\mathbb{R}$ that may be interpreted as the probability distribution for measurements of A in the state ψ.

Axiom 3 provides motivation for the idea that two unit vectors that differ by a constant represent the same physical state. If $\psi_2 = c\psi_1$ with $|c| = 1$, then for any operator A, we have

$$\langle \psi_2, A\psi_2 \rangle = \langle c\psi_1, Ac\psi_1 \rangle = |c|^2 \langle \psi_1, A\psi_1 \rangle = \langle \psi_1, A\psi_1 \rangle.$$

Thus, the expectation values of all observables are the same in the state ψ_2 as in the state ψ_1.

Notation 3.10 *If A is a self-adjoint operator on $\mathbf{H}$ and $\psi \in \mathbf{H}$ is a unit vector, the expectation value of A in the state ψ is denoted $\langle A \rangle_\psi$ and is defined (in light of Axiom 3) to be*

$$\langle A \rangle_\psi = \langle \psi, A\psi \rangle. \tag{3.17}$$

Proposition 3.11 (Eigenvectors) *If a quantum system is in a state described by a unit vector $\psi \in \mathbf{H}$ and for some quantum observable $\hat{f}$ we have $\hat{f}\psi = \lambda\psi$ for some $\lambda \in \mathbb{R}$, then*

$$E(f^m) = \left\langle (\hat{f})^m \right\rangle_\psi = \lambda^m \tag{3.18}$$

for all positive integers m. The unique probability measure consistent with this condition is the one in which f has the definite value λ, with probability one.

What the proposition means is that if ψ is an eigenvector for $\hat{f}$, then measurements of f for a particle in the state ψ are not actually random, but rather always give the answer of λ. If $\hat{f}\psi = \lambda\psi$, then $\langle \psi, (\hat{f})^m \psi \rangle = \lambda^m \langle \psi, \psi \rangle = \lambda^m$. Thus, by (3.15), we want to find a probability measure μ on $\mathbb{R}$ such that

$$\int_{\mathbb{R}} x^m \, d\mu = \lambda^m, \tag{3.19}$$

for all non-negative integers m. The proposition is claiming that there is one and only one such measure, namely the δ-measure at the point λ.

Because $\hat{f}$ is assumed to be self-adjoint and therefore symmetric, Proposition 3.4 thus tells us that the every eigenvalue for $\hat{f}$ is real.

Proof. The relation (3.18) follows from (3.15) and the fact that $\hat{f}\psi = \lambda\psi$. Meanwhile, if μ is the δ-measure at λ, then certainly (3.19) holds. Meanwhile, since the mth moment grows only exponentially with m, even the most elementary uniqueness results for the moment problem show that the δ-measure is the *only* measure with these moments. (See, e.g., Theorem 8.1 in Chap. 4 of [18].) ∎

If, more generally, the state of the system is a *linear combination* of eigenvectors for $\hat{f}$, measurements of f will no longer be deterministic.

Example 3.12 *Suppose $\hat{f}$ has an orthonormal basis $\{e_j\}$ of eigenvectors with distinct (real) eigenvalues λ_j. Suppose also that ψ is a unit vector in* **H** *with the expansion*

$$\psi = \sum_{j=1}^{\infty} a_j e_j. \tag{3.20}$$

Then for a measurement in the state ψ of the observable f, the observed value of f will always be one of the numbers λ_j. Furthermore, the probability of observing the value λ_j is given by

$$\mathrm{Prob}\{f = \lambda_j\} = |a_j|^2. \tag{3.21}$$

Assuming that ψ is in the domain of $(\hat{f})^m$, it is easy to verify that the probabilities in (3.21) are consistent with the expectation values given in Axiom 3. After all, if ψ is given as in (3.20), then we can readily calculate that $\langle \psi, (\hat{f})^m \psi \rangle$ equals $\sum |a_j|^2 \lambda_j^m$, which is nothing but the mth moment associated with the probability distribution in (3.21). In general, we cannot quite *derive* (3.21) from Axiom 3, since the uniqueness results for the moment problem might not apply. Nevertheless, (3.21) is the most natural candidate for the probabilities, and we will assume that this formula holds.

It is not difficult to extend Example 3.12 to the case where the eigenvalues are not distinct: For *any* sequence $\{\lambda_j\}$ of eigenvalues, the probability of observing some value λ will be the sum of $|a_j|^2$ over all those values of j for which $\lambda_j = \lambda$. For any self-adjoint operator A, the spectral theorem implies that A has *either* an orthonormal basis of eigenvectors *or* some continuous analog thereof. In particular, given a self-adjoint operator A and a unit vector $\psi \in \mathbf{H}$, the spectral theorem will give us a probability measure μ_ψ^A on $\mathbb{R}$ that we interpret as describing the probabilities for a measurement of A in the state ψ. See Proposition 7.17 in the bounded case and Definition 10.7 in the unbounded case.

Axiom 4 *Suppose a quantum system is initially in a state ψ and that a measurement of an observable f is performed. If the result of the measurement is the number $\lambda \in \mathbb{R}$, then immediately after the measurement, the system will be in a state ψ' that satisfies*

$$\hat{f}\psi' = \lambda\psi'.$$

The passage from ψ to ψ' is called the collapse of the wave function. Here $\hat{f}$ is the self-adjoint operator associated with f by Axiom 2.

Let us assume again that $\hat{f}$ has an orthonormal basis of eigenvectors $\{e_j\}$ with distinct eigenvalues λ_j. Then we can say, more specifically, that if we observe the value λ_j in a measurement of $\hat{f}$ (and we *will* always observe one of the λ_j's) then $\psi' = e_j$. That is, the measurement "collapses" the wave function by throwing away all the components of ψ in the direction of the e_k's, except the one with $k = j$.

This idea of the collapse of the wave function has generated an enormous amount of discussion and controversy. One way to look at the situation is to think that the wave function ψ is not actually the state of the system—although we continue to use the standard physics term, "state." Rather, the wave function is the thing that encodes the *probabilities* for the state of the system. The collapse of the wave function is then something similar to a conditional probability; the probabilities for future measurements of the system should be consistent with the outcome of the measurement we just made. Paul Dirac has described the collapse of the wave function as being not a discontinuous change in the state of the system, but a discontinuous change in our *knowledge of* the state of the system.

In any case, Axiom 4 guarantees the following reasonable principle: If we measure f and then measure f again a very short time later, the result of the second measurement will agree with the result of the first measurement. Thus, immediately after the first measurement, the probabilities for a second measurement of f are not those associated with the vector ψ, but rather those associated with the state ψ'. (Since ψ' is an eigenvector for $\hat{f}$ with eigenvalue λ, Proposition 3.11 tells us that measurements of f in the state ψ' always give the value of λ.)

Note that Axiom 4 only tells us something about the state of the system *immediately* after a measurement. Following the measurement, the state of the system will evolve in time in the usual way (Sect. 3.7). A significant time after the measurement, then, the system will probably no longer be in the state ψ'.

Let us conclude this section by considering an example of how one makes a measurement of a real-world physical system, namely, the hydrogen atom. The Hamiltonian operator $\hat{H}$ for a hydrogen atom has negative eigenvalues of the form

$$-\frac{R}{n^2},\tag{3.22}$$

where R is the *Rydberg constant* and $n = 1, 2, 3, \ldots$ These energies will be derived in Chap. 18. Negative eigenvalues are of greater interest than positive ones, because negative eigenvalues describes states where the electron is bound to the nucleus. If an electron is placed into a state having energy $-R/n_1^2$, with $n_1 > 1$, it will eventually "decay" into a state with lower energy, say, $-R/n_2^2$, with $n_2 < n_1$. (The most readily observed cases are those with $n_2 = 2$ and $n_2 = 1$.) In the process of decaying, the electron emits a photon, with the energy of the photon being equal to the change in energy of the electron, namely,

$$E_{\text{photon}} = \frac{R}{n_2^2} - \frac{R}{n_1^2}.\tag{3.23}$$

Meanwhile, the frequency of the photon is proportional to its energy. Thus, by observing the frequency of the emitted photon, one can determine the change in energy of the electron and thus determine the values of n_1 and n_2.

A general "bound state" of the hydrogen atom (a state in which the electron is bound to the nucleus), will be a linear combination of eigenvectors for $\hat{H}$ with various different eigenvalues of the form (3.22). To measure the energy of the electron, we simply wait for the electron to decay into a lower-energy state and emit a photon, observe the frequency of the photon, and work backwards to the energy of the electron. If we consider many "identically prepared" electrons, all having the same wave function that is a linear combination of eigenvectors, we will observe many different frequencies for the emitted photons, and thus many different energies for the electron. The *probabilities* for the observed energies of the electron will follow the principle spelled out in Example 3.12.

In basic probability theory, if Y is a random variable then the *variance* σ^2 of Y is computed as

$$\sigma^2 = E\left[(Y - E(Y))^2\right],$$

where E denotes the mean or expectation value of a random variable. The *standard deviation* $\sigma := \sqrt{\sigma^2}$ is a measure of the "typical" deviation from the mean $E(X)$. Observe that the variance may be computed as

$$\begin{aligned}
\sigma^2 &= E\left[Y^2 - 2E(Y)Y + E(Y)^2\right] \\
&= E(Y^2) - 2E(Y)^2 + E(Y)^2 \\
&= E(Y^2) - E(Y)^2.
\end{aligned}\tag{3.24}$$

Definition 3.13 *If A is a self-adjoint operator on a Hilbert space* **H** *and ψ is a unit vector in* **H**, *let $\Delta_\psi A$ denote the standard deviation associated with measurements of A in the state ψ, which is computed as*

$$(\Delta_\psi A)^2 = \left\langle (A - \langle A \rangle_\psi I)^2 \right\rangle_\psi$$
$$= \langle A^2 \rangle_\psi - \left(\langle A \rangle_\psi \right)^2.$$

We refer to $\Delta_\psi A$ as the **uncertainty** *of A in the state ψ.*

For any single observable A, it is possible to choose ψ so that $\Delta_\psi A$ is as small as we like. In Chap. 12, however, we will see that when two observables A and B do not commute, then $\Delta_\psi A$ and $\Delta_\psi B$ cannot both be made arbitrarily small for the same ψ. In particular, we will derive there the famous *Heisenberg uncertainty principle*, which states that

$$(\Delta_\psi X)(\Delta_\psi P) \geq \frac{\hbar}{2},$$

for all ψ for which $\Delta_\psi X$ and $\Delta_\psi P$ are defined.

3.7   Time-Evolution in Quantum Theory

3.7.1   The Schrödinger Equation

Up to now, we have been considering the wave function ψ at a fixed time. We now consider the way in which the wave function evolves in time. Recall that in the Hamiltonian formulation of classical mechanics (Sect. 2.5), the time-evolution of the system is governed by the Hamiltonian (energy) function H, through Hamilton's equations. According to Axiom 2, there is a corresponding self-adjoint linear operator $\hat{H}$ on the quantum Hilbert space **H**, which we call the *Hamiltonian operator* for the system. See Sect. 3.7.4 for an example.

Recall that we motivated the definition of the momentum operator by the *de Broglie hypothesis*, $p = \hbar k$, where k is the spatial frequency of the wave function. We can similarly motivate the time-evolution in quantum mechanics by a similar relation between the energy and the temporal frequency of our wave function:

$$E = \hbar \omega. \tag{3.25}$$

This relationship between energy and temporal frequency is nothing but the relationship proposed by Planck in his model of blackbody radiation (Sect. 1.1.3). Suppose that a wave function ψ_0 has definite energy E, meaning that ψ_0 is an eigenvector for $\hat{H}$ with eigenvalue E. Then (3.25) means that the time-dependence of the wave function should be purely at frequency $\omega = E/\hbar$. That is to say, if the state of the system at time $t = 0$ is ψ_0, then the state of the system at any other time t should be

$$\psi(t) = e^{-i\omega t}\psi_0 = e^{-iEt/\hbar}\psi_0. \tag{3.26}$$

We can rewrite (3.26) as a differential equation:

$$\frac{d\psi}{dt} = -\frac{iE}{\hbar}\psi = \frac{E}{i\hbar}\psi. \tag{3.27}$$

Note that we are taking "temporal frequency ω" to mean that the time-dependence is of the form $e^{-i\omega t}$, whereas we took "spatial frequency k" to mean that the space-dependence is of the form e^{ikx}, with no minus sign in the exponent. This curious convention is convenient when we look at pure exponential solutions to the free Schrödinger equation (Chap. 4) of the form $\exp[i(kx - \omega t)]$, which describes a solution moving to the right with speed ω/k.

Equation (3.27) tells us the time-evolution for a particle that is initially in a state of definite energy, that is, an eigenvector for the Hamiltonian operator. A natural way to generalize this equation is to recognize that $E\psi$ is nothing but $\hat{H}\psi$, since ψ is just a multiple of ψ_0, which is an eigenvector for $\hat{H}$ with eigenvalue E. Replacing E by $\hat{H}$ in (3.27) leads to the following general prescription for the time-evolution of a quantum system.

Axiom 5 *The time-evolution of the wave function ψ in a quantum system is given by the Schrödinger equation,*

$$\frac{d\psi}{dt} = \frac{1}{i\hbar}\hat{H}\psi. \tag{3.28}$$

Here $\hat{H}$ is the operator corresponding to the classical Hamiltonian H by means of Axiom 2.

Although both Hamilton's equations and the Schrödinger equation involve a Hamiltonian, the two equations otherwise do not seem parallel. Of course, since quantum mechanics is not classical mechanics, we should not expect the two theories to have the same time-evolution. Nevertheless, we might hope to see *some* similarities between the time-evolution of a classical system and that of the corresponding quantum system. Such a similarity can be seen when we consider how the expectation values of observables evolve in quantum mechanics.

Proposition 3.14 *Suppose $\psi(t)$ is a solution of the Schrödinger equation and A is a self-adjoint operator on $\mathbf{H}$. Assuming certain natural domain conditions hold, we have*

$$\frac{d}{dt}\langle A\rangle_{\psi(t)} = \left\langle \frac{1}{i\hbar}[A, \hat{H}]\right\rangle_{\psi(t)}, \tag{3.29}$$

where $\langle A \rangle_\psi$ is as in Notation 3.10 and where $[\cdot, \cdot]$ denotes the commutator, *defined as*

$$[A, B] = AB - BA.$$

Equation (3.29) should be compared to the way a function f on the classical phase space evolves in time along a solution of Hamilton's equations: $df/dt = \{f, H\}$. We see, then, that the commutator of operators (divided by $i\hbar$) plays a role in quantum mechanics similar to the role of the Poisson bracket in classical mechanics.

Proof. Let $\psi(t)$ be a solution to the Schrödinger equation and let us compute at first without worrying about domains of the operators involved. If we use the product rule (Exercise 1) for differentiation of the inner product, we obtain

$$\frac{d}{dt} \langle \psi(t), A\psi(t) \rangle = \left\langle \frac{d\psi}{dt}, A\psi \right\rangle + \left\langle \psi, A\frac{d\psi}{dt} \right\rangle$$

$$= \frac{i}{\hbar} \left\langle \hat{H}\psi, A\psi \right\rangle - \frac{i}{\hbar} \left\langle \psi, A\hat{H}\psi \right\rangle$$

$$= \frac{1}{i\hbar} \left\langle \psi, [A, \hat{H}]\psi \right\rangle,$$

where in the last step we have used the self-adjointness of $\hat{H}$ to move it to the other side of the inner product. Recall that we are following the convention of putting the complex conjugate on the first factor in the inner product, which accounts for the plus sign in the first term on the second line. Rewriting this using Notation 3.10 gives the desired result.

If A and $\hat{H}$ are (as usual) unbounded operators, then the preceding calculation is not completely rigorous. Since, however, we are deferring a detailed examination of issues of unbounded operators until Chap. 9, let us simply state the conditions needed for the calculation to be valid. For every $t \in \mathbb{R}$, we need to have $\psi(t) \in \mathrm{Dom}(A) \cap \mathrm{Dom}(\hat{H})$, we need $A\psi(t) \in \mathrm{Dom}(\hat{H})$, and we need $\hat{H}\psi(t) \in \mathrm{Dom}(A)$. (These conditions are needed for $[A, \hat{H}]\psi(t)$ to be defined.) In addition, we need $A\psi(t)$ to be a continuous path in $\mathbf{H}$. ∎

Note that to see interesting behavior in the time-evolution of a quantum system, there has to be *noncommutativity* present. If all the physically interesting operators A commuted with the Hamiltonian operator $\hat{H}$, then $[\hat{H}, A]$ would be zero and the expectation values of these operators would be constant in time. Noncommutativity of the basic operators is therefore an essential property of quantum mechanics. In the case of a particle in $\mathbb{R}^1$, noncommutativity is built into the commutation relation for X and P, given in Proposition 3.8.

Although it is not reasonable to have *all* physically interesting operators commute with $\hat{H}$, there may be *some* operators with this property. If $[A, \hat{H}] = 0$, then the expectation value of A (and, indeed, all the moments of A) is independent of time along any solution of the Schrödinger equation.

We may therefore call such an operator A a *conserved quantity* (or *constant of motion*). Just as in the classical setting, conserved quantities (when we can find them) are helpful in understanding how to solve the Schrödinger equation.

Proposition 3.14 suggests that the map

$$(A, B) \longmapsto \frac{1}{i\hbar}[A, B],$$

where A and B are self-adjoint operators, plays a role similar to that of the Poisson bracket in classical mechanics. This analogy is supported by the following list of elementary properties of the commutator, which should be compared to the properties of the Poisson bracket listed in Proposition 2.23.

Proposition 3.15 *For any vector space V over $\mathbb{C}$ and linear operators A, B, and C on V, the following relations hold.*

1. $[A, B + \alpha C] = [A, B] + \alpha[A, C]$ *for all* $\alpha \in \mathbb{C}$

2. $[B, A] = -[A, B]$

3. $[A, BC] = [A, B]C + B[A, C]$

4. $[A, [B, C]] = [[A, B], C] + [B, [A, C]]$

Property 4 is equivalent to the *Jacobi identity*,

$$[A, [B, C]] + [B, [C, A]] + [C, [A, B]] = 0, \tag{3.30}$$

as can easily be seen using the skew-symmetry of the commutator.

Proof. The first two properties of the commutator are obvious, and the third is easily verified by writing things out. Property 4 can also be proved by writing things out, but it is slightly messier. Each of the three double commutators on the left-hand side of (3.30) generates four terms, for a total of 12 terms. Each term has the operators A, B, and C multiplied together in some order. It is a straightforward but unenlightening calculation to verify that each of the six possible orderings of A, B, and C occurs twice, with opposite signs. ∎

If A and B are bounded *self-adjoint* operators on some Hilbert space, then it is straightforward to check that $(1/(i\hbar))[A, B]$ is again self-adjoint (Exercise 3). If A and B are unbounded self-adjoint operators, then the operator $(1/(i\hbar))[A, B]$ will be self-adjoint under suitable assumptions on the domains of A and B.

Proposition 3.16 *If $\phi(t)$ and $\psi(t)$ are solutions to the Schrödinger equation (3.28), the quantity $\langle \phi(t), \psi(t) \rangle$ is independent of t. In particular, $\|\psi(t)\|$ is independent of t, for any solution $\psi(t)$ of the Schrödinger equation.*

Proof. Using again the product rule, we have

$$\frac{d}{dt}\langle\phi(t),\psi(t)\rangle = \left\langle\frac{1}{i\hbar}\hat{H}\phi(t),\psi(t)\right\rangle + \left\langle\phi(t),\frac{1}{i\hbar}\hat{H}\psi(t)\right\rangle$$

$$= -\frac{1}{i\hbar}\left\langle\hat{H}\phi(t),\psi(t)\right\rangle + \frac{1}{i\hbar}\left\langle\phi(t),\hat{H}\psi(t)\right\rangle$$

Since $\hat{H}$ is self-adjoint, we can move $\hat{H}$ to the other side of the inner product and the derivative is equal to 0. ∎

3.7.2 Solving the Schrödinger Equation by Exponentiation

The Schrödinger equation is an example of a equation of the form

$$\frac{dv}{dt} = Av, \tag{3.31}$$

where A is a linear operator on a Hilbert space. (In the Schrödinger case, we have $A = -(i/\hbar)\hat{H}$.) Let us think of (3.31) in the case where the Hilbert space is the finite-dimensional space $\mathbb{C}^n$. In that case, we can think of A as an $n \times n$ matrix, in which case (3.31) is the sort of equation encountered in the elementary theory of ordinary differential equations. The solution of this system (in the finite-dimensional case) can be expressed as

$$v(t) = e^{tA}v_0,$$

where the matrix exponential e^{tA} is defined by a convergent power series and where $v_0 = v(0)$ is the initial condition. If A is diagonalizable, then the exponential can by computed by using a basis of eigenvectors. (See Sect. 16.4 for more information.)

The Schrödinger equation simply replaces $\mathbb{C}^n$ by a Hilbert space **H** and the matrix A by the linear operator $-(i/\hbar)\hat{H}$.

Claim 3.17 *Suppose $\hat{H}$ is a self-adjoint operator on* **H**. *If a reasonable meaning can be given to the expression $e^{-it\hat{H}/\hbar}$, then the Schrödinger equation can be solved by setting*

$$\psi(t) = e^{-it\hat{H}/\hbar}\psi_0. \tag{3.32}$$

To see why the claim should be true, we expect that we can differentiate the operator-valued expression $e^{-it\hat{H}/\hbar}$ with respect to t as we would in the finite-dimensional case. The differentiation, then, would pull down a factor of $-i\hat{H}/\hbar$, which would indicate that $\psi(t)$ indeed solves the Schrödinger equation. Furthermore, when $t = 0$, $e^{-it\hat{H}/\hbar}$ should be equal to I, so that $\psi(0)$ is indeed ψ_0.

If $\hat{H}$ is a bounded operator (which is rarely the case), then the exponential $e^{-it\hat{H}/\hbar}$ can be defined by a convergent power series, precisely as in the finite-dimensional case. In that case, Claim 3.17 is an easily proved theorem.

In the more typical case where $\hat{H}$ is unbounded, convergence of the series for the exponential is a rather delicate matter, and it is better instead to use the spectral theorem. We leave a general discussion of the spectral theorem to Chaps. 7 and 10, and here consider only the case of a pure point spectrum. A (possibly unbounded) self-adjoint operator $\hat{H}$ is said to have a pure point spectrum if there exists an orthonormal basis $\{e_j\}$ for $\mathbf{H}$ consisting of eigenvectors for $\hat{H}$. If $\hat{H}e_j = E_j e_j$ for some $E_j \in \mathbb{R}$, then the exponential can be defined by requiring that

$$e^{-it\hat{H}/\hbar}e_j = e^{-itE_j/\hbar}e_j. \tag{3.33}$$

The operator $e^{-it\hat{H}/\hbar}$ is unitary and thus bounded; it is the unique bounded operator on $\mathbf{H}$ satisfying (3.33).

It is not precisely true that every self-adjoint operator has an orthonormal basis of eigenvectors, even if the operator is bounded. Nevertheless, given a self-adjoint operator A, the spectral theorem tells us that there is a decomposition of $\mathbf{H}$ into "generalized eigenspaces" for A. It is, however, a bit complicated to state the precise sense of this decomposition, especially in the case of unbounded operators. Still, Claim 3.17 allows us to identify one goal for the spectral theorem: Whatever the spectral theorem says, it ought to allow us to make sense of the expression e^{iaA}, for any self-adjoint operator A and real number a. This goal will indeed be realized, in the bounded case in Chap. 7 and in the unbounded case in Chap. 10.

We should add two points of clarification regarding the expression (3.32). First, in writing (3.32), we have not "really" solved the Schrödinger equation. For this expression to be useful, we need to compute $e^{-it\hat{H}/\hbar}$ in some relatively explicit way. If, for example, we can actually compute an orthonormal basis of eigenvectors for $\hat{H}$, then in light of (3.33), we are on our way to understanding the behavior of the operator $e^{-it\hat{H}/\hbar}$. Second, although $\hat{H}$ is an unbounded operator, which is not defined on all of $\mathbf{H}$ but only on a dense subspace, the operator $e^{-it\hat{H}/\hbar}$ is unitary and defined on all of $\mathbf{H}$. Thus, the right-hand side of (3.32) makes sense for any ψ_0 in $\mathbf{H}$. Nevertheless, we cannot expect that $e^{-it\hat{H}/\hbar}\psi_0$ actually solves the Schrödinger equation (in the natural Hilbert space sense) unless ψ_0 belongs to the domain of $\hat{H}$. (See Lemma 10.17 in Sect. 10.2.)

3.7.3 Eigenvectors and the Time-Independent Schrödinger Equation

As we saw in the preceding section, eigenvectors for the Hamiltonian operator are of great importance in solving the Schrödinger equation. In light of this fact, we make the following definition.

Definition 3.18 *If $\hat{H}$ is the Hamiltonian operator for a quantum system, the eigenvector equation*

$$\hat{H}\psi = E\psi, \quad E \in \mathbb{R}, \tag{3.34}$$

is called the **time-independent Schrödinger equation**.

As always in eigenvector equations, we are trying to determine both the numbers E for which (3.34) has a nonzero solution (the eigenvalues) and the corresponding vectors ψ (the eigenvectors). When quantum texts speak of "solving," say, the quantum harmonic oscillator, what they usually mean is finding all of the solutions to the time-independent Schrödinger equation. (See, e.g., Chaps. 5 and 11.) If ψ is a solution to the time-independent Schrödinger equation, then the solution to the time-*dependent* Schrödinger equation with initial condition ψ is simply $\psi(t) = e^{-itE/\hbar}\psi$. Since $\psi(t)$ is just a constant multiple of ψ, we see that $\psi(t)$ represents the same physical state as ψ. Thus, a solution to the time-independent Schrödinger equation is sometimes called a *stationary state*.

3.7.4   The Schrödinger Equation in $\mathbb{R}^1$

Let us now consider the simplest example for the Hamiltonian operator $\hat{H}$. For a particle moving in $\mathbb{R}^1$, recall (Sect. 3.5) that we have identified the position operator X as being multiplication by x and the momentum operator as $P = -i\hbar \, d/dx$. The classical Hamiltonian for such a particle is typically taken to be of the form $H(x,p) = p^2/(2m) + V(x)$, where V is the potential energy function. In that case, we may reasonably take

$$\hat{H} = \frac{P^2}{2m} + V(X).$$

Here the operator $V(X)$ is simply multiplication by the potential energy function $V(x)$. (This operator may also be thought of as the function V applied to the operator X in the sense of the functional calculus coming from the spectral theorem.) We see, then, that

$$\hat{H}\psi(x) = -\frac{\hbar^2}{2m}\frac{d^2\psi}{dx^2} + V(x)\psi(x). \tag{3.35}$$

An operator of the form (3.35), or an analogously defined operator in higher dimensions, is referred to as a *Schrödinger operator*. (The term *Hamiltonian operator* refers more generally to whatever operator governs the time-evolution of a quantum system, regardless of its form.)

If our Hamiltonian is of the form given in (3.35), then the time-dependent Schrödinger equation takes the form

$$\frac{\partial\psi(x,t)}{\partial t} = \frac{i\hbar}{2m}\frac{\partial^2\psi(x,t)}{\partial x^2} - \frac{i}{\hbar}V(x)\psi(x,t), \tag{3.36}$$

which is a linear partial differential equation. By contrast, Newton's equation for a particle in $\mathbb{R}^1$ is a typically nonlinear ordinary differential equation.

For a particle in $\mathbb{R}^1$, the time-independent Schrödinger equation is an ordinary differential equation, one that is linear but that has nonconstant coefficients, unless V happens to be constant. For simple examples of the potential function V, there are relatively standard methods of ordinary differential equations that can be brought to bear on the time-independent Schrödinger equation.

3.7.5 Time-Evolution of the Expected Position and Expected Momentum

Since a quantum particle does not have a fixed position or momentum, it does not make sense to ask whether the particle satisfies Newton's equation. It does, however, make sense to ask whether the *expected values* of the position and momentum satisfy Newton's equation (in the form of Hamilton's equations).

Proposition 3.19 *Suppose $\psi(t)$ is a solution to the Schrödinger equation (3.36) for a sufficiently nice potential V and for a sufficiently nice initial condition $\psi(0) = \psi_0$. Then the expected position and expected momentum in the state $\psi(t)$ satisfy*

$$\frac{d}{dt} \langle X \rangle_{\psi(t)} = \frac{1}{m} \langle P \rangle_{\psi(t)} \tag{3.37}$$

$$\frac{d}{dt} \langle P \rangle_{\psi(t)} = - \langle V'(X) \rangle_{\psi(t)}. \tag{3.38}$$

The assumptions in the proposition are there for two reasons: First, to ensure that $\hat{H}$ is actually a self-adjoint operator (see Sect. 9.9) and second, to ensure that the domain assumptions in Proposition 3.14 are satisfied. If we assume, for example, that $V(x)$ is a bounded-below polynomial in x and that ψ_0 belongs to the Schwartz space (A.15), then both of these concerns will be taken care of. Once these technicalities are addressed, the proof of Proposition 3.19 is a straightforward application of Proposition 3.14; see Exercise 4. Note that (3.37) says that in a certain sense, the velocity of a quantum particle is $1/m$ times the momentum, just as in the classical case.

At first glance, it might appear that the pair $(\langle X \rangle_{\psi(t)}, \langle P \rangle_{\psi(t)})$ is a solution to Hamilton's equations, and indeed (3.37) is precisely what Hamilton's equations require. To get a solution to Hamilton's equations, however, we would need the right-hand side of (3.38) to equal $-V'(\langle X \rangle_{\psi(t)})$. But in general,

$$\langle V'(X) \rangle_{\psi} \neq V'(\langle X \rangle_{\psi}).$$

Consider, for example, the case $V'(x) = x^3 + x^2$. If ψ is an even function, then $\langle X \rangle_{\psi} = 0$ and so $V'(\langle X \rangle_{\psi}) = 0$. But $\langle X^3 + X^2 \rangle_{\psi}$ will not be zero, because the X^3 term will be zero and the X^2 term will be positive. We conclude, then, that $\langle X \rangle_{\psi(t)}$ and $\langle P \rangle_{\psi(t)}$ usually *do not* evolve along solutions to Hamilton's equations.

There is, however, one case in which $\langle V'(X) \rangle_\psi$ coincides with $V'(\langle X \rangle_\psi)$, and that is the case in which V is quadratic, in which case V' is linear. In that case we have

$$\langle V'(X) \rangle_\psi = \langle aX + bI \rangle_\psi = a \langle X \rangle_\psi + b = V'(\langle X \rangle_\psi).$$

Thus, the expected position and expected momentum *do* follow classical trajectories in the case of a quadratic potential. It is not surprising that this case is special in quantum mechanics, since it is also special in classical mechanics; this is the case in which Newton's law is a linear differential equation.

Although the expected position and expected momentum do not (in general) exactly follow classical trajectories, they will do so approximately under certain conditions. If the wave function $\psi(x)$ is concentrated mostly near a single point $x = x_0$, then $\langle V'(X) \rangle_\psi$ and $V'(\langle X \rangle_\psi)$ will both be approximately equal to $V'(x_0)$. In that case, the expected position and expected momentum of the particle will *approximately* follow a classical trajectory, at least for as long as the wave function remains concentrated near a single point.

3.8   The Heisenberg Picture

The "Heisenberg picture" of quantum mechanics is based on Heisenberg's matrix model of quantum mechanics (Sect. 1.3). In the Heisenberg picture, one thinks of the *operators* (quantum observables) as evolving in time, while the vectors in the Hilbert space (quantum states) remain independent of time. This is to be contrasted with the approach to quantum mechanics we have been using up to now (the "Schrödinger picture"), in which the observables are independent of time and the states evolve in time.

Definition 3.20 *In the Heisenberg picture, each self-adjoint operator A evolves in time according to the operator-valued differential equation*

$$\frac{dA(t)}{dt} = \frac{1}{i\hbar}[A(t), \hat{H}], \tag{3.39}$$

where $\hat{H}$ is the Hamiltonian operator of the system, and where $[\cdot, \cdot]$ is the commutator, given by $[A, B] = AB - BA$.

Note that since $\hat{H}$ commutes with itself, the operator $\hat{H}$ remains constant in time, even in the Heisenberg picture. This observation is the quantum counterpart to the fact that the classical Hamiltonian H remains constant along a solution of Hamilton's equations.

Given the self-adjoint operator $\hat{H}$, the spectral theorem will give us a way to construct a family of unitary operators $e^{-it\hat{H}/\hbar}$, $t \in \mathbb{R}$, and this family of operators computes the time-evolution of states in the Schrödinger picture (Sect. 3.7.2). It is easy to check (at least formally) that the solution to (3.39) can be expressed as

$$A(t) = e^{it\hat{H}/\hbar} A e^{-it\hat{H}/\hbar}. \tag{3.40}$$

Now, if ψ is the state of the system (now considered to be independent of time), then the expectation of $A(t)$ in the state ψ is defined to be $\langle A(t) \rangle_\psi = \langle \psi, A(t)\psi \rangle$. We may then compute that

$$\langle A(t) \rangle_\psi = \left\langle \psi, e^{it\hat{H}/\hbar} A e^{-it\hat{H}/\hbar}\psi \right\rangle$$
$$= \left\langle e^{-it\hat{H}/\hbar}\psi, A e^{-it\hat{H}/\hbar}\psi \right\rangle$$
$$= \langle \psi(t), A\psi(t) \rangle,$$

where $\psi(t)$ is time-evolved state of the system in the Schrödinger picture. Here, we have used that the adjoint of $e^{it\hat{H}/\hbar}$ is $e^{-it\hat{H}/\hbar}$, which is formally clear and which is a consequence of the spectral theorem.

Note that in the Schrödinger picture, $\langle \psi(t), A\psi(t) \rangle$ is the expectation value of A in the state $\psi(t)$. We conclude, then, that the Heisenberg picture and the Schrödinger picture give rise to precisely the same expectation values for observables as a function of time, and are therefore physically equivalent. Although we will work primarily with the Schrödinger picture of quantum mechanics, the Heisenberg picture is also important, for example, in quantum field theory.

Proposition 3.21 *Suppose $\hat{H} = P^2/(2m) + V(X)$, where V is a bounded-below polynomial. Then for any $t \in \mathbb{R}$ we have*

$$\hat{H} = \frac{1}{2m}\left(P(t)\right)^2 + V(X(t)). \tag{3.41}$$

Note that since $[\hat{H}, \hat{H}] = 0$, the Hamiltonian $\hat{H}$ is independent of time, even in the Heisenberg picture. Thus, the right-hand side of (3.41) is actually independent of t, even though $P(t)$ and $X(t)$ depend on t. Equation (3.41) holds also for sufficiently nice nonpolynomial functions V, but some limiting argument would be required in the proof. The assumption that V be bounded below is to ensure that $\hat{H}$ is actually an (essentially) self-adjoint operator; compare Sect. 9.10.

Lemma 3.22 *Suppose A is a self-adjoint operator on $\mathbf{H}$ and that $A(\cdot)$ is a solution to (3.39) with $A(0) = A$. Then for any positive integer m, the map*

$$t \mapsto (A(t))^m$$

is also a solution to (3.39).

That is to say, the time-evolution of the mth power of A is the same as the mth power of the time-evolution of A; that is, $A^m(t) = (A(t))^m$.

Proof. If we use (3.40), then the result holds because

$$e^{it\hat{H}/\hbar} A^m e^{-it\hat{H}/\hbar} = e^{it\hat{H}/\hbar} A e^{-it\hat{H}/\hbar} e^{it\hat{H}/\hbar} A e^{-it\hat{H}/\hbar} \cdots e^{it\hat{H}/\hbar} A e^{-it\hat{H}/\hbar}$$

$$= \left(e^{it\hat{H}/\hbar} A e^{-it\hat{H}/\hbar} \right)^m.$$

It is also easy to check that $A(t)^m$ satisfies the differential equation (3.39). ∎

With this lemma in hand, it is easy to prove the proposition.

Proof of Proposition 3.21. On the one hand, since $[\hat{H}, \hat{H}] = 0$, the time-evolved operator $\hat{H}(t)$ is simply equal to $\hat{H}$. On the other hand, if we time-evolve $P^2/(2m) + V(X)$ using Lemma 3.22, we obtain the expression on the right-hand side of (3.41). ∎

Proposition 3.23 *Suppose the Hamiltonian of a quantum system is as in Proposition 3.21. Then the operators $X(t)$ and $P(t)$ defined by (3.39) satisfy the following operator-valued differential equation:*

$$\frac{dX}{dt} = \frac{1}{m} P(t)$$

$$\frac{dP}{dt} = -V'(X(t)). \tag{3.42}$$

Proof. See Exercise 7. ∎

Proposition 3.23 means that the operator-valued functions $X(t)$ and $P(t)$ satisfy the operator analogs of the classical equations of motion $dx/dt = p(t)/m$ and $dp/dt = -V'(x(t))$. Nevertheless, the *expectation values* of $X(t)$ and $P(t)$ *do not* satisfy the ordinary equations of motion, as we have already seen by calculating in the Schrödinger picture. If we take expectation values in the system (3.42), we get the same answer as in Proposition 3.19, namely,

$$\frac{d}{dt} \langle X(t) \rangle_\psi = \frac{1}{m} \langle P(t) \rangle_\psi$$

$$\frac{d}{dt} \langle P(t) \rangle_\psi = - \langle V'(X(t)) \rangle_\psi.$$

These are *not* the classical equations of motion, unless the expectation value of the operator $V'(X(t))$ coincides with V' applied to the expectation value of $X(t)$, which is usually not the case.

3.9   Example: A Particle in a Box

Let us consider quantum mechanics in one space dimension for a particle that is confined to move in a "box," which we describe as the interval $0 \leq x \leq L$. Our goal is to find all of the eigenvectors and eigenvalues of the Schrödinger operator, that is, to find solutions of the time-independent Schrödinger equation $\hat{H}\psi = E\psi$. In solving this equation, we may think of the constraint to the box as follows. Imagine a particle moving in $\mathbb{R}^1$ in the presence of a potential V that is 0 for x between 0 and L and takes some very large constant value C on the rest of the real line. Classically, this would mean that the particle has to have very high energy (greater than C) to escape from the box. Quantum mechanically, if we have a solution of the time-independent Schrödinger equation $\hat{H}\psi = E\psi$ for this potential (with $E \ll C$), then we expect ψ to decay rapidly for x outside of the box. (We will see this behavior explicitly in Chap. 5.) In the limit as C tends to infinity, we expect solutions of the time-independent Schrödinger equation to be zero outside the box and to *tend to zero as we approach the ends of the box.*

The upshot of this discussion is that we are looking for smooth functions ψ on $[0, L]$ that satisfy the differential equation

$$-\frac{\hbar^2}{2m}\frac{d^2\psi}{dx^2} = E\psi(x), \quad 0 \leq x \leq L \tag{3.43}$$

and the boundary conditions

$$\psi(0) = \psi(L) = 0. \tag{3.44}$$

For $E > 0$, the solution space to (3.43) will be the span of two complex exponentials, or equivalently a sine and a cosine function:

$$\psi(x) = a\sin\left(\frac{\sqrt{2mE}}{\hbar}x\right) + b\cos\left(\frac{\sqrt{2mE}}{\hbar}x\right). \tag{3.45}$$

If we now impose the boundary condition $\psi(0) = 0$, we get that $b = 0$, leaving only the sine term. If we then impose the condition $\psi(L) = 0$, we will obtain $a = 0$—which would mean that ψ is identically zero—*unless*

$$\sin\left(\frac{\sqrt{2mE}}{\hbar}L\right) = 0. \tag{3.46}$$

Since we are interested in solutions to (3.43) where ψ is not identically zero, we want (3.46) to hold. Thus, the argument of sine function must be an integer multiple of π. This condition imposes a restriction on the value of E, namely that E should be of the form

$$E_j := \frac{j^2\pi^2\hbar^2}{2mL^2}, \tag{3.47}$$

for some positive integer j.

It is a simple exercise (Exercise 8) to verify that for $E \leq 0$, the only solution to (3.43) satisfying the boundary conditions (3.44) is the one with ψ identically zero.

Proposition 3.24 *The following functions are solutions to (3.43) satisfying the boundary conditions (3.44):*

$$\psi_j(x) = \sqrt{\frac{2}{L}} \sin\left(\frac{j\pi x}{L}\right), \quad j = 1, 2, 3, \ldots,$$

and the corresponding eigenvalues E_j are given by (3.47). The functions ψ_j form an orthonormal basis for the Hilbert space $L^2([0, L])$.

Proof. We have already verified the equation and eigenvalue for each ψ_j. It is a simple computation to verify that the ψ_j's are orthonormal, and the elementary theory of Fourier series (Fourier sine series, in this case) shows that the ψ_j's form an orthonormal *basis* for $L^2([0, L])$. ∎

The Hamiltonian operator for this problem (in which $V = 0$ inside the box) is given by

$$\hat{H}\psi = -\frac{\hbar^2}{2m}\frac{d^2\psi}{dx^2}.$$

This operator is an unbounded operator and is not defined on the whole Hilbert space $L^2([0, L])$, but only on a dense subspace $\mathrm{Dom}(\hat{H}) \subset L^2([0, L])$. The domain of $\hat{H}$ should be chosen in such a way that $\hat{H}$ is essentially self-adjoint and, thus, symmetric (Sect. 3.2), meaning that

$$\left\langle \phi, \hat{H}\psi \right\rangle = \left\langle \hat{H}\phi, \psi \right\rangle \tag{3.48}$$

for all ϕ, ψ in $\mathrm{Dom}(\hat{H})$. For (3.48) to hold, ϕ and ψ must satisfy appropriate boundary conditions, which will allow the boundary terms in the integration by parts to be zero. (See Exercise 9.)

Mathematically, then, it is necessary to impose some boundary conditions in order for $\hat{H}$ to be an essentially self-adjoint operator. The particular choice of boundary conditions (3.44) is based on the idea of approximating the box by a very large "confining" potential outside the box. See Chap. 9 for an extensive discussion of domain issues for unbounded operator.

3.10   Quantum Mechanics for a Particle in $\mathbb{R}^n$

Up to this point, we have been considering a quantum particle moving in $\mathbb{R}^1$. It is straightforward, however, to generalize to a quantum particle moving in $\mathbb{R}^n$. The Hilbert space for a particle in $\mathbb{R}^n$ is $L^2(\mathbb{R}^n)$, rather than $L^2(\mathbb{R})$. Instead of single position operator, we have n such operators, given by

$$X_j\psi(\mathbf{x}) = x_j\psi(\mathbf{x}), \quad j = 1, \ldots, n.$$

Similarly, we have n momentum operators, given by

$$P_j\psi(\mathbf{x}) = -i\hbar\frac{\partial\psi}{\partial x_j}.$$

As in the $\mathbb{R}^1$ case, X_j does not commute with P_j but satisfies $[X_j, P_j] = i\hbar I$. On the other hand, X_j commutes with X_k and P_j commutes with P_k. Furthermore, X_j commutes with P_k for $j \neq k$. These formulas are referred to as the *canonical commutation relations*.

Proposition 3.25 (Canonical Commutation Relations) *The position and momentum operators satisfy*

$$\frac{1}{i\hbar}[X_j, X_k] = 0$$

$$\frac{1}{i\hbar}[P_j, P_k] = 0$$

$$\frac{1}{i\hbar}[X_j, P_k] = \delta_{jk} I \tag{3.49}$$

for all $1 \leq j, k \leq n$.

These relations are the quantum counterparts of the Poisson bracket relations among the position and momentum *functions* in classical mechanics. Specifically, the role of the Poisson bracket in Proposition 2.24 is played in Proposition 3.25 by the quantity $(1/(i\hbar))[\cdot, \cdot]$.

If the classical Hamiltonian for a particle in $\mathbb{R}^n$ is of the usual form (kinetic energy plus potential energy), then we may analogously define the Hamiltonian operator to be of the form

$$\hat{H} = \sum_{j=1}^{n} \frac{P_j^2}{2m} + V(\mathbf{X}), \tag{3.50}$$

where $V(\mathbf{X})$ denotes the result of applying the function V to the commuting family of operators $\mathbf{X} = (X_1, \ldots, X_n)$. It it natural to identify $V(\mathbf{X})$ with the operator of multiplication by the function $V(\mathbf{x})$. In that case, we may write $\hat{H}$ more explicitly as

$$\hat{H}\psi(\mathbf{x}) = -\frac{\hbar}{2m}\Delta\psi(\mathbf{x}) + V(\mathbf{x})\psi(\mathbf{x}),$$

where Δ is the Laplacian, given by

$$\Delta = \sum_{j=1}^{n} \frac{\partial^2}{\partial x_j^2}.$$

We refer to an operator of the form (3.50) as a Schrödinger operator.

We may also introduce angular momentum operators defined by analogy to the classical angular momentum functions.

Definition 3.26 *For each pair (j, k) with $1 \leq j, k \leq n$, define the **angular momentum operator** $\hat{J}_{jk}$ by the formula*

$$\hat{J}_{jk} = X_j P_k - X_k P_j.$$

As in the classical case, we have $\hat{J}_{jk} = 0$ when $j = k$. When $j \neq k$, X_j and P_k commute, so the order of the factors in the definition of $\hat{J}_{jk}$ is not important. Explicitly, we have

$$\hat{J}_{jk} = -i\hbar \left(x_j \frac{\partial}{\partial x_k} - x_k \frac{\partial}{\partial x_j} \right).$$

The operator in parentheses is the angular derivative $(\partial/\partial\theta)$ in the (x_j, x_k) plane.

When $n = 3$, it is customary to use the quantum counterpart of the classical angular momentum *vector*, namely,

$$\hat{J}_1 := X_2 P_3 - X_3 P_2; \quad \hat{J}_2 := X_3 P_1 - X_1 P_3; \quad \hat{J}_3 := X_1 P_2 - X_2 P_1. \quad (3.51)$$

When $n = 3$, every $\hat{J}_{jk}$ with $j \neq k$ is one of the above three operators or the negative thereof.

3.11 Systems of Multiple Particles

Suppose now we have a system of N quantum particles moving in $\mathbb{R}^n$. If the particles are all of different types (e.g., one electron and one proton), then the Hilbert space for this system is $L^2(\mathbb{R}^{nN})$. That is, the wave function ψ of the system is a function of variables $\mathbf{x}^1, \mathbf{x}^2, \ldots, \mathbf{x}^N$, with each $\mathbf{x}^j$ belonging to $\mathbb{R}^n$. If we normalize ψ to be a unit vector in $L^2(\mathbb{R}^{nN})$, then $|\psi(\mathbf{x}^1, \mathbf{x}^2, \ldots, \mathbf{x}^N)|^2$ is to be interpreted as the joint probability distribution for the positions of the N particles.

We may introduce position operators X_k^j (the kth component of the position of the jth particle) and momentum operators P_k^j in obvious analogy to the definition for a single particle. The typical Hamiltonian operator for such a system is then

$$\hat{H}\psi(\mathbf{x}^1, \ldots, \mathbf{x}^N) = -\sum_{j=1}^{N} \frac{\hbar^2}{2m_j} \Delta_j \psi(\mathbf{x}^1, \ldots, \mathbf{x}^N) + V(\mathbf{x}^1, \ldots, \mathbf{x}^N)\psi(\mathbf{x}),$$

where m_j is the mass of the jth particle. Here Δ_j means the Laplacian with respect to the variable $\mathbf{x}^j \in \mathbb{R}^n$, with the other variables fixed.

As we will see in Chap. 19, the Hilbert space for a composite system, made up of various subsystems, is typically taken to be the (Hilbert) *tensor product* of the individual Hilbert spaces. In the present context, we may think of our system of being made up of N subsystems, each being one of the individual particles. Fortunately, there is a natural isomorphism (Proposition 19.12) between $L^2(\mathbb{R}^{nN})$ and the tensor product of N copies of $\mathbb{R}^n$, so that the approach we are taking here is consistent with the general philosophy.

If the particles in question are identical (say, all electrons), then there is an additional complication to the description of the Hilbert space for the system. In standard quantum theory, we are supposed to believe that "identical particles are indistinguishable." What this means is that the wave function should have the property that if we interchange, say, $\mathbf{x}^1$ with $\mathbf{x}^2$, then the new wave function should represent the same physical state as the original wave function. Recalling that two unit vectors in the quantum Hilbert space represent the same physical state if and only if they differ by a constant of absolute value 1, this means we should have

$$\psi(\mathbf{x}^2, \mathbf{x}^1, \mathbf{x}^3, \ldots, \mathbf{x}^N) = u\psi(\mathbf{x}^1, \mathbf{x}^2, \mathbf{x}^3, \ldots, \mathbf{x}^N),$$

for some constant u with $|u| = 1$. Applying this rule twice gives that ψ is $u^2\psi$, so evidently u must be either 1 or -1.

Particles in quantum mechanics are grouped into two types, according to whether the constant u in the previous paragraph is 1 or -1. Particles with $u = 1$ are called *bosons* and particles with $u = -1$ are called *fermions*. Whether a particle is a boson or a fermion is determined by the *spin* of the particle, a concept that we have not yet introduced. Nevertheless, we can say that particles without spin are bosons. For a collection of N identical spinless particles moving in $\mathbb{R}^3$, the proper Hilbert space is the *symmetric subspace* of $L^2(\mathbb{R}^{3N})$, that is, the space of functions in $L^2(\mathbb{R}^{3N})$ that are invariant under arbitrary permutations of the variables. We will have more to say about spin and systems of identical particles in Chaps. 17 and 19.

3.12   Physics Notation

In quantum mechanics, physicists almost invariably use the *Dirac notation* (or *bra-ket notation*) introduced by Dirac in 1939 [5]. This notation is made up of Notations 3.27–3.29 below. In this section, we explore the Dirac notation along with a few other notational differences between the mathematics and physics literature.

Before proceeding it is important to point out that when using Dirac notation, it is essential that the complex conjugate in the inner product should go on the *first* factor.

Notation 3.27 *A vector ψ in $\mathbf{H}$ is referred to as a **ket** and is denoted $|\psi\rangle$. A continuous linear functional on $\mathbf{H}$ is called a **bra**. For any $\phi \in \mathbf{H}$, let $\langle\phi|$ denote the bra given by*

$$\langle\phi|\,(\psi) = \langle\phi, \psi\rangle.$$

That is to say, $\langle\phi|$ is the "inner product with ϕ" functional. The bracket (or bra-ket) of two vectors $\phi, \psi \in \mathbf{H}$ is the result of applying the bra $\langle\phi|$ to the ket $|\psi\rangle$, namely the inner product of the ϕ and ψ, denoted $\langle\phi|\psi\rangle$.

If A is an operator on $\mathbf{H}$ and ϕ is a vector in $\mathbf{H}$, then we can form the linear functional $\langle \phi | A$, i.e., the linear map $\psi \mapsto \langle \phi | A\psi \rangle$. Physicists generally write an expression of this form as

$$\langle \phi | A | \psi \rangle .$$

This notation emphasizes that there are two different ways of thinking of this quantity. We may think of $\langle \phi | A | \psi \rangle$ either as the linear functional $\langle \phi | A$ applied to the vector $| \psi \rangle$, or as the linear functional $\langle \phi |$ applied to the vector $A | \psi \rangle$.

Notation 3.28 *For any ϕ and ψ in $\mathbf{H}$, the expression $|\phi\rangle\langle\psi|$ denotes the linear operator on $\mathbf{H}$ given by*

$$(|\phi\rangle\langle\psi|)(\chi) = |\phi\rangle\langle\psi|\chi\rangle = \langle\psi|\chi\rangle\,|\phi\rangle .$$

That is, in mathematics notation, $|\phi\rangle\langle\psi|$ is the operator sending χ to $\langle\psi,\chi\rangle\,\phi$.

The operator $|\phi\rangle\langle\psi|$ associates to each (ket) vector $|\chi\rangle$ a new vector in the only way that makes notational sense: We interpret $|\phi\rangle\langle\psi||\chi\rangle$ as the vector $|\phi\rangle$ multiplied by the scalar $\langle\psi|\chi\rangle$.

Notation 3.29 *Given a family of vectors in $\mathbf{H}$ labeled by, say, three indices n, l, and m, rather than denoting these vectors as $|\psi_{n,l,m}\rangle$, a physicist will denote them simply as $|n, l, m\rangle$.*

This notation is not without its pitfalls. If we have two different sets of vectors labeled by the same set of indices, a mathematician can simply label them as $\phi_{n,l,m}$ and $\psi_{n,l,m}$, but the physicist has a problem.

As an example of the Dirac notation, suppose that an operator $\hat{H}$ has an orthonormal basis of eigenvectors ψ_n. A physicist would express the decomposition of a general vector in terms of this basis as

$$I = \sum_n |n\rangle\langle n| , \tag{3.52}$$

where ψ_n is represented simply as $|n\rangle$ and where $|n\rangle\langle n|$ is (given that $|n\rangle$ is a unit vector) the orthogonal projection onto the one-dimensional subspace spanned by the vector $|n\rangle$.

Notation 3.30 *In the physics literature, the complex conjugate of a complex number z is denoted as z^*, rather than $\bar{z}$, as in the mathematics literature. What a mathematician calls the adjoint of an operator and denotes by A^*, a physicist calls the Hermitian conjugate of A and denotes by $A^\dagger$. Physicists refer to self-adjoint operators as Hermitian.*

We may express the concept of an adjoint (or Hermitian conjugate) of an operator using Dirac notation, as follows. If A is a bounded operator on $\mathbf{H}$, then $A^\dagger$ is the unique bounded operator such that

$$\langle \psi | A = \langle A^\dagger \psi | .$$

One peculiarity of the physics literature on quantum mechanics is a conspicuous failure of most articles to state what the Hilbert space is. Rather than starting by defining the Hilbert space in which they are working, physicists generally start by writing down the commutation relations that hold among various operators on the space. Thus, for example, a physicist might begin with position and momentum operators X and P, satisfying $[X, P] = i\hbar I$, without ever specifying what space these operators are operating on. The justification for this omission is, presumably, the Stone–von Neumann theorem, which asserts that (provided the operators satisfy the expected "exponentiated" relations) there is, up to unitary equivalence, only one Hilbert space with operators satisfying these relations and on which the operators act irreducibly. (See Chap. 14 for a precise statement of the result.) It is, nevertheless, disconcerting for a mathematician to encounter an entire paper full of computations involving certain operators, without any specification of what space these operators are operating on, let alone *how* the operators act on the space.

This practice among physicists represents something of a role reversal. In the setting of linear algebra, for example, a mathematician might say, "Let V be a n-dimensional vector space over $\mathbb{R}$." If a physicist says, "Oh, so it's $\mathbb{R}^n$," the mathematician will reply, "No, no, you don't have to choose a basis." By contrast, in quantum mechanics, it is the physicist who does not want to choose a particular realization of the space. A physicist will simply write down the commutation relations between, say, X and P. If pressed, the physicist might say that he is working in an irreducible representation of those relations. If a mathematician then says, "Oh, so it's $L^2(\mathbb{R})$," the physicist will reply, "No, no, there is no preferred realization."

Notation 3.31 *Given an irreducible representation of the canonical commutation relations, and given a vector ψ in the corresponding Hilbert space, a physicist will speak of the position wave function $\psi(x)$, defined by*

$$\psi(x) = \langle x|\psi\rangle . \qquad (3.53)$$

Here, $\langle x|$ is the bra associated with the ket $|x\rangle$, where $|x\rangle$ is supposed to be an eigenvector for the position operator with eigenvalue x.

See, again, Chap. 14 for the precise notion of "irreducible representation of the canonical commutation relations." One may similarly define the *momentum wave function* by taking the inner product of ψ with the eigenvectors of the momentum operator, which are also non-normalizable. See Sect. 6.6 for details.

A mathematician might find Notation 3.31 objectionable on the grounds that the operator X does not actually have any eigenvectors. After all, it is harmless, in view of the Stone–von Neumann theorem, to work in the "Schrödinger representation," in which our Hilbert space is $L^2(\mathbb{R})$ and the position operator X is just multiplication by x. Given a number x_0, there is no nonzero element ψ of $L^2(\mathbb{R})$ for which $X\psi = x_0\psi$. After all, any ψ satisfying this equation would have to be supported at the point $x = x_0$, in which case ψ would equal zero almost everywhere and would be the zero element of $L^2(\mathbb{R})$. A physicist, on the other hand, would say that the desired eigenfunction is $\psi(x) = \delta(x - x_0)$, where δ is the Dirac delta-"function." The fact that $\delta(x - x_0)$ is not actually *in* the Hilbert space $L^2(\mathbb{R})$ does not concern the physicist; it is simply a "non-normalizable state." The mathematical theory of such non-normalizable states comes under the heading "generalized eigenvectors." See Sect. 6.6 for a discussion of this issue in the case of the eigenvectors of the momentum operator.

A more subtle issue regarding the "position eigenvectors" is that each eigenvector is unique only up to multiplication by a constant. If one wants the momentum operator to act on the position wave function, as defined by (3.53), in the usual way, one must make a consistent choice of normalization of the eigenvectors of the position operators. Specifically, one should choose the constants in such a way that the exponentiated momentum operator $\exp(iaP/\hbar)$ maps $|x\rangle$ to $|x + a\rangle$.

3.13   Exercises

1. Suppose that $\phi(t)$ and $\psi(t)$ are differentiable functions with values in a Hilbert space $\mathbf{H}$, meaning that the limit

$$\frac{d\phi}{dt} := \lim_{h \to 0} \frac{\phi(t + h) - \phi(t)}{h}$$

exists in the norm topology of $\mathbf{H}$ for each t, and similarly for $\psi(t)$. Show that

$$\frac{d}{dt}\langle \phi(t), \psi(t)\rangle = \left\langle \frac{d\phi}{dt}, \psi(t) \right\rangle + \left\langle \phi(t), \frac{d\psi}{dt} \right\rangle.$$

2. Suppose A and B are operators on a *finite-dimensional* Hilbert space and suppose that $AB - BA = cI$ for some constant c. Show that $c = 0$.

   Note: This shows that the commutation relations in (3.8) are a purely infinite-dimensional phenomenon.

3. If A is a bounded operator on a Hilbert space $\mathbf{H}$, then there exists a unique bounded operator A^* on $\mathbf{H}$ satisfying $\langle \phi, A\psi\rangle = \langle A^*\phi, \psi\rangle$ for all ϕ and ψ in $\mathbf{H}$. (Appendix A.4.3.) The operator A^* is called the *adjoint* of A, and A is called *self-adjoint* if $A^* = A$.

   (a) Show that for any bounded operator A and constant $c \in \mathbb{C}$, we have $(cA)^* = \bar{c}A^*$, where $\bar{c}$ is the complex conjugate of c.

(b) Show that if A and B are self-adjoint, then the operator

$$\frac{1}{i\hbar}[A, B]$$

is also self-adjoint.

4. Verify Proposition 3.19 using Proposition 3.14. Note that the operator $V'(X)$ means simply the operator of multiplication by the function $V'(x)$.

5. Suppose that ψ is a unit vector in $L^2(\mathbb{R})$ such that the functions $x\psi(x)$ and $x^2\psi(x)$ also belong to $L^2(\mathbb{R})$. Show that

$$\langle X^2 \rangle_\psi > \left(\langle X \rangle_\psi \right)^2 .$$

   Hint: Consider the integral

$$\int_{-\infty}^{\infty} (x - a)^2 |\psi(x)|^2 \, dx,$$

where $a = \langle X \rangle_\psi$.

6. Consider the Hamiltonian $\hat{H}$ for a quantum harmonic oscillator, given by

$$\hat{H} = -\frac{\hbar^2}{2m}\frac{d^2}{dx^2} + \frac{k}{2}x^2,$$

where k is the spring constant of the oscillator. Show that the function

$$\psi_0(x) = \exp\left\{ -\frac{\sqrt{km}}{2\hbar}x^2 \right\}$$

is an eigenvector for $\hat{H}$ with eigenvalue $\hbar\omega/2$, where $\omega := \sqrt{k/m}$ is the classical frequency of the oscillator.

   Note: We will explore the eigenvectors and eigenvalues of $\hat{H}$ in detail in Chap. 11.

7. Prove Proposition 3.23.

   Hint: Show that $[P(t), \hat{H}] = ([P, \hat{H}])(t)$ and $[X(t), \hat{H}] = ([X, \hat{H}])(t)$.

8. (a) Find the general solution to (3.43), where E is a negative real number. Show that the only such solution that satisfies the boundary conditions (3.44) is identically zero.

   (b) Establish the same result as in Part (a) for $E = 0$.

9. (a) Suppose ϕ and ψ are smooth functions on $[0, L]$ satisfying the boundary conditions (3.44). Using integration by parts, show that

$$\left\langle \phi, \hat{H}\psi \right\rangle = \left\langle \hat{H}\phi, \psi \right\rangle,$$

where $\hat{H} = -(\hbar^2/2m)\, d^2/dx^2$ and where

$$\langle \phi, \psi \rangle = \int_0^L \overline{\phi(x)}\psi(x)\ dx.$$

(b) Show that the result of Part (a) fails if ϕ and ψ are arbitrary smooth functions (not satisfying the boundary conditions).

10. Let $\hat{J}_1$, $\hat{J}_2$, and $\hat{J}_3$ be the angular momentum operators for a particle moving in $\mathbb{R}^3$. Using the canonical commutation relations (Proposition 3.25), show that these operators satisfy the commutation relations

$$\frac{1}{i\hbar}[\hat{J}_1, \hat{J}_2] = \hat{J}_3; \quad \frac{1}{i\hbar}[\hat{J}_2, \hat{J}_3] = \hat{J}_1; \quad \frac{1}{i\hbar}[\hat{J}_3, \hat{J}_1] = \hat{J}_2.$$

This is the quantum mechanical counterpart to Exercise 19 in the previous chapter.

4

The Free Schrödinger Equation

In this chapter, we consider various methods of solving the free Schrödinger equation in one space dimension. Here "free" means that there is no force acting on the particle, so that we may take the potential V to be identically zero. Thus, the free Schrödinger equation is

$$\frac{\partial \psi}{\partial t} = \frac{i\hbar}{2m}\frac{\partial^2 \psi}{\partial x^2},\tag{4.1}$$

subject to an initial condition of the form

$$\psi(x,0) = \psi_0(x).$$

We will identify some key features of solutions to this equation, such as the "spread of the wave packet" and the distinction between "phase velocity" and "group velocity." In particular, the notion of group velocity will confirm our expectation that a particle of momentum p should travel with velocity $v = p/m$.

Before attempting to solve the free Schrödinger equation, let us make a simple observation about the time evolution of the expected values of the position and momentum. If we apply Proposition 3.19 in the case that V is identically equal to zero, we have

$$\frac{d}{dt}\langle X \rangle_{\psi(t)} = \frac{1}{m}\langle P \rangle_{\psi(t)}$$

$$\frac{d}{dt}\langle P \rangle_{\psi(t)} = 0.$$

B.C. Hall, *Quantum Theory for Mathematicians*, Graduate Texts in Mathematics 267, DOI 10.1007/978-1-4614-7116-5_4, © Springer Science+Business Media New York 2013

Thus, the expectation value of P is independent of time, which then means that the expectation value of X is linear in time:

$$\langle X \rangle_{\psi(t)} = \langle X \rangle_{\psi_0} + \frac{t}{m} \langle P \rangle_{\psi_0}$$

$$\langle P \rangle_{\psi(t)} = \langle P \rangle_{\psi_0} .$$

Thus, the free Schrödinger equation is one of the special cases in which the expected values of the position and momentum exactly follow classical trajectories (and those classical trajectories are very simple in the case $V \equiv 0$).

4.1   Solution by Means of the Fourier Transform

We look for solutions of the free Schrödinger equation on $\mathbb{R}^1$ of the form

$$\psi(x,t) = e^{i(kx - \omega(k)t)}, \tag{4.2}$$

where k is the frequency in space and $\omega(k)$ is the frequency in time, which is an as-yet-undetermined function of k. (Of course, such a solution is not square-integrable in x for a fixed t, but we will find our way back to square-integrable solutions eventually.) Plugging this into (4.1) easily gives the formula for ω as a function of k:

$$\omega(k) = \frac{\hbar k^2}{2m}. \tag{4.3}$$

A formula of this sort, expressing the temporal frequency ω as a function of the spatial frequency k in a solution of some partial differential equation, is called a *dispersion relation*.

Observe that (4.2) can be written as

$$\psi(x,t) = \exp\left[ik\left(x - \frac{\omega(k)}{k}t \right) \right]. \tag{4.4}$$

Now, replacing a function $f(x)$ by $f(x - a)$ has the effect of shifting f to the right by a. Thus, the time-evolution has the effect of shifting the initial function to the right by an amount equal to $(\omega(k)/k)t$. This means that the function $\psi(x,t)$ is moving to the right with speed $\omega(k)/k$. This speed, for reasons that will be clearer in Sect. 4.3, is called the *phase velocity*.

The phase velocity, then, is the speed at which a pure exponential solution of our equation (the free Schrödinger equation) propagates. We compute the phase velocity as $\omega(k)/k = \hbar k/(2m)$. Now, we have said that a wave function of the form e^{ikx} represents a particle with momentum $p = \hbar k$. We thus arrive at the following curious conclusion.

Proposition 4.1 *The phase velocity of a particle with momentum $p = \hbar k$ is*

$$phase\ velocity = \frac{\omega(k)}{k} = \frac{\hbar k}{2m} = \frac{p}{2m}.$$

This velocity is half the velocity of a classical particle of momentum p.

Proposition 4.1 might make us think that our basic relation $p = \hbar k$ is off by a factor of 2. We will see, however, that the phase velocity, that is, the velocity of a pure exponential solution, is not the "real" velocity of a particle with momentum p. The real velocity is the "group velocity," which will turn out to be, as expected, p/m.

Leaving aside for now the question of the velocity, let us build up a general solution to (4.1) from solutions of the form (4.2). We make use of the Fourier transform, discussed in Appendix A.3. We can then express the solution to the free Schrödinger equation, for "nice" initial conditions, as a "superposition" of these pure exponential solutions.

Proposition 4.2 *Suppose that ψ_0 is a "nice" function, for example, a Schwartz function (Definition A.15). Let $\hat{\psi}_0$ denote the Fourier transform of ψ_0 and define $\psi(x,t)$ by*

$$\psi(x,t) = \frac{1}{\sqrt{2\pi}} \int_{-\infty}^{\infty} \hat{\psi}_0(k) e^{i(kx-\omega(k)t)} \, dk, \tag{4.5}$$

where $\omega(k)$ is defined by (4.3). Then $\psi(x,t)$ solves the free Schrödinger equation with initial condition ψ_0.

The assumption that ψ be a Schwartz function is stronger than necessary. The reader is invited to trace through the argument and find suitable weaker conditions.

Proof. Since the Fourier transform of a Schwartz function is a Schwartz function, $\hat{\psi}_0(k)$ will decay faster than $1/k^4$ as k tends to $\pm\infty$. Meanwhile, by integrating the derivative of the function e^{ikx}, we obtain the estimate

$$\left| \frac{e^{ik(x+h)} - e^{ikx}}{h} \right| \leq |k|.$$

We can then apply dominated convergence, using $|k| \left| \hat{\psi}_0(k) \right|$ as our dominating function, to move a derivative with respect to x under the integral sign in the formula for $\psi(x,t)$. This derivative pulls down a factor of ik inside the integral. The decay of $\hat{\psi}_0$ allows us to repeat this argument to move a second derivative with respect inside the integral. We can also move a derivative with respect to t inside the integral, by a similar argument.

Since $\exp\{i(kx - \omega(k)t)\}$ satisfies the Schrödinger equation for each fixed k, differentiation under the integral shows that $\psi(x,t)$ satisfies the Schrödinger equation as well. The Fourier inversion formula shows that $\psi(x,0) = \psi_0(x)$. ∎

Proposition 4.3 *If* $\psi(x,t)$ *is as in Proposition 4.2, then the Fourier transform of* $\psi(x,t)$*, with respect to* x *with* t *fixed, is given by*

$$\hat{\psi}(k,t) = \hat{\psi}_0(k)\exp\left[-i\frac{\hbar k^2 t}{2m}\right]. \qquad (4.6)$$

Proof. We can write (4.5) as

$$\psi(x,t) = \frac{1}{\sqrt{2\pi}}\int_{-\infty}^{\infty} e^{ikx}\left[\hat{\psi}_0(k)e^{-i\omega(k)t}\right]\,dk.$$

By the uniqueness of the Fourier decomposition (i.e., the injectivity of the inverse Fourier transform, which follows from the Plancherel formula), the Fourier transform of $\psi(x,t)$ (with respect to x) must be the function in square brackets. Putting in the expression (4.3) for $\omega(k)$ establishes the desired result. ∎

Now, the Fourier transform is a unitary map from $L^2(\mathbb{R})$ onto $L^2(\mathbb{R})$. Thus, for *any* ψ_0 in $L^2(\mathbb{R})$, $\hat{\psi}_0$ also belongs to $L^2(\mathbb{R})$. Since the quantity multiplying $\hat{\psi}_0(k)$ in (4.6) has absolute value 1, the right-hand side of (4.6) is a well-defined square-integrable function of k, for any ψ_0 in $L^2(\mathbb{R})$, which has a well-defined inverse Fourier transform in $L^2(\mathbb{R})$.

Definition 4.4 *For any* $\psi_0 \in L^2(\mathbb{R})$*, define, for each* $t \in \mathbb{R}$*,* $\psi(x,t)$ *to be the unique element of* $L^2(\mathbb{R})$ *that has a Fourier transform (with respect to* x*) given by (4.6).*

Definition 4.4 defines a time-evolution for arbitrary initial conditions in $L^2(\mathbb{R})$. For general $\psi_0 \in L^2(\mathbb{R})$, however, $\psi(x,t)$ may not satisfy the Schrödinger equation in the classical, pointwise sense, simply because $\psi(x,t)$ may fail to be differentiable, either in x or in t. Nevertheless, $\psi(x,t)$, as defined by Definition 4.4, always satisfies the Schrödinger equation in the weak (distributional) sense. See Exercise 1.

4.2   Solution as a Convolution

According to Proposition 4.3, we see that the Fourier transform of the time-t wave function is the product of the Fourier transform of ψ_0 and the function $\exp[-it\hbar k^2/(2m)]$. According to Proposition A.21, the inverse Fourier transform of a product of two sufficiently nice functions is $1/\sqrt{2\pi}$ times the *convolution* of the two separate inverse Fourier transforms. Here the convolution $\phi * \psi$ of two functions ϕ and ψ is defined to be

$$(\phi * \psi)(x) = \int_{-\infty}^{\infty} \phi(x-y)\psi(y)\,dy,$$

whenever the integral is convergent for all x.

Formally, then, we ought to have

$$\psi(x,t) = \psi_0 * K_t, \tag{4.7}$$

where

$$K_t = \frac{1}{\sqrt{2\pi}} \mathcal{F}^{-1} \left\{ \exp\left[-i\frac{\hbar k^2 t}{2m} \right] \right\}.$$

The problem with is idea is that the function $\exp[-it\hbar k^2/(2m)]$ is not a "nice" function in the usual sense. Certainly, this function is not the Fourier transform of some function in $L^1(\mathbb{R}) \cap L^2(\mathbb{R})$, because if it were, then the function would have to tend to zero at infinity (Proposition A.14). Therefore, we cannot directly apply Proposition A.21, even if ψ_0 is in $L^1(\mathbb{R}) \cap L^2(\mathbb{R})$.

Fortunately, the desired inverse Fourier transform can be computed as a convergent improper integral (Exercise 2), with the following result:

$$K_t(x) := \frac{1}{2\pi} \int_{-\infty}^{\infty} e^{ikx} \exp\left[-i\frac{\hbar k^2 t}{2m} \right] dk = \sqrt{\frac{m}{i2\pi\hbar t}} \exp\left\{ i\frac{mx^2}{2t\hbar} \right\}. \tag{4.8}$$

Here, the square root is the one with positive real part. The function K_t is called the *fundamental solution* of the free Schrödinger equation. (See Fig. 4.1.) This function does indeed satisfy the free Schrödinger equation, as we can easily verify by direct differentiation.

The preceding discussion should make the following result plausible.

Theorem 4.5 *Suppose $\psi_0 \in L^2(\mathbb{R}) \cap L^1(\mathbb{R})$. Then $\psi(x,t)$, as defined by (4.5), may be computed for all $t \neq 0$ as*

$$\psi(x,t) = \sqrt{\frac{m}{2\pi i t \hbar}} \int_{-\infty}^{\infty} \exp\left\{ i\frac{m}{2t\hbar}(x-y)^2 \right\} \psi_0(y) \, dy.$$

The expression for $\psi(x,t)$ is $(2\pi)^{-1/2} K_t * \psi_0$, where K_t is as in (4.8).

Proof. For any set $E \subset \mathbb{R}$, let 1_E denote the indicator function of E, that is, the function that is 1 on E and 0 elsewhere. Then $K_t 1_{[-n,n]}$ belongs to $L^1(\mathbb{R}) \cap L^2(\mathbb{R})$ for any positive integer n. By Proposition A.21, then, we have

$$\mathcal{F}\left((K_t 1_{[-n,n]}) * \psi_0 \right) = \sqrt{2\pi} \mathcal{F}(K_t 1_{[-n,n]}) \mathcal{F}(\psi_0). \tag{4.9}$$

Because ψ_0 is in $L^1(\mathbb{R})$, it is easy to see that $K_t 1_{[-n,n]} * \psi_0$ converges pointwise to $K_t * \psi_0$. On the other hand, using the argument in Exercise 2, we can see that $\mathcal{F}(K_t 1_{[-n,n]})$ is bounded by a constant independent of n and converges pointwise to the function

$$\frac{1}{\sqrt{2\pi}} \exp\left[-i\frac{\hbar k^2 t}{2m} \right]. \tag{4.10}$$

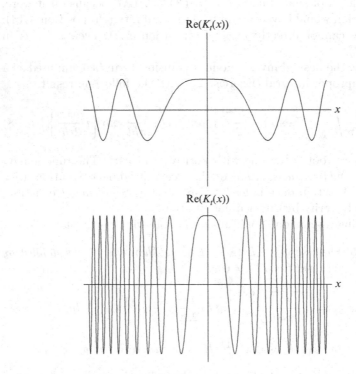

FIGURE 4.1. The real part of $K_t(x)$, for $t = 1$ (*top*) and $t = 0.2$ (*bottom*).

Equation (4.10) is enough to show that the right-hand side of (4.9) converges in $L^2(\mathbb{R})$ to the function

$$\exp\left[-i\frac{\hbar k^2 t}{2m}\right]\hat{\psi}_0(k).$$

By the Plancherel theorem, $K_t 1_{[-n,n]} * \psi_0$ must also be converging in $L^2(\mathbb{R})$, and the L^2 limit must coincide with the pointwise limit, which is $K_t * \psi_0$. Thus, taking limits on both sides of (4.9) shows that the Fourier transform of $K_t * \psi_0$ is what we want it to be. $\blacksquare$

In general, to be considered the *fundamental* solution of a certain equation, a function should converge to a Dirac δ-function (Example A.26), in the distribution sense, as t tends to zero. Since $|K_t(x)|$ is independent of x for each t, it might seem doubtful that K_t has this property. On the other hand, we can see $K_t(x)$ oscillates very rapidly except near $x = 0$. (See Fig. 4.1.) This oscillation causes the integral of $K_t(x)$ against some nice function $\psi(x)$ to be small, except for the part of the integral near $x = 0$. Indeed, because the Fourier transform of K_t converges to the constant function $1/\sqrt{2\pi}$ (which is what we get by formally taking the Fourier transform of the δ-function) as t tends to zero, it is not hard to show that K_t does, in fact, converge to a δ-function. The details of this verification are left to the reader.

4.3 Propagation of the Wave Packet: First Approach

Let us consider the Schrödinger equation in $\mathbb{R}^1$ with an initial condition ψ_0 that is a "wave packet," meaning a complex exponential multiplied by some function that localizes ψ_0 in space. Specifically, we take

$$\psi_0(x) = e^{ip_0 x/\hbar} A_0(x), \qquad (4.11)$$

where A_0 is some real, positive function and p_0 is a nonzero real number. (The case $p_0 = 0$ should be treated separately.) We also assume that A_0 is "slowly varying" compared to $e^{ip_0 x/\hbar}$, meaning that A_0 is approximately constant over many periods of the function $e^{ip_0 x/\hbar}$. (We will give a more precise meaning to the "slowly varying" condition shortly.) Thus, if we look at $\psi_0(x)$ on a distance scale of a small number of periods of the function $e^{ip_0 x/\hbar}$, then ψ_0 will look like a constant times $e^{ip_0 x/\hbar}$, which, as we have seen, represents a particle with momentum p_0. We expect, then, that the wave function ψ_0 represents a particle with momentum approximately equal to p_0.

Let us now try to solve the free Schrödinger equation in terms of the amplitude and phase of the wave function. We write

$$\psi(x,t) = A(x,t)e^{i\theta(x,t)}$$

where A and θ are real-valued functions. If we plug this expression for ψ into the free Schrödinger equation and then cancel a factor of $e^{i\theta(x,t)}$ from every term, we obtain the equation

$$\frac{\partial A}{\partial t} + i\frac{\partial \theta}{\partial t}A = \frac{i\hbar}{2m}\frac{\partial^2 A}{\partial x^2} - \frac{\hbar}{m}\frac{\partial A}{\partial x}\frac{\partial \theta}{\partial x} - \frac{i\hbar}{2m}A\left(\frac{\partial \theta}{\partial x}\right)^2 - \frac{\hbar}{2m}A\frac{\partial^2 \theta}{\partial x^2}. \quad (4.12)$$

Since A and θ are real-valued, we may separately equate the real and imaginary parts of (4.12), giving

$$\frac{\partial A}{\partial t} = -\frac{\hbar}{m}\frac{\partial A}{\partial x}\frac{\partial \theta}{\partial x} - \frac{\hbar}{2m}A\frac{\partial^2 \theta}{\partial x^2} \quad (4.13)$$

and (after dividing the imaginary part of (4.12) by A)

$$\frac{\partial \theta}{\partial t} = \frac{\hbar}{2m}\frac{1}{A}\frac{\partial^2 A}{\partial x^2} - \frac{\hbar}{2m}\left(\frac{\partial \theta}{\partial x}\right)^2. \quad (4.14)$$

Any solution to this system of partial differential equations will yield a solution $\psi(x,t) = A(x,t)e^{i\theta(x,t)}$ to the free Schrödinger equation.

Since we are assuming A is "slowly varying" compared to θ, it is reasonable to think that the first term on the right-hand side of (4.14) will be small compared to the second term. That is to say, we interpret the slowly varying condition to mean

$$\frac{1}{A}\frac{\partial^2 A}{\partial x^2} \ll \left(\frac{\partial \theta}{\partial x}\right)^2, \quad (4.15)$$

where the symbol $\ll$ means "much smaller than." We will take initial conditions such that (4.15) holds at $t = 0$, and then we will assume that (4.15) continues to hold at least for small positive times. We may then (to first approximation) drop the first term on the right-hand side of (4.14), giving the following simplified version of (4.14):

$$\frac{\partial \theta}{\partial t} = -\frac{\hbar}{2m}\left(\frac{\partial \theta}{\partial x}\right)^2. \quad (4.16)$$

We now look for a solution to the pair of equations (4.13) and (4.16) with initial conditions corresponding to (4.11).

Proposition 4.6 *A solution to the approximate equations (4.13) and (4.16) with initial condition $\theta(x,0) = p_0 x/\hbar$ is given by*

$$\theta(x,t) = \frac{p_0}{\hbar}\left(x - \frac{p_0}{2m}t\right) \quad (4.17)$$

and

$$A(x,t) = A_0\left(x - \frac{p_0}{m}t\right). \quad (4.18)$$

This yields an approximate solution to the free Schrödinger equation given by

$$\psi(x,t) = A_0\left(x - \frac{p_0}{m}t\right)\exp\left[i\frac{p_0}{\hbar}\left(x - \frac{p_0}{2m}t\right)\right]. \tag{4.19}$$

Note from (4.17) and (4.18) that if the "slowly varying" condition (4.15) holds at time 0, it will continue to hold for all positive times in our approximate solution.

Proof. Although (4.16) is a nonlinear equation, we can find a solution to it with the simple initial conditions $\theta(x,0) = p_0 x/\hbar$, namely,

$$\begin{aligned}\theta(x,t) &= \frac{p_0 x}{\hbar} - \frac{p_0^2}{2m\hbar}t \\ &= \frac{p_0}{\hbar}\left(x - \frac{p_0}{2m}t\right).\end{aligned} \tag{4.20}$$

Since $\partial\theta/\partial x = p_0/\hbar$ and $\partial^2\theta/\partial x^2 = 0$, if we plug (4.20) back into (4.13) we obtain

$$\frac{\partial A}{\partial t} = -\frac{p_0}{m}\frac{\partial A}{\partial x}.$$

The (presumably unique) solution to this linear equation with initial condition $A(x,0) = A_0(x)$ is

$$A(x,t) = A_0\left(x - \frac{p_0}{m}t\right), \tag{4.21}$$

as claimed. ∎

We hope that the solution (4.19) to the system of equations (4.13) and (4.16) is a close approximation to the solution to the original pair of equations (4.13) and (4.14)—assuming, of course, that A_0 is slowly varying compared to $\theta_0(x) = p_0 x/\hbar$. It is not especially easy to estimate directly how rapidly solutions to (4.13) and (4.16) diverge from solutions to (4.13) and (4.14). We will therefore leave an estimate of the error in our approximation until the next section, where we will obtain the same approximate solution by a different method.

Note that a function of the form $f(x,t) = \phi(x-vt)$ is moving to the right with constant velocity v. (If v is negative, then, of course, this means the function is moving to the left.) Observe that both the amplitude $A(x,t)$ and the phase $\exp\{i\theta(x,t)\}$ are of this form, but with two *different* velocities.

Conclusion 4.7 *In the approximate solution (4.19) to the free Schrödinger equation, the amplitude $A(x,t)$ is moving with velocity p_0/m, whereas the phase $\theta(x,t)$ is moving with velocity $p_0/(2m)$. These two velocities are called the group velocity and the phase velocity, respectively:*

$$phase\ velocity = \frac{p_0}{2m}$$
$$group\ velocity = \frac{p_0}{m}.$$

Note that the formula for the phase velocity agrees with the one given previously in Sect. 4.1, the velocity of propagation of a pure exponential solution to the free Schrödinger equation. Indeed, nothing prevents us from taking $A_0 \equiv 1$, in which case the left-hand side of (4.15) is actually identically zero, so that a solution to (4.13) and (4.16) is actually a solution to (4.13) and (4.14).

Which of the velocities is the "real" velocity of the particle? The answer is: the group velocity. After all, the probability distribution for the particle's position is determined by the *amplitude* of the wave function and is unaffected by the phase. It is the amplitude that determines (as much as it can be determined) where the particle is. Thus, the true velocity of the particle should be the velocity at which the amplitude propagates. Figure 4.2 shows the propagation of the real part of a wave packet, with the motion of a single peak indicated by the shaded region. The phase velocity determines the speed at which the individual peaks in the real part of ψ move, whereas the group velocity determines the speed of the packet as a whole. Since the peak we are tracking lags well behind the motion of the whole packet, we see that the phase velocity is smaller than the group velocity.

We should expect that solutions to our approximate equations (4.13) and (4.16) will diverge slowly over time from solutions to the free Schrödinger equation (4.13) and (4.14). For sufficiently long times, there may be a significant difference between approximate and true solutions. This expectation is confirmed in Sect. 4.5, where we investigate the spread of the wave packet, a phenomenon that is not seen in our approximation.

4.4   Propagation of the Wave Packet: Second Approach

We have seen that the general solution of the free Schrödinger equation can be obtained by means of the Fourier transform as

$$\psi(x,t) = \frac{1}{\sqrt{2\pi}} \int_{-\infty}^{\infty} \hat{\psi}_0(k) \exp\left[i\left(kx - \omega(k)t\right)\right] dk, \qquad (4.22)$$

where

$$\omega(k) = \frac{\hbar k^2}{2m}. \qquad (4.23)$$

Let us assume that ψ_0 has approximate momentum equal to p_0. Thus, we expect that $\hat{\psi}_0(k)$ will be concentrated near $k_0 := p_0/\hbar$. If that is the case, then only the values of k close to k_0 are important. For k close to k_0, we use the first-order Taylor expansion

$$\omega(k) \approx \omega(k_0) + \omega'(k_0)(k - k_0), \qquad (4.24)$$

where for now we do not put in the explicit formula for $\omega'(k_0)$.

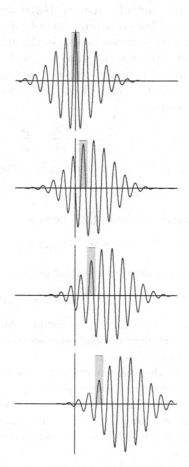

FIGURE 4.2. Propagation of Re[ψ], with motion of a single peak shaded.

Inserting (4.24) into (4.22), we get two factors that are independent of k and come outside the integral, leaving us with

$$\psi(x,t) \approx \frac{1}{\sqrt{2\pi}} e^{i\omega'(k_0)k_0 t} e^{-i\omega(k_0)t} \int_{-\infty}^{\infty} \hat{\psi}_0(k) \exp\left[ik(x - \omega'(k_0)t)\right] \, dk$$

$$= e^{i\omega'(k_0)k_0 t} e^{-i\omega(k_0)t} \psi_0(x - \omega'(k_0)t). \tag{4.25}$$

Note that the factors in front of $\psi_0(x - \omega'(k_0)t)$ are simply constants, that is, independent of x. These constants do not affect the "state" of the system, in that we have said that two vectors in the quantum Hilbert space that differ by a constant represent the same physical state. Ignoring these constants, we are left with the factor of $\psi_0(x - \omega'(k_0)t)$, which is simply shifting to the right at speed $\omega'(k_0)$. Thus, the (approximate) velocity at which our wave packet is moving is

$$\text{velocity} \approx \omega'(k_0) = \frac{\hbar k_0}{m} = \frac{p_0}{m}.$$

Let us consider the special case in which ψ_0 is of the form

$$\psi_0(x) = e^{ik_0 x} A_0(x),$$

where A_0 is real and positive. Then (4.25) becomes

$$e^{i\omega'(k_0)k_0 t} e^{-i\omega(k_0)t} e^{ik_0(x - \omega'(k_0)t)} A_0(x - \omega'(k_0)t).$$

After canceling the terms involving $\omega'(k_0)k_0 t$ in the exponent, we obtain

$$\psi(x,t) \approx e^{i(k_0 x - \omega(k_0)t)} A_0(x - \omega'(k_0)t).$$

Recalling that $p_0 = \hbar k_0$ and putting in the formula for ω, we see that this approximation to $\psi(x,t)$ is precisely the same as the one we obtained, by a different method, in Proposition 4.6.

As in Sect. 4.3, we see that the velocity at which a pure exponential solution of the free Schrödinger equation propagates [namely, $\omega(k_0)/k_0 = \hbar k_0/(2m)$] is not the same as the velocity at which the overall wave packet propagates. Rather, as seen in (4.25), the wave packet propagates at a velocity given by $\omega'(k_0) = \hbar k_0/m$. We may summarize this conclusion in the following proposition.

Proposition 4.8 *The speed at which a pure exponential solution of the free Schrödinger equation propagates is*

$$phase \ velocity = \frac{\omega(k_0)}{k_0} = \frac{\hbar k_0}{2m} = \frac{p_0}{2m}.$$

By contrast, the (approximate) speed at which the wave packet propagates is

$$group \ velocity = \left.\frac{d\omega}{dk}\right|_{k=k_0} = \frac{\hbar k_0}{m} = \frac{p_0}{m}.$$

The disadvantage of the method we used in Sect. 4.3 is that it does not easily yield estimates on how big an error there is in our approximation. In the current section, however, we can estimate the error by comparing the Fourier transforms of the exact solution and the approximate solution. Our error estimate will involve a quantity κ defined as follows:

$$\kappa = \left[\int_{-\infty}^{\infty} \left| \hat{\psi}_0(k) \right|^2 (k - k_0)^4 \, dk \right]^{1/4}. \tag{4.26}$$

The quantity κ is, roughly, half the width of the interval around k_0 on which most of $\hat{\psi}(k)$ is concentrated. If, for example, $\hat{\psi}$ is supported in the interval $[k_0 - \varepsilon, k_0 + \varepsilon]$, then $\kappa \leq \varepsilon$, assuming that ψ—and therefore $\hat{\psi}$—is a unit vector. (A more common measure of concentration would replace $(k - k_0)^4$ by $(k - k_0)^2$ and the fourth root of the integral by the square root. But the "quartic" measure of concentration in (4.26) is the one that arises in estimating the error of our approximations in this section.)

Proposition 4.9 *Let $\psi(x, t)$ be the exact solution to the free Schrödinger equation with initial condition ψ_0, and let $\phi(x, t)$ be the approximate solution given by the right-hand side of (4.25). Then the following L^2 estimate holds:*

$$\| \psi(x, t) - \phi(x, t) \|_{L^2(\mathbb{R})} \leq \frac{|t| \, \hbar \kappa^2}{2m} = |t| \, \omega(\kappa), \tag{4.27}$$

where the L^2 norm is with respect to x with t fixed and where $\omega(\cdot)$ is defined by (4.23).

Equation (4.27) means that the L^2 norm of the error will be small, provided that

$$|t| \ll \frac{1}{\omega(\kappa)}.$$

If κ is much smaller than k_0, then $1/\omega(\kappa)$ will be much larger than $1/\omega(k_0)$. That means that the timescale on which the true and approximate solutions diverge will be long compared to the timescale on which our approximate solution is oscillating.

Proof. Let $\hat{\psi}(k, t)$ and $\hat{\phi}(k, t)$ denote the Fourier transforms of ϕ and ψ with respect to x, with t fixed. From (4.22) we can read off that

$$\hat{\psi}(k, t) = e^{-i\omega(k)t} \hat{\psi}_0(k).$$

Meanwhile, $\hat{\phi}(k, t)$ is obtained from $\hat{\psi}(k, t)$ by replacing $\omega(k)$ by the right-hand side of (4.24). Now, direct calculation shows that

$$\omega(k) - (\omega(k_0) + \omega'(k_0)(k - k_0)) = \frac{\hbar}{2m}(k - k_0)^2.$$

From this expression and the elementary estimate $\left|e^{i\theta} - e^{i\phi}\right| \le |\theta - \phi|$, we obtain

$$\left|\hat{\psi}(k,t) - \hat{\phi}(k,t)\right| \le \frac{|t|\,\hbar}{2m}(k - k_0)^2 \left|\hat{\psi}_0(k)\right|. \qquad (4.28)$$

The estimate (4.27) then follows by the Plancherel theorem and the definition of κ. ∎

For a more detailed version of the approach used in this section, see Sect. 5.6 of [30].

4.5   Spread of the Wave Packet

We use the uncertainty (Definition 3.13) $\Delta_\psi X$ in the position of the particle as a measure of the "width" of $\psi(x)$ as a function of x. At the level of approximation considered in the previous two sections, the uncertainty in the position of a free particle is independent of time. After all, in the approximate solution (4.19), the amplitude of the wave function simply shifts to the right at a speed equal to the group velocity, without changing shape. A more precise calculation, however, shows that after sufficiently long times, the wave packet spreads out in space. (Exercise 7 gives an idea of the time scale on which this spread takes place.)

We can compute the time-evolution of the uncertainty in the particle's position without having to solve the full Schrödinger equation, by using Proposition 3.14 from Chap. 3. We start by observing that for a free particle, our Hamiltonian is simply $P^2/(2m)$, which commutes with P. It follows that the expected value and uncertainty for the particle's momentum (and, indeed, the entire probability distribution of the momentum) are independent of time. Meanwhile, to compute the time-dependence of $\langle X \rangle$ and $\langle X^2 \rangle$, we use Proposition 3.14 along with the commutation relation $[X, P] = i\hbar I$ (Proposition 3.8).

Proposition 4.10 *For a wave function $\psi(x,t)$ evolving according to the free Schrödinger equation on $\mathbb{R}^1$, the expectation values for X and X^2 evolve as follows:*

$$\langle X \rangle_{\psi(t)} = \langle X \rangle_{\psi_0} + \frac{t}{m}\langle P \rangle_{\psi_0}$$

and

$$\langle X^2 \rangle_{\psi(t)} = \langle X^2 \rangle_{\psi_0} + \frac{t}{m}\langle XP + PX \rangle_{\psi_0} + \frac{t^2}{m^2}\langle P^2 \rangle_{\psi(0)}.$$

These relations imply the following result:

$$\left(\Delta_{\psi(t)}X\right)^2$$
$$= \frac{t^2}{m^2}\left(\Delta_{\psi_0}P\right)^2 + \frac{t}{m}\left(\langle XP + PX \rangle_{\psi_0} - 2\langle X \rangle_{\psi_0}\langle P \rangle_{\psi_0}\right) + \left(\Delta_{\psi_0}X\right)^2.$$

For a unit vector ψ_0 in $L^2(\mathbb{R})$, the uncertainty $\Delta_{\psi_0} P$ in the momentum cannot be zero, because the uncertainty would be zero only if ψ_0 is an eigenvector for the momentum operator. But the eigenvectors for P are the functions of the form e^{ikx}, which are not in $L^2(\mathbb{R})$. Thus, the leading coefficient in the expression for $(\Delta_{\psi(t)} X)^2$ is never zero, and thus $\Delta_{\psi(t)} X$ tends to infinity as t tends to infinity.

Proof. We compute that

$$\begin{aligned}[P^2, X] &= P^2 X - PXP + PXP - XP^2 \\ &= P[P, X] + [P, X]P \\ &= -2i\hbar P.\end{aligned}$$

Thus (as we have already noted in Sect. 3.7.5),

$$\frac{d}{dt}\langle X \rangle_{\psi(t)} = \left\langle \frac{i}{\hbar}(-2i\hbar P) \right\rangle_{\psi(t)} = \frac{\langle P \rangle_{\psi(t)}}{m} = \frac{\langle P \rangle_{\psi_0}}{m}, \qquad (4.29)$$

where we have used in the last equality that the expected momentum is independent of time. Since the derivative of $\langle X \rangle_{\psi(t)}$ is constant, $\langle X \rangle_{\psi(t)}$ itself is a linear function of t, which gives the first result in the proposition.

Meanwhile, a little algebra shows that

$$\begin{aligned}[P^2, X^2] &= P[P, X]X + [P, X]PX + XP[P, X] + X[X, P]P \\ &= -2i\hbar(PX + XP),\end{aligned}$$

and

$$[P^2, PX + XP] = P[P^2, X] + [P^2, X]P = -4i\hbar P^2.$$

Thus

$$\frac{d}{dt}\langle X^2 \rangle_{\psi(t)} = \frac{i}{2m\hbar}\langle [P^2, X^2] \rangle_{\psi(t)} = \frac{1}{m}\langle XP + PX \rangle_{\psi(t)}$$

and

$$\begin{aligned}\frac{d^2}{dt^2}\langle X^2 \rangle_{\psi(t)} &= \frac{i}{\hbar}\frac{1}{m}\frac{1}{2m}\langle [P^2, XP + PX] \rangle_{\psi(t)} \\ &= \frac{2}{m^2}\langle P^2 \rangle_{\psi(t)} = \frac{2}{m^2}\langle P^2 \rangle_{\psi_0}.\end{aligned}$$

Since the second derivative of $\langle X^2 \rangle_{\psi(t)}$ is independent of t, $\langle X^2 \rangle_{\psi(t)}$ itself is a quadratic polynomial in t, the coefficients of which are determined by the value of $\langle X \rangle_{\psi(t)}$ and its first two time-derivatives at $t = 0$. This leads to the second result in the proposition. The last result follows by direct calculation. ∎

4.6 Exercises

1. A locally integrable function $\psi(x,t)$ satisfies the free Schrödinger equation in the weak (or distributional) sense if for each smooth compactly supported function χ, we have

$$\int_{\mathbb{R}^2} \psi(x,t) \left[\frac{\partial \chi}{\partial t} + \frac{i\hbar}{2m} \frac{\partial^2 \chi}{\partial x^2} \right] \, dx \, dt = 0. \qquad (4.30)$$

[One obtains (4.30) by assuming $\partial\psi/\partial t - (i\hbar/2m)\partial^2\psi/\partial x^2$ is zero, integrating against $\chi(x,t)$, and then formally integrating by parts.]

(a) Show that if $\psi(x,t)$ is smooth as a function of x and t then ψ satisfies the free Schrödinger equation in the pointwise sense if and only if ψ satisfies the free Schrödinger equation in the weak sense.

Hint: Proposition A.23 may be useful.

(b) For any $\psi_0 \in L^2(\mathbb{R})$, define $\psi(x,t)$ by Definition 4.4. Show that ψ satisfies the free Schrödinger equation in the weak sense.

First show that the function ψ_A given by

$$\psi_A(x,t) = \frac{1}{\sqrt{2\pi}} \int_{-A}^{A} \hat{\psi}_0(k) e^{i(kx-\omega(k)t)} \, dk$$

satisfies the free Schrödinger equation in the weak sense, for each A.

2. (a) Show that for any $a \in \mathbb{C}$ with $\mathrm{Re}(a) > 0$,

$$\left(\int_{-\infty}^{\infty} e^{-x^2/(2a)} \, dx \right)^2 = \int_{\mathbb{R}^2} e^{-(x^2+y^2)/(2a)} \, dx \, dy$$

$$= 2\pi a,$$

where the integral over $\mathbb{R}^2$ can be evaluated using polar coordinates. Conclude that

$$\int_{-\infty}^{\infty} e^{-x^2/(2a)} \, dx = \sqrt{2\pi a}, \qquad (4.31)$$

where the square root is the one with positive real part.

(b) Show that for all $A, B > 0$ we have

$$\int_A^B e^{-x^2/(2a)} \, dx = -\frac{a}{x} e^{-x^2/(2a)} \Big|_A^B + \int_A^B \frac{a}{x^2} e^{-x^2/(2a)} \, dx$$

for any nonzero complex number a. Using this, show that the integral in (4.31) is convergent for all nonzero a with $\mathrm{Re}\, a \geq 0$, provided the integral is interpreted as an improper integral (i.e., the limit as A tends to infinity of an integral from $-A$ to A).

(c) Now show that the result of Part (a) is valid also for nonzero values of a with $\operatorname{Re} a = 0$.

Hint: Given $\beta \neq 0$, show that the (improper) integral from A to ∞ of $\exp[-x^2/(2(\alpha + i\beta))]$ is small for large A, uniformly in $\alpha \in [0, 1]$.

(d) Show that

$$\frac{1}{2\pi} \int_{-\infty}^{\infty} e^{ikx} e^{-it\hbar k^2/(2m)} \, dk = \sqrt{\frac{m}{2\pi i \hbar t}} e^{imx^2/(2t\hbar)},$$

where the integral is interpreted as an improper integral and the square root is the one with positive real part.

3. Suppose ϕ is a Schwartz function (Definition A.15) and ψ belongs to $L^2(\mathbb{R})$. Show that the convolution $\phi * \psi$ is smooth (infinitely differentiable).

4. Consider the *heat equation* for a function $\psi(x, t)$, given by

$$\frac{\partial \psi}{\partial t} = \alpha \frac{\partial^2 \psi}{\partial x^2},$$

where α is a constant, subject to the initial condition $\psi(x, 0) = \psi_0(x)$.

(a) Derive a differential equation for $\hat{\psi}(k, t)$, the Fourier transform of a solution of the heat equation with respect to x, with t fixed, assuming that $\psi(x, t)$ is a "nice" function of x for each t. Solve this equation subject to the initial condition $\hat{\psi}(k, 0) = \hat{\psi}_0(k)$.

(b) Obtain an expression for the solution to the heat equation as a convolution of ψ_0 with a "fundamental solution" to the heat equation.

Note: As we will discuss in Chap. 20, the heat equation can be thought of as a sort of "imaginary time" version of the free Schrödinger equation.

5. Suppose we take an initial condition in the free Schrödinger equation with initial phase given by $\theta_0(x) = p_0 x/\hbar$ and initial amplitude given by $A_0(x)$, as in (4.11). Suppose also that the initial amplitude is of the form

$$A_0(x) = \exp\left\{ -\frac{1}{2} \left(\frac{x - x_0}{L} \right)^2 \right\}.$$

Note that A_0 is centered around the point x_0 and that the parameter L is a measure of the "width" in space of our initial wave packet. A function of the form $\psi_0(x) = e^{ip_0 x/\hbar} A_0(x)$, with A_0 as above, is called a *Gaussian wave packet*.

Compute the quantity

$$\frac{1}{\left(\frac{\partial \theta_0}{\partial x}\right)^2} \left(\frac{1}{A_0} \frac{\partial^2 A_0}{\partial x^2}\right). \tag{4.32}$$

Assuming that $\hbar$ is small compared to Lp_0, show that (4.32) is small, except at points where our initial wave packet is very small.

Note: This shows that our "slowly varying" assumption (4.15) is reasonable for the case of Gaussian wave packets.

6. The *Klein–Gordon equation*, a proposed relativistic alternative to the Schrödinger equation, is the equation

$$\frac{1}{c^2} \frac{\partial^2 \psi}{\partial t^2} = \frac{\partial^2 \psi}{\partial x^2} - \frac{m^2 c^2}{\hbar^2} \psi,$$

where $m > 0$ is the mass of the particle and c is the speed of light.

(a) Obtain the dispersion relation for the Klein–Gordon equation, that is, the expression for $\omega(k)$ that makes the function $\exp[i(kx - \omega(k)t]$ a solution to the Klein–Gordon equation.

(b) Show that the *phase velocity* $\omega(k)/k$ satisfies $|\omega(k)/k| > c$, that the *group velocity* $d\omega(k)/dk$ satisfies $|d\omega/dk| < c$, and that

$$\text{(phase velocity)(group velocity)} = c^2.$$

Note: Since the Klein–Gordon equation is second order in time, there will be two possible values for $\omega(k)$ for each k, one positive and one negative. The results of Part (b) hold for both of the two "branches" of $\omega(k)$.

7. Consider the uncertainty $\Delta_{\psi(t)} X$ of a wave function $\psi(t)$ evolving according to the free Schrödinger equation. Show that

$$\left|\frac{d}{dt}\left(\Delta_{\psi(t)} X\right)\right| \leq \frac{\Delta_{\psi_0} P}{m} \tag{4.33}$$

for all t and that

$$\lim_{t \to +\infty} \frac{d}{dt}\left(\Delta_{\psi(t)} X\right) = \frac{\Delta_{\psi_0} P}{m}.$$

Note: By comparison,

$$\frac{d}{dt}\langle X \rangle_{\psi(t)} = \frac{\langle P \rangle_{\psi_0}}{m}. \tag{4.34}$$

If $\hat{\psi}_0(k)$ is concentrated in a sufficiently small region around a nonzero number $k_0 = p_0/\hbar$, then $\Delta_{\psi_0} P$ will be small compared to $\langle P \rangle_{\psi_0}$. In that case, by comparing (4.33) to (4.34), we see that the rate at which the wave packet spreads out is small compared to the rate at which the wave packet moves.

5

A Particle in a Square Well

5.1 The Time-Independent Schrödinger Equation

It is difficult to solve the time-dependent Schrödinger equation explicitly, even in relatively simple cases. (Even for the free Schrödinger equation, we made do in Chap. 4 with solutions that are either approximate or that involve an integral that is not explicitly evaluated.) Usually, then, one analyzes the time-independent Schrödinger equation (the eigenvector equation for $\hat{H}$) and then attempts to infer something about the time-dependent problem from the results. There are a number of problems, including the harmonic oscillator and the hydrogen atom, in which the time-independent Schrödinger equation can be solved explicitly.

In this section, we will consider a simple but instructive example, which can be solved by elementary methods. We consider the time-independent Schrödinger equation in $\mathbb{R}^1$, with a potential of the form

$$V(x) = \begin{cases} -C, & -A \leq x \leq A \\ 0, & |x| > A \end{cases}, \qquad (5.1)$$

where A and C are positive constants. The region $-A \leq x \leq A$ is the "square well" for the potential (Fig. 5.1).

Let us think first for a moment about the behavior of a *classical* particle in a square well. If we think of V as the limit of a sequence of potentials that change linearly from -1 to 0 in a small interval around ± 1, we may expect the following behavior for a particle in a square well. If the energy of the particle is negative, then the particle must be in the well. In that

B.C. Hall, *Quantum Theory for Mathematicians*, Graduate Texts
in Mathematics 267, DOI 10.1007/978-1-4614-7116-5_5,
© Springer Science+Business Media New York 2013

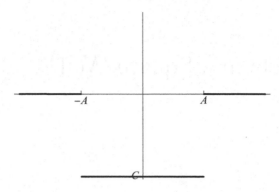

FIGURE 5.1. A square well potential.

case, it will move with constant speed until it hits the edge of the well, at which point it will reflect instantaneously off the wall and move with the same speed in the opposite direction. If the energy of the particle is positive, it will move always in the same direction, with speed equal to one constant when it is not in the well and speed equal to a different constant when it is in the well.

In the quantum case, we will be interested mainly in eigenvectors for the Schrödinger operator with negative eigenvalues ($E < 0$). Of course, on the quantum side of things, energy eigenvectors do not change in time, except for an overall phase factor. Nevertheless, since the classical particle with $E < 0$ spends the same amount of time in each part of the well, we may expect that the quantum particle will have approximately equal *probability* of being found in each part of the well. This expectation will be fulfilled for "highly excited states," such as the one in Fig. 5.7. For the quantum particle, however, there is a small but nonzero probability of finding the particle outside the well, which is impossible classically.

Our goal is to study the time-independent Schrödinger equation, that is, the eigenvalue equation

$$-\frac{\hbar^2}{2m}\frac{d^2\psi}{dx^2} + V(x)\psi(x) = E\psi(x), \tag{5.2}$$

where both the eigenvalues E and the associated eigenvectors ψ (or "eigenfunctions," in physics terminology) are as yet unknown. As a second-order linear ordinary differential equation, this equation always has (for any value of E) a two-dimensional solution space. We are, however, looking for solutions that lie in the quantum Hilbert space $L^2(\mathbb{R})$. We will see there are actually only a finitely many E's, all of them with $E < 0$, for which (5.2) has a nonzero solution in $L^2(\mathbb{R})$. In this case, then, the Schrödinger operator $\hat{H}$ has a *discrete spectrum* below zero and a *continuous spectrum* above zero.

5.2 Domain Questions and the Matching Conditions

Before starting to solve (5.2), we must give some heed to the unbounded nature of the Hamiltonian operator. The Schrödinger operator

$$\hat{H} = -\frac{\hbar^2}{2m}\frac{d^2}{dx^2} + V(X)$$

on the left-hand side of (5.2) is an *unbounded* operator, meaning that there is no constant C such that $\|\hat{H}\psi\| \le C\|\psi\|$, where $\|\cdot\|$ is the L^2 norm. On the other hand, we want to define $\hat{H}$ in such a way that it is self-adjoint. But according to Corollary 9.9, a self-adjoint operator that is defined on the whole Hilbert space must be bounded.

We conclude, then, that $\hat{H}$ is not going to be defined on the entire Hilbert space $L^2(\mathbb{R})$, but only on a dense subspace thereof. In practical terms, saying that $\hat{H}$ is not defined on the whole Hilbert space means simply that for many functions ψ in $L^2(\mathbb{R})$, the second derivative $d^2\psi/dx^2$ does not exist, or exists but fails to be in L^2. (In our example, the potential V is bounded, and so $V\psi$ will always be in L^2 provided that ψ is in L^2.)

Since the potential V for a square well is bounded, the domain of the Hamiltonian $\hat{H} = P^2/(2m) + V(X)$ is the same as the domain of the kinetic energy operator $P^2/(2m) = -(\hbar^2/2m)d^2/dx^2$. As we will see in Sect. 9.7, the domain of the kinetic energy operator may be described as the space of L^2 functions ψ for which $d^2\psi/dx^2$, computed in the weak or distributional sense (Appendix A.3.3), again belongs to $L^2(\mathbb{R})$. This condition is equivalent to the statement that there exists some L^2 function ϕ such that ψ is the second integral of ϕ (for some choice of the constants of integration).

Meanwhile, since our potential is piecewise constant, any solution ψ to (5.2) will be smooth except possibly at the transition points $x = \pm A$, and both ψ and ψ' will have left and right limits at A and $-A$. Indeed, on each of the intervals $(-\infty, -A)$, $(-A, A)$, and (A, ∞), any solution to (5.2) will be simply a linear combination of (real or complex) exponentials. For functions of this sort, it is not hard to see when we are in the domain of $\hat{H}$.

Proposition 5.1 *Suppose ψ is smooth on each of the intervals $(-\infty, -A)$, $(-A, A)$, and (A, ∞). Then ψ belongs to the domain of $\hat{H}$ [with potential function given by (5.1)] if and only if the (1) ψ and $d\psi/dx$ are continuous at $x = \pm A$, and (2) $d^2\psi/dx^2$ belongs to $L^2(\mathbb{R})$.*

Proof. Suppose first that ψ satisfies the conditions (1) and (2). Then it is not hard to see (Exercise 1) that the second derivative of ψ in the distribution sense is simply the function $d^2\psi/dx^2$, computed in the ordinary pointwise sense for $x \neq \pm A$. (The second derivative may not exist at $x = \pm A$, but we simply leave $d^2\psi/dx^2$ undefined at these two points, which form a set of measure zero.) Thus, $d^2\psi/dx^2$, computed in the distribution sense, is an element of $L^2(\mathbb{R})$.

On the other hand, if either ψ of ψ' has a discontinuity at $x = A$ or at $x = -A$, then (Exercise 1 again) the distributional derivative will contain either a multiple of a δ-function of a multiple of the derivative of δ-function at one of these points. But neither a δ-function nor the derivative of δ-function is a square-integrable function. ∎

Let us think about what the continuity condition on ψ and $d\psi/dx$ means in practical terms. Since V is constant on $(-\infty, -A)$, we can easily solve (5.2) on that interval, obtaining a two-dimensional solution space. Once we choose a solution from this solution space, then the values of ψ and $d\psi/dx$ as x approaches $-A$ from the left will serve as the initial conditions for solving (5.2) on $(-A, A)$. Thus, the requirement of continuity for ψ and $d\psi/dx$ serve as a "matching condition" between the solution on $(-\infty, -A)$ and the solution on $(-A, A)$. We cannot just separately pick any solution to (5.2) on $(-\infty, -A)$ and any solution on $(-A, A)$; at the boundary, the values of ψ and $d\psi/dx$ must match. (This same matching condition appears in elementary treatments of ordinary differential equations with discontinuous coefficients.)

Once we pick a solution on $(-\infty, -A)$ we get a unique solution on $(-A, A)$—and then the values of ψ and $d\psi/dx$ as we approach A from the left will serve as the initial conditions for solving (5.2) on (A, ∞). The conclusion is that once we pick a solution to (5.2) on $(-\infty, -A)$ (from the two-dimensional solution space), we have no additional choices to make; the differential equation along with the matching conditions give a unique way to extend the solution from $(-\infty, -A)$ to the whole real line.

5.3   Finding Square-Integrable Solutions

If $E > 0$, then any solution to (5.2) will be a combination of two complex exponentials in the range $x < -A$; such a function cannot be square-integrable unless it is identically zero. If, however, we take ψ to be identically zero in the region $x < -A$, then our continuity condition requires that ψ and $d\psi/dx$ approach 0 as x approaches $-A$ from the right. Thus, the matching conditions at $-A$ force the solution to be identically zero in $[-A, A]$ as well. Finally, by matching across $x = A$, we get an identically zero solution on $[A, \infty)$. Thus, for $E > 0$, any solution to (5.2) satisfying the continuity conditions in Proposition 5.1 must be identically zero. A similar analysis applies when $E = 0$, where the solutions to (5.2) on $(-\infty, A]$ would be of the form $c_1 + c_2 x$, which is square-integrable only if $c_1 = c_2 = 0$.

The conclusion, then, is that to have a chance to get a solution to (5.2) that is square-integrable and in the domain of $\hat{H}$, we must take $E < 0$. For $E < 0$, the solution to (5.2) on $(-\infty, -A)$ will be a linear combination of the two exponentials $\exp(\alpha x)$ and $\exp(-\alpha x)$, where

$$\alpha = \frac{\sqrt{2m\,|E|}}{\hbar}. \tag{5.3}$$

For ψ to be square-integrable over $(-\infty, -A)$, the coefficient of $\exp(-\alpha x)$ must be zero, since this term grows exponentially as x tends to $-\infty$. Thus, the value of ψ on $(-\infty, -A)$ must be $c\exp(\alpha x)$. Once we choose a value for c, we get a unique solution on $(-A, A)$ by matching ψ and ψ' across $x = -A$. We then get a unique solution on (A, ∞) by matching across $x = A$. The solution on (A, ∞) will be again be a linear combination of $\exp(\alpha x)$ and $\exp(-\alpha x)$. For ψ to be in L^2, we need the coefficient of $\exp(\alpha x)$ on (A, ∞) to be zero. We have no choice, however, about what ψ is on (A, ∞); the coefficient of $\exp(\alpha x)$ either comes out to be zero or it does not.

The conclusion, then, is that for any $E < 0$, there is a unique (up to a constant) solution to (5.2) that is square-integrable on the interval $(-\infty, -A)$. This solution then gives rise to a unique solution on $(-A, A)$ and then to a unique solution on (A, ∞), up to a constant. Unless we are lucky, the solution on (A, ∞) will grow exponentially and thus fail to be in L^2. Therefore, in most cases there will be no nonzero solution to (5.2) that satisfies the continuity condition and is square-integrable over the whole real line. The hope is that for *certain special values of* E, we will be able to find a solution that decays exponentially *both* on $(-\infty, -A)$ and on (A, ∞), in which case the solution will belong to $L^2(\mathbb{R})$.

It can be shown (Exercise 6) that there are no nonzero square-integrable solutions with $E \le -C$. Therefore, any square-integrable solutions to (5.2) that may exist must come from the range $-C < E < 0$. To analyze this range, let us rewrite the time-independent Schrödinger equation by dividing through by $-\hbar^2/(2m)$, yielding the equation

$$\frac{d^2\psi}{dx^2} = \begin{cases} \varepsilon\psi & |x| > A \\ -(c - \varepsilon)\psi & |x| < A \end{cases}. \tag{5.4}$$

where

$$\varepsilon = -\frac{2mE}{\hbar^2}$$

$$c = \frac{2mC}{\hbar^2}. \tag{5.5}$$

Note that although E is assumed to be negative, we have normalized ε to be positive; the condition $-C < E < 0$ corresponds to $0 < \varepsilon < c$.

Because our potential function V is even, it is easy to see that for any solution ψ to (5.4), the even and odd parts of ψ are also solutions. We can, therefore, analyze even solutions and odd solutions separately. We begin with the even case. For $x < -A$, every solution to (5.4) that is square-integrable over $(-\infty, A)$ is of the form

$$\psi(x) = ae^{\sqrt{\varepsilon}x}, \quad x \le -A. \tag{5.6}$$

Since we assume that ψ is even, we then have

$$\psi(x) = ae^{-\sqrt{\varepsilon}x}, \quad x \ge A. \tag{5.7}$$

Meanwhile, for $-A < x < A$, every even solution is of the form

$$\psi(x) = b\cos\left(\sqrt{c - \varepsilon}x\right). \tag{5.8}$$

Proposition 5.2 *Let ψ be the function defined in (5.6)–(5.8). Then there exist nonzero constants a and b so that ψ belongs to the domain of $\hat{H}$ if and only if the following matching condition holds:*

$$\sqrt{\varepsilon} = \sqrt{c - \varepsilon}\tan\left(\sqrt{c - \varepsilon}A\right). \tag{5.9}$$

Proof. Clearly both ψ and $d^2\psi/dx^2$ belong to $L^2(\mathbb{R})$. Thus, in light of Proposition 5.1, we need only ensure that $\psi(x)$ and $\psi'(x)$ are continuous at $x = \pm A$. Since the exponential functions are never zero, we may always ensure that ψ itself is continuous by taking any value we like for b and then choosing a appropriately Once ψ has been made to be continuous, ψ' will be continuous provided that $\psi'(x)/\psi(x)$ has the same value as we approach $\pm A$ from inside the well or from the outside. To obtain the condition (5.9), we compute ψ'/ψ from (5.6) and then from (5.8), evaluate both quantities at $x = -A$, and then equate the two values of ψ'/ψ. Because we have made our solution an even function, we get the same matching condition at $x = A$ as at $x = -A$.

Now, in deriving (5.9), we implicitly assumed that ψ is nonzero at $x = \pm A$. We do not, however, get any nonzero solutions in which $\psi(\pm A) = 0$. After all, at points where the cosine function in (5.8) is zero, its derivative is nonzero. But no choice of the constant in front of the exponentials (5.6) and (5.7) will produce a function that is zero but has a derivative that is nonzero. ∎

Proposition 5.3 *For all positive values of c and A, there exists at least one $\varepsilon \in (0, c)$ such that (5.9) holds.*

Proof. Case 1: $\sqrt{c}A < \pi/2$. In this case, as ε varies between 0 and c, the left-hand side of (5.9) will vary between 0 and some positive number, whereas the right-hand side of (5.9) will vary between some positive number and 0. By the intermediate value theorem, there must exist $\varepsilon \in (0, c)$ for which (5.9) holds. See Fig. 5.2.

Case 2: $\sqrt{c}A \geq \pi/2$. In this case, there is $\varepsilon_0 \in [0, c]$ for which $\sqrt{c - \varepsilon_0}A = \pi/2$. As ε decreases from c to ε_0, the right-hand side of (5.9) will vary from 0 to $+\infty$. Thus, for ε slightly larger than ε_0, the right-hand side of (5.9) will be larger than the left-hand side. By the intermediate value theorem, there must exist $\varepsilon \in (\varepsilon_0, c)$ for which (5.9) holds. See Fig. 5.3 for a case $\sqrt{c}A$ slightly larger than $\pi/2$ and Fig. 5.4 for a case with $\sqrt{c}A$ much larger than $\pi/2$. ∎

Note that if $\sqrt{c}A$ is much larger than $\pi/2$, then there will be multiple solutions of (5.9), as can be seen in Fig. 5.4.

We have found, then, at least one solution ψ to (5.4) that satisfies the matching condition and for which both ψ and ψ'' decay exponentially at infinity. Since this ψ belongs to the domain of $\hat{H}$, we have established the following result.

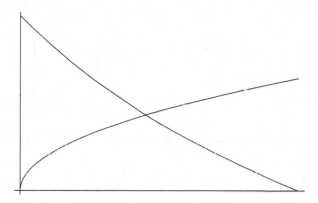

FIGURE 5.2. Solving the matching condition, Case 1.

Proposition 5.4 *For any positive values of A and C, there exists at least one value of E in the range $-C < E < 0$ for which (5.2) has a nonzero solution in the domain of $\hat{H}$, given by the formula*

$$\psi(x) = \begin{cases} \cos\left(\sqrt{c - \varepsilon}x\right) & -A \leq x \leq A \\ \cos\left(\sqrt{c - \varepsilon}A\right)\exp[-\sqrt{\varepsilon}(|x| - A)] & |x| \geq A \end{cases},$$

where c and ε are defined in (5.5) and where ε satisfies (5.9).

In Proposition 5.4, we have *not* normalized ψ to be a unit vector in $L^2(\mathbb{R})$, but rather have normalized ψ to equal 1 at the origin. In Figs. 5.5–5.7, we plot our eigenfunction in several different cases. In Fig. 5.5, we have a "shallow" well, with $\sqrt{c}A = 1$. In that case, we obtain only one even eigenvector, which is the *ground state* of the system (i.e., the eigenvector with the smallest eigenvalue). Next, we consider a "deep" well, with $\sqrt{c}A = 30$. For this well, the ground state is shown in Fig. 5.5 and an "excited state"

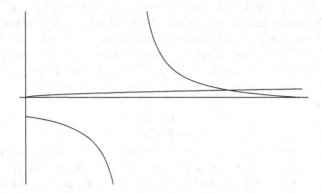

FIGURE 5.3. Solving the matching condition, Case 2a.

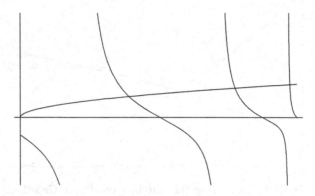

FIGURE 5.4. Solving the matching conditions, Case 2b.

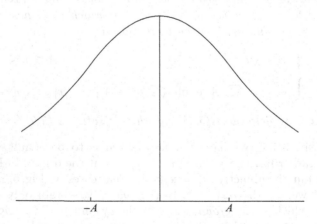

FIGURE 5.5. Ground state for a shallow potential well.

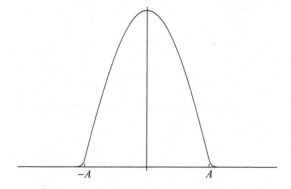

FIGURE 5.6. Ground state for a deep potential well.

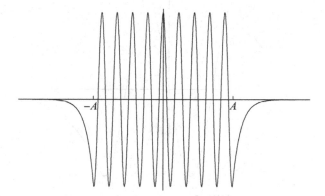

FIGURE 5.7. Excited state for a deep potential well.

(i.e., an eigenvector with an eigenvalue that is not the smallest) is shown in Fig. 5.7.

Note that in the shallow well, the ground state extends quite a bit beyond the interval $[-A, A]$, whereas in the deep well, the ground state goes to zero very quickly as soon as we move outside the well. On the other hand, the excited state in Fig. 5.7 extends comparatively far outside the well.

It is straightforward to adapt the preceding analysis to the odd case. The matching condition (5.9) is replaced by

$$\sqrt{\varepsilon} = -\sqrt{c - \varepsilon} \cot \left(\sqrt{c - \varepsilon} A \right) \tag{5.10}$$

(Exercise 2) and the formula for the eigenvectors is now

$$\psi(x) = \begin{cases} \sin \left(\sqrt{c - \varepsilon} x \right) & -A \leq x \leq A \\ \pm \sin \left(\sqrt{c - \varepsilon} A \right) \exp[-\sqrt{\varepsilon}(|x| - A)] & |x| \geq A \end{cases},$$

where we take the $+$ sign for $x > A$ and the $-$ sign for $x < -A$.

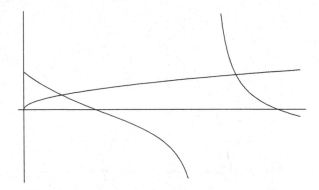

FIGURE 5.8. Matching condition for odd solutions.

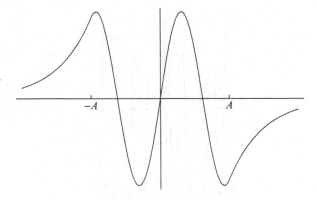

FIGURE 5.9. An odd solution.

If $\sqrt{c}A < \pi/2$, then the matching condition (5.10) will have no solutions, since the right-hand side of (5.10) will be negative for all $\varepsilon \in (0, c)$. For large values of $\sqrt{c}A$, there will be several solutions to (5.10). A typical matching scenario and an associated eigenfunction are plotted in Figs. 5.8 and 5.9.

5.4   Tunneling and the Classically Forbidden Region

Let us now briefly compare the classical situation to the quantum one. Classically, if a particle has energy E, then since the kinetic energy $p^2/(2m)$ is always non-negative, the particle simply cannot be located at a point x with $V(x) > E$. Thus, the region $V(x) \le E$ may be called the "classically allowed" region and the region $V(x) > E$ the "classically forbidden" region. In the case of a square well potential (5.1), if $-C < E < 0$, then the "well" itself (i.e., the region with $-A \le x \le A$) is the classically allowed region and the outside of the well (i.e., the region with $|x| > A$) is the classically forbidden region.

Quantum mechanically, if $\hat{H}\psi = E\psi$, then the particle has a definite value for the energy, namely E. We see, however, that such a particle has a nonzero probability of being located in the classically forbidden region. Note that although the wave function is not zero in the classically forbidden region, it does decay exponentially with the distance from the classically allowed region. That is to say, the quantum particle can penetrate some distance into the classically forbidden region. Note, however, that if E is much less than zero—i.e., ε is large—then a state with $\hat{H}\psi = E\psi$ will decay very rapidly outside the well (like $\exp[-\sqrt{\varepsilon}(|x| - A)]$).

More generally, we can think about the time-dependent Schrödinger equation for a particle with energy *approximately* equal to E. If we require that the energy be *exactly* equal to E, then there is no interesting time-dependence, since the solution to the time-dependent Schrödinger equation is simply a constant time ψ_0. We can, however, think of a particle where the uncertainty in the energy is nonzero but small. Suppose such a particle is traveling through a region with $V < E$ and then approaches a region with $V > E$ (a "potential barrier"). Classically, the particle would just reflect off of this barrier and go back in the other direction. Quantum mechanically, though, it is possible for the particle to "tunnel" through the potential barrier and come out the other side. That is to say, at some later time, there will be some non-negligible portion of the wave function on the far side of the barrier.

5.5 Discrete and Continuous Spectrum

Our analysis of the eigenvector equation (5.2) for $-C < E < 0$ shows that there are only finitely many values of E in this range for which we get square-integrable solutions. It is not hard to analyze the case $E \leq -C$ with the result that all nonzero solutions grow exponentially in at least one direction (Exercise 6). Meanwhile, for $E > 0$, any solution to (5.2) on $(-\infty, -A)$ has sinusoidal behavior and is not square-integrable unless it is identically zero, in which case (by our matching condition) the solution must be zero everywhere.

The upshot is that we obtain only finitely many square-integrable solutions to (5.2), up to multiplying each solution by a constant. Clearly, then, the "true" eigenvectors for $\hat{H}$ [i.e., the ones that actually belong to the Hilbert space $L^2(\mathbb{R})$] cannot form an orthonormal basis for $L^2(\mathbb{R})$. Nevertheless, the spectral theorem (Chap. 7) provides something like a orthonormal-basis decomposition of elements of $L^2(\mathbb{R})$ in terms of the solutions to (5.2). A general element ψ of $L^2(\mathbb{R})$ will be a sum of two terms. The first term is a linear combination of the true (L^2) eigenvectors for

$\hat{H}$, which have $E < 0$. The second term is a continuous superposition (i.e., an integral) of the non–square-integrable "generalized eigenvectors" with $E > 0$.

In Chap. 9, we will introduce the notion of the *spectrum* of a (possibly unbounded) self-adjoint operator A. We will see that a number λ belongs to the spectrum of A if for all $\varepsilon > 0$ there exists a unit vector ψ in the domain of A for which $\|A\psi - \lambda\psi\| < \varepsilon$. In the case of the Hamiltonian operator $\hat{H}$ with a square well potential, it is not hard to show that every real number E with $E \geq 0$ belongs to the spectrum of $\hat{H}$ (Exercise 4.).

It can be shown that if a number $E < 0$ is not an eigenvalue (i.e., if there are no nonzero L^2 solutions to $\hat{H}\psi = E\psi$), then E is not an element of the spectrum of $\hat{H}$. This result is hinted at by Exercise 5. Thus, the spectrum of $\hat{H}$ consists of a finite number of points in $(-C, 0)$ (at least one), together with the whole half line $[0, \infty)$.

5.6 Exercises

1. (a) Suppose ψ is a smooth function on each of the intervals $(-\infty, -A)$, $(-A, A)$, and (A, ∞) and that both ψ and ψ' are continuous at $x = A$ and at $x = -A$. Show that for any smooth function χ with compact support, we have

$$\int_{-\infty}^{\infty} \chi''(x)\psi(x) \, dx = \int_{-\infty}^{\infty} \chi(x)\psi''(x) \, dx, \qquad (5.11)$$

where we leave $\psi''(x)$ undefined at $x = \pm A$ if the second derivative does not exist at those points. (In light of Definition A.28, (5.11) means that the second derivative of ψ, in the distribution sense, is simply the function ψ''.)

Hint: Choose some interval $[-R, R]$ with $R > A$ containing the support of χ. Now use integration by parts separately on each of the intervals $[-R, -A]$, $[-A, A]$, and $[A, R]$, paying careful attention to the boundary terms.

   (b) Suppose now that ψ is a smooth function on each of the intervals $(-\infty, -A)$, $(-A, A)$, and (A, ∞), and that both ψ and ψ' have left and right limits at $x = \pm A$, but that, say, ψ' has a discontinuity at $x = -A$. Show that (5.11) has to be modified by adding a nonzero multiple of $\chi(-A)$ to the right-hand side.

2. Verify the matching condition (5.10) for odd solutions of the time-independent Schrödinger equation.

3. Let ω be a nonzero real number and consider a function of the form

$$\psi(x) = a \cos(\omega x) + b \sin(\omega x),$$

for real numbers a and b. If a and b are not both zero, show that for any $A \in \mathbb{R}$, we have

$$\lim_{B \to +\infty} \int_A^B \psi(x)^2 \, dx = +\infty.$$

4. Let f be a C^∞ function on the interval $(0, 1)$ with the property that $f(x) = 1$ for $0 < x < 1/3$ and $f(x) = 0$ for $2/3 < x < 1$. Then define a family of "cutoff" functions χ_n on $\mathbb{R}$ by the formula

$$\chi_n(x) = \begin{cases} 0 & |x| \geq n+1 \\ 1 & |x| \leq n \\ f(-x-n) & -(n+1) < x < -n \\ f(x-n) & n < x < n+1 \end{cases}.$$

Given any $E > 0$, let ψ be a nonzero solution to (5.2) for which $\psi(x)$ and $\psi'(x)$ are continuous at $x = \pm A$. Let $\psi_n = \psi \chi_n$. Show that ψ_n belongs to the domain of $\hat{H}$ and that

$$\lim_{n \to \infty} \frac{\left\| \hat{H}\psi_n - E\psi_n \right\|}{\|\psi_n\|} = 0.$$

Note: As we will see in Chap. 9, this implies that every real number E with $E > 0$ belongs to the *spectrum* of the operator $\hat{H}$.

Hint: In estimating $\|\psi_n\|$, it may be helpful to apply Exercise 3 to the real and imaginary parts of ψ outside the well.

5. Suppose $E < 0$ and suppose that there exists no nonzero square-integrable solutions to (5.2) for which ψ and ψ' are continuous. Let ψ be a nonzero solution of (5.2) for which $\psi(x)$ and $\psi'(x)$ are continuous at $x = \pm A$ and let ψ_n be as in Exercise 4. Show that

$$\frac{\left\| \hat{H}\psi_n - E\psi_n \right\|}{\|\psi_n\|}$$

does not tend to zero as n tends to infinity.

6. (a) Show that for $E < -C$, there are no nonzero square-integrable solutions to (5.2) for which ψ and ψ' are continuous.

   (b) Obtain the result of Part (a) when $E = -C$.

   Hint: Analyze the even and odd cases separately.

7. Let the *ground state* for a particle in a square well denote the eigenvector with the lowest (most negative) eigenvalue, which corresponds to the largest value for ε.

(a) Show that the ground state is always an even function. That is to say, show that the largest value of ε satisfying (5.9) is always larger than any solution to (5.10).

(b) Show that the ground state is a nowhere-zero function.

6
Perspectives on the Spectral Theorem

6.1 The Difficulties with the Infinite-Dimensional Case

Suppose A is a self-adjoint $n \times n$ matrix, meaning that $A_{kj} = \overline{A_{jk}}$ for all $1 \leq j, k \leq n$. Then a standard result in linear algebra asserts that there exist an orthonormal basis $\{\mathbf{v}_j\}_{j=1}^n$ for $\mathbb{C}^n$ and real numbers $\lambda_1, \ldots, \lambda_n$ such that $A\mathbf{v}_j = \lambda_j \mathbf{v}_j$. (See Theorem 18 in Chap. 8 of [24] and Exercise 4 in Chap. 7.)

We may state the same result in basis-independent language as follows. Suppose $\mathbf{H}$ is a finite-dimensional Hilbert space and A is a self-adjoint linear operator on $\mathbf{H}$, meaning that $\langle \phi, A\psi \rangle = \langle A\phi, \psi \rangle$ for all $\phi, \psi \in \mathbf{H}$. Then there exists an orthonormal basis of $\mathbf{H}$ consisting of eigenvectors for A with real eigenvalues.

Since there is a standard notion of orthonormal bases for general Hilbert spaces, we might hope that a similar result would hold for self-adjoint operators on infinite-dimensional Hilbert spaces. Simple examples, however, show that a self-adjoint operator may not have *any* eigenvectors. Consider, for example, $\mathbf{H} = L^2([0,1])$ and an operator A on $\mathbf{H}$ defined by

$$(A\psi)(x) = x\psi(x). \tag{6.1}$$

Then A satisfies $\langle \phi, A\psi \rangle = \langle A\phi, \psi \rangle$ for all $\phi, \psi \in L^2([0,1])$, and yet A has no eigenvectors. After all, if $x\psi(x) = \lambda\psi(x)$, then ψ would have to be supported on the set where $x = \lambda$, which is a set of measure zero. Thus, only the zero element of $L^2([0,1])$ satisfies $A\psi = \lambda\psi$.

B.C. Hall, *Quantum Theory for Mathematicians*, Graduate Texts in Mathematics 267, DOI 10.1007/978-1-4614-7116-5_6, © Springer Science+Business Media New York 2013

Now, a physicist would say that the operator A in (6.1) *does* have eigenvectors, namely the distributions $\delta(x - \lambda)$. (See Appendix A.3.3.) These distributions indeed satisfy $x\delta(x - \lambda) = \lambda\delta(x - \lambda)$, but they do not belong to the Hilbert space $L^2([0,1])$. Such "eigenvectors," which belong to some larger space than **H**, are known as *generalized eigenvectors*. Even though these generalized eigenvectors are not actually in the Hilbert space, we may hope that there is some sense in which they form something like a orthonormal basis. See Sect. 6.6 for an example of how such a "basis" might function.

Let us mention in passing that our simple expectation of a true orthonormal basis of eigenvectors *is* realized for compact self-adjoint operators, where an operator A on **H** is said to be compact if the image under A of every bounded set in **H** has compact closure; see Theorem VI.16 in Volume I of [34]. The operators of interest in quantum mechanics, however, are not compact. (Of course, even if a self-adjoint operator is not compact, it *might* still have an orthonormal basis of eigenvectors, as, e.g., in the case of the Hamiltonian operator for a harmonic oscillator. See Chap. 11.)

Meanwhile, there is another serious difficulty that arises with self-adjoint operators in the infinite-dimensional case. Most of the self-adjoint operators A of quantum mechanics are *unbounded* operators, meaning that there is no constant C such that $\|A\psi\| \leq C\|\psi\|$ for all ψ. Suppose, for example, that A is the position operator X on $L^2(\mathbb{R})$, given by $(X\psi)(x) = x\psi(x)$. If 1_E denotes the indicator function of E (the function that is 1 on E and 0 elsewhere), then it is apparent that

$$\|X1_{[n,n+1]}\| \geq n\,\|1_{[n,n+1]}\|$$

for every positive integer n, and, thus, X cannot be bounded. Now, using the closed graph theorem and elementary results from Sect. 9.3, it can be shown that if A is defined on all of **H** and satisfies $\langle \phi, A\psi \rangle = \langle A\phi, \psi \rangle$ for all $\phi, \psi \in$ **H**, then A must be bounded. (See Corollary 9.9.) Thus, if A is unbounded and self-adjoint, *it cannot be defined on all of* **H**.

We define, then, an "unbounded operator on **H** " to be a linear operator from a *dense subspace* of **H**—known as the domain of A—to **H**. The notion of self-adjointness for such operators is more complicated than in the bounded case. The obvious condition, that $\langle \phi, A\psi \rangle$ should equal $\langle A\phi, \psi \rangle$ for all ϕ and ψ in the domain of A, is not the "right" condition. Specifically, that condition is not sufficient to guarantee that the spectral theorem applies to A. Rather, for any unbounded operator A, we will define the adjoint A^* of A, which will be an unbounded operator with its own domain. An unbounded operator is then defined to be self-adjoint if the domains of A and A^* are the same *and* A and A^* agree on their common domain. That is to say, self-adjointness means not only that A and A^* agree whenever they are both defined, but also that the *domains* of A and A^* agree.

6.2   The Goals of Spectral Theory

Before getting into the details of the spectral theory, let us think for a moment about what it is we want the spectral theorem to do for us. In the first place, we would like the spectral theorem to allow us to apply various functions to an operator. We saw, for example, that the time-dependent Schrödinger equation can be "solved" by setting $\psi(t) = \exp\{-it\hat{H}/\hbar\}\psi_0$. Because the Hamiltonian operator $\hat{H}$ is unbounded, it is not convenient to use power series to define the exponential. If, however, $\hat{H}$ has a true orthonormal basis $\{e_k\}$ of eigenvectors with corresponding eigenvalues λ_n, then we can define $\exp\{-it\hat{H}/\hbar\}$ to be the unique bounded operator with the property that

$$e^{-it\hat{H}/\hbar}e_k = e^{-it\lambda_k/\hbar}e_k$$

for all k.

In cases where $\hat{H}$ does not have a true orthonormal basis of eigenvectors, we would like the spectral theorem to provide a "functional calculus" for $\hat{H}$, that is, a system for applying functions (including exponentials) to $\hat{H}$. This functional calculus should have properties similar to what we have in the case of a true orthonormal basis of eigenvectors.

In the second place, we would like the spectral theorem to provide a probability distribution for the result of measuring a self-adjoint operator A. Let us recall how measurement probabilities work in the case that A has a true orthonormal basis $\{e_j\}$ of eigenvectors with eigenvalues λ_j. Building on Example 3.12, we may compute the probabilities in such a case as follows. Given any Borel set E of $\mathbb{R}$, let V_E be the closed span of all the eigenvectors for A with eigenvalues in E, and let P_E be the orthogonal projection onto V_E. Then for any unit vector ψ, we have

$$\text{prob}_\psi(A \in E) = \langle \psi, P_E \psi \rangle. \tag{6.2}$$

In particular, if the eigenvalues are distinct and ψ decomposes as $\psi = \sum_j c_j e_j$, the probability of observing the value λ_j will be $|c_j|^2$ (as in Example 3.12), since $P_{\{\lambda_j\}}$ is just the projection onto e_j.

In cases where A does not have a true orthonormal basis of eigenvectors, we would like the spectral theorem to provide a family of projection operators P_E, one for each Borel subset $E \subset \mathbb{R}$, which will allow us to define probabilities as in (6.2). We will call these projection operators *spectral projections* and the associated subspaces V_E *spectral subspaces*. (Thus, P_E is the orthogonal projection onto V_E.) Intuitively, V_E may be thought of as the closed span of all the generalized eigenvectors with eigenvalues in E.

In the first version of the spectral theorem, both these goals will be achieved, with the spectral projections being provided by a *projection-valued measure* and the functional calculus being provided by integration with respect to this measure. Although having (generalized) eigenvectors for a self-adjoint operator is, from a practical standpoint, of secondary importance, we provide a framework for understanding such eigenvectors, using the concept of a *direct integral*. The second version of the spectral theorem decomposes the Hilbert space **H** as a direct integral, with respect to a certain measure μ, of generalized eigenspaces for a self-adjoint operator A. The generalized eigenspace for a particular eigenvalue λ will not actually be a subspace of **H**, unless $\mu(\{\lambda\}) > 0$. Thus, the notion of a direct integral gives a rigorous meaning to the notion of "eigenvectors" that are not actually in the Hilbert space.

6.3   A Guide to Reading

Although the portion of this book devoted to spectral theory is unavoidably technical in places, it has been designed so that the reader can take in as much or as little as desired. The reader who is willing to take things on faith can simply take in the examples of the position and momentum operators in Sects. 6.4 and 6.6 and accept these as prototypes of how the spectral theorem works. The reader who wants more details can find the statement of the spectral theorem for bounded operators, in two different forms, in Chap. 7, and can find the basics of unbounded self-adjoint operators in Chap. 9. Finally, the reader who wants a complete treatment of the subject can find full proofs of the spectral theorem in both forms, first for bounded operators in Chap. 8, and then for unbounded operators in Chap. 10.

6.4   The Position Operator

As our first example, let us consider the position operator X, given by $(X\psi)(x) = x\psi(x)$, acting on the Hilbert space $\mathbf{H} = L^2(\mathbb{R})$. As for the similar operator in Sect. 6.1, X has no true eigenvectors, that is, no eigenvectors that are actually in **H**. If we think that the *generalized* eigenvectors for X are the distributions $\delta(x-\lambda)$, $\lambda \in \mathbb{R}$, then we may make an educated guess that the spectral subspace V_E should consist of those functions that "supported" on E, that is, those that are zero almost everywhere on the complement of E. (A superposition of the "functions" $\delta(x-\lambda)$, with $\lambda \in E$, should be a function supported on E.)

The spectral projection P_E is then the orthogonal projection onto V_E, which may be computed as

$$P_E\psi = 1_E\psi,$$

where 1_E is the indicator function of E. In that case, we have, following (6.2),

$$\mathrm{prob}_\psi\left(X \in E\right) = \langle\psi, P_E\psi\rangle = \int_E |\psi(x)|^2 \, dx.$$

This formula is just what we would have expected from our discussion in Chap. 3, where we claimed that the probability distribution for the position of the particle is $|\psi(x)|^2$.

Meanwhile, let us consider the functional calculus for X. If $f(\lambda) = \lambda^m$, then $f(X)$ should be just the mth power of X, which is multiplication by x^m. It seems reasonable, then, to think that for any function f, we should define $f(X)$ to be simply multiplication by $f(x)$. In particular, the operator e^{iaX} should be simply multiplication by e^{iax}, which is a bounded operator on $L^2(\mathbb{R})$.

6.5   Multiplication Operators

Since the position operator acts simply as multiplication by the function x, it is straightforward to find the spectral subspaces and also to construct the functional calculus for X. We may consider multiplication operators in a more general setting. If $\mathbf{H} = L^2(X, \mu)$ and h is a real-valued measurable function on X, then we may define the multiplication operator M_h on $L^2(X, \mu)$ by

$$M_h\psi = h\psi.$$

We can then construct spectral subspaces as

$$V_E = \{\psi \,|\, \psi \text{ is supported on } h^{-1}(E)\}$$

and define a *functional calculus* by

$$f(A) = \text{multiplication by } f \circ h.$$

One form of spectral theorem may now be stated simply as follows: *A self-adjoint operator A on a separable Hilbert space is unitarily equivalent to a multiplication operator.* That is to say, there is some σ-finite measure space (X, μ) and some measurable function h on X such that A is unitarily equivalent to multiplication by h. (See Theorem 7.20.) Although this version of the spectral theorem is compellingly easy to state, there is slight modification of it, involving direct integrals, that is in some ways even better. See Sect. 7.3 for more information.

6.6   The Momentum Operator

Let us now see how the spectral theorem works out in the case of the momentum operator, $P = -i\hbar \, d/dx$ on $L^2(\mathbb{R})$. The "eigenvectors" for P are the functions e^{ikx}, $k \in \mathbb{R}$, with the corresponding eigenvalues being $\hbar k$. Although the functions e^{ikx} are not in $L^2(\mathbb{R})$, the Fourier transform shows that any function in $L^2(\mathbb{R})$ can be expanded as a *superposition* (i.e., continuous version of a linear combination) of these functions. (See Appendix A.3.2.) Indeed, the Fourier transform is very much like the decomposition of a vector in an orthonormal basis, in that the Fourier coefficients $\hat{\psi}(k)$ can be expressed in terms of the "inner product" of a function ψ with e^{ikx}:

$$\hat{\psi}(k) = (2\pi)^{-1/2} \int_{-\infty}^{\infty} e^{-ikx} \psi(x) \, dx = (2\pi)^{-1/2} \left\langle e^{ikx}, \psi \right\rangle_{L^2(\mathbb{R})},$$

if we ignore the fact that e^{ikx} is not actually in L^2.

Indeed, physicists frequently understand the Fourier transform by asserting that the functions $e^{ikx}/\sqrt{2\pi}$ form an "orthonormal basis in the continuous sense" for $L^2(\mathbb{R})$. Orthonormality in the continuous sense is supposed to mean that one replaces the usual Kronecker delta in the definition of an orthonormal set by the Dirac δ-*function*

$$\left\langle \frac{e^{ikx}}{\sqrt{2\pi}}, \frac{e^{ilx}}{\sqrt{2\pi}} \right\rangle_{L^2(\mathbb{R})} = \delta(k - l), \tag{6.3}$$

where δ is supposed to satisfy

$$\int_{-\infty}^{\infty} f(k)\delta(k - l) \, dk = f(l)$$

for all continuous functions f. (Rigorously, $\delta(k - l)$ is a distribution; see Appendix A.3.3.)

To give some rigorous meaning to (6.3), note that although the inner product of e^{ikx} and e^{ilx} is not defined, we may approximate this inner product by the expression

$$\frac{1}{2\pi} \int_{-A}^{A} e^{-ikx} e^{ilx} \, dx = \frac{1}{2\pi} \frac{e^{-i(k-l)x}}{-i(k-l)} \Bigg|_{-A}^{A} = \frac{A}{\pi} \frac{\sin\left[A(k-l)\right]}{A(k-l)}.$$

It is possible to show that the above function, viewed as a function of k for fixed A and l, behaves like $\delta(k - l)$ in the limit as A tends to infinity. That is to say, for all sufficiently nice functions ψ, we have

$$\lim_{A \to \infty} \int_{-\infty}^{\infty} \psi(k) \frac{A}{\pi} \frac{\sin\left[A(k-l)\right]}{A(k-l)} dk = \psi(l). \tag{6.4}$$

Here is a heuristic argument for (6.4). By making the change of variable $k' = k - l$, we may reduce the general problem to the case $l = 0$. If we then make the change of variable $\kappa = Ak$, the desired result is equivalent to

$$\lim_{A \to +\infty} \int_{-\infty}^{\infty} \frac{1}{\pi} \frac{\sin \kappa}{\kappa} f\left(\frac{\kappa}{A}\right) \, d\kappa = f(0). \tag{6.5}$$

Now, *if* we can bring the limit inside the integral, $f(\kappa/A)$ will tend to $f(0)$ as A tends to infinity. Since the rest of the integrand on the right-hand side of (6.5) is already independent of A, the result would then follow if we could show that

$$\int_{-\infty}^{\infty} \frac{1}{\pi} \frac{\sin \kappa}{\kappa} \, d\kappa = 1. \tag{6.6}$$

Even though the integral in (6.6) is not absolutely convergent, it is a convergent improper integral. The value of the integral can be obtained by the method of contour integration (or the method of consulting a table of integrals), and indeed (6.6) holds. Since (6.3) is, in any case, only a heuristic way of thinking about the Fourier transform, we will not take the time to develop a rigorous version of the preceding argument.

It is possible to derive, at least formally, many of the standard properties of the Fourier transform by using (6.3), just as one can obtain properties of Fourier *series* by using the orthonormality of the functions $e^{2\pi i n x}$ in $L^2([0,1])$. More importantly, the Fourier transform is precisely the unitary transformation that changes the momentum operator into a multiplication operator. To see this property of the Fourier transform more clearly, we introduce a simple rescaling of it.

Definition 6.1 *For any $\psi \in L^2(\mathbb{R})$, define $\tilde{\psi}$ by*

$$\tilde{\psi}(p) = \frac{1}{\sqrt{\hbar}} \hat{\psi}\left(\frac{p}{\hbar}\right),$$

so that

$$\tilde{\psi}(p) = \frac{1}{\sqrt{2\pi\hbar}} \int_{-\infty}^{\infty} e^{-ipx/\hbar} \psi(x) \, dx.$$

*The function $\tilde{\psi}(p)$ is the **momentum wave function** associated with ψ.*

By the Plancherel theorem (Theorem A.19) and a change of variable, if ψ is a unit vector, then so is $\hat{\psi}$ and also $\tilde{\psi}$. For any unit vector ψ, we interpret $|\tilde{\psi}(p)|^2$ as the probability density for the momentum of the particle, just as $|\psi(x)|^2$ is the probability distribution of the position of the particle. Using Proposition A.17, we may readily verify that for nice enough ψ, we have

$$\widetilde{P\psi}(p) = p\tilde{\psi}(p). \tag{6.7}$$

Equation (6.7) means that the unitary map $\psi \to \tilde{\psi}$ turns the momentum operator P into multiplication by p. That is to say, the spectral theorem, in its "multiplication operator" form, is accomplished in this case by the Fourier transform (scaled as in Definition 6.1).

In terms of the momentum wave function, we may define spectral projections and a functional calculus for P, just as in Sect. 6.5. For any Borel set $E \subset \mathbb{R}$, we may define a projection P_E to be the orthogonal projection onto to the space of functions ψ for which $\tilde{\psi}(p)$ is zero almost everywhere outside of E. If f is any bounded measurable function on $\mathbb{R}$, we can define an operator $f(P)$ by defining $f(P)\psi$ to be the unique element of $L^2(\mathbb{R})$ for which

$$\widetilde{f(P)\psi}(p) = f(p)\tilde{\psi}(p).$$

7

The Spectral Theorem for Bounded Self-Adjoint Operators: Statements

In the present chapter, we will consider the spectral theorem for *bounded* self-adjoint operators, leaving a discussion of unbounded operators to Chaps. 9 and 10. The proofs of the main theorems (two different versions of the spectral theorem) are moderately long and are deferred to Chap. 8. After some elementary definitions and results in Sect. 7.1, we come to the main results in Sects. 7.2 and 7.3. Throughout the chapter, $\mathbf{H}$ will, as usual, denote a separable Hilbert space over $\mathbb{C}$.

7.1 Elementary Properties of Bounded Operators

As usual, we will let $\mathbf{H}$ denote a separable complex Hilbert space. Recall from Appendix A.3.4 that a linear operator A on $\mathbf{H}$ is said to be *bounded* if the *operator norm* of A,

$$\|A\| := \sup_{\psi \in \mathbf{H} \setminus \{0\}} \frac{\|A\psi\|}{\|\psi\|} \tag{7.1}$$

is finite. The space of bounded operators on $\mathbf{H}$ forms a Banach space under the operator norm, and we have the inequality

$$\|AB\| \leq \|A\| \, \|B\| \tag{7.2}$$

for all bounded operators A and B.

Definition 7.1 *The Banach space of bounded operators on* $\mathbf{H}$, *with respect to the operator norm (7.1), is denoted* $\mathcal{B}(\mathbf{H})$.

B.C. Hall, *Quantum Theory for Mathematicians*, Graduate Texts in Mathematics 267, DOI 10.1007/978-1-4614-7116-5_7, © Springer Science+Business Media New York 2013

Recall (Appendix A.4.3) that for any $A \in \mathcal{B}(\mathbf{H})$ there is a unique operator $A^* \in \mathcal{B}(\mathbf{H})$, called the adjoint of A, such that

$$\langle \phi, A\psi \rangle = \langle A^*\phi, \psi \rangle$$

for all $\phi, \psi \in \mathbf{H}$. An operator $A \in \mathcal{B}(\mathbf{H})$ is called *self-adjoint* if $A^* = A$. We say that $A \in \mathcal{B}(\mathbf{H})$ is *non-negative* if

$$\langle \psi, A\psi \rangle \geq 0 \qquad (7.3)$$

for all $\psi \in \mathbf{H}$.

Proposition 7.2 *For all $A \in \mathcal{B}(\mathbf{H})$, we have*

$$\|A^*\| = \|A\|$$

and

$$\|A^*A\| = \|A\|^2.$$

In particular, if A is self-adjoint, we have the useful result that $\|A^2\| = \|A\|^2$.

Proof. The operator norm of A can also be computed as

$$\|A\| = \sup_{\|\psi\|=1} \|A\psi\|.$$

Furthermore, for any vector $\phi \in \mathbf{H}$, $\|\phi\| = \sup_{\|\chi\|=1} |\langle \chi, \phi \rangle|$. (Inequality one direction is by the Cauchy–Schwarz inequality, and inequality the other direction is by taking χ to be a multiple of ϕ.) Thus,

$$\|A\| = \sup_{\|\phi\|=\|\psi\|=1} |\langle \phi, A\psi \rangle|.$$

From this, we get

$$\begin{aligned}
\|A^*\| &= \sup_{\|\phi\|=\|\psi\|=1} |\langle \phi, A^*\psi \rangle| \\
&= \sup_{\|\phi\|=\|\psi\|=1} |\langle A\phi, \psi \rangle| \\
&= \sup_{\|\phi\|=\|\psi\|=1} |\langle \psi, A\phi \rangle| \\
&= \|A\|.
\end{aligned}$$

Meanwhile, $\|A^*A\| \leq \|A^*\| \|A\| = \|A\|^2$. On the other hand,

$$\begin{aligned}
\|A^*A\| &= \sup_{\|\phi\|=\|\psi\|=1} |\langle \phi, A^*A\psi \rangle| \\
&= \sup_{\|\phi\|=\|\psi\|=1} |\langle A\phi, A\psi \rangle| \\
&\geq \sup_{\|\psi\|=1} |\langle A\psi, A\psi \rangle| \\
&= \|A\|^2,
\end{aligned}$$

which establishes the inequality in the other order. ∎

We now record an elementary but very useful result.

Proposition 7.3 *For all $A \in \mathcal{B}(\mathbf{H})$, we have*

$$[\text{Range}(A)]^{\perp} = \ker(A^*),$$

where for any $B \in \mathcal{B}(\mathbf{H})$, $\ker(B)$ denotes the kernel of B.

Proof. Suppose first that ψ belongs to $[\text{Range}(A)]^{\perp}$. Then for all $\phi \in \mathbf{H}$, we have

$$0 = \langle \psi, A\phi \rangle = \langle A^*\psi, \phi \rangle. \tag{7.4}$$

This implies that $A^*\psi = 0$ and thus that $\psi \in \ker(A^*)$. Conversely, suppose $\psi \in \ker(A^*)$. Then for all $\phi \in \mathbf{H}$, (7.4) holds (reading the equation from right to left). This shows that ψ is orthogonal to every element of the form $A\phi$, meaning that $\psi \in [\text{Range}(A)]^{\perp}$. ∎

Next, we define the *spectrum* of a bounded operator, which plays the same role as the set of eigenvalues in the finite-dimensional case.

Definition 7.4 *For $A \in \mathcal{B}(\mathbf{H})$, the **resolvent set** of A, denoted $\rho(A)$ is the set of all $\lambda \in \mathbb{C}$ such that the operator $(A - \lambda I)$ has a bounded inverse. The **spectrum** of A, denoted by $\sigma(A)$, is the complement in $\mathbb{C}$ of the resolvent set. For λ in the resolvent set of A, the operator $(A - \lambda I)^{-1}$ is called the **resolvent** of A at λ.*

Saying that $(A - \lambda I)$ has a bounded inverse means that there exists a bounded operator B such that

$$(A - \lambda I)B = B(A - \lambda I) = I.$$

If A is bounded and $A - \lambda I$ is one-to-one and maps $\mathbf{H}$ onto $\mathbf{H}$, then it follows from the closed graph theorem (Theorem A.39) that the inverse map must be bounded. Thus, the resolvent set of A can alternatively be described as the set of $\lambda \in \mathbb{C}$ for which $A - \lambda I$ is one-to-one and onto.

Proposition 7.5 *For all $A \in \mathcal{B}(\mathbf{H})$, the following results hold.*

1. *The spectrum $\sigma(A)$ of A is a closed, bounded, and nonempty subset of $\mathbb{C}$.*

2. *If $|\lambda| > \|A\|$, then λ is in the resolvent set of A.*

Lemma 7.6 *Suppose $X \in \mathcal{B}(\mathbf{H})$ satisfies $\|X\| < 1$. Then the operator $I - X$ is invertible, with the inverse given by the following convergent series in $\mathcal{B}(\mathbf{H})$:*

$$(I - X)^{-1} = I + X + X^2 + X^3 + \cdots. \tag{7.5}$$

Proof. As a consequence of (7.2), we have $\|X^m\| \leq \|X\|^m$. The (geometric) series on the right-hand side of (7.5) is therefore absolutely convergent and thus convergent in the Banach space $\mathcal{B}(\mathbf{H})$ (Appendix A.3.4). If we multiply this series on either side by $(I - X)$, everything will cancel except I, showing that the sum of the series is the inverse of $(I - X)$. ∎

Proof of Proposition 7.5. For any nonzero $\lambda \in \mathbb{C}$, consider the operator

$$A - \lambda I = -\lambda \left(I - \frac{A}{\lambda} \right).$$

If $|\lambda| > \|A\|$, then $\|A/\lambda\| < 1$, and $I - A/\lambda$ is invertible by the lemma. It then follows that $A - \lambda I$ is invertible, with

$$(A - \lambda I)^{-1} = -\frac{1}{\lambda} \left(I + \frac{A}{\lambda} + \frac{A^2}{\lambda^2} + \cdots \right). \tag{7.6}$$

Thus, λ is in the resolvent set of A. This establishes Point 2 in the proposition and shows that $\sigma(A)$ is bounded.

Suppose now that $\lambda_0 \in \mathbb{C}$ is in the resolvent set of A. Then for another number $\lambda \in \mathbb{C}$, we have

$$A - \lambda I = A - \lambda_0 I - (\lambda - \lambda_0)I$$
$$= (A - \lambda_0 I)\,(I - (\lambda - \lambda_0)\,(A - \lambda_0 I)^{-1}). \tag{7.7}$$

Thus, if

$$|\lambda - \lambda_0| < \frac{1}{\|(A - \lambda_0 I)^{-1}\|},$$

both factors on the right-hand side of (7.7) will be invertible, so that $A - \lambda I$ is also invertible. Thus, the resolvent set of A is open and the spectrum is closed.

To show that $\sigma(A)$ is nonempty, note that $A - \lambda I$ may be computed as follows:

$$(A - \lambda I)^{-1} = (I - (\lambda - \lambda_0)(A - \lambda_0 I)^{-1})^{-1}(A - \lambda_0 I)^{-1}$$
$$= \left(\sum_{m=0}^{\infty} (\lambda - \lambda_0)^m ((A - \lambda_0 I)^{-1})^m \right) (A - \lambda_0 I)^{-1}. \tag{7.8}$$

Thus, near any point λ_0 in the resolvent set of A, the resolvent $(A - \lambda I)^{-1}$ can be computed by the locally convergent series (7.8) in powers of $\lambda - \lambda_0$, with the coefficients of the series being elements of $\mathcal{B}(\mathbf{H})$. For any $\phi, \psi \in \mathbf{H}$, the map

$$\lambda \mapsto \langle \phi, (A - \lambda I)^{-1} \psi \rangle \tag{7.9}$$

will be given by a locally convergent power series with coefficients in $\mathbb{C}$, meaning that the function (7.9) is a holomorphic function on the resolvent set of A. Furthermore, from (7.6) we can see that $\left\|(A - \lambda I)^{-1}\right\|$ tends to zero as $|\lambda|$ tends to infinity, and so also does the right-hand side of (7.9).

If $\sigma(A)$ were the empty set, the function (7.9) would be holomorphic on all of $\mathbb{C}$ and tending to zero at infinity. By Liouville's theorem, the right-hand side of (7.9) would have to be identically zero for all ϕ and ψ, which would mean that $(A - \lambda I)^{-1}$ is the zero operator. But since $(A - \lambda I)(A - \lambda I)^{-1} = I$, the operator $(A - \lambda I)^{-1}$ cannot be zero. ∎

If $A\psi = \lambda\psi$ for some $\lambda \in \mathbb{C}$ and some nonzero $\psi \in \mathbf{H}$, then $(A - \lambda I)$ has a nonzero kernel and so λ is in the spectrum of A. Thus, any eigenvalue for A is contained in the spectrum of A. In the infinite-dimensional case, however, the converse is not true: A point in the spectrum may not be an eigenvalue for A. Nevertheless, for a bounded *self-adjoint* operator A, the spectrum of A may be described in a way that is not too far removed from what we have in the finite-dimensional case.

Proposition 7.7 *If $A \in \mathcal{B}(\mathbf{H})$ is self-adjoint, then the following results hold.*

1. *The spectrum of A is contained in the real line.*

2. *A number $\lambda \in \mathbb{R}$ belongs to the spectrum of A if and only if there exists a sequence ψ_n of nonzero vectors in $\mathbf{H}$ such that*

$$\lim_{n \to \infty} \frac{\|A\psi_n - \lambda\psi_n\|}{\|\psi_n\|} = 0. \tag{7.10}$$

Condition 2 in the proposition says that $\lambda \in \mathbb{R}$ belongs to the spectrum if and only if λ is "almost an eigenvalue," meaning that there exists $\psi \neq 0$ for which $A\psi$ is equal to $\lambda\psi$ plus an error that is small compared to the size of ψ.

Lemma 7.8 *If $A \in \mathcal{B}(\mathbf{H})$ is self-adjoint, then for all $\lambda = a + ib \in \mathbb{C}$, we have*

$$\langle (A - \lambda I)\psi, (A - \lambda I)\psi \rangle \geq b^2 \langle \psi, \psi \rangle. \tag{7.11}$$

Proof. We compute that

$$\begin{aligned}
\langle (A &- (a + ib)I)\psi, (A - (a + ib)I)\psi \rangle \\
&= \langle (A - aI)\psi, (A - aI)\psi \rangle + ib \langle \psi, (A - aI)\psi \rangle \\
&\quad - ib \langle (A - aI)\psi, \psi \rangle + b^2 \langle \psi, \psi \rangle. \tag{7.12}
\end{aligned}$$

Since A is self-adjoint, so is $A - aI$, from which we see that the second and third terms on the right-hand side of (7.12) cancel, leaving us with

$$\langle (A - \lambda I)\psi, (A - \lambda I)\psi \rangle = \langle (A - aI)\psi, (A - aI)\psi \rangle + b^2 \langle \psi, \psi \rangle,$$

from which the desired inequality follows. ∎

Proof of Proposition 7.7. For Point 1, we need to show that any complex number $\lambda = a + ib$ with $b \neq 0$ belongs to the resolvent set of A. Since $b \neq 0$, (7.11) shows that $A - \lambda I$ is injective. Meanwhile, by Proposition 7.3, $\text{Range}(A - \lambda I)^{\perp} = \ker(A - \bar{\lambda}I)$. Since $\bar{\lambda}$ also has nonzero imaginary part, $A - \bar{\lambda}I$ is injective, and so the range of $A - \lambda I$ is dense in $\mathbf{H}$. To show that the range is all of $\mathbf{H}$, consider any $\phi \in \mathbf{H}$ and choose a sequence $\phi_n = (A - \lambda I)\psi_n$ in $\text{Range}(A - \lambda I)$ with $\phi_n \to \phi$. Applying (7.11) with ψ replaced by $\psi_n - \psi_m$ shows that $\langle \psi_n \rangle$ is a Cauchy sequence. Thus, $\psi_n \to \psi$ for some $\psi \in \mathbf{H}$. Since A is bounded,

$$(A - \lambda I)\psi = \lim_{n \to \infty} (A - \lambda I)\psi_n = \lim_{n \to \infty} \phi_n = \phi.$$

We conclude, then, that $A - \lambda I$ is one-to-one and onto. The inverse operator $(A - \lambda I)^{-1}$ is bounded, by (7.11) (or by the closed graph theorem).

For Point 2, assume there exists a sequence as in (7.10), and suppose that $A - \lambda I$ had an inverse. Letting $\phi_n = (A - \lambda I)\psi_n$, we have $\psi_n = (A - \lambda I)^{-1}\phi_n$ and so (7.10) says that

$$\lim_{n \to \infty} \frac{\|\phi_n\|}{\|(A - \lambda I)^{-1}\phi_n\|} = 0,$$

which shows that $(A - \lambda I)^{-1}$ is actually unbounded. Thus, $A - \lambda I$ cannot have a *bounded* inverse.

Conversely, if, for some $\lambda \in \mathbb{R}$, no such sequence exists, then there exists some $\varepsilon > 0$ such that

$$\|(A - \lambda I)\psi\| \geq \varepsilon \|\psi\| \tag{7.13}$$

for all $\psi \in \mathbf{H}$. Then $A - \lambda I$ is injective and Proposition 7.3 tells us that the range of the self-adjoint operator $A - \lambda I$ is dense in $\mathbf{H}$. Arguing as in the preceding paragraphs with (7.13) in place of (7.11), we can see that the range of $A - \lambda I$ is also closed, hence all of $\mathbf{H}$. This shows that $A - \lambda I$ has an inverse. ∎

Example 7.9 *Let $\mathbf{H} = L^2([0,1])$ and let A be the operator on $\mathbf{H}$ defined by*

$$(A\psi)(x) = x\psi(x).$$

Then this operator is bounded and self-adjoint, and its spectrum is given by

$$\sigma(A) = [0,1].$$

As we have already noted in Sect. 6.1, the operator A does not have any (true) eigenvectors.

Proof. It is apparent that $\|A\psi\| \leq \|\psi\|$ and that $\langle \phi, A\psi \rangle = \langle A\phi, \psi \rangle$ for all $\phi, \psi \in \mathbf{H}$, so that A is bounded and self-adjoint. Given $\lambda \in (0,1)$, consider the functions $\psi_n := 1_{[\lambda, \lambda + 1/n]}$, which satisfy $\|\psi_n\|^2 = 1/n$. On the other hand, since $|x - \lambda| \leq 1/n$ on $[\lambda, \lambda + 1/n]$, we have

$$\|(A - \lambda I)\psi_n\|^2 \leq 1/n^3.$$

Thus, by Proposition 7.7, λ belongs to the spectrum of A. Since this holds for all $\lambda \in (0,1)$ and the spectrum of A is closed, $\sigma(A) \supset [0,1]$.

Meanwhile, if $\lambda \notin [0,1]$, then the function $1/(x - \lambda)$ is bounded on $[0,1]$, and so $A - \lambda I$ has a bounded inverse, consisting of multiplication by $1/(x - \lambda)$. Thus, $\sigma(A) = [0,1]$. ∎

7.2 Spectral Theorem for Bounded Self-Adjoint Operators, I

7.2.1 Spectral Subspaces

Given a bounded (for now) self-adjoint operator A, we hope to associate with each Borel set $E \subset \sigma(A)$ a closed subspace V_E of $\mathbf{H}$, where we think intuitively that V_E is the closed span of the generalized eigenvectors for A with eigenvalues in E. [We could do this more generally for any $E \subset \mathbb{R}$, but we do not expect any contribution from $\mathbb{R} \setminus \sigma(A)$.] We would expect the collection of these subspaces to have the following properties.

1. $V_{\sigma(A)} = \mathbf{H}$ and $V_\varnothing = \{0\}$.

2. If E and F are disjoint, then $V_E \perp V_F$.

3. For any E and F, $V_{E \cap F} = V_E \cap V_F$.

4. If $E_1, E_2, \ldots$ are disjoint and $E = \cup_j E_j$, then

$$V_E = \bigoplus_j V_{E_j}.$$

5. For any E, V_E is invariant under A.

6. If $E \subset [\lambda_0 - \varepsilon, \lambda_0 + \varepsilon]$ and $\psi \in V_E$, then

$$\|(A - \lambda_0 I)\psi\| \le \varepsilon \|\psi\|.$$

The condition $V_{\sigma(A)} = \mathbf{H}$ captures the idea that our generalized eigenvectors should span $\mathbf{H}$, while Property 2 captures the idea that our generalized eigenvectors should have some sort of orthogonality for distinct eigenvalues, even if they are not actually in the Hilbert space. In Property 4, there may be infinitely many of the E_j's, in which case, the direct sum is in the Hilbert space sense (Definition A.45). Properties 5 and 6 capture the idea that V_E is made up of generalized eigenvectors for A with eigenvalues in E.

7.2.2 Projection-Valued Measures

It is convenient to describe closed subspaces of a Hilbert space $\mathbf{H}$ in terms of the associated orthogonal projection operators. Recall (Proposition A.57) that, given a closed subspace V of $\mathbf{H}$, there exists a unique bounded operator P that equals the identity on V and equals zero on the orthogonal complement $V^\perp$ of V. This operator is called the orthogonal projection onto V and satisfies $P^2 = P$ and $P^* = P$. The following definition expresses the first four properties of our spectral subspaces—the ones that do not involve the operator A—in terms of the corresponding orthogonal projections. Since those properties are similar to those of a measure, we use the term *projection-valued measure*.

Definition 7.10 *Let X be a set and Ω a σ-algebra in X. A map $\mu : \Omega \to \mathcal{B}(\mathbf{H})$ is called a* **projection-valued measure** *if the following properties are satisfied.*

1. *For each $E \in \Omega$, $\mu(E)$ is an orthogonal projection.*

2. *$\mu(\varnothing) = 0$ and $\mu(X) = I$.*

3. *If $E_1, E_2, E_3, \ldots$ in Ω are disjoint, then for all $v \in \mathbf{H}$, we have*

$$\mu\left(\bigcup_{j=1}^{\infty} E_j\right) v = \sum_{j=1}^{\infty} \mu(E_j) v,$$

where the convergence of the sum is in the norm topology on $\mathbf{H}$.

4. *For all $E_1, E_2 \in \Omega$, we have $\mu(E_1 \cap E_2) = \mu(E_1)\mu(E_2)$.*

Note that if E_1 and E_2 are disjoint, then Properties 2 and 4 tell us that $\mu(E_1)\mu(E_2) = 0$, from which it follows (Exercise 10) that the range of $\mu(E_1)$ and the range of $\mu(E_2)$ are perpendicular. It is then not hard to verify that $\mu(E_1)\mu(E_2)$ is the projection onto the intersection of the ranges of $\mu(E_1)$ and $\mu(E_2)$ (Exercise 11). Thus, if we define, for each $E \in \Omega$, a closed subspace $V_E := \text{Range}(\mu(E))$, then the collection of V_E's satisfy the first four properties that we anticipated for spectral subspaces.

In the next subsection, we will associate a projection-valued measure μ^A with each bounded self-adjoint operator A. In that case, the projection $\mu^A(E)$ will be thought of as a projection onto the spectral subspace corresponding to E. We are about to introduce the notion of operator-valued integration with respect to a projection-valued measure. In the case of the projection-valued measure μ^A associated with A, this operator-valued integral will be the *functional calculus* for A.

Observe that, for any projection-valued measure μ and $\psi \in \mathbf{H}$, we can form an ordinary (positive) real-valued measure μ_ψ by setting

$$\mu_\psi(E) = \langle \psi, \mu(E)\psi \rangle \qquad (7.14)$$

for all $E \in \Omega$. This observation provides a link between integration with respect to a projection-valued measure and integration with respect to an ordinary measure.

Proposition 7.11 (Operator-Valued Integration) *Let Ω be a σ-algebra in a set X and let $\mu : \Omega \to \mathcal{B}(\mathbf{H})$ be a projection-valued measure. Then there exists a unique linear map, denoted $f \mapsto \int_\Omega f \, d\mu$, from the space of bounded, measurable, complex-valued functions on Ω into $\mathcal{B}(\mathbf{H})$ with the property that*

$$\left\langle \psi, \left(\int_X f \, d\mu \right) \psi \right\rangle = \int_X f \, d\mu_\psi \qquad (7.15)$$

for all f and all $\psi \in \mathbf{H}$, where μ_ψ is given by (7.14). This integral has the following additional properties.

1. *For all $E \in \Omega$, we have*

$$\int_X 1_E \, d\mu = \mu(E).$$

   In particular, the integral of the constant function 1 is I.

2. *For all f, we have*

$$\left\| \int_X f \, d\mu \right\| \leq \sup_{\lambda \in X} |f(\lambda)|. \qquad (7.16)$$

3. *Integration is multiplicative: For all f and g, we have*

$$\int_X fg \, d\mu = \left(\int_X f \, d\mu \right) \left(\int_X g \, d\mu \right). \qquad (7.17)$$

4. *For all f, we have*

$$\int_X \bar{f} \, d\mu = \left(\int_X f \, d\mu \right)^*.$$

In particular, if f is real-valued, then $\int_X f \, d\mu$ is self-adjoint.

By Property 1 and linearity, integration with respect to μ has the expected behavior on simple functions. It then follows from Property 2 that the integral of an arbitrary bounded measurable function f can be computed as follows. Take a sequence s_n of simple functions converging uniformly to f; the integral of f is then the limit, in the operator norm topology, of the integral of the s_n's.

Although the multiplicative property of the integral may seem surprising at first, observe that for any $E_1, E_2 \in \Omega$, Property 3 in Definition 7.10 tells us that

$$\left(\int_X 1_{E_1} \, d\mu \right) \left(\int_X 1_{E_2} \, d\mu \right) = \mu(E_1)\mu(E_2) = \mu(E_1 \cap E_2)$$

$$= \int_X 1_{E_1} \cdot 1_{E_2} \, d\mu.$$

Thus, multiplicativity of the integral at the level of indicator functions is built into the definition of a projection-valued measure.

If one wanted to make a real-valued measure for which the corresponding integral was multiplicative, then since $1_E \cdot 1_E = 1_E$, the integral of 1_E—namely, $\mu(E)$—would have to satisfy $\mu(E)^2 = \mu(E)$. This would mean that $\mu(E)$ is 0 or 1 for all E. For such measures, one would indeed obtain multiplicativity of the integral, but measures with this property are not very interesting. For operator-valued measures, we can have *interesting* examples where the integral is multiplicative, simply because there are many more idempotents (elements A with $A^2 = A$) in $\mathcal{B}(\mathbf{H})$ than in $\mathbb{R}$.

Proof of Proposition 7.11. Given a projection-valued measure μ and a bounded measurable function f on X, define a map $Q_f : \mathbf{H} \to \mathbb{C}$ by

$$Q_f(\psi) = \int_X f \, d\mu_\psi,$$

where μ_ψ is given by (7.14). If f is an indicator function, then $Q_f(\psi) = \langle \psi, \mu(E)\psi \rangle$ is a bounded quadratic form. (See Definition A.60.) It is straightforward to show, passing from indicator functions to simple functions and then to general functions, that for any bounded measurable f, Q_f is a bounded quadratic form, with

$$|Q_f(\psi)| \le \left(\sup_{\lambda \in X} |f(\lambda)| \right) \|\psi\|^2. \tag{7.18}$$

It then follows from Proposition A.63 that there is a unique bounded operator A_f such that

$$Q_f(\psi) = \langle \psi, A_f \psi \rangle$$

for all $\psi \in \mathbf{H}$. We set $\int_X f \, d\mu = A_f$. From the way A_f is defined, it satisfies (7.15). The uniqueness of the linear map $f \mapsto \int_X f \, d\mu$ follows from the uniqueness in Proposition A.63.

If $f = 1_E$, then $Q_f(\psi) = \mu_\psi(E) = \langle \psi, \mu(E)\psi \rangle$, in which case the unique associated operator A_f is $\mu(E)$. This establishes Property 1. Property 2 follows from (7.18).

For Property 3, we have already observed that multiplicativity of the integral, at the level of indicator functions, is built into the definition of a projection-valued measure. Since both sides of (7.17) are bilinear in (ϕ, ψ), we have (7.17) for simple functions. Using Property 2, we can then obtain (7.17) for all bounded measurable functions by taking limits.

Finally, if f is real valued, then $Q_f(\psi)$ will be real for all $\psi \in \mathbf{H}$. Thus, by Proposition A.63, the associated operator A_f will be self-adjoint. Property 4 then follows by linearity. ∎

7.2.3 The Spectral Theorem

We are ready to state one version of the spectral theorem for bounded self-adjoint operators.

Theorem 7.12 (Spectral Theorem, First Form) *If $A \in \mathcal{B}(\mathbf{H})$ is self-adjoint, then there exists a unique projection-valued measure μ^A on the Borel σ-algebra in $\sigma(A)$, with values in projections on $\mathbf{H}$, such that*

$$\int_{\sigma(A)} \lambda \, d\mu^A(\lambda) = A. \tag{7.19}$$

Since the spectrum $\sigma(A)$ of A is bounded, the function $f(\lambda) := \lambda$ is bounded on $\sigma(A)$. The proof of this theorem is given in Chap. 8.

Definition 7.13 (Functional Calculus) *If $A \in \mathcal{B}(\mathbf{H})$ is self-adjoint and $f : \sigma(A) \to \mathbb{C}$ is a bounded measurable function, define an operator $f(A)$ by setting*

$$f(A) = \int_{\sigma(A)} f(\lambda) \, d\mu^A(\lambda),$$

where μ^A is the projection-valued measure in Theorem 7.12.

We may extend the projection-valued measure μ^A from $\sigma(A)$ to all of $\mathbb{R}$ by assigning measure 0 to $\mathbb{R} \setminus \sigma(A)$. Then, roughly speaking, $f(A)$ is the operator that is equal to $f(\lambda)I$ on the range of the projection operator $\mu^A([\lambda, \lambda + d\lambda))$.

Since the integral with respect to μ^A is multiplicative, it follows from (7.19) that if $f(\lambda) = \lambda^m$ for some positive integer m, then $f(A)$ is the mth power of A. Further, since the series $e^{a\lambda} = \sum_{m=0}^{\infty} (a\lambda)^m / m!$ converges uniformly on the compact set $\sigma(A)$, the operator e^{aA} (computed using the functional calculus for the function $f(\lambda) = e^{u\lambda}$) may be computed as a power series.

Definition 7.14 (Spectral Subspaces) *For $A \in \mathcal{B}(\mathbf{H})$, let μ^A be the associated projection-valued measure, extended to be a measure on $\mathbb{R}$ by setting $\mu^A(\mathbb{R} \setminus \sigma(A)) = 0$. Then for each Borel set $E \subset \mathbb{R}$, define the* **spectral subspace** V_E *of $\mathbf{H}$ by*

$$V_E = \mathrm{Range}(\mu^A(E)).$$

The definition of a projection-valued measure implies that these spectral subspaces satisfy the first four properties listed in Sect. 7.2.1. We now show that (7.19) implies the remaining two properties we anticipated for the spectral subspaces.

Proposition 7.15 *If $A \in \mathcal{B}(\mathbf{H})$ is self-adjoint, the spectral subspaces associated with A have the following properties.*

1. Each spectral subspace V_E is invariant under A.

2. If $E \subset [\lambda_0 - \varepsilon, \lambda_0 + \varepsilon]$ then for all $\psi \in V_E$, we have

$$\|(A - \lambda_0 I)\psi\| \le \varepsilon \|\psi\|.$$

3. The spectrum of $A|_{V_E}$ is contained in the closure of E.

4. If λ_0 is in the spectrum of A, then for every neighborhood U of λ_0, we have $V_U \ne \{0\}$, or, equivalently, $\mu(U) \ne 0$.

Proof. For Point 1, observe that for any bounded measurable functions f and g on $\sigma(A)$, the operators $f(A)$ and $g(A)$ commute, since the product in either order is equal to the integral of the function $fg = gf$ with respect to μ^A. In particular, A, which is the integral of the function $f(\lambda) = \lambda$, commutes with $\mu^A(E)$, which is the integral of the function 1_E. Thus, given a vector $\mu^A(E)\phi$ in the range of $\mu^A(E)$, we have

$$A\mu^A(E)\phi = \mu^A(E)A\phi,$$

which is again in the range of $\mu^A(E)$, establishing the invariance of the spectral subspace.

For Point 2, suppose that $\psi \in V_E$, where $E \subset [\lambda_0 - \varepsilon, \lambda_0 + \varepsilon]$. Then ψ is in the range of $\mu^A(E)$, and so

$$(A - \lambda_0 I)\psi = (A - \lambda_0 I)\mu^A(E)\psi.$$

But $\mu^A(E) = 1_E(A)$ and $A - \lambda_0 I = f(A)$, where $f(\lambda) = \lambda - \lambda_0$. By the multiplicativity of the integral, then,

$$(A - \lambda_0 I)\psi = (f1_E)(A)\psi.$$

But $|f(\lambda)1_E(\lambda)| \le \varepsilon$ and so by (7.16), the operator $(f1_E)(A)$ has norm at most ε.

For Point 3, if λ_0 is not in $\bar{E}$, then the function $g(\lambda) := 1_E(\lambda)(1/(\lambda - \lambda_0))$ is bounded. Thus, $g(A)$ is a bounded operator and

$$g(A)(A - \lambda_0 I) = (A - \lambda_0 I)g(A) = 1_E(A).$$

This shows that the restriction to V_E of $g(A)$ is the inverse of the restriction to V_E of A. Thus, λ_0 is not in the spectrum of $A|_{V_E}$.

For Point 4, fix $\lambda_0 \in \sigma(A)$ and suppose for some $\varepsilon > 0$, we have $\mu((\lambda_0 - \varepsilon, \lambda_0 + \varepsilon)) = 0$. Consider, then, the bounded function f defined by

$$f(\lambda) = \begin{cases} \frac{1}{\lambda - \lambda_0} & |\lambda - \lambda_0| \ge \varepsilon \\ 0 & |\lambda - \lambda_0| < \varepsilon \end{cases}.$$

Since $f(\lambda) \cdot (\lambda - \lambda_0)$ equals 1 except on $(\lambda_0 - \varepsilon, \lambda_0 + \varepsilon)$, the equation $f(\lambda) \cdot (\lambda - \lambda_0) = 1$ holds μ-almost everywhere. Thus, the integral of this function coincides with the integral of the constant function 1, which is I. Since the integral is multiplicative, we see that

$$f(A)(A - \lambda_0 I) = (A - \lambda_0 I)f(A) = I,$$

showing that the bounded operator $f(A)$ is the inverse of $(A - \lambda_0 I)$. This contradicts the assumption that $\lambda_0 \in \sigma(A)$. ∎

Proposition 7.16 *If $A \in \mathcal{B}(\mathbf{H})$ is self-adjoint and $B \in \mathcal{B}(\mathbf{H})$ commutes with A, the following results hold.*

1. *For all bounded measurable functions f on $\sigma(A)$, the operator $f(A)$ commutes with B.*

2. *Each spectral subspace for A is invariant under B.*

The proof of this proposition is deferred until Chap. 8. We conclude this section by fulfilling (at least for bounded self-adjoint operators) one of the goals of the spectral theorem, namely to give a probability measure describing the probabilities for measurements of a self-adjoint operator A in the state ψ.

Proposition 7.17 *Suppose $A \in \mathcal{B}(\mathbf{H})$ is self-adjoint and $\psi \in \mathbf{H}$ is a unit vector. Then there exists a unique probability measure μ_ψ^A on $\mathbb{R}$ such that*

$$\int_{\mathbb{R}} \lambda^m \, d\mu_\psi^A(\lambda) = \langle \psi, A^m \psi \rangle$$

for all non-negative integers m.

We will prove a version of Proposition 7.17 for unbounded self-adjoint operators in Chap. 9. In the unbounded case, however, we will not obtain uniqueness of the probability measure, even if ψ is in the domain of A^m for all m. Even in the unbounded case, however, the spectral theorem provides a *canonical* choice of the probability measure.
Proof. We define a measure μ_ψ^A on $\sigma(A)$ as in Sect. 7.2.2 by

$$\mu_\psi^A(E) = \langle \psi, \mu^A(E)\psi \rangle.$$

The properties of integration with respect to μ^A then tell us that

$$\langle \psi, A^m \psi \rangle = \left\langle \psi, \left(\int_{\sigma(A)} \lambda^m \, d\mu^A(\lambda) \right) \psi \right\rangle = \int_{\sigma(A)} \lambda^m \, d\mu_\psi^A(\lambda).$$

We then extend μ_ψ^A to $\mathbb{R}$ by setting it equal to zero on $\mathbb{R} \setminus \sigma(A)$, establishing the existence of the desired probability measure on $\mathbb{R}$. Since

$$|\langle \psi, A^m \psi \rangle| \leq \|\psi\|^2 \, \|A^m\| \leq \|\psi\|^2 \, \|A\|^m,$$

the moments grow only exponentially with m. Thus, standard uniqueness results for the moment problem (e.g., Theorem 8.1 in Chap. 4 of [18]) give the uniqueness of μ_ψ^A. ∎

7.3   Spectral Theorem for Bounded Self-Adjoint Operators, II

As we have already noted in Sect. 6.5, one version of the spectral theorem asserts that every self-adjoint operator is unitarily equivalent to a multiplication operator. In the case of a bounded self-adjoint operator A, on a separable Hilbert space $\mathbf{H}$, this result means that A is unitarily equivalent to the operator M_h on $L^2(X, \mu)$, where (X, μ) is a σ-finite measure space, h is a measurable, real-valued function, and M_h is the operator of multiplication by h:

$$(M_h\psi)(\lambda) = h(\lambda)\psi(\lambda).$$

Although the "multiplication operator" form of the spectral theorem (Theorem 7.20) has the advantage of being easy to state, there is an even better version involving the concept of a direct integral. It is straightforward to extend the notion of an L^2 space to an L^2 space with values in a Hilbert space $\mathbf{H}$. In a direct integral, we extend the concept one step further, by allowing the Hilbert space to depend on the point. We begin with a measure space (X, μ) and then have one Hilbert space $\mathbf{H}_\lambda$ for each λ in X. An element of the direct integral is a function s on X such that $s(\lambda)$ belongs to $\mathbf{H}_\lambda$ for each $\lambda \in X$. Given a *real-valued* measurable function h on X, it makes sense to multiply an element s of the direct integral by h.

The direct integral form of the spectral theorem says a bounded self-adjoint operator A is unitarily equivalent to a multiplication operator on a direct integral. By extending multiplication operators to the more general setting of direct integrals (instead of just ordinary L^2 spaces), we gain several benefits. First, the set X and the function h become canonical: The set X is simply the spectrum of A and the function h is simply $h(\lambda) = \lambda$. Second, the direct integral approach carries with it a notion of "generalized eigenvectors," since the space $\mathbf{H}_\lambda$ can be thought of as the space of generalized eigenvectors with eigenvalue λ. (The spaces $\mathbf{H}_\lambda$ are not, in general, contained in the direct integral Hilbert space. Thus, direct integrals give a rigorous meaning to the idea of "eigenvectors" that are not in the Hilbert space on which the operator acts.) Third, the direct integral approach gives a simple way to classify self-adjoint operators up to unitary equivalence: Two self-adjoint operators are unitarily equivalent if and only if their direct integral representations are equivalent in a natural sense (Proposition 7.24).

If one really wants the simplicity of the (ordinary) multiplication operator version of the spectral theorem, it is a simple matter to prove this result using precisely the same methods as in the proof of the direct integral version. (See Theorem 7.20.) Nevertheless, the direct integral version is, arguably, the most definitive version of the spectral theorem for a single self-adjoint operator.

We turn now to the definition of a direct integral. Suppose μ is a σ-finite measure on a σ-algebra Ω of sets in X. Suppose also that for each $\lambda \in X$, we have a separable Hilbert space $\mathbf{H}_\lambda$ with inner product $\langle \cdot, \cdot \rangle_\lambda$. We want to define the direct integral of the $\mathbf{H}_\lambda$'s with respect to μ. Elements of the direct integral will be *sections* s, meaning that s is a function on X with values in the union of the $\mathbf{H}_\lambda$'s, having the property that

$$s(\lambda) \in \mathbf{H}_\lambda$$

for each λ in X. We would like to define the norm of a section s by the formula

$$\|s\|^2 = \int_X \langle s(\lambda), s(\lambda) \rangle_\lambda \, d\mu(\lambda),$$

provided that the integral on the right-hand side is finite. The inner product of two sections s_1 and s_2 (with finite norm) should then be given by the formula

$$\langle s_1, s_2 \rangle := \int_X \langle s_1(\lambda), s_2(\lambda) \rangle_\lambda \, d\mu(\lambda).$$

The problem with this description of the norm and inner product on the direct integral is that we have not said anything about measurability. As things stand, it does not make sense to ask whether a section s is measurable, since the space in which $s(\lambda)$ takes its values is different for each λ. We must, therefore, introduce some additional structure that gives rise to a notion of measurability. (The measurability issue is a technicality that can be ignored on a first reading.)

One way to address the measurability issue is to choose a *simultaneous orthonormal basis* for each of the Hilbert spaces $\mathbf{H}_\lambda$. To deal with the possibility that different spaces can have different dimensions, we slightly modify the concept of an orthonormal basis. We say that a family $\{e_j\}$ of vectors is an orthonormal basis for a Hilbert space $\mathbf{H}$ if $\langle e_j, e_k \rangle = 0$ for $j \neq k$, the norm of each e_j is *either* 0 or 1, and the closure of the span of the e_j's is all of $\mathbf{H}$. This just means that we allow some of the vectors in our basis to be zero, with the nonzero vectors forming an orthonormal basis in the usual sense.

We now define a simultaneous orthonormal basis for a family $\{\mathbf{H}_\lambda\}$ of separable Hilbert spaces to be a collection $\{e_j(\cdot)\}_{j=1}^\infty$ of sections with the property that for each λ, $\{e_j(\lambda)\}_{j=1}^\infty$ is an orthonormal basis for $\mathbf{H}_\lambda$. Provided that the function $\lambda \mapsto \dim \mathbf{H}_\lambda$ is a measurable function from X into $[0, \infty]$, it is possible to choose a simultaneous orthonormal basis $\{e_j(\cdot)\}$ such that $\langle e_j(\lambda), e_k(\lambda) \rangle$ is measurable for all j and k. Having chosen a simultaneous orthonormal basis with this property, we *define* a section s to be measurable if the function

$$\lambda \mapsto \langle e_j(\lambda), s(\lambda) \rangle_\lambda$$

is a measurable complex-valued function for each j. Our assumption on the e_j's means that the e_j's themselves are measurable sections.

We refer to a choice of simultaneous orthonormal basis, chosen so that $\langle e_j(\lambda), e_k(\lambda) \rangle$ is measurable, as a *measurability structure* on the collection of $\mathbf{H}_\lambda$'s. Given two measurable sections s_1 and s_2, the function

$$\lambda \mapsto \langle s_1(\lambda), s_2(\lambda) \rangle_\lambda = \sum_{j=1}^{\infty} \langle s_1(\lambda), e_j(\lambda) \rangle_\lambda \langle e_j(\lambda), s_2(\lambda) \rangle_\lambda$$

is also measurable.

Definition 7.18 *Suppose the following structures are given: (1) a σ-finite measure space (X, Ω, μ), (2) a collection $\{\mathbf{H}_\lambda\}_{\lambda \in X}$ of separable Hilbert spaces for which the dimension function is measurable, and (3) a measurability structure on $\{\mathbf{H}_\lambda\}_{\lambda \in X}$. Then the **direct integral** of the $\mathbf{H}_\lambda$'s with respect to μ, denoted*

$$\int_X^{\oplus} \mathbf{H}_\lambda \, d\mu(\lambda),$$

is the space of equivalence classes of almost-everywhere-equal measurable sections s for which

$$\|s\|^2 := \int_X \langle s(\lambda), s(\lambda) \rangle_\lambda \, d\mu(\lambda) < \infty.$$

The inner product $\langle s_1, s_2 \rangle$ of two such sections s_1 and s_2 is given by the formula

$$\langle s_1, s_2 \rangle := \int_X \langle s_1(\lambda), s_2(\lambda) \rangle_\lambda \, d\mu(\lambda).$$

To see that the integral defining the inner product of two finite-norm sections is finite, note that $|\langle s_1(\lambda), s_2(\lambda) \rangle_\lambda| \leq \|s_1(\lambda)\|_\lambda \|s_2(\lambda)\|_\lambda$. By assumption, $\|s_j(\lambda)\|_\lambda$ is a square-integrable function of λ for $j = 1, 2$, and the product of two square-integrable functions is integrable. Thus, the integrand in the definition of $\langle s_1, s_2 \rangle$ is also integrable. It is not hard to show, using an argument similar to the proof of completeness of L^2 spaces, that a direct integral of Hilbert spaces is a Hilbert space.

Let us think of two important special cases of the direct integral construction. First, if each of the $\mathbf{H}_\lambda$'s is simply $\mathbb{C}$, then the direct integral (with the obvious measurability structure) is simply $L^2(X, \mu)$. Second, suppose that $X = \{\lambda_1, \lambda_2, \ldots\}$ is countable, Ω is the σ-algebra of all subsets of X, and μ is the counting measure on X. Then the direct integral is the *Hilbert space direct sum* (Definition A.45).

Given a direct integral, suppose we have some $\lambda_0 \in X$ for which $\{\lambda_0\}$ is measurable and such that $c := \mu(\{\lambda_0\}) > 0$. Then we can embed $\mathbf{H}_{\lambda_0}$ isometrically into the direct integral by mapping each $\psi \in \mathbf{H}_{\lambda_0}$ to the section s given by

$$s(\lambda) = \begin{cases} \frac{1}{\sqrt{c}}\psi, & \lambda = \lambda_0 \\ 0, & \lambda \neq \lambda_0 \end{cases}.$$

Even if $\mu(\{\lambda_0\}) = 0$, we may still think that $\mathbf{H}_{\lambda_0}$ is a sort of "generalized subspace" of the direct integral.

Theorem 7.19 (Spectral Theorem, Second Form) *If $A \in \mathcal{B}(\mathbf{H})$ is self-adjoint, then there exists a σ-finite measure μ on $\sigma(A)$, a direct integral*

$$\int_{\sigma(A)}^{\oplus} \mathbf{H}_\lambda \, d\mu(\lambda),$$

and a unitary map U between $\mathbf{H}$ and the direct integral such that

$$\left[U A U^{-1}(s) \right](\lambda) = \lambda s(\lambda) \tag{7.20}$$

for all sections s in the direct integral.

The proof of Theorem 7.19 is given in the next chapter, along with the proof of our first version of the spectral theorem. In the meantime, let us think about what this version of the spectral theorem is saying. We may think that the unitary map U is an identification of our original Hilbert space $\mathbf{H}$ with a certain direct integral over the spectrum of A. Under this identification, the self-adjoint operator A becomes the operator of multiplication by λ, that is, the map sending the section $s(\lambda)$ to $\lambda s(\lambda)$. Roughly speaking, then, the operator A acts (under our identification) as λI on each space $\mathbf{H}_\lambda$. Thus, we may think of $\mathbf{H}_\lambda$ as being something like an "eigenspace" for A, for each element λ of the spectrum of A. Of course, unless $\mu(\{\lambda\}) > 0$, the Hilbert space $\mathbf{H}_\lambda$ is not actually contained in $\mathbf{H}$. Nevertheless, we may think of elements of a given $\mathbf{H}_\lambda$ as "generalized eigenvectors" for the operator A.

The direct integral formulation of the spectral theorem leads readily to a classification result for bounded self-adjoint operators. See Proposition 7.24 later in this section. Meanwhile, as we noted earlier in this section, the method of proof for Theorem 7.19 also yields a version of the spectral theorem involving multiplication operators on ordinary L^2 spaces.

Theorem 7.20 (Spectral Theorem, Multiplication Operator Form) *Suppose $A \in \mathcal{B}(\mathbf{H})$ is self-adjoint. Then there exists a σ-finite measure space (X, μ), a bounded, measurable, real-valued function h on X, and a unitary map $U : \mathbf{H} \to L^2(X, \mu)$ such that*

$$[U A U^{-1}(\psi)](\lambda) = h(\lambda)\psi(\lambda)$$

for all $\psi \in L^2(X, \mu)$.

We return now to a discussion of the direct integral version of the spectral theorem. This version gives a simple description of the functional calculus.

Proposition 7.21 *Suppose $A \in \mathcal{B}(\mathbf{H})$ is self-adjoint and U is a unitary map as in Theorem 7.19. Then for any bounded measurable function f on $\sigma(A)$, we have*

$$[Uf(A)U^{-1}(s)](\lambda) = f(\lambda)s(\lambda).$$

Thus, roughly speaking, $f(A)$ is defined to be $f(\lambda)I$ on each "generalized eigenspace" $\mathbf{H}_\lambda$. Proposition 7.21 follows directly from (7.20) if f is a polynomial; the result for continuous f then follows by taking uniform limits. The result for general f is then easily established by using the limiting arguments of Chap. 8, especially Exercise 3.

Let us now consider what sort of uniqueness there should be in the second version of the spectral theorem. There is a "trivial" source of nonuniqueness coming from the possibility that some of the $\mathbf{H}_\lambda$'s may have dimension 0. Let E_0 denote the set of λ for which $\dim \mathbf{H}_\lambda = 0$. Even if $\mu(E_0) > 0$, the set E_0 makes no contribution to the norm of a section, since every section is automatically zero on E_0. Thus, we may define a new measure $\tilde{\mu}$ by setting $\tilde{\mu}(E) = \mu(E \cap E_0^c)$, so that $\tilde{\mu}$ agrees with μ on E_0^c but is zero on E_0. Then the direct integrals of the $\mathbf{H}_\lambda$'s with respect to μ and with respect to $\tilde{\mu}$ are "indistinguishable." Thus, we can always modify a direct integral so as to assume that $\dim \mathbf{H}_\lambda > 0$ for almost every λ.

Meanwhile, unlike the projection-valued measure μ^A in Theorem 7.12, the measure μ in Theorem 7.19 is not unique, but only unique up to equivalence, where two σ-finite measures on a given measurable space are equivalent if they have precisely the same sets of measure zero. For a given measure μ, the Hilbert spaces $\mathbf{H}_\lambda$ are unique only up to unitary equivalence, meaning that only the *dimension* of the spaces is uniquely determined. Even the dimension of $\mathbf{H}_\lambda$ is uniquely determined only up to a set of μ-measure zero. As it turns out, the sources of nonuniqueness in this paragraph and the previous paragraph are all that exist.

Proposition 7.22 (Uniqueness in Theorem 7.19) *Suppose $A \in \mathcal{B}(\mathbf{H})$ is self-adjoint and consider two different direct integrals as in Theorem 7.19, one with measure $\mu^{(1)}$ and Hilbert spaces $\mathbf{H}_\lambda^{(1)}$ and the other with measure $\mu^{(2)}$ and Hilbert spaces $\mathbf{H}_\lambda^{(2)}$. If $\dim \mathbf{H}_\lambda^{(j)} > 0$ for $\mu^{(j)}$-almost every λ $(j = 1, 2)$, then $\mu^{(1)}$ and $\mu^{(2)}$ are mutually absolutely continuous and*

$$\dim \mathbf{H}_\lambda^{(1)} = \dim \mathbf{H}_\lambda^{(2)}$$

for $\mu^{(j)}$-almost every λ $(j = 1, 2)$.

See the end of the next chapter for a sketch of the proof of this uniqueness result.

Theorem 7.19 should be thought of as a refinement of our earlier form (Theorem 7.12) of the spectral theorem, in the sense that we can easily recover Theorem 7.12 from Theorem 7.19. In the setting of Theorem 7.19, and given a measurable set $E \subset \sigma(A)$, let V_E denote the space of (equivalence classes) of sections s that are supported on E, that is, for which $s(\lambda) = 0$ for μ-almost every λ in E^c. This is easily seen to be a closed subspace. Let P_E denote the orthogonal projection onto V_E, and define

$$\mu^A(E) = U^{-1}P_E U. \tag{7.21}$$

It is straightforward to check that μ^A is a projection-valued measure on $\sigma(A)$, with values in $\mathcal{B}(\mathbf{H})$, and that $\int_{\sigma(A)} \lambda \, d\mu^A(\lambda) = A$.

Note that both versions of the spectral theorem for A involve a measure, the first, denoted μ^A, being a projection-valued measure, and the second, denoted μ, being an ordinary measure with values in the non-negative real numbers. The following result shows the relationship between the two measures.

Proposition 7.23 *Suppose $A \in \mathcal{B}(\mathbf{H})$ is self-adjoint, μ^A is the projection-valued measure given by Theorem 7.12 and μ is a real-valued measure as in Theorem 7.19. If $\dim \mathbf{H}_\lambda > 0$ for μ-almost every λ, then for any Borel set $E \subset \sigma(A)$, $\mu^A(E) = 0$ if and only if $\mu(E) = 0$.*

Of course, the 0 in the expression $\mu^A(E) = 0$ is the zero *operator*, whereas the 0 in the expression $\mu(E) = 0$ is the *number* 0. Nevertheless, we may think of Proposition 7.23 as saying that μ^A and μ are equivalent in the usual measure-theoretic sense, having precisely the same sets of measure zero.

Proof. As we have remarked, given a direct integral as in Theorem 7.19, we can construct a projection-valued measure by means of (7.21), and this projection-valued measure satisfies $\int_{\sigma(A)} \lambda \, d\mu^A(\lambda) = A$. This projection-valued measure must coincide with the one in Theorem 7.12, by the uniqueness in that theorem.

Now, if $\mu(E) = 0$, then any section supported on E is zero almost everywhere and thus represents the zero element of the direct integral. In that case, $V_E = 0$ and so $\mu^A(E) = 0$ by (7.21). In the other direction, suppose $\mu(E) > 0$. Since μ is σ-finite, E will contain a measurable subset F such that $0 < \mu(F) < \infty$. Then let s be the section given by

$$s(\lambda) = \sum_{j=1}^{\infty} \frac{1}{2^j} e_j(\lambda)$$

for $\lambda \in F$ and $s(\lambda) = 0$ for $\lambda \in F^c$, where $\{e_j(\cdot)\}$ is our measurability structure for the direct integral. Then

$$\langle s(\lambda), e_j(\lambda) \rangle_\lambda = \frac{1}{2^j} \langle e_j(\lambda), e_j(\lambda) \rangle_\lambda 1_F(\lambda),$$

which is a measurable function of λ for all j, so that s is measurable. Since we assume that $\mathbf{H}_\lambda$ has nonzero dimension for μ-almost every λ, s will be nonzero almost everywhere on F and thus will have positive norm. The norm of s is finite because $\|s(\lambda)\| \leq 1$ and F has finite measure. Thus, $V_E \neq 0$ and $\mu^A(E) \neq 0$. $\blacksquare$

We say that self-adjoint operators A_1 and A_2 on Hilbert spaces $\mathbf{H}_1$ and $\mathbf{H}_2$ are *unitarily equivalent* if there exists a unitary map $U : \mathbf{H}_1 \to \mathbf{H}_2$ such that

$$A_2 = U A_1 U^{-1}.$$

Using Proposition 7.22, we can give a classification of bounded self-adjoint operators on separable Hilbert spaces up to unitary equivalence. For a given bounded self-adjoint operator A, we call the function $\lambda \mapsto \dim \mathbf{H}_\lambda$ the *multiplicity function* for A. It is well defined (independent of the choice of direct integral decomposition) up to a set of measure zero. It turns out that bounded self-adjoint operators are characterized, up to unitary equivalence, by the spectrum of A as a set, the equivalence class of the measure μ in Theorem 7.19, and the multiplicity function.

Proposition 7.24 *Suppose A_1 and A_2 are bounded self-adjoint operators on separable Hilbert spaces $\mathbf{H}_1$ and $\mathbf{H}_2$, respectively. Choose direct integral representations for A_1 and A_2 as in Theorem 7.19, with the associated measures μ_1 and μ_2 chosen so that $\dim \mathbf{H}_\lambda > 0$ for μ_j-almost every λ ($j = 1, 2$). Then A_1 and A_2 are unitarily equivalent if and only if the following conditions are satisfied.*

1. *$\sigma(A_1) = \sigma(A_2)$.*

2. *The measures μ_1 and μ_2 are mutually absolutely continuous.*

3. *The multiplicity functions of A_1 and A_2 coincide up to a set of measure zero.*

See Exercise 12 for a proof of this result.

7.4   Exercises

1. Suppose A and B are *commuting* linear operators on a nonzero finite-dimensional vector space.

   (a) Show that each eigenspace for A is invariant under B.

   (b) Show that A and B have at least one simultaneous eigenvector, that is, a nonzero vector v with $Av = \lambda v$ and $Bv = \mu v$, for some constants $\lambda, \mu \in \mathbb{C}$.

2. Suppose that $A \in \mathcal{B}(\mathbf{H})$ is *normal*, meaning that $AA^* = A^*A$. Suppose that for some $\psi \in \mathbf{H}$ and $\lambda \in \mathbb{C}$ we have $A\psi = \lambda\psi$. Show that $A^*\psi = \bar{\lambda}\psi$.

   Hint: Compute $\|(A^* - \bar{\lambda})\psi\|$.

3. Suppose a closed subspace V of $\mathbf{H}$ is invariant under a bounded operator A, meaning that $A\psi \in V$ for all $\psi \in V$. Show that the orthogonal complement $V^{\perp}$ of V is invariant under A^*.

4. (a) Suppose that $\mathbf{H}$ is a finite-dimensional Hilbert space over $\mathbb{C}$ and A is a normal linear operator on $\mathbf{H}$ in the sense of Exercise 2. Show that there exists an orthonormal basis for V consisting of simultaneous eigenvectors for A and A^*.

   Hint: Use Exercises 1 and 3.

   (b) Suppose A is a linear operator on a finite-dimensional Hilbert space $\mathbf{H}$ over $\mathbb{C}$ and suppose there exists an orthonormal basis for V consisting of eigenvectors of A. Show that A commutes with A^*.

5. Suppose $A \in \mathcal{B}(\mathbf{H})$ has an inverse A^{-1} in $\mathcal{B}(\mathbf{H})$. Show that $(A^{-1})^* A^* = A^*(A^{-1})^* = I$. Conclude that A^* is invertible and $(A^*)^{-1} = (A^{-1})^*$.

6. Suppose U is a unitary operator on $\mathbf{H}$ (Definition A.55). Show that the spectrum of U is contained in the unit circle.

   Hint: By writing $U - \lambda I$ as $(-\lambda)(I - U/\lambda)$ or as $U(I - \lambda U^{-1})$, show that any λ with $|\lambda| \neq 1$ is in the resolvent set of λ.

7. Suppose that $A \in \mathcal{B}(\mathbf{H})$ is self-adjoint and non-negative, that is, that A satisfies (7.3). Show that the spectrum of A is contained in the interval $[0, \infty)$.

   Note: Conversely, if $A \in \mathcal{B}(\mathbf{H})$ is self-adjoint and $\sigma(A) \subset [0, \infty)$, then A is non-negative. See Exercise 2 in Chap. 8.

8. Suppose $A \in \mathcal{B}(\mathbf{H})$ is invertible. Show that there exists $\varepsilon > 0$ such that for all $B \in \mathcal{B}(\mathbf{H})$ with $\|B - A\| < \varepsilon$, B is also invertible.

   Hint: Use a power series argument as in the proof of Proposition 7.5.

9. Assume $A \in \mathcal{B}(\mathbf{H})$ is self-adjoint.

   (a) Suppose $\lambda_0 \in \mathbb{C}$ is a point in the resolvent set of A. Show that

   $$\left\|(A - \lambda_0 I)^{-1}\right\| = \frac{1}{d(\lambda_0, \sigma(A))},$$

   where $d(\lambda_0, \sigma(A)) = \inf_{\lambda \in \sigma(A)} |\lambda - \lambda_0|$.

   Hint: Think of $(A - \lambda_0 I)^{-1}$ as a function of A in the sense of the functional calculus for A.

   (b) Given $\lambda_0 \in \mathbb{C}$, suppose that there exists some nonzero $\psi \in \mathbf{H}$ such that

   $$\|A\psi - \lambda_0\psi\| < \varepsilon \|\psi\|.$$

   Show that there exists $\lambda \in \sigma(A)$ such that $|\lambda - \lambda_0| < \varepsilon$.

10. Suppose V_1 and V_2 are two closed subspaces of $\mathbf{H}$, with associated orthogonal projections P_1 and P_2. Show that V_1 and V_2 are orthogonal if and only if $P_1 P_2 = 0$.

11. Suppose μ is a projection-valued measure on (X, Ω). Show that for any $E_1, E_2 \in \Omega$, $\mu(E_1)\mu(E_2)$ is the projection onto the closed subspace $\mathrm{Range}(\mu(E_1)) \cap \mathrm{Range}(\mu(E_2))$.

    Hint: Write E_1 as $E_1 = (E_1 \cap E_2) \cup (E_1 \backslash E_2)$ and use Exercise 10.

12. Prove Proposition 7.24.

    Hint: Use Proposition 7.22 and the Radon–Nikodym theorem (Theorem A.6).

8

The Spectral Theorem for Bounded Self-Adjoint Operators: Proofs

In this chapter we give proofs of all versions of the spectral theorem stated in the previous chapter.

8.1 Proof of the Spectral Theorem, First Version

A proof of the spectral theorem, in its projection-valued measure form, can be obtained in two main stages. The first stage of the proof is to define a *continuous* functional calculus, meaning we associate with each continuous function f on $\sigma(A)$ an operator $f(A)$. The map $f \mapsto f(A)$ should have the property that if f is the function $f(\lambda) = \lambda^m$, then $f(A) = A^m$. The continuous functional calculus is then constructed by approximating continuous functions on $\sigma(A)$ by polynomials. The Stone–Weierstrass theorem tells us that polynomials are dense in the continuous functions on $\sigma(A)$; it remains only to show that if a sequence p_n of polynomials converges uniformly to some continuous function f on $\sigma(A)$, then the operators $p_n(A)$ converge to some operator, which we will then call $f(A)$.

The second stage of the proof is to show that the continuous functional calculus can be represented as integration against a projection-valued measure. This result is just an operator-valued version of the Riesz representation theorem from measure theory (Theorem 8.5). Indeed, we will see that this operator-valued version of the Riesz representation theorem can be reduced to the usual form of the theorem.

B.C. Hall, *Quantum Theory for Mathematicians*, Graduate Texts in Mathematics 267, DOI 10.1007/978-1-4614-7116-5_8,
© Springer Science+Business Media New York 2013

8.1.1   Stage 1: The Continuous Functional Calculus

We begin by defining, for any $A \in \mathcal{B}(\mathbf{H})$, the *spectral radius* $R(A)$ by

$$R(A) = \sup_{\lambda \in \sigma(A)} |\lambda|.$$

(By Propositions 7.5 and 7.7, $\sigma(A)$ is a nonempty, bounded subset of $\mathbb{R}$.) According to Point 2 of Proposition 7.5, we have

$$R(A) \leq \|A\|$$

for any $A \in \mathcal{B}(\mathbf{H})$. In general, $\|A\|$ can be much bigger than $R(A)$. For example, if A is a nilpotent matrix, then $R(A) = 0$ but $\|A\|$ can be arbitrarily large.

Lemma 8.1 *If $A \in \mathcal{B}(\mathbf{H})$ is self-adjoint, the norm and the spectral radius of A are equal:*
$$\|A\| = R(A).$$

In preparation for the proof, we determine the radius of convergence of the power series for the resolvent given in the proof of Proposition 7.5. According to Proposition 7.2, we have

$$\|A^* A\| = \|A\|^2$$

for any $A \in \mathcal{B}(\mathbf{H})$. If A is self-adjoint, we obtain

$$\|A^2\| = \|A\|^2.$$

Iterating this relation gives

$$\left\| A^{2^n} \right\| = \|A\|^{2^n} \tag{8.1}$$

for all n.

Consider, for a bounded self-adjoint operator A, the following formal expression for the resolvent of A:

$$(A - \lambda I)^{-1} = -\frac{1}{\lambda} \left(I - \frac{A}{\lambda} \right)^{-1}$$

$$= -\sum_{m=0}^{\infty} \frac{A^m}{\lambda^{m+1}}. \tag{8.2}$$

If $|\lambda| > \|A\|$, then the proof of Proposition 7.5 shows that the series (8.2) converges in the operator norm topology and that the sum of the series is indeed the inverse of $(A - \lambda I)$. If, on the other hand, $|\lambda| \leq \|A\|$, it follows from (8.1) that the norms of the terms in (8.2) do not tend to zero, and so the series cannot converge in the operator norm topology. We may say, then, that the series (8.2) has radius of convergence equal to $\|A\|$.

Proof of Lemma 8.1. We know that $R(A) \leq \|A\|$. To show that $R(A) = \|A\|$, we wish to argue that $(A - \lambda I)^{-1}$ is a holomorphic operator-valued function of λ on the set $|\lambda| > R(A)$, and therefore the Laurent series of $(A - \lambda I)^{-1}$ must converge for $|\lambda| > R(A)$. But the Laurent series of $(A - \lambda I)^{-1}$ is just the series in (8.2), and we have shown that the series diverges when $|\lambda| \leq \|A\|$. This would be a contradiction if $R(A)$ were less than $\|A\|$.

To flesh out the argument, recall the formula (7.8) in the proof of Proposition 7.5 for the resolvent of A.

That formula expresses the map $\lambda \mapsto (A - \lambda I)^{-1}$ as a convergent power series in powers of $\lambda - \lambda_0$, near any point λ_0 in the resolvent set of A. It follows that for any bounded linear functional $\xi \in \mathcal{B}(\mathbf{H})^*$, the complex-valued function

$$\lambda \mapsto \xi((A - \lambda I)^{-1})$$

is holomorphic on the resolvent set of A. This function has a unique Laurent series, which is given by applying ξ term by term to (8.2). The series will converge on the largest annulus contained in the resolvent set of A, namely the set of λ with $|\lambda| > R(A)$.

Convergence of (8.2) means that $\left|\xi(A^m/\lambda^{m+1})\right|$ is bounded as function of m, for each ξ and each λ with $|\lambda| > R(A)$. Thus, by (a corollary of) the uniform boundedness principle (Appendix A.3.4), the set $\{A^m/\lambda^{m+1}\}_{m=0}^{\infty}$ is bounded in the Banach space $\mathcal{B}(\mathbf{H})$, for all λ with $|\lambda| > R(A)$. In particular, for each λ with $|\lambda| > R(A)$, there is a constant C such that

$$\frac{\|A^{2^n}\|}{|\lambda|^{2^n}} = \frac{\|A\|^{2^n}}{|\lambda|^{2^n}} \leq C.$$

If $\|A\|$ were greater than $R(A)$, this inequality would be false for λ satisfying $R(A) < |\lambda| < \|A\|$. ∎

The next key step in Stage 1 of the proof is to understand how the spectrum of a self-adjoint operator transforms under application of a polynomial.

Lemma 8.2 (Spectral Mapping Theorem) *For all $A \in \mathcal{B}(\mathbf{H})$ and all polynomials p, we have*

$$\sigma(p(A)) = p(\sigma(A)).$$

That is to say, the spectrum of $p(A)$ consists precisely of the numbers of the form $p(\lambda)$, with λ in the spectrum of A.

Proof. The result is trivial if p is constant. When $\deg p \geq 1$, let p given by

$$p(z) = a_n z^n + a_{n-1} z^{n-1} + \cdots + a_0$$

be an arbitrary polynomial. We first show that $p(\sigma(A)) \subset \sigma(p(A))$. Suppose, then, that $\lambda \in \sigma(A)$. Observe that

$$p(A) - p(\lambda)I = a_n(A^n - \lambda^n I) + a_{n-1}(A^{n-1} - \lambda^{n-1}I) + \cdots + a_0 I - a_0 I.$$

Now,

$$A^k - \lambda^k I = (A - \lambda I)(A^{k-1} + \lambda A^{k-2} + \lambda^2 A^{k-3} + \cdots + \lambda^{k-1}I).$$

Thus, we can pull out a factor of $(A - \lambda I)$ from each nonzero term in $p(A) - p(\lambda)I$, giving

$$p(A) - p(\lambda)I = (A - \lambda I)q(A)$$

where q is a polynomial (depending on λ). Since, by assumption, $A - \lambda I$ is not invertible, and since $(A - \lambda I)$ commutes with $q(A)$, $(A - \lambda I)q(A)$ cannot be invertible (Exercise 1). This shows that $p(\lambda)$ belongs to the spectrum of $p(A)$.

We now show that $\sigma(p(A)) \subset p(\sigma(A))$. Suppose, then, that $\gamma \in \sigma(p(A))$. Since $\mathbb{C}$ is algebraically closed, we can factor the polynomial $p(z) - \gamma$, as a function of z, as

$$p(z) - \gamma = c(z - b_1)(z - b_2) \cdots (z - b_n). \tag{8.3}$$

Thus,

$$p(A) - \gamma I = c(A - b_1 I)(A - b_2 I) \cdots (A - b_n I).$$

Since $p(A) - \gamma I$ is assumed to be noninvertible, there must be some j such that $(A - b_j I)$ is noninvertible, that is, for which $b_j \in \sigma(A)$. Then (8.3) tells us that $p(b_j) - \gamma = 0$, meaning that $\gamma = p(b_j)$. Thus, γ is of the form $p(\lambda)$ for some $\lambda (= b_j)$ in $\sigma(A)$. $\blacksquare$

The last step in Stage 1 of our proof is to apply the Stone–Weierstrass theorem to show that polynomials are dense in $\mathcal{C}(\sigma(A); \mathbb{R})$ (the space of continuous, real-valued functions on $\sigma(A)$) with respect to the supremum norm.

Proposition 8.3 *Suppose $A \in \mathcal{B}(\mathbf{H})$ is self-adjoint. Then there exists a unique bounded linear map from $\mathcal{C}(\sigma(A); \mathbb{R})$ into $\mathcal{B}(\mathbf{H})$, denoted by $f \mapsto f(A)$, such that when $f(\lambda) = \lambda^m$, we have $f(A) = A^m$. The map $f \mapsto f(A)$, $f \in \mathcal{C}(\sigma(A); \mathbb{R})$, is called the (real-valued) functional calculus for A.*

Proof. Note that if A is self-adjoint, then $p(A)$ is self-adjoint provided that p is a *real-valued* polynomial (i.e., one where all the coefficients are real numbers). Thus, combining the spectral mapping theorem with the equality of the norm and spectral radius, we have the following: If A is a self-adjoint operator and p is a real-valued polynomial, then

$$\|p(A)\| = \sup_{\lambda \in \sigma(A)} |p(\lambda)|. \tag{8.4}$$

Thus, the map $p \to p(A)$ is an isometric linear map from the space of polynomials on $\sigma(A)$ (with the supremum norm) into the space of bounded operators on $\mathbf{H}$.

According to the Stone–Weierstrass theorem polynomials are dense in $C(\sigma(A); \mathbb{R})$. Thus, by the BLT theorem (Theorem A.36), we can extend the map $p \mapsto p(A)$ uniquely to a bounded linear map of $C(\sigma(A); \mathbb{R})$ into $\mathcal{B}(\mathbf{H})$.
∎

Proposition 8.4 *If $A \in \mathcal{B}(\mathbf{H})$ is self-adjoint, the (real-valued) continuous functional calculus for A, mapping $C(\sigma(A); \mathbb{R})$ into $\mathcal{B}(\mathbf{H})$, has the following properties.*

1. **Multiplicativity:** *For all f, g, we have*

$$(fg)(A) = f(A)g(A),$$

   where fg denotes the pointwise product of f and g.

2. **Self-adjointness:** *For all f, the operator $f(A)$ is self-adjoint.*

3. **Non-negativity:** *For all f, if f is non-negative, then $f(A)$ is a non-negative operator.*

4. **Norm and spectrum properties:** *For all f, we have*

$$\|f(A)\| = \sup_{\lambda \in \sigma(A)} |f(\lambda)| \tag{8.5}$$

   and

$$\sigma(f(A)) = \{f(\lambda) \mid \lambda \in \sigma(A)\}. \tag{8.6}$$

Proof. Point 1 holds for polynomials and thus, by taking limits, for all $f \in C(\sigma(A); \mathbb{R})$. Furthermore, if p is a real-valued polynomial and A is self-adjoint, then $p(A)$ is self adjoint. From this, we get Point 2 by taking limits. If $f \in C(\sigma(A); \mathbb{R})$ is non-negative, then $f = g^2$, where $g = \sqrt{f}$ is real-valued. Thus, $g(A)$ is self-adjoint and for all $\psi \in \mathbf{H}$, Point 1 tells us that

$$\langle \psi, f(A)\psi \rangle = \langle \psi, g(A)^2 \psi \rangle = \langle g(A)\psi, g(A)\psi \rangle \geq 0, \tag{8.7}$$

which establishes Point 3. We have already established (8.5) in (8.4) for polynomials; the result for general $f \in C(\sigma(A); \mathbb{R})$ follows by taking limits.

To establish (8.6), suppose first that $\lambda_0 \in \mathbb{C}$ is not in the range of f. Then the function $g(\lambda) := 1/(f(\lambda) - \lambda_0)$ is continuous on $\sigma(A)$ and the operator $g(A)$ will be the inverse of $f(A) - \lambda_0 I$, showing that λ_0 is not in the spectrum of $f(A)$.

In the other direction, suppose that $\lambda_0 = f(\mu)$ for some $\mu \in \sigma(A)$; we want to show that $f(\mu) \in \sigma(f(A))$. Suppose now that $f(A) - f(\mu)I$ were invertible and choose a sequence p_n of polynomials converging uniformly to f on $\sigma(A)$. By Exercise 8 in Chap. 7, any operator sufficiently close to $f(A) - f(\mu)I$ in the operator norm topology would also be invertible. In particular, $p_n(A) - p_n(\mu)I$ would have to be invertible for all sufficiently large n, contradicting the spectral mapping theorem. ∎

8.1.2 Stage 2: An Operator-Valued Riesz Representation Theorem

We turn now to Stage 2 of the proof of the spectral theorem. We will make use of the Riesz representation theorem from measure theory (*not* the result about continuous linear functionals on a Hilbert space). The following form of this result is sufficient for our purposes.

Theorem 8.5 (Riesz Representation Theorem) *Let X be a compact metric space and let $\mathcal{C}(X; \mathbb{R})$ denote the space of continuous, real-valued functions on X. Suppose $\Lambda : \mathcal{C}(X; \mathbb{R}) \to \mathbb{R}$ is a linear functional with the property that $\Lambda(f)$ is non-negative whenever all the values of f are non-negative. Then there exists a unique (real-valued, positive) measure μ on the Borel σ-algebra in X for which*

$$\Lambda(f) = \int_X f \, d\mu$$

for all $f \in \mathcal{C}(X; \mathbb{R})$.

See pp. 353–354 of Volume I of [34] for a short proof in the case in which X is a compact subset of $\mathbb{R}$, which is all we really require. For the full result stated above, see Theorems 7.2 and 7.8 in [12]. Observe that μ is a finite measure, with $\mu(X) = \Lambda(\mathbf{1})$, where $\mathbf{1}$ is the constant function.

Given a bounded self-adjoint operator $A \in \mathcal{B}(\mathbf{H})$, we have constructed, in the previous subsection, a continuous functional calculus for A. This calculus is a map, denoted $f \mapsto f(A)$, from $\mathcal{C}(\sigma(A); \mathbb{R})$ into $\mathcal{B}(\mathbf{H})$. If $f \in \mathcal{C}(\sigma(A); \mathbb{R})$ is non-negative, then (Point 3 of Proposition 8.4) $f(A)$ is a non-negative operator. Thus, given $\psi \in \mathbf{H}$, if we define a linear functional Λ_ψ on $\mathcal{C}(\sigma(A); \mathbb{R})$ by the formula

$$\Lambda_\psi(f) = \langle \psi, f(A)\psi \rangle \,,$$

Λ_ψ will satisfy the hypotheses of the Riesz representation theorem. Thus, for each $\psi \in \mathbf{H}$, we obtain a unique measure μ_ψ such that

$$\langle \psi, f(A)\psi \rangle = \int_{\sigma(A)} f(\lambda) \, d\mu_\psi(\lambda) \tag{8.8}$$

for all $f \in \mathcal{C}(\sigma(A); \mathbb{R})$. Note that

$$\mu_\psi(\sigma(A)) = \Lambda_\psi(\mathbf{1}) = \|\psi\|^2 \,. \tag{8.9}$$

Definition 8.6 *If f is a bounded measurable (complex-valued) function on $\sigma(A)$, define a map $Q_f : \mathbf{H} \to \mathbb{C}$ by the formula*

$$Q_f(\psi) = \int_{\sigma(A)} f(\lambda) \, d\mu_\psi(\lambda),$$

where μ_ψ is the measure in (8.8).

If f happens to be real valued and continuous, then $Q_f(\psi)$ is equal $\langle \psi, f(A)\psi \rangle$, in which case Q_f is a bounded quadratic form. (See Definition A.60 and Example A.62.) It turns out that Q_f is a bounded quadratic form for any bounded measurable f, in which case Proposition A.63 allows us to associate with Q_f a bounded operator, which we denote by $f(A)$. Once the relevant properties of $f(A)$ are established, we will construct the desired projection-valued measure by setting $\mu^A(E) = 1_E(A)$.

Proposition 8.7 *For any bounded measurable function f on $\sigma(A)$, the map Q_f in Definition 8.6 is a bounded quadratic form.*

Proof. Let $\mathcal{F}$ denote the space of all bounded, Borel-measurable functions f for which Q_f is a quadratic form. Then $\mathcal{F}$ is a vector space and contains $\mathcal{C}(\sigma(A); \mathbb{R})$. Furthermore, $\mathcal{F}$ is closed under uniformly bounded pointwise limits, because $Q_f(\psi)$ is continuous with respect to such limits, by dominated convergence. Standard measure-theoretic techniques (Exercise 3) then show that $\mathcal{F}$ is the space of all bounded Borel-measurable functions on X.

Meanwhile, it follows from (8.9) that

$$|Q_f(\psi)| \leq \sup_{\lambda \subset \sigma(A)} |f(\lambda)| \, \|\psi\|^2,$$

showing that Q_f is always a *bounded* quadratic form. ∎

Definition 8.8 *For a bounded measurable function f on $\sigma(A)$, let $f(A)$ be the operator associated to the quadratic form Q_f by Proposition A.63. This means that $f(A)$ is the unique operator such that*

$$\langle \psi, f(A)\psi \rangle = Q_f(\psi) = \int_{\sigma(A)} f \, d\mu_\psi$$

for all $\psi \in \mathbf{H}$.

Observe that if f is real valued, then $Q_f(\psi)$ is real for all $\psi \in \mathbf{H}$, which means (Proposition A.63) that the associated operator $f(A)$ is self-adjoint. We will shortly associate with A a projection-valued measure μ^A, and we will show that $f(A)$, as given by Definition 8.8, agrees with $f(A)$ as given by $\int_{\sigma(A)} f(\lambda) \, d\mu^A(\lambda)$. [See (8.10) and compare Definition 7.13.]

Proposition 8.9 *For any two bounded measurable functions f and g, we have*

$$(fg)(A) = f(A)g(A).$$

Proof. Let $\mathcal{F}_1$ denote the space of bounded measurable functions f such that $(fg)(A) = f(A)g(A)$ for all $g \in \mathcal{C}(\sigma(A); \mathbb{R})$. Then $\mathcal{F}_1$ is a vector space and contains $\mathcal{C}(\sigma(A); \mathbb{R})$. We have already noted that dominated convergence guarantees that the map $f \mapsto Q_f(\psi)$, $\psi \in \mathbf{H}$, is continuous under uniformly bounded pointwise convergence. By the polarization identity (Proposition A.59), the same is true for the map $f \mapsto L_f(\phi, \psi)$, where L_f is the sesquilinear form associated to Q_f. Now, by the polarization identity, f will be in $\mathcal{F}_1$ provided that

$$\langle \psi, (fg)(A)\psi \rangle = \langle \psi, f(A)g(A)\psi \rangle$$

or, equivalently,

$$Q_{fg}(\psi) = L_f(\psi, g(A)\psi)$$

for all $\psi \in \mathbf{H}$ and all $g \in \mathcal{C}(\sigma(A); \mathbb{R})$. From this, we can see that $\mathcal{F}_1$ is closed under uniformly bounded pointwise limits. Thus, by Exercise 3, $\mathcal{F}_1$ consists of all bounded, Borel-measurable functions.

We now let $\mathcal{F}_2$ denote the space of all bounded, Borel-measurable functions f such that $(fg)(A) = f(A)g(A)$ for all bounded Borel-measurable functions g. Our result for $\mathcal{F}_1$ shows that $\mathcal{F}_2$ contains $\mathcal{C}(\sigma(A); \mathbb{R})$. Thus, the same argument as for $\mathcal{F}_1$ shows that $\mathcal{F}_2$ consists of all bounded, Borel-measurable functions. ∎

Theorem 8.10 *Suppose $A \in \mathcal{B}(\mathbf{H})$ is self-adjoint. For any measurable set $E \subset \sigma(A)$, define an operator $\mu^A(E)$ by*

$$\mu^A(E) = 1_E(A),$$

where $1_E(A)$ is given by Definition 8.8. Then μ^A is a projection-valued measure on $\sigma(A)$ and satisfies

$$\int_{\sigma(A)} \lambda \, d\mu^A(\lambda) = A.$$

Theorem 8.10 establishes the existence of the projection-valued measure in our first version of the spectral theorem (Theorem 7.12).

Proof. Since 1_E is real-valued and satisfies $1_E \cdot 1_E = 1_E$, Proposition 8.4 tells us that $1_E(A)$ is self-adjoint and satisfies $1_E(A)^2 = 1_E(A)$. Thus, $\mu^A(E)$ is an orthogonal projection (Proposition A.57), for any measurable set $E \subset X$. If E_1 and E_2 are measurable sets, then $1_{E_1 \cap E_2} = 1_{E_1} \cdot 1_{E_2}$ and so

$$\mu^A(E_1 \cap E_2) = \mu^A(E_1)\mu^A(E_2).$$

If $E_1, E_2, \ldots$ are disjoint measurable sets, then $\mu^A(E_j)\mu^A(E_k) = \mu^A(\varnothing) = 0$, for $j \neq k$, and so the ranges of the projections $\mu^A(E_j)$ and $\mu^A(E_k)$ are orthogonal. It then follows by an elementary argument that, for all $\psi \in \mathbf{H}$, we have

$$\sum_{j=1}^{\infty} \mu^A(E_j)\psi = P\psi,$$

where the sum converges in the norm topology of $\mathbf{H}$ and where P is the orthogonal projection onto the smallest closed subspace containing the range of $\mu^A(E_j)$ for every j. On the other hand, if $E := \cup_{j=1}^{\infty} E_j$, then the sequence $f_N := \sum_{j=1}^{N} 1_{E_j}$ is uniformly bounded (by 1) and converges pointwise to 1_E. Thus, using again dominated convergence in (8.8),

$$\lim_{N \to \infty} \left\langle \psi, \sum_{j=1}^{N} 1_{E_j}(A)\psi \right\rangle = \langle \psi, 1_E(A)\psi \rangle .$$

It follows that $1_E(A)$ coincides with P, which establishes the desired countable additivity for μ^A.

Finally, if $f = 1_E$ for some Borel set E, then

$$\int_{\sigma(A)} f(\lambda) \, d\mu^A(\lambda) = f(A), \tag{8.10}$$

where $f(A)$ is given by Definition 8.8. [The integral is equal to $\mu^A(E)$, which is, by definition, equal to $1_E(A)$.] The equality (8.10) then holds for simple functions by linearity and for all bounded, Borel-measurable functions by taking limits. In particular, if $f(\lambda) = \lambda$, then the integral of f against μ^A agrees with $f(A)$ as defined in Definition 8.8, which agrees with $f(A)$ as defined in the continuous functional calculus, which in turn agrees with $f(A)$ as defined for polynomials—namely, $f(A) = A$. This means that

$$\int_{\sigma(A)} \lambda \, d\mu^A(\lambda) = A$$

as desired. $\blacksquare$

We have now completed the existence of the projection-valued measure μ^A in Theorem 7.12. The uniqueness of μ^A is left as an exercise (Exercise 4). We close this section by proving Proposition 7.16, which states that if a bounded operator B commutes with a bounded self-adjoint operator A, then B commutes with $f(A)$, for all bounded, Borel-measurable functions f on $\sigma(A)$.

Proof of Proposition 7.16. If B commutes with A, then B commutes with $p(A)$, for any polynomial p. Thus, by taking limits as in the construction of the continuous functional calculus, B will commute with $f(A)$ for any continuous real-valued function f on $\sigma(A)$. We now let $\mathcal{F}$ denote the space of all bounded, Borel-measurable functions f on $\sigma(A)$ for which $f(A)$ commutes with B, so that $C(\sigma(A); \mathbb{R})$.

To show that a bounded measurable f belongs to $\mathcal{F}$, it suffices to show that for all $\phi, \psi \in \mathbf{H}$ we have $\langle \phi, f(A)B\psi \rangle = \langle \phi, Bf(A)\psi \rangle$, or, equivalently, $\langle \phi, f(A)B\psi \rangle = \langle B^*\phi, f(A)\psi \rangle$. That is, we want

$$L_f(\phi, B\psi) = L_f(B^*\phi, \psi).$$

But we have seen that for fixed vectors $\psi_1, \psi_2 \in \mathbf{H}$, the map $f \mapsto L_f(\psi_1, \psi_2)$ is continuous under uniformly bounded pointwise limits. Thus, $\mathcal{F}$ is closed under such limits, which implies (Exercise 3) that $\mathcal{F}$ contains all bounded, Borel-measurable functions. ∎

8.2   Proof of the Spectral Theorem, Second Version

We now turn to the proof of Theorem 7.19. As in the proof of Theorem 7.12, we will make use of continuous functional calculus for a bounded self-adjoint operator A and the Riesz representation theorem. We begin by establishing the special case in which A has a **cyclic vector**, that is, a vector ψ with the property that the vectors $A^k\psi$, $k = 0, 1, 2, \ldots$, span a dense subspace of $\mathbf{H}$. In that case, the direct integral will be simply an L^2 space (i.e., the Hilbert spaces $\mathbf{H}_\lambda$ are equal to $\mathbb{C}$ for all λ). Thus, in this special case, the direct integral and multiplication operator versions of the spectral theorem coincide.

Lemma 8.11 *Suppose $A \in \mathcal{B}(\mathbf{H})$ is self-adjoint and ψ is a cyclic vector for A. Let μ_ψ be the unique measure on $\sigma(A)$, given by Theorem 8.5, for which*

$$\langle \psi, f(A)\psi \rangle = \int_{\sigma(A)} f(\lambda) \, d\mu_\psi(\lambda) \tag{8.11}$$

for all $f \in \mathcal{C}(\sigma(A); \mathbb{R})$. Then there exists a unitary map

$$U : \mathbf{H} \to L^2(\sigma(A), \mu_\psi)$$

such that

$$\left[UAU^{-1}\phi \right](\lambda) = \lambda\phi(\lambda)$$

for all $\phi \in L^2(\sigma(A), \mu_\psi)$.

Proof. We start by defining U on the complex vector space of vectors of the form $p(A)\psi$, where p is a complex-valued polynomial, as follows:

$$U[p(A)\psi] = p.$$

To show that U is well defined, write p as $p = p_1 + ip_2$, where p_1 and p_2 are real-valued polynomials. Since $p_1(A)$ and $p_2(A)$ are self-adjoint and commuting, we obtain

$$\langle p(A)\psi, p(A)\psi \rangle = \langle \psi, \left[p_1(A)^2 + p_2(A)^2 \right] \psi \rangle$$

$$= \int_{\sigma(A)} \left[p_1(\lambda)^2 + p_2(\lambda)^2 \right] \, d\mu_\psi(\lambda), \qquad (8.12)$$

by canceling cross terms and applying (8.11). Thus, if $p(A)\psi = 0$ in $\mathbf{H}$, then $p(\lambda) = 0$ for μ_ψ-almost every λ in $\sigma(A)$, so that p represents the zero element of $L^2(\sigma(A), \mu_\psi)$.

Equation (8.12) shows also that the map U is isometric on its initial domain. This initial domain is dense in $\mathbf{H}$ since it contains the vectors $A^k \psi$ and ψ is cyclic. Thus, the BLT theorem (Theorem A.36) tells us that U extends uniquely to an isometric map of $\mathbf{H}$ into $L^2(\sigma(A), \mu_\psi)$. Since polynomials are dense in $L^2(\sigma(A), \mu_\psi)$ (by the Stone–Weierstrass theorem and Theorem A.10), U actually is unitary.

Now, since U takes $A^k \psi$ to the function $\lambda \mapsto \lambda^k$ in $L^2(\sigma(A), \mu_\psi)$, we have that $UAU^{-1}(\lambda^k) = \lambda^{k+1}$. Thus,

$$[UAU^{-1}p](\lambda) = \lambda p(\lambda)$$

for all polynomials p. Since polynomials are dense in $L^2(\sigma(A), \mu_\psi)$, we have $[UAU^{-1}\phi](\lambda) = \lambda \phi(\lambda)$ for all $\phi \in L^2(\sigma(A), \mu_\psi)$, as claimed. ∎

Lemma 8.12 *Suppose $A \in \mathcal{B}(\mathbf{H})$ is self-adjoint and μ^A is the associated projection-valued measure on $\sigma(A)$, as in Theorem 8.10. Then there exists a non-negative real-valued measure μ on $\sigma(A)$ such that for all Borel sets $E \subset \sigma(A)$, we have $\mu^A(E) = 0$ if and only if $\mu(E) = 0$.*

Proof. Let $\{e_j\}$ be an orthonormal basis for $\mathbf{H}$ and let μ_{e_j} be the associated real-valued measures, given by $\mu_{e_j}(E) = \langle e_j, \mu^A(E)e_j \rangle$. Then $\mu_{e_j}(\sigma(A)) = \langle e_j, Ie_j \rangle = 1$ for all j. Thus, the formula

$$\mu := \sum_j \frac{1}{j^2} \mu_{e_j}$$

defines a finite measure on $\sigma(A)$. Given some Borel set $E \subset \sigma(A)$, if $\mu^A(E) = 0$, then $\mu_{e_j}(E) = 0$ for all j and so $\mu(E) = 0$. Conversely, if $\mu(E) = 0$, then

$$0 = \langle e_j, \mu^A(E)e_j \rangle = \langle \mu^A(E)e_j, \mu^A(E)e_j \rangle$$

for all j, since $\mu^A(E)$ is self-adjoint and $\mu^A(E)^2 = \mu^A(E)$. Thus, $\mu^A(E)e_j = 0$ for all j, which means that $\mu^A(E) = 0$. ∎

Lemma 8.13 *If $A \in \mathcal{B}(\mathbf{H})$ is self-adjoint, then $\mathbf{H}$ can be decomposed as an orthogonal direct sum of closed nonzero subspaces W_j, where each W_j is invariant under A and where the restriction of A to W_j has a cyclic vector ψ_j. The number of W_j's is either finite or countably infinite.*

Proof. Recall our standing assumption that $\mathbf{H}$ is separable, and let $\{\phi_j\}$ be a countable dense subset of $\mathbf{H}$. Let W_1 be the closed subspace of $\mathbf{H}$ spanned by $\phi_1, A\phi_1, A^2\phi_1, \ldots$. Then W_1 is invariant under A and $\psi_1 := \phi_1$ is a cyclic vector for $A|_{W_1}$. If $W_1 = \mathbf{H}$ then we are done. If not, let j be the smallest number such that ϕ_j is not contained in W_1. Let ψ_2 be the orthogonal projection of ϕ_j onto the orthogonal complement of W_1, and let W_2 be the closed span of $\psi_2, A\psi_2, A^2\psi_2, \ldots$. Then W_2 is invariant under A and ψ_2 is a cyclic vector for $A|_{W_2}$. Furthermore, since A is self-adjoint and leaves W_1 invariant, it also leaves $W_1^\perp$ invariant, which means that $A^k\psi_2$ is orthogonal to W_1 for all k, so that W_2 is orthogonal to W_1.

If, now, $W_1 \oplus W_2 = \mathbf{H}$, we are done. If not, we let k be the smallest number such that ϕ_k is not in $W_1 \oplus W_2$ and we let ψ_3 be the projection of ϕ_k onto the orthogonal complement of $W_1 \oplus W_2$, and so on. Continuing on in this way, we obtain an orthogonal collection of closed subspaces that are invariant under A, each of which has a cyclic vector. Either the process terminates with finitely many of these subspaces spanning $\mathbf{H}$, or we get an infinite family. In the latter case, each ϕ_j belongs to the span of the W_j's and hence the (Hilbert space) direct sum of the W_j's is all of $\mathbf{H}$. $\blacksquare$

We are now ready for the proof of our second form of the spectral theorem.

Proof of Theorem 7.19. Let $\{W_j, \psi_j\}$ be as in Lemma 8.13, and let A_j denote the restriction of A to W_j, which is a bounded self-adjoint operator on the Hilbert space W_j. For each A_j, we can obtain a unitary map U_j as in Lemma 8.11, and we wish to piece these maps together for different values of j to obtain a direct integral decomposition for all of $\mathbf{H}$. To facilitate piecing the maps together, we will modify the U_j's so that they all map to L^2 spaces over a subset of $\sigma(A)$ with respect to the *same measure* μ.

If we apply Lemma 8.11 to A_j, we get a unitary map

$$U_j : W_j \to L^2(\sigma(A_j), \mu_{\psi_j})$$

such that $U_j A U_j^{-1}$ is the operator of multiplication by λ. Here, μ_{ψ_j} is the measure on $\sigma(A_j)$ given by $\mu_{\psi_j}(E) = \langle \psi_j, \mu^{A_j}(E)\psi_j \rangle$. Now, according to Exercise 5, the spectrum of A_j is contained in the spectrum of A. Furthermore, if E is a measurable subset of $\sigma(A_j) \subset \sigma(A)$, then 1_E may be thought of as a measurable function either on $\sigma(A_j)$ or on $\sigma(A)$. Exercise 5 tells us that $1_E(A_j)$, as defined by the functional calculus for A_j, coincides with the restriction to W_j of $1_E(A)$. Thus, if $1_E(A) = 0$ then $1_E(A_j) = 0$ as well. Equivalently, if $\mu^A(E) = 0$ then $\mu^{A_j}(E) = 0$, where μ^{A_j} is the projection-valued measure associated to the self-adjoint operator A_j.

Let us now choose a measure μ as in Lemma 8.12. Any set of measure zero for μ is a set of measure zero for μ^A and thus also for μ^{A_j} and then for μ_{ψ_j}. Thus, if we extend μ_{ψ_j} to a measure on $\sigma(A)$ by making it zero on $\sigma(A) \setminus \sigma(A_j)$, we have that μ_{ψ_j} is absolutely continuous with respect to μ.

By the Radon–Nikodym theorem (Theorem A.6), each μ_{ψ_j} has a density ρ_j with respect to μ, and this density is nonzero μ_{ψ_j}-almost everywhere.

Now, the map

$$f \mapsto \rho_j^{1/2} f$$

is easily seen to be a unitary map of $L^2(\sigma(A_j), \mu_{\psi_j})$ to $L^2(\sigma(A_j), \mu)$. Thus, we can define a unitary map

$$\tilde{U}_j : W_j \to L^2(\sigma(A_j), \mu)$$

by setting

$$(\tilde{U}_j \psi)(\lambda) = \rho_j(\lambda)^{1/2} (U_j \psi)(\lambda).$$

Since multiplication by $(\rho_j)^{1/2}$ commutes with multiplication by λ, we have

$$\left(\tilde{U}_j A_j \tilde{U}_j^{-1} \right)(\psi)(\lambda) = \lambda \psi(\lambda).$$

Now, $L^2(\sigma(A_j), \mu)$ can be thought of as a direct integral over $\sigma(A)$ with respect to μ, where we take $\mathbf{H}_\lambda^j = \mathbb{C}$ for $\lambda \in \sigma(A_j)$ and we take $\mathbf{H}_\lambda^j = \{0\}$ if $\lambda \in \sigma(A_j)^c$. We now define another direct integral over $\sigma(A)$ in which the Hilbert spaces $\mathbf{H}_\lambda$, $\lambda \in \sigma(A)$, are defined by

$$\mathbf{H}_\lambda = \bigoplus_j \mathbf{H}_\lambda^j.$$

Here the measurable structure on the direct integral is defined by setting

$$e_j(\lambda) = \begin{cases} (0, 0, \ldots, 1, 0, 0, \ldots), & \lambda \in E_j \\ (0, 0, \ldots, 0, 0, 0, \ldots), & \lambda \in E_j^c \end{cases},$$

where the 1 is in the jth slot. Since each $\mathbf{H}_\lambda$ is a direct sum of the $\mathbf{H}_\lambda^j$'s, the direct integral of the $\mathbf{H}_\lambda$'s is the Hilbert space direct sum of the direct integral of the $\mathbf{H}_\lambda^j$'s, which is just $L^2(\sigma(A_j), \mu)$.

Meanwhile, $\mathbf{H}$ is the direct sum of the W_j's, and we have unitary maps $\tilde{U}_j$ of W_j to $L^2(\sigma(A_j), \mu)$ such that $\tilde{U}_j A \tilde{U}_j^{-1}$ is just multiplication by λ on $L^2(E_j, \mu)$. Thus, we can assemble the $\tilde{U}_j$'s into a single unitary map U of $\mathbf{H}$ to the integral of the $\mathbf{H}_\lambda$'s, and we will have UAU^{-1} equal to multiplication by λ, as desired. ∎

In the interest of brevity, we will not give a complete proof of Proposition 7.22 (uniqueness in Theorem 7.19), but only indicate the main ideas. To establish the equivalence of $\mu^{(1)}$ and $\mu^{(2)}$, we observe that both measures have the same sets of measure zero as the projection-valued measure μ^A (Proposition 7.23). Meanwhile, if we have two different direct integrals, each unitarily equivalent to $\mathbf{H}$ as in (7.20), then there will be a unitary map V between the two direct integrals that commutes with the operator $s(\lambda) \mapsto \lambda s(\lambda)$. Using an argument similar to that in Exercise 7, we can show that there must be bounded maps $V_\lambda : \mathbf{H}_\lambda^{(1)} \to \mathbf{H}_\lambda^{(2)}$ such that $(Vs)(\lambda) = V_\lambda s(\lambda)$ for almost every λ. Then we argue that the only way V can be unitary is if V_λ is unitary for almost every λ. This implies that $\dim \mathbf{H}_\lambda^{(1)} = \dim \mathbf{H}_\lambda^{(2)}$ for almost every λ.

Finally, we briefly indicate the proof of the multiplication operator form of the spectral theorem.

Proof of Theorem 7.20. Let W_j be as in Lemma 8.13 and let A_j be the restriction of A to W_j. By the proof of Theorem 7.19, each A_j is unitarily equivalent to multiplication by λ on the Hilbert space $L^2(\sigma(A_j), \mu_j)$, for some finite measure μ_j on $\sigma(A_j)$. Let X be the disjoint union of the sets $\sigma(A_j)$, let μ be the sum of the measures μ_j, and let h be the function whose restriction to each $\sigma(A_j)$ is the function $\lambda \mapsto \lambda$. Then $L^2(X, \mu)$ is the orthogonal direct sum of the Hilbert spaces $L^2(\sigma(A_j), \mu_j)$, which means that $L^2(X, \mu)$ may be identified unitarily with $\mathbf{H} = \oplus W_j$ in an obvious way. Under this identification, the operator A corresponds to multiplication by h. ∎

8.3 Exercises

1. (a) Suppose $A, B \in \mathcal{B}(\mathbf{H})$ commute and A is not invertible. Show that AB is not invertible.

    Hint: First show that if AB were invertible, then A would have both a left inverse and a right inverse. Then show that the left inverse and right inverse would need to be equal.

   (b) Show that the result of Part (a) is false if we omit the assumption that A and B commute.

2. (a) Suppose $A \in \mathcal{B}(\mathbf{H})$ is self-adjoint and $\sigma(A) \subset [0, \infty)$. Show that A has a self-adjoint square root in $\mathcal{B}(\mathbf{H})$ and therefore that A is a non-negative operator (i.e., $\langle \psi, A\psi \rangle \geq 0$ for all $\psi \in \mathbf{H}$).

   (b) Give an example of a bounded operator A on a Hilbert space such that $\sigma(A) \subset [0, \infty)$ but A is not non-negative.

3. Let X be a compact metric space and let $\mathcal{C}(X; \mathbb{R})$ denote the space of continuous real-valued functions on X. Suppose that $\mathcal{F}$ is a set of bounded, measurable, complex-valued functions on X with the following properties: (1) $\mathcal{F}$ is a complex vector space, (2) $\mathcal{F}$ contains $\mathcal{C}(X; \mathbb{R})$, and (3) $\mathcal{F}$ is closed under pointwise limits of uniformly bounded sequences. (A sequence f_n is uniformly bounded if there exists a constant C such that $|f_n(x)| \leq C$ for all n and x).

   (a) Let $\mathcal{L}_0$ denote the collection of those measurable sets E for which 1_E is a uniformly bounded limit of a sequence of continuous functions. Show that $\mathcal{L}_0$ is an algebra and contains all open sets in X.

(b) Let $\mathcal{L}_1$ denote the collection of all measurable sets in E for which 1_E belongs to $\mathcal{F}$. Using the monotone class lemma (Theorem A.8), show that $\mathcal{L}_1$ consists of all Borel sets in X.

(c) Show that $\mathcal{F}$ consists of all bounded, Borel-measurable functions on X.

4. Suppose $A \in \mathcal{B}(\mathbf{H})$ is self-adjoint μ^A and ν^A are two projection-valued measures on $\sigma(A)$ such that

$$\int_{\sigma(A)} \lambda \, d\mu^A(\lambda) = \int_{\sigma(A)} \lambda \, d\nu^A(\lambda) = A.$$

Show that integration with respect to μ^A agrees with integration with respect to ν^A, first on polynomials, then on continuous functions, and finally on bounded measurable functions. Conclude that $\mu^A = \nu^A$.

Hint: Use Exercise 17.

5. Suppose $A \in \mathcal{B}(\mathbf{H})$ is self-adjoint operator and V is a closed subspace of $\mathbf{H}$ that is invariant under A.

(a) Using Proposition 7.7, show that the spectrum of the restriction to V of A is contained in the spectrum of A.

(b) Suppose now that f is a bounded measurable function on $\sigma(A)$, which means that f is also a function on $\sigma(A|_V) \subset \sigma(A)$. Show that V is invariant under $f(A)$ and that

$$f(A)|_V = f(A|_V),$$

where the operator on the right-hand side is defined by the measurable functional calculus for the bounded self-adjoint operator $A|_V$.

6. Suppose $A \in \mathcal{B}(\mathbf{H})$ is self-adjoint and ψ is an eigenvector for A, that is, a nonzero vector with $A\psi = \lambda\psi$ for some $\lambda \in \mathbb{R}$. Show that for any bounded measurable function f on $\sigma(A)$ we have

$$f(A)\psi = f(\lambda)\psi.$$

Hint: Use Exercise 5.

7. Suppose $K \subset \mathbb{R}$ is a compact set and μ is a finite measure on K. Let A be the bounded operator on $L^2(K, \mu)$ given by

$$(A\psi)(\lambda) = \lambda\psi(\lambda).$$

Now suppose that B is a bounded operator on $L^2(K, \mu)$ that commutes with A.

(a) Let $\phi = B1$, where 1 denotes the constant function, so that $\phi \in L^2(K, \mu)$. Show that for all continuous functions ψ on K, we have $B\psi = \phi\psi$.

(b) Using Exercise 3, show that for all bounded, Borel-measurable functions ψ on K, we have $B\psi = \phi\psi$.

(c) Show that ϕ is essentially bounded (i.e., bounded outside a set of μ-measure zero). Conclude that $B\psi = \phi\psi$ for all $\psi \in L^2(K, \mu)$.

8. If $A \in \mathcal{B}(\mathbf{H})$ is self-adjoint, define $U(t) \in \mathcal{B}(\mathbf{H})$ by $U(t) = \exp\{itA\}$ for each $t \in \mathbb{R}$, where the exponential is defined by the functional calculus for A.

(a) Show that $U(t)$ is unitary for all t and that $U(s)U(t) = U(s + t)$. (A family of operators with this property is called a *one-parameter unitary group*.)

(b) Show that the map $t \mapsto U(t)$ is continuous in the operator norm topology.

(c) Give an example of a one-parameter unitary group on a Hilbert space that is not continuous in the operator norm topology.

See Sect. 10.2 for more on one-parameter unitary groups.

9

Unbounded Self-Adjoint Operators

9.1 Introduction

Recall that most of the operators of quantum mechanics, including those representing position, momentum, and energy, are not defined on the entirety of the relevant Hilbert space, but only on a dense subspace thereof. In the case of the position operator, for example, given $\psi \in L^2(\mathbb{R})$, the function $X\psi(x) = x\psi(x)$ could easily fail to be in $L^2(\mathbb{R})$. Nevertheless, the space of ψ's in $L^2(\mathbb{R})$ for which $x\psi(x)$ is again in $L^2(\mathbb{R})$ is a dense subspace of $L^2(\mathbb{R})$. A closely related property of these operators is that they are not bounded, meaning that there is no constant C such that

$$\|A\psi\| \le C \|\psi\|$$

for all ψ for which A is defined. Because our operators are unbounded, we cannot use the BLT (bounded linear transformation) theorem to extend them to the whole Hilbert space.

In this chapter and the following one, we are going to study unbounded operators defined on dense subspaces of a Hilbert space $\mathbf{H}$. We will introduce the "correct" notion of self-adjointness for unbounded operators, namely the one for which the spectral theorem holds. As it turns out, the obvious candidate for a definition of self-adjointness, namely that $\langle \phi, A\psi \rangle = \langle A\phi, \psi \rangle$ for all ϕ and ψ in the domain of A, is *not* the correct one. Rather, for any unbounded operator A, we will define another unbounded operator A^*, the adjoint of A, with its own naturally defined domain. Then A is

B.C. Hall, *Quantum Theory for Mathematicians*, Graduate Texts
in Mathematics 267, DOI 10.1007/978-1-4614-7116-5_9,
© Springer Science+Business Media New York 2013

said to be self-adjoint if A^* and A are the same operators *with the same domain.*

In the present chapter, we give the definition of an unbounded self-adjoint operator, along with conditions for self-adjointness and several examples and counterexamples. We defer a discussion of the spectral theorem itself until Chap. 10. The statement of the spectral theorem (either in terms of projection-valued measures or in terms of direct integrals) is essentially the same as in the bounded case, with only a few modifications to deal with the domain of the operator.

Although this chapter is rather technical, a reader who is willing to accept some things on faith may wish simply to read the definitions of self-adjoint and essentially self-adjoint operators in Sect. 9.2, and then skip to the statements of Theorem 9.21 and Corollary 9.22 in Sect. 9.5. As in previous chapters, $\mathbf{H}$ will denote a separable Hilbert space over $\mathbb{C}$.

9.2   Adjoint and Closure of an Unbounded Operator

Recall that we briefly introduced unbounded operators in Sect. 3.2. According to Definition 3.1, an *unbounded operator* A on $\mathbf{H}$ is a linear map of some dense subspace $\mathrm{Dom}(A) \subset \mathbf{H}$ (the domain of A) into $\mathbf{H}$. As in Sect. 3.2, "unbounded" means "not necessarily bounded," meaning that we permit the case in which $\mathrm{Dom}(A) = \mathbf{H}$ and A is bounded.

Now, if A *is* bounded, then for any ϕ, the linear functional

$$\langle \phi, A \cdot \rangle$$

is bounded. Thus, by the Riesz theorem (Theorem A.52), there is a unique χ such that

$$\langle \phi, A \cdot \rangle = \langle \chi, \cdot \rangle .$$

We then define the adjoint A^* of A by setting $A^*\phi$ equal to χ. (See Sect. A.4.)

If A is unbounded, then $\langle \phi, A \cdot \rangle$ is not *necessarily* bounded, but may be bounded for certain vectors ϕ. If $\langle \phi, A \cdot \rangle$ does happen to be bounded, for some $\phi \in \mathbf{H}$, then the BLT theorem (Theorem A.36) says that this linear functional has a unique bounded extension from $\mathrm{Dom}(A)$ to all $\mathbf{H}$. The Riesz theorem then tells us that there is a unique χ such that this linear functional is "inner product with χ." This line of reasoning leads to the following definition, which was already introduced briefly in Sect. 3.2.

Definition 9.1 *Suppose A is an operator defined on a dense subspace* $\mathrm{Dom}(A) \subset \mathbf{H}$. *Let* $\mathrm{Dom}(A^*)$ *to be the space of all* $\phi \in \mathbf{H}$ *for which the linear functional*

$$\psi \mapsto \langle \phi, A\psi \rangle , \quad \psi \in \mathrm{Dom}(A),$$

is bounded. For $\phi \in \mathrm{Dom}(A^*)$, define $A^*\phi$ to be the unique vector such that $\langle \phi, A\psi \rangle = \langle A^*\phi, \psi \rangle$ for all $\psi \in \mathrm{Dom}(A)$.

Saying that $\langle \phi, A\cdot \rangle$ is bounded means, explicitly, that there exists a constant C such that $|\langle \phi, A\psi \rangle| \leq C \|\psi\|$ for all $\psi \in \mathrm{Dom}(A)$. As in the bounded case, the operator A^* is linear on its domain, and is called the *adjoint* of A.

Another way to think about the definition of A^* is as follows. Given a vector ϕ, if there exists a vector χ such that $\langle \phi, A\psi \rangle = \langle \chi, \psi \rangle$ for all $\psi \in \mathrm{Dom}(A)$, then ϕ belongs to $\mathrm{Dom}(A^*)$ and $A^*\phi = \chi$. By the Riesz theorem, such a χ will exist if and only if $\langle \phi, A\cdot \rangle$ is bounded, which means this way of thinking about A^* is equivalent to Definition 9.1.

Given a densely defined operator A, the adjoint A^* of A could fail to be densely defined. This situation, however, is a pathology that does not usually occur for operators of interest in applications.

Definition 9.2 *An unbounded operator A on $\mathbf{H}$ is **symmetric** if*

$$\langle \phi, A\psi \rangle = \langle A\phi, \psi \rangle \tag{9.1}$$

for all $\phi, \psi \in \mathrm{Dom}(A)$.

As we will see shortly, if A is symmetric, then A^* is an extension of A, in the sense of the following definition.

Definition 9.3 *An unbounded operator A is an **extension** of an unbounded operator B if $\mathrm{Dom}(A) \supset \mathrm{Dom}(B)$ and $A = B$ on $\mathrm{Dom}(B)$.*

If A is an extension of B, then very likely A is given by the same "formula" as B. If $\mathbf{H} = L^2(\mathbb{R})$, for example, both operators might be given by the formula $-i\hbar \, d/dx$ on their respective domains. Nevertheless, if $\mathrm{Dom}(A) \neq \mathrm{Dom}(B)$, then A is still a different operator from B.

Proposition 9.4 *An unbounded operator A is symmetric if and only if A^* is an extension of A.*

Proof. If A is symmetric, then for all $\phi \in \mathrm{Dom}(A)$, (9.1) and the Cauchy–Schwarz inequality show that

$$|\langle \phi, A\psi \rangle| \leq \|A\phi\| \, \|\psi\| \, ,$$

showing that $\phi \in \mathrm{Dom}(A^*)$. In that case, (9.1) shows that the unique vector $A^*\phi$ for which $\langle \phi, A\psi \rangle = \langle A^*\phi, \psi \rangle$ is nothing but $A\phi$, which means that A^* agrees with A on $\mathrm{Dom}(A)$.

In the other direction, if A^* is an extension of A, then for each $\phi \in \mathrm{Dom}(A)$, we have

$$\langle \phi, A\psi \rangle = \langle A^*\phi, \psi \rangle = \langle A\phi, \psi \rangle \, ,$$

for all $\psi \in \mathrm{Dom}(A)$, which shows that A is symmetric. ∎

We come now to the key definition of this section, that of self-adjointness. This notion constitutes the hypothesis of the spectral theorem for unbounded operators.

Definition 9.5 *An unbounded operator A on $\mathbf{H}$ is* **self-adjoint** *if*

$$\mathrm{Dom}(A^*) = \mathrm{Dom}(A)$$

and $A^*\phi = A\phi$ for all $\phi \in \mathrm{Dom}(A)$.

We may reformulate the definition of self-adjointness by saying that A is self-adjoint if A^* is equal to A, provided that equality of unbounded operators is understood to include *equality of domains*. Every self-adjoint operator is symmetric (by Proposition 9.4), but there exist many operators that are symmetric without being self-adjoint. In light of Proposition 9.4, a symmetric operator is self-adjoint if and only if $\mathrm{Dom}(A^*) = \mathrm{Dom}(A)$. In trying to show that a symmetric operator is self-adjoint, the difficulty lies in showing that $\mathrm{Dom}(A^*)$ is no bigger than $\mathrm{Dom}(A)$.

Definition 9.6 *An unbounded operator A on $\mathbf{H}$ is said to be* **closed** *if the graph of A is a closed subset of $\mathbf{H} \times \mathbf{H}$. An unbounded operator A on $\mathbf{H}$ is said to be* **closable** *if the closure in $\mathbf{H} \times \mathbf{H}$ of the graph of A is the graph of a function. If A is closable, then the closure A^{cl} of A is the operator with graph equal to the closure of the graph of A.*

To be more explicit, an operator A is closed if and only if the following condition holds: Suppose a sequence ψ_n belongs to $\mathrm{Dom}(A)$ and suppose that there exist vectors ψ and ϕ in $\mathbf{H}$ with $\psi_n \to \psi$ and $A\psi_n \to \phi$. Then ψ belongs to $\mathrm{Dom}(A)$ and $A\psi = \phi$. Regarding closability, an operator A is *not* closable if there exist two elements in the closure of the graph of A of the form (ϕ, ψ) and (ϕ, χ), with $\psi \neq \chi$. Another way of putting it is to say that an operator A is closable if there exists some closed extension of it, in which case the closure of A is the smallest closed extension of A.

The notion of the closure of a (closable) operator is useful because it sweeps away some of the arbitrariness in the choice of a domain of an operator. If we consider, for example, the operator $A = -i\hbar\, d/dx$ as an unbounded operator on $L^2(\mathbb{R})$, there are many different reasonable choices for $\mathrm{Dom}(A)$, including (1) the space of C^∞ functions of compact support, (2) the Schwartz space (Definition A.15), and (3) the space of continuously differentiable functions ψ for which both ψ and ψ' belong to $L^2(\mathbb{R})$. As it turns out, each of these three choices for $\mathrm{Dom}(A)$ leads to the same operator A^{cl}. Note that we are not claiming that *every* choice for $\mathrm{Dom}(A)$ leads to the same closure; nevertheless, it is often the case that *many* reasonable choices do lead to the same closure.

Definition 9.7 *An unbounded operator A on $\mathbf{H}$ is said to be* **essentially self-adjoint** *if A is symmetric and closable and A^{cl} is self-adjoint.*

Actually, as we shall see in the next section, a symmetric operator is always closable. Many symmetric operators fail to be even essentially self-adjoint. We will see examples of such operators in Sects. 9.6 and 9.10. Section 9.5 gives some reasonably simple criteria for determining when a symmetric operator is essentially self-adjoint.

9.3   Elementary Properties of Adjoints and Closed Operators

In this section, we spell out some of the most basic and useful properties of adjoints and closures of unbounded operators. In Sect. 9.5, we will draw on these results to prove some more substantial results. In what follows, if we say that two operators "coincide," it means that they *have the same domain* and that they are equal on that common domain.

Proposition 9.8   *1. If A is an unbounded operator on* **H**, *then the graph of the operator A^* (which may or may not be densely defined) is closed in* **H** $\times$ **H**.

2. A symmetric operator is always closable.

Proof. Suppose ψ_n is a sequence in the domain of A^* that converges to some $\psi \in$ **H**. Suppose also that $A^*\psi_n$ converges to some $\phi \in$ **H**. Then $\langle \psi_n, A \cdot \rangle = \langle A^*\psi_n, \cdot \rangle$ and for any $\chi \in \mathrm{Dom}(A)$, we have

$$\langle \psi, A\chi \rangle = \lim_{n \to \infty} \langle \psi_n, A\chi \rangle = \lim_{n \to \infty} \langle A^*\psi_n, \chi \rangle = \langle \phi, \chi \rangle .$$

This shows that ψ belongs to the domain of A^* and that $A^*\psi = \phi$, establishing that the graph of A^* is closed.

If A is symmetric, A^* is an extension of A. Since, as we have just proved, A^* is closed, A has a closed extension and is therefore closable. ∎

Corollary 9.9 *If A is a symmetric operator with* $\mathrm{Dom}(A) =$ **H**, *then A is bounded.*

Proof. Since A is symmetric, it is closable by Proposition 9.8. But since the domain of A is already all of **H**, the closure of A must coincide with A itself. (The closure of A always agrees with A on $\mathrm{Dom}(A)$, which in this case is all of **H**.) Thus, A is a closed operator defined on all of **H**, and the closed graph theorem (Theorem A.39) implies that A is bounded. ∎

Proposition 9.10 *If A is a closable operator on* **H**, *then the adjoint of A^{cl} coincides with the adjoint of A.*

Proof. Suppose that for some $\psi \in$ **H** there exists a ϕ such that $\langle \psi, A^{cl}\chi \rangle = \langle \phi, \chi \rangle$ for all $\chi \in \mathrm{Dom}(A^{cl})$. Since A^{cl} is an extension of A, it follows that $\langle \psi, A\chi \rangle = \langle \phi, \chi \rangle$ for all $\chi \in \mathrm{Dom}(A)$. This shows that $\mathrm{Dom}(A^*) \supset \mathrm{Dom}((A^{cl})^*)$ and that A^* agrees with $(A^{cl})^*$ on $\mathrm{Dom}((A^{cl})^*)$.

In the other direction, suppose for some $\psi \in \mathbf{H}$ there exists a ϕ such that $\langle \psi, A\chi \rangle = \langle \phi, \chi \rangle$ for all $\chi \in \mathrm{Dom}(A)$. Suppose now $\xi \in \mathrm{Dom}(A^{cl})$ with $A^{cl}\xi = \eta$. Then there exists a sequence χ_n in $\mathrm{Dom}(A)$ with $\chi_n \to \xi$ and $A\chi_n \to \eta$, and we have

$$\langle \psi, A\chi_n \rangle = \langle \phi, \chi_n \rangle$$

for all n. Letting n tend to infinity, we obtain $\langle \psi, \eta \rangle = \langle \phi, \xi \rangle$, or $\langle \psi, A^{cl}\xi \rangle = \langle \phi, \xi \rangle$. This shows that $\psi \in \mathrm{Dom}((A^{cl})^*)$ and $A^{cl}\psi = \phi$. Thus, $\mathrm{Dom}(A^*) \subset \mathrm{Dom}((A^{cl})^*)$. ∎

Proposition 9.11 *If A is essentially self-adjoint, then A^{cl} is the unique self-adjoint extension of A.*

Proof. Suppose B is a self-adjoint extension of A. Since $B = B^*$, B is closed and is, therefore, an extension of A^{cl}. It then follows from the definition of the adjoint that $\mathrm{Dom}(B^*) \subset \mathrm{Dom}(A^{cl})$. Thus, we have

$$\mathrm{Dom}(B^*) \subset \mathrm{Dom}(A^{cl}) \subset \mathrm{Dom}(B).$$

Since B is self-adjoint, all three of the above sets must be equal, so actually $B = A^{cl}$. ∎

Proposition 9.12 *If A is an unbounded operator on $\mathbf{H}$, then*

$$(\mathrm{Range}(A))^{\perp} = \ker(A^*).$$

Proof. First assume that $\psi \in (\mathrm{Range}(A))^{\perp}$. Then for all $\phi \in \mathrm{Dom}(A)$ we have

$$\langle \psi, A\phi \rangle = 0.$$

That is to say, the linear functional $\langle \psi, A\cdot \rangle$ is bounded—in fact, zero—on $\mathrm{Dom}(A)$. Thus, from the definition of the adjoint, we conclude that $\psi \in \mathrm{Dom}(A^*)$ and $A^*\psi = 0$.

Meanwhile, suppose that ψ is in $\mathrm{Dom}(A^*)$ and that $A^*\psi = 0$. The only way this can happen is if the linear functional $\langle \psi, A\cdot \rangle$ is zero on $\mathrm{Dom}(A)$, which means that ψ is orthogonal to the image of A. ∎

Proposition 9.13 *Suppose A is an unbounded operator on $\mathbf{H}$ and that B is a bounded operator defined on all of $\mathbf{H}$. Let $A + B$ denote the operator with $\mathrm{Dom}(A + B) = \mathrm{Dom}(A)$ and given by $(A + B)\psi = A\psi + B\psi$ for all $\psi \in \mathrm{Dom}(A)$. Then $(A+B)^*$ has the same domain as A^* and $(A+B)^*\psi = A^*\psi + B^*\psi$ for all $\psi \in \mathrm{Dom}(A^*)$.*

In particular, the sum of an unbounded self-adjoint operator and a bounded self-adjoint operator (defined on all of $\mathbf{H}$) is self-adjoint on the domain of the unbounded operator.

Proof. See Exercise 3. ∎

The sum of two unbounded self-adjoint operators is not, in general, self-adjoint. See Sect. 9.9 for more information about this issue.

Proposition 9.14 *Let A be a closed operator and λ an element of $\mathbb{C}$. Suppose that there exists $\varepsilon > 0$ such that*

$$\|(A - \lambda I)\psi\| \geq \varepsilon \|\psi\| \tag{9.2}$$

for all A in $\mathrm{Dom}(A)$. Then the range of $A - \lambda I$ is a closed subspace of $\mathbf{H}$.

Here, we take the domain of the operator $A - \lambda I$ to coincide with the domain of A, as in Proposition 9.13.

Proof. Assume that ϕ_n is a sequence in the range of $A - \lambda I$ converging to some ϕ. Then $\phi_n = (A - \lambda I)\psi_n$, for some sequence ψ_n in $\mathrm{Dom}(A)$. Applying (9.2) with $\psi = \psi_n - \psi_m$ shows that $\|\psi_n - \psi_m\| \leq (1/\varepsilon) \|\phi_n - \phi_m\|$. This means that ψ_n is Cauchy and thus convergent to some vector ψ. Since $\psi_n \to \psi$ and $(A - \lambda I)\psi_n = \phi_n \to \phi$, we have that

$$A\psi_n = \lambda\psi_n + \phi_n \to \lambda\psi + \phi.$$

Thus, by the definition of a closed operator, $\psi \in \mathrm{Dom}(A)$ and $A\psi = \lambda\psi + \phi$. This means that $(A - \lambda I)\psi = \phi$ and so the range of $A - \lambda I$ is closed. ∎

We conclude this section with a simple example for which we can compute the adjoint and closure explicitly.

Example 9.15 *Let $\langle e_j \rangle$ be an orthonormal basis for $\mathbf{H}$ and let $\langle \lambda_j \rangle$ be an arbitrary sequence of real numbers. Define an operator A on $\mathbf{H}$ with $\mathrm{Dom}(A)$ equal to the space of finite linear combinations of the e_j's, with A itself defined by*

$$Ae_j = \lambda_j e_j.$$

Then A is symmetric and closable and $\mathrm{Dom}(A^*) - \mathrm{Dom}(A^{cl}) = V$, where

$$V = \left\{ \psi = \sum_j a_j e_j \,\middle|\, \sum_j (1 + \lambda_j^2) |a_j|^2 < \infty \right\}. \tag{9.3}$$

For any $\psi = \sum_j a_j e_j$ in V, we have

$$A^*\psi = A^{cl}\psi = \sum_j a_j \lambda_j e_j. \tag{9.4}$$

Thus, $(A^{cl})^* = A^* = A^{cl}$, showing that A is essentially self-adjoint.

Proof. Note that for any sequence $\langle a_j \rangle$ of coefficients satisfying the condition on the right-hand side of (9.3), we have $\sum_j |a_j|^2 < \infty$ and, thus, the sum $\sum_j a_j e_j$ converges in **H**. Suppose first that $\phi = \sum_j a_j e_j$ belongs V. Then for any $\psi = \sum_j b_j e_j$ (finite sum) in the domain of A we have

$$\langle \phi, A\psi \rangle = \sum_j \overline{a_j} \lambda_j b_j$$

and so by the Cauchy–Schwarz inequality,

$$|\langle \phi, A\psi \rangle| \leq \left(\sum_j \lambda_j^2 |a_j|^2 \right)^{1/2} \|\psi\|.$$

Thus, $\langle \phi, A\cdot \rangle$ is a bounded linear functional, showing that $\phi \in \text{Dom}(A^*)$. Furthermore, it is apparent that $\langle \phi, A\psi \rangle = \langle \chi, \psi \rangle$ for all $\psi \in \text{Dom}(A)$, where $\chi = \sum_j a_j \lambda_j e_j$.

Meanwhile, suppose $\phi = \sum_j a_j e_j$ belongs to the domain of A^*, and consider $\psi_N := \sum_{j=1}^N \lambda_j a_j e_j$ in $\text{Dom}(A)$. Then

$$|\langle \phi, A\psi_N \rangle| = \sum_{j=1}^N \lambda_j^2 |a_j|^2 = \left(\sum_{j=1}^N \lambda_j^2 |a_j|^2 \right)^{1/2} \|\psi_N\|.$$

Since $\phi \in \text{Dom}(A^*)$, the functional $\langle \phi, A\cdot \rangle$ is bounded, and so $\sum_{j=1}^N \lambda_j^2 |a_j|^2$ must be bounded, independent of N, and so $\sum_j \lambda_j^2 |a_j|^2 < \infty$. Since ϕ belongs to **H**, we have also that $\sum_j |a_j|^2 < \infty$, showing that ϕ is in V.

Turning now to the closure of A, it is apparent that A is symmetric and thus closable, by Proposition 9.8. Suppose $\psi = \sum_j a_j e_j$ belongs to V and consider $\psi_N := \sum_{j=1}^N a_j e_j$. Clearly, ψ_N converges to ψ. Furthermore, since $\psi \in V$, we see that $A\psi_N$ converges to the vector $\sum_j a_j \lambda_j e_j$. This shows that $\psi \in \text{Dom}(A^{cl})$ and that $A^{cl}\psi = \sum_j a_j \lambda_j e_j$. Thus, each element of V belongs to $\text{Dom}(A^{cl})$ and A^{cl} is given on V by (9.4).

Now, the space V forms a Hilbert space with respect to the norm given by

$$\|\psi\|_V^2 = \sum_j (1 + \lambda_j^2) |a_j|^2,$$

where $\psi = \sum_j a_j e_j$. [To establish completeness of V with respect to this norm, note that V can be identified isometrically with $L^2(\mathbb{N})$ with respect to the measure μ for which $\mu(\{j\}) = 1 + \lambda_j^2$.] Suppose, now, that we have a sequence $\langle \psi^m \rangle$ in $\text{Dom}(A)$ for which both $\langle \psi_m \rangle$ and $\langle A\psi_m \rangle$ are convergent. Then $\langle \psi^m \rangle$ forms a Cauchy sequence in V which converges to some element ψ of V. Since $\|\psi\|_\mathbf{H} \leq \|\psi\|_V$ for all $\psi \in \text{Dom}(A)$, we see that ψ^m also converges in **H** to $\psi \in V$. This shows that each element of $\text{Dom}(A^{cl})$ belongs to V. ∎

9.4   The Spectrum of an Unbounded Operator

Recall that if A is a bounded operator, then a number $\lambda \in \mathbb{C}$ belongs to the *resolvent set* of A if the operator $A - \lambda I$ has a bounded inverse, and λ belongs to the *spectrum* of A if $A - \lambda I$ does not have a bounded inverse. For an unbounded operator A, we will say that a number $\lambda \in \mathbb{C}$ is in the resolvent set of A if $A - \lambda I$ has a bounded inverse. That is, even though A is unbounded, for λ to be in the resolvent set of A, there must be a *bounded* inverse to $A - \lambda I$; otherwise, λ is in the spectrum of A. We make this characterization more precise in the following definition.

Definition 9.16 *Suppose A is an unbounded operator on* **H**. *A number $\lambda \in \mathbb{C}$ belongs to the* **resolvent set** *of A if there exists a bounded operator B with the following properties: (1) For all $\psi \in$ **H**, $B\psi$ belongs to* $\mathrm{Dom}(A)$ *and $(A - \lambda I)B\psi = \psi$, and (2) for all $\psi \in \mathrm{Dom}(A)$ we have $B(A - \lambda I)\psi = \psi$.*
If no such bounded operator B exists, then λ belongs to the **spectrum** *of A.*

Note that we are implicitly taking $\mathrm{Dom}(A - \lambda I)$ to equal $\mathrm{Dom}(A)$, as in Proposition 9.13. As in the bounded case, even if A is self-adjoint, points λ in the spectrum of A are not necessarily eigenvalues; that is, there does not necessarily exist a nonzero $\psi \in \mathrm{Dom}(A)$ with $A\psi = \lambda\psi$. On the other hand, if $A\psi = \lambda\psi$ for some $\psi \in \mathrm{Dom}(A)$, then $A - \lambda I$ is not injective and thus λ certainly does belong to the spectrum of A.

Theorem 9.17 *If A is an unbounded self-adjoint operator on* **H**, *the spectrum of A is contained in the real line.*

If A is symmetric but not self-adjoint, then the spectrum of A must contain points not in the real line. Indeed, Theorem 9.21 will show that at least one of $(A - iI)$ and $(A + iI)$ must fail to be surjective, and thus at least one of the numbers i and $-i$ is in the spectrum of A. Nevertheless, a symmetric operator cannot have nonreal eigenvalues, as we showed already in Proposition 3.4.

Proof. Consider a complex number $\lambda = a + ib$ with $b \neq 0$. Since A is symmetric, the proof of Lemma 7.8 applies, giving

$$\langle (A - \lambda I)\psi, (A - \lambda I)\psi \rangle \geq b^2 \langle \psi, \psi \rangle \tag{9.5}$$

for all $\psi \in \mathrm{Dom}(A)$. This shows that $(A - \lambda I)$ is injective.

Meanwhile, applying Propositions 9.12 and 9.13 with $B = -\lambda I$ we see that

$$(\mathrm{Range}(A - \lambda I))^{\perp} = \ker((A - \lambda I)^*) = \ker(A^* - \bar{\lambda} I) = \ker(A - \bar{\lambda} I).$$

Since $\bar{\lambda}$ again has nonzero imaginary part, $A - \bar{\lambda} I$ is also injective, showing that $\mathrm{Range}(A - \lambda I)$ is dense in **H**. Since $A = A^*$ is closed, (9.5) allows us to apply Proposition 9.14 to show that $\mathrm{Range}(A - \lambda I)$ is closed, hence all of **H**.

We have shown, then, that $(A - \lambda I)$ maps $\mathrm{Dom}(A)$ injectively onto $\mathbf{H}$. It follows from (9.5) (or the closed graph theorem) that the inverse operator is bounded, so that λ is in the resolvent set of A. ∎

Our next result shows that the spectrum of an unbounded self-adjoint operator has properties similar to that of a bounded self-adjoint operator.

Proposition 9.18 *If A is an unbounded self-adjoint operator on $\mathbf{H}$, then the following hold.*

1. *A number $\lambda \in \mathbb{R}$ belongs to the spectrum of A if and only if there exists a sequence ψ_n of nonzero vectors in $\mathrm{Dom}(A)$ such that*

$$\lim_{n \to \infty} \frac{\|(A - \lambda I)\psi_n\|}{\|\psi_n\|} = 0. \qquad (9.6)$$

2. *The spectrum $\sigma(A)$ of A is a closed subset of $\mathbb{R}$.*

Although the spectrum of a bounded self-adjoint operator is a bounded subset of $\mathbb{R}$, the spectrum of an unbounded self-adjoint operator will be unbounded. Indeed, it can be shown (using the spectral theorem) that if a self-adjoint operator has bounded spectrum, then the operator must be bounded.

Proof. For Point 1, if a sequence as in (9.6) existed, then as in the proof of Proposition 7.7, $A - \lambda I$ could not have a *bounded* inverse, so λ must be in the spectrum of A. Conversely, suppose no such sequence exists. Then there is some $\varepsilon > 0$ such that

$$\|(A - \lambda I)\psi\| \geq \varepsilon \|\psi\| \qquad (9.7)$$

for all $\psi \in \mathrm{Dom}(A)$. This means that $A - \lambda I$ is injective and that, by Proposition 9.14, the range of $A - \lambda I$ is closed. But

$$(A - \lambda I)^* = A^* - \lambda I = A - \lambda I$$

and $A - \lambda I$ is injective, so by Proposition 9.12, the range of $A - \lambda I$ is all of $\mathbf{H}$. This means $A - \lambda I$ has an inverse, which is bounded by (9.7). Thus λ is not in the spectrum of A.

Point 2 is left as an exercise (Exercise 4). ∎

Definition 9.19 *Let A be an unbounded operator on $\mathbf{H}$. Then A is **non-negative** if $\langle \psi, A\psi \rangle \geq 0$ for all $\psi \in \mathrm{Dom}(A)$ and A is **bounded below** by $c \in \mathbb{R}$ if $\langle \psi, A\psi \rangle \geq c \|\psi\|^2$ for all $\psi \in \mathrm{Dom}(A)$.*

Proposition 9.20 *Let A be an unbounded self-adjoint operator on $\mathbf{H}$. If A is non-negative, then the spectrum of A is contained in $[0, \infty)$. More generally, if A is bounded below by c, then the spectrum of A is contained in $[c, \infty)$.*

We will eventually see, using the spectral theorem for unbounded self-adjoint operators, that the converse to Proposition 9.20 also holds: If the spectrum of a self-adjoint operator A is contained in $[0, \infty)$, then A is non-negative, and if the spectrum of A is contained in $[c, \infty)$, then A is bounded below by c. These results follow easily, for example, from the form of the spectral theorem in Theorem 10.9.

Proof. Suppose A is bounded below by c and λ is a point in the spectrum of A. If ψ_n be a sequence as in Point 1 of Proposition 9.18, with the ψ_n's normalized to be unit vectors, then

$$\lim_{n \to \infty} |\langle \psi_n, (A - \lambda I)\psi_n \rangle| \leq \lim_{n \to \infty} \|(A - \lambda I)\psi_n\| = 0.$$

On the other hand, $A = \lambda I + (A - \lambda I)$, and so

$$\langle \psi_n, A\psi_n \rangle = \lambda + \langle \psi_n, (A - \lambda I)\psi_n \rangle.$$

Thus, $\langle \psi_n, A\psi_n \rangle$ converges to λ $(= \lambda \langle \psi_n, \psi_n \rangle)$ as n tends to infinity. Since A is bounded below by c, we must have $\lambda \geq c$. This establishes the result for operators bounded below by c. Specializing to $c = 0$ gives the result for non-negative operators. ∎

9.5 Conditions for Self-Adjointness and Essential Self-Adjointness

In this section, we give criteria for determining whether a symmetric operator is self-adjoint or essentially self-adjoint. See also Sect. 10.2 for the connection between self-adjoint operators and one-parameter unitary groups.

Theorem 9.21 *If A is a symmetric operator on* **H**, *then A is essentially self-adjoint if and only if* Range$(A - iI)$ *and* Range$(A + iI)$ *are dense subspaces of* **H**.

Using Proposition 9.12, we can reformulate this result as follows.

Corollary 9.22 *If A is a symmetric operator on* **H**, *then A is essentially self-adjoint if and only if the operators $A^* + iI$ and $A^* - iI$ are injective on* Dom(A^*).

As Exercise 11 shows, it is possible to have one of the operators $A^* + iI$ and $A^* - iI$ be injective and the other fail to be injective.

Proof of Theorem 9.21. Assume first that A is essentially self-adjoint, so that A^{cl} is self-adjoint. Then $A^* = (A^{cl})^* = A^{cl}$, and so

$$[\text{Range}(A - iI)]^{\perp} = \ker(A^* + iI) = \ker(A^{cl} + iI) = \{0\},$$

by Theorem 9.17, and similarly for the range of $A + iI$.

Conversely, assume A is symmetric and that $A - iI$ and $A + iI$ both have dense range. Since $(A^{cl})^* = A^*$ is a closed extension of A, it is also an extension of A^{cl}, showing that A^{cl} is symmetric. We may then apply Lemma 7.8—the proof of which requires only symmetry—to the operator A^{cl} with $\lambda = i$, giving

$$\left\| (A^{cl} - iI)\psi \right\|^2 \geq \|\psi\|^2 \tag{9.8}$$

and showing that $A^{cl} - iI$ is injective. Since the range of $A - iI$ is dense, the range of $A^{cl} - iI$ is certainly also dense. But since A^{cl} is closed, (9.8) and Proposition 9.14 tell us that the range of $A^{cl} - iI$ is closed, hence all of $\mathbf{H}$. Similar reasoning shows that the range of $A^{cl} + iI$ is also all of $\mathbf{H}$.

Now, by Proposition 9.13, $(A^{cl} - iI)^* = (A^{cl})^* + iI$, which is an extension of $A^{cl} + iI$. Suppose $(A^{cl})^* + iI$ is a *proper* extension of $A^{cl} + iI$, that is, that the domain of $(A^{cl})^* + iI$ is strictly bigger than the domain of $A^{cl} + iI$. Then since $A^{cl} + iI$ already maps *onto* $\mathbf{H}$, $(A^{cl})^* + iI$ cannot be injective. Thus, the operator

$$(A^{cl})^* + iI = A^* + iI = (A - iI)^*$$

must have a nontrivial kernel. Then by Proposition 9.12, Range$(A - iI)$ is not dense, contradicting our assumptions.

We conclude, therefore, that $(A^{cl})^* + iI$ is *not* a proper extension of $A^{cl} + iI$, i.e., that $(A^{cl})^* + iI = A^{cl} + iI$ (with equality of domains). This, by Proposition 9.13, means that $(A^{cl})^* = A^*$ (with equality of domains), which is what we are trying to prove. ∎

Proposition 9.23 *If A is a symmetric operator on $\mathbf{H}$, then A is self-adjoint if and only if*

$$\text{Range}(A - iI) = \text{Range}(A + iI) = \mathbf{H}.$$

Proof. Suppose first that A is self-adjoint. Then by Theorem 9.21, the ranges of $A - iI$ and $A + iI$ are dense in $\mathbf{H}$. On the other hand,

$$\|(A - iI)\psi\|^2 \geq \|\psi\|^2, \tag{9.9}$$

by (the proof of) Lemma 7.8, with $\lambda = i$. Since, also, $A = A^*$ is closed, Proposition 9.14 tells us that the range of $A - iI$ is closed, hence all of $\mathbf{H}$. A similar argument shows that the range of $A + iI$ is all of $\mathbf{H}$.

Conversely, suppose that the ranges of $A - iI$ and $A + iI$ are all of $\mathbf{H}$. Then A is essentially self-adjoint by Theorem 9.21, so that A^* is self-adjoint. Since $A - iI$ already maps *onto* $\mathbf{H}$, if A^* were a nontrivial extension of A, then $A^* - iI$ could not be injective. But (9.9), with A replaced by A^*, shows that $A^* - iI$ is injective. Thus, $A = A^*$ and so A is self-adjoint. ∎

In the case that A is positive-semidefinite (i.e., $\langle \psi, A\psi \rangle \geq 0$ for all $\psi \in$ Dom(A)), there is another self-adjointness condition, the proof of which is very similar to that of Theorem 9.22.

Theorem 9.24 *Suppose that A is a symmetric operator on $\mathbf{H}$ and that $\langle \psi, A\psi \rangle \geq 0$ for all $\psi \in \mathrm{Dom}(A)$. Then A is essentially self-adjoint if and only if $A + I$ has dense range. Equivalently, A is essentially self-adjoint if and only if $A^* + I$ is injective.*

Proof. Assume first that A is essentially self-adjoint. Then $(A + I)^* = A^* + I = A^{cl} + I$. It is easily seen that A^{cl} is also positive definite, and so

$$\langle \psi, (A^{cl} + I)\psi \rangle = \langle \psi, \psi \rangle + \langle \psi, A^{cl}\psi \rangle \geq \langle \psi, \psi \rangle \qquad (9.10)$$

Thus, $A^{cl} + I = (A + I)^*$ is injective. Thus, the range of $A + I$ is dense, by Proposition 9.12.

Now assume that $A + I$ has dense range. By (9.10), $A^{cl} + I$ is injective and by (9.10) and Proposition 9.14, the range of $A^{cl} + I$ is closed, hence all of $\mathbf{H}$. Assume $\mathrm{Dom}(A^*)$ is strictly larger than $\mathrm{Dom}(A^{cl})$. Then because $A^{cl} + I$ is already surjective, $A^* + I$ (which has a domain equal to the domain of A^*) cannot be injective. Thus, $A^* + I = (A + I)^*$ has a nontrivial kernel, which means that the range of $A + I$ is not dense. This is a contradiction, and so the domain of A^* must actually be equal to the domain of A^{cl}. Since A and so also A^{cl} are symmetric, this means that A^{cl} is self-adjoint. ∎

Example 9.25 *Suppose that A is a symmetric operator on $\mathbf{H}$ that has an orthonormal basis of eigenvectors. That is to say, suppose there is an orthonormal basis $\{e_j\}$ for $\mathbf{H}$ such that for each j, we have $e_j \in \mathrm{Dom}(A)$ and $Ae_j = \lambda_j e_j$ for some real number λ_j. Then A is essentially self-adjoint.*

This result is a strengthening of Example 9.15, in that we do not assume that the domain of A is equal to the space of finite linear combinations of the e_j's.

Proof. For any j, $(A - iI)e_j = (\lambda_j - i)e_j$. Since λ_j is real, we have a nonzero multiple of e_j belonging to $\mathrm{Range}(A - iI)$, for each j. This shows that $\mathrm{Range}(A - iI)$ is dense, and similarly for $\mathrm{Range}(A + iI)$. ∎

Example 9.26 *Suppose $\mathbf{H}$ is a Hilbert space direct sum of a sequence of separable Hilbert spaces $\mathbf{H}_j$:*

$$\mathbf{H} = \bigoplus_{j=1}^{\infty} \mathbf{H}_j.$$

Suppose also that A_j is a bounded self-adjoint operator on $\mathbf{H}_j$, for each j. Define a subspace V of $\mathbf{H}$ by

$$V = \left\{ \psi = (\psi_1, \psi_2, \ldots) \,\middle|\, \sum_{j=1}^{\infty} \left(\|\psi_j\|_j^2 + \|A_j\psi_j\|_j^2 \right) < \infty \right\}.$$

Suppose now that A is a symmetric operator on $\mathbf{H}$ whose domain contains the finite direct sum of the $\mathbf{H}_j$'s and such that $A|_{\mathbf{H}_j} = A_j$. Then A is *essentially self-adjoint*, $\text{Dom}(A^{cl}) = \text{Dom}(A^*) = V$, *and*

$$A^{cl}\psi = A^*\psi = (A_1\psi_1, A_2\psi_2, \ldots) \tag{9.11}$$

for all $\psi = (\psi_1, \psi_2, \ldots)$ *in* V.

See Definition A.45 for the definition of the Hilbert direct sum and the finite direct sum of a sequence of Hilbert spaces. Example 9.25 is the special case of Example 9.26 in which each $\mathbf{H}_j$ has dimension 1. This result will be useful to us in Chap. 10.

Proof. Since A_j is self-adjoint, the ranges of $A_j - iI$ and $A_j + iI$ are dense in $\mathbf{H}_j$. Thus, the closure of the range of $A - iI$ contains each $\mathbf{H}_j$ and is therefore dense in $\mathbf{H}$, and similarly for $A + iI$. This shows that A is essentially self-adjoint.

It remains to show that the domain of $A^* = A^{cl}$ is V. Let W denote the finite direct sum of the $\mathbf{H}_j$'s. By the argument in the previous paragraph, $A|_W$ is essentially self-adjoint. Then A^* is a symmetric extension of $(A|_W)^*$, which must coincide with $(A|_W)^*$. Thus, it suffices to consider the case $\text{Dom}(A) = W$.

If we assume that $\text{Dom}(A) = W$, we can compute the adjoint of A by the argument in Example 9.15. If $\phi \in V$, then the Cauchy–Schwarz inequality shows that the linear functional $\langle \phi, A \cdot \rangle$ is bounded and that $A^*\phi$ is as (9.11). On the other hand, if $\langle \phi, A \cdot \rangle$ is bounded, where $\phi = (\phi_1, \phi_2, \ldots)$, take

$$\psi_N = (\phi_1, \phi_2, \ldots, \phi_N, 0, 0, \ldots).$$

Then, as in the proof of Example 9.15, the only way we can have $|\langle \phi, A\psi_N \rangle| \leq C \|\psi_N\|$ is if ϕ belongs to V. ∎

9.6   A Counterexample

In this section, we will examine an elementary example of an operator that is symmetric but not essentially self-adjoint. Our example will be essentially the momentum operator on a finite interval, with "wrong" boundary conditions. (A more sophisticated example is given in Sect. 9.10.) We take our Hilbert space to be $L^2([0, 1])$.

Proposition 9.27 *Let* $\text{Dom}(A) \subset L^2([0, 1])$ *be the space of continuously differentiable functions* f *on* $[0, 1]$ *satisfying*

$$\psi(0) = \psi(1) = 0.$$

For $\psi \in \text{Dom}(A)$, *define*

$$A\psi = -i\hbar \frac{d\psi}{dx}.$$

Then A *is symmetric but not essentially self-adjoint.*

We can understand the failure of essential self-adjointness of A in practical terms as a failure of the spectral theorem. The eigenvector equation $A\psi = \lambda\psi$ for $\lambda \in \mathbb{R}$ is a first-order ordinary differential equation, whose general solution is $\psi(x) = ce^{i\lambda x}$, where c is a constant. The only way such a function can satisfy the boundary conditions $\psi(0) = \psi(1) = 0$ is if $c = 0$, in which case ψ is the zero vector. Thus, A has no eigenvectors. Furthermore, taking the closure of A does not help, because, as the proof will show, the boundary conditions survive taking the closure.

Proof of symmetry. Using integration by parts we see that for all ϕ and ψ in $\mathrm{Dom}(A)$ we have

$$\int_0^1 \overline{\phi(x)}\frac{d\psi}{dx}\ dx = \overline{\phi(1)}\psi(1) - \overline{\phi(0)}\psi(0) - \int_0^1 \overline{\frac{d\phi}{dx}}\psi(x)\ dx. \qquad (9.12)$$

Since we assume ϕ and ψ are in $\mathrm{Dom}(A)$, the boundary terms are zero and we get

$$\left\langle \phi, \frac{d\psi}{dx} \right\rangle_{L^2([0,1])} = -\left\langle \frac{d\phi}{dx}, \psi \right\rangle_{L^2([0,1])}.$$

Because there is a conjugate in one side of the inner product but not the other, it follows that

$$\left\langle \phi, -i\hbar\frac{d\psi}{dx} \right\rangle_{L^2([0,1])} = \left\langle -i\hbar\frac{d\phi}{dx}, \psi \right\rangle_{L^2([0,1])},$$

as claimed. ∎

We now consider A^{cl} and $A^* = (A^{cl})^*$. We will see that there are elements of the domain of the adjoint that are not in the domain of the closure.

Lemma 9.28 *If ϕ is a continuously differentiable function on $[0,1]$, then $\phi \in \mathrm{Dom}(A^*)$ and $A^*\phi = -i\hbar\ d\phi/dx$.*

Proof. If ϕ is continuously differentiable, then for any ψ in $\mathrm{Dom}(A)$, we may integrate by parts as in (9.12). Since ψ is zero at both ends of the interval, the boundary terms vanish and we obtain

$$\langle \phi, A\psi \rangle = i\hbar \int_0^1 \overline{\frac{d\phi}{dx}}\psi(x)\ dx$$

$$= \int_0^1 \overline{\left(-i\hbar\frac{d\phi}{dx}\right)}\psi(x)\ dx \qquad (9.13)$$

Since $d\phi/dx$ is continuous and hence in $L^2([0,1])$, we see that (9.13) is a continuous linear functional, as a function of ψ with fixed ϕ. Thus, ψ is in the domain of A^*, and $A^*\phi = -i\ d\phi/dx$. ∎

Proof of Proposition 9.27. Suppose ψ is in the domain of A^{cl}. Then there exist ψ_n in $\mathrm{Dom}(A)$ such that ψ_n converges to ψ and $A\psi_n$ converges to some $\chi \in L^2([0,1])$. Since the derivatives of the ψ_n's are converging in L^2, the ψ_n's themselves must be converging uniformly, as can be shown by writing each ψ_n as the integral of its derivative. (See Exercise 10.) It follows that every element of $\mathrm{Dom}(A^{cl})$ is continuous and vanishes at both ends of the interval. On the other hand, $\mathrm{Dom}(A^*)$ contains all smooth functions, including many that do not vanish at the ends of the interval. Thus, A^{cl} and $(A^{cl})^* = A^*$ do not have the same domains. ∎

It follows from Lemma 9.28 that *every* complex number λ belongs to the spectrum of A^{cl}. See Exercise 9.

The reason that A fails to be essentially self-adjoint is that we impose *too many* boundary conditions on functions in the domain of A, which results in there being *too few* boundary conditions (in this case, no boundary conditions at all) on functions in the domain of A^*. In this example, A^* is given by the same *formula* as A ($-id/dx$ in both cases), but the domain of A^* is bigger than the domain of A^{cl}.

Suppose we define another operator B, still given by the formula $-i\,d/dx$, but with the domain of B to be the space of continuously differentiable functions ψ with $\psi(0) = \psi(1)$. If we integrate by parts as in (9.12), the boundary terms will cancel, showing that B is symmetric. Meanwhile, the functions $\psi_n(x) := e^{2\pi inx}$, $n \in \mathbb{Z}$, form an orthonormal basis for $L^2([0,1])$ consisting of eigenvectors for B, with real eigenvalues $\lambda_n = 2\pi n$. Thus, by Example 9.25, B is essentially self-adjoint.

9.7   An Example

We now give an example of an operator that *is* essentially self-adjoint. Let $C_c^\infty(\mathbb{R})$ denote the space of smooth, compactly supported functions on $\mathbb{R}$.

Proposition 9.29 *Let P be the densely defined operator with $\mathrm{Dom}(P) = C_c^\infty(\mathbb{R}) \subset L^2(\mathbb{R})$ and given by $P\psi = -i\hbar\,d\psi/dx$. Then P is essentially self-adjoint.*

Proof. Our strategy is to apply Corollary 9.22. Since P is symmetric, we expect that P^* will be given by the formula $-i\hbar\,d/dx$, on some suitable domain inside $L^2(\mathbb{R})$. Thus, if $\psi \in \ker(P^* + iI)$, this should mean that $-i\hbar\,d\psi/dx = -i\psi$, or $d\psi/dx = (1/\hbar)\psi(x)$, which ought to imply that $\psi(x) = ce^{x/\hbar}$, for some constant c. Since $ce^{x/\hbar}$ belongs to $L^2(\mathbb{R})$ only if $c = 0$, we hope to conclude that $\psi = 0$.

To say that $\psi \in L^2(\mathbb{R})$ belongs to the kernel of $P^* + iI$ means that ψ belongs to $\mathrm{Dom}(P^*)$ and that $P^*\psi = -i\psi$. This holds if and only if

$$-i\hbar \int_{\mathbb{R}} \overline{\frac{d\chi}{dx}}\psi(x)\,dx = i \int_{\mathbb{R}} \overline{\chi(x)}\psi(x)\,dx$$

for all $\chi \in C_c^\infty(\mathbb{R})$. For any $\xi \in C_c^\infty(\mathbb{R})$, if we take $\chi(x) = \xi(x)e^{-x/\hbar}$ and combine the integrals into one, we get

$$0 = -i \int_{\mathbb{R}} \left[\hbar e^{-x/\hbar} \overline{\frac{d\xi}{dx}} - e^{-x/\hbar} \overline{\xi(x)} + e^{-x/\hbar} \overline{\xi(x)} \right] \psi(x) \, dx$$

$$= -i\hbar \int_{\mathbb{R}} \overline{\frac{d\xi}{dx}} e^{-x/\hbar} \psi(x) \, dx. \tag{9.14}$$

Now, (9.14) says that the derivative of $e^{-x/\hbar}\psi(x)$ in the weak or distributional sense is zero. (See Proposition A.29 in Appendix A.3.3.) Thus, by the remarks immediately following Proposition A.5, we must have $e^{-x/\hbar}\psi(x) = c$ for some c, meaning that $\psi(x) = ce^{x/\hbar}$. Since we also assume that ψ belongs to $\text{Dom}(P^*) \subset L^2(\mathbb{R})$, we must have $c = 0$, so that ψ is the zero element of $L^2(\mathbb{R})$.

We have shown, then, that only 0 belongs to the kernel of $P^* + iI$. A similar argument with i replaced by $-i$ and $e^{x/\hbar}$ by $e^{-x/\hbar}$ shows that only 0 belongs to the kernel of $P^* - iI$. Thus, by Corollary 9.22, P is essentially self-adjoint. ∎

9.8 The Basic Operators of Quantum Mechanics

In this section, we consider several of the unbounded self-adjoint operators that arise in quantum mechanics. We find natural domains of self- adjointness for the position, momentum, kinetic energy, and potential energy operators. Since Schrödinger operators are more complicated to analyze, we postpone a discussion of them until the next section. We begin with the potential energy operator.

Proposition 9.30 *Suppose $V : \mathbb{R}^n \to \mathbb{R}$ is a measurable function. Let $V(\mathbf{X})$ be the unbounded operator with domain*

$$\text{Dom}(V(\mathbf{X})) = \left\{ \psi \in L^2(\mathbb{R}^n) \, \middle| \, V(\mathbf{x})\psi(\mathbf{x}) \in L^2(\mathbb{R}^n) \right\}$$

and given by

$$[V(\mathbf{X})\psi](\mathbf{x}) = V(\mathbf{x})\psi(\mathbf{x}).$$

Then $\text{Dom}(V(\mathbf{X}))$ is dense in $L^2(\mathbb{R}^n)$ and $V(\mathbf{X})$ is self-adjoint on this domain.

Proof. Define a subset E_m of $\mathbb{R}^n$ by

$$E_m = \{\mathbf{x} \in \mathbb{R}^n \,|\, |V(\mathbf{x})| < m\},$$

so that $\cup_m E_m = \mathbb{R}^n$. Then for any $\psi \in L^2(\mathbb{R}^n)$, the function $\psi 1_{E_m}$ belongs to $\text{Dom}(V(\mathbf{X}))$. On the other hand, using dominated convergence, we have $\psi 1_{E_m} \to \psi$ as $m \to \infty$, establishing that $\text{Dom}(V(\mathbf{X}))$ is dense.

Since V is real-valued, it is easy to see that $V(\mathbf{X})$ is symmetric on $\mathrm{Dom}(V(\mathbf{X}))$. Thus, $V(\mathbf{X})^*$ is an extension of $V(\mathbf{X})$.

Meanwhile, suppose $\phi \in \mathrm{Dom}(V(\mathbf{X})^*)$, meaning that

$$\psi \mapsto \int_X \overline{\phi(x)} V(x) \psi(x) \, dx, \quad \psi \in \mathrm{Dom}(V(\mathbf{X})) \tag{9.15}$$

is a bounded linear functional. This linear functional has a unique bounded extension to L^2 and, thus, Thus, there exists a unique $\chi \in L^2(\mathbb{R}^n)$ such that

$$\int_X \overline{\psi(x)} V(x) \phi(x) \, dx = \int_X \overline{\chi(x)} \phi(x) \, dx, \tag{9.16}$$

or

$$\int_X \left[\overline{\psi(x)} V(x) - \overline{\chi(x)} \right] \phi(x) \, dx = 0$$

for all $\phi \in \mathrm{Dom}(V(\mathbf{X}))$.

Taking $\phi = (\psi V - \chi) 1_{E_m}$, we see that $\psi V - \chi$ is zero almost everywhere on E_m, for all m, hence zero almost everywhere on $\mathbb{R}^n$. Thus, ψV is equal to χ as an element of $L^2(\mathbb{R}^n)$. This shows that $\psi \in \mathrm{Dom}(V(\mathbf{X}))$. Thus, actually, $\mathrm{Dom}(V(\mathbf{X})^*) = \mathrm{Dom}(V(\mathbf{X}))$. Since we have already shown that $V(\mathbf{X})^*$ is an extension of $V(\mathbf{X})$, we conclude that $V(\mathbf{X})$ is self-adjoint on $\mathrm{Dom}(V(\mathbf{X}))$. $\blacksquare$

If we specialize the preceding proposition to the case $V(\mathbf{x}) = x_j$, we obtain the following result about the position operator.

Corollary 9.31 *The position operator X_j is self-adjoint on the domain*

$$\mathrm{Dom}(X_j) = \left\{ \psi \in L^2(\mathbb{R}^n) \, \middle| \, x_j \psi(\mathbf{x}) \in L^2(\mathbb{R}^n) \right\}.$$

We now turn to consideration of the momentum operator. Since the Fourier transform converts $\partial/\partial x_j$ into multiplication by ik_j (Proposition A.17) we can use the preceding results on multiplication operators to obtain a natural domain on which the momentum operator is self-adjoint.

Proposition 9.32 *For each $j = 1, 2, \ldots, n$, define a domain $\mathrm{Dom}(P_j) \subset L^2(\mathbb{R}^n)$ as follows:*

$$\mathrm{Dom}(P_j) = \left\{ \psi \in L^2(\mathbb{R}^n) \, \middle| \, k_j \hat{\psi}(\mathbf{k}) \in L^2(\mathbb{R}^n) \right\},$$

where $\hat{\psi}$ is the Fourier transform of ψ. Define P_j on this domain by

$$P_j \psi = \mathcal{F}^{-1}(\hbar k_j \hat{\psi}(\mathbf{k})).$$

Then P_j is self-adjoint on $\mathrm{Dom}(P_j)$.

The domain $\mathrm{Dom}(P_j)$ of P_j can also be described as the set of all $\psi \in L^2(\mathbb{R}^n)$ such that $\partial\psi/\partial x_j$, computed in the distribution sense, belongs to $L^2(\mathbb{R}^n)$. For any $\psi \in \mathrm{Dom}(P_j)$, we have $P_j \psi = -i\hbar \partial\psi/\partial x_j$, where $\partial\psi/\partial x_j$ is computed in the distribution sense.

Saying that the distributional derivative of ψ belongs to $L^2(\mathbb{R}^n)$ means (Proposition A.29) that there exists a (unique) ϕ in $L^2(\mathbb{R}^n)$ such that

$$-\left\langle \frac{\partial \chi}{\partial x_j}, \psi \right\rangle = \langle \chi, \phi \rangle$$

for all $\chi \in C_c^\infty(\mathbb{R}^n)$. If ψ is continuously differentiable, then the distributional derivative of ψ coincides with the ordinary derivative of ψ. Thus, if $\psi \in L^2(\mathbb{R}^n)$ is continuously differentiable, then ψ belongs to $\mathrm{Dom}(P_j)$ if and only if $\partial\psi/\partial x_j$, computed in the pointwise sense, belongs to $L^2(\mathbb{R}^n)$, in which case $P_j\psi = -i\hbar\partial\psi/\partial x_j$. On the other hand, if $\psi \in \mathrm{Dom}(P_j)$, it is not necessarily the case that ψ is *continuously* differentiable.

In the case $n = 1$, the domain of P_1 certainly contains $C_c^\infty(\mathbb{R})$, since each element ψ of $C_c^\infty(\mathbb{R})$ is a Schwartz function (Definition A.15), so that $\hat{\psi}$ is also a Schwartz function, in which case $k\hat{\psi}(k)$ belongs to $L^2(\mathbb{R})$. Now, as shown in Sect. 9.7, the operator $-i\hbar d/dx$ is essentially self-adjoint on $C_c^\infty(\mathbb{R})$, which means that this operator has a unique self-adjoint extension. This self-adjoint extension must, therefore, agree with the operator P_1 in the $n = 1$ case of Proposition 9.32.

Lemma 9.33 *Suppose $\psi \in L^2(\mathbb{R}^n)$ has the property that $\partial\psi/\partial x_j$, computed in the distribution sense, is equal to an L^2 function ϕ. Then $\hat{\phi}(\mathbf{k}) = ik_j\hat{\psi}(\mathbf{k})$, showing that $k_j\hat{\psi}(\mathbf{k})$ belongs to $L^2(\mathbb{R}^n)$.*

Conversely, suppose $\psi \in L^2(\mathbb{R}^n)$ has the property that $k_j\hat{\psi}(\mathbf{k})$ belongs to $L^2(\mathbb{R}^n)$. Then $\partial\psi/\partial x_j$, computed in the distribution sense, is equal to the L^2 function $\mathcal{F}^{-1}(ik_j\mathcal{F}(\psi))$.

Proof. Suppose $\partial\psi/\partial x_j$, computed in the distribution sense, is equal to the L^2 function ϕ (see Definition A.28). Then by the unitarity of the Fourier transform (Theorem A.19) and its behavior with respect to differentiation (Proposition A.17), we have

$$\langle \chi, \phi \rangle = -\left\langle \frac{\partial \chi}{\partial x_j}, \psi \right\rangle$$
$$= -\langle ik_j\mathcal{F}(\chi), \mathcal{F}(\psi) \rangle,$$

for all $\chi \in C_c^\infty(\mathbb{R})$. Thus,

$$\langle \mathcal{F}(\chi), \mathcal{F}(\phi) \rangle = -\langle ik_j\mathcal{F}(\chi), \mathcal{F}(\psi) \rangle, \quad \chi \in C_c^\infty(\mathbb{R}).$$

Writing this equality out as an integral, we have

$$\int_{\mathbb{R}^n} \overline{\hat{\chi}(\mathbf{k})}\hat{\phi}(\mathbf{k}) \, d\mathbf{k} = -\int_{\mathbb{R}^n} \overline{ik_j\hat{\chi}(\mathbf{k})}\hat{\psi}(\mathbf{k}) \, d\mathbf{k}$$
$$= \int_{\mathbb{R}^n} \overline{\hat{\chi}(\mathbf{k})}ik_j\hat{\psi}(\mathbf{k}) \, d\mathbf{k} \qquad (9.17)$$

for all $\chi \in C_c^\infty(\mathbb{R}^n)$.

We now claim that because (9.17) holds for *all* $\chi \in C_c^\infty(\mathbb{R}^n)$, we must have $\hat{\phi}(\mathbf{k}) = ik_j\hat{\psi}(\mathbf{k})$ for almost every $\mathbf{k}$. Using the Stone–Weierstrass theorem and Theorem A.10, it is not hard to show that the space of smooth functions with support in $[a, b]$ is dense in $L^2([a, b])$, for all $a < b \in \mathbb{R}$. Since both $\hat{\phi}$ and $k_j\hat{\psi}(\mathbf{k})$ are locally square-integrable, we see that these two functions are equal almost everywhere on $[a, b]$, for all $a < b \in \mathbb{R}$, and hence equal almost everywhere on $\mathbb{R}$.

Since $\hat{\phi}$ is globally square-integrable, so is $k_j\hat{\psi}(\mathbf{k})$. Furthermore, by the injectivity of the L^2 Fourier transform, we have

$$\frac{\partial \psi}{\partial x_j} = \phi = \mathcal{F}^{-1}(ik_j\mathcal{F}(\psi))$$

as claimed.

The argument for the second part of the lemma is similar and left as an exercise (Exercise 12). ∎

Proof of Proposition 9.32. By Proposition 9.30, the operator of multiplication by k_j is an unbounded self-adjoint operator on $L^2(\mathbb{R}^n)$, with domain equal to the set of ϕ for which $k_j\phi(\mathbf{k})$ belongs to $L^2(\mathbb{R}^n)$. It then follows from the unitarity of the Fourier transform that $P_j = \hbar\mathcal{F}^{-1}M_{k_j}\mathcal{F}$ is self-adjoint on $\mathcal{F}^{-1}(\text{Dom}(M_{k_j}))$, where M_{k_j} denotes multiplication by k_j.

The second characterization of $\text{Dom}(P_j)$ follows from Lemma 9.33. ∎

Proposition 9.34 *Define a domain* $\text{Dom}(\Delta)$ *as follows:*

$$\text{Dom}(\Delta) = \left\{ \psi \in L^2(\mathbb{R}^n) \,\middle|\, |\mathbf{k}|^2 \,\hat{\psi}(\mathbf{k}) \in L^2(\mathbb{R}^n) \right\}.$$

Define Δ *on this domain by the expression*

$$\Delta\psi = -\mathcal{F}^{-1}(|\mathbf{k}|^2 \,\hat{\psi}(\mathbf{k})), \tag{9.18}$$

where $\hat{\psi}$ *is the Fourier transform of* ψ *and* $\mathcal{F}^{-1}$ *is the inverse Fourier. Then* Δ *is self-adjoint on* $\text{Dom}(\Delta)$.

The domain $\text{Dom}(\Delta)$ *may also be described as the set of all* $\psi \in L^2(\mathbb{R}^n)$ *such that* $\Delta\psi$, *computed in the distribution sense, belongs to* $L^2(\mathbb{R}^n)$. *If* $\psi \in \text{Dom}(\Delta)$, *then* $\Delta\psi$ *as defined by (9.18) agrees with* $\Delta\psi$ *computed in the distribution sense.*

The proof of Proposition 9.34 is extremely similar to that of Proposition 9.32 and is omitted. Of course, the kinetic energy operator $-\hbar^2\Delta/(2m)$ is also self-adjoint on the same domain as Δ. It is easy to see from (9.18) and the unitarity of the Fourier transform that $-\hbar^2\Delta/(2m)$ is non-negative, that is, that

$$\left\langle \psi, -\frac{\hbar^2}{2m}\Delta\psi \right\rangle \geq 0$$

for all $\psi \in \text{Dom}(\Delta)$.

Using the same reasoning as in Sects. 9.6 and 9.7, it is not hard to show that the operators P_j and Δ are essentially self-adjoint on $C_c^\infty(\mathbb{R}^n)$. See Exercise 16.

Care must be exercised in applying Proposition 9.34. Although the function

$$\psi(\mathbf{x}) := \frac{1}{|\mathbf{x}|}$$

is harmonic on $\mathbb{R}^3 \setminus \{0\}$, the Laplacian over $\mathbb{R}^3$ of ψ in the distribution sense is *not* zero (Exercise 13). (It can be shown, by carefully analyzing the calculation in the proof of Proposition 9.35, that $\Delta\psi$ is a nonzero multiple of a δ-function.) This example shows that if a function ψ has a singularity, calculating the Laplacian of ψ away from the singularity may not give the correct distributional Laplacian of ψ. For example, the function ϕ in $L^2(\mathbb{R}^3)$ given by

$$\phi(\mathbf{x}) := \frac{e^{-|\mathbf{x}|^2}}{|\mathbf{x}|} \tag{9.19}$$

is not in $\mathrm{Dom}(\Delta)$, even though both ϕ and $\Delta\phi$ are (by direct computation) square-integrable over $\mathbb{R}^3 \setminus \{0\}$. Indeed, when $n \leq 3$, every element of $\mathrm{Dom}(\Delta)$ is continuous (Exercise 14).

Proposition 9.35 *Suppose $\psi(\mathbf{x}) = g(\mathbf{x})f(|\mathbf{x}|)$, where g is a smooth function on $\mathbb{R}^n$ and f is a smooth function on $(0, \infty)$. Suppose also that f satisfies*

$$\lim_{r \to 0^+} r^{n-1} f(r) = 0$$

$$\lim_{r \to 0^+} r^{n-1} f'(r) = 0.$$

If both ψ and $\Delta\psi$ are square-integrable over $\mathbb{R}^n \setminus \{0\}$, then ψ belongs to $\mathrm{Dom}(\Delta)$.

Note that the second condition in the proposition fails if $n = 3$ and $f(r) = 1/r$. We will make use of this result in Chap. 18.

Proof. To apply Proposition 9.34, we need to compute $\langle \psi, \Delta\chi \rangle$, for each $\chi \in C_c^\infty(\mathbb{R}^n)$. We choose a large cube C, centered at the origin and such that the support of χ is contained in the interior of C. Then we consider the integral of $\bar{\psi}(\partial^2\chi/\partial x_j^2)$ over $C \setminus C_\varepsilon$, where C_ε is a cube centered at the origin and having side-length ε. We evaluate the x_j-integral first and we integrate by parts twice. For "good" values of the remaining variables, x_j ranges over all of C, in which case there are no boundary terms to worry about. For "bad" values of the remaining variables, we get two kinds of boundary terms, one involving $\bar{\psi}(\partial\chi/\partial x_j)$ and one involving $(\partial\bar{\psi}/\partial x_j)\chi$, in both cases integrated over two opposite faces of C_ε.

Now,

$$\frac{\partial\psi}{\partial x_j} = \frac{\partial g}{\partial x_j} f(|\mathbf{x}|) + g(\mathbf{x})\frac{df}{dr}\frac{x_j}{r}.$$

Since the area of the faces of the cube is ε^{n-1}, the assumption on f will cause the boundary terms to disappear in the limit as ε tends to zero. Furthermore, both ψ and $\Delta\psi$ are in $L^2(\mathbb{R}^n)$ and thus in $L^1(C)$, where in the case of $\Delta\psi$, we simply leave the value at the origin (which is a set of measure zero) undefined. Thus, integrals of $\bar{\psi}\Delta\chi$ and $(\Delta\bar{\psi})\chi$ over $C\backslash C_\varepsilon$ will converge to integrals over C. Since the boundary terms vanish in the limit, we are left with

$$\langle \psi, \Delta\chi \rangle = \langle \Delta\psi, \chi \rangle.$$

Thus, the distributional Laplacian of ψ is simply integration against the "pointwise" Laplacian, ignoring the origin. Proposition 9.34 then tells us that $\psi \in \mathrm{Dom}(\Delta)$. ∎

9.9   Sums of Self-Adjoint Operators

In the previous section, we have succeeded in defining the Laplacian Δ, and hence also the kinetic energy operator $-\hbar^2\Delta/(2m)$, as a self-adjoint operator on a natural dense domain in $L^2(\mathbb{R}^n)$. We have also defined the potential energy operator $V(\mathbf{X})$ as a self-adjoint operator on a different dense domain, for any measurable function $V : \mathbb{R}^n \to \mathbb{R}$. To obtain the Schrödinger operator $-\hbar^2\Delta/(2m)+V(\mathbf{X})$, we "merely" have to make sense of the sum of two unbounded self-adjoint operators. This task, however, turns out to be more difficult than might be expected. In particular, if V is a highly singular function, then $-\hbar^2\Delta/(2m) + V(\mathbf{X})$ may fail to be self-adjoint or essentially self-adjoint on any natural domain.

Definition 9.36 *If A and B are unbounded operators on $\mathbf{H}$, then $A + B$ is the operator with domain*

$$\mathrm{Dom}(A + B) := \mathrm{Dom}(A) \cap \mathrm{Dom}(B)$$

and given by $(A + B)\psi = A\psi + B\psi$.

The sum of two unbounded self-adjoint operators A and B may fail to be self-adjoint or even essentially self-adjoint. [If, however, B is bounded with $\mathrm{Dom}(B) = \mathbf{H}$, then Proposition 9.13 shows that $A + B$ is self-adjoint on $\mathrm{Dom}(A) \cap \mathrm{Dom}(B) = \mathrm{Dom}(A)$.] For one thing, if A and B are unbounded, then $\mathrm{Dom}(A) \cap \mathrm{Dom}(B)$ may fail to be dense in $\mathbf{H}$. But even if $\mathrm{Dom}(A) \cap \mathrm{Dom}(B)$ is dense in $\mathbf{H}$, it can easily happen that $A + B$ is not essentially self-adjoint on this domain. (See, for example, Sect. 9.10.) Many things that are simple for bounded self-adjoint operators becomes complicated when dealing with unbounded self-adjoint operators!

In this section, we examine criteria on a function V under which the Schrödinger operator

$$\hat{H} = -\frac{\hbar^2}{2m}\Delta + V$$

is self-adjoint or essentially self-adjoint on some natural domain inside $L^2(\mathbb{R}^n)$.

Theorem 9.37 (Kato–Rellich Theorem) *Suppose that A and B are unbounded self-adjoint operators on $\mathbf{H}$. Suppose that $\mathrm{Dom}(A) \subset \mathrm{Dom}(B)$ and that there exist positive constants a and b with $a < 1$ such that*

$$\|B\psi\| \le a\,\|A\psi\| + b\,\|\psi\| \tag{9.20}$$

for all $\psi \in \mathrm{Dom}(A)$. Then $A + B$ is self-adjoint on $\mathrm{Dom}(A)$ and essentially self-adjoint on any subspace of $\mathrm{Dom}(A)$ on which A is essentially self-adjoint. Furthermore, if A is non-negative, then the spectrum of $A + B$ is bounded below by $-b/(1 - a)$.

Note that since we assume $\mathrm{Dom}(B) \supset \mathrm{Dom}(A)$, the natural domain for $A + B$ is $\mathrm{Dom}(A) \cap \mathrm{Dom}(B) = \mathrm{Dom}(A)$. An operator B satisfying (9.20) is said to be *relatively bounded* with respect to A, with relative bound a.

Proof. We use the trivial variant of Theorem 9.21 given in Exercise 8. Choose a positive real number μ large enough that $a + b/\mu < 1$, which is possible because we assume $a < 1$. Then for any $\psi \in \mathrm{Dom}(A)$, we have

$$(A + B + i\mu I)\psi = \left(B(A + i\mu I)^{-1} + I\right)(A + i\mu I)\psi. \tag{9.21}$$

For any $\psi \in \mathbf{H}$, we compute that

$$\left\|B(A + i\mu I)^{-1}\psi\right\| \le a\left\|A(A + i\mu I)^{-1}\psi\right\| + b\left\|(A + i\mu I)^{-1}\psi\right\|$$
$$\le \left(a + \frac{b}{\mu}\right)\|\psi\|. \tag{9.22}$$

Here we have made use of the estimates

$$\left\|A(A + i\mu I)^{-1}\right\| < 1, \quad \left\|(A + i\mu I)^{-1}\right\| < \frac{1}{\mu},$$

both of which are elementary (Exercise 17).

If C denotes the operator $B(A + i\mu I)^{-1}$, (9.22) tells us that $\|C\| < (a + b/\mu) < 1$. Thus, by Lemma 7.6, $C + I$ is invertible. Furthermore, since A is self-adjoint, $A + i\mu I$ maps $\mathrm{Dom}(A)$ *onto* $\mathbf{H}$. Thus, (9.21) tells us that $A + B + i\mu I$ also maps $\mathrm{Dom}(A)$ onto $\mathbf{H}$. The same argument shows that $A + B - i\mu I$ maps $\mathrm{Dom}(A)$ onto $\mathbf{H}$ and we conclude, by Exercise 8, that $A + B$ is self-adjoint on $\mathrm{Dom}(A)$.

Suppose, in addition, that A is non-negative. Let us replace $i\mu$ by $\lambda > 0$, in (9.21). Calculating as in (9.22), using the estimates in Exercise 18, we obtain that

$$\left\|B(A + \lambda I)^{-1}\psi\right\| \le \left(a + \frac{b}{\lambda}\right)\|\psi\|$$

for all $\psi \in \mathbf{H}$. If $\lambda > b/(1 - a)$, then $a + b/\lambda < 1$, and by the above argument, $\mathrm{Range}(A + B + \lambda I) = \mathbf{H}$. Furthermore, since $A + B + \lambda I$ is self-adjoint, Proposition 9.12 tells us that $\ker(A + B + \lambda I) = \{0\}$. This shows that $A + B + \lambda I$ is invertible and $-\lambda$ is in the resolvent set of $A + B$. We conclude, then, that the spectrum of $A+B$ is contained in $[-b/(1-a), +\infty)$.

The last part of the theorem, concerning essential self-adjointness, is left as an exercise (Exercise 19). ∎

Theorem 9.38 *Suppose n is at most 3 and $V : \mathbb{R}^n \to \mathbb{R}$ is a measurable function that can be decomposed as a sum of two real-valued, measurable functions V_1 and V_2, with V_1 belonging to $L^2(\mathbb{R}^n)$ and V_2 being bounded. Then the Schrödinger operator $-\hbar^2\Delta/(2m)+V(\mathbf{X})$ is self-adjoint on $\mathrm{Dom}(\Delta)$. Furthermore, $-\hbar^2\Delta/(2m) + V(\mathbf{X})$ is bounded below.*

Implicit in the statement of the theorem is that $\mathrm{Dom}(V(\mathbf{X}))$, as given in Proposition 9.30, contains $\mathrm{Dom}(\Delta)$. A result similar to Theorem 9.38 in $\mathbb{R}^n$, $n \geq 4$, but the condition that V_1 belongs to $L^2(\mathbb{R}^n)$ is replaced by the condition that V_1 belongs to $L^p(\mathbb{R}^n)$ for some $p > n/2$. See Theorem X.20 in Volume II of [34].

Proof. We apply the Kato–Rellich theorem with $A = -\hbar^2\Delta/2m$ and $B = V(\mathbf{X})$. Assume $\psi \in \mathrm{Dom}(\Delta)$ and fix some $\varepsilon > 0$. By Exercise 14, there exists a constant c_ε such that

$$|\psi(\mathbf{x})| \leq \varepsilon \|\Delta\psi\| + c_\varepsilon \|\psi\|$$

for all $\mathbf{x} \in \mathbb{R}^n$. Thus, if V is as in the theorem and $\psi \in \mathrm{Dom}(\Delta)$,

$$\|V\psi\| \leq \sup |\psi(\mathbf{x})| \|V_1\| + \sup |V_2(\mathbf{x})| \|\psi\|$$
$$\leq \varepsilon \|V_1\| \|\Delta\psi\| + (c_\varepsilon \|V_1\| + \sup |V_2(\mathbf{x})|) \|\psi\|.$$

This shows that $\mathrm{Dom}(V(\mathbf{X})) \supset \mathrm{Dom}(\Delta)$. Since ε is arbitrary, we can arrange for the constant in front of $\|\Delta\psi\|$ to be less than one and the Kato–Rellich theorem applies. ∎

Theorem 9.39 *Suppose n is at most 3 and $V : \mathbb{R}^n \to \mathbb{R}$ is a measurable function that can be decomposed as a sum of three real-valued, measurable functions V_1, V_2, and V_3, with V_1 belonging to $L^2(\mathbb{R}^n)$, V_2 being bounded, and V_3 being non-negative and locally square-integrable. Then the Schrödinger operator $-\hbar^2\Delta/(2m) + V(\mathbf{X})$ is essentially self-adjoint on $C_c^\infty(\mathbb{R}^n)$.*

The proof of this result would take us too far afield and is omitted. See Theorem X.29 in Volume II of [34]. Note that we assume only that V_3 is non-negative and *locally* square-integrable; V_3 can tend to $+\infty$ arbitrarily fast at infinity. Again, the same result applies in $\mathbb{R}^n$, $n \geq 4$, if the condition on V_1 is replaced by the assumption that $V_1 \in L^p(\mathbb{R}^n)$ for some $p > n/2$.

Proposition 9.40 *Fix $\mathbf{a}$ and $\mathbf{b}$ in $\mathbb{R}^n$ and let $\mathbf{a} \cdot \mathbf{X} + \mathbf{b} \cdot \mathbf{P}$ denote the operator given by*

$$(\mathbf{a} \cdot \mathbf{X} + \mathbf{b} \cdot \mathbf{P})\psi(\mathbf{x}) = (\mathbf{a} \cdot \mathbf{x})\psi(\mathbf{x}) - i\hbar \sum_{j=1}^n b_j \frac{\partial \psi}{\partial x_j}.$$

Then $\mathbf{a} \cdot \mathbf{X} + \mathbf{b} \cdot \mathbf{P}$ *is essentially self-adjoint on* $C_c^\infty(\mathbb{R}^n)$.

Proof. We use the same strategy as in Sect. 9.7, namely we explicitly solve the equation $A^*\psi = \pm i\psi$ and find that there are no nonzero, square-integrable solutions.

The case $\mathbf{b} = 0$ is not hard to analyze and is left as an exercise (Exercise 20). Assume, then, that $\mathbf{b} \neq 0$. By making a rotational change of variables, we can assume that $\mathbf{b} = \alpha \mathbf{e}_1$ and $\mathbf{a} = \beta \mathbf{e}_1 + \gamma \mathbf{e}_2$, so that

$$(A\psi)(\mathbf{x}) = (\beta x_1 + \gamma x_2)\psi(\mathbf{x}) - i\hbar\alpha\frac{\partial\psi}{\partial x_1}. \tag{9.23}$$

(If $n = 1$, the γx_2 term is not present.) As in the proof of Proposition 9.29, the adjoint A^* of A will be given by the same formula as A, with $\mathrm{Dom}(A^*)$ consisting of those elements ψ of $L^2(\mathbb{R}^n)$ for which the right-hand side of (9.23), computed in the distributional sense, belongs to $L^2(\mathbb{R}^n)$.

We now apply the criterion for essential self-adjointness in Corollary 9.22. We need to show that the equations $A^*\psi = i\psi$ and $A^*\psi = -i\psi$ have no nonzero solutions in $\mathrm{Dom}(A^*)$. After rewriting the equation $A^*\psi = i\psi$ as

$$\frac{\partial\psi}{\partial x_1} = -\frac{i}{\hbar\alpha}(\beta x_1 + \gamma x_2)\psi(\mathbf{x}) - \frac{1}{\hbar\alpha}\psi(\mathbf{x}), \tag{9.24}$$

we can easily find the general distributional solution as

$$\psi(\mathbf{x}) = c(x_2,\dots,x_n)\exp\left\{-\frac{i\beta}{2\alpha\hbar}x_1^2 - \frac{i\gamma}{\alpha\hbar}x_1 x_2 - \frac{1}{\alpha\hbar}x_1\right\}. \tag{9.25}$$

[It is easily verified that if we let ϕ equal ψ divided by the exponential on the right-hand side of (9.25), then ϕ satisfies $\partial\phi/\partial x_1 = 0$ in the distributional sense. Exercise 21 then tells us that ϕ must be a function of $x_2,\dots,x_n$.] Since the exponential factor is never square integrable as a function of x_1 with x_2 fixed, the only way that ψ can be square integrable is if c is zero for almost every value of $(x_2,\dots,x_n)$, in which case ψ is the zero element of $L^2(\mathbb{R}^n)$. A similar argument shows that the equation $A^*\psi = -i\psi$ has no nonzero solutions. ∎

9.10   Another Counterexample

In this section, we will show that the Schrödinger operator $\hat{H} = P^2/(2m) - X^4$ is *not* essentially self-adjoint on $C_c^\infty(\mathbb{R})$, even though $\hat{H}$ is certainly symmetric. By contrast, $P^2/(2m) + X^4$ is essentially self-adjoint, by Theorem 9.39. The operator $P^2/(2m) - X^4$ is a more serious counterexample than the one in Sect. 12.2, in that it does not involve any obviously incorrect choice of boundary conditions. On the other hand, it should not be surprising that something goes "wrong" in a quantum system with a potential equal to $-x^4$. After all, a classical system with this potential has trajectories that go to infinity in finite time (see Exercise 4 in Chap. 2).

To show that $\hat{H}$ is not essentially self-adjoint, we will show that the adjoint $\hat{H}^*$ is not symmetric. Suppose ψ is a C^∞ function such that both ψ and the function

$$-\frac{\hbar^2}{2m}\psi''(x) - x^4\psi(x) \tag{9.26}$$

belong to $L^2(\mathbb{R})$. Using integration by parts, as in the proof of Lemma 9.28, we can see that ψ is in the domain of $\hat{H}^*$ and $\hat{H}^*\psi$ is the function in (9.26). We will construct an approximate eigenvector $\psi \in \mathrm{Dom}(\hat{H}^*)$ for $\hat{H}^*$ with an imaginary eigenvalue $i\alpha$, which will show that $\hat{H}^*$ is not symmetric and thus $\hat{H}$ is not essentially self-adjoint.

Theorem 9.41 *Define an operator $\hat{H}$ with $\mathrm{Dom}(\hat{H}) = C_c^\infty(\mathbb{R})$ by the formula*

$$\hat{H} = -\frac{\hbar^2}{2m}\frac{d^2}{dx^2} - x^4.$$

Then $\hat{H}$ is not essentially self-adjoint.

In preparation for the proof, let us define a function $p(x)$ on $\mathbb{R}$ such that

$$\frac{p(x)^2}{2m} - x^4 = i\alpha,$$

that is,

$$p(x) = \sqrt{2m}\sqrt{x^4 + i\alpha}. \tag{9.27}$$

Here we take the square root that is in the first quadrant. The function $p(x)$ represents "the momentum of a classical particle with energy $i\alpha$."

Lemma 9.42 *If ψ_α is given by*

$$\psi_\alpha(x) = \frac{1}{\sqrt{p(x)}}\exp\left\{\frac{i}{\hbar}\int_0^x p(y)\,dy\right\}, \tag{9.28}$$

then ψ_α belongs to $L^2(\mathbb{R})$ and the function

$$-\frac{\hbar^2}{2m}\frac{d^2\psi_\alpha}{dx^2} - x^4\psi_\alpha \tag{9.29}$$

also belongs to $L^2(\mathbb{R})$. Furthermore, we have

$$\left[-\frac{\hbar^2}{2m}\frac{d^2}{dx^2} - x^4 - i\alpha\right]\psi_\alpha(x) = -\frac{\hbar^2}{2m}\psi_\alpha(x)m_\alpha(x),$$

where

$$m_\alpha(x) = \frac{5}{4}\frac{x^6}{(x^4 + i\alpha)^2} - 3\frac{x^2}{(x^4 + i\alpha)}.$$

It will be apparent from the proof that the two terms in (9.29) are *not* separately in $L^2(\mathbb{R})$. The motivation for the definition of ψ_α comes from the WKB approximation (Chap. 15) with a complex value for the energy.
Proof. Let us consider the integral of p,

$$\int_0^x p(y)\, dy = \sqrt{2m} \int_0^x \sqrt{y^4 + i\alpha}\, dy.$$

Using the power series for $(1+x)^a$ we see that for large y,

$$\sqrt{y^4 + i\alpha} = y^2\sqrt{1 + i\alpha/y^4} = y^2\left(1 + \frac{i\alpha}{2y^4} + O\left(\frac{1}{y^8}\right)\right).$$

From this estimate, it is easy to see that the imaginary part of $\int_0^x p(y)\, dy$ remains bounded as x tends to $\pm\infty$. It follows that the exponential in the definition of ψ is bounded, from which it is easy to see that ψ is square integrable.

Now, using the formula for the second derivative of a product, we obtain

$$-\hbar^2 \frac{d^2}{dx^2}\psi_\alpha = \left[\frac{p(x)^2}{\sqrt{p(x)}} - i\hbar \frac{p'(x)}{\sqrt{p(x)}} - 2\hbar^2\left(-\frac{1}{2}\frac{p'(x)}{p(x)^{3/2}}\right)\frac{ip(x)}{\hbar}\right.$$

$$\left. -\hbar^2 \frac{d^2}{dx^2}\frac{1}{\sqrt{p(x)}}\right] \exp\left\{\frac{i}{\hbar}\int_0^x p(y)\, dy\right\}. \tag{9.30}$$

The factor of $1/\sqrt{p(x)}$ in the definition of ψ_α was chosen precisely so that the second and third terms in square brackets will cancel. If we replace $p^2(x)$ in the numerator of the first term by $2m(x^4 + i\alpha)$, we obtain

$$-\frac{\hbar^2}{2m}\psi_\alpha''(x) - x^4\psi_\alpha - i\alpha\psi_\alpha = -\frac{\hbar^2}{2m}\left(\frac{d^2}{dx^2}p(x)^{-1/2}\right)\exp\left\{\frac{i}{\hbar}\int_0^x p(y)\, dy\right\}.$$

It is then an elementary calculation to show that

$$\frac{d^2}{dx^2}p(x)^{-1/2} = p(x)^{-1/2}\left[\frac{5}{4}(x^4 + i\alpha)^{-2}x^6 - 3(x^4 + i\alpha)^{-1}x^2\right],$$

from which the lemma follows. ∎
Proof of Theorem 9.41. If $\hat{H}$ were essentially self-adjoint, $\hat{H}^*$ (which would coincide with $\hat{H}^{cl}$) would be self-adjoint and, in particular, symmetric. If this were the case, we would have, by the proof of Lemma 7.8,

$$\left\langle (\hat{H}^* - i\alpha I)\psi, (\hat{H}^* - i\alpha I)\psi \right\rangle \geq \alpha^2 \langle \psi, \psi \rangle \tag{9.31}$$

for all $\psi \in \mathrm{Dom}(\hat{H}^*)$ and $\alpha \in \mathbb{R}$. But if ψ_α is the function in Lemma 9.42, the discussion preceding Theorem 9.41 shows that ψ_α belongs to $\mathrm{Dom}(\hat{H}^*)$.

Furthermore, it is easily verified that there is a constant C such that $|m_\alpha(x)| \leq C$ for all $\alpha \geq 1$ and $x \in \mathbb{R}$. Thus, for all sufficiently large α, we have

$$\left\| (\hat{H}^* - i\alpha I)\psi_\alpha \right\|^2 \leq \frac{\hbar^4}{4m^2} C^2 \left\| \psi_\alpha \right\|^2 < \alpha^2 \left\| \psi_\alpha \right\|^2,$$

contradicting (9.31). ∎

See Exercise 22 for a more explicit approach to showing that $\hat{H}^*$ is not symmetric.

9.11 Exercises

1. Show that an unbounded operator A fails to be closable if and only if the closure of the graph of A contains an element of the form $(0, \psi)$ with $\psi \neq 0$.

2. Define an unbounded operator A on $L^2([0,1])$ with domain $\mathrm{Dom}(A) = C([0,1])$ by
$$Af = f(0)\mathbf{1},$$
where $\mathbf{1}$ is the constant function. Show that A is not closable.

3. Prove Proposition 9.13.

4. Suppose that A is an unbounded self-adjoint operator on $\mathbf{H}$ and that numbers λ_n in $\sigma(A)$ converge to some $\lambda \in \mathbb{R}$. Using Point 1 of Proposition 9.18, show that $\lambda \in \sigma(A)$.

5. Suppose A is a closed operator on $\mathbf{H}$. Show that the kernel of A is a closed subspace of $\mathbf{H}$.

6. Suppose A is a closed operator on $\mathbf{H}$. Define a norm $\|\cdot\|_1$ on $\mathrm{Dom}(A)$ by
$$\|\psi\|_1 = \|\psi\| + \|A\psi\|.$$
Show that $\mathrm{Dom}(A)$ is a Banach space with respect to $\|\cdot\|_1$.

7. Let A be an unbounded operator on $\mathbf{H}$.

   (a) Show that if A is symmetric, then A^{cl} is also symmetric.

   (b) Show that if B is an extension of A, then A^* is an extension of B^*.

   (c) Suppose A is self-adjoint and B is an extension of A. Show that if B is symmetric, then $\mathrm{Dom}(A) = \mathrm{Dom}(B)$. (That is to say, a self-adjoint operator has no proper symmetric extensions.)

8. Fix a positive real number μ.

    (a) Show that a symmetric operator A is self-adjoint if and only if $\mathrm{Range}(A + i\mu I)$ and $\mathrm{Range}(A - i\mu I)$ are equal to $\mathbf{H}$.

    (b) Show that a symmetric operator A is essentially self-adjoint if and only if $\mathrm{Range}(A + i\mu I)$ and $\mathrm{Range}(A - i\mu I)$ are dense in $\mathbf{H}$.

9. Let A be the operator considered in Sect. 9.6. Using Lemma 9.28, show that for each $\lambda \in \mathbb{C}$, there exists $\psi \in \mathrm{Dom}(A^*)$ with $A^*\psi = \lambda\psi$. Conclude that each $\lambda \in \mathbb{C}$ belongs to the spectrum of A^{cl}.

    Hint: Recall that $(A^{cl})^* = A^*$.

10. Let A be the operator considered in Sect. 9.6 and suppose ψ is in the domain of A^{cl}. Then there exists a sequence ψ_n in $\mathrm{Dom}(A)$ such that ψ_n converges to ψ in $L^2([0,1])$ and such that $A\psi_n$ converges to some χ in $L^2([0,1])$.

    (a) Show that

$$\psi_n(x) = \left\langle 1_{[0,x]}, \frac{d\psi_n}{dx} \right\rangle = i \left\langle 1_{[0,x]}, A\psi_n \right\rangle$$

    for all $x \in [0,1]$.

    (b) Show that ψ_n converges *uniformly* to the function

$$\psi(x) = i \left\langle 1_{[0,x]}, \chi \right\rangle.$$

    (c) Conclude that ψ is continuous and satisfies $\psi(0) = \psi(1) = 0$.

11. Take $\mathbf{H} = L^2((0, \infty))$ and let A be the operator $-i\, d/dx$, with $\mathrm{Dom}(A)$ consisting of those smooth functions that are supported on a compact subset of $(0, \infty)$. (Such a function is, in particular, zero on $(0, \varepsilon)$ for some $\varepsilon > 0$.) Show that A is symmetric and that $A^* + iI$ is injective but that $A^* - iI$ is not injective.

    Hint: Imitate the arguments in the proof of Propositions 9.27 and 9.29.

12. Prove the second part of Lemma 9.33.

13. Let χ be a smooth, radial function on $\mathbb{R}^3$ such that for $|\mathbf{x}| < 1$ we have $\chi(\mathbf{x}) = 1$, for $|\mathbf{x}| > 2$ we have $\chi(\mathbf{x}) = 0$, and for $1 < |\mathbf{x}| < 2$, we have $\partial\chi/\partial r < 0$. Show that

$$\int_{\mathbb{R}^3} \frac{1}{|\mathbf{x}|} \Delta\chi(\mathbf{x})\, d\mathbf{x} < 0,$$

which shows that the Laplacian of $1/|\mathbf{x}|$, in the distribution sense, is not zero.

Hint: Let $E = C_1 \backslash C_2$, where C_1 is a cube centered at the origin with side length 3 and where C_2 is a cube centered at the origin with side length 1/2. Then E contains the support of $\Delta\chi$. Using integration by parts on E, show that

$$\int_{\mathbb{R}^3} \frac{1}{|\mathbf{x}|} \Delta\chi(\mathbf{x}) \, d\mathbf{x} = -\int_{\mathbb{R}^3} \nabla\left(\frac{1}{|\mathbf{x}|}\right) \cdot \nabla\chi(\mathbf{x}) \, d\mathbf{x}.$$

14. Let $\mathrm{Dom}(\Delta) \subset L^2(\mathbb{R}^n)$ denote the domain of the Laplacian, as given in Proposition 9.34, and assume $n \leq 3$.

   (a) Show that each $\psi \in \mathrm{Dom}(\Delta)$ is continuous and that there exists constants c_1 and c_2 such that

   $$|\psi(\mathbf{x})| \leq c_1 \|\psi\| + c_2 \left\| |\mathbf{k}|^{9/5} \left| \hat{\psi}(\mathbf{k}) \right| \right\|,$$

   for all $\psi \in \mathrm{Dom}(\Delta)$.

   Hint: Show that $\hat{\psi}$ is in L^1 by expressing $\hat{\psi}$ as the product of two L^2 functions.

   (b) Show that for any $\varepsilon > 0$, there exists a constant c_ε such that

   $$|\psi(\mathbf{x})| \leq c_\varepsilon \|\psi\| + \varepsilon \|\Delta\psi\|$$

   for all $\psi \in \mathrm{Dom}(\Delta)$.

15. Recall the definitions of $\mathrm{Dom}(P_j)$ and $\mathrm{Dom}(\Delta)$ in Sect. 9.8. Let $\mathrm{Dom}(P_j^2)$ be the set of all ψ belonging to $\mathrm{Dom}(P_j)$ such that $P_j\psi$ again belongs to $\mathrm{Dom}(P_j)$. Show that

$$\bigcap_{j=1}^n \mathrm{Dom}(P_j^2) = \mathrm{Dom}(\Delta).$$

16. Let Q_j denote the restriction to $C_c^\infty(\mathbb{R}^n)$ of the momentum operator P_j. Show that $\mathrm{Dom}(Q_j^*) = \mathrm{Dom}(P_j)$. Conclude that Q_j is essentially self-adjoint.

17. Let A be an unbounded self-adjoint operator on $\mathbf{H}$ and let μ be a nonzero real number.

   (a) Show that $\left\|(A + i\mu I)^{-1}\right\| \leq 1/|\mu|$. Note that $(A + i\mu I)^{-1}$ exists, by Theorem 9.17.

   (b) Show that for all $\psi \in \mathbf{H}$,

   $$\|\psi\|^2 = \left\|A(A + i\mu I)^{-1}\psi\right\|^2 + \mu^2 \left\|(A + i\mu I)^{-1}\psi\right\|^2.$$

   Conclude that $\left\|A(A + i\mu I)^{-1}\right\| \leq 1$.

18. Let A be an unbounded self-adjoint operator on $\mathbf{H}$. Suppose A is non-negative (Definition 9.19) and let λ be a positive real number.

   (a) Show that $\left\|(A + \lambda I)^{-1}\right\| \leq 1/\lambda$.
   (b) Show that for all $\psi \in \mathbf{H}$,

   $$\|\psi\|^2 \geq \left\|A(A + \lambda I)^{-1}\psi\right\|^2 + \lambda^2 \left\|(A + \lambda I)^{-1}\psi\right\|^2.$$

   Conclude that $\left\|A(A + \lambda I)^{-1}\right\| < 1$.

19. Prove the last part of Theorem 9.37, concerning domains of essential self-adjointness.

   Hint: If A is self-adjoint on $\mathrm{Dom}(A)$ and $V \subset \mathrm{Dom}(A)$ is a dense subspace of $\mathbf{H}$, then A is essentially self-adjoint on V if and only if the closure of $A|_V$ is equal to A.

20. Let A be the operator $\mathbf{b} \cdot \mathbf{X}$ on the domain $C_c^\infty(\mathbb{R}^n)$, for some $\mathbf{b} \in \mathbb{R}^n$.

   (a) Using the definition of the adjoint of an unbounded operator, show that $\mathrm{Dom}(A^*)$ consists of all those ψ in $L^2(\mathbb{R}^n)$ for which the function $(\mathbf{b} \cdot \mathbf{x})\psi(\mathbf{x})$ again belongs to $L^2(\mathbb{R}^n)$.
   (b) Using Proposition 9.30, show that A is essentially self-adjoint.

21. (a) Show that a function $\phi \in C_c^\infty(\mathbb{R}^n)$ can be expressed as $\phi = \partial\chi/\partial x_1$ for some $\chi \in C_c^\infty(\mathbb{R}^n)$ if and only if ϕ satisfies

   $$\int_{-\infty}^{\infty} \phi(x_1, x_2, \ldots, x_n) \, dx_1 = 0$$

   for all $(x_2, \ldots, x_n)$.

   (b) Fix a function $\gamma \in C_c^\infty(\mathbb{R})$ such that $\int_{-\infty}^{\infty} \gamma(x) \, dx = 1$. Show that any $\phi \in C_c^\infty(\mathbb{R}^n)$ can be expressed as

   $$\phi(\mathbf{x}) = f(x_2, \ldots, x_n)\gamma(x_1) + \frac{\partial\chi}{\partial x_1}$$

   for some $\chi \in C_c^\infty(\mathbb{R}^n)$, where f is the element of $C_c^\infty(\mathbb{R}^{n-1})$ given by

   $$f(x_2, \ldots, x_n) = \int_{-\infty}^{\infty} \phi(x_1, x_2, \ldots, x_n) \, dx_1.$$

   (c) Suppose T is a distribution on $\mathbb{R}^n$ with the property that

   $$\frac{\partial T}{\partial x_1} = 0.$$

Define a distribution c on $\mathbb{R}^{n-1}$ by the formula

$$c(f) = T(f(x_2, \ldots, x_n)\gamma(x_1)).$$

Show that for all $\phi \in C_c^\infty(\mathbb{R}^n)$ we have

$$T(\phi) = c(\tilde{\phi}),$$

where $\tilde{\phi} \in C_c^\infty(\mathbb{R}^{n-1})$ is given by

$$\tilde{\phi}(x_2, \ldots, x_n) = \int_{\mathbb{R}} \phi(x_1, x_2, \ldots, x_n) \, dx_1.$$

22. Let $\hat{H}$ denote the Schrödinger operator in Theorem 9.41 and let ψ_α be the function defined in Lemma 9.42.

(a) Show that

$$\left\langle \psi_\alpha, \hat{H}^*\psi_\alpha \right\rangle - \left\langle \hat{H}^*\psi_\alpha, \psi_\alpha \right\rangle$$
$$= -\frac{\hbar^2}{2m} \lim_{A \to \infty} \left[\overline{\psi_\alpha(x)}\psi'_\alpha(x) \Big|_{-A}^{A} - \overline{\psi'_\alpha(x)}\psi_\alpha(x) \Big|_{-A}^{A} \right].$$

(b) Now show by direct calculation that $\left\langle \psi, \hat{H}^*\psi \right\rangle \neq \left\langle \hat{H}^*\psi, \psi \right\rangle$.

10

The Spectral Theorem for Unbounded Self-Adjoint Operators

This chapter gives statements and proofs of the spectral theorem for unbounded self-adjoint operators, in the same forms as in the bounded case, in terms of projection-valued measures, in terms of direct integrals, and in terms of multiplication operators. The proof reduces the spectral theorem for an unbounded self-adjoint operator A to spectral theorem for the *bounded* operator $U := (A + iI)(A - iI)^{-1}$ (Sect. 10.4). This bounded operator is, however, not self-adjoint but rather unitary. Thus, before coming to the proof of the spectral theorem for unbounded self-adjoint operators, we prove (Sect. 10.3) the spectral theorem for bounded *normal* operators, those that commute with their adjoints. (A unitary operator U certainly commutes with its adjoint $U^* = U^{-1}$.) The proof for a bounded normal operator B is the same as for bounded self-adjoint operators, except for the step in which we approximate continuous functions on $\sigma(B)$ by polynomials. Since $\sigma(B)$ is not necessarily contained in $\mathbb{R}$, we need to use the complex version of the Stone–Weierstrass theorem, which requires us to consider polynomials in λ *and* $\bar\lambda$. We must then prove a strengthened version of the spectral mapping theorem before proceeding along the lines of the proof for bounded self-adjoint operators.

In Sect. 10.2, we discuss Stone's theorem, which gives a one-to-one correspondence between strongly continuous one-parameter unitary groups and self-adjoint operators. One direction of Stone's theorem follows from the spectral theorem, that is, from the functional calculus that results from the spectral theorem.

B.C. Hall, *Quantum Theory for Mathematicians*, Graduate Texts
in Mathematics 267, DOI 10.1007/978-1-4614-7116-5_10,
© Springer Science+Business Media New York 2013

10.1 Statements of the Spectral Theorem

The statement of the spectral theorem—in any of the forms that we have considered—is almost the same for unbounded self-adjoint operators as for bounded ones. The only difference is that the statement of the theorem in the unbounded case has to contain some description of the domain of the operator.

Recall that if μ is a projection-valued measure on (X, Ω) with values in $\mathcal{B}(\mathbf{H})$ and ψ is an element of $\mathbf{H}$, then we can construct a non-negative, real-valued measure μ_ψ from μ by setting $\mu_\psi(E) = \langle \psi, \mu(E)\psi \rangle$, for each measurable set E. To motivate the following definition, consider integration of a bounded measurable function f against a projection-valued measure μ. Since the integral is multiplicative and complex-conjugation of a function corresponds to adjoint of the operator, we have

$$\left\langle \left(\int_X f \, d\mu \right) \psi, \left(\int_X f \, d\mu \right) \psi \right\rangle = \left\langle \psi, \left(\int_X \bar{f} f \, d\mu \right) \psi \right\rangle$$

$$= \int_X |f|^2 \, d\mu_\psi. \tag{10.1}$$

Suppose, now, that f is an *unbounded* measurable function on X and we wish to define $\int_X f \, d\mu$, which will presumably be an unbounded operator. It seems reasonable to define the domain of f to be the set of ψ for which the right-hand side of (10.1) is finite.

Proposition 10.1 *Suppose μ is a projection-valued measure on (X, Ω) with values in $\mathcal{B}(\mathbf{H})$ and $f : X \to \mathbb{C}$ is a measurable function (not necessarily bounded). Define a subspace W_f of $\mathbf{H}$ by*

$$W_f = \left\{ \psi \in \mathbf{H} \,\middle|\, \int_X |f(\lambda)|^2 \, d\mu_\psi(\lambda) < \infty \right\}. \tag{10.2}$$

Then there exists a unique unbounded operator on $\mathbf{H}$ with domain W_f—which is denoted by $\int_X f \, d\mu$—with the property that

$$\left\langle \psi, \left(\int_X f \, d\mu \right) \psi \right\rangle = \int_X f(\lambda) \, d\mu_\psi(\lambda)$$

for all ψ in W_f. This operator satisfies (10.1) for all $\psi \in W_f$.

Note that since μ_ψ is a finite measure for all ψ, if f is bounded then the domain of $\int_X f \, d\mu$ is all of $\mathbf{H}$. Thus, in the bounded case, the definition of $\int_X f \, d\mu$ in Proposition 10.1 agrees with our earlier definition (in Chap. 7) of the integral. This means, in particular, that if f is a bounded function, $\int_X f \, d\mu$ is a bounded operator. Proposition 10.1 follows immediately from the following result.

Proposition 10.2 *Let f be a measurable function on X and let W_f be as in (10.2). Then the following results hold.*

1. *The space W_f is a dense subspace of $\mathbf{H}$ and the map $Q_f : W_f \to \mathbb{C}$ given by*

$$Q_f(\psi) = \int_X f(\lambda) \, d\mu_\psi(\lambda)$$

   is a quadratic form on W_f.

2. *If L_f is the associated sesquilinear form on W_f, we have*

$$|L_f(\phi, \psi)| \leq \|\phi\| \, \|f\|_{L^2(X, \mu_\psi)} \tag{10.3}$$

   for all $\phi, \psi \in W_f$.

3. *For each $\psi \in W_f$, there is a unique $\chi \in \mathbf{H}$ such that $L_f(\phi, \psi) = \langle \phi, \chi \rangle$ for all $\phi \in W_f$. Furthermore, the map $\psi \mapsto \chi$ is linear and for all $\psi \in W_f$, we have*

$$\|\chi\|^2 = \int_X |f|^2 \, d\mu_\psi \tag{10.4}$$

Proof. It is easy to see that W_f is closed under scalar multiplication. To show that it is closed under addition, note that since $\mu(E)$ is self-adjoint and satisfies $\mu(E)^2 = \mu(E)$, we have

$$\begin{aligned}
\mu_{\phi+\psi}(E) &= \|\mu(E)(\phi + \psi)\|^2 \\
&\leq (\|\mu(E)\phi\| + \|\mu(E)\psi\|)^2 \\
&\leq 2\|\mu(E)\phi\|^2 + 2\|\mu(E)\psi\|^2 \\
&= 2\mu_\phi(E) + 2\mu_\psi(E),
\end{aligned}$$

where in the third line we have use the elementary inequality $(x + y)^2 \leq 2x^2 + 2y^2$.

To show that W_f is dense in $\mathbf{H}$, let $E_n = \{x \in X \mid |f(x)| < n\}$. If $\psi \in \mathrm{Range}(\mu(E_n))$, then $\mu_\psi(E_n^c) = 0$, and, thus,

$$\int_X |f|^2 \, d\mu_\psi = \int_{E_n} |f|^2 \, d\mu_\psi \leq n^2 \mu_\psi(E_n) < \infty, \tag{10.5}$$

showing that ψ belongs to W_f. Since also $\cup_n E_n = X$, the union of the ranges of the $\mu(E_n)$'s is dense and contained in W_f.

If f is bounded, Q_f may be computed as

$$Q_f(\psi) = \left\langle \psi, \left(\int_X f \, d\mu\right)\psi \right\rangle, \quad \psi \in \mathbf{H},$$

where $\int_X f\, d\mu$ is as in Chap. 7. Thus, Q_f is a quadratic form for which the associated sesquilinear form is

$$L_f(\phi, \psi) = \left\langle \phi, \left(\int_X f\, d\mu \right) \psi \right\rangle, \quad \phi, \psi \in \mathbf{H}.$$

This form satisfies

$$|L_f(\phi, \psi)| \le \|\phi\| \left\| \left(\int_X f\, d\mu \right) \psi \right\|$$
$$= \|\phi\| \|f\|_{L^2(X, \mu_\psi)}, \tag{10.6}$$

for all $\phi, \psi \in \mathbf{H}$, where in the second line we have used (10.1).

If f is unbounded and ψ belongs to W_f, let $f_n = f 1_{E_n}$. Then $Q_f(\psi) = \lim_{n\to\infty} Q_{f_n}(\psi)$, by monotone convergence, in which case, it is easy to see that Q_f is still a quadratic form and that (10.6) still holds for all $\phi \in \mathbf{H}$. From (10.6), we see that for each $\psi \in W_f$, the conjugate-linear functional $\phi \mapsto L_f(\phi, \psi)$ is bounded. Thus, by (the complex-conjugate of) the Riesz theorem, there is a unique vector χ such that $L_f(\phi, \psi) = \langle \phi, \chi \rangle$. Furthermore, (10.6) tells us that $\|\chi\| \le \|f\|_{L^2(X, \mu_\psi)}$. Conversely, since $L_f(\phi, \psi) = \langle \phi, \chi \rangle$, (10.6) is an equality when $\phi = \chi$, showing that $\|\chi\| \ge \|f\|_{L^2(X, \mu_\psi)}$. Finally, the map $\psi \mapsto \chi$ is linear because $L_f(\phi, \psi)$ is linear in ψ. ∎

Proposition 10.3 *If f is a real-valued, measurable function on X, then $\int_X f\, d\mu$ is self-adjoint on W_f.*

Proof. Let $A_f = \int_X f\, d\mu$. Define subsets F_n of X by

$$F_n = \{x \in X \mid n - 1 \le |f(x)| < n\},$$

so that X is the disjoint union of the F_n's, and let $W^n = \mathrm{Range}(\mu(F_n))$. As in the proof of Proposition 10.2, any $\psi \in W^n$ is in W_f, and the quadratic form Q_f is bounded on W^n [compare (10.5)]. Furthermore, if $\phi \in (W^n)^\perp$ and $\psi \in W^n$, it is straightforward to check that $\mu_{\phi+\psi} = \mu_\phi + \mu_\psi$ and so

$$Q_f(\phi + \psi) = Q_f(\phi) + Q_f(\psi). \tag{10.7}$$

From (10.7), we obtain, by the polarization identity,

$$\langle \phi, A_f \psi \rangle = L_f(\phi, \psi) = 0.$$

This shows that $A_f \psi$ belongs to $(W^n)^{\perp\perp} = W^n$.

We conclude that A_f maps W^n boundedly to itself. Indeed, the restriction to W^n of A_f coincides with the restriction to W^n of the bounded operator obtained by integrating $f 1_{F_n}$ with respect to μ (compare the quadratic forms). Furthermore, since Q_f is real-valued, the restriction of A_f to W^n is self-adjoint (Proposition A.63).

Now, $\mathbf{H}$ is the orthogonal direct sum of the W^n's, meaning that $\mathbf{H}$ may be identified with the set of infinite sequences $(\psi_1, \psi_2, \psi_3, \ldots)$ with $\psi_n \in W^n$ and such that

$$\sum_{n=1}^{\infty} \|\psi_n\|^2 < \infty.$$

If A_n denotes the restriction of A_f to W^n, then under this decomposition of $\mathbf{H}$, we have

$$W_f = \left\{ \psi \in \mathbf{H} \,\middle|\, \sum_{n=1}^{\infty} \|A_n \psi_n\|^2 < \infty \right\}$$

$$= \left\{ \psi = (\psi_1, \psi_2, \ldots) \,\middle|\, \sum_{n=1}^{\infty} \left(\|\psi_n\|^2 + \|A_n \psi_n\|^2 \right) < \infty \right\}. \tag{10.8}$$

To verify (10.8), we note that

$$\int_X |f|^2 \, d\mu_\psi = \sum_{n=1}^{\infty} \int_{W_n} |f|^2 \, d\mu_\psi = \sum_{n=1}^{\infty} \|A_n \psi_n\|^2. \tag{10.9}$$

The first equality is by monotone convergence and the second holds because $\mu_\psi = \mu_{\psi_n}$ on W^n. In particular, the first quantity in (10.9) is finite if and only if the last quantity if finite.

By a similar argument, for $\psi \in W_f$, we have

$$Q_f(\psi) = \int_X f(\lambda) \, d\mu_\psi(\lambda) = \sum_{n=1}^{\infty} \langle \psi_n, A_n \psi_n \rangle,$$

from which it follows that

$$L_f(\phi, \psi) = \sum_{n=1}^{\infty} \langle \phi_n, A_n \psi_n \rangle$$

for all $\phi, \psi \in W_f$. From this we see that $A_f \psi$ is the vector represented by the sequence $(A_1 \psi_1, A_2 \psi_2, \ldots)$. It then follows from Example 9.26 that A_f is self-adjoint. ∎

Theorem 10.4 (Spectral Theorem, First Form) *Suppose A is a self-adjoint operator on $\mathbf{H}$. Then there is a unique projection-valued measure μ^A on $\sigma(A)$ with values in $\mathcal{B}(\mathbf{H})$ such that*

$$\int_{\sigma(A)} \lambda \, d\mu^A(\lambda) = A. \tag{10.10}$$

Since the spectrum of A is typically an unbounded set, the function $f(\lambda) = \lambda$ is an unbounded function on $\sigma(A)$. Note also that the equality in (10.10) includes, as always, equality of domains. That is, the domain of the integral on the left-hand side, namely the space W_f in Proposition 10.1, coincides with $\text{Dom}(A)$. The proof of this theorem is given in Sect. 10.4.

Definition 10.5 (Functional Calculus) *For any measurable function f on $\sigma(A)$, define a (possibly unbounded) operator, denoted $f(A)$, by*

$$f(A) = \int_{\sigma(A)} f(\lambda) \, d\mu^A(\lambda).$$

As usual, we can extend the projection-valued measure μ^A from $\sigma(A)$ to $\mathbb{R}$ by setting μ^A equal to zero on the complement of $\sigma(A)$.

Definition 10.6 (Spectral Subspaces) *If A is a self-adjoint operator on $\mathbf{H}$, then for any Borel set $E \subset \mathbb{R}$, define the **spectral subspace** V_E of $\mathbf{H}$ by*

$$V_E = \mathrm{Range}(\mu^A(E)).$$

Definition 10.7 (Measurement Probabilities) *If A is a self-adjoint operator on $\mathbf{H}$, then for any unit vector $\psi \in \mathbf{H}$, define a probability measure μ_ψ^A on $\mathbb{R}$ by the formula*

$$\mu_\psi^A(E) = \langle \psi, \mu^A(E)\psi \rangle.$$

If the operator A represents some observable in quantum mechanics, then we interpret μ_ψ^A to be the probability distribution for the result of measuring A in the state ψ.

Proposition 10.8 *Let A be a self-adjoint operator on $\mathbf{H}$. Then the spectral subspaces V_E associated to A have the following properties.*

1. *If E is a bounded subset of $\mathbb{R}$, then $V_E \subset \mathrm{Dom}(A)$, V_E is invariant under A, and the restriction of A to V_E is bounded.*

2. *If E is contained in $(\lambda_0 - \varepsilon, \lambda_0 + \varepsilon)$, then for all $\psi \in V_E$, we have*

$$\|(A - \lambda_0 I)\psi\| \leq \varepsilon \|\psi\|.$$

Proof. Point 1 holds because the function $f(\lambda) = \lambda$ is bounded on E. (See the proof of Proposition 10.3.) Point 2 then holds because, as in the proof of Proposition 10.3, the restriction of A to V_E coincides with the restriction to V_E of the operator $f(A)$, where $f(\lambda) = \lambda 1_E(\lambda)$. ∎

Theorem 10.9 (Spectral Theorem, Second Form) *Suppose A is a self-adjoint operator on $\mathbf{H}$. Then there is a σ-finite measure μ on $\sigma(A)$, a direct integral*

$$\int_{\sigma(A)}^{\oplus} \mathbf{H}_\lambda \, d\mu(\lambda),$$

and a unitary map U from $\mathbf{H}$ to the direct integral such that:

$$U(\mathrm{Dom}(A)) = \left\{ s \in \int_{\sigma(A)}^{\oplus} \mathbf{H}_\lambda \, d\mu(\lambda) \,\middle|\, \int_{\sigma(A)} \|\lambda s(\lambda)\|_\lambda^2 \, d\mu(\lambda) < \infty \right\}$$

and such that

$$\left(UAU^{-1}(s)\right)(\lambda) = \lambda s(\lambda)$$

for all $s \in U(\mathrm{Dom}(A))$.

Theorem 10.10 (Spectral Theorem, Multiplication Operator Form)
Suppose A is a self-adjoint operator on $\mathbf{H}$. Then there is a σ-finite measure space (X, μ), a measurable, real-valued function h on X, and a unitary map $U : \mathbf{H} \to L^2(X, \mu)$ such that

$$U(\mathrm{Dom}(A)) = \left\{ \psi \in L^2(X, \mu) \,\middle|\, h\psi \in L^2(X, \mu) \right\}$$

and such that

$$(UAU^{-1}(\psi))(x) = h(x)\psi(x)$$

for all $\psi \in U(\mathrm{Dom}(A))$.

These theorems are also proved in Sect. 10.4.

10.2   Stone's Theorem and One-Parameter Unitary Groups

In this section we explore the notion of one-parameter unitary groups and their connection to self-adjoint operators. We assume here the spectral theorem, the proof of which (in Sect. 10.4) does not use any results from this section.

Definition 10.11 *A **one-parameter unitary group** on $\mathbf{H}$ is a family $U(t)$, $t \in \mathbb{R}$, of unitary operators with the property that $U(0) = I$ and that $U(s+t) = U(s)U(t)$ for all $s, t \in \mathbb{R}$. A one-parameter unitary group is said to be **strongly continuous** if*

$$\lim_{s \to t} \|U(t)\psi - U(s)\psi\| = 0 \tag{10.11}$$

for all $\psi \in \mathbf{H}$ and all $t \in \mathbb{R}$.

Almost all one-parameter unitary groups arising in applications are strongly continuous.

Example 10.12 *Let $\mathbf{H} = L^2(\mathbb{R}^n)$ and let $U_{\mathbf{a}}(t)$ be the translation operator given by*

$$(U_{\mathbf{a}}(t)\psi)(\mathbf{x}) = \psi(\mathbf{x} + t\mathbf{a}). \tag{10.12}$$

Then $U(\cdot)$ is a strongly continuous one-parameter unitary group.

Proof. It is easy to see that $U_{\mathbf{a}}(\cdot)$ is a one-parameter unitary group. To see that $U_{\mathbf{a}}(\cdot)$ is strongly continuous, consider first the case in which ψ is continuous and compactly supported. Since a continuous function on a compact metric space is automatically uniformly continuous, it follows that $\psi(\mathbf{x}+t\mathbf{a})$ tends uniformly to $\psi(\mathbf{x})$ as t tends to zero. Since also the support of ψ is compact and thus of finite measure, it follows that $\psi(\mathbf{x}+t\mathbf{a})$ tends to $\psi(\mathbf{x})$ in $L^2(\mathbb{R}^n)$ as t tends to zero.

Now, the space $C_c(\mathbb{R}^n)$ of continuous functions of compact support is dense in $L^2(\mathbb{R}^n)$ (Theorem A.10). Thus, given $\varepsilon > 0$ and $\psi \in L^2(\mathbb{R}^n)$, we can find $\phi \in C_c(\mathbb{R}^n)$ such that $\|\psi - \phi\|_{L^2(\mathbb{R})} < \varepsilon/3$. Then choose δ so that $\|U_{\mathbf{a}}(a)\phi - \phi\| < \varepsilon/3$ whenever $|a| < \delta$. Then given $t \in \mathbb{R}$, if $|t - s| < \delta$, we have

$$\|U_{\mathbf{a}}(t)\psi - U_{\mathbf{a}}(s)\psi\|$$
$$\leq \|U_{\mathbf{a}}(t)\psi - U_{\mathbf{a}}(t)\phi\| + \|U_{\mathbf{a}}(t)\phi - U_{\mathbf{a}}(s)\phi\| + \|U_{\mathbf{a}}(s)\phi - U_{\mathbf{a}}(s)\psi\|$$
$$= \|U_{\mathbf{a}}(t)(\psi - \phi)\| + \|U_{\mathbf{a}}(s)(U_{\mathbf{a}}(t - s)\phi - \phi)\| + \|U_{\mathbf{a}}(s)(\phi - \psi)\|. \quad (10.13)$$

Since $U_{\mathbf{a}}(t)$ and $U_{\mathbf{a}}(s)$ are unitary, we can see that each of the terms on the last line of (10.13) is less than $\varepsilon/3$. ∎

Note that for $\mathbf{a} \neq 0$ the unitary group $U_{\mathbf{a}}(\cdot)$ in Example 10.12 is *not* continuous in the operator norm topology. After all, given any $\varepsilon \neq 0$, we can take a nonzero element ψ of $L^2(\mathbb{R}^n)$ that is supported in a very small ball around the origin. Then $U_{\mathbf{a}}(\varepsilon)\psi$ is orthogonal to ψ and has the same norm as ψ, so that

$$\|U_{\mathbf{a}}(\varepsilon)\psi - U_{\mathbf{a}}(0)\psi\| = \|U_{\mathbf{a}}(\varepsilon)\psi - \psi\| = \sqrt{2}\,\|\psi\|.$$

Thus, $\|U_{\mathbf{a}}(\varepsilon) - U_{\mathbf{a}}(0)\| \geq \sqrt{2}$ for all $\varepsilon \neq 0$.

Definition 10.13 *If $U(\cdot)$ is a strongly continuous one-parameter unitary group, the **infinitesimal generator** of $U(\cdot)$ is the operator A given by*

$$A\psi = \lim_{t \to 0} \frac{1}{i} \frac{U(t)\psi - \psi}{t}, \quad (10.14)$$

with $\mathrm{Dom}(A)$ consisting of the set of $\psi \in \mathbf{H}$ for which the limit in (10.14) exists in the norm topology on $\mathbf{H}$.

The following result shows that we can construct a strongly continuous one-parameter unitary group from any self-adjoint operator A by setting $U(t) = e^{iAt}$. Furthermore, the original operator A is precisely the infinitesimal generator of $U(t)$.

Proposition 10.14 *Suppose A is a self-adjoint operator on $\mathbf{H}$ and let $U(\cdot)$ be defined by*

$$U(t) = e^{itA},$$

where the operator e^{itA} is defined by the functional calculus for A. Then the following hold.

1. *$U(\cdot)$ is a strongly continuous one-parameter unitary group.*

2. *For all $\psi \in \mathrm{Dom}(A)$, we have*

$$A\psi = \lim_{t \to 0} \frac{1}{i} \frac{U(t)\psi - \psi}{t},$$

   where the limit is in the norm topology on **H**.

3. *For all $\psi \in$* **H**, *if the limit*

$$\lim_{t \to 0} \frac{1}{i} \frac{U(t)\psi - \psi}{t}$$

   exists in the norm topology on **H**, *then $\psi \in \mathrm{Dom}(A)$ and the limit is equal to $A\psi$.*

Proof. Since $\sigma(A) \subset \mathbb{R}$, the function $f(\lambda) := e^{it\lambda}$ is bounded on $\sigma(A)$ and satisfies $f(\lambda)\overline{f(\lambda)} = 1$ for all $\lambda \in \sigma(A)$. Thus, the operator $f(A)$ is bounded and satisfies

$$f(A)f(A)^* = f(A)^*f(A) = I,$$

which shows that $f(A) = e^{itA}$ is unitary. The multiplicativity of the functional calculus then tells us that $U(\cdot)$ is a one-parameter unitary group. To see that $U(t)$ is strongly continuous, note that

$$\|U(t)\psi - U(s)\psi\|^2 = \langle \psi, (U(t)^* - U(s)^*)(U(t) - U(s))\psi \rangle$$
$$= \int_{-\infty}^{\infty} \left| e^{it\lambda} - e^{is\lambda} \right|^2 d\mu_\psi^A(\lambda). \qquad (10.15)$$

The integral on the right-hand side of (10.15) tends to zero as s approaches t, by dominated convergence.

For Point 2, from recall from Theorem 10.4 that $A = \int_{-\infty}^{\infty} \lambda \, d\mu^A(\lambda)$, and take $\psi \in \mathrm{Dom}(A)$. Then, by (10.4), we have

$$\left\| \frac{1}{i} \frac{U(t)\psi - \psi}{t} - A\psi \right\|^2 = \int_{-\infty}^{\infty} \left| \frac{1}{i} \frac{e^{it\lambda} - 1}{t} - \lambda \right|^2 d\mu_\psi^A(\lambda). \qquad (10.16)$$

If we write the function $e^{it\lambda} - 1$ as the integral of its derivative with respect to λ, starting at $\lambda = 0$, we can see that $\left| (e^{it\lambda} - 1)/t \right| \leq \lambda$. Meanwhile, since ψ is in the domain of the operator $A = \int_{-\infty}^{\infty} \lambda \, d\mu^A(\lambda)$, we have $\int_{-\infty}^{\infty} \lambda^2 \, d\mu_\psi^A(\lambda) < \infty$. Thus, we may apply dominated convergence, with $4\lambda^2$ as our dominating function, to show that the right-hand side of (10.16) tends to zero as t tends to zero.

For Point 3, let B be the infinitesimal generator of $U(\cdot)$. If ϕ and ψ belong to $\mathrm{Dom}(B)$, then

$$
\begin{aligned}
\langle \phi, B\psi \rangle &= \lim_{t \to 0} \left\langle \phi, \frac{1}{i} \frac{U(t)\psi - \psi}{t} \right\rangle \\
&= \lim_{t \to 0} \left\langle -\frac{1}{i} \frac{U(t)^*\phi - \phi}{t}, \psi \right\rangle \\
&= \lim_{t \to 0} \left\langle \frac{1}{i} \frac{U(-t)\phi - \phi}{(-t)}, \psi \right\rangle \\
&= \langle B\phi, \psi \rangle .
\end{aligned}
$$

Thus, B is symmetric. On the other hand, Point 2 shows that B is an extension of A, so by Exercise 7 in Chap. 9, $B = A$ (with equality of domain). ∎

Theorem 10.15 (Stone's Theorem) *Suppose $U(\cdot)$ is a strongly continuous one-parameter unitary group on $\mathbf{H}$. Then the infinitesimal generator A of $U(\cdot)$ is densely defined and self-adjoint, and $U(t) = e^{itA}$ for all $t \in \mathbb{R}$.*

If $U(\cdot)$ is a strongly continuous one-parameter unitary group, then $U(\cdot)$ is continuous in the operator norm topology if and only if the infinitesimal generator of $U(\cdot)$ is a bounded operator (Exercise 1). As Example 10.12 suggests, most one-parameter unitary groups that arise in applications are *not* continuous in the operator norm topology.

Before giving the proof of Stone's theorem, let us work out the generator of the group in Example 10.12.

Example 10.16 *If $U_{\mathbf{a}}(\cdot)$, $\mathbf{a} \in \mathbb{R}^n$, is the strongly continuous one-parameter unitary group in Example 10.12, then each $\psi \in C_c^\infty(\mathbb{R}^n)$ is in the domain of the infinitesimal generator A of $U_{\mathbf{a}}(\cdot)$ and for all such ψ, we have*

$$
A\psi = -i \sum_j a_j \frac{\partial \psi}{\partial x_j}. \tag{10.17}
$$

Furthermore, A is essentially self-adjoint on $C_c^\infty(\mathbb{R}^n)$.

Proof. The formula for the infinitesimal generator is easy to establish for ψ in $C_c^\infty(\mathbb{R}^n)$. The essential self-adjointness of A is a special case of Proposition 13.5 (the proof of which is similar to the proof of Proposition 9.29). ∎

We now establish two intermediate results before coming to the proof of Stone's theorem.

Lemma 10.17 *Let $U(\cdot)$ be a strongly continuous one-parameter unitary group and let A be its infinitesimal generator. If $\psi \in \mathrm{Dom}(A)$, then for all $t \in \mathbb{R}$, the vector $U(t)\psi$ belongs to $\mathrm{Dom}(A)$ and*

$$
\lim_{h \to 0} \frac{U(t+h)\psi - U(t)\psi}{h} = iU(t)A\psi = iAU(t)\psi. \tag{10.18}
$$

Note that Lemma 10.17 tells us that the curve $\psi(t) := U(t)\psi_0$ in **H** satisfies the differential equation

$$\frac{d\psi}{dt} = iA\psi(t)$$

in the natural Hilbert space sense, provided that ψ_0 belongs to $\mathrm{Dom}(A)$. This result, together with Proposition 10.14, tells us that if $\psi_0 \in \mathrm{Dom}(\hat{H})$, then the curve $\psi(t) := e^{-it\hat{H}/\hbar}\psi_0$ indeed solves the Schrödinger equation in the Hilbert space sense.

Proof. We compute that

$$\frac{U(t+h)\psi - U(t)\psi}{h} = U(t)\frac{[U(h)\psi - \psi]}{h}. \tag{10.19}$$

Since $\psi \in \mathrm{Dom}(A)$, the limit as h tends to zero of (10.19) exists and is equal to $iU(t)A\psi$. On the other hand,

$$\frac{U(t+h)\psi - U(t)\psi}{h} = \frac{U(h)(U(t)\psi) - (U(t)\psi)}{h}.$$

Thus, the limit as h tends to zero of (10.19) is, by the definition of A, equal to $iA(U(t)\psi)$. This shows that $U(t)\psi$ is in the domain of A and establishes the second equality in (10.18). ■

Lemma 10.18 *For any strongly continuous one-parameter unitary group $U(\cdot)$, the infinitesimal generator A is densely defined.*

Proof. Given any continuous function f of compact support, define an operator B_f by setting

$$B_f = \int_{-\infty}^{\infty} f(\tau)U(\tau)\, d\tau.$$

Here, the operator-valued integral is the unique bounded operator such that

$$\langle \phi, B_f\psi \rangle = \int_{-\infty}^{\infty} f(\tau)\langle \phi, U(\tau)\psi \rangle\, d\tau. \tag{10.20}$$

[It is easy to see that right-hand side of (10.20) defines a bounded sesquilinear form, for each fixed $f \in C_c^{\infty}(\mathbb{R})$.]

Using the group property of $U(\cdot)$, we see that

$$U(t)B_f\psi - B_f\psi = \int_{-\infty}^{\infty} [f(\tau)U(\tau+t)\psi - f(\tau)U(\tau)\psi]\, d\tau$$

$$= \int_{-\infty}^{\infty} [f(\tau-t) - f(\tau)]U(\tau)\psi\, d\tau,$$

where in the second line, we have made a change of variable in the first term in the integral. From this, we easily obtain that

$$\lim_{t \to 0} \frac{U(t)B_f\psi - B_f\psi}{t} = -\int_{-\infty}^{\infty} f'(\tau)U(\tau)\psi \, d\tau.$$

This shows that $B_f\psi$ is in the domain of A for all $\psi \in \mathbf{H}$ and $f \in C_c^{\infty}(\mathbb{R})$.

Now choose a sequence $f_n \in C_c^{\infty}(\mathbb{R})$ such that f_n is non-negative and supported in the interval $[-1/n, 1/n]$ and such that $\int_{-\infty}^{\infty} f_n(\tau) \, d\tau = 1$. Then for any $\psi \in \mathbf{H}$, we have

$$B_{f_n}\psi - \psi = \int_{-\infty}^{\infty} f_n(\tau)[U_n(\tau)\psi - \psi] \, d\tau,$$

so that

$$\|B_{f_n}\psi - \psi\| \le \int_{-\infty}^{\infty} f_n(\tau) \|U(\tau)\psi - \psi\| \, d\tau$$

$$\le \sup_{-1/n \le \tau \le 1/n} \|U(\tau)\psi - \psi\|.$$

Since $U(\cdot)$ is strongly continuous, we see that $B_{f_n}\psi$ converges to ψ as $n \to \infty$. Thus, every element of $\mathbf{H}$ can be approximated by vectors in the domain of A. ∎

Proof of Theorem 10.15. Suppose $U(\cdot)$ is a strongly continuous one-parameter unitary group and A is its infinitesimal generator. By Lemma 10.18, A is densely defined. As shown in the proof of Proposition 10.14, A (denoted by B in that proof) is symmetric.

Next, we show that A is essentially self-adjoint. Suppose now that ψ belongs to the kernel of $A^* - iI$, i.e., $A^*\psi = i\psi$. Given $\phi \in \text{Dom}(A)$, set $y(t) = \langle U(t)\phi, \psi \rangle$, so that $|y(t)| \le \|\phi\| \|\psi\|$. On the other hand, we expect that $U(t) = e^{iAt}$, so that $U(t)^*$ should be e^{-iA^*t}. Thus, $y(t)$ should (formally) be equal to $\langle \phi, e^t\psi \rangle$. If this is correct, then since $y(t)$ is a bounded function of t, we must have $\langle \phi, \psi \rangle = 0$. Thus, ψ would be orthogonal to every element of a dense subspace of $\mathbf{H}$, showing that $\psi = 0$. We could then similarly argue that $\ker(A^* + iI) = \{0\}$, which would show that A is essentially self-adjoint.

To make the argument rigorous, we apply Lemma 10.17, giving

$$\frac{d}{dt} \langle U(t)\phi, \psi \rangle = \langle iAU(t)\phi, \psi \rangle = \langle iU(t)\phi, A^*\psi \rangle$$

$$= \langle iU(t)\phi, i\psi \rangle = \langle U(t)\phi, \psi \rangle.$$

Thus, the function $y(t) := \langle U(t)\phi, \psi \rangle$ satisfies the ordinary differential equation $dy/dt = y$. The unique solution to this equation is $y(t) = y(0)e^t$. Since y is bounded, we must have $0 = y(0) = \langle \phi, \psi \rangle$ for all $\phi \in \text{Dom}(A)$, which implies that $\psi = 0$. Thus, $\ker(A^* - iI) = \{0\}$, and by a similar argument $\ker(A^* + iI) = \{0\}$. This shows (Corollary 9.22) that A is essentially self-adjoint.

We can now construct a strongly continuous unitary group $V(\cdot)$ by setting $V(t) = e^{iA^{cl}t}$. To show that $V(\cdot) = U(\cdot)$, take $\psi \in \mathrm{Dom}(A) \subset \mathrm{Dom}(A^{cl})$ and set $w(t) = U(t)\psi - V(t)\psi$. By Proposition 10.14, the infinitesimal generator of $V(\cdot)$ is A^{cl}. Thus, applying Lemma 10.17 to both $U(\cdot)$ and $V(\cdot)$, we have

$$\frac{d}{dt}w(t) = iAU(t)\psi - iAV(t)\psi$$
$$= iAw(t),$$

where the limit defining dw/dt is taken in the norm topology on $\mathbf{H}$. Thus,

$$\frac{d}{dt}\|w(t)\|^2 = \langle iAw(t), w(t)\rangle + \langle w(t), iAw(t)\rangle$$
$$= -i\langle Aw(t), w(t)\rangle + i\langle w(t), Aw(t)\rangle$$
$$= 0,$$

because A is symmetric. Since also $w(0) = 0$, we conclude that $w(t) = 0$ for all t. Thus, $U(\cdot)$ and $V(\cdot)$ agree on a dense subspace and hence on all of $\mathbf{H}$.

We now know that $U(t) = e^{iA^{cl}t}$. It then follows from Points 2 and 3 of Proposition 10.14 that the infinitesimal generator of $U(\cdot)$ (namely A) is precisely A^{cl}. That is, $A = A^{cl}$ and $U(t) = e^{iAt}$. Furthermore, we have already shown that A is essentially self-adjoint and we now know that $A = A^{cl}$, so A is actually self-adjoint. Finally, if B is any self-adjoint operator for which $U(t) = e^{iBt}$, then by Proposition 10.14, B must be the infinitesimal generator of $U(\cdot)$, i.e., $B = A$. ∎

10.3   The Spectral Theorem for Bounded Normal Operators

We are going to prove the spectral theorem for an unbounded self-adjoint operator by reducing it to the spectral theorem for a bounded operator. The reduction, however, will not be to a bounded *self-adjoint* operator, but rather to a *unitary* operator. Although we proved the spectral theorem only for bounded self-adjoint operators, the theorem applies more generally to bounded *normal* operators. (See Exercise 4 in Chap. 7 for the matrix case.)

Definition 10.19 *A bounded operator A on $\mathbf{H}$ is **normal** if A commutes with its adjoint: $AA^* = A^*A$.*

Every bounded self-adjoint operator is obviously normal. Other examples of normal operators are skew-self-adjoint operators ($A^* = -A$) and unitary operators $(UU^* = U^*U = I)$. The spectrum of a bounded normal operator need not be contained in $\mathbb{R}$, but can be an arbitrary closed, bounded, nonempty subset of $\mathbb{C}$. On the other hand, if U is unitary, then the spectrum of U is contained in the unit circle (Exercise 6 in Chap. 7).

In this section, we consider the spectral theorem for a bounded normal operator A. The statements of the two versions of the theorem are precisely the same as in the self-adjoint case, except that $\sigma(A)$ is no longer necessarily contained in the real line. Almost all of the proofs of these results are the same as in the self-adjoint case; we will, therefore, consider only those steps where some modification in the argument is required.

Theorem 10.20 *Suppose $A \in \mathcal{B}(\mathbf{H})$ is normal. Then there exists a unique projection-valued measure μ^A on the Borel σ-algebra in $\sigma(A)$, with values in $\mathcal{B}(\mathbf{H})$, such that*

$$\int_{\sigma(A)} \lambda \, d\mu^A(\lambda) = A.$$

Furthermore, for any measurable set $E \subset \sigma(A)$, $\mathrm{Range}(\mu^A(E))$ is invariant under A and A^*.

Once we have the projection-valued measure μ^A, we can define a *functional calculus* for A, as in the self-adjoint case, by setting

$$f(A) = \int_{\sigma(A)} f(\lambda) \, d\mu^A(\lambda)$$

for any bounded measurable function f on $\sigma(A)$.

We can also define *spectral subspaces*, as in the self-adjoint case, by setting

$$V_E := \mathrm{Range}(\mu^A(E))$$

for each Borel set $E \subset \sigma(A)$. These spectral subspaces have precisely the same properties (with the same proofs) as in Proposition 7.15, with the following two exceptions. First, the assertion that V_E is invariant under A should be replaced by the assertion that V_E is invariant under A and A^*. Second, in Point 2 of the proposition, the condition $E \subset [\lambda_0 - \varepsilon, \lambda_0 + \varepsilon]$ should be replaced by $E \subset \overline{D}(\lambda_0, \varepsilon)$, where $D(z, r)$ denotes the disk of radius r in $\mathbb{C}$ centered at z.

Meanwhile, the spectral theorem in its direct integral and multiplication operator versions also holds for a bounded normal operator A. The statements are identical to the self-adjoint case, except that we no longer assume $\sigma(A) \subset \mathbb{R}$ and we no longer assume that the function h in the multiplication operator version is real valued.

Let us recall the two stages in the proof of the spectral theorem (first version) for bounded self-adjoint operators. The first stage is the construction of the continuous functional calculus. The steps in this construction are (1) the equality of the norm and spectral radius for self-adjoint operators,

(2) the spectral mapping theorem, and (3) the Stone–Weierstrass theorem. The second stage is a sort of operator-valued Riesz representation theorem, which we prove by reducing it to the ordinary Riesz representation theorem using quadratic forms. In generalizing from bounded self-adjoint to bounded normal operators, the second stage of the proof is precisely the same as in the self-adjoint case. In the first stage, however, there are some additional ideas needed in each step of the argument.

There is a relatively simple argument that reduces the equality of norm and spectral radius for normal operators to the self-adjoint case. Meanwhile, since the spectral mapping theorem, as stated in Chap. 8, already holds for arbitrary bounded operators, it *appears* that no change is needed in this step. We must think, however, about the proper notion of "polynomial." For a general normal operator A, the spectrum of A is not contained in $\mathbb{R}$, and, thus, powers of λ are complex-valued functions on $\sigma(A)$. We must, therefore, use the complex-valued version of the Stone–Weierstrass theorem (Appendix A.3.1), which requires that our algebra of functions be closed under complex-conjugation. This means that we need to consider polynomials in λ *and* $\bar{\lambda}$, that is, linear combinations of functions of the form $\lambda^m \bar{\lambda}^n$.

What we need, then, is a form of the spectral mapping theorem that applies to this sort of polynomial. On the operator side, the natural counterpart to the complex conjugate of a function is the adjoint of an operator. Thus, applying the function $\lambda^m \bar{\lambda}^n$ to a normal operator A should give $A^m (A^*)^n$. The desired "spectral mapping theorem" is then the following: If p is a polynomial in two variables, and A is a bounded normal operator, then

$$\sigma(p(A, A^*)) = \left\{ p(\lambda, \bar{\lambda}) \,\middle|\, \lambda \in \sigma(A) \right\}. \tag{10.21}$$

This statement is true (Theorem 10.23), but its proof is not nearly as simple as the proof of the ordinary spectral mapping theorem. One way to prove (10.21) is to use the theory of commutative C^*-algebras, as in [33]. (See Theorem 11.19 in [33] along with the assertion on p. 321 that the spectrum of an element is independent of the algebra containing that element.) Another approach is the direct argument found in Bernau [3], which uses no fancy machinery but which is long and not easily motivated. A third approach is to use the spectral theorem for bounded self-adjoint operators to help us prove (10.21); this is the approach we will follow.

We begin with the equality of norm and spectral radius and then turn to (10.21).

Proposition 10.21 *If $A \in \mathcal{B}(\mathbf{H})$ is normal, then*

$$\|A\| = R(A).$$

Lemma 10.22 *If A and B are commuting elements of $\mathcal{B}(\mathbf{H})$, then*

$$R(AB) \le R(A)R(B).$$

Proof. If A is any bounded operator, the proof of Lemma 8.1 shows that for any real number T with $T > R(A)$, we have

$$\lim_{m \to \infty} \frac{\|A^m\|}{T^m} = 0.$$

If A and B are two commuting bounded operators and S and T are two real numbers, with $S > R(A)$ and $T > R(B)$, then

$$\frac{\|(AB)^m\|}{S^m T^m} = \frac{\|A^m B^m\|}{S^m T^m} \leq \frac{\|A^m\| \|B^m\|}{S^m T^m}.$$

Thus,

$$\lim_{m \to \infty} \frac{\|(AB)^m\|}{S^m T^m} = 0. \tag{10.22}$$

Meanwhile, if we apply the expression for the resolvent in the proof of Lemma 8.1 to AB, we obtain

$$(AB - \lambda)^{-1} = -\sum_{m=0}^{\infty} \frac{A^m B^m}{\lambda^{m+1}}, \tag{10.23}$$

since A and B commute. For any λ_1 with $|\lambda_1| > R(A)R(B)$, take λ_2 with $|\lambda_1| > |\lambda_2| > R(A)R(B)$. The terms in (10.23) with $\lambda = \lambda_2$ tend to zero by (10.22), which means that (10.23) converges with $\lambda = \lambda_1$. Thus, λ_1 is in the resolvent set of AB. ∎

Proof of Proposition 10.21. For any bounded operator, $\|A\| \geq R(A)$ (Proposition 7.5). To get the inequality in the other direction, recall (Proposition 7.2) that $\|A\|^2 = \|A^*A\|$. Note also that A^*A is self-adjoint, since its adjoint is $A^*A^{**} = A^*A$. Thus, if A and A^* commute, we have

$$\|A\|^2 = \|A^*A\| = R(A^*A) \leq R(A^*)R(A)$$
$$\leq \|A^*\| R(A) = \|A\| R(A).$$

Here we have used Lemmas 8.1 and 10.22 and the general inequality between norm and spectral radius. Dividing by $\|A\|$ gives $\|A\| \leq R(A)$, unless $\|A\| = 0$, in which case the desired inequality is trivially satisfied. ∎

Theorem 10.23 *If $A \in \mathcal{B}(\mathbf{H})$ is normal, then for any polynomial p in two variables, we have*

$$\sigma\left(p(A, A^*)\right) = \left\{ p(\lambda, \bar{\lambda}) \big| \lambda \in \sigma(A) \right\}.$$

If, for example, $p(\lambda, \bar{\lambda}) = \lambda^2 \bar{\lambda}^3$, then $p(A, A^*) = A^2 (A^*)^3$. Note that since A and A^* are assumed to commute, the map sending the polynomial $p(\lambda, \bar{\lambda})$ to $p(A, A^*)$ is an algebra homomorphism. That is to say, $(pq)(A, A^*) = p(A, A^*)q(A, A^*)$. This would not be the case if A did not commute with A^*.

We begin by proving Theorem 10.23 in the case that A is a normal *matrix*. Although the matrix case is quite simple, it provides an outline for our assault on the general result.

Proof of Theorem 10.23 in the Matrix Case. For matrices, the spectrum is nothing but the set of eigenvalues. If A commutes with A^*, then for any $\lambda \in \mathbb{C}$,

$$\langle (A^* - \bar{\lambda}I)\psi, (A^* - \bar{\lambda}I)\psi \rangle = \langle \psi, (A - \lambda I)(A^* - \bar{\lambda}I)\psi \rangle$$
$$= \langle \psi, (A^* - \bar{\lambda}I)(A - \lambda I)\psi \rangle$$
$$= \langle (A - \lambda I)\psi, (A - \lambda I)\psi \rangle \qquad (10.24)$$

Thus, if ψ is an eigenvalue for A with eigenvalue λ, ψ is automatically an eigenvalue for A^* with eigenvalue $\bar{\lambda}$. It then easily follows that ψ is an eigenvector for $p(A, A^*)$ with eigenvalue $p(\lambda, \bar{\lambda})$.

In the other direction, suppose μ is an eigenvalue for $p(A, A^*)$ and let W denote the μ-eigenspace for $p(A, A^*)$. Since A and A^* commute with each other, they also commute with $p(A, A^*)$. Thus, A and A^* preserve W, as is easily verified, and the operator $A|_W$ will have some eigenvector ψ with eigenvalue λ. Since $A\psi = \lambda\psi$, then, as in (10.24), $A^*\psi = \bar{\lambda}\psi$ and so

$$p(A, A^*)\psi = p(\lambda, \bar{\lambda})\psi.$$

Since also $p(A, A^*)\psi = \mu\psi$, by assumption, we have $\mu = p(\lambda, \bar{\lambda})$, where λ is an eigenvalue for A. ∎

We now attempt to run the same argument for a bounded normal operator on $\mathbf{H}$, replacing "eigenvector" with "almost eigenvector," where ψ is an ε-almost eigenvector for ψ if $\|(A - \lambda I)\psi\|$ is less than $\varepsilon \|\psi\|$. The main difficulty with this approach is that for a given eigenvalue λ, the set of ε-almost eigenvectors is not a vector space. To surmount this difficulty, we will use the spectral theorem for the self-adjoint operator B^*B, where $B = p(A, A^*) - \mu I$, with $\mu \in \sigma(p(A, A^*))$. We will construct a spectral subspace W for B^*B such that W is invariant under A and A^* and such that each element of W is an ε-almost eigenvector for $p(A, A^*)$ with eigenvalue μ. (Note, however, that we are not claiming that W contains *all* the ε-almost eigenvectors for $p(A, A^*)$.)

Definition 10.24 *If $A \in \mathcal{B}(\mathbf{H})$, then an ε-almost eigenvector for A with eigenvalue $\lambda \in \mathbb{C}$ is a nonzero vector $\psi \in \mathbf{H}$ such that*

$$\|(A - \lambda I)\psi\| < \varepsilon \|\psi\|.$$

We now establish three lemmas about almost eigenvectors, the last of which makes use of the spectral theorem for bounded self-adjoint operators. With these lemmas in hand, we will have a clear path to imitate the proof of the matrix case of Theorem 10.23.

Lemma 10.25 *Suppose $A \in \mathcal{B}(\mathbf{H})$ is normal.*

1. *If ψ is an ε-almost eigenvector for A with eigenvalue λ, then ψ is an ε-almost eigenvector for A^* with eigenvalue $\bar{\lambda}$.*

2. *A number $\lambda \in \mathbb{C}$ belongs to $\sigma(A)$ if and only if for all $\varepsilon > 0$, there exists an ε-almost eigenvector with eigenvalue λ.*

Proof. Point 1 follows immediately from (10.24), which holds for bounded normal operators, not just matrices. For Point 2, suppose that an ε-almost eigenvector with eigenvalue λ exists for all $\varepsilon > 0$. Then $A - \lambda I$ cannot have a *bounded* inverse, and so $\lambda \in \sigma(A)$. In the other direction, if there is some $\varepsilon > 0$ for which no ε-almost eigenvector exists, then

$$\|(A - \lambda I)\psi\| \geq \varepsilon \|\psi\| \tag{10.25}$$

for all $\psi \in \mathbf{H}$, showing that $A - \lambda I$ is injective. By (10.24), the same inequality hods with $A - \lambda I$ replaced by $A^* - \bar{\lambda} I$. Thus, $A^* - \bar{\lambda} I$ is injective, so by Proposition 7.3, the range of $A - \lambda I$ is dense in $\mathbf{H}$. Using (10.25) as in the proof of Proposition 7.7, it is easily seen that the range of $A - \lambda I$ is also closed, hence all of $\mathbf{H}$. Thus, $(A - \lambda I)$ is invertible and the inverse is bounded, by (10.25). ∎

Lemma 10.26 *Suppose $A \in \mathcal{B}(\mathbf{H})$ is normal. Then for each polynomial p in two variables and each number $\lambda \in \mathbb{C}$, there is a constant C such that if ψ is an ε-almost eigenvector for A with eigenvalue λ, then ψ is a $(C\varepsilon)$-almost eigenvector for $p(A, A^*)$ with eigenvalue $p(\lambda, \bar{\lambda})$.*

Proof. We decompose $p(A, A^*) - p(\lambda, \bar{\lambda})I$ into a linear combination of terms of the form $A^k (A^*)^l - \lambda^k \bar{\lambda}^l$ and we estimate such terms by induction on $k + l$. If $k = 1$ and $l = 0$, there is nothing to prove, and if $k = 0$ and $l = 1$, we use (10.24). Assume now that we have established the desired result for $k + l = N$ and consider a case with $k + l = N + 1$. If $k > 0$, we write

$$\left(A^k (A^*)^l - \lambda^k \bar{\lambda}^l\right) \psi = A^{k-1}(A^*)^l (A - \lambda I) \psi$$
$$+ \lambda \left(A^{k-1}(A^*)^l - \lambda^{k-1} \bar{\lambda}^l I\right) \psi. \tag{10.26}$$

Since ψ is an ε-almost eigenvector and A and A^* are bounded, the norm of the first term on the right-hand side of (10.26) is at most $c_1 \varepsilon$. By induction, the norm of the second term on the right-hand side of (10.26) is at most $|\lambda| c_2 \varepsilon$. Thus, the norm of the left-hand side of (10.26) is at most $(c_1 + |\lambda| c_2)\varepsilon$. A similar analysis holds if $k = 0$, in which case $l > 0$. ∎

Lemma 10.27 *Let $A \in \mathcal{B}(\mathbf{H})$ be normal, let p be a polynomial in two variables, and let μ be an element of the spectrum of $p(A, A^*)$. Then for all $\varepsilon > 0$, there exists a nonzero closed subspace W^ε of $\mathbf{H}$ such that W^ε is invariant under A and A^* and such that every nonzero element of W^ε is an ε-almost eigenvector for $p(A, A^*)$ with eigenvalue μ.*

Proof. Fix some μ in the spectrum of $p(A, A^*)$ and let $B = p(A, A^*) - \mu I$. Then B is normal and 0 belongs to the spectrum of B. Using Point 2 of Lemma 10.25 and Lemma 10.26, we see that 0 belongs to the spectrum of the self-adjoint operator B^*B. We apply the spectral theorem to B^*B and we let W^ε be the spectral subspace for B^*B corresponding to the interval $(-\varepsilon^2, \varepsilon^2)$. By Proposition 7.15, W^ε is nonzero and invariant under B^*B, and the restriction of B^*B to W^ε has norm at most ε^2. Thus, for all $\psi \in W^\varepsilon$ we have

$$\langle B\psi, B\psi \rangle = \langle \psi, B^*B\psi \rangle \le \|\psi\| \, \|B^*B\psi\| \le \varepsilon^2 \|\psi\|^2 .$$

Since $B = p(A, A^*) - \mu I$, this shows that every nonzero element of W^ε is an ε-almost eigenvector for $p(A, A^*)$ with eigenvalue μ. Furthermore, A and A^* commute with B^*B and thus they preserve each spectral subspace of B^*B (Proposition 7.16) including W^ε. ∎

Proof of Theorem 10.23. Suppose first that λ belongs to the spectrum of A. By Point 2 of Lemma 10.25, A has ε-almost eigenvalues with eigenvalue λ for every $\varepsilon > 0$. Lemma 10.26 then shows that $p(A, A^*)$ has $(C\varepsilon)$-almost eigenvectors with eigenvalue $p(\lambda, \bar{\lambda})$ for every $\varepsilon > 0$, which shows that $p(\lambda, \bar{\lambda})$ is in the spectrum of $p(A, A^*)$.

In the other direction, suppose that μ is in the spectrum of $p(A, A^*)$. For any $\varepsilon > 0$, we consider the nonzero subspace W^ε in Lemma 10.27, which is invariant under A and A^*. The restriction of A to W^ε is again a normal operator (Exercise 8), and $A|_{W^\varepsilon}$ has nonempty spectrum (Proposition 7.5). If we fix some $\lambda \in \sigma(A|_{W^\varepsilon})$, Lemma 10.25 tells us that there exists an ε-almost eigenvector ψ for A in W^ε. By Lemma 10.26, ψ is a $(C\varepsilon)$-almost eigenvector for $p(A, A^*)$ with eigenvalue $p(\lambda, \bar{\lambda})$. Meanwhile, since $\psi \in W^\varepsilon$, the same vector ψ is also an ε-almost eigenvector for $p(A, A^*)$ with eigenvalue μ. It then is easy to see (Exercise 10) that

$$\left| \mu - p(\lambda, \bar{\lambda}) \right| < C\varepsilon + \varepsilon. \tag{10.27}$$

Since (10.27) holds for all $\varepsilon > 0$, we can find a sequence λ_n of points in $\sigma(A)$ such that $p(\lambda_n, \bar{\lambda}_n) \to \mu$. Since $\sigma(A)$ is compact, we can pass to a subsequence of the λ_n's that is convergent to some $\lambda \in \sigma(A)$, and this λ will satisfy $p(\lambda, \bar{\lambda}) = \mu$. ∎

Combining Theorem 10.23 with the equality of the norm and spectral radius for normal operators (Proposition 10.21), we have the following result. If $A \in \mathcal{B}(\mathbf{H})$ is normal and p is a polynomial in two variables, then

$$\|p(A, A^*)\| = \sup_{\lambda \in \sigma(A)} \left| p(\lambda, \bar{\lambda}) \right| .$$

The map $p \mapsto p(A, A^*)$ has the property that $\bar{p}(A, A^*) = (p(A, A^*))^*$, where the polynomial $\bar{p}$ is the complex-conjugate of p. In particular, if p takes only real values on $\sigma(A)$, then $p(A, A^*)$ is self-adjoint.

By the complex-valued version of the Stone–Weierstrass theorem (A.12), polynomials in λ and $\bar{\lambda}$ are dense in $\mathcal{C}(\sigma(A); \mathbb{C})$, the space of continuous complex-valued functions on $\sigma(A)$. Thus, the BLT theorem (Theorem A.36) tells that we can extend the map $p \mapsto p(A, A^*)$ to an isometric map of $\mathcal{C}(\sigma(A); \mathbb{C})$ into $\mathcal{B}(\mathbf{H})$. This extension, which we call the *continuous functional calculus* for A, has all the same properties as in the self-adjoint case.

Now that the continuous functional calculus for normal operators has been established, the proof of the spectral theorem—in any of its various versions—proceeds exactly as in the self-adjoint case. There is no need, then, to repeat the arguments given in Chap. 8.

10.4 Proof of the Spectral Theorem for Unbounded Self-Adjoint Operators

To prove the spectral theorem for an unbounded self-adjoint operator A, we will construct from A a certain unitary (and thus normal) operator U. We then apply the spectral theorem for bounded normal operators to U and translate this result into the desired result for A. To motivate the construction of U, consider the function

$$C(x) := \frac{x+i}{x-i}, \quad x \in \mathbb{R}. \tag{10.28}$$

It is a simple matter to check that C maps $\mathbb{R}$ injectively onto $S^1 \backslash \{1\}$, with inverse given by

$$D(u) := i\frac{u+1}{u-1}, \quad u \in S^1 \backslash \{1\}. \tag{10.29}$$

Furthermore, we have $\lim_{x \to \pm\infty} C(x) = 1$. The function $C(x)$ in (10.28) is the simplest bounded, injective function one can define on $\mathbb{R}$.

We wish to apply the map C to a self-adjoint operator A. If A is bounded and self-adjoint, it is straightforward to check that the operator $(A+iI)(A-iI)^{-1}$ is unitary (Exercise 5). Even in the unbounded case, it is possible to make sense of the operator $U := C(A)$, and we can recover A from U, by (essentially) applying D. The operator U is unitary and is known as the *Cayley transform* of A.

Recall that if A is self-adjoint, then i is in the resolvent set of A and the operator $(A - iI)^{-1}$ maps $\mathbf{H}$ into $\mathrm{Dom}(A)$.

Theorem 10.28 (Cayley Transform) *If A is a self-adjoint operator on* $\mathbf{H}$*, let U be the operator defined by*

$$U\psi = (A + iI)(A - iI)^{-1}\psi.$$

Then the following results hold.

1. *The operator U is a unitary operator on* **H**.

2. *The operator $U - I$ is injective.*

3. *The range of the operator $U - I$ is equal to* $\mathrm{Dom}(A)$ *and for all* $\psi \in$ $\mathrm{Range}(U - I)$ *we have*

$$A\psi = i(U + I)(U - I)^{-1}\psi. \tag{10.30}$$

According to Point 2, $U - I$ is injective, while according to Point 3, the range of $U - I$ is $\mathrm{Dom}(A)$. Thus, in (10.30), the expression $(U - I)^{-1}$ refers to the inverse of the one-to-one and onto map $U - I : \mathbf{H} \to \mathrm{Dom}(A)$. We are *not* claiming that 1 is in the resolvent set of U. That is to say, $(U - I)^{-1}$ is not a bounded operator, unless $\mathrm{Dom}(A) = \mathbf{H}$, which occurs only if A is bounded.

Proof. The resolvent operator $(A - iI)^{-1}$ must be injective, because

$$(A - iI)(A - iI)^{-1}\psi = \psi$$

for all $\psi \in \mathbf{H}$. Furthermore, $(A - iI)^{-1}$ maps **H** *onto* $\mathrm{Dom}(A)$, because

$$\psi = (A - iI)^{-1}(A - iI)\psi$$

for all $\psi \in \mathrm{Dom}(A)$. Since $-i$ is also in the resolvent set of A, similar reasoning shows that $A + iI$ maps $\mathrm{Dom}(A)$ injectively onto **H**. Thus, U is the composition of one operator that maps **H** injectively onto $\mathrm{Dom}(A)$ and another operator that maps $\mathrm{Dom}(A)$ injectively onto **H**, so that U maps **H** injectively onto **H**.

Now, for any $\phi \in \mathrm{Dom}(A)$ we have

$$\langle (A + iI)\phi, (A + iI)\phi \rangle = \langle A\phi, A\phi \rangle + \langle \phi, \phi \rangle$$
$$= \langle (A - iI)\phi, (A - iI)\phi \rangle ,$$

because of a familiar cancellation of cross terms. Thus, applying this with $\phi = (A - iI)^{-1}\psi$ shows that for any $\psi \in \mathbf{H}$, we have

$$\langle (A + iI)(A - iI)^{-1}\psi, (A + iI)(A - iI)^{-1}\psi \rangle$$
$$= \langle (A - iI)(A - iI)^{-1}\psi, (A - iI)(A - iI)^{-1}\psi \rangle$$
$$= \langle \psi, \psi \rangle .$$

Thus, U is one-to-one and onto and preserves norms and is therefore unitary.

For Point 2, observe that for any $\psi \in \mathbf{H}$, we have

$$(A + iI)(A - iI)^{-1}\psi = ((A - iI) + 2iI)(A - iI)^{-1}\psi$$
$$= \psi + 2i(A - iI)^{-1}\psi. \tag{10.31}$$

Thus, since $(A - iI)^{-1}$ is injective, we cannot have $U\psi = \psi$ unless $\psi = 0$.
Finally, for Point 3, (10.31) says that

$$U - I = 2i(A - iI)^{-1}, \tag{10.32}$$

which means (by the reasoning at the start of the proof) that the range of $U - I$ is $\mathrm{Dom}(A)$. For $\psi \in \mathrm{Dom}(A)$, we then have

$$
\begin{aligned}
(U + I)(U - I)^{-1}\psi &= \frac{1}{2i}(U + I)(A - iI)\psi \\
&= \frac{1}{2i}\left[(A + iI) + (A - iI)\right]\psi \\
&= \frac{1}{i}A\psi,
\end{aligned}
$$

which establishes Point 3. ∎

We may apply the spectral theorem for bounded normal operators to associate a projection-valued measure μ^U to U. We will then transfer this measure from $S^1\backslash\{0\}$ to $\mathbb{R}$ by means of the map D in (10.29) to obtain the desired projection-valued measure μ^A for A.

Proposition 10.29 *Let A be a self-adjoint operator on $\mathbf{H}$, let U be the unitary operator in Theorem 10.28, and let $D : S^1\backslash\{0\} \to \mathbb{R}$ be as in (10.29). Then*

$$A = D(U), \tag{10.33}$$

where $D(U)$ is defined by the functional calculus for U.

More precisely, $D(U) = \int_{\sigma(U)} D(\lambda)\, d\mu^U(\lambda)$, where μ^U is the projection-valued measure associated to U by the spectral theorem for bounded normal operators. Note that by Point 2 of Theorem 10.28, 1 is not an eigenvalue for U and thus $\mu^U(\{1\}) = 0$. Thus, D is an almost-everywhere-defined function on $\sigma(U)$, even if $1 \in \sigma(A)$. As always, the equality in (10.33) includes equality of domains, where the domain of $\int_{\sigma(U)} D\, d\mu^U$ is the space W_D in Proposition 10.1.

Proposition 10.29 should certainly be plausible in light of the previously established formula (10.30) for A in terms of U.

Proof. Suppose E is a Borel subset of $S^1\backslash\{0\}$ such that the closure of E does not contain 1, and let $V_E = \mathrm{Range}(\mu^U(E))$ be the associated spectral subspace. Then the spectrum of $U|_E$ is contained in $\bar{E}$, which means that the functions $u \mapsto D(u)$ and $u \mapsto 1/(u-1)$ are bounded on $\sigma(U|_{V_E})$. Now, by comparing the quadratic forms, we can see that $D(U)|_{V_E} = D(U|_{V_E})$. Then by the multiplicativity of the functional calculus for U on bounded functions, we have

$$D(U)\psi = i(U + I)(U - I)^{-1}\psi$$

for all $\psi \in V_E$. Thus, by Point 3 of Theorem 10.28, $D(U)$ agrees with A on V_E.

Meanwhile, if we decompose $S^1 \backslash \{0\}$ as the disjoint union of sets E_n for which $\bar{E}_n$ does not contain 1, then $\mathbf{H}$ is the Hilbert space direct sum of the subspaces V_{E_n}. Now, A and (by Proposition 10.3) $D(U)$ are both self-adjoint. Furthermore, these operators agree on the finite direct sum of the V_{E_n}'s and they are essentially self-adjoint on this finite sum, by Example 9.26. Thus, A and $D(U)$ must be equal (with equality of domain). ∎

Theorem 10.30 *Define a projection-valued measure μ^A on $\mathbb{R}$ by*

$$\mu^A(E) = \mu^U(C(E)). \tag{10.34}$$

Then

$$A = \int_{\mathbb{R}} \lambda \, d\mu^A(\lambda), \tag{10.35}$$

where μ^U is the projection-valued measure coming from the spectral theorem for the bounded normal operator U and C is the map defined in (10.28).

Proof. If for any $\psi \in \mathbf{H}$, we define $\mu_\psi^U(E) = \langle \psi, \mu^U \psi \rangle$ and similarly define μ_ψ^A, then we have

$$\mu_\psi^A(E) = \mu_\psi^U(C(E)).$$

By the abstract change of variables theorem from measure theory, we have

$$\int_{\mathbb{R}} \lambda^2 \, d\mu_\psi^A(\lambda) = \int_{S^1 \backslash \{0\}} D(u)^2 \, d\mu_\psi^U(u), \tag{10.36}$$

since D is the inverse map to C. Thus, the two operators in (10.35) have the same domain. Furthermore, if we replace λ^2 by λ and $D(u)^2$ by $D(u)$ in (10.36), we see that the operators in (10.35) are also equal. ∎

Proof of Theorem 10.4. The existence of the desired projection-valued measure μ^A is the content of Theorem 10.30. To establish uniqueness, suppose ν^A is a projection-valued measure on $\sigma(A)$ such that $\int \lambda \, d\nu^A(\lambda) = A$. Consider then the operator $C(A)$ as defined by integration of the function $c(\lambda)$ against ν^A. Arguing as in the proof of Proposition 10.29, we can see that $C(A)$, computed in this fashion, coincides with the operator $U = C(A)$ defined as the product of $(A + iI)$ and $(A - iI)^{-1}$.

Now define a projection-valued measure ν^U on S^1 by setting $\nu^U(E) = \nu^A(C^{-1}(E))$. Then as in the proof of Theorem 10.30, we have $\int_{S^1} u \, d\nu^U (u) = U$. The uniqueness part of the spectral theorem for U (Theorem 10.20) then tells us that $\nu^U = \mu^U$, from which it follows that $\nu^A = \mu^A$. ∎

Proof of Theorem 10.9. By the direct-integral form of the spectral theorem for $U = C(A)$, there is a family of Hilbert spaces $\mathbf{H}_\lambda$, $\lambda \in \sigma(U) \subset S^1$, and a positive, real-valued measure μ on $\sigma(U)$ such that $\mathbf{H}$ is unitarily equivalent to $\int_{\sigma(U)} \mathbf{H}_\lambda \, d\mu$, in such a way that the operator U corresponds to the map $s(\lambda) \mapsto \lambda s(\lambda)$. Since 1 is not an eigenvalue for U, either $\mathbf{H}_1 = \{0\}$ or $\mu(\{1\}) = 0$. Either way, $\mathbf{H}_1$ is "negligible" in the direct integral. We can then define a family of Hilbert spaces $\mathbf{K}_\lambda := \mathbf{H}_{C(\lambda)}$, for $\lambda \in \sigma(A) \subset \mathbb{R}$, and a measure ν on $\sigma(A)$ given by $\nu(E) = \mu(C(E))$. We may then form the direct integral $\int_{\sigma(A)} \mathbf{K}_\lambda \, d\nu$. This direct integral is unitarily equivalent in an obvious way to $\int_{\sigma(U)} \mathbf{H}_\lambda \, d\mu$. We wish to show, then, that $\int_{\sigma(A)} \mathbf{K}_\lambda \, d\nu$ is unitarily equivalent to $\mathbf{H}$ in such a way that the operator A corresponds to the (unbounded) operator mapping $s(\lambda)$ to $\lambda s(\lambda)$. Since the argument is similar to that in the proof of Theorem 10.4, we omit the details.

As in the proof of Theorem 10.4, the uniqueness in Theorem 10.9 can be reduced to the uniqueness for the direct-integral form of the spectral theorem for U. ∎

The proof of the multiplication operator form of the spectral theorem for unbounded operators is similar to the preceding proofs and is omitted.

10.5   Exercises

1. (a) If A is a bounded self-adjoint operator, show that $U(t) := e^{iAt}$ is continuous in the operator norm topology.

    (b) Using the spectral theorem, show that if A is a self-adjoint operator and $\sigma(A)$ is a bounded subset of $\mathbb{R}$, then A is bounded.

    (c) Suppose A is a self-adjoint operator that is not bounded. Show that $U(t) := e^{iAt}$ is not continuous in the operator norm topology.

    Hint: Consider ψ in a spectral subspace of the form $V_{(\lambda_0 - \varepsilon, \lambda_0 + \varepsilon)}$, where λ_0 is a point in $\sigma(A)$ with $|\lambda_0|$ large.

2. Let P_j be the unbounded self-adjoint operator defined in Sect. 9.8. Show that the one-parameter unitary group e^{itP_j} generated by P_j is given by
$$(e^{itP_j} \psi)(\mathbf{x}) = \psi(\mathbf{x} + t\hbar \mathbf{e}_j)$$
for all $\psi \in L^2(\mathbb{R}^n)$, where $\mathbf{e}_j$ is the jth element of the standard basis for $\mathbb{R}^n$.

    Hint: First determine the Fourier transform of $e^{itP_j} \psi$, using Proposition 9.32.

3. If A is an unbounded self-adjoint operator on $\mathbf{H}$, let us say that a family $\psi(t)$ of elements of $\mathbf{H}$ satisfies the equation
$$\frac{d\psi}{dt} = iA\psi(t) \tag{10.37}$$

in the strong sense if each $\psi(t)$ belongs to $\mathrm{Dom}(A)$ and

$$\lim_{h \to 0} \left\| \frac{\psi(t+h) - \psi(t)}{h} - iA\psi(t) \right\| = 0$$

for every $t \in \mathbb{R}$. If we define $\psi(t)$ by $\psi(t) = e^{itA}\psi_0$, for some $\psi_0 \in \mathbf{H}$, show that $\psi(t)$ satisfies (10.37) in the strong sense if and only if ψ_0 belongs to $\mathrm{Dom}(A)$.

4. Suppose A is an unbounded self-adjoint operator and suppose that there exists a number $\gamma \in \mathbb{R}$ and a nonzero vector $\psi \in \mathrm{Dom}(A)$ such that

$$\|A\psi - \gamma\psi\| < \varepsilon \|\psi\|$$

for some $\varepsilon > 0$. Show that there exists a number $\tilde{\gamma}$ in the spectrum of A such that $|\gamma - \tilde{\gamma}| < \varepsilon$.

Hint: If no such $\tilde{\gamma}$ existed, the function $f(\lambda) := 1/|\lambda - \gamma|$ would satisfy $|f(\lambda)| \leq 1/\varepsilon$ for all $\lambda \in \sigma(A)$. Consider, then, the operator $f(A)$, which is nothing but $(A - \gamma I)^{-1}$.

5. If A is a bounded self-adjoint operator, show that the operator $C(A)$ given by

$$C(A) = (A + iI)(A - iI)^{-1}$$

is unitary and that 1 is in the resolvent set of $C(A)$. Show also that A can be recovered from $C(A)$ by the formula

$$A = i(C(A) + I)(C(A) - I)^{-1}.$$

6. Show that Lemma 10.22 is false if we do not assume that A and B commute.

7. Let A be a normal matrix and p a polynomial in two variables. Show by example that an eigenvector for $p(A, A^*)$ is not necessarily an eigenvector for A.

Note: Nevertheless, the proof of the matrix case of Theorem 10.23 shows that if μ is an eigenvalue for $p(A, A^*)$, then there exists *some* eigenvector for $p(A, A^*)$ with eigenvalue μ that is also an eigenvector for A.

8. Suppose $A \in \mathcal{B}(\mathbf{H})$ and W is a closed subspace of $\mathbf{H}$ that is invariant under A and A^*.

(a) Show that $(A|_W)^* = A^*|_W$.

(b) Show that if A is normal, the restriction of A to W is normal.

9. (a) Suppose that $\mathbf{H}$ is finite dimensional, A is a normal operator on $\mathbf{H}$, and W is a subspace of $\mathbf{H}$ that is invariant under A. Show that W is invariant under A^*.

   (b) Show by example that the result of Part (a) is false if $\mathbf{H}$ is infinite dimensional.

10. Given $A \in \mathcal{B}(\mathbf{H})$, suppose that the same vector ψ is an ε-almost eigenvector for A with eigenvalue λ and a δ-almost eigenvector for A with eigenvalue μ. Show that $|\lambda - \mu| < \varepsilon + \delta$.

11

The Harmonic Oscillator

11.1 The Role of the Harmonic Oscillator

The harmonic oscillator is an important model for various reasons. In solid-state physics, for example, a crystal is modeled as a large number of coupled harmonic oscillators. Using the notion of "normal modes," this model is then transformed into independent one-dimensional harmonic oscillators with different frequencies. In the quantum mechanical setting, the excitations of the different normal modes are called *phonons*.

A free quantum field theory is similarly modeled as a family of coupled harmonic oscillators, except that in the field theory setting we have infinitely many of the oscillators. Even interacting quantum field theories are often described using the harmonic oscillator raising and lowering operators, which are referred to as creation and annihilation operators in the context of field theory.

Our approach to analyzing the harmonic oscillator also introduces the algebraic approach to quantum mechanics, in which algebra (commutation relations between various operators) substantially replaces analysis (differential equations) as the way to solve quantum systems. Most of the effort in analyzing the harmonic oscillator occurs in the algebraic section (Sect. 11.2), with the remaining analytic issues being taken care of in Sects. 11.3 and 11.4.

B.C. Hall, *Quantum Theory for Mathematicians*, Graduate Texts
in Mathematics 267, DOI 10.1007/978-1-4614-7116-5_11,
© Springer Science+Business Media New York 2013

11.2   The Algebraic Approach

In this section we will derive as much information as possible about the Hamiltonian operator for a quantum harmonic oscillator using only the commutation relation between the position and momentum operators,

$$[X, P] = i\hbar I. \tag{11.1}$$

Here, as usual, $[\cdot, \cdot]$ denotes the commutator, given by $[A, B] = AB - BA$. We consider, then, a harmonic oscillator with Hamiltonian given by

$$\hat{H} = \frac{P^2}{2m} + \frac{k}{2}X^2, \tag{11.2}$$

where k is a positive constant. Our goal is to see what we can say about the eigenvectors and eigenvalues of $\hat{H}$ using only the fact that X and P are self-adjoint operators satisfying (11.1), without making use of the actual formulas for these operators.

To be honest, we are actually assuming certain domain conditions regarding the operators X and P, in addition to the commutation relation (11.1), namely that the vectors ψ_n in Theorem 11.2 are actually in the domain of X and P (and thus, also, in the domain of the raising and lowering operators). In this section, we follow the usual physics practice of assuming that all the vectors we work with are in the domain of all the relevant operators. This assumption will turn out to be correct in the case we are actually considering, in which X and P are the usual position and momentum operators on $L^2(\mathbb{R})$. (See Sect. 11.4.) It is a more complicated matter to work out the domain conditions that must be imposed on two self-adjoint operators satisfying (11.1) in order for the argument of the present section to be valid. We will come back to this issue in Chap. 14.

Following, again, the convention in the physics literature, we now eliminate the spring constant k in favor of the frequency $\omega = \sqrt{k/m}$ of the corresponding classical harmonic oscillator. [Solutions to Hamilton's equations with classical Hamiltonian $H(x, p)$ equal to $p^2/(2m) + kx^2/2$ are sinusoidal with frequency $\sqrt{k/m}$.] Replacing k by $m\omega^2$, we may rewrite (11.2) as

$$\hat{H} = \frac{1}{2m}\left(P^2 + (m\omega X)^2\right). \tag{11.3}$$

We now introduce the *lowering operator* a, given by

$$a = \frac{m\omega X + iP}{\sqrt{2\hbar m\omega}} \tag{11.4}$$

and its adjoint a^*, the *raising* operator," given by

$$a^* = \frac{m\omega X - iP}{\sqrt{2\hbar m\omega}}. \tag{11.5}$$

The reason for the terminology "raising" and "lowering" is that these operators raise and lower the eigenvalue for the Hamiltonian, as we will see shortly. In the context of quantum field theory, operators very much like a and a^* are called *creation operators* and *annihilation operators*, respectively, because they map from the n-particle space to either the $(n+1)$-particle space or the $(n-1)$-particle space, thus "creating" or "annihilating" a particle.

In the world of noncommuting operators, $(A-B)(A+B)$ does not equal $A^2 - B^2$; rather,

$$(A - B)(A + B) = A^2 - B^2 + [A, B].$$

Thus, if we compute a^*a using (11.1) we get

$$a^*a = \frac{1}{2\hbar m\omega} \left((m\omega X)^2 + P^2 + im\omega [X, P] \right)$$

$$= \frac{1}{\hbar\omega} \frac{1}{2m} \left(P^2 + (m\omega X)^2 \right) - \frac{1}{2} I.$$

From this we obtain

$$\hat{H} = \hbar\omega \left(a^*a + \frac{1}{2} I \right).$$

The $\frac{1}{2} I$ on the right-hand side of this expression should be thought of as a "quantum correction," in that there would be no such term in the analogous formula for the *classical* Hamiltonian.

It suffices to work out the spectral properties (eigenvectors and eigenvalues) of a^*a. To get back to $\hat{H}$, we keep the same eigenvectors and simply add $1/2$ to the eigenvalues and then multiply by $\hbar\omega$. We compute that

$$[a, a^*] = \frac{1}{2\hbar m\omega} \left([m\omega X, -iP] + [iP, m\omega X] \right)$$

$$= \frac{1}{2\hbar m\omega} \left(\hbar m\omega I + \hbar m\omega I \right)$$

$$= I. \tag{11.6}$$

From this, it is easy to compute that

$$[a, a^*a] = a \tag{11.7}$$

$$[a^*, a^*a] = -a^*. \tag{11.8}$$

Now, a^*a is self-adjoint (or, at the least, symmetric) because $(a^*a)^* = a^*a^{**} = a^*a$. This operator is also non-negative, because

$$\langle \psi, a^*a\psi \rangle = \langle a\psi, a\psi \rangle \geq 0$$

for all ψ. We now come to a key computation, which demonstrates the utility of the operators a and a^*.

Proposition 11.1 *Suppose that ψ is an eigenvector for a^*a with eigenvalue λ. Then*

$$a^*a(a\psi) = (\lambda - 1)a\psi$$
$$a^*a(a^*\psi) = (\lambda + 1)a^*\psi.$$

Thus, either $a\psi$ is zero or $a\psi$ is an eigenvector for a^*a with eigenvalue $\lambda - 1$. Similarly, either $a^*\psi$ is zero or $a^*\psi$ is an eigenvector for a^*a with eigenvalue $\lambda + 1$. That is to say, the operators a^* and a raise and lower the eigenvalues of a^*a, respectively.

Proof. Using the commutation relation (11.7), we find that

$$a^*a(a\psi) = (a(a^*a) - a)\,\psi = (\lambda - 1)a\psi.$$

A similar calculation applies to $a^*\psi$, using (11.8). ∎

If ψ is an eigenvector for a^*a with eigenvalue λ, then

$$\lambda \langle \psi, \psi \rangle = \langle \psi, a^*a\psi \rangle = \langle a\psi, a\psi \rangle \geq 0,$$

which means that $\lambda \geq 0$. Let us assume that a^*a has at least one eigenvector ψ, with eigenvalue λ, which we expect since a^*a is self-adjoint. Since a lowers the eigenvalue of a^*a, if we apply a repeatedly to ψ, we must eventually get zero. After all, if $a^n\psi$ were always nonzero, these vectors would be, for large n, eigenvectors for a^*a with negative eigenvalue, which we have seen is impossible.

It follows that there exists some $N \geq 0$ such that $a^N\psi \neq 0$ but $a^{N+1}\psi = 0$. If we define ψ_0 by

$$\psi_0 := a^N\psi,$$

then $a\psi_0 = 0$, which means that $a^*a\psi_0 = 0$. Thus, ψ_0 is an eigenvector for a^*a with eigenvalue 0. (It follows that the original eigenvalue λ must have been equal to the non-negative integer N.)

The conclusion is this: Provided that a^*a has at least one eigenvector ψ, we can find a nonzero vector ψ_0 such that

$$a\psi_0 = a^*a\psi_0 = 0.$$

Since a^*a cannot have negative eigenvalues, we may call ψ_0 a "ground state" for a^*a, that is, an eigenvector with lowest possible eigenvalue. We may then apply the raising operator a^* repeatedly to ψ_0 to obtain eigenvectors for a^*a with positive eigenvalues.

Theorem 11.2 *If ψ_0 is a unit vector with the property that $a\psi_0 = 0$, then the vectors*

$$\psi_n := (a^*)^n\psi_0, \quad n \geq 0,$$

satisfy the following relations for all $n, m \geq 0$:

$$a^* \psi_n = \psi_{n+1}$$
$$a^* a \psi_n = n \psi_n$$
$$\langle \psi_n, \psi_m \rangle = n! \delta_{n,m}$$
$$a \psi_{n+1} = (n+1) \psi_n.$$

Let us think for a moment about what this is saying. We have an orthogonal "chain" of eigenvectors for $a^* a$ with eigenvalues $0, 1, 2, \ldots$, with the norm of ψ_n equal to $\sqrt{n!}$. The raising operator a^* shifts us up the chain, while the lowering operator a shifts us down the chain (up to a constant). In particular, the "ground state" ψ_0 is annihilated by a. Thus, we have a complete understanding of how a and a^* act on this chain of eigenvectors for $a^* a$.

Proof. The first result is the definition of ψ_{n+1} and the second follows from Proposition 11.1 and the fact that $a^* a \psi_0 = 0$. For the third result, if $n \neq m$, we use the general result that eigenvectors for a self-adjoint operator (in our case, $a^* a$) with distinct eigenvalues are orthogonal. (This result actually applies to operators that are only symmetric.)

If $n = m$, we work by induction. For $n = 0$, $\langle \psi_0, \psi_0 \rangle = 1$ is assumed. If we assume $\langle \psi_n, \psi_n \rangle = n!$, we compute that

$$
\begin{aligned}
\langle \psi_{n+1}, \psi_{n+1} \rangle &= \langle a^* \psi_n, a^* \psi_n \rangle \\
&= \langle \psi_n, a a^* \psi_n \rangle \\
&= \langle \psi_n, (a^* a + 1) \psi_n \rangle \\
&= (n+1) \langle \psi_n, \psi_n \rangle \\
&= (n+1)!.
\end{aligned}
$$

Finally, we compute that

$$a \psi_{n+1} = a a^* \psi_n = (a^* a + I) \psi_n = (n+1) \psi_n,$$

which establishes the last claimed result. ∎

It is now reasonable to ask whether the vectors $\{\psi_n\}_{n=0}^{\infty}$ form an orthonormal *basis* for the quantum Hilbert space. Suppose this is *not* the case. If we then let V denote the closed span of the ψ_n's, V will be invariant under both a and a^*. Thus, by elementary linear algebra, the orthogonal complement $V^{\perp}$ of V will also be invariant under the adjoint operators a^* and a, and therefore also under $a^* a$. Therefore, we can begin our analysis anew in $V^{\perp}$, with the result that we will obtain a new ground state $\phi_0 \in V^{\perp}$ (satisfying $a \phi_0 = 0$) that is orthogonal to the original ground state ψ_0. If, then, the closed span of the ψ_n's is *not* the whole Hilbert space, there will exist at least two independent solutions of the equation $a \psi = 0$. To put this claim the other way around, if it turns out that there is only one solution (up to a constant) of $a\psi = 0$, then we expect that the vectors obtained by applying a^* repeatedly to the solution will form an orthogonal *basis* for our Hilbert space. (Because we are glossing over various technical issues having to do with the domains of various operators, this conclusion should not be regarded as completely rigorous.)

11.3 The Analytic Approach

In the preceding section, we analyzed the eigenvectors of the operator a^*a as much as possible using only the commutation relation $[a, a^*] = I$, which follows from the underlying commutation relation $[X, P] = i\hbar I$. To progress further, we must recall the actual formula for the operators a and a^*.

To simplify our analysis, let us introduce the following natural scale of distance for our problem:

$$D := \sqrt{\frac{\hbar}{m\omega}}.$$

We then introduce a normalized position variable, measured in units of D,

$$\tilde{x} := \frac{x}{D}, \tag{11.9}$$

so that

$$\frac{d}{d\tilde{x}} = \sqrt{\frac{\hbar}{m\omega}} \frac{d}{dx}.$$

A calculation gives the following simple expressions for the raising and lowering operators:

$$a = \frac{1}{\sqrt{2}}\left(\tilde{x} + \frac{d}{d\tilde{x}}\right)$$

$$a^* = \frac{1}{\sqrt{2}}\left(\tilde{x} - \frac{d}{d\tilde{x}}\right). \tag{11.10}$$

Note that the constants m, ω, and $\hbar$ have conveniently disappeared from the formulas.

Given the expression in (11.10), we can easily solve the (first-order, linear) equation $a\psi_0 = 0$ as

$$\psi_0(\tilde{x}) = Ce^{-\tilde{x}^2/2}. \tag{11.11}$$

If we take C to be positive, then our normalization condition determines its value to be $\sqrt{\pi}/D$, by Proposition A.22. (The normalization condition is that the integral of $|\psi_0|^2$ with respect to dx—not $d\tilde{x}$—should be 1.) We obtain, then,

$$\psi_0(x) = \sqrt{\frac{\pi m\omega}{\hbar}} \exp\left\{-\frac{m\omega}{2\hbar}x^2\right\}. \tag{11.12}$$

It remains only to apply a^* repeatedly to ψ_0 to get the "excited states" ψ_n.

Theorem 11.3 *The ground state ψ_0 of the harmonic oscillator is given by (11.12). The excited states ψ_n are given by*

$$\psi_n = H_n \, \psi_0 \tag{11.13}$$

where H_n is a polynomial of degree n given inductively by the formulas

$$H_0(\tilde{x}) = 1$$

$$H_{n+1}(\tilde{x}) = \frac{1}{\sqrt{2}} \left(2\tilde{x} H_n(\tilde{x}) - \frac{dH_n(\tilde{x})}{d\tilde{x}} \right).$$

Here, $\tilde{x}$ is the normalized position variable given by (11.9).

The polynomials H_n are essentially (modulo various normalization conventions) the *Hermite polynomials*.

Proof. When $n = 0$, (11.13) reduces to $\psi_0 = \psi_0$. Assuming that (11.13) holds for some n, we compute ψ_{n+1} as

$$\psi_{n+1} = a^* \psi_n = \frac{1}{\sqrt{2}} \left(\tilde{x} H_n(\tilde{x}) C e^{-\tilde{x}^2/2} - \frac{d}{d\tilde{x}} \left[H_n(\tilde{x}) C e^{-\tilde{x}^2/2} \right] \right)$$

$$= \frac{1}{\sqrt{2}} \left(2\tilde{x} H_n(\tilde{x}) - \frac{dH_n}{d\tilde{x}} \right) C e^{-\tilde{x}^2/2} = H_{n+1}(\tilde{x}) \psi_0(\tilde{x}),$$

as claimed. ∎

Figure 11.1 shows the ground state of the harmonic oscillator, along with the excited states with $n = 5$ and $n = 30$. Each eigenfunction is plotted as a function of the normalized position variable $\tilde{x}$. In each case, the shaded region indicates the extent of the classically allowed region, that is, the range in which a classical particle with energy E_n can move. Note that each wave function decays rapidly outside the classically allowed region. In the last image, we can see that frequency of oscillation of the wave function is greatest in the middle of the classically allowed region, while the amplitude of the wave function is greatest near the ends of the classically allowed region. Intuitively, these properties of the wave function reflect that a classical particle with energy E_n has largest momentum in the middle of the classically allowed region (where the potential is smallest) and that the classical particle spends more time at the ends of the classically allowed region, since it is moving slowest there. Further development of this sort of reasoning may be found in Chap. 15.

11.4   Domain Conditions and Completeness

Although the analysis in Sect. 11.2 is typical of what is found in physics texts, it is not completely rigorous from a mathematician's point of view.

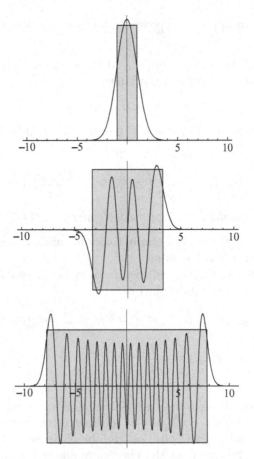

FIGURE 11.1. Harmonic oscillator eigenvectors with $n = 0$, $n = 5$, and $n = 30$. In each case, the classically allowed region is shaded.

The main problem is that the lowing operator a, the raising operator a^*, and the product operator a^*a are all unbounded operators. The difficulty in working with unbounded operators is that one constantly has to check that a vector is in the domain of the relevant operator before applying that operator. For example, suppose we have a vector ψ_0 in the domain of a and satisfying $a\psi_0 = 0$. We wish to apply the raising operator a^* to ψ_0 and we then want to argue that

$$a^*a(a^*\psi_0) = a^*\psi_0.$$

This is easy enough to verify (as we did in the previous section) *provided* that all vectors are in the domain of the relevant operators. But how do we know that ψ_0 is in the domain of a^*? And even if it is, how do we know that $a^*\psi_0$ is in the domain of a^*a?

These concerns are not just theoretical. Consider a general pair of operators A and B satisfying $[A, B] = i\hbar I$. If we try to analyze an operator of the form $\alpha A^2 + \beta B^2$, for $\alpha, \beta > 0$, by the methods of Sect. 11.2, things can easily go awry, as the counterexample in Sect. 12.2 demonstrates. Fortunately, in the case of the ordinary position and momentum operators, the putative eigenfunctions ψ_n for a^*a in Theorem 11.3 are very nice functions, in the form of a polynomial times a Gaussian. Thus, there is no difficulty in verifying that these functions are in the domain of any finite product of creation and annihilation operators. It follows that if a and a^* are given in terms of the usual position and momentum operators and ψ_0 given by (11.12), the relations in Theorem 11.2 indeed hold.

In particular, we can see that the ψ_n's form an orthogonal *set* of functions in $L^2(\mathbb{R})$. Showing that they form an orthogonal *basis* is also not terribly difficult.

Theorem 11.4 *The functions*

$$\psi_n(x) = H_n(\tilde{x})\psi_0(\tilde{x})$$

$$= H_n\left(\sqrt{\frac{m\omega}{\hbar}}x\right)\sqrt[4]{\frac{\pi m\omega}{\hbar}}\exp\left\{-\frac{m\omega}{2\hbar}x^2\right\}$$

form an orthogonal basis *for the Hilbert space* $L^2(\mathbb{R})$.

The following result is the key to the proof.

Lemma 11.5 *For all $\alpha \in \mathbb{C}$, the partial sums of the series*

$$\sum_{n=0}^{\infty}\frac{\alpha^n\tilde{x}^n}{n!}e^{-\tilde{x}^2/2}$$

converge in $L^2(\mathbb{R})$ *to the function* $e^{\alpha\tilde{x}}e^{-\tilde{x}^2/2}$.

Proof. We need to show that

$$\left\|e^{\alpha\tilde{x}}e^{-\tilde{x}^2/2} - \sum_{n=0}^{N}\frac{\alpha^n\tilde{x}^n}{n!}e^{-\tilde{x}^2/2}\right\|^2 = \int\left|\sum_{n=N+1}^{\infty}\frac{\alpha^n\tilde{x}^n}{n!}e^{-\tilde{x}^2/2}\right|^2 d\tilde{x} \quad (11.14)$$

tends to zero as N tends to infinity. The integrand on the right-hand side of (11.14) tends to zero pointwise. If we can find a suitable dominating function, we can use dominated convergence to conclude that the integral also tends to zero. We see that

$$\left|\sum_{n=N+1}^{\infty}\frac{\alpha^n\tilde{x}^n}{n!}e^{-\tilde{x}^2/2}\right|^2 \leq \left(\sum_{n=0}^{\infty}\frac{|\alpha\tilde{x}|^n}{n!}e^{-\tilde{x}^2/2}\right)^2$$

$$= e^{2|\alpha||\tilde{x}|}e^{-\tilde{x}^2}.$$

Since this last function certainly has finite integral, dominated convergence applies and we are done. ∎

Proof of Theorem 11.4. It is easily seen that the raising and lowering operators map the Schwartz space $\mathcal{S}(\mathbb{R})$ (Definition A.15) into itself. Furthermore, it is easy to verify (Exercise 1) that

$$\left\langle \frac{d\phi}{dx}, \psi \right\rangle = \left\langle \phi, \frac{d\psi}{dx} \right\rangle,$$

for all $\phi, \psi \in \mathcal{S}(\mathbb{R})$. From this, we can easily verify that for all $\phi, \psi \in \mathcal{S}(\mathbb{R})$,

$$\langle \phi, a\psi \rangle = \langle a^*\phi, \psi \rangle$$

and so also

$$\langle \phi, a^*a\psi \rangle = \langle a^*a\phi, \psi \rangle.$$

It is evident that both the ground state ψ_0 and all the excited states ψ_n occurring in Theorem 11.4 belong to $\mathcal{S}(\mathbb{R})$. Thus, the proof of Theorem 11.2 is indeed valid. We conclude, then, that the ψ_n's form an orthogonal *set* of vectors in $L^2(\mathbb{R})$ and that they are eigenvectors for $\hat{H}$ with the indicated eigenvalues.

It remains to show that the ψ_n's form an orthogonal *basis* for $L^2(\mathbb{R})$. Let V denote the space of finite linear combinations of the ψ_n's. Since H_n is a polynomial of degree n, it is easily seen that V consists precisely functions of the form

$$\psi(\tilde{x}) = p(\tilde{x})e^{-\tilde{x}^2/2},$$

where p is a polynomial.

Lemma 11.5 then shows that $e^{ik\tilde{x}}e^{-\tilde{x}^2/2}$ belongs to the L^2-closure of V for all $k \in \mathbb{R}$. Thus, if ψ is orthogonal to every element of $\bar{V}$, we have

$$\int_{\mathbb{R}} e^{-ik\tilde{x}}e^{-\tilde{x}^2/2}\psi(\tilde{x})\,d\tilde{x} = 0 \tag{11.15}$$

for all k. Now, since $e^{-\tilde{x}^2/2}$ belongs to $L^\infty(\mathbb{R}) \cap L^2(\mathbb{R})$ and ψ belongs to $L^2(\mathbb{R})$, their product belongs to $L^2(\mathbb{R}) \cap L^1(\mathbb{R})$. Thus, (11.15) tells us that the L^2 Fourier transform of $e^{-\tilde{x}^2/2}\psi(\tilde{x})$ is identically zero. Thus, $e^{-\tilde{x}^2/2}\psi(\tilde{x})$ must be the zero element of $L^2(\mathbb{R})$, by the Plancherel theorem, and so $\psi(\tilde{x}) = 0$ almost everywhere. This shows that $V^\perp = \{0\}$, meaning that V is dense in $L^2(\mathbb{R})$. ∎

11.5   Exercises

1. Show that for any Schwartz functions ϕ and ψ, we have

$$\langle \phi, a\psi \rangle = \langle a^*\phi, \psi \rangle,$$

as expected.

Hint: Use integration by parts on the interval $[-A, A]$ and show that the boundary terms tend to zero as A tends to infinity.

2. Show that the polynomials H_n satisfy the following relations:

$$H_{n-1}(y) = \frac{1}{n\sqrt{2}}H'_n(y)$$

and

$$H_{n+1}(y) = \frac{1}{\sqrt{2}}\left(2yH_n(y) - n\sqrt{2}H_{n-1}(y)\right).$$

Hint: Start with the relation $a\psi_n = n\psi_{n-1}$.

3. Establish the following *Rodrigues formula* for the polynomials H_n:

$$H_n(y) = (-1)^n 2^{-n/2} \frac{\left(\frac{d}{dy}\right)^n e^{-y^2}}{e^{-y^2}}.$$

4. In this exercise, we prove the following claim: The polynomial H_n has n distinct real zeros and the zeros of H_n "interlace" with the zeros of H_{n-1}, meaning that there is exactly one zero of H_{n-1} between each pair of consecutive zeros of H_n.

   (a) Verify the claim for H_1 and H_0.

   (b) Assume, inductively, that H_n and H_{n-1} have distinct real zeros and that the zeros interlace. Show that H_{n-1} alternates in sign at consecutive zeros of H_n. Then show that H_{n+1} and H_{n-1} have opposite signs at each zero of H_n, so that H_{n+1} also alternates in sign at consecutive zeros of H_n. Conclude that H_{n+1} must have at least one zero between each pair of consecutive zeros of H_n.
   Hint: Use Exercise 2.

   (c) Show that H_{n+1} and H_{n-1} have the same sign near $\pm\infty$ but opposite signs at the largest and smallest zeros of H_n. Conclude that H_{n+1} has at least one zero below the smallest zero of H_n and at least one zero above the largest zero of H_n.

   (d) Conclude that H_{n+1} has $n+1$ real zeros that interlace with the zeros of H_n.

5. Let $\tilde{\psi}_n = \psi_n / \|\psi_n\|$ be the normalized nth excited state.

   (a) Let $\tilde{X} = X/D$, where $D = (\hbar/m\omega)^{1/2}$. Show that

   $$\left\langle \tilde{X}^2 \right\rangle_{\tilde{\psi}_n} = n + \frac{1}{2}.$$

   Hint: Express $\tilde{X}$ in terms of a and a^*, using (11.10), and then use Theorem 11.2.

(b) Show that

$$\langle X \rangle_{\tilde{\psi}_n} = 0$$

$$\Delta_{\tilde{\psi}_n} X = \left(\frac{\hbar(n + 1/2)}{m\omega} \right)^{1/2}.$$

(c) If T and V denote the kinetic energy and potential energy terms, respectively, in (11.3), show that

$$\langle T \rangle_{\tilde{\psi}_n} = \langle V \rangle_{\tilde{\psi}_n} = \frac{1}{2} \hbar\omega \left(n + \frac{1}{2} \right).$$

12

The Uncertainty Principle

In this chapter, we will continue our investigation of the consequences of the commutation relations among the position and momentum operators. We will mostly consider a particle in $\mathbb{R}^1$, where we have

$$[X, P] = i\hbar I. \tag{12.1}$$

We have already seen that much of the analysis of the Hamiltonian $\hat{H}$ for the quantum harmonic oscillator (given by $c_1 P^2 + c_2 X^2$) can be carried out using only the commutation relation (12.1). There are two other main results that can be derived from these commutation relations: the Heisenberg uncertainty principle and the Stone–von Neumann theorem. The uncertainty principle states that the product of the uncertainty in X and the uncertainty in P cannot be smaller than $\hbar/2$. The Stone-von Neumann theorem, meanwhile, states that *any* two self-adjoint operators A and B satisfying $[A, B] = i\hbar I$ "look like" several copies of the standard position and momentum operators acting on $L^2(\mathbb{R})$. Both results are true only under certain technical domain conditions, which we will need to examine carefully. We discuss the uncertainty principle in this chapter and the Stone–von Neumann theorem in the next chapter.

The uncertainty principle states that for all ψ in $L^2(\mathbb{R})$ satisfying certain domain conditions, we have

$$(\Delta_\psi X)(\Delta_\psi P) \geq \frac{\hbar}{2},$$

where, for any observable A, we let $\Delta_\psi A$ denote the "uncertainty" in measurements of A in the state ψ (Definition 3.13). This means that one cannot

B.C. Hall, *Quantum Theory for Mathematicians*, Graduate Texts in Mathematics 267, DOI 10.1007/978-1-4614-7116-5_12, © Springer Science+Business Media New York 2013

make both the uncertainty in position and the uncertainty in momentum arbitrarily small in the same state ψ.

Although we can easily make $\Delta_\psi X$ as small as we want simply be taking ψ to be supported in a small interval, if we do that, $\Delta_\psi P$ will be large. Similarly, we can make $\Delta_\psi P$ as small as we like, by taking the momentum wave function $\tilde{\psi}(p)$ (Sect. 6.6) to be supported in a small interval, but then $\Delta_\psi X$ will get large. In the idealized limit in which the position wave function is concentrated at a single point, $\psi(x)$ would be a multiple of $\delta(x - a)$ for some a, in which case, the momentum wave function $\tilde{\psi}(p)$ would be a multiple of $e^{-ipa/\hbar}$. In that case, $|\tilde{\psi}(p)|^2$ is constant, meaning that the momentum wave function is completely spread out over the whole real line.

This uncertainty principle may be interpreted as saying that it is impossible to simultaneously measure the position and momentum of a quantum particle. After all, we have said (Axiom 4) that if we perform a measurement of an observable A with a discrete spectrum, then immediately after the measurement the state ψ of the system should be an eigenvector for A. If A has a continuous spectrum, this principle is replaced by the requirement that after the measurement, the uncertainty in A should very small. If we could measure both the position and the momentum of the particle simultaneously with arbitrary precision, then after the measurement, *both* ΔX and ΔP would have to be very small, violating the uncertainty principle.

Now, on the scale of everyday life, Planck's constant is very small. If, for example, we measure mass in units of grams, distance in units of centimeters, and time in units of seconds, then $\hbar$ has the numerical value of 1.054×10^{-27}. Thus, on "macroscopic" scales of energy and momentum, it is possible for the uncertainties in position and momentum both to be very small. But on the atomic scale, the uncertainty principle puts a substantial limitation on how localized the position and momentum of a particle can be.

In Sect. 12.1, we prove a version of the uncertainty principle for *any* two operators A and B satisfying $[A, B] = i\hbar I$, under a seemingly innocuous assumption on the domains of the operators involved. In Sect. 12.2, however, we see that the domain assumptions are not so innocuous after all. In that section, we encounter two operators satisfying $[A, B] = i\hbar I$ on a dense subspace of the Hilbert space, along with a vector ψ such that the uncertainty in A is finite and the uncertainty in B is zero. The existence of such a vector is surely contrary to the spirit of the uncertainty principle, even though it does not violate the version of the uncertainty principle proved in Sect. 12.1. (The vector ψ in Sect. 12.2 does not satisfy the domain assumptions of Theorem 12.4.) Finally, in Sect. 12.3, we show that for the usual position and momentum operators on $L^2(\mathbb{R})$, no such counterexamples occur: If $\Delta_\psi X$ and $\Delta_\psi P$ are both defined, then $(\Delta_\psi X)(\Delta_\psi P) \geq \hbar/2$.

12.1 Uncertainty Principle, First Version

In this section, it is essential that we make sure that all vectors are in the domains of the various operators we want to apply to these vectors. With this concern in mind, we make the following definition. (Compare Definition 9.36.)

Definition 12.1 *If A and B are unbounded operators on* **H**, *define AB to be the operator with domain*

$$\mathrm{Dom}(AB) = \{\psi \in \mathrm{Dom}(B) \,|\, B\psi \in \mathrm{Dom}(A)\}$$

and given by $(AB)\psi = A(B\psi)$.

Even if $\mathrm{Dom}(A)$ and $\mathrm{Dom}(B)$ are dense in **H**, it could happen that $\mathrm{Dom}(AB)$ is not dense in **H**.

Recall (Definition 3.13) that the uncertainty of a symmetric operator A in a state ψ is defined to be

$$(\Delta_\psi A)^2 = \left\langle \left(A - \langle A\rangle_\psi I\right)^2 \right\rangle_\psi. \tag{12.2}$$

As written, this definition requires that ψ belong to the domain of $(A - \langle A\rangle_\psi I)^2$, which is the same as the domain of A^2. However, since we assume that A is symmetric, then $\langle A\rangle_\psi = \langle \psi, A\psi\rangle$ is real, so that $A - \langle A\rangle_\psi I$ is again symmetric. Thus, (12.2) can be rewritten as

$$(\Delta_\psi A)^2 = \left\langle (A - \langle A\rangle_\psi I)\psi, (A - \langle A\rangle_\psi I)\psi \right\rangle.$$

Having written the uncertainty in this way, it is natural to extend the definition of uncertainty to vectors that belong only to $\mathrm{Dom}(A)$, as follows.

Definition 12.2 *If A is a symmetric operator on* **H**, *then for all unit vectors* ψ *in* $\mathrm{Dom}(A)$, *the* **uncertainty** $\Delta_\psi A$ *of A in the state* ψ *is given by*

$$(\Delta_\psi A)^2 = \left\langle (A - \langle A\rangle_\psi I)\psi, (A - \langle A\rangle_\psi I)\psi \right\rangle. \tag{12.3}$$

By expanding out the right-hand side of (12.3), we see that the uncertainty may also be computed as

$$(\Delta_\psi A)^2 = \langle A\psi, A\psi\rangle - (\langle \psi, A\psi\rangle)^2.$$

[Compare (3.24).] Of course, if ψ happens to be in the domain of A^2, then Definition 12.2 agrees with (12.2).

Proposition 12.3 *If A is a symmetric operator on* **H**, *then for all unit vectors* $\psi \in \mathrm{Dom}(A)$, *we have* $\Delta_\psi A = 0$ *if and only if* ψ *is an eigenvector for A.*

Proof. If $\Delta_\psi A = 0$, then from (12.3), we see that $(A - \langle A \rangle_\psi I)\psi = 0$, meaning that ψ is an eigenvector for A with eigenvalue $\langle A \rangle_\psi$. Conversely, if $A\psi = \lambda\psi$ for some λ, then $\langle \psi, A\psi \rangle = \lambda \langle \psi, \psi \rangle = \lambda$. Thus, $(A - \langle A \rangle_\psi I)\psi = 0$, which, by (12.3), means that $\Delta_\psi A = 0$. ∎

As discussed in the introduction to this chapter, we expect that immediately after a measurement of an observable A, the state of the system will have very small uncertainty for A. Indeed, if A has discrete spectrum, we expect that the state of the system will be an eigenvector for A. Even in the case of a continuous spectrum, we expect that the uncertainty in A can be made as small as one wishes, by making more and more precise measurements. Suppose now that one wishes to observe simultaneously two (or more) different observables, represented by operators A and B. In the case of a discrete spectrum, the system after the measurement should be *simultaneously* an eigenvector for A and an eigenvector for B. In the case where A and B commute, this idea is reasonable. There is a version of the spectral theorem for commuting self-adjoint operators; in the case of discrete spectrum, it says that two commuting self-adjoint operators have an orthonormal basis of simultaneous eigenvectors with real eigenvalues. (In the case of unbounded operators, there are, as usual, technical domain conditions in defining what it means for two self-adjoint operators to commute.)

In the case where A and B do not commute, they do not need to have any simultaneous eigenvectors. Certainly, A and B cannot have an orthonormal basis of simultaneous eigenvectors, or they *would* in fact commute. The lack of simultaneous eigenvectors suggests, then, that it is simply not possible to make a simultaneous measurement of two self-adjoint operators unless they commute. In standard physics terminology, the quantities A and B are said to be "incommensurable," meaning not capable of being measured at the same time. (See Exercise 2 for a classification of the simultaneous eigenvectors of a representative pair of noncommuting operators.)

In the case of a continuous spectrum, the notion of an eigenvector is replaced by the notion of a state with very small uncertainty for the relevant operator. In light of our discussion of simultaneous eigenvectors, we may expect that for noncommuting operators, it may be difficult to find states where the uncertainties of both operators are small. This expectation is realized in the following version of the uncertainty principle.

Theorem 12.4 *Suppose A and B are symmetric operators and ψ is a unit vector belonging to* $\mathrm{Dom}(AB) \cap \mathrm{Dom}(BA)$. *Then*

$$(\Delta_\psi A)^2 (\Delta_\psi B)^2 \geq \frac{1}{4} \left| \langle [A, B] \rangle_\psi \right|^2. \tag{12.4}$$

Note that if $\psi \in \mathrm{Dom}(AB)$ then in particular, $\psi \in \mathrm{Dom}(B)$, and if $\psi \in \mathrm{Dom}(BA)$ then $\psi \in \mathrm{Dom}(A)$. Thus, the assumptions on ψ are sufficient to guarantee that $\Delta_\psi A$ and $\Delta_\psi B$ make sense as in Definition 12.2.

Proof. Define operators A' and B' by $A' := A - \langle \psi, A\psi \rangle I$ and $B' := B - \langle \psi, B\psi \rangle I$. (We use the same domains for A' and B' as for A and B, and it is easily verified that A' and B' are still symmetric on those domains.) Then by the Cauchy–Schwarz inequality, we obtain

$$\langle A'\psi, A'\psi \rangle \langle B'\psi, B'\psi \rangle \geq |\langle A'\psi, B'\psi \rangle|^2 \tag{12.5}$$

$$\geq |\text{Im} \langle A'\psi, B'\psi \rangle|^2 \tag{12.6}$$

$$= \frac{1}{4} |\langle A'\psi, B'\psi \rangle - \langle B'\psi, A'\psi \rangle|^2. \tag{12.7}$$

The assumptions on ψ guarantee that $B\psi \in \text{Dom}(A)$ and hence also that $B'\psi \in \text{Dom}(A')$, and similarly with A' and B' reversed. Since A' and B' are symmetric, we may rewrite (12.7) as

$$\langle A'\psi, A'\psi \rangle \langle B'\psi, B'\psi \rangle \geq \frac{1}{4} |\langle \psi, A'B'\psi \rangle - \langle \psi, B'A'\psi \rangle|^2$$

$$= \frac{1}{4} |\langle \psi, [A', B']\psi \rangle|^2.$$

Now, since the identity operator commutes with everything, the commutator of A' and B' is the same as the commutator of A and B. Furthermore, $\langle A'\psi, A'\psi \rangle$ is nothing but $(\Delta_\psi A)^2$ and similarly for B. Thus, we obtain

$$(\Delta_\psi A)^2 (\Delta_\psi B)^2 \geq \frac{1}{4} |\langle \psi, [A, B]\psi \rangle|^2,$$

which is what we wanted to prove. ∎

We now specialize Theorem 12.4 to the case in which the commutator is $i\hbar I$ and take the square root of both sides.

Corollary 12.5 *Suppose A and B are symmetric operators satisfying*

$$[A, B] = i\hbar I$$

on $\text{Dom}(AB) \cap \text{Dom}(BA)$. *Then if* $\psi \in \text{Dom}(AB) \cap \text{Dom}(BA)$ *is a unit vector, we have*

$$(\Delta_\psi A)(\Delta_\psi B) \geq \frac{\hbar}{2}. \tag{12.8}$$

In particular, for all unit vectors $\psi \in L^2(\mathbb{R})$ *in* $\text{Dom}(XP) \cap \text{Dom}(PX)$, *we have*

$$(\Delta_\psi X)(\Delta_\psi P) \geq \frac{\hbar}{2}. \tag{12.9}$$

Note that the factor of $\hbar$ appearing on the right-hand side of (12.8) is really just $|\langle \psi, [A, B]\psi \rangle|$. Since, however, ψ is a unit vector and $[A, B] = i\hbar I$, ψ drops out of the right-hand side of our inequality. We see then that both sides of (12.9) make sense whenever $\Delta_\psi X$ and $\Delta_\psi P$ make sense, namely, whenever ψ belongs to $\text{Dom}(X)$ and to $\text{Dom}(P)$. (Recall Definition 12.2.)

On the other hand, the *proof* that we have given for (12.9) requires ψ to be in both $\mathrm{Dom}(XP)$ and $\mathrm{Dom}(PX)$. Nevertheless, it is natural to ask whether (12.9) holds for all ψ in $\mathrm{Dom}(X) \cap \mathrm{Dom}(P)$. We may similarly ask whether (12.8) holds for all ψ in $\mathrm{Dom}(A) \cap \mathrm{Dom}(B)$. As we will see in Sects. 12.2 and 12.3, the answer to the first question is yes and the answer to the second question is no.

Meanwhile, it is of interest to investigate "minimum uncertainty states," that is, states ψ for which the inequality (12.4) is an equality.

Proposition 12.6 *If A and B are symmetric and ψ is a unit vector in* $\mathrm{Dom}(AB) \cap \mathrm{Dom}(BA)$, *equality holds in (12.4) if and only if one of the following holds: (1) ψ is an eigenvector for A, (2) ψ is an eigenvector for B, or (3) ψ is an eigenvector for an operator of the form*

$$A - i\gamma B$$

for some nonzero real number γ.

In the case $A = X$ and $B = P$, we will consider examples where equality holds in Sect. 12.4.

Proof. To get equality in (12.4), we must have equality in both (12.5) and (12.6). Equality in (12.5) occurs if and only if $A'\psi = 0$ or $B'\psi = 0$ or $A'\psi = cB'\psi$ for some nonzero constant c. If $A'\psi$ is zero, ψ is an eigenvector for A with eigenvalue $\langle A\rangle_\psi$. In that case, equality holds in (12.6) as well. Conversely, if ψ is an eigenvector for A with some eigenvalue λ, then $\langle A\rangle_\psi = \lambda$ and $A'\psi = 0$. Similarly, $B'\psi = 0$ if and only if ψ is an eigenvector for B.

Meanwhile, suppose $A'\psi$ and $B'\psi$ are nonzero and $A'\psi = cB'\psi$, so that equality holds in (12.5). Then equality holds (12.6) if and only if $c = i\gamma$ for some nonzero $\gamma \in \mathbb{R}$. Thus, when $A'\psi$ and $B'\psi$ are nonzero, we get equality in (12.4) if and only if

$$A'\psi = i\gamma B'\psi \tag{12.10}$$

for some nonzero real number γ. Recalling the definition of A' and B', (12.10) says that

$$(A - \langle \psi, A\psi\rangle\, I)\psi = i\gamma(B - \langle \psi, B\psi\rangle\, I)\psi \tag{12.11}$$

or

$$(A - i\gamma B)\psi = \lambda\psi, \tag{12.12}$$

where $\lambda = \langle \psi, A\psi\rangle - i\gamma\,\langle \psi, B\psi\rangle$.

Thus, if (12.11) holds, ψ is an eigenvector of $A - i\gamma B$. Conversely, if ψ is an eigenvector for $A - i\gamma B$ with some eigenvalue $\lambda = c + id$ in $\mathbb{C}$, then

$$(c + id)\,\|\psi\|^2 = \langle \psi, (A - i\gamma B)\psi\rangle = \langle \psi, A\psi\rangle - i\gamma\,\langle \psi, B\psi\rangle. \tag{12.13}$$

Since A and B are assumed to be symmetric and ψ is a unit vector, we may equate real and imaginary parts in (12.13) to obtain

$$c = \langle \psi, A\psi\rangle\,; \quad d = -\gamma\,\langle \psi, B\psi\rangle\,.$$

From this we can see that (12.11) and (12.10) hold, and thus equality holds in (12.4). ∎

12.2 A Counterexample

In this section, we consider the Hilbert space $L^2[-1,1]$. As our "position" operator, we use the usual formula,

$$A\psi(x) = x\psi(x).$$

Note that A is a *bounded* operator, because we restrict x to the bounded interval $[-1,1]$. As such, A is defined (and self-adjoint) on the whole Hilbert space $L^2(\mathbb{R})$. As our "momentum" operator, we again use the usual formula,

$$B = -i\hbar\frac{d}{dx}.$$

As the domain of B we will take the space of continuously differentiable functions ψ on $[-1,1]$ satisfying the *periodic boundary condition*,

$$\psi(-1) = \psi(1). \tag{12.14}$$

To verify that B is symmetric, note that for any C^1 functions ϕ and ψ, we have

$$\int_{-1}^{1} \overline{\phi(x)}\frac{d\psi}{dx} \, dx = \overline{\phi(1)}\psi(1) - \overline{\phi(-1)}\psi(-1) - \int_{-1}^{1} \overline{\frac{d\phi}{dx}}\psi(x) \, dx.$$

If both ϕ and ψ satisfy the periodic boundary condition (12.14), the boundary terms cancel out to zero. This shows that the operator d/dx is skew-symmetric on $\mathrm{Dom}(B)$, from which it follows that $-i\hbar d/dx$ is symmetric on $\mathrm{Dom}(B)$. Actually, since the functions

$$\psi_n(x) := \frac{1}{\sqrt{2}}e^{\pi i n x}, \quad n \in \mathbb{Z}, \tag{12.15}$$

constitute an orthonormal basis of eigenvectors for B with real eigenvalues, B is essentially self-adjoint, by Example 9.25.

Now, for all $\psi \in \mathrm{Dom}(AB) \cap \mathrm{Dom}(BA)$ we have, by direct calculation,

$$AB\psi - BA\psi = i\hbar\psi, \tag{12.16}$$

just as for the usual position and momentum operators. Furthermore, $\mathrm{Dom}(AB) \cap \mathrm{Dom}(BA)$ is dense in $\mathbf{H}$, since it contains all continuously differentiable functions ψ such that $\psi(0) = \psi(1) = 0$. Consider, now, the function $\psi_n(x)$ in (12.15), for some integer n. Clearly, ψ_n is in the domain of B, since $B\psi_n$ is just a multiple of ψ_n. Since ψ_n is an eigenvector for B, the uncertainty of B in the state ψ_n is zero! Meanwhile, since A is bounded, the uncertainty of A is well defined and finite. Thus, $\Delta_{\psi_n} A$ and $\Delta_{\psi_n} B$ are both unambiguously defined and

$$(\Delta_{\psi_n} A)(\Delta_{\psi_n} B) = 0. \tag{12.17}$$

How can (12.17) hold? Is it not, in light of (12.16), a violation of (12.8) in Corollary 12.5? The answer is no, for the reason that ψ_n does not satisfy the domain assumptions in that corollary. Specifically, $A\psi_n$ is not in the domain of B, since $A\psi_n$ is does not satisfy the periodic boundary condition in the definition of $\mathrm{Dom}(B)$. Thus, ψ_n does not belong to $\mathrm{Dom}(BA)$.

Although it does not contradict Corollary 12.5, (12.17) certainly violates the spirit of the uncertainty principle. In the next section, we will show that no such strange counterexamples occur for the usual position and momentum operators.

12.3   Uncertainty Principle, Second Version

In this section, we will see that if A and B are taken to be the usual position and momentum operators X and P, the uncertainty principle holds whenever $\Delta_\psi X$ and $\Delta_\psi P$ are defined. We continue to use Definition 12.2 for the definition of the uncertainty in any operator, in which case, for $\Delta_\psi X$ and $\Delta_\psi P$ to be defined, we require only that ψ belong to $\mathrm{Dom}(X)$ and $\mathrm{Dom}(P)$.

We are now ready to formulate the strong version of the uncertainty principle.

Theorem 12.7 *Suppose ψ is a unit vector in $L^2(\mathbb{R})$ belonging to $\mathrm{Dom}(X)\cap \mathrm{Dom}(P)$. Then*

$$(\Delta_\psi X)(\Delta_\psi P) \geq \frac{\hbar}{2}, \tag{12.18}$$

where $\Delta_\psi X$ and $\Delta_\psi P$ are given by Definition 12.2.

Proof. According to Stone's theorem and Example 10.16, the operator P is $\hbar$ times the infinitesimal generator of the group $U(\cdot)$ of translations. That is to say, for all $\psi \in \mathrm{Dom}(P)$, we have

$$(P\psi)(x) = -i\hbar \lim_{a \to 0} \frac{\psi(x + a) - \psi(x)}{a},$$

where the limit is in the L^2 norm sense. Thus,

$$\langle X\psi, P\psi \rangle = \lim_{a \to 0} \left\langle X\psi, -i\hbar \left(\frac{\psi(x+a) - \psi(x)}{a} \right) \right\rangle$$

$$= \lim_{a \to 0} \left(\frac{1}{a} \langle x\psi(x), -i\hbar\psi(x+a) \rangle + \frac{i\hbar}{a} \langle X\psi, \psi \rangle \right)$$

$$= \lim_{a \to 0} \left(\frac{1}{a} \langle i\hbar(y-a)\psi(y-a), \psi(y) \rangle + \frac{i\hbar}{a} \langle X\psi, \psi \rangle \right),$$

where in the last step we have made the change of variable $y = x + a$.

If we rename the variable of integration back to x, we get

$$\langle X\psi, P\psi \rangle$$

$$= \lim_{a \to 0} \left(\left\langle i\hbar X \left(\frac{\psi(x-a) - \psi(x)}{a} \right), \psi(x) \right\rangle + i\hbar \langle \psi(x-a), \psi(x) \rangle \right)$$

$$= \lim_{a \to 0} \left(\left\langle i\hbar \left(\frac{\psi(x-a) - \psi(x)}{a} \right), X\psi(x) \right\rangle + i\hbar \langle \psi(x-a), \psi(x) \rangle \right)$$

$$= \langle P\psi, X\psi \rangle + i\hbar \langle \psi, \psi \rangle. \tag{12.19}$$

In the second equality, we have used that X is symmetric and that (check) if $\psi \in \mathrm{Dom}(X)$, then $\psi(x-a) \in \mathrm{Dom}(X)$ for each fixed a. In the last equality, we get a minus sign from having $\psi(x-a) - \psi(x)$ rather than $\psi(x+a) - \psi(x)$, and we use that translation is strongly continuous.

It should be noted that (12.19) is precisely what we would get by formally moving X to the right-hand side of the inner product, using the commutation relation $XP - PX = i\hbar I$, and then moving P to the left-hand side of the inner product. But to make *that* calculation rigorous, we would need to assume that ψ is in the domain of XP and the domain of PX. In (12.19), on the other hand, we have obtained the desired conclusion assuming only that ψ is in the domain of X and in the domain of P.

Having obtained (12.19), we can easily verify that for any real constants α and β, we have

$$\langle (X - \alpha I)\psi, (P - \beta I)\psi \rangle = \langle (P - \beta I)\psi, (X - \alpha I)\psi \rangle + i\hbar \langle \psi, \psi \rangle. \tag{12.20}$$

Solving (12.20) for $\langle \psi, \psi \rangle$ gives

$$\langle \psi, \psi \rangle = \frac{1}{i\hbar} \left(\langle (X - \alpha I)\psi, (P - \beta I)\psi \rangle - \langle (P - \beta I)\psi, (X - \alpha I)\psi \rangle \right)$$

$$= \frac{2}{\hbar} \mathrm{Im} \langle (X - \alpha I)\psi, (P - \beta I)\psi \rangle$$

$$\leq \frac{2}{\hbar} \| (X - \alpha I)\psi \| \, \| (P - \beta I)\psi \|, \tag{12.21}$$

by the Cauchy–Schwarz inequality. If ψ is a unit vector and we take $\alpha = \langle X \rangle_\psi$, and $\beta = \langle P \rangle_\psi$, then $\| (X - \alpha I)\psi \|^2 = (\Delta_\psi X)^2$ and $\| (P - \beta I)\psi \|^2 = (\Delta_\psi P)^2$. Thus, we get

$$1 \leq \frac{2}{\hbar}(\Delta_\psi X)(\Delta_\psi P),$$

which is equivalent to what we want to prove. ∎

We know from Sect. 12.2 that the strong form of the uncertainty principle does not hold if X and P are replaced by two arbitrary operators satisfying $AB - BA = ihI$ on $\text{Dom}(AB) \cap \text{Dom}(BA)$, even if $\text{Dom}(AB) \cap \text{Dom}(BA)$ is dense in **H**. Nevertheless, if we look carefully at the proof of Theorem 12.7, we can see what assumptions we would need on A and B to make the proof go through in a more general setting.

Theorem 12.8 *Suppose A and B are self-adjoint operators on* **H**. *Suppose that for all $a \in \mathbb{R}$ and $\psi \in \text{Dom}(A)$, we have that $e^{iaB}\psi$ belongs to $\text{Dom}(A)$ and that*

$$Ae^{iaB}\psi = e^{iaB}A\psi - \hbar a e^{iaB}\psi. \qquad (12.22)$$

Then for all unit vectors ψ in $\text{Dom}(A) \cap \text{Dom}(B)$, we have

$$(\Delta_\psi A)(\Delta_\psi B) \geq \frac{\hbar}{2},$$

where $\Delta_\psi A$ and $\Delta_\psi B$ are defined by Definition 12.2.

The relation

$$e^{iaB}A = Ae^{iaB} + \hbar a e^{iaB}, \quad a \in \mathbb{R}, \qquad (12.23)$$

which holds on $\text{Dom}(A)$, is a "semi-exponentiated" form of the canonical commutation relations. As shown in Exercise 6, there is a *formal* argument (ignoring domain issues) that the commutation relations $[A, B] = i\hbar I$ ought to imply the relations (12.22). Nevertheless, as Exercise 7 shows, this formal argument does not always give the correct conclusion. In Sect. 14.2, we will encounter a "fully exponentiated" form of the canonical commutation relations, in which both A and B are exponentiated.
Proof. See Exercise 5. ∎

Corollary 12.9 *For any $j = 1, \ldots n$ and any unit vector $\psi \in L^2(\mathbb{R}^n)$ with $\psi \in \text{Dom}(X_j) \cap \text{Dom}(P_j)$, we have*

$$(\Delta_\psi X_j)(\Delta_\psi P_j) \geq \frac{\hbar}{2}.$$

Proof. In the case that $A = X_j$ and $B = P_j$, we have $(e^{iaB/\hbar}\psi)(\mathbf{x}) = \psi(\mathbf{x} + a\mathbf{e}_j)$, by Exercise 2 in Chap. 10. Thus, in this case, (12.22) says that

$$(x_j + a)\psi(\mathbf{x} + a\mathbf{e}_j) = x_j\psi(\mathbf{x} + a\mathbf{e}_j) + a\psi(\mathbf{x} + a\mathbf{e}_j),$$

which is true. ∎

12.4 Minimum Uncertainty States

In this section, we look at the states that give equality in the uncertainty principle. Such states are known as minimum uncertainty states or *coherent states*. As in the general setting of Proposition 12.6, the condition for a equality is an eigenvector condition. That is to say, even though in Theorem 12.7, we allow ψ's that are not $\mathrm{Dom}(XP) \cap \mathrm{Dom}(PX)$, we do not get any new minimum uncertainty states by this weakening of our domain assumptions.

Proposition 12.10 *A unit vector $\psi \in \mathrm{Dom}(X) \cap \mathrm{Dom}(P)$ satisfies*

$$(\Delta_\psi X)(\Delta_\psi P) = \frac{\hbar}{2}$$

if and only if ψ satisfies

$$(X + i\delta P)\psi = \lambda\psi \tag{12.24}$$

for some nonzero real number δ and some complex number λ.

For convenience, we have made the substitution $\delta = -\gamma$ in (12.24) relative to Proposition 12.6.

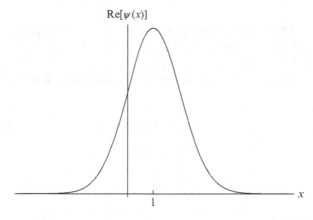

Re[$\psi(x)$]

FIGURE 12.1. Minimum uncertainty state with $\langle X \rangle = 1$, $\langle P \rangle = 0$, and $\Delta X = 1/2$.

Proof. All the relations in the proof of Theorem 12.7 are equalities, except for the inequality in the last line of (12.21). Equality will hold in that line if and only if one of $(X - \alpha I)\psi$ and $(P - \beta I)\psi$ is zero or $(P - \beta I)\psi$ is a pure-imaginary multiple of $(X - \alpha I)\psi$. Now, if ψ is a unit vector in $L^2(\mathbb{R})$, then neither ψ nor the Fourier transform of ψ can be supported at a single point; thus, neither $(X - \alpha I)\psi$ nor $(P - \beta I)\psi$ can be zero. We are left, then, with the condition that

$$(X - \alpha I)\psi = i\gamma(P - \beta I)\psi, \tag{12.25}$$

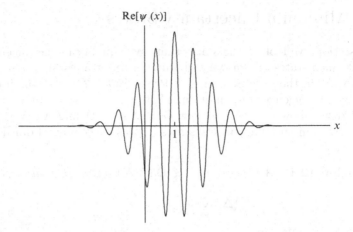

FIGURE 12.2. Minimum uncertainty state with $\langle X \rangle = 1$, $\langle P \rangle = 10$, and $\Delta X = 1/2$.

where γ is a nonzero real number, $\alpha = \langle A \rangle_\psi$ and $\beta = \langle B \rangle_\psi$. As in the proof of Proposition 12.6, (12.25) is equivalent to the assertion that ψ is an eigenvector for the operator $X - i\gamma P$. Letting $\delta = -\gamma$ gives the desired result. ∎

Proposition 12.11 *If the parameter δ in (12.24) is negative, there are no nonzero solutions to (12.24). If the parameter δ is positive, there exists a unique (up to multiplication by a constant) solution $\psi_{\delta,\lambda}$ to (12.24) for every complex number λ. The function $\psi_{\delta,\lambda}$ has the following additional properties*

$$\langle X \rangle = \operatorname{Re} \lambda$$
$$\langle P \rangle = \frac{1}{\delta} \operatorname{Im} \lambda$$
$$\frac{\Delta X}{\Delta P} = \delta.$$

Explicitly, we have

$$\psi_{\delta,\lambda}(x) = c_1 \exp\left\{ -\frac{(x-\lambda)^2}{2\delta\hbar} \right\}$$
$$= c_2 \exp\left\{ -\frac{(x - \langle X \rangle)^2}{2\delta\hbar} \right\} \exp\left\{ \frac{i \langle P \rangle x}{\hbar} \right\},$$

where all expectation values are taken in the state $\psi_{\delta,\lambda}$.

Note that among states with $(\Delta X)(\Delta P) = \hbar/2$, we can arrange for $\Delta X / \Delta P$ to be any positive real number, and once we have chosen $\Delta X / \Delta P$, we can then arrange for $\langle X \rangle$ and $\langle P \rangle$ to be any two real numbers. On the

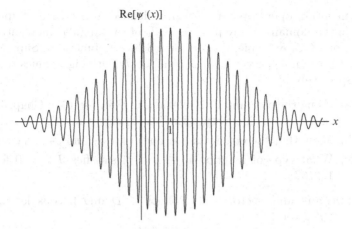

FIGURE 12.3. Minimum uncertainty state with $\langle X \rangle = 1$, $\langle P \rangle = 20$, and $\Delta X = 1$.

other hand, once $\Delta X / \Delta P$ and $\langle X \rangle$ and $\langle P \rangle$ have been specified, there is a unique quantum state with $(\Delta X)(\Delta P) = \hbar/2$. In Figs. 12.1–12.3, we have plotted the real part of $\psi_{\delta,\lambda}$ for several different values of the parameters, in a system of units for which $\hbar = 1$.

Proof. The equation $(X + i\delta P)\psi = \lambda\psi$ amounts to

$$x\psi + \delta\hbar \frac{d\psi}{dx} = \lambda\psi(x), \qquad (12.26)$$

where ψ is assumed to be in the domain of P, so that the distributional derivative of ψ is an L^2 function. If ψ were smooth, then the unique solution to (12.26) would be the function $\psi_{\delta,\lambda}$ given in the proposition, which is square-integrable if and only if $\delta > 0$. Even (12.26) is only assumed to hold in the distribution sense, the argument in the proof of Proposition 9.29 (with $e^{-x/\hbar}\psi(x)$ replaced by $\exp[(x-\lambda)^2/(2\delta\hbar)]\psi(x)$) shows that there are no additional solutions. The formulas for $\langle X \rangle$, $\langle P \rangle$, and $\Delta X / \Delta P$ can be computed either by tracing through the arguments in the proof of Theorem 12.7 or by direct calculation with the formula for $\psi_{\delta,\lambda}$. ∎

12.5 Exercises

1. Let α be a positive real number. Show that the following "additive" version of the uncertainty principle holds for all unit vectors $\psi \in \mathrm{Dom}(X) \cap \mathrm{Dom}(P)$:

$$\alpha\Delta_\psi X + \frac{1}{\alpha}\Delta_\psi P \geq \sqrt{2\hbar}.$$

2. In this exercise, we classify the simultaneous eigenvectors of the *noncommuting* operators $\hat{J}_1$ and $\hat{J}_2$. Let $\hat{J}_1$, $\hat{J}_2$, and $\hat{J}_3$ denote the angular

momentum operators on $L^2(\mathbb{R}^3)$ as defined in Sect. 3.10. Suppose ψ is in the domain of any product $\hat{J}_j \hat{J}_k$ of two angular momentum operators. (For example, ψ could be a Schwartz function.) Suppose also that ψ is an eigenvector for $\hat{J}_1$ and for $\hat{J}_2$ with eigenvalues α and β, respectively.

   (a) Using the commutation relations in Exercise 10 in Chap. 3, show that ψ is an eigenvector for $\hat{J}_3$ with eigenvalue 0.

   (b) Show that the eigenvalues α and β for $\hat{J}_1$ and $\hat{J}_2$ must be zero.

   (c) What type of function $\psi \in L^2(\mathbb{R}^3)$ satisfies $\hat{J}_j \psi = 0$ for $j = 1, 2, 3$?

3. Given any unit vector $\psi \in \mathrm{Dom}(X) \cap \mathrm{Dom}(P)$, consider another vector ϕ given by

$$\phi(x) = e^{ibx/\hbar}\psi(x - a).$$

Show that ϕ is a unit vector belonging to $\mathrm{Dom}(X) \cap \mathrm{Dom}(P)$ and that

$$\langle X \rangle_\phi = \langle X \rangle_\psi + a$$
$$\Delta_\phi X = \Delta_\psi X$$

and

$$\langle P \rangle_\phi = \langle P \rangle_\psi + b$$
$$\Delta_\phi P = \Delta_\psi P.$$

4. We have seen that a unit vector $\psi \in \mathrm{Dom}(X) \cap \mathrm{Dom}(P)$ is a minimum uncertainty state [i.e., $(\Delta_\psi X)(\Delta_\psi P) = \hbar/2$] if and only if there exists some $\delta > 0$ such that ψ is an eigenvector of the operator $X + i\delta P$. In that case, ψ is also an eigenvector for any operator of the form $c(X + i\delta P)$, with c being a nonzero constant. Consider, then, some fixed $\delta > 0$ and define an operator a by the formula

$$a = \frac{\frac{1}{\delta}(X + i\delta P)}{\sqrt{2\hbar/\delta}}.$$

Then a is just the annihilation operator, as defined in Chap. 11, for a harmonic oscillator with $m\omega = 1/\delta$. Thus, a and its adjoint a^* satisfy the relation $[a, a^*] = I$, and we have the "chain" of eigenvectors $\psi_n \in L^2(\mathbb{R})$ satisfying the properties listed in Theorem 11.2.

   (a) For any $\lambda \in \mathbb{C}$, find constants c_n so that the vector

$$\phi_\lambda := \sum_{n=0}^{\infty} c_n \psi_n$$

is an eigenvector for a with eigenvalue λ. Show that the resulting series converges in **H**.

(b) Let ϕ_λ denote the eigenvector obtained in Part (a), normalized so that $c_0 = 1$. Show that

$$\phi_\lambda = e^{\lambda a^*}\phi_0,$$

where the exponential is defined by

$$e^{\lambda a^*}\phi_0 = \sum_{n=0}^{\infty} \frac{\lambda^n}{n!}(a^*)^n\phi_0.$$

with convergence in $L^2(\mathbb{R})$.

5. Prove Theorem 12.8, following the outline of the proof of Theorem 12.7. Recall from Sect. 10.2 that $B/\hbar$ is the infinitesimal generator of the one-parameter unitary group $U(a) := e^{iaB/\hbar}$.

6. If X and Y are bounded operators, we may define $\mathrm{ad}_X(Y) = [X, Y]$, where $[X, Y] = XY - YX$. Thus, say, $(\mathrm{ad}_X)^3(Y) = [X, [X, [X, Y]]]$. It is not hard to show that for any bounded operators Y and X, we have

$$e^X Y e^{-X} = e^{\mathrm{ad}_X}(Y)$$
$$= Y + [X, Y] + \frac{[X, [X, Y]]}{2!} + \frac{[X, [X, [X, Y]]]}{3!} + \cdots.$$
$$(12.27)$$

(See Proposition 2.25 and Exercise 2.19 of [21].)

Suppose A and B are unbounded self-adjoint operators satisfying $[A, B] = i\hbar I$ on $\mathrm{Dom}(AB) \cap \mathrm{Dom}(BA)$. Show that *if* we could apply (12.27) with $X = iaB/\hbar$ and $Y = A$ (even though X and Y are unbounded), then A and B would satisfy (12.22).

7. Let A be the operator in Sect. 12.2, and let B be the unique self-adjoint extension of the operator B in that section. Show that the operators $X = iaB/\hbar$ and $Y = A$ do *not* satisfy (12.27).

Note: This result shows the hazards involved formally applying results for bounded operators to unbounded operators.

Hint: Show that the unitary operators $U(a) := \exp(iaB/\hbar)$ consist of "translation with wrap around," first on the eigenvectors of B and then on the whole Hilbert space.

13

Quantization Schemes for Euclidean Space

13.1 Ordering Ambiguities

One of the axioms of quantum mechanics states, "To each real-valued function f on the classical phase space there is associated a self-adjoint operator $\hat{f}$ on the quantum Hilbert space." The attentive reader will note that we have not, up to this point, given a general procedure for constructing $\hat{f}$ from f. If we call $\hat{f}$ the *quantization* of f, then we have only discussed the quantizations of a few very special classical observables, such as position, momentum, and energy.

Let us now think about what would go into quantizing a (more-or-less) general observable. Let us consider for simplicity a particle moving in $\mathbb{R}^1$ and let us assume that quantizations of x and p are the usual position and momentum operators X and P. What should the quantization of, say, xp be? Classically, xp and px are the same, but quantum mechanically, XP does not equal PX. Furthermore, neither XP nor PX is self-adjoint, because $(XP)^* = P^*X^* = PX$, and $PX \neq XP$. In this case, then, a reasonable candidate for the quantization would be

$$\widehat{xp} = \frac{1}{2}(XP + PX).$$

The significance of this simple example is that the failure of commutativity among quantum operators creates an ambiguity in the quantization process. It does not make sense to simply "replace x by X and p by P everywhere in the formula," since the ordering of position and momentum makes no difference on the classical side, but it does on the quantum

B.C. Hall, *Quantum Theory for Mathematicians*, Graduate Texts in Mathematics 267, DOI 10.1007/978-1-4614-7116-5_13,
© Springer Science+Business Media New York 2013

side. Up to this point, we have not really had to confront this ambiguity, because of the special form of the observables we have quantized. The Hamiltonian, for example, is typically of the form $H(x,p) = p^2/(2m) + V(x)$. Since each term contains only x or only p, it is natural to quantize H to $\hat{H} = P^2/(2m) + V(X)$, where $V(X)$ may be defined by the functional calculus or simply as multiplication by $V(x)$. In defining the angular momentum operators, we do encounter products of position and momentum, but never of the same component of position and momentum. For a particle in $\mathbb{R}^2$, for example, we have, $J = x_1 p_2 - x_2 p_1$. On the quantum side, X_1 commutes with P_2 and X_2 with P_2, and thus there is no ambiguity: $X_1 P_2 - X_2 P_1$ is the same as $P_2 X_1 - P_1 X_2$.

When we turn to the quantization of a general observable, however, we must confront the ordering ambiguity directly. Groenewold's theorem (Sect. 13.4) suggests that there is no single "perfect" quantization scheme. Nevertheless, there is one that is generally acknowledged as having the best properties, the Weyl quantization, and we spend most of our time with that particular scheme. Other quantization schemes do also play a role in physics, however; Wick-ordered quantization, notably, plays an important role in quantum field theory. (In quantum field theory, the replacement of certain Weyl-quantized operators with their Wick-quantized counterparts is interpreted as a type of renormalization.)

13.2   Some Common Quantization Schemes

In this section, we consider several of the most commonly used quantization schemes. For simplicity, we limit our attention to systems with one degree of freedom and to classical observables that are polynomials in x and p. (We consider the Weyl quantization in greater generality in Sect. 13.3.) Furthermore, we resolve in this section not to worry about domain questions and simply to use $C_c^\infty(\mathbb{R})$ as the domain for all of our operators. Thus, in this section, equality of operators means equality as maps of $C_c^\infty(\mathbb{R})$ to itself. It should be noted that the operators of the sort we will be considering may very well fail to be essentially self-adjoint, even if they are symmetric. Section 9.10 shows, for example, that the operator $P^2 - cX^4$, for $c > 0$, is not essentially self-adjoint on $C_c^\infty(\mathbb{R})$. We follow the terminology of harmonic analysis by referring to a classical symbol f as the *symbol* of its quantization $\hat{f}$. Once we have discussed each quantization scheme briefly, we will formalize the definitions of all the schemes in Definition 13.1.

The simplest approach to quantization is to choose, once and for all, which to put first, the position or the momentum operators. We may, for example, choose to put the momentum operators to the right, acting first, and the position operators to the left, acting second. In this approach, a polynomial in x and p will quantize to a differential operator in "standard form," with all the derivatives acting first, followed by multiplication operators. In harmonic analysis, there is a method for extending this quantization scheme to more-or-less arbitrary symbols, f. For a general (nonpolynomial) symbol f, the resulting operator $\hat{f}$ is known as a *pseudodifferential operator*.

A serious drawback of the pseudodifferential quantization is that even when the symbol f is real-valued, the operator $\hat{f}$ it produces is typically not self-adjoint (or even symmetric). If, for example, $f(x,p) = xp$, then the associated operator is XP, the adjoint of which is PX, which is not equal to XP. The simplest way to fix this problem is to symmetrize the operator by taking half the sum of the operator and its adjoint.

The Weyl quantization, meanwhile, takes more seriously the possibility of different orderings of X and P, by considering *all* possible orderings. Thus, in quantizing, say, x^2p^2, the Weyl quantization will give

$$\frac{1}{6}(X^2P^2 + XPXP + XP^2X + PX^2P + PXPX + P^2X^2).$$

For a general monomial, the Weyl quantization similarly averages all the possible orderings of the position and momentum operators.

For Wick-ordered and anti-Wick-ordered quantization, we no longer regard the position and momentum operators as the "basic" operators, but rather the creation and annihilation operators. Specifically, given any positive real number α, we introduce complex coordinates on the classical phase space by

$$z = x - i\alpha p$$
$$\bar{z} = x + i\alpha p. \tag{13.1}$$

(Although it would seem more natural to define z to be $x + i\alpha p$, this choice would lead to problems later, especially with the Segal–Bargmann transform.) We then consider the corresponding quantum operators, which we call the raising and lowering operators:

$$a^* = X - i\alpha P$$
$$a = X + i\alpha P. \tag{13.2}$$

In comparing these operators to the ones defined in the context of the harmonic oscillator, we should think of α as corresponding to $1/(m\omega)$. Even with this identification, however, the operators in (13.2) differ by a constant from the raising and lowering operators of Chap. 11. [The overall normalization of the raising and lowering operators is not important in this context, provided that we are consistent in the normalization between (13.1) and (13.2).] In particular, the commutator of a and a^* is not I but rather $2\alpha\hbar I$.

In Wick-ordered quantization, we begin by expressing the classical observable f in terms of z and $\bar{z}$ rather than in terms of x and p. When we quantize, we put all the lowering operators (coming from the factors of $\bar{z}$ in f) to the right, acting first, and the raising operators (coming from the factors of z in f) to the left, acting second. This approach to quantization is useful in quantum field theory, where letting the lowering operators act first can cause certain otherwise ill-defined expressions to become well defined. In anti-Wick-ordered quantization, we do the reverse, putting the raising operators to the right, acting first. Although anti-Wick-ordered quantization seems singular in the context of quantum field theory, in systems with finitely many degrees of freedom, it is actually better behaved than Wick-ordered quantization.

Definition 13.1 *Define several different quantization schemes for symbols that are polynomials in x and p as follows. Each scheme is uniquely determined—as a map from polynomials on $\mathbb{R}^2$ into operators on $C_c^\infty(\mathbb{R})$—by the indicated formulas.*

1. **Pseudodifferential operator quantization:**

$$Q(x^j p^k) = X^j P^k.$$

2. **Symmetrized pseudodifferential operator quantization:**

$$Q(x^j p^k) = \frac{1}{2}(X^j P^k + P^k X^j).$$

3. **Weyl quantization:**

$$Q(x^j p^k) = \frac{1}{(j+k)!} \sum_{\sigma \in S_{j+k}} \sigma\left(X, X, \ldots, X, P, P, \ldots, P\right),$$

   where for any operators $A_1, A_2, \ldots, A_n$ and any $\sigma \in S_n$, we define

$$\sigma(A_1, A_2, \ldots, A_n) = A_{\sigma(1)} A_{\sigma(2)} \cdots A_{\sigma(n)}. \tag{13.3}$$

4. **Wick-ordered quantization** *with parameter α:*

$$Q((x + i\alpha p)^j (x - i\alpha p)^k) = (X - i\alpha P)^k (X + i\alpha P)^j, \quad \alpha > 0.$$

5. **Anti-Wick-ordered quantization** *with parameter α:*

$$Q((x + i\alpha p)^j (x - i\alpha p)^k) = (X + i\alpha P)^j (X - i\alpha P)^k, \quad \alpha > 0.$$

In applications, the most useful quantization schemes are the Wick-ordered, anti-Wick-ordered, and Weyl schemes. All of the quantization schemes in Definition 13.1 except the pseudodifferential operator quantization have the property of mapping real-valued polynomials to symmetric operators on $C_c^\infty(\mathbb{R})$. (See Exercise 3 in the case of the Wick- and anti-Wick-ordered quantizations.)

In comparing the different quantization schemes, it is important to recognize that two different expressions may describe the same operator. We may calculate, for example, that

$$\frac{1}{2}(XP^2 + P^2X) = \frac{1}{2}(PXP + [X, P]P + PXP - P[X, P])$$
$$= PXP,$$

since $[X, P]$ is a multiple of the identity and thus commutes with P. As a result, we can eliminate the PXP term in the Weyl quantization of xp^2, with the result that

$$Q_{\text{Weyl}}(xp^2) = \frac{1}{3}(XP^2 + PXP + P^2X) = \frac{1}{2}(XP^2 + P^2X), \quad (13.4)$$

which coincides, in this very special case, with the symmetrized pseudodifferential quantization of xp^2.

Example 13.2 *If $f(x, p) = x^2$, then the Weyl, Wick-ordered and anti-Wick-ordered quantizations of f are as follows:*

$$Q_{\text{Weyl}}(x^2) = X^2$$
$$Q_{\text{Wick}}(x^2) = X^2 - \frac{1}{2}\alpha h I$$
$$Q_{\text{anti-Wick}}(x^2) = X^2 + \frac{1}{2}\alpha h I.$$

Proof. The value for $Q_{\text{Weyl}}(x^2)$ is apparent. To compute the Wick- and anti-Wick-ordered quantizations, we first write x as $(z + \bar{z})/2$, so that

$$x^2 = \frac{(z + \bar{z})^2}{4} = \frac{1}{4}(z^2 + 2z\bar{z} + \bar{z}^2).$$

Thus, we have, for example,

$$Q_{\text{Wick}}(x^2) = \frac{1}{4}\left((X - i\alpha P)^2 + 2(X - i\alpha P)(X + i\alpha P) + (X + i\alpha P)^2\right).$$

When we expand this expression out, the P^2 terms cancel, and the XP and PX terms from $(X - i\alpha P)^2$ will cancel with the XP and PX terms from $(X + i\alpha P)^2$. Thus, we will be left with X^2 terms and the XP and PX terms from the cross-term above:

$$Q_{\text{Wick}}(x^2) = \frac{1}{4}\left(4X^2 + 2i\alpha[X, P]\right).$$

Using the commutation relation between X and P gives the desired result. The calculation of $Q_{\mathrm{antiWick}}(x^2)$ is identical except that the order of the factors in the cross-term is reversed, which gives the opposite sign for the $[X, P]$ term. ∎

Proposition 13.3 *The Weyl quantization—viewed as a linear map of the space of polynomials on $\mathbb{R}^2$ into operators on $C_c^\infty(\mathbb{R})$—is uniquely characterized by the following identity:*

$$Q_{\mathrm{Weyl}}((ax + bp)^j) = (aX + bP)^j \tag{13.5}$$

for all non-negative integers j and all $a, b \in \mathbb{C}$.

Proof. The Weyl quantization is easily seen to satisfy the identity

$$Q_{\mathrm{Weyl}}((a_1 x + b_1 p) \cdots (a_j x + b_j p))$$
$$= \frac{1}{j!} \sum_{\sigma \in S_j} \sigma(a_1 X + b_1 P, \ldots, a_j X + b_j P), \tag{13.6}$$

for all sequences $a_1, \ldots, a_j$ and $b_1, \ldots, b_j$ of complex numbers, where the expression $\sigma(\cdot, \cdot, \ldots, \cdot)$ is defined by (13.3). Specializing to the case where all the a_j's are equal to a and all the b_j's are equal to b gives (13.5). Conversely, suppose that Q is any linear map of polynomials into operators on $C_c^\infty(\mathbb{R})$ satisfying $Q((ax + bp)^j) = (aX + bP)^j$ for all a, b, and j. For each j, let V_j denote the space of homogeneous polynomials f of degree j such that $Q(f) = Q_{\mathrm{Weyl}}(f)$. Then V_j contains all polynomials of the form $(ax + bp)^j$, and thus, by Exercise 1, V_j consists of all homogeneous polynomials of degree j, so that $Q = Q_{\mathrm{Weyl}}$. ∎

Proposition 13.4 *The Weyl quantization satisfies*

$$Q_{\mathrm{Weyl}}(xg) = Q_{\mathrm{Weyl}}(x)Q_{\mathrm{Weyl}}(g) - \frac{i\hbar}{2}Q_{\mathrm{Weyl}}\left(\frac{\partial g}{\partial p}\right) \tag{13.7}$$

$$= Q_{\mathrm{Weyl}}(g)Q_{\mathrm{Weyl}}(x) + \frac{i\hbar}{2}Q_{\mathrm{Weyl}}\left(\frac{\partial g}{\partial p}\right) \tag{13.8}$$

and

$$Q_{\mathrm{Weyl}}(pg) = Q_{\mathrm{Weyl}}(p)Q_{\mathrm{Weyl}}(g) + \frac{i\hbar}{2}Q_{\mathrm{Weyl}}\left(\frac{\partial g}{\partial x}\right) \tag{13.9}$$

$$= Q_{\mathrm{Weyl}}(g)Q_{\mathrm{Weyl}}(p) - \frac{i\hbar}{2}Q_{\mathrm{Weyl}}\left(\frac{\partial g}{\partial x}\right) \tag{13.10}$$

for all polynomials g in x and p.

It should be noted that the formulas for the Weyl quantization in Proposition 13.4 may not give the same "expression" for $Q_{\mathrm{Weyl}}(f)$ as does Definition 13.1, but it does give the same operator. [Compare (13.4).]

Proof. Suppose $A = (a_1 X + b_1 P)$ and $B = (a_2 X + b_2 P)$. Then $[A, B]$ is a multiple of I, from which we can easily verify that

$$AB^j = B^k AB^{j-k} + k[A, B]B^{j-1},$$

for $0 \leq k \leq j$. If we sum this relation over k and divide by $j + 1$, we obtain

$$AB^j = \frac{1}{j+1} \sum_{k=0}^{j} B^k AB^{j-k} + \frac{1}{j+1} \frac{j(j+1)}{2}[A, B]B^{j-1}. \tag{13.11}$$

Now, A is the Weyl quantization of $(a_1 X + b_1 p)$ and B^j is the Weyl quantization of $(a_2 x + b_2 p)^j$, and both terms on the right-hand side of (13.11) are easily recognized as Weyl quantizations. Thus, after rearranging the terms and evaluating the commutator, (13.11) becomes,

$$Q_{\text{Weyl}}((a_1 x + b_1 p)(a_2 x + b_2 p)^j)$$
$$= Q_{\text{Weyl}}(a_1 x + b_1 p)Q_{\text{Weyl}}((a_2 x + b_2 p)^j)$$
$$- i\hbar \frac{j}{2}(a_1 b_2 - a_2 b_1)Q_{\text{Weyl}}((a_1 x + b_1 p)^{j-1}). \tag{13.12}$$

Meanwhile, if we run the same argument starting with $B^j A$ we obtain a similar result:

$$Q_{\text{Weyl}}((a_1 x + b_1 p)(a_2 x + b_2 p)^j)$$
$$= Q_{\text{Weyl}}((a_2 x + b_2 p)^j)Q_{\text{Weyl}}(a_1 x + b_1 p)$$
$$+ i\hbar \frac{j}{2}(a_1 b_2 - a_2 b_1)Q_{\text{Weyl}}((a_1 x + b_1 p)^{j-1}). \tag{13.13}$$

If we specialize to the case $(a_1, b_1) = (1, 0)$ and $(a_2, b_2) = (a, b)$, we get

$$Q_{\text{Weyl}}(x(ax + bp)^j) - Q_{\text{Weyl}}(x)Q_{\text{Weyl}}((ax + bp)^j)$$
$$- i\hbar \frac{j}{2}bQ_{\text{Weyl}}((ax + bp)^{j-1}), \tag{13.14}$$

where the last term on the right-hand side of (13.14) is $-i\hbar/2$ times the Weyl quantization of $\partial(ax + bp)^j/\partial p$. Thus, (13.14) is precisely (13.7) in the case $g(x, p) = (ax + bp)^j$. We can then see from Exercise 1 that (13.7) hold for all polynomials g. The proofs of (13.8), (13.9), and (13.10) are similar. ∎

13.3   The Weyl Quantization for $\mathbb{R}^{2n}$

In this section, we study the Weyl quantization on a much larger class of symbols (i.e., classical observables) than the polynomial symbols considered in the previous section. We also generalize from symbols defined on $\mathbb{R}^2$ to symbols defined on $\mathbb{R}^{2n}$.

13.3.1   Heuristics

It is a straightforward matter to extent the Weyl quantization on polynomials from $\mathbb{R}^2$ to $\mathbb{R}^{2n}$. This extended quantization will satisfy

$$Q_{\text{Weyl}}((\mathbf{a} \cdot \mathbf{p} + \mathbf{b} \cdot \mathbf{p})^j) = (\mathbf{a} \cdot \mathbf{X} + \mathbf{b} \cdot \mathbf{P})^j \qquad (13.15)$$

for all $\mathbf{a}, \mathbf{b} \in \mathbb{R}^n$ and all non-negative integers j, as in Proposition 13.3 in the $n = 1$ case. Suppose we wish to extend Q_{Weyl} to certain nonpolynomial symbols, starting with complex exponentials. If we multiply (13.15) by $(i)^j/j!$ and sum on j, we would expect to have

$$Q_{\text{Weyl}}\left(e^{i(\mathbf{a} \cdot \mathbf{x} + \mathbf{b} \cdot \mathbf{p})}\right) = e^{i(\mathbf{a} \cdot \mathbf{X} + \mathbf{b} \cdot \mathbf{P})}. \qquad (13.16)$$

Now, if f is any sufficiently nice function on $\mathbb{R}^{2n}$, we can expand f as an integral involving functions of the form $\exp(i(\mathbf{a} \cdot \mathbf{x} + \mathbf{b} \cdot \mathbf{p}))$, by using the Fourier transform:

$$f(\mathbf{x}, \mathbf{p}) = (2\pi)^{-n} \int_{\mathbb{R}^{2n}} \hat{f}(\mathbf{a}, \mathbf{b}) e^{i(\mathbf{a} \cdot \mathbf{x} + \mathbf{b} \cdot \mathbf{p})} \, d\mathbf{a} \, d\mathbf{b},$$

where $\hat{f}$ is the Fourier transform of f. In light of (13.16), it is then natural to define

$$Q_{\text{Weyl}}(f) = (2\pi)^{-n} \int_{\mathbb{R}^{2n}} \hat{f}(\mathbf{a}, \mathbf{b}) e^{i(\mathbf{a} \cdot \mathbf{X} + \mathbf{b} \cdot \mathbf{P})} \, d\mathbf{a} \, d\mathbf{b}. \qquad (13.17)$$

Before proceeding, let us pause for a moment to compute the operator $\exp(i(\mathbf{a} \cdot \mathbf{X} + \mathbf{b} \cdot \mathbf{P}))$. If A and B are bounded operators that commute with their commutator (i.e., such that $[A, [A, B]] = [B, [A, B]] = 0$), then

$$e^{A+B} = e^{-[A,B]/2} e^A e^B. \qquad (13.18)$$

(See Theorem 14.1, which is proved in Sect. 3.1 of [21]. Equation (13.18) is a special case of the Baker–Campbell–Hausdorff Formula.) If we *formally* apply (13.18) with $A = i\mathbf{a} \cdot \mathbf{X}$ and $B = i\mathbf{b} \cdot \mathbf{P}$ (even though these are unbounded operators), we obtain

$$e^{i(\mathbf{a} \cdot \mathbf{X} + \mathbf{b} \cdot \mathbf{P})} = e^{i\hbar(\mathbf{a} \cdot \mathbf{b})/2} e^{i\mathbf{a} \cdot \mathbf{X}} e^{i\mathbf{b} \cdot \mathbf{P}}. \qquad (13.19)$$

Meanwhile, by Example 10.16 in Sect. 10.2, we know that

$$(e^{i\mathbf{b} \cdot \mathbf{P}} \psi)(\mathbf{x}) = \psi(\mathbf{x} + \hbar \mathbf{b}).$$

Thus, we may reasonably hope that

$$\left(e^{i(\mathbf{a} \cdot \mathbf{X} + \mathbf{b} \cdot \mathbf{P})} \psi\right)(\mathbf{x}) = e^{i\hbar(\mathbf{a} \cdot \mathbf{b})/2} e^{i\mathbf{a} \cdot \mathbf{x}} \psi(\mathbf{x} + \hbar \mathbf{b}). \qquad (13.20)$$

In general, we get incorrect results if we formally apply results for bounded operators to operators that are unbounded. In this case, however, the result of the formal calculation is correct. The simplest way to prove this is to replace $\mathbf{a}$ and $\mathbf{b}$ by $t\mathbf{a}$ and $t\mathbf{b}$ on the right-hand side of (13.19) and to check that the result is a strongly continuous one-parameter unitary group.

Proposition 13.5 *For all* **a** *and* **b** *in* $\mathbb{R}^n$, *the operators* $U_{\mathbf{a},\mathbf{b}}(t)$ *on* $L^2(\mathbb{R}^n)$ *given by*

$$(U_{\mathbf{a},\mathbf{b}}(t)\psi)(\mathbf{x}) = e^{it^2\hbar(\mathbf{a}\cdot\mathbf{b})/2}e^{it\mathbf{a}\cdot\mathbf{x}}\psi(\mathbf{x} + t\hbar\mathbf{b}) \tag{13.21}$$

form a strongly continuous one-parameter unitary group. The infinitesimal generator of this group coincides with $\mathbf{a}\cdot\mathbf{X} + \mathbf{b}\cdot\mathbf{P}$ *on* $C_c^\infty(\mathbb{R}^n)$ *and is essentially self-adjoint on this domain. Thus, if* $\mathbf{a}\cdot\mathbf{X} + \mathbf{b}\cdot\mathbf{P}$ *denotes the unique self-adjoint extension of the infinitesimal generator on* $C_c^\infty(\mathbb{R}^n)$, *it follows from Stone's theorem that*

$$e^{it(\mathbf{a}\cdot\mathbf{X}+\mathbf{b}\cdot\mathbf{P})} = e^{it^2\hbar(\mathbf{a}\cdot\mathbf{b})/2}e^{it\mathbf{a}\cdot\mathbf{X}}e^{it\mathbf{b}\cdot\mathbf{P}}$$

for all $t \in \mathbb{R}$. *In particular, (13.19) and (13.20) hold.*

Proof. It is apparent that $U_{\mathbf{a},\mathbf{b}}$ is unitary for each **a** and **b**, and it is a simple direct computation to show that it is indeed a unitary group. Strong continuity is proved in the usual way using a dense subspace, as in the proof of Example 10.12. When ψ is in $C_c^\infty(\mathbb{R}^n)$, it is easy to differentiate the right-hand side of (13.21) with respect to t at $t = 0$ to obtain the formula for the infinitesimal generator. Finally, the essential self-adjointness of $\mathbf{a}\cdot\mathbf{X} + \mathbf{b}\cdot\mathbf{P}$ on $C_c^\infty(\mathbb{R}^n)$ is precisely the content of Proposition 9.40. ∎

With the computation of the operator $e^{i(\mathbf{a}\cdot\mathbf{X}+\mathbf{b}\cdot\mathbf{P})}$ in hand, we return to our analysis of the proposed formula (13.17) for the general Weyl quantization. If the Fourier transform of f is in $L^1(\mathbb{R}^{2n})$, we can regard the right-hand side of (13.17) as an absolutely convergent "Bochner" integral with values in the Banach space $\mathcal{B}(\mathbf{H})$. For our purposes, however, it is more convenient to think of operators on $L^2(\mathbb{R}^n)$ as integral operators and to write down a formula for the integral kernel of $Q_{\text{Weyl}}(f)$ in terms of f itself. (But see Exercise 7.)

At a formal level, the operator mapping ψ to $e^{i\hbar(\mathbf{a}\cdot\mathbf{b})/2}e^{i\mathbf{a}\cdot\mathbf{x}}\psi(\mathbf{x} + \hbar\mathbf{b})$ may be thought of as an "integral" operator, with integral kernel given by

$$e^{i\hbar(\mathbf{a}\cdot\mathbf{b})/2}e^{i\mathbf{a}\cdot\mathbf{x}}\delta_n(\mathbf{x} + \hbar\mathbf{b} - \mathbf{y}), \tag{13.22}$$

where δ_n is an n-dimensional delta-function (the n-dimensional analog of the distribution in Example A.26). Thus, it should be possible to obtain the integral kernel of $Q_{\text{Weyl}}(f)$ by integrating the preceding expression against $\hat{f}(\mathbf{a}, \mathbf{b})$. To evaluate the resulting integral, we make the change of variable $\mathbf{c} = \hbar\mathbf{b}$, in which case we obtain

$$(2\pi\hbar)^{-n}\int_{\mathbb{R}^n}\int_{\mathbb{R}^n} e^{i(\mathbf{a}\cdot\beta)/2}e^{i\mathbf{a}\cdot\mathbf{x}}\delta_n(\mathbf{x} + \mathbf{c} - \mathbf{y})\hat{f}(\mathbf{a}, \mathbf{c}/\hbar)\, d\mathbf{c}\, d\mathbf{a}$$

$$= (2\pi\hbar)^{-n}\int_{\mathbb{R}^n} e^{i(\mathbf{a}\cdot(\mathbf{y}-\mathbf{x}))/2}e^{i\mathbf{a}\cdot\mathbf{x}}\hat{f}(\mathbf{a}, (\mathbf{y} - \mathbf{x})/\hbar)\, d\mathbf{a}$$

$$= \hbar^{-n}(2\pi)^{-n/2}\left[(2\pi)^{-n/2}\int_{\mathbb{R}^n} e^{i\mathbf{a}\cdot(\mathbf{x}+\mathbf{y})/2}\hat{f}(\mathbf{a}, (\mathbf{y} - \mathbf{x})/\hbar)\, d\mathbf{a}\right]. \tag{13.23}$$

We may recognize the integral in square brackets in the last line of (13.23) as undoing the Fourier transform of f in the $\mathbf{x}$-variable, leaving us with the partial Fourier transform of f in the $\mathbf{p}$ variable, evaluated at the points $(\mathbf{x} + \mathbf{y})/2, (\mathbf{y} - \mathbf{x})/\hbar$. (The partial Fourier transform means the ordinary Fourier transform with respect to one of the variables, with the other variable fixed.) Thus, we expect that $Q_{\text{Weyl}}(f)$ should be the integral operator with integral kernel κ_f given by

$$\kappa_f(\mathbf{x}, \mathbf{y}) = (2\pi\hbar)^{-n} \int_{\mathbb{R}^n} f((\mathbf{x} + \mathbf{y})/2, \mathbf{p}) e^{-i(\mathbf{y} - \mathbf{x}) \cdot \mathbf{p}/\hbar} \, d\mathbf{p}. \qquad (13.24)$$

13.3.2 The L^2 Theory

With the preceding calculations as motivation, we now define $Q_{\text{Weyl}}(f)$ to be the integral operator with kernel κ_f, beginning with the case in which f belongs to $L^2(\mathbb{R}^{2n})$. The resulting operators will turn out to be Hilbert–Schmidt operators on $L^2(\mathbb{R}^n)$.

If $\mathbf{H}$ is a Hilbert space and $A \in \mathcal{B}(\mathbf{H})$ is a non-negative self-adjoint operator on $\mathbf{H}$, then it can be shown that A has a well-defined (but possibly infinite) *trace*. What this means is that the value of

$$\text{trace}(A) := \sum_j \langle e_j, A e_j \rangle$$

is the same for each orthonormal basis $\{e_j\}$ of $\mathbf{H}$. Note that since A is a non-negative operator, $\langle e_j, A e_j \rangle$ is a non-negative real number, so that the sum is always defined, but may have the value $+\infty$.

Now, if A is any bounded operator, then A^*A is self-adjoint and non-negative. We say that A is *Hilbert–Schmidt* if

$$\text{trace}(A^*A) < \infty.$$

Given two Hilbert–Schmidt operators A and B, it can be shown that A^*B is a trace-class operator, meaning that the sum

$$\text{trace}(A^*B) := \sum_{j=1}^{\infty} \langle e_j, A^* B e_j \rangle$$

is absolutely convergent and the value of the sum is independent of the choice of orthonormal basis. We define the *Hilbert–Schmidt inner product* of A and B and the associated *Hilbert–Schmidt norm* of A by

$$\langle A, B \rangle_{\text{HS}} := \text{trace}(A^*B)$$
$$\|A\|_{\text{HS}} := \sqrt{\text{trace}(A^*A)}.$$

It can be shown that the space of Hilbert–Schmidt operators on $\mathbf{H}$ forms a Hilbert space with respect to the Hilbert–Schmidt inner product.

(See Sect. 19.2 for more details.) We denote the space of Hilbert–Schmidt operators on $\mathbf{H}$ by $HS(\mathbf{H})$.

We will make use of the following standard (and elementary) result characterizing Hilbert–Schmidt operators on $L^2(\mathbb{R}^n)$ in terms of integral operators. (See, for example, Theorem VI.23 in Volume I of [34].)

Proposition 13.6 *If κ is in $L^2(\mathbb{R}^n \times \mathbb{R}^n)$ then for every $\psi \in L^2(\mathbb{R}^n)$, the integral*

$$A_\kappa(\psi)(\mathbf{x}) := \int_{\mathbb{R}^n} \kappa(\mathbf{x}, \mathbf{y})\psi(\mathbf{y}) \, d\mathbf{y} \tag{13.25}$$

is absolutely convergent for almost every $\mathbf{x} \in \mathbb{R}^n$, and $A_\kappa(\psi)$ also belongs to $L^2(\mathbb{R}^n)$. Furthermore, the operator A_κ is a Hilbert–Schmidt operator on $L^2(\mathbb{R}^n)$ and

$$\|A_\kappa\|_{HS} = \|\kappa\|_{L^2(\mathbb{R}^n \times \mathbb{R}^n)}.$$

Conversely, for any Hilbert–Schmidt operator A on $L^2(\mathbb{R}^n)$, there exists a unique $\kappa \in L^2(\mathbb{R}^n \times \mathbb{R}^n)$ such that $A = A_\kappa$.

We are now ready, using discussion in Sect. 13.3.1 as motivation, to define the Weyl quantization of L^2 symbols.

Definition 13.7 *For all $f \in L^2(\mathbb{R}^{2n})$, define $\kappa_f : \mathbb{R}^{2n} \to \mathbb{C}$ by*

$$\kappa_f(\mathbf{x}, \mathbf{y}) = (2\pi\hbar)^{-n} \int_{\mathbb{R}^n} f((\mathbf{x}+\mathbf{y})/2, \mathbf{p})e^{-i(\mathbf{y}-\mathbf{x})\cdot\mathbf{p}/\hbar} \, d\mathbf{p}, \tag{13.26}$$

*and define the **Weyl quantization** of f, as an operator on $L^2(\mathbb{R}^n)$, by*

$$Q_{\text{Weyl}}(f) = A_{\kappa_f},$$

where A_{κ_f} is defined by (13.25).

The integral in (13.26) is not necessarily absolutely convergent, and should be understood as computing a partial Fourier transform. Thus, we should, strictly speaking, replace the right-hand side of (13.26) with

$$\lim_{R \to \infty} (2\pi\hbar)^{-n} \int_{|\mathbf{p}| \le R} f((\mathbf{x}+\mathbf{y})/2, \mathbf{p})e^{-i(\mathbf{y}-\mathbf{x})\cdot\mathbf{p}/\hbar} \, d\mathbf{p}, \tag{13.27}$$

where the limit is in the norm topology of $L^2(\mathbb{R}^{2n})$. [The partial Fourier transform maps the Schwartz space $\mathcal{S}(\mathbb{R}^{2n})$ to itself. By Fubini's theorem and the Plancherel formula for $\mathbb{R}^n$, the partial Fourier transform is an L^2-isometry and extends to a unitary map of $L^2(\mathbb{R}^{2n})$ to itself. This unitary map can be computed by the usual formula on functions in $L^1 \cap L^2$ and can be computed by the limiting formula similar to (13.27) in general.]

In words, we may describe the procedure for computing κ_f at a point $(\mathbf{x}^1, \mathbf{x}^2)$ in $\mathbb{R}^{2n}$ as follows. First, compute the partial Fourier transform $\mathcal{F}_{\mathbf{p}}$

of $f(\mathbf{x}, \mathbf{p})$ in the $\mathbf{p}$-variable, resulting in the function $(\mathcal{F}_\mathbf{p} f)(\mathbf{x}, \xi)$. Then evaluate $\mathcal{F}_\mathbf{p} f$ at the point $\mathbf{x} = (\mathbf{x}^1 + \mathbf{x}^2)/2$, $\xi = (\mathbf{x}^2 - \mathbf{x}^1)/\hbar$. Finally, multiply the result by $\hbar^{-n}(2\pi)^{-n/2}$ to get

$$\kappa_f(\mathbf{x}^1, \mathbf{x}^2) = \hbar^{-n}(2\pi)^{-n/2}(\mathcal{F}_\mathbf{p} f)((\mathbf{x}^1 + \mathbf{x}^2)/2, (\mathbf{x}^2 - \mathbf{x}^1)/\hbar). \quad (13.28)$$

Theorem 13.8 *The map Q_{Weyl} is a constant multiple of a unitary map of $L^2(\mathbb{R}^{2n})$ onto $\mathrm{HS}(L^2(\mathbb{R}^n))$. The inverse map $Q_{\mathrm{Weyl}}^{-1} : \mathrm{HS}(L^2(\mathbb{R}^n)) \to L^2(\mathbb{R}^{2n})$ is given by*

$$Q_{\mathrm{Weyl}}^{-1}(A)(\mathbf{x}, \mathbf{p}) = \hbar^n \int_{\mathbb{R}^n} \kappa(\mathbf{x} - \hbar\mathbf{b}/2, \mathbf{x} + \hbar\mathbf{b}/2) e^{i\mathbf{b} \cdot \mathbf{p}} \, d\mathbf{b},$$

where κ is the integral kernel of A as in Proposition 13.6.

Furthermore, for all $f \in L^2(\mathbb{R}^{2n})$, we have $Q_{\mathrm{Weyl}}(\bar{f}) = Q_{\mathrm{Weyl}}(f)^*$; in particular, $Q_{\mathrm{Weyl}}(f)$ is self-adjoint if f is real valued.

Properly speaking, the integral in the theorem should be understood as an L^2 limit, as in (13.27). The fact that Q_{Weyl} is unitary (up to a constant) tells us that for an appropriate constant c, the operators $ce^{i(\mathbf{a} \cdot \mathbf{X} + \mathbf{b} \cdot \mathbf{P})}$ form an "orthonormal basis in the continuous sense" for the Hilbert space $\mathrm{HS}(L^2(\mathbb{R}^n))$. (Compare Sect. 6.6.)

It is possible, using the same formulas, to extend the notion of Weyl quantization to symbols belonging the space of tempered distributions, that is, the space of continuous linear functionals on $\mathcal{S}(\mathbb{R}^{2n})$. We will not, however, develop this construction here. See [11] for more information.

Proof. Proposition 13.6 gives a unitary identification of $\mathrm{HS}(L^2(\mathbb{R}^n))$ with $L^2(\mathbb{R}^n \times \mathbb{R}^n)$. Thus, it suffices to show that the map $f \mapsto \kappa_f$ is a multiple of a unitary map. This result holds because the partial Fourier transform is a unitary map of $L^2(\mathbb{R}^{2n})$ to itself and composition with an invertible linear map is a constant multiple of a unitary map. The inverse of the map $f \mapsto \kappa_f$ is obtained by inverting the linear map and undoing the partial Fourier transform. Finally, it is apparent from (13.26) that

$$\kappa_{\bar{f}}(\mathbf{x}, \mathbf{y}) = \overline{\kappa_f(\mathbf{y}, \mathbf{x})}.$$

This, along with Exercise 6, shows that $Q_{\mathrm{Weyl}}(\bar{f}) = Q_{\mathrm{Weyl}}(f)^*$. ∎

13.3.3   The Composition Formula

If f and g are L^2 functions on $\mathbb{R}^{2n}$, then $Q_{\mathrm{Weyl}}(f)$ and $Q_{\mathrm{Weyl}}(g)$ are Hilbert–Schmidt operators, in which case their product is again Hilbert–Schmidt. (Indeed, the product of a Hilbert–Schmidt operator and a bounded operator is always Hilbert–Schmidt.) Thus, since Q_{Weyl} is a bijection of $L^2(\mathbb{R}^{2n})$ with $\mathrm{HS}(L^2(\mathbb{R}^n))$, there is a unique L^2 function, which we denote by $f \star g$, such that

$$Q_{\mathrm{Weyl}}(f) Q_{\mathrm{Weyl}}(g) = Q_{\mathrm{Weyl}}(f \star g). \quad (13.29)$$

(Of course, the operator $\star$, like the Weyl quantization itself, depends on $\hbar$, but we suppress this dependence in the notation.)

Proposition 13.9 *The Moyal product $f \star g$ may be characterized in terms of the Fourier transform as*

$$\widehat{(f \star g)}(\mathbf{a}, \mathbf{b}) = (2\pi)^{-n} \iint e^{-i\hbar(\mathbf{a} \cdot \mathbf{b}' - \mathbf{b} \cdot \mathbf{a}')/2}$$
$$\times \hat{f}(\mathbf{a} - \mathbf{a}', \mathbf{b} - \mathbf{b}')\hat{g}(\mathbf{a}', \mathbf{b}') \, d\mathbf{a}' \, d\mathbf{b}',$$

where both integrals are over $\mathbb{R}^n$.

Note that if we set $\hbar = 0$ in the above formula, $\widehat{f \star g}$ reduces to $(2\pi)^{-n}$ times the convolution of $\hat{f}$ and $\hat{g}$, which is nothing but the Fourier transform of fg. It is thus not difficult to show (Exercise 10) that

$$\lim_{\hbar \to 0^+} f \star g = fg.$$

That is to say, the Moyal product $f \star g$ is a "deformation" of the ordinary pointwise product of functions on $\mathbb{R}^{2n}$. More generally, the Moyal product can be expanded in an asymptotic expansion in powers of $\hbar$, as explained in Sect. 2.3 of [11]. This expansion terminates in the case that f and g are both polynomials.

Proof. It is, of course, possible to obtain this formula using kernel functions. It is, however, easier to work with the (13.17), which can be shown (Exercise 7) to give the same result as Definition 13.7 when f is a Schwartz function. We assume standard properties of the Bochner integral for functions with values in a Banach space [in our case, $\mathcal{B}(\mathbf{H})$], which are similar to those of the Lebesgue integral. (See, for example, Sect. V.5 of [46].)

We have, then,

$$Q_{\text{Weyl}}(f)Q_{\text{Weyl}}(g) = (2\pi)^{-n} \iint \hat{f}(\mathbf{a}, \mathbf{b})e^{i(\mathbf{a} \cdot \mathbf{X} + \mathbf{b} \cdot \mathbf{P})} \, d\mathbf{a} \, d\mathbf{b}$$
$$\times (2\pi)^{-n} \iint \hat{g}(\mathbf{a}', \mathbf{b}')e^{i(\mathbf{a}' \cdot \mathbf{X} + \mathbf{b}' \cdot \mathbf{P})} \, d\mathbf{a}' \, d\mathbf{b}'. \quad (13.30)$$

Now, it is an easy calculation to verify, using Proposition 13.5, that

$$e^{i(\mathbf{a} \cdot \mathbf{X} + \mathbf{b} \cdot \mathbf{P})}e^{i(\mathbf{a}' \cdot \mathbf{X} + \mathbf{b}' \cdot \mathbf{P})} = e^{-i\hbar(\mathbf{a} \cdot \mathbf{b}' - \mathbf{b} \cdot \mathbf{a}')/2}e^{i((\mathbf{a} + \mathbf{a}') \cdot \mathbf{X} + (\mathbf{b} + \mathbf{b}') \cdot \mathbf{P})}, \quad (13.31)$$

which is what one obtains by formally applying the special case of the Baker–Campbell–Hausdorff formula in (13.18). Thus, we may combine the integrals in (13.30) to obtain

$$Q_{\text{Weyl}}(f)Q_{\text{Weyl}}(g) = (2\pi)^{-2n} \iiiint e^{-i\hbar(\mathbf{a} \cdot \mathbf{b}' - \mathbf{b} \cdot \mathbf{a}')/2}e^{i((\mathbf{a} + \mathbf{a}') \cdot \mathbf{X} + (\mathbf{b} + \mathbf{b}') \cdot \mathbf{P})}$$
$$\times \hat{f}(\mathbf{a}, \mathbf{b})\hat{g}(\mathbf{a}', \mathbf{b}') \, d\mathbf{a} \, d\mathbf{b} \, d\mathbf{a}' \, d\mathbf{b}'.$$

By introducing new variables $\mathbf{c} = \mathbf{a} + \mathbf{a}'$ and $\mathbf{d} = \mathbf{b} + \mathbf{b}'$ in the $\mathbf{a}$ and $\mathbf{b}$ integrals and reversing the order of integration, we obtain, after simplifying the exponent,

$$
\begin{aligned}
&Q_{\text{Weyl}}(f)Q_{\text{Weyl}}(g) \\
&= (2\pi)^{-n} \iint \left[(2\pi)^{-n} \iint e^{-i\hbar(\mathbf{c}\cdot\mathbf{b}' - \mathbf{d}\cdot\mathbf{a}')/2} \right. \\
&\quad \left. \times \hat{f}(\mathbf{c} - \mathbf{a}', \mathbf{d} - \mathbf{b}')\hat{g}(\mathbf{a}', \mathbf{b}') \, d\mathbf{a}' \, d\mathbf{b}' \right] e^{i(\mathbf{c}\cdot\mathbf{X} + \mathbf{d}\cdot\mathbf{P})} \, d\mathbf{c} \, d\mathbf{d}.
\end{aligned}
$$

From this and (13.17), we see that $Q_{\text{Weyl}}(f)Q_{\text{Weyl}}(g)$ is the Weyl quantization of the function whose Fourier transform is the quantity in square brackets above, which is what we wanted to show. ∎

Proposition 13.10 *The Moyal product $f \star g$ extends to a continuous map of $L^2(\mathbb{R}^{2n}) \times L^2(\mathbb{R}^{2n})$ into $L^2(\mathbb{R}^{2n})$ and the composition formula (13.29) holds for all f and g in $L^2(\mathbb{R}^{2n})$.*

Proof. A standard inequality asserts that for any two Hilbert–Schmidt operators A and B, we have

$$
\|AB\|_{\text{HS}} \le \|A\|_{\text{HS}} \|B\|_{\text{HS}}.
$$

It follows that the product map $(A, B) \mapsto AB$ is a continuous map of $\text{HS}(L^2(\mathbb{R}^n)) \times \text{HS}(L^2(\mathbb{R}^n))$ to $\text{HS}(L^2(\mathbb{R}^n))$. Meanwhile, the Weyl quantization is a constant multiple of a unitary map from $L^2(\mathbb{R}^{2n})$ to $\text{HS}(L^2(\mathbb{R}^n))$. For Schwartz functions f and g, the Moyal product is nothing but

$$
f \star g = Q_{\text{Weyl}}^{-1}(Q_{\text{Weyl}}(f)Q_{\text{Weyl}}(g)). \tag{13.32}
$$

The right-hand side of (13.32) provides the desired continuous extension of $f \star g$. Clearly, the composition formula (13.29) holds for this extension. ∎

13.3.4 Commutation Relations

In quantum mechanics, the commutator of two operators (divided by $i\hbar$) plays a role similar to that of the Poisson bracket in classical mechanics. Thus, we may naturally ask: To what extent does the Weyl quantization (or any other quantization scheme) map Poisson brackets to commutators? The short answer is: Not always. Indeed, as we will see in Sect. 13.4, no "reasonable" quantization scheme can give an exact correspondence between $\{f, g\}$ on the classical side and $[A, B]/(i\hbar)$ on the quantum side. Nevertheless, such an exact correspondence does hold for various special classes of symbols. If we consider, for example, the class of symbols that depend only on $\mathbf{x}$ and not on $\mathbf{p}$, then on the classical side, all such functions Poisson commute. The Weyl quantization maps such functions $f(\mathbf{x})$ to the operator of multiplication by $f(\mathbf{x})$, and thus the quantizations of any two such functions commute. A more interesting (in particular, noncommutative) example is as follows.

Proposition 13.11 *Suppose f is a polynomial in $\mathbf{x}$ and $\mathbf{p}$ of degree at most 2 and g is an arbitrary polynomial in $\mathbf{x}$ and $\mathbf{p}$. Then*

$$\frac{1}{i\hbar}[Q_{\mathrm{Weyl}}(f), Q_{\mathrm{Weyl}}(g)] = Q_{\mathrm{Weyl}}(\{f, g\}), \tag{13.33}$$

where $\{f, g\}$ is the Poisson bracket of f and g.

Here, we define the Weyl quantization by the obvious n-variable extension of Definition 13.1, and we regard all operators as operating simply on $C_c^\infty(\mathbb{R}^n)$. See Exercise 8 for another class of symbols on which (13.33) holds. Although the requirement that g be a polynomial can be relaxed, we will not attempt to obtain the optimal version of the result.

Proof. For notational simplicity, we abbreviate $Q_{\mathrm{Weyl}}(f)$ to $Q(f)$ for the duration of the proof. If f has degree zero, then both sides of the desired equality are zero. Turning to case in which f has degree 1, we use the n-variable extension of Proposition 13.4, the proof of which is essentially the same as the 1-variable result. The result is as follows:

$$Q(x_j g) = Q(x_j)Q(g) - \frac{i\hbar}{2}Q\left(\frac{\partial g}{\partial p_j}\right)$$

$$= Q(g)Q(x_j) + \frac{i\hbar}{2}Q\left(\frac{\partial g}{\partial p_j}\right).$$

By subtracting these two formulas and rearranging, we get

$$\frac{1}{i\hbar}[Q(x_j), Q(g)] = Q\left(\frac{\partial g}{\partial p_j}\right) = Q(\{x_j, g\}).$$

A very similar argument establishes the desired result when $f = p_j$ and thus for all homogeneous polynomials of degree 1.

Suppose now that f_1 and f_2 are homogeneous polynomials of degree 1 in $\mathbf{x}$ and $\mathbf{p}$. Then it follows easily from Proposition 13.4 that for any polynomial h, we have

$$Q(f_j h) = \frac{1}{2}(Q(f_j)Q(h) + Q(h)Q(f_j)), \quad j = 1, 2. \tag{13.34}$$

In particular, we have

$$Q(f_1 f_2) = \frac{1}{2}(Q(f_1)Q(f_2) + Q(f_2)Q(f_1)). \tag{13.35}$$

Using (13.35) and the product rule for commutators (Proposition 3.15), we have

$$\frac{1}{i\hbar}[Q(f_1 f_2), Q(g)]$$

$$= \frac{1}{2i\hbar}([Q(f_1), Q(g)]Q(f_2) + Q(f_1)[Q(f_2), Q(g)]$$

$$+ [Q(f_2), Q(g)]Q(f_1) + Q(f_2)[Q(f_1), Q(g)]).$$

Using the degree-1 case of the result we are trying to prove, along with (13.34), we get

$$
\begin{aligned}
\frac{1}{i\hbar}[Q(f_1 f_2), Q(g)] &= \frac{1}{2}(Q(\{f_1, g\})Q(f_2) + Q(f_1)Q(\{f_2, g\}) \\
&\quad + Q(\{f_2, g\})Q(f_1) + Q(f_2)Q(\{f_1, g\})) \\
&= Q(f_2\{f_1, g\}) + Q(f_1\{f_2, g\}) \\
&= Q(\{f_1 f_2, g\}),
\end{aligned} \tag{13.36}
$$

where in the last equality we have used the product rule for the Poisson bracket. We have now established the desired result when f is a homogeneous polynomial of degree 0, 1, or 2. ∎

At first glance, it appears that one could extend the result to the case where f has degree 3, by considering three homogenous polynomials f_1, f_2, and f_3 of degree 1 and symmetrizing as in (13.35). The argument breaks down, however, because the $Q(f_j)$'s do not commute. The $Q(f_j)$'s will not always occur in the correct order to allow us to pull the f_j's back inside the Weyl quantization, the way we did in (13.36) in the degree-2 case. Indeed, an elementary but tedious calculations shows that

$$
\frac{1}{i\hbar}[Q_{\text{Weyl}}(x^2 p), Q_{\text{Weyl}}(xp^2)] = 3X^2 P^2 - 6i\hbar XP - \hbar^2 I,
$$

whereas

$$
Q_{\text{Weyl}}(\{x^2 p, xp^2\}) = 3X^2 P^2 - 6i\hbar XP - \frac{3}{2}\hbar^2 I,
$$

so that the two expressions differ by $\hbar^2 I/2$.

We conclude this section with a brief glimpse of an important "equivariance" property of the Weyl quantization. Note that the Poisson bracket of two real valued homogeneous polynomials of degree 2 is again real valued and homogeneous of degree 2. The space of real homogeneous polynomials of degree 2 thus forms a Lie algebra (Sect. 16.3) with respect to the Poisson bracket. This Lie algebra is naturally isomorphic to the Lie algebra $\mathsf{sp}(n; \mathbb{R})$ of Lie group $\mathsf{Sp}(n; \mathbb{R})$, the real symplectic group. This group is the group of invertible linear transformations that preserve a skew-symmetric form on $\mathbb{R}^{2n}$. See Chap. 16 for information about Lie groups and their Lie algebras.

If we apply Proposition 13.11 in the case in which both f and g are homogeneous of degree 2, we see that the map $\pi(f) := Q_{\text{Weyl}}(f)$ is a representation of $\mathsf{sp}(n; \mathbb{R})$ in the space of skew-symmetric operators on $L^2(\mathbb{R}^n)$. It can be shown that associated to this representation of $\mathsf{sp}(n; \mathbb{R})$ there is a projective unitary representation Π of the group $\mathsf{Sp}(n; \mathbb{R})$, known as the *metaplectic representation*. (See, again, Chap. 16 for definitions.) Proposition 13.11 is the infinitesimal version of the following equivariance property of the Weyl quantization: For all $A \in \mathsf{Sp}(n; \mathbb{R})$ and all $f \in L^2(\mathbb{R}^{2n})$, we have

$$
Q_{\text{Weyl}}(f \circ A^{-1}) = \Pi(A)Q_{\text{Weyl}}(f)\Pi(A)^{-1}.
$$

See Theorem 2.15 and Chap. 4 of [11] [where our $\Pi(A)$ corresponds to $\mu((A^*)^{-1})$ in Folland's notation] for this result and much more about the metaplectic representation.

13.4   The "No Go" Theorem of Groenewold

In Sect. 13.3.4, we noted that the Weyl quantization on polynomials satisfies

$$\frac{1}{i\hbar}[Q_{\text{Weyl}}(f), Q_{\text{Weyl}}(g)] = Q_{\text{Weyl}}(\{f, g\}), \tag{13.37}$$

provided that f is a polynomial of degree 2, but not in general. One might think that the failure of (13.37) represents a shortcoming in the definition of the Weyl quantization, which could be remedied by an alternative definition. In this section, however, we will see that no quantization scheme that maps x_j and p_j to the usual position and momentum operators X_j and P_j can satisfy (13.37) for general polynomials in **x** and **p**. This sort of nonexistence result, of a construct satisfying seemingly natural and desirable conditions, is referred to in the physics literature as a "no go" theorem.

In light of this result, one might think that perhaps the position and momentum operators should be defined differently, possibly with an accompanying change in the choice of the quantum Hilbert space. Indeed, there *is* a map Q that satisfies (13.37) for all f and g, namely the prequantization map described in Sect. 23.3. The prequantization map accomplishes this feat by drastically enlarging the quantum Hilbert space, from $L^2(\mathbb{R}^n)$ to $L^2(\mathbb{R}^{2n})$. The Hilbert space $L^2(\mathbb{R}^{2n})$ is considered to be "too big" from a physical standpoint, which explains why the map Q is only "prequantization" rather than "quantization." (The prequantization map has a number of other undesirable features that are described in Sect. 23.3.) If one imposes a natural "smallness" assumption on the quantum Hilbert space (irreducibility under the action of the position and momentum operators), then the Stone–von Neumann theorem will tell us that (modulo certain technical domain assumptions) *any* choice of position and momentum operators satisfying the canonical commutation relations is unitarily equivalent to the usual ones.

The upshot of the discussion in the two preceding paragraphs is that there is no physically reasonable quantization scheme that satisfies (13.37) for all (polynomial) functions f and g.

We turn, now, to Groenewold's "no go" theorem. We need to make domain assumptions, so that it makes sense to compute the commutators of the quantized operators. The simplest approach is to assume that the quantization $Q(f)$ of any polynomial f will be in the algebra generated by the X's and P's, and thus that $Q(f)$ will be a differential operator with polynomial coefficients. There is a variant of this result, known as van

Hove's theorem, that proves a similar "no go" result under a more general assumption about the form of the quantized operators. See [15] for a rigorous proof of van Hove's theorem.

Definition 13.12 *For any $k \geq 0$, let $\mathcal{P}_k$ denote the space of homogeneous polynomials of degree k and let $\mathcal{P}_{\leq k}$ denote the space of all polynomials of degree at most k.*

Theorem 13.13 (Groenewold's Theorem) *Let $\mathcal{D}(\mathbb{R}^n)$ denote the space of differential operators on $\mathbb{R}^n$ with polynomial coefficients. There does not exist a linear map $Q : \mathcal{P}_{\leq 4} \to \mathcal{D}(\mathbb{R}^n)$ with the following properties.*

1. $Q(1) = I$.

2. $Q(x_j) = X_j$ and $Q(p_j) = P_j$.

3. For all f and g in $\mathcal{P}_{\leq 3}$, we have

$$Q(\{f,g\}) = \frac{1}{i\hbar}[Q(f), Q(g)]. \tag{13.38}$$

Note that in Property 3 of the theorem, we assume that f and g belong to $\mathcal{P}_{\leq 3}$ rather than $\mathcal{P}_{\leq 4}$. This assumption guarantees that $\{f, g\}$ belongs to $\mathcal{P}_{\leq 4}$, so that the left-hand side of (13.38) is defined.

Our strategy in proving Groenewold's theorem is the following. We know (Proposition 13.11) that the Weyl quantization satisfies (13.38) if f has degree at most 2 and g has degree at most 3. Using this result, we can show that any map Q satisfying the properties in Theorem 13.13 must coincide with the Weyl quantization on $\mathcal{P}_{\leq 3}$. We then identify a polynomial $f \in \mathcal{P}_4$ that can be expressed as a Poisson bracket in two different ways, $f = \{g, h\} = \{g', h'\}$, with g, h, g', and h' in $\mathcal{P}_3$. Upon calculating that $[Q_{\text{Weyl}}(g), Q_{\text{Weyl}}(h)]$ does not coincide with $[Q_{\text{Weyl}}(g'), Q_{\text{Weyl}}(h')]$, we will have a contradiction.

The proof will consist of several lemmas, followed by the *coup de grâce*.

Lemma 13.14 *Consider an element A of $\mathcal{D}(\mathbb{R}^n)$ expressed as*

$$A = \sum_{\mathbf{k}} f_{\mathbf{k}}(\mathbf{x}) \left(\frac{\partial}{\partial \mathbf{x}}\right)^{\mathbf{k}},$$

where $\mathbf{k}$ ranges over multi-indices, where the $f_{\mathbf{k}}$'s are polynomials, and where only finitely many of the $f_{\mathbf{k}}$'s are nonzero. Then A is the zero operator on $C_c^\infty(\mathbb{R}^n)$ only if each of the $f_{\mathbf{k}}$'s is zero.

Proof. For each multi-index $\mathbf{k}$, let $|\mathbf{k}| = k_1 + \cdots + k_n$. Suppose not all the $f_{\mathbf{k}}$'s are zero, let N be the smallest non-negative integer for which $f_{\mathbf{k}}$ is nonzero for some $\mathbf{k}$ with $|\mathbf{k}| = N$, and let $\mathbf{k}_0$ be some multi-index with

$|\mathbf{k}_0| = N$ and $f_{\mathbf{k}_0} \neq 0$. Let us apply A to a function g that is equal, in a neighborhood of the origin, to $\mathbf{x}^{\mathbf{k}_0}$. Then all the terms in Ag other than the $f_{\mathbf{k}_0}$ term will be zero in a neighborhood of the origin, whereas the $f_{\mathbf{k}_0}$ term will be a nonzero constant in a neighborhood of the origin. Thus, A is not the zero operator. ∎

Lemma 13.15 *If A belongs to $\mathcal{D}(\mathbb{R}^n)$ and A commutes with X_j and P_j for all $j = 1, \ldots, n$, then $A = cI$ for some $c \in \mathbb{C}$.*

Proof. We may easily prove by induction that

$$\left(\frac{\partial}{\partial x_j}\right)^k (x_j g(\mathbf{x})) = k \left(\frac{\partial}{\partial x_j}\right)^{k-1} g(\mathbf{x}) + x_j \left(\frac{\partial}{\partial x_j}\right)^k g(\mathbf{x})$$

for any polynomial g. Thus, for any multi-index $\mathbf{k}$, we have

$$\left[f(\mathbf{x}) \left(\frac{\partial}{\partial \mathbf{x}}\right)^{\mathbf{k}}, X_j \right] = k_j f(\mathbf{x}) \left(\frac{\partial}{\partial \mathbf{x}}\right)^{\mathbf{k}-\mathbf{e}_j}. \tag{13.39}$$

Suppose A is a nonzero element of $\mathcal{D}(\mathbb{R}^n)$ that commutes with each X_j. If $\deg(A) = M$, consider a nonzero term in A of degree M:

$$f_{\mathbf{k}_0}(\mathbf{x}) \left(\frac{\partial}{\partial \mathbf{x}}\right)^{\mathbf{k}_0}, \quad |\mathbf{k}_0| = M, \ f_{\mathbf{k}_0} \neq 0.$$

If $M > 0$, we can pick some j such that the jth entry of $\mathbf{k}_0$ is nonzero. By (13.39) and our assumption on A, we have

$$0 = [A, X_j] = (\mathbf{k}_0)_j f_{\mathbf{k}_0}(\mathbf{x}) \left(\frac{\partial}{\partial \mathbf{x}}\right)^{\mathbf{k}_0-\mathbf{e}_j} + \text{ other terms},$$

where the other terms involve multi-indices of the form $\mathbf{k} - \mathbf{e}_j$, with $\mathbf{k} \neq \mathbf{k}_0$. Thus, by Lemma 13.14, $[A, X_j]$ is not the zero operator.

We see, then, that any $A \in \mathcal{D}(\mathbb{R}^n)$ that commutes with each X_j must be of degree zero; that is, A must simply be multiplication by some polynomial $f(\mathbf{x})$. If, in addition, A commutes with each P_j, then

$$0 = [f(\mathbf{x}), P_j] = i\hbar \frac{\partial f}{\partial x_j}(\mathbf{x}).$$

Thus, actually, f must be constant and A is a multiple of the identity operator. ∎

Lemma 13.16 *For any $f \in \mathcal{P}_2$, there exist $g_1, \ldots, g_j$ and $h_1, \ldots, h_j$ in $\mathcal{P}_2$ such that*

$$f = \{g_1, h_1\} + \cdots + \{g_j, h_j\}.$$

Furthermore, for any $f' \in \mathcal{P}_3$, there exist elements $g'_1, \ldots, g'_k$ of $\mathcal{P}_3$ and $h'_1, \ldots, h'_k$ of $\mathcal{P}_2$ such that

$$f' = \{g'_1, h'_1\} + \cdots + \{g'_k, h'_k\}.$$

Proof. See Exercise 12. ■

Lemma 13.17 *If Q satisfies the conditions in Theorem 13.13, then Q coincides with Q_{Weyl} on $\mathcal{P}_{\leq 3}$.*

Proof. Our argument leans heavily on Proposition 13.11. Note that, by assumption, Q coincides with Q_{Weyl} on $\mathcal{P}_{\leq 1}$. For $f \in \mathcal{P}_2$, let us write $Q(f)$ as

$$Q(f) = Q_{\text{Weyl}}(f) + A_f.$$

For any $g \in \mathcal{P}_{\leq 1}$, we have, by (13.38) and Proposition 13.11,

$$
\begin{aligned}
Q(\{f,g\}) &= \frac{1}{i\hbar}[Q(f), Q(g)] \\
&= \frac{1}{i\hbar}[Q_{\text{Weyl}}(f), Q_{\text{Weyl}}(g)] + \frac{1}{i\hbar}[A_f, Q_{\text{Weyl}}(g)] \\
&= Q_{\text{Weyl}}(\{f,g\}) + \frac{1}{i\hbar}[A_f, Q_{\text{Weyl}}(g)] \\
&= Q(\{f,g\}) + \frac{1}{i\hbar}[A_f, Q_{\text{Weyl}}(g)],
\end{aligned}
\tag{13.40}
$$

since $\{f,g\} \in \mathcal{P}_{\leq 1}$. Thus, $[A_f, Q_{\text{Weyl}}(g)] = 0$ for every $g \in \mathcal{P}_1$, and so, by Lemma 13.15, we must have $A_f = c_f I$ for some constant c_f.

Now, if h is in $\mathcal{P}_2$, we have, by the just-established result and Proposition 13.11,

$$
\begin{aligned}
Q(\{f,h\}) &= \frac{1}{i\hbar}[Q(f), Q(h)] \\
&= \frac{1}{i\hbar}[Q_{\text{Weyl}}(f) + c_f I, Q_{\text{Weyl}}(h) + c_h I] \\
&= \frac{1}{i\hbar}[Q_{\text{Weyl}}(f), Q_{\text{Weyl}}(h)] \\
&= Q_{\text{Weyl}}(\{f,h\}).
\end{aligned}
\tag{13.41}
$$

That is to say, Q and Q_{Weyl} agree on elements of $\mathcal{P}_2$ of the form $\{f,h\}$, for $f, h \in \mathcal{P}_2$. Thus, by Lemma 13.16, Q and Q_{Weyl} agree on all of $\mathcal{P}_2$, and so on all of $\mathcal{P}_{\leq 2}$.

We now use the $\mathcal{P}_{\leq 2}$ case of the lemma to establish the $\mathcal{P}_3$ case. Given $f \in \mathcal{P}_3$, we write $Q(f) = Q_{\text{Weyl}}(f) + B_f$. Given $g \in \mathcal{P}_{\leq 1}$, we have $\{f,g\} \in \mathcal{P}_{\leq 2}$. Thus, we may argue as in (13.40), applying the just-established $\mathcal{P}_{\leq 2}$ case of the lemma to $\{f,g\}$ in the last step. The conclusion is that $[B_f, Q(g)] = 0$ for all $f \in \mathcal{P}_{\leq 2}$ and thus, by Lemma 13.15, that $B_f = d_f I$ for some constant d_f. Meanwhile, if $h \in \mathcal{P}_2$, we argue as in (13.41), but with c_f replaced by d_f and with c_h now known to be zero. The conclusion is that Q agrees with Q_{Weyl} for all elements of $\mathcal{P}_3$ of the form $\{f,h\}$ with $f \in \mathcal{P}_3$ and $h \in \mathcal{P}_2$, and thus, by Lemma 13.16, for all elements of $\mathcal{P}_3$. ■

Proof of Theorem 13.13. Assume, toward a contradiction, that a map Q as in the theorem exists. Let f be the polynomial given by

$$f(\mathbf{x}, \mathbf{p}) = x_1^2 p_1^2.$$

We observe that f can be written in two different ways as a Poisson bracket:

$$x_1^2 p_1^2 = \frac{1}{9}\{x_1^3, p_1^3\} = \frac{1}{3}\{x_1^2 p_1, x_1 p_1^2\}.$$

Thus, by Lemma 13.17, we must have

$$\frac{1}{9}[Q_{\text{Weyl}}(x_1^3), Q_{\text{Weyl}}(p_1^3)] = i\hbar Q(x_1^2 p_1^2)$$

$$= \frac{1}{3}[Q_{\text{Weyl}}(x_1^2 p_1), Q_{\text{Weyl}}(x_1 p_1^2)].$$

On the other hand, if we apply both commutators to the constant function $\mathbf{1}$ (or to a function equal to 1 in a neighborhood of the origin), we obtain

$$\frac{1}{9}[Q_{\text{Weyl}}(x_1^3), Q_{\text{Weyl}}(p_1^3)]\mathbf{1} = \frac{1}{9}(X_1^3 P_1^3 - P_1^3 X_1^3)\mathbf{1}$$

$$= -\frac{1}{9}(-i\hbar)^3 6 \cdot \mathbf{1}.$$

Meanwhile, if we compute the quantizations as in (13.4) and then drop all terms involving $P_1 \mathbf{1}$, we obtain (after a small computation)

$$\frac{1}{3}[Q_{\text{Weyl}}(x_1^2 p_1), Q_{\text{Weyl}}(x_1 p_1^2)]\mathbf{1} = \frac{1}{12}(X_1^2 P_1^3 X_1 + P_1 X_1^2 P_1^2 X_1)\mathbf{1}$$

$$- \frac{1}{12}(X_1 P_1^3 X_1^2 + P_1^2 X_1 P_1 X_1^2)\mathbf{1}$$

$$= -\frac{1}{12} P_1^2 X_1 P_1 X_1^2 \mathbf{1}$$

$$= -\frac{1}{12}(-i\hbar)^3 4 \cdot \mathbf{1}.$$

Since $6/9$ does not equal $4/12$, we have a contradiction. ∎

13.5   Exercises

1. Let $\mathcal{P}_j$ denote the space of complex-valued homogeneous polynomials on $\mathbb{R}^2$ of degree j. Then $\mathcal{P}_j$ is a complex vector space of dimension $j+1$, which we may identify with $\mathbb{C}^{j+1}$ using the obvious basis for $\mathcal{P}_j$. Let V_j denote the complex subspace of $\mathcal{P}_j$ spanned by polynomials of the form $(ax + bp)^j$, with $a, b \in \mathbb{C}$. Show that $V_j = \mathcal{P}_j$.

   Hint: Since every subspace of $\mathbb{C}^{j+1}$ is (topologically) closed, if $\gamma(t)$ is a smooth curve in V_j, the derivative $\gamma'(t)$ will also lie in V_j.

2. Show that symmetrized pseudodifferential operator quantization of x^2p^2 is equal to $Q_{\mathrm{Weyl}}(x^2p^2) - \hbar^2/2$.

3. Show that Wick-ordered and anti-Wick-ordered quantizations map real-valued polynomials to symmetric operators on $C_c^\infty(\mathbb{R})$.

   Hint: Compare the values of each quantization scheme on $z^k\bar{z}^l$ and on $\overline{(z^k\bar{z}^l)}$.

4. Consider a classical harmonic oscillator with Hamiltonian

$$H(x,p) = \frac{p^2}{2m} + \frac{1}{2}m\omega^2 x^2 = \frac{1}{2}m\omega^2\left(x^2 + \left(\frac{p}{m\omega}\right)^2\right),$$

   where ω is the frequency of the oscillator. Consider the Wick- and anti-Wick-ordered quantizations with parameter $\alpha = 1/(m\omega)$. Show that

$$Q_{\mathrm{Wick}}(H) = Q_{\mathrm{Weyl}}(H) - \frac{1}{2}\hbar\omega$$

$$Q_{\mathrm{anti-Wick}}(H) = Q_{\mathrm{Weyl}}(H) + \frac{1}{2}\hbar\omega.$$

5. Let $U_{\mathbf{a},\mathbf{b}}(t)$ be as in Proposition 13.5. Show by direct calculation that these operators form a one-parameter unitary group.

6. Given $\kappa \in L^2(\mathbb{R}^n \times \mathbb{R}^n)$, let A_κ denote the associated integral operator on $L^2(\mathbb{R}^n)$, as in Proposition 13.6. Show that the adjoint A^* of A is also an integral operator, with integral kernel κ' given by

$$\kappa'(\mathbf{x},\mathbf{y}) = \overline{\kappa(\mathbf{y},\mathbf{x})}.$$

7. Suppose that $f \in L^2(\mathbb{R}^{2n})$ and that $\hat{f} \in L^1(\mathbb{R}^{2n})$. Then the right-hand side of (13.17) may be understood as an absolutely convergent "Bochner" integral with values in the Banach space $\mathcal{B}(L^2(\mathbb{R}^n))$. Show that $Q_{\mathrm{Weyl}}(f)$ as defined by (13.17) coincides with $Q_{\mathrm{Weyl}}(f)$ as defined in Definition 13.7.

   Hint: The Bochner integral commutes with applying a bounded linear functional. Use this result with the linear functional $\Lambda_{\phi,\psi}(A) := \langle \phi, A\psi \rangle$ on $\mathcal{B}(L^2(\mathbb{R}^n))$. Then use the expression in (13.23) for κ_f, which follows from Definition 13.7 by applying a partial Fourier transform.

8. (a) Show that for any polynomial f in one variable, we have

$$Q_{\mathrm{Weyl}}(f(x)p) = f(X)P - \frac{i\hbar}{2}f'(X).$$

(b) Show that for any two polynomials f and g, the Poisson bracket $\{f(x)p, g(x)p\}$ is of the form $h(x)p$ for some polynomial h.

(c) Show that for any two polynomials f and g, we have

$$\frac{1}{i\hbar}[Q_{\text{Weyl}}(f(x)p), Q_{\text{Weyl}}(g(x)p)] = Q_{\text{Weyl}}(\{f(x)p, g(x)p\}).$$

9. (a) Given ϕ and ψ in $L^2(\mathbb{R}^n)$, let $|\phi\rangle\langle\psi|$ be the operator defined in Notation 3.28. Show that $|\phi\rangle\langle\psi|$ can be expressed as an integral operator as in Proposition 13.6 and determine the associated integral kernel κ.

   (b) For $\sigma > 0$, let $\psi_\sigma \in L^2(\mathbb{R}^n)$ be given by the expression

   $$\psi_\sigma(\mathbf{x}) = (\pi\sigma)^{-n/4}e^{-|\mathbf{x}|^2/(2\sigma)}.$$

   Using Proposition A.22, show that ψ_σ is a unit vector in $L^2(\mathbb{R}^n)$ and that the Weyl symbol of the corresponding one-dimensional projection operator $|\psi_\sigma\rangle\langle\psi_\sigma|$ is given by

   $$Q_{\text{Weyl}}^{-1}(|\psi_\sigma\rangle\langle\psi_\sigma|) = 2^n e^{-|\mathbf{x}|^2/\sigma}e^{-\sigma|\mathbf{p}|^2/\hbar^2}.$$

   Note: If we give σ the value $\hbar/(m\omega)$, the Gaussian function ψ_α may be thought of as the ground state for an n-dimensional harmonic oscillator. (Compare the functions in Theorem 11.3.) The computation in this exercise plays an important role in the proof of the Stone–von Neumann theorem in Chap. 14.8.

10. If f and g are Schwartz functions on $\mathbb{R}^{2n}$, show that $\widehat{f \star g}$ converges in the L^1 norm to $(2\pi)^{-n}\hat{f} * \hat{g}$, where $*$ denotes convolution. Conclude that $f \star g$ converges uniformly to fg as $\hbar$ tends to zero.

11. Suppose that $f(\mathbf{p}, \mathbf{q})$ is a homogeneous polynomial of degree 2. Show that for each t, the Hamiltonian flow Φ_t associated with f is a linear map of $\mathbb{R}^{2n}$ to itself.

12. Prove Lemma 13.16.

    Hint: Let $g_1 \in \mathcal{P}_2$ be given by

    $$g_1(\mathbf{x}, \mathbf{p}) = \sum_{j=1}^{n} x_j p_j.$$

    Show that for any monomial of the form $\mathbf{x}^\mathbf{j}\mathbf{p}^\mathbf{k}$, we have $\{g_1, \mathbf{x}^\mathbf{j}\mathbf{p}^\mathbf{k}\} = (|\mathbf{k}| - |\mathbf{j}|)\mathbf{x}^\mathbf{j}\mathbf{p}^\mathbf{k}$. Thus, most of the standard basis elements f for $\mathcal{P}_2$ and all of the standard basis elements f for $\mathcal{P}_3$ can be obtained as nonzero multiples of $\{g_1, f\}$.

14

The Stone–von Neumann Theorem

The Stone–von Neumann theorem is a uniqueness theorem for operators satisfying the canonical commutation relations. Suppose A and B are two self-adjoint operators on $\mathbf{H}$ satisfying $[A, B] = i\hbar I$. Suppose also that A and B act irreducibly on $\mathbf{H}$, meaning that the only closed subspaces of $\mathbf{H}$ invariant under A and B are $\{0\}$ and $\mathbf{H}$. Then provided that certain technical assumptions hold (the exponentiated commutation relations), we will conclude that A and B are unitarily equivalent to the usual position and momentum operators X and P. That is, there is a unitary operator $U : \mathbf{H} \to L^2(\mathbb{R})$ such that $UAU^{-1} = X$ and $UBU^{-1} = P$. If $\mathbf{H}$ is not irreducible, then it decomposes as a direct sum of invariant subspaces V_l for A and B, and the restrictions of A and B to each V_l are unitarily equivalent to the usual X and P.

We begin this chapter with a heuristic argument for the Stone–von Neumann theorem, an argument that glosses over certain (essential but technical) domain issues. Then we introduce the exponentiated commutation relations, which should be thought of as a sort of mild strengthening of the ordinary canonical commutation relations. Finally, we give a precise statement of the theorem and provide a proof.

14.1 A Heuristic Argument

Suppose that A and B are any two (possibly unbounded) self-adjoint operators on a separable Hilbert space $\mathbf{H}$ satisfying $[A, B] = i\hbar I$. What we

B.C. Hall, *Quantum Theory for Mathematicians*, Graduate Texts in Mathematics 267, DOI 10.1007/978-1-4614-7116-5_14,
© Springer Science+Business Media New York 2013

would *like* to conclude is that **H** looks like a Hilbert space direct sum of closed subspaces V_l that are invariant under A and B, and such that each V_l is unitarily equivalent to $L^2(\mathbb{R})$ in a way that turns the operators A and B into the standard position and momentum operators X and P. That is to say, we hope to find unitary maps $U_l : V_l \to L^2(\mathbb{R})$ such that

$$U_l A U_l^{-1} = X$$
$$U_l B U_l^{-1} = P.$$

This conclusion is, however, not quite correct, for reasons having to do with the domains of the relevant operators. Nevertheless, let us consider a heuristic argument for this conclusion. We start by forming a lowering operator α and a raising operator α^* by analogy to the definitions of a and a^* in Chap. 11:

$$\alpha = \frac{m\omega A + iB}{\sqrt{2\hbar m\omega}}; \quad \alpha^* = \frac{m\omega A - iB}{\sqrt{2\hbar m\omega}}.$$

Then we look at the kernel W of the lowering operator α, which will be a closed subspace of **H**, provided that α is a closed operator. The elements of W may be thought of as "ground states" for the operator $\alpha^*\alpha$. Choose an orthonormal basis $\{\phi_0^l\}$ for W and define vectors

$$\phi_m^l := (\alpha^*)^m \phi_0^l.$$

It is not hard to show that for $l \neq l'$, ϕ_m^l is orthogonal to $\phi_{m'}^{l'}$ for all m and m'. Let V_l denote the closed span of the vectors ψ_m^l, $m = 0, 1, 2, \ldots$.

Using the calculation in Sect. 11.2, we can see that the way α and α^* act on each chain (the vectors ψ_m^l with l fixed and m varying) is precisely the same as the way the standard lowering and raising operators a and a^* act on the chain of eigenvectors for a^*a. Thus, for each l, we can construct a unitary map U_l from V_l to $L^2(\mathbb{R})$ by mapping the vectors ϕ_m^l in V_j to the vectors ψ_m in $L^2(\mathbb{R})$ described in Theorems 11.3 and 11.4. (In particular, the vector $\psi_0 \in L^2(\mathbb{R})$ is the ground state for the harmonic oscillator, which is a Gaussian.) Since the formula for how α and α^* act is the same as the formula for how a and a^* act, U_l will "intertwine" α with a and α^* with a and a^*, meaning that $U_l\alpha = aU_l$, and similarly for α^* and a^*. It follows that U_l also intertwines A with X and B with P.

It remains only to argue (heuristically) that the spaces V_l fill up the whole Hilbert space **H**. Clearly, the span V of the V_l's is invariant under both α and α^*. Thus, the orthogonal complement $V^\perp$ of V is invariant under the adjoints α^* and α. If $V^\perp$ is not zero, then arguing as in Chap. 11, there should be a ground state in $V^\perp$, that is a nonzero vector annihilated by α. This vector would be orthogonal to all the ϕ_0^l's, contradicting the assumption that the ϕ_0^l's form an orthonormal basis for the kernel of α.

The preceding heuristic argument cannot be completely rigorous, however, since the counterexample in Sect. 12.2 gives a pair of operators A

and B that satisfy the canonical commutation relations but are clearly not unitarily equivalent to the usual position and momentum operators. After all, the "position" operator A in that section is a bounded operator, which cannot be unitarily equivalent to the usual position operator.

What goes wrong is, as usual, a matter of domain considerations. Setting m, $\hbar$, and ω equal to 1, we can look for a vector ϕ_0 that is annihilated by the operator

$$\alpha = \frac{1}{\sqrt{2}}(A + iB) = \frac{1}{\sqrt{2}}\left(x + \frac{d}{dx}\right).$$

By the same argument as in Chap. 11, ϕ_0 must be a constant multiple of the function $e^{-x^2/2}$. The function $\phi_1 := \alpha^* \phi_0$ is then a multiple of $xe^{-x^2/2}$. The problem is that ϕ_1 is not in the domain of α^*. After all, ϕ_1 does not satisfy the periodic boundary condition $\psi(-1) = \psi(1)$ that defines the domain of B. Thus, we cannot continue to apply α^* to obtain an orthogonal chain of vectors and the entire argument breaks down.

What we need, then, is some additional condition that will distinguish between the "good" cases of the canonical commutation relations and the "bad" cases. One possibility for this additional condition is the exponentiated form of the canonical commutation relations, which are discussed in the following section. Our rigorous proof (Sect. 14.3) of the Stone–von Neumann theorem will follow the same outline as the heuristic argument in this section, except that the unbounded operators α and α^* will be replaced by certain bounded operators, constructed by an analog of the Weyl quantization.

14.2 The Exponentiated Commutation Relations

If A is a bounded operator on a Hilbert space $\mathbf{H}$, we may define the *exponential* of A, denoted either e^A or $\exp(A)$, by the power series

$$e^A = \sum_{m=0}^{\infty} \frac{A^m}{m!},$$

where $A^0 = I$. A standard power series argument shows that *if* $A, B \in \mathcal{B}(\mathbf{H})$ commute, then

$$e^{A+B} = e^A e^B, \quad [A, B] = 0. \tag{14.1}$$

(See Exercise 6 in Chap. 16.) Even when A and B do not commute, there is a formula, called the *Baker–Campbell–Hausdorff formula*, that expresses $e^A e^B$, for sufficiently small A and B, in the form

$$e^A e^B = \exp\left\{A + B + \frac{1}{2}[A, B] + \frac{1}{12}[A, [A, B]] + \cdots\right\},$$

where the terms indicated by $\cdots$ are iterated commutators involving A and B. (See Chap. 3 of [21] for more information.) A very special case of this formula is obtained in the case where A and B commute with their commutator, so that all higher commutators are zero.

Theorem 14.1 *Suppose $A, B \in \mathcal{B}(\mathbf{H})$ commute with their commutator, that is,*

$$[A, [A, B]] = [B, [A, B]] = 0.$$

Then

$$e^A e^B = e^{A+B+\frac{1}{2}[A,B]}.$$

This relation may also be written as

$$e^{A+B} = e^{-\frac{1}{2}[A,B]} e^A e^B.$$

Note that in this special case of the Baker–Campbell–Hausdorff formula, no smallness assumption is imposed on A and B.

Proof. We will prove that

$$e^{tA} e^{tB} = e^{t(A+B) + \frac{t^2}{2}[A,B]}, \tag{14.2}$$

which reduces to the desired result at $t = 1$. Since $[A, B]$ commutes with everything in sight, we can use (14.1) to split the exponential on the right-hand side of (14.2) into two and then move the factor involving $[A, B]$ to the other side. Thus, (14.2) is equivalent to the relation

$$e^{tA} e^{tB} e^{-t^2 [A,B]/2} = e^{t(A+B)}. \tag{14.3}$$

Let $\alpha(t)$ denote the left-hand side of (14.3). We will show that $\alpha(t)$ satisfies a simple differential equation, which may be solved explicitly to obtain $\alpha(t) = e^{t(A+B)}$.

Using term-by-term differentiation, it is easy to verify that

$$\frac{d}{dt} e^{tC} = C e^{tC} = e^{tC} C$$

for any $C \in \mathcal{B}(\mathbf{H})$, and that

$$\frac{d}{dt} e^{-t^2 [A,B]/2} = e^{-t^2 [A,B]/2} (-t[A, B]).$$

We may then differentiate $\alpha(t)$ using the product rule, which is proved the same way as in the scalar case, giving

$$\frac{d\alpha}{dt} = e^{tA} A e^{tB} e^{-t^2 [A,B]/2} + e^{tA} e^{tB} B e^{-t^2 [A,B]/2}$$

$$+ e^{tA} e^{tB} e^{-t^2 [A,B]/2} (-t[A, B]).$$

To simplify our expression for $d\alpha/dt$, we need an intermediate result. By the product rule

$$\frac{d}{dt}e^{-tB}Ae^{tB} = e^{-tB}[A,B]e^{tB} = [A,B], \tag{14.4}$$

because B—and, thus, e^B—commutes with $[A,B]$. Noting that $e^{-tB}Ae^{tB} = A$ when $t = 0$, we may integrate (14.4) to get

$$e^{-tB}Ae^{tB} = A + t[A,B]. \tag{14.5}$$

(The difference of the two sides of (14.5) has derivative zero, so by Part (a) of Exercise 2, the two sides are equal up to a constant, which is seen to be zero by evaluating at $t = 0$.)

Using (14.5), we obtain

$$e^{tA}Ae^{tB} = e^{tA}e^{tB}(e^{-tB}Ae^{tB}) = e^{tA}e^{tB}(A + t[A,B]).$$

Moreover, since everything commutes with $[A,B]$, we may commute anything we want past $e^{-t^2[A,B]/2}$. Thus,

$$\frac{d\alpha}{dt} = \alpha(t)(A + t[A,B] + B - t[A,B])$$
$$= \alpha(t)(A + B).$$

Now, according to Exercise 2, the unique solution to the differential equation $d\alpha/dt = \alpha(t)(A+B)$ is $\alpha(t) = \alpha(0)e^{t(A+B)}$. Since $\alpha(0) = I$, we obtain the desired result (14.3). ∎

Suppose, now, that A and B are *unbounded* self-adjoint operators satisfying

$$[A,B] = i\hbar I, \tag{14.6}$$

where the exponentials e^{isA} and e^{itB} are defined by means of the spectral theorem. If we *formally* apply Theorem 14.1 to isA and itB (even these operators are unbounded), we obtain

$$e^{i(sA+tB)} = e^{ist\hbar/2}e^{isA}e^{itB} = e^{-ist\hbar/2}e^{itB}e^{isA}$$

so that

$$e^{isA}e^{itB} = e^{-ist\hbar}e^{itB}e^{isA}. \tag{14.7}$$

It is essential to emphasize that the conclusion (14.7) is only formal, since it assumes that results for bounded operators carry over to unbounded operators, which is false in general. Nevertheless, we may hope that in "good" cases, self-adjoint operators satisfying (14.6) will also satisfy (14.7).

Extending the preceding discussion to the case of several degrees of freedom in an obvious way, we are led to the following definition.

Definition 14.2 *If $A_1, \ldots, A_n$ and $B_1, \ldots, B_n$ are possibly unbounded self-adjoint operators on* **H**, *the A's and B's satisfy the* **exponentiated commutation relations** *if the following relations hold for all $1 \leq j, k \leq n$ and $s, t \in \mathbb{R}$:*

$$e^{isA_j} e^{itA_k} = e^{itA_k} e^{isA_j}$$

$$e^{isB_j} e^{itB_k} = e^{itB_k} e^{isB_j}$$

$$e^{isA_j} e^{itB_k} = e^{-ist\hbar\delta_{jk}} e^{itB_k} e^{isA_j}.$$

The operators e^{isA_j} and e^{itB_k} are defined by the spectral theorem for unbounded self-adjoint operators, and they are unitary operators, defined on all of **H**. Thus, when we say that the exponentiated commutation relations hold, we mean that they hold on the entire Hilbert space **H**.

Notation 14.3 *Suppose operators $A_1, \ldots, A_n$ and $B_1, \ldots, B_n$ satisfy the exponentiated commutation relations. Then for all* **a** *and* **b** *in $\mathbb{R}^n$, let $e^{i(\mathbf{a} \cdot \mathbf{A} + \mathbf{b} \cdot \mathbf{B})}$ denote the unitary operator given by*

$$e^{i(\mathbf{a} \cdot \mathbf{A} + \mathbf{b} \cdot \mathbf{B})} = e^{i\hbar(\mathbf{a} \cdot \mathbf{b})/2} e^{ia_1 A_1} \cdots e^{ia_n A_n} e^{ib_1 B_1} \cdots e^{ib_n B_n}. \qquad (14.8)$$

Equation (14.8) is nothing but what we obtain by formally applying Theorem 14.1 to the operators $i\mathbf{a} \cdot \mathbf{A}$ and $i\mathbf{b} \cdot \mathbf{B}$ and then further splitting the exponentials by formally applying (14.1). The notation may be further justified by checking (Exercise 4) that the operators

$$U_{\mathbf{a},\mathbf{b}}(t) := e^{it^2\hbar(\mathbf{a} \cdot \mathbf{b})/2} e^{ita_1 A_1} \cdots e^{ita_n A_n} e^{itb_1 B_1} \cdots e^{itb_n B_n} \qquad (14.9)$$

form a strongly continuous one-parameter unitary group. If we then *define* $\mathbf{a} \cdot \mathbf{A} + \mathbf{b} \cdot \mathbf{B}$ as the infinitesimal generator (Sect. 10.2) of $U_{\mathbf{a},\mathbf{b}}$, the relation (14.8) will indeed hold. Using the definition of $e^{i(\mathbf{a} \cdot \mathbf{A} + \mathbf{b} \cdot \mathbf{B})}$ and the exponentiated commutation relations, a simple calculation shows that

$$e^{i(\mathbf{a} \cdot \mathbf{A} + \mathbf{b} \cdot \mathbf{B})} e^{i(\mathbf{a}' \cdot \mathbf{A} + \mathbf{b}' \cdot \mathbf{B})} = e^{-i\hbar(\mathbf{a} \cdot \mathbf{b}' - \mathbf{b} \cdot \mathbf{a}')/2} e^{i((\mathbf{a}+\mathbf{a}') \cdot \mathbf{A} + (\mathbf{b}+\mathbf{b}') \cdot \mathbf{B})}. \quad (14.10)$$

In particular, $e^{-i(\mathbf{a} \cdot \mathbf{A} + \mathbf{b} \cdot \mathbf{B})}$ is the inverse of $e^{i(\mathbf{a} \cdot \mathbf{A} + \mathbf{b} \cdot \mathbf{B})}$, as the notation suggests.

The following examples show that in the good case (the usual position and momentum operators on $L^2(\mathbb{R}^n)$), the exponentiated commutation relations do hold, where as in the bad case (the counterexample in Sect. 12.2), they do not.

Example 14.4 *Let A_j be the usual position operator X_j acting on $L^2(\mathbb{R}^n)$ and let B_j be the usual momentum operator P_j. Then the A's and B's satisfy the exponentiated commutation relations.*

Proof. Since X_j is just multiplication by x_j, it is easily verified that e^{isX_j} is just multiplication by e^{isx_j}. Meanwhile, the exponentiated momentum operators satisfy (Example 10.16)

$$(e^{itP_j}\psi)(\mathbf{x}) = \psi(\mathbf{x} + t\hbar\mathbf{e}_j).$$

It is then evident that e^{isX_j} commutes with e^{itX_k} and that e^{isP_j} commutes with e^{itP_k}. We may also compute that

$$(e^{itP_k}e^{isX_j}\psi)(\mathbf{x}) = e^{is(\mathbf{x}+t\hbar\mathbf{e}_k)_j}\psi(\mathbf{x} + t\hbar\mathbf{e}_k)$$
$$= e^{ist\hbar\delta_{jk}}(e^{isX_j}e^{itP_k}\psi)(\mathbf{x}),$$

which is what we wanted to prove. ∎

Example 14.5 *Let A be the operator in Sect. 12.2 and let B be the (unique self-adjoint extension of) the operator in that section. Then A and B do not satisfy the exponentiated commutation relations.*

Proof. The operator A is multiplication by x, and so the operator e^{isA} is just multiplication by e^{isx}. Meanwhile, the operator B is $-i\hbar \, d/dx$, with periodic boundary conditions. We will now demonstrate that e^{itB} consists of "translation with wraparound." Specifically, for any $a \in \mathbb{R}$ and $\psi \in L^2([-1,1])$, let us define $S_a\psi \in L^2([-1,1])$ by

$$(S_a\psi)(x) = \psi(x + a - 2m_{x,a}),$$

where m_x is the unique integer such that

$$-1 \le x + a - 2m_{x,a} < 1.$$

It is easy to check that S_a is a unitary map of $L^2([0,1])$ for each $a \in \mathbb{R}$. We then claim that

$$e^{itB} = S_{\hbar t}. \tag{14.11}$$

To verify the correctness of (14.11), observe that B has an orthonormal basis of eigenvectors, namely the functions $\psi_n(x) := e^{\pi i n x}$, $n \in \mathbb{Z}$, with the corresponding eigenvalues being $\pi n\hbar$. Thus, if we compute e^{itB} by means of the spectral theorem, we have

$$e^{itB}\psi_n = e^{\pi i n t\hbar}\psi_n.$$

On the other hand,

$$(S_a\psi_n)(x)(e^{\pi i n x}) = e^{\pi i n(x+a-2m_{x,a})}$$
$$= e^{-2\pi i n m_{x,a}}e^{\pi i n a}e^{\pi i n x}$$
$$= e^{\pi i n a}\psi_n(x),$$

showing that e^{itB} and $S_{\hbar t}$ agree on each of the functions ψ_n, $n \in \mathbb{Z}$, and thus on all of $L^2([-1,1])$.

Having computed both e^{isA} and e^{itB}, we may now easily see that these operators do not satisfy the exponentiated commutation relations. We have, for example, that

$$e^{isA}e^{itB}\mathbf{1} = e^{isx},$$

whereas

$$e^{itB}e^{isA}\mathbf{1} = e^{is(x+t\hbar-2m_{x,a})}.$$

The function $e^{is(x+t\hbar-2m_{x,a})}$ is not equal to $e^{ist\hbar}e^{isx}$ but rather to

$$e^{ist\hbar}e^{isx}e^{-2ism_{x,a}},$$

where $e^{-2ism_{x,a}}$ is not always equal to 1. ∎

14.3 The Theorem

We give two versions of the Stone–von Neumann theorem, one for general operators satisfying the exponentiated commutation relations and one for the special case where the operators act irreducibly.

Definition 14.6 *Operators $A_1, \ldots, A_n$ and $B_1, \ldots, B_n$ satisfying the exponentiated commutation relations are said to act **irreducibly** on $\mathbf{H}$ if the only closed subspaces of $\mathbf{H}$ that are invariant under every e^{itA_j} and every e^{itB_j} are $\{0\}$ and $\mathbf{H}$.*

Proposition 14.7 *The usual position and momentum operators act irreducibly on $L^2(\mathbb{R}^n)$.*

We delay the proof of this result until near the end of this section.

Theorem 14.8 (Stone–von Neumann Theorem) *Suppose $A_1, \ldots, A_n$ and $B_1, \ldots, B_n$ are self-adjoint operators on $\mathbf{H}$ satisfying the exponentiated commutation relations. Then $\mathbf{H}$ can be decomposed as an orthogonal direct sum of closed subspaces $\{V_l\}$ with the following properties. First, each V_l is invariant under e^{itA_j} and e^{itB_j} for all j and t. Second, there exist unitary operators $U_l : V_l \to L^2(\mathbb{R}^n)$ such that*

$$U_l e^{itA_j} U_l^{-1} = e^{itX_j}$$

and

$$U_l e^{itB_j} U_l^{-1} = e^{itP_j}$$

for all j and t.

If, in addition, the A's and B's act irreducibly on $\mathbf{H}$, then there exists a single unitary map $U : \mathbf{H} \to L^2(\mathbb{R}^n)$ such that

$$U e^{itA_j} U^{-1} = e^{itX_j}$$

and

$$U e^{itB_j} U^{-1} = e^{itP_j},$$

for all t. The map U is unique up to multiplication by a constant of absolute value 1.

The preceding results can be expressed in terms of the Heisenberg group; see Exercise 6.

Our strategy (as in von Neumann's 1931 paper [41]) in proving Theorem 14.8 is to follow the outline of the heuristic argument in Sect. 14.1, but replacing the unbounded raising and lowering operators by the bounded operators $e^{i(\mathbf{a}\cdot\mathbf{A}+\mathbf{b}\cdot\mathbf{B})}$ in Notation 14.3. If we define $\phi_0 \in L^2(\mathbb{R}^n)$ by

$$\phi_0(\mathbf{x}) = (\pi\sigma)^{-n/4} e^{-|\mathbf{x}|^2/(2\sigma)}, \tag{14.12}$$

for some $\sigma > 0$, then ϕ_0 is a unit vector, which we may think of as the ground state of an n-dimensional harmonic oscillator with frequency $\omega = \hbar/(m\sigma)$. We can easily compute the Weyl symbol of the projection $|\phi_0\rangle\langle\phi_0|$ onto ϕ_0 as follows:

$$f_0(\mathbf{x},\mathbf{p}) := Q_{\text{Weyl}}^{-1}(|\phi_0\rangle\langle\phi_0|) = 2^n e^{-|\mathbf{x}|^2/\sigma} e^{-\sigma|\mathbf{p}|^2/\hbar^2}. \tag{14.13}$$

(See Exercise 9 in Chap. 13).

We may define a generalized Weyl quantization Q for $\mathbf{H}$ by using the operators $e^{i(\mathbf{a}\cdot\mathbf{A}+\mathbf{b}\cdot\mathbf{B})}$ in place of the operators $e^{i(\mathbf{a}\cdot\mathbf{X}+\mathbf{b}\cdot\mathbf{P})}$ in (13.17). We will show that the operator $P := Q(f_0)$ is an orthogonal projection, and we will take $W := \text{Range}(P)$ as our space of ground states in $\mathbf{H}$. A crucial result will be that the projection P is nonzero and, indeed, that the restriction of P to any nonzero subspace invariant under the $e^{i(\mathbf{a}\cdot\mathbf{A}+\mathbf{b}\cdot\mathbf{B})}$'s is nonzero.

If $\{\psi^l\}$ is an orthonormal basis for W, consider the vectors

$$\psi_{\mathbf{a},\mathbf{b}}^l := e^{i(\mathbf{a}\cdot\mathbf{A}+\mathbf{b}\cdot\mathbf{B})}\psi^l.$$

We will show that these vectors are orthogonal for different values of l, and that for fixed l, the inner product of two such vectors is the same as in the $L^2(\mathbb{R}^n)$ case. Thus, if V_l denotes the closed span of the $\psi_{\mathbf{a},\mathbf{b}}^l$'s with l fixed and $\mathbf{a}$ and $\mathbf{b}$ varying, we can construct a unitary map from V_l to $L^2(\mathbb{R}^n)$ that intertwines the operators $e^{i(\mathbf{a}\cdot\mathbf{A}+\mathbf{b}\cdot\mathbf{B})}$ with the operators $e^{i(\mathbf{a}\cdot\mathbf{X}+\mathbf{b}\cdot\mathbf{P})}$. The sum of the V_l's must be all of $\mathbf{H}$, for if not, the orthogonal complement Y of the span would be invariant under the $e^{i(\mathbf{a}\cdot\mathbf{A}+\mathbf{b}\cdot\mathbf{B})}$'s. Thus, the restriction of P to Y would be nonzero, implying that there are elements of $W := \text{Range}(P)$ orthogonal to every ψ^l, contradicting the assumption that the ψ^l's span W.

The rest of this section will flesh out the argument sketched in the preceding paragraphs.

Definition 14.9 *Suppose self-adjoint operators $A_1, \ldots, A_n$ and $B_1, \ldots, B_n$ satisfy the exponentiated commutation relations on* **H**. *For any $f \in \mathcal{S}(\mathbb{R}^{2n})$, define $Q(f) \in \mathcal{B}(\mathbf{H})$ by the formula*

$$Q(f) = (2\pi)^{-n} \int_{\mathbb{R}^{2n}} \hat{f}(\mathbf{a}, \mathbf{b}) e^{i(\mathbf{a} \cdot \mathbf{A} + \mathbf{b} \cdot \mathbf{B})} \, d\mathbf{a} \, d\mathbf{b},$$

where $\hat{f}$ is the Fourier transform of f and where $e^{i(\mathbf{a} \cdot \mathbf{A} + \mathbf{b} \cdot \mathbf{B})}$ is as in Notation 14.3. The integral is a Bochner integral with values in the Banach space $\mathcal{B}(\mathbf{H})$.

We will assume the following standard properties of the Bochner integral (Sect. V.5 of [46]). First, any continuous function $f : \mathbb{R}^{2n} \to \mathcal{B}(\mathbf{H})$ for which $\int \|f(x)\| \, dx < \infty$ has a well-defined Bochner integral. Second, the Bochner integral commutes with applying bounded linear transformations. Third, a version of Fubini's theorem holds.

Proposition 14.10 *For any operators satisfying the exponentiated commutation relations, the associated map Q in Definition 14.9 has the following properties.*

  1. *If $f \in \mathcal{S}(\mathbb{R}^{2n})$ is real valued, $Q(f)$ is self-adjoint.*

  2. *For all $\mathbf{a}$ and $\mathbf{b}$ in $\mathbb{R}^n$ and $f \in \mathcal{S}(\mathbb{R}^n)$, we have*

  $$e^{i(\mathbf{a} \cdot \mathbf{A} + \mathbf{b} \cdot \mathbf{B})} Q(f) = Q(f')$$
  $$Q(f) e^{i(\mathbf{a} \cdot \mathbf{A} + \mathbf{b} \cdot \mathbf{B})} = Q(f''),$$

  where f' and f'' are the functions with Fourier transforms given by

  $$\widehat{f'}(\mathbf{a}', \mathbf{b}') = e^{i\hbar(\mathbf{a}' \cdot \mathbf{b} - \mathbf{a} \cdot \mathbf{b}')/2} \, \hat{f}(\mathbf{a}' - \mathbf{a}, \mathbf{b}' - \mathbf{b})$$
  $$\widehat{f''}(\mathbf{a}', \mathbf{b}') = e^{-i\hbar(\mathbf{a}' \cdot \mathbf{b} - \mathbf{a} \cdot \mathbf{b}')/2} \hat{f}(\mathbf{a}' - \mathbf{a}, \mathbf{b}' - \mathbf{b})$$

  3. *For all f and g in $\mathcal{S}(\mathbb{R}^{2n})$, we have*

  $$Q(f) Q(g) = Q(f \star g),$$

  where $\star$ is the Moyal product described in Proposition 13.9.

  4. *For all $f \in \mathcal{S}(\mathbb{R}^n)$, if $Q(f) = 0$ then $f = 0$.*

Using both parts of Point 2 of the theorem, we can see that for all $\mathbf{a}, \mathbf{b} \in \mathbb{R}^n$, we have

$$e^{-i(\mathbf{a} \cdot \mathbf{A} + \mathbf{b} \cdot \mathbf{B})} Q(f) e^{i(\mathbf{a} \cdot \mathbf{A} + \mathbf{b} \cdot \mathbf{B})} = Q(g),$$

where

$$\hat{g}(\mathbf{a}', \mathbf{b}') = e^{i\hbar(\mathbf{a}' \cdot \mathbf{b} - \mathbf{a} \cdot \mathbf{b}')} \hat{f}(\mathbf{a}', \mathbf{b}'). \tag{14.14}$$

Proof. For Point 1, we can re-express $Q(f)$ as

$$(2\pi)^{-n} \int_{\mathbb{R}^{2n}} \frac{1}{2} \left[\hat{f}(\mathbf{a}, \mathbf{b}) e^{i(\mathbf{a} \cdot \mathbf{A} + \mathbf{b} \cdot \mathbf{B})} + \hat{f}(-\mathbf{a}, -\mathbf{b}) e^{-i(\mathbf{a} \cdot \mathbf{A} + \mathbf{b} \cdot \mathbf{B})} \right] \, d\mathbf{a} \, d\mathbf{b},$$

since the change of variable $\mathbf{a}' = -\mathbf{a}$, $\mathbf{b}' = -\mathbf{b}$ brings the second term equal to the first term. If f is real valued, then $\hat{f}(-\mathbf{a}, -\mathbf{b})$ is the conjugate of $\hat{f}(\mathbf{a}, \mathbf{b})$, so that the expression in square brackets in the integral is self-adjoint for each $(\mathbf{a}, \mathbf{b})$.

For the first part of Point 2, we use (14.10) to obtain

$$e^{i(\mathbf{a} \cdot \mathbf{A} + \mathbf{b} \cdot \mathbf{B})} Q(f)$$
$$= (2\pi)^{-n} \int_{\mathbb{R}^{2n}} e^{-i\hbar(\mathbf{a} \cdot \mathbf{b}' - \mathbf{b} \cdot \mathbf{a}')/2} \hat{f}(\mathbf{a}', \mathbf{b}') e^{i((\mathbf{a}+\mathbf{a}') \cdot \mathbf{A} + (\mathbf{b}+\mathbf{b}') \cdot \mathbf{B})} \, d\mathbf{a}' \, d\mathbf{b}'.$$

Making the change of variables $\mathbf{a}'' = \mathbf{a}' + \mathbf{a}$ and $\mathbf{b}'' = \mathbf{b}' + \mathbf{b}$ and simplifying gives the desired result. The proof of the second part of Point 2 is similar.

The proof of Point 3 is precisely the same as the proof of Proposition 13.9, which relies only on the exponentiated commutation relations.

For Point 4, suppose that $Q(f) = 0$ for some $f \in \mathcal{S}(\mathbb{R}^{2n})$. Then for all $\phi, \psi \in \mathbf{H}$ and all $\mathbf{a}, \mathbf{b} \in \mathbb{R}^n$, we have

$$0 = \left\langle e^{i(\mathbf{a} \cdot \mathbf{A} + \mathbf{b} \cdot \mathbf{B})} \phi, Q(f) e^{i(\mathbf{a} \cdot \mathbf{A} + \mathbf{b} \cdot \mathbf{B})} \psi \right\rangle$$
$$= \left\langle \phi, e^{-i(\mathbf{a} \cdot \mathbf{A} + \mathbf{b} \cdot \mathbf{B})} Q(f) e^{i(\mathbf{a} \cdot \mathbf{A} + \mathbf{b} \cdot \mathbf{B})} \psi \right\rangle$$
$$= \langle \phi, Q(g) \psi \rangle$$

where g is as in (14.14). Thus,

$$0 = \int e^{i\hbar(\mathbf{a}' \cdot \mathbf{b} - \mathbf{a} \cdot \mathbf{b}')} \hat{f}(\mathbf{a}', \mathbf{b}') \left\langle \phi, e^{i(\mathbf{a}' \cdot \mathbf{A} + \mathbf{b}' \cdot \mathbf{B})} \psi \right\rangle \, d\mathbf{a}' \, d\mathbf{b}' \qquad (14.15)$$

for all ϕ, ψ and $\mathbf{a}, \mathbf{b}$. But (14.15) is just computing the inverse Fourier transform of the function $\hat{f}(\mathbf{a}', \mathbf{b}') \langle \phi, e^{i(\mathbf{a}' \cdot \mathbf{A} + \mathbf{b}' \cdot \mathbf{B})} \psi \rangle$, evaluated at the point $(-\mathbf{a}, \mathbf{b})$. By the Fourier inversion formula, then, this function must be zero for almost every pair $(\mathbf{a}', \mathbf{b}')$. Now, the function $\langle \phi, e^{i(\mathbf{a}' \cdot \mathbf{A} + \mathbf{b}' \cdot \mathbf{B})} \psi \rangle$ is a continuous function of $(\mathbf{a}, \mathbf{b})$ and by taking $\phi = e^{i(\mathbf{a}_0 \cdot \mathbf{A} + \mathbf{b}_0 \cdot \mathbf{B})} \psi$, it can be made to be nonzero at any given point $(\mathbf{a}_0, \mathbf{b}_0)$ in $\mathbb{R}^{2n}$, and thus also in a neighborhood of that point. Thus, actually, $\hat{f}$ is identically zero and so also is f. ∎

Lemma 14.11 *Let f_0 be the function on $\mathbb{R}^{2n}$ given by*

$$f_0(\mathbf{x}, \mathbf{p}) = 2^n e^{-|\mathbf{x}|^2/\sigma} e^{-\sigma|\mathbf{p}|^2/\hbar^2},$$

where σ is a fixed positive number. Then for all $\mathbf{a}, \mathbf{b} \in \mathbb{R}^n$, we have

$$Q(f_0) e^{i(\mathbf{a} \cdot \mathbf{A} + \mathbf{b} \cdot \mathbf{B})} Q(f_0) = e^{-\sigma|\mathbf{a}|^2/4} e^{-\hbar^2|\mathbf{b}|^2/(4\sigma)} Q(f_0). \qquad (14.16)$$

In particular,

$$Q(f_0)^2 = Q(f_0).$$

Proof. By Proposition 14.10, (14.16) is equivalent to the assertion that

$$f_0 \star f_0' = e^{-\sigma|\mathbf{a}|^2/4} e^{-\hbar^2|\mathbf{b}|^2/(4\sigma)} f_0. \tag{14.17}$$

Now, it is certainly possible to establish (14.17) by direct computation from the definitions of f_0' and $\star$; all the integrals involved will be Gaussian integrals, which can be evaluated by means of Proposition A.22. This approach, however, is both painful and unilluminating. A more sensible approach is to observe that is suffices to verify (14.16) for the ordinary Weyl quantization on $L^2(\mathbb{R}^n)$. After all, (14.16) is equivalent to (14.17), which in turn is equivalent to the identity

$$\begin{aligned}
Q_{\text{Weyl}}(f_0) &e^{i(\mathbf{a}\cdot\mathbf{X}+\mathbf{b}\cdot\mathbf{P})} Q_{\text{Weyl}}(f_0) \\
&= e^{-\sigma|\mathbf{a}|^2/4} e^{-\hbar^2|\mathbf{b}|^2/(4\sigma)} Q_{\text{Weyl}}(f_0), \tag{14.18}
\end{aligned}$$

by applying Proposition 14.10 in the case $Q = Q_{\text{Weyl}}$.

Now, by Exercise 9 in Chap. 13, $Q_{\text{Weyl}}(f_0)$ is the one-dimensional projection $|\phi_0\rangle\langle\phi_0|$, where $\phi_0(\mathbf{x}) = (\pi\alpha)^{-n/4} e^{-|\mathbf{x}|^2/(2\sigma)}$. Thus,

$$\begin{aligned}
Q_{\text{Weyl}}(f_0) e^{i(\mathbf{a}\cdot\mathbf{A}+\mathbf{b}\cdot\mathbf{B})} Q_{\text{Weyl}}(f_0) &= |\phi_0\rangle\langle\phi_0| \, e^{i(\mathbf{a}\cdot\mathbf{X}+\mathbf{b}\cdot\mathbf{P})} \, |\phi_0\rangle\langle\phi_0| \\
&= c\,|\phi_0\rangle\langle\phi_0|, \tag{14.19}
\end{aligned}$$

where

$$c = \langle\phi_0|\, e^{i(\mathbf{a}\cdot\mathbf{X}+\mathbf{b}\cdot\mathbf{P})} \,|\phi_0\rangle.$$

To compute c, we use (13.20), which gives

$$c = (\pi\alpha)^{-n/2} e^{i\hbar(\mathbf{a}\cdot\mathbf{b})/2} \int_{\mathbb{R}^n} e^{-|\mathbf{x}|^2/(2\sigma)} e^{i\mathbf{a}\cdot\mathbf{x}} e^{-|\mathbf{x}+\hbar\mathbf{b}|^2/(2\sigma)} \, d\mathbf{x}. \tag{14.20}$$

The integral in (14.20) can be computed by expanding $|\mathbf{x} + \hbar\mathbf{b}|^2$, collecting terms in the exponent, and applying Proposition A.22. The result, after a bit of algebra, is

$$c = e^{-\sigma|\mathbf{a}|^2/4} e^{-\hbar|\mathbf{b}|^2/(4\sigma)},$$

which gives (14.18). ∎

We now prove the claimed irreducibility of the usual position and momentum operators.

Proof of Proposition 14.7. Given operators $A_1, \ldots, A_n$ and $B_1, \ldots, B_n$ satisfying the exponentiated commutation relations, consider the operator $Q(f_0)$, where f_0 is as in (14.13). According to Lemma 14.11, $Q(f_0)^2 = Q(f_0)$. Since also f_0 is real valued, $Q(f_0)$ is self-adjoint and thus an orthogonal projection. Suppose that the range of the orthogonal projection $Q(f_0)$ is one-dimensional. We then claim that the A's and B's act irreducibly. If not, there would exist a nontrivial closed subspace V that is invariant under each of the operators $e^{i(\mathbf{a}\cdot\mathbf{A}+\mathbf{b}\cdot\mathbf{B})}$. Then the nonzero subspace $V^\perp$ would also be invariant under each of the operators $(e^{i(\mathbf{a}\cdot\mathbf{A}+\mathbf{b}\cdot\mathbf{B})})^* = e^{-i(\mathbf{a}\cdot\mathbf{A}+\mathbf{b}\cdot\mathbf{B})}$. Thus, the exponentiated commutation relations are satisfied in both V and $V^\perp$, with the A's and B's being the infinitesimal generators of the restrictions of e^{itA_j} and e^{itB_j} to each subspace.

It follows that the restriction of $Q(f_0)$ to each of these subspaces may be thought of as the generalized Weyl quantizations for V and $V^\perp$ of the function f_0. Applying Point 4 of Proposition 14.10 to V and to $V^\perp$, we conclude that the restrictions of $Q(f_0)$ to V and to $V^\perp$ are nonzero. Thus, both V and $V^\perp$ will contain nonzero elements of $\text{Range}(Q(f_0))$, contradicting our assumption that $\text{Range}(Q(f_0))$ is one dimensional.

In case of $L^2(\mathbb{R}^n)$, we have $Q_{\text{Weyl}}(f_0) = |\phi_0\rangle\langle\phi_0|$, where ϕ_0 is given by (14.12), which clearly has a one-dimensional range. Thus, the usual position and momentum operators act irreducibly on $L^2(\mathbb{R}^n)$. ∎

We are finally ready for the proof of the Stone–von Neumann theorem.

Proof of Theorem 14.8. Let $W = \text{Range}(Q(f_0))$, where f_0 is given by (14.13) for some fixed $\sigma > 0$. For $\phi, \psi \in W$, we can use (14.10), Lemma 14.11, and the fact that $Q(f_0)$ is the identity on W to obtain

$$\left\langle e^{i(\mathbf{a}\cdot\mathbf{A}+\mathbf{b}\cdot\mathbf{B})}\phi, e^{i(\mathbf{a}'\cdot\mathbf{A}+\mathbf{b}'\cdot\mathbf{B})}\psi \right\rangle$$

$$= \left\langle Q(f_0)\psi, e^{-i(\mathbf{a}\cdot\mathbf{A}+\mathbf{b}\cdot\mathbf{B})}e^{i(\mathbf{a}'\cdot\mathbf{A}+\mathbf{b}'\cdot\mathbf{B})}Q(f_0)\psi \right\rangle$$

$$= e^{i\hbar(\mathbf{a}\cdot\mathbf{b}'-\mathbf{b}\cdot\mathbf{a}')/2}\left\langle \phi, Q(f_0)e^{i((\mathbf{a}'-\mathbf{a})\cdot\mathbf{A}+(\mathbf{b}'-\mathbf{b})\cdot\mathbf{B})}Q(f_0)\psi \right\rangle$$

$$= e^{i\hbar(\mathbf{a}\cdot\mathbf{b}'-\mathbf{b}\cdot\mathbf{a}')/2}e^{-\sigma|\mathbf{a}'-\mathbf{a}|^2/4}e^{-\hbar^2|\mathbf{b}'-\mathbf{b}|^2/(4\sigma)}\left\langle \phi, \psi \right\rangle. \qquad (14.21)$$

Now let $\{\psi^l\}$ be an orthonormal basis for W and define vectors $\psi^l_{\mathbf{a},\mathbf{b}}$, $\mathbf{a}, \mathbf{b} \in \mathbb{R}^n$, by

$$\psi^l_{\mathbf{a},\mathbf{b}} = e^{i(\mathbf{a}\cdot\mathbf{A}+\mathbf{b}\cdot\mathbf{B})}\psi^l.$$

By (14.21), $\psi^l_{\mathbf{a},\mathbf{b}}$ is orthogonal to $\psi^{l'}_{\mathbf{a}',\mathbf{b}'}$ whenever $l \neq l'$. Furthermore,

$$\langle\psi^l_{\mathbf{a},\mathbf{b}}, \psi^l_{\mathbf{a}',\mathbf{b}'}\rangle = e^{i\hbar(\mathbf{a}\cdot\mathbf{b}'-\mathbf{b}\cdot\mathbf{a}')/2}e^{-\sigma|\mathbf{a}'-\mathbf{a}|^2/4}e^{-\hbar^2|\mathbf{b}'-\mathbf{b}|^2/(4\sigma)}, \qquad (14.22)$$

where the right-hand side of (14.22) is "universal," that is, independent of l and independent of the particular Hilbert space in which we are working.

Let V_l be the closed span of the vectors $\psi^l_{\mathbf{a},\mathbf{b}}$ with l fixed and $\mathbf{a}, \mathbf{b}$ varying. We may define a map $U_l : V_l \to L^2(\mathbb{R}^n)$ by requiring that

$$U_l\left(\sum_{j=1}^N \alpha_j\psi^l_{\mathbf{a}_j,\mathbf{b}_j}\right) = \sum_{j=1}^N \alpha_j\phi_{\mathbf{a}_j,\mathbf{b}_j},$$

for every sequence $\mathbf{a}_1, \ldots, \mathbf{a}_N$ and $\mathbf{b}_1, \ldots, \mathbf{b}_N$ of vectors, where

$$\phi_{\mathbf{a},\mathbf{b}} = e^{i(\mathbf{a}\cdot\mathbf{X}+\mathbf{b}\cdot\mathbf{P})}\phi_0.$$

This map is isometric by (14.22) on linear combinations of the $\psi_{\mathbf{a,b}}^l$'s and thus extends uniquely to an isometric map of V_l into $L^2(\mathbb{R}^n)$. [In particular, U_l is well defined: If some linear combination of $\psi_{\mathbf{a,b}}^l$'s is zero, then this linear combination has norm zero and so its image under U_l also has norm zero and is thus zero in $L^2(\mathbb{R}^n)$.]

Now, V_l is invariant under the operators $e^{i(\mathbf{a}\cdot\mathbf{A}+\mathbf{b}\cdot\mathbf{B})}$ by (14.10), and, similarly, the image of V_l under U_l is invariant under the operators $e^{i(\mathbf{a}\cdot\mathbf{X}+\mathbf{b}\cdot\mathbf{P})}$. By the irreducibility of $L^2(\mathbb{R}^n)$ (Proposition 14.7), we conclude that V_l maps onto $L^2(\mathbb{R}^n)$ and is, therefore, unitary. Furthermore, using (14.10) and the analogous expression (13.31) for the position and momentum operators, it is easy to check that each U_l intertwines $e^{i(\mathbf{a}\cdot\mathbf{A}+\mathbf{b}\cdot\mathbf{B})}$ with $e^{i(\mathbf{a}\cdot\mathbf{A}+\mathbf{b}\cdot\mathbf{B})}$, for all $\mathbf{a}, \mathbf{b} \in \mathbb{R}^n$. In particular, taking either $\mathbf{a} = t\mathbf{e}_j$ and $\mathbf{b} = 0$ or $\mathbf{a} = 0$ and $\mathbf{b} = t\mathbf{e}_j$ we see that U_l intertwines e^{itA_j} with e^{itX_j}. Similarly, U_l intertwines e^{itB_j} with e^{itP_j}.

We now argue that the Hilbert space direct sum of the orthogonal subspaces V_l is all of $\mathbf{H}$. If not, then as in the proof of Proposition 14.7, the orthogonal complement Y of this sum would be invariant under the operators $e^{i(\mathbf{a}\cdot\mathbf{A}+\mathbf{b}\cdot\mathbf{B})}$ and thus also under the operator $Q(f_0)$. Furthermore, as in the proof of Proposition 14.7, the restriction of $Q(f_0)$ to Y would be nonzero. Thus, there would exist elements of $W = \text{Range}(Q(f_0))$ orthogonal to each ψ^l, contradicting the assumption that the ψ^l's span W.

It remains only to address the irreducible case. If the A's and B's act irreducibly, then there can be only one subspace, $V_1 = \mathbf{H}$, which means that W must be one dimensional. Any unitary map $U : \mathbf{H} \to L^2(\mathbb{R}^n)$ that intertwines each operator $e^{i(\mathbf{a}\cdot\mathbf{A}+\mathbf{b}\cdot\mathbf{B})}$ with $e^{i(\mathbf{a}\cdot\mathbf{X}+\mathbf{b}\cdot\mathbf{P})}$ must also intertwine each operator of the form $Q(f)$ with $Q_{\text{Weyl}}(f)$. It follows that U must map the one-dimensional subspace W unitarily onto the one-dimensional range of $Q_{\text{Weyl}}(f_0) = |\phi_0\rangle\langle\phi_0|$. Thus, the restriction of U to W is unique up to a constant of absolute value 1. But the reasoning leading to the existence of U shows that U is determined by its action on W, so the entire map U is unique up to a constant. ∎

14.4　The Segal–Bargmann Space

A simple example of the Stone–von Neumann theorem is provided by the Hilbert space $\mathbf{H} := L^2(\mathbb{R}^n)$, together with the operators $A_j := P_j$, and $B_j := -X_j$. In that case (Exercise 3), the unitary map U in the Stone–von Neumann theorem will simply be a scaled version of the Fourier transform, as in Definition 6.1. To obtain a more interesting example, we construct a Hilbert space consisting of holomorphic functions on $\mathbb{C}^n$.

14.4.1   The Raising and Lowering Operators

A smooth function on $F : \mathbb{C}^n \to \mathbb{C}$ is said to be *holomorphic* if it is holomorphic as a function of z_j with the other z_k's fixed. Equivalently, F is holomorphic if $\partial F/\partial \bar{z}_j = 0$, where

$$\frac{\partial}{\partial \bar{z}_j} = \frac{1}{2}\left(\frac{\partial}{\partial x_j} + i\frac{\partial}{\partial y_j}\right).$$

The operator

$$\frac{\partial}{\partial z_j} := \frac{1}{2}\left(\frac{\partial}{\partial x_j} - i\frac{\partial}{\partial y_j}\right)$$

preserves the space of holomorphic functions on $\mathbb{C}^n$.

Considered the operators z_j (i.e., multiplication by z_j) and $\hbar\, \partial/\partial z_j$, acting on the space of holomorphic functions on $\mathbb{C}^n$. Fock [9] observed that these operators satisfy the following commutation relations:

$$[z_j, z_k] = \left[\hbar\frac{\partial}{\partial z_j}, \hbar\frac{\partial}{\partial z_k}\right] = 0$$

$$\left[\hbar\frac{\partial}{\partial z_j}, z_k\right] = \hbar\delta_{jk}I. \tag{14.23}$$

These are essentially the same commutation relations as the raising and lowering operators considered in Sect. 11.2. Specifically, (14.23) are the relations that would be satisfied by the natural higher-dimensional analogs of the operators a and a^* in that section if we omitted the factor of $\sqrt{\hbar}$ in the denominator in (11.4) and (11.5).

Now, if we wish to interpret the operators z_j and $\hbar\, \partial/\partial z_j$ as raising and lowering operators, then we should look for an inner product on the space of holomorphic functions that would make these two operators adjoints of each other. After all, the analysis in Chap. 11 strongly depends on the assumption that a and a^* are adjoints of each other. In the early 1960s, Segal [36] and Bargmann [2] identified such an inner product. Once we have described this Segal–Bargmann inner product, we will construct self-adjoint "position" and "momentum" operators as appropriate linear combinations of z_j and $\hbar\, \partial/\partial z_j$. We will then verify the exponentiated commutation relations and irreducibility, allowing us to apply the Stone–von Neumann theorem.

We look for an L^2 inner product with respect to a measure having a positive density with respect to the Lebesgue measure on $\mathbb{C}^n$.

Lemma 14.12 *Suppose that μ is a smooth, strictly positive density on $\mathbb{C}^n$ and that F and G are sufficiently nice (but not necessarily holomorphic)* *functions on* $\mathbb{C}^n$. *Then*

$$\int_{\mathbb{C}^n} \overline{F(\mathbf{z})} \frac{\partial G}{\partial z_j} \mu(\mathbf{z}) \; d\mathbf{z}$$

$$= -\int_{\mathbb{C}^n} \overline{\frac{\partial F}{\partial \bar{z}_j}} G(\mathbf{z}) \mu(\mathbf{z}) \; d\mathbf{z} - \int_{\mathbb{C}^n} \overline{\frac{\partial \log \mu}{\partial \bar{z}_j} F(\mathbf{z})} \; G(\mathbf{z}) \; d\mathbf{z}, \qquad (14.24)$$

where $d\mathbf{z}$ *denotes the* $2n$-*dimensional Lebesgue measure on* $\mathbb{C}^n \cong \mathbb{R}^{2n}$.

Equation (14.24) tells us that

$$\left(\frac{\partial}{\partial z_j} \right)^* = -\frac{\partial}{\partial \bar{z}_j} - \frac{\partial \log \mu}{\partial \bar{z}_j},$$

where the adjoint is computed with respect to the inner product for the Hilbert space $L^2(\mathbb{C}^n, \mu)$. If we restrict the adjoint operator $(\partial/\partial z_j)^*$ to the space of holomorphic functions, then the $\partial/\partial \bar{z}_j$ term is zero, by the definition of a holomorphic function.

Proof. Let us approximate the integral over $\mathbb{C}^n$ on the left-hand side of (14.24) by an integral over a large cube. By performing either the x_j-integral or the y_j-integral first, we can integrate by parts to push the derivatives with respect to x_j or y_j off of G and onto the product of $\bar{F}$ and μ (with a minus sign). The boundary term in the integration by parts will involve the function $\overline{F(\mathbf{z})}G(\mathbf{z})\mu(\mathbf{z})$ integrated over two opposite faces of the cube. If this function tends to zero sufficiently rapidly at infinity, the boundary terms will vanish in the limit. In that case, we obtain

$$\int_{\mathbb{C}^n} \overline{F(\mathbf{z})} \frac{\partial G}{\partial z_j} \mu(\mathbf{z}) \; d\mathbf{z}$$

$$= -\int_{\mathbb{C}^n} \left(\frac{\partial}{\partial z_j} \overline{F(\mathbf{z})} \right) G(\mathbf{z}) \mu(\mathbf{z}) \; d\mathbf{z} - \int_{\mathbb{C}^n} \overline{F(\mathbf{z})} G(\mathbf{z}) \frac{\partial \mu}{\partial z_j} \; d\mathbf{z},$$

provided that all three of the above integrals are absolutely convergent. Since $\partial \bar{F}/\partial z_j = \overline{\partial F/\partial \bar{z}_j}$ and

$$\frac{\partial \mu}{\partial z_j} = \frac{\partial \log \mu}{\partial z_j} \mu = \overline{\frac{\partial \log \mu}{\partial \bar{z}_j}} \mu,$$

we obtain (14.24). ∎

We now look for a density $\mu_\hbar$ for which $\partial \log \mu / \partial \bar{z}_j = -z_j/\hbar$. In that case, the adjoint operator $(\partial/\partial z_j)^*$ preserves the holomorphic subspace of $L^2(\mathbb{C}^n, \mu_\hbar)$ and is given on this subspace by multiplication by $z_j/\hbar$.

Lemma 14.13 *Specialize Lemma 14.12 to the case in which* F *and* G *are holomorphic polynomials and* μ *is the density* $\mu_\hbar$ *given by*

$$\mu_\hbar(\mathbf{z}) = \frac{1}{(\pi\hbar)^n} e^{-|\mathbf{z}|^2/\hbar}. \qquad (14.25)$$

Then we have

$$\int_{\mathbb{C}^n} \overline{F(\mathbf{z})} \frac{\partial G}{\partial z_j} \mu_\hbar(\mathbf{z}) \ d\mathbf{z} = \frac{1}{\hbar} \int_{\mathbb{C}^n} \overline{z_j F(\mathbf{z})} G(\mathbf{z}) \mu_\hbar(\mathbf{z}) \ d\mathbf{z}. \tag{14.26}$$

Proof. In the case that F and G are holomorphic polynomials, $\partial F/\partial \bar{z}_j = 0$, so the first term on the right-hand side of (14.24) is zero. Furthermore, $\bar{F}G\mu$ decreases rapidly at infinity and so the boundary terms vanish in this case. Finally, we may compute $\partial \log \mu_\hbar / \partial \bar{z}_j$ as $-z_j/\hbar$, giving (14.26). $\blacksquare$

Definition 14.14 *The **Segal–Bargmann space**, denoted $\mathcal{H}L^2(\mathbb{C}^n, \mu_\hbar)$ is the space of holomorphic functions F on $\mathbb{C}^n$ for which*

$$\|F\|_\hbar := \left(\int_{\mathbb{C}^n} |F(\mathbf{z})|^2 \ \mu_\hbar(\mathbf{z}) \ d\mathbf{z} \right)^{1/2} < \infty,$$

*where $\mu_\hbar$ is as in (14.25). Define **raising** and **lowering operators** a_j^* and a_j on $\mathcal{H}L^2(\mathbb{C}^n, \mu_\hbar)$ by*

$$a_j^* = z_j$$
$$a_j = \hbar \frac{\partial}{\partial z_j},$$

with the domain of a_j and a_j^* consisting of the space of holomorphic polynomials.

In light of Lemma 14.13, the operators a_j and a_j^* satisfy

$$\langle F, a_j G \rangle_{\mathcal{H}L^2(\mathbb{C}^n, \mu_\hbar)} - \langle a_j^* F, G \rangle_{\mathcal{H}L^2(\mathbb{C}^n, \mu_\hbar)}$$

for all holomorphic polynomials F and G, thus justifying the notation a_j^* for the raising operator. The space $\mathcal{H}L^2(\mathbb{C}^n, \mu_\hbar)$ is also sometimes called the *Fock space*. It should be noted, however, that in quantum field theory, the term Fock space also refers to a different (but related) space—the completion of the tensor algebra over a fixed Hilbert space.

Proposition 14.15 *The Segal–Bargmann space is complete with respect to the norm $\|\cdot\|_\hbar$ and forms a Hilbert space with respect to the associated inner product,*

$$\langle F, G \rangle_\hbar := \int_{\mathbb{C}^n} \overline{F(\mathbf{z})} G(\mathbf{z}) \mu_\hbar(\mathbf{z}) \ d\mathbf{z}.$$

Furthermore, the space of holomorphic polynomials forms a dense subspace of the Segal–Bargmann space.

Note that elements of $\mathcal{H}L^2(\mathbb{C}^n, \mu_\hbar)$ are actual functions on $\mathbb{C}^n$, *not* equivalence classes of functions. Nevertheless, we can regard $\mathcal{H}L^2(\mathbb{C}^n, \mu_\hbar)$ as a subspace of $L^2(\mathbb{C}^n, \mu_\hbar)$, since each equivalence class of almost-everywhere equal functions contains at most one holomorphic representative.

Proof. Given any $\mathbf{z}_0 \in \mathbb{C}^n$ and $R > 0$, let $P_{\mathbf{z}_0,R}$ denote the polydisk given by

$$P_{\mathbf{z}_0} = \left\{ \mathbf{z} \in \mathbb{C}^n \,\middle|\, |z_j - (\mathbf{z}_0)_j| < R, \ j = 1, \dots, n \right\}.$$

Using a power-series argument, it is easy to show that the value of a holomorphic function F at $\mathbf{z}_0$ is equal to the average of F over $P_{\mathbf{z}_0,R}$. We can then multiply and divide by $\mu_\hbar$ to obtain

$$F(\mathbf{z}_0) = \frac{1}{(\pi R^2)^n} \int_{P_{\mathbf{z}_0,R}} \frac{1}{\mu_\hbar(\mathbf{z})} F(\mathbf{z}) \, \mu_\hbar(\mathbf{z}) \, d\mathbf{z}.$$

The Cauchy–Schwarz inequality then tells us that

$$|F(\mathbf{z}_0)|$$
$$\leq \frac{1}{(\pi R^2)^n} \left(\sup_{\mathbf{z} \in P_{\mathbf{z}_0,R}} \frac{1}{\mu_\hbar(\mathbf{z})} \right) \left\| \mathbf{1}_{P_{\mathbf{z}_0,R}} \right\|_{L^2(\mathbb{C}^n, \mu_\hbar)} \|F\|_{L^2(\mathbb{C}^n, \mu_\hbar)} . \qquad (14.27)$$

This inequality tells us that pointwise evaluation [the map $F \mapsto F(\mathbf{z}_0)$] is a bounded linear functional on the Segal–Bargmann space.

Suppose now that F_n is a sequence of holomorphic functions such that F_n converges in $L^2(\mathbb{C}^n, \mu_\hbar)$ to some F. Using (14.27), we can easily show that F_n converges to F uniformly on compact sets, which implies that F is also holomorphic. This shows that the holomorphic subspace of $L^2(\mathbb{C}^n, \mu_\hbar)$ is closed and hence is a Hilbert space.

To show the denseness of polynomials, consider some $F \in \mathcal{H}L^2(\mathbb{C}^n, \mu_\hbar)$ and let

$$F(\mathbf{z}) = \sum_{\mathbf{n}} a_{\mathbf{n}} \mathbf{z}^{\mathbf{n}} \qquad (14.28)$$

be the Taylor expansion of F, where $\mathbf{n}$ ranges over all multi-indices. This series converges to F uniformly on compact subsets of $\mathbb{C}^n$. We claim that the terms in (14.28) are orthogonal. To see this, use Fubini's theorem to perform the integration of $\overline{\mathbf{z}^{\mathbf{n}}}$ against $\mathbf{z}^{\mathbf{m}}$ one variable at a time. Using polar coordinates in each copy of $\mathbb{C}$, we can see that we will get zero if the power of z_j in $\mathbf{z}^{\mathbf{n}}$ is not the same as the power of z_j in $\mathbf{z}^{\mathbf{m}}$.

Since it is orthogonal, the series in (14.28) will converge in $L^2(\mathbb{C}^n, \mu_\hbar)$ provided that the sum of the squares of the norms of the terms is finite. If $P_{0,R}$ is a sequence of polydisks of increasing radius centered at the origin, the argument in the preceding paragraph shows that the terms in (14.28) are orthogonal in $L^2(P_{0,R}, \mu_\hbar)$. Since the series converges uniformly on $P_{0,R}$, we can then interchange sum and integral to obtain

$$\sum_{\mathbf{n}} |a_{\mathbf{n}}|^2 \|\mathbf{z}^{\mathbf{n}}\|^2_{L^2(P_{0,R}, \mu_\hbar)} = \|F\|^2_{L^2(P_{0,R}, \mu_\hbar)} .$$

By applying monotone convergence to both the sum over $\mathbf{n}$ and the integrals over $P_{0,R}$, we may let R tend to infinity to obtain

$$\sum_{\mathbf{n}} |a_{\mathbf{n}}|^2 \, \|\mathbf{z}^{\mathbf{n}}\|^2_{L^2(\mathbb{C}^n,\mu_\hbar)} = \|F\|^2_{L^2(\mathbb{C}^n,\mu_\hbar)} < \infty.$$

Thus, the series in (14.28) converges in $L^2(\mathbb{C}^n,\mu_\hbar)$ and this L^2 limit must coincide with the pointwise limit, namely F itself. $\blacksquare$

14.4.2 The Exponentiated Commutation Relations

To apply the Stone–von Neumann theorem to the Segal–Bargmann space, we define self-adjoint "position" and "momentum" operators as follows:

$$A_j = \frac{1}{\sqrt{2}}\left(z_j + \hbar \frac{\partial}{\partial z_j}\right)$$

$$B_j = \frac{i}{\sqrt{2}}\left(z_j - \hbar \frac{\partial}{\partial z_j}\right).$$

We will identify one-parameter unitary groups having (extensions of) these operators as their infinitesimal generators, which will show (by Stone's theorem) that the generators are indeed self-adjoint on suitable domains. We will then verify the exponentiated commutation relations and check irreducibility.

Let us compute heuristically and then check that our results are correct. If we formally apply Theorem 14.1 to the (unbounded) operators $\sum \bar{a}_j z_j$ and $-\hbar \sum a_j \partial/\partial z_j$, we obtain

$$\exp\left\{\sum_{j=1}^n \left(-\bar{a}_j z_j + \hbar a_j \frac{\partial}{\partial z_j}\right)\right\}$$

$$= \exp\left\{-\frac{1}{2}\hbar |\mathbf{a}|^2\right\} \exp\left\{-\sum_{j=1}^n a_j z_j\right\} \exp\left\{\hbar \sum_{j=1}^n a_j \frac{\partial}{\partial z_j}\right\}. \tag{14.29}$$

This calculation suggests that we define operators $T_{\mathbf{a}}$ by the formula

$$(T_{\mathbf{a}}F)(\mathbf{z}) = e^{-\hbar |\mathbf{a}|^2/2} e^{-\bar{\mathbf{a}}\cdot\mathbf{z}} F(\mathbf{z} + \hbar \mathbf{a}), \quad \mathbf{a} \in \mathbb{C}^n, \tag{14.30}$$

where for any $\mathbf{a}, \mathbf{b} \in \mathbb{C}^n$, we define $\mathbf{a}\cdot\mathbf{b} = \sum_j a_j b_j$ (no complex conjugates). Since the exponent on the left-hand side of (14.29) is skew-self-adjoint (the difference of an operator and its adjoint), we expect the operators $T_{\mathbf{a}}$ to be unitary. For suitable choices of $\mathbf{a}$, the operator on the left-hand side of (14.29) will become the one-parameter group generated by A_j or B_j.

Theorem 14.16 *For each* $\mathbf{a} \in \mathbb{C}^n$, *the operator* $T_{\mathbf{a}}$ *defined by (14.30) is a unitary operator on the Segal–Bargmann space, and the map* $\mathbf{a} \mapsto T_{\mathbf{a}}$ *is strongly continuous. These operators satisfy*

$$T_{\mathbf{a}}T_{\mathbf{b}} = e^{i\hbar \operatorname{Im}(\bar{\mathbf{a}}\cdot\mathbf{b})} T_{\mathbf{a}+\mathbf{b}}. \tag{14.31}$$

In particular, for each j, *the maps*

$$U_j(t) := T_{it\mathbf{e}_j/\sqrt{2}}; \quad V_j(t) := T_{t\mathbf{e}_j/\sqrt{2}}$$

are strongly continuous one-parameter unitary groups. The infinitesimal generators A_j *and* B_j *of these groups satisfy the exponentiated commutation relations.*

For any $F \in \operatorname{Dom}(A_j)$, *we have*

$$(A_j F)(\mathbf{z}) = \frac{1}{\sqrt{2}} \left(z_j F(\mathbf{z}) + \hbar \frac{\partial F}{\partial z_j} \right)$$

and for any $F \in \operatorname{Dom}(B_j)$, *we have*

$$(B_j F)(\mathbf{z}) = \frac{i}{\sqrt{2}} \left(z_j F(\mathbf{z}) - \hbar \frac{\partial F}{\partial z_j} \right).$$

Furthermore, the domains of A_j *and* B_j *contain all holomorphic polynomials.*

Finally, the operators A_j *and* B_j *act irreducibly on the Segal–Bargmann space, in the sense of Definition 14.6.*

Proof. It is evident that $T_{\mathbf{a}}F(\mathbf{z})$ is holomorphic as a function of $\mathbf{z}$ for each fixed $\mathbf{a}$. Meanwhile, for any $F \in \mathcal{H}L^2(\mathbb{C}^n, \mu_\hbar)$, we have

$$\|T_{\mathbf{a}}F\|_{L^2(\mathbb{C}^n,\mu_\hbar)}^2 = (\pi\hbar)^{-n} \int_{\mathbb{C}^n} e^{-\hbar|\mathbf{a}|^2} e^{-2\operatorname{Re}(\bar{\mathbf{a}}\cdot\mathbf{z})} |F(\mathbf{z}+\hbar\mathbf{a})|^2 e^{-|\mathbf{z}|^2/\hbar} \, d\mathbf{z}$$

$$= (\pi\hbar)^{-n} \int_{\mathbb{C}^n} e^{-|\mathbf{z}+\hbar\mathbf{a}|^2/\hbar} |F(\mathbf{z}+\hbar\mathbf{a})|^2 \, d\mathbf{z}$$

$$= \|F\|_{L^2(\mathbb{C}^n,\mu_\hbar)}^2,$$

showing that $T_{\mathbf{a}}$ is isometric. The formula for $T_{\mathbf{a}}T_{\mathbf{b}}$ follows from direct computation (Exercise 7), and from this formula we see that $T_{\mathbf{a}}T_{-\mathbf{a}} = I$, which shows that $T_{\mathbf{a}}$ is surjective and thus unitary. The strong continuity of $T_{\mathbf{a}}$ is easily verified on polynomials (Exercise 8), which are dense in the $\mathcal{H}L^2(\mathbb{C}^n, \mu_\hbar)$.

It easily follows from (14.31) that $U_j(\cdot)$ and $V_j(\cdot)$ are one-parameter unitary groups, and also that (the infinitesimal generators of) these unitary groups satisfy the exponentiated commutation relations. If F is in the domain of the infinitesimal generator of $U_j(\cdot)$, the limit

$$(A_j F)(\mathbf{z}) := \frac{1}{i} \lim_{t\to 0} \frac{1}{t} \left[e^{-\hbar t^2/4} e^{itz_j/\sqrt{2}} F(\mathbf{z}+it\hbar\mathbf{e}_j/\sqrt{2}) - F(\mathbf{z}) \right] \tag{14.32}$$

must exist in $L^2(\mathbb{C}^n, \mu_\hbar)$. The L^2 limit coincides with the easily computed pointwise limit, giving

$$A_j F(\mathbf{z}) = \frac{1}{i}\left(\frac{i}{\sqrt{2}} z_j F(\mathbf{z}) + \frac{i\hbar}{\sqrt{2}}\frac{\partial F}{\partial z_j}\right),$$

as claimed. If F is a polynomial, it is easily shown, using dominated convergence, that the limit in (14.32) exists in $L^2(\mathbb{C}^n, \mu_\hbar)$. The analysis of B_j is similar.

Finally, we address irreducibility. If the A_j's and B_j's did not act irreducibly, then in the application of the Stone–von Neumann theorem to $\mathcal{H}L^2(\mathbb{C}^n, \mu_\hbar)$, there would exist at least two subspaces V_l. Thus, there would exist at least two linearly independent vectors F_l such that for all j, we have that F_l is in the domain of A_j and B_j and

$$0 = (A_j + iB_j)F_l = \frac{2\hbar}{\sqrt{2}}\frac{\partial F_l}{\partial z_j}.$$

(Take F_l to be the preimage under U_l of the function ϕ_0 in (14.12), with $\sigma = \hbar$.) This would mean that each F_l is constant, contradicting the assumption that the F_l's are linearly independent. ∎

14.4.3   The Reproducing Kernel

According to (14.27), evaluation of $F \in \mathcal{H}L^2(\mathbb{C}^n, \mu_\hbar)$ at a fixed point $\mathbf{z}$ is a continuous linear functional. Thus, this linear functional can be written as the inner product with a unique element $\chi_\mathbf{z}$ of $\mathcal{H}L^2(\mathbb{C}^n, \mu_\hbar)$, which we now compute. The vector $\chi_\mathbf{z}$ is called the *coherent state* with parameter $\mathbf{z}$.

Proposition 14.17 *For all $F \in \mathcal{H}L^2(\mathbb{C}^n, \mu_\hbar)$, we have*

$$F(\mathbf{z}) - \int_{\mathbb{C}^n} e^{\mathbf{z}\cdot\bar{\mathbf{w}}/\hbar} F(\mathbf{w})\mu_\hbar(\mathbf{w})\; d\mathbf{w}. \tag{14.33}$$

The function $e^{\mathbf{z}\cdot\bar{\mathbf{w}}/\hbar}$ is called the *reproducing kernel* for $\mathcal{H}L^2(\mathbb{C}^n, \mu_\hbar)$, since integration against this kernel simply gives back (or "reproduces") the function F. Of course, the relation (14.33) holds only for *holomorphic* functions in $L^2(\mathbb{C}^n, \mu_\hbar)$. Equation (14.33) can be rewritten as

$$F(\mathbf{z}) = \langle \chi_\mathbf{z}, F \rangle_{\mathcal{H}L^2(\mathbb{C}^n, \mu_\hbar)},$$

where

$$\chi_\mathbf{z}(\mathbf{w}) = e^{\bar{\mathbf{z}}\cdot\mathbf{w}/\hbar}.$$

Proof. We begin by establishing the result in the case $\mathbf{z} = 0$. We have already established, in the proof of Proposition 14.15, that the Taylor series of F converges to F in $\mathcal{H}L^2(\mathbb{C}^n, \mu_\hbar)$, and the distinct monomials in this series are orthogonal. Thus, when computing $\langle 1, F\rangle_{\mathcal{H}L^2(\mathbb{C}^n, \mu_\hbar)}$, only the constant term in the expansion of F survives, giving

$$\langle 1, F\rangle_{\mathcal{H}L^2(\mathbb{C}^n, \mu_\hbar)} = F(0)\,\langle 1, 1\rangle_{\mathcal{H}L^2(\mathbb{C}^n, \mu_\hbar)} = F(0), \tag{14.34}$$

since $\mu_\hbar$ is a probability measure. But this relation is precisely the $\mathbf{z} = 0$ case of (14.33).

Let us now apply (14.34) to $T_\mathbf{a}F$, where $T_\mathbf{a}$ is the unitary operator in (14.30). According to Theorem 14.16, $T_\mathbf{a}$ is unitary with inverse equal to $T_{-\mathbf{a}}$, giving

$$(T_\mathbf{a}F)(0) = \langle 1, T_\mathbf{a}F\rangle_{\mathcal{H}L^2(\mathbb{C}^n, \mu_\hbar)} = \langle T_{-\mathbf{a}}1, F\rangle_{\mathcal{H}L^2(\mathbb{C}^n, \mu_\hbar)}.$$

Writing this relation out using $\mathbf{w}$ as our variable of integration gives

$$e^{-\hbar|\mathbf{a}|^2/2}F(\hbar\mathbf{a}) = \int \overline{e^{-\hbar|\mathbf{a}|^2/2}e^{\bar{\mathbf{a}}\cdot\mathbf{w}}}F(\mathbf{w})\mu_\hbar(\mathbf{w})\ dw.$$

Setting $\mathbf{a} = \mathbf{z}/\hbar$ and simplifying gives the desired result. ∎

14.4.4   The Segal–Bargmann Transform

Since the operators A_j and B_j in Theorem 14.16 satisfy the exponentiated commutation relations and act irreducibly on $\mathcal{H}L^2(\mathbb{C}^n, \mu_\hbar)$, the second part of the Stone–von Neumann theorem tells us that there is a unitary map $U : \mathcal{H}L^2(\mathbb{C}^n, \mu_\hbar) \to L^2(\mathbb{R}^n)$, unique up to a constant, that intertwines these operator with the usual position and momentum operators. The inverse map $V : L^2(\mathbb{R}^n) \to \mathcal{H}L^2(\mathbb{C}^n, \mu_\hbar)$ is called the *Segal–Bargmann transform*.

Theorem 14.18 *Let V be the inverse of the map $U : \mathcal{H}L^2(\mathbb{C}^n, \mu_\hbar) \to L^2(\mathbb{R}^n)$ given by the Stone–von Neumann theorem, normalized so that V takes the function $\phi_0 \in L^2(\mathbb{R}^n)$ in (14.12) (with $\sigma = \hbar$) to the constant function $1 \in \mathcal{H}L^2(\mathbb{C}^n, \mu_\hbar)$. Then V may be computed as follows:*

$$(V\psi)(\mathbf{z}) = (\pi\hbar)^{-n/4}\int_{\mathbb{R}^n} \exp\left\{-\frac{1}{2\hbar}\left(\mathbf{z}\cdot\mathbf{z} - 2\sqrt{2}\mathbf{z}\cdot\mathbf{x} + \mathbf{x}\cdot\mathbf{x}\right)\right\}\psi(\mathbf{x})\ dx.$$

Recall that we define $\mathbf{a}\cdot\mathbf{b} = \sum_j a_jb_j$ for all $\mathbf{a}, \mathbf{b} \in \mathbb{C}^n$, with no complex conjugates in the definition. In particular, the integrand in the formula for $V\psi$ is a holomorphic function of $\mathbf{z}$, for each fixed $\mathbf{x}$.

Note that the value of $(V\psi)(\mathbf{z})$ at $\mathbf{z} = \mathbf{0}$ is simply the inner product of ψ with the ground state function ϕ_0, with $\sigma = \hbar$. The proof of Theorem 14.18 will show that the value of $(V\psi)(\mathbf{z})$ at an arbitrary $\mathbf{z}$ is a certain constant $c_\mathbf{z}$ times the inner product of ψ with a *phase space translate* of ϕ_0, that is, a vector of the form $e^{i\mathbf{a}\cdot\mathbf{X}}e^{i\mathbf{b}\cdot\mathbf{P}}\phi_0$. [See (14.36).] According to (the obvious higher-dimensional counterpart to) Proposition 12.11, ϕ_0 is a minimum uncertainty state, meaning that equality is achieved in Corollary 12.9 for each

j. Thus, by (the obvious higher-dimensional counterpart to) Exercise 3 in Chap. 12, each state of the form $e^{i\mathbf{a}\cdot\mathbf{X}}e^{i\mathbf{b}\cdot\mathbf{P}}\phi_0$ is also a minimum uncertainty state.

Proof. By the unitarity of V and the $\mathbf{z} = 0$ case of Proposition 14.17, we have

$$\langle\phi_0,\psi\rangle_{L^2(\mathbb{R}^n)} = \langle V\phi_0, V\psi\rangle_{\mathcal{H}L^2(\mathbb{C}^n,\mu_\hbar)} = \langle 1, V\psi\rangle_{\mathcal{H}L^2(\mathbb{C}^n,\mu_\hbar)} = (V\psi)(0).$$

Thus, the value of $V\psi$ at 0 is just the inner product of ψ with ϕ_0. More generally,

$$\begin{aligned}
\langle e^{-i\mathbf{a}\cdot\mathbf{X}}e^{-i\mathbf{b}\cdot\mathbf{P}}\phi_0,\psi\rangle &= \langle\phi_0, e^{i\mathbf{b}\cdot\mathbf{P}}e^{i\mathbf{a}\cdot\mathbf{X}}\psi\rangle \\
&= \langle V\phi_0, Ve^{i\mathbf{b}\cdot\mathbf{P}}e^{i\mathbf{a}\cdot\mathbf{X}}\psi\rangle \\
&= \langle 1, e^{i\mathbf{b}\cdot\mathbf{B}}e^{i\mathbf{a}\cdot\mathbf{A}}V\psi\rangle \\
&= (e^{i\mathbf{b}\cdot\mathbf{B}}e^{i\mathbf{a}\cdot\mathbf{A}}V\psi)(0), \qquad (14.35)
\end{aligned}$$

where $e^{i\mathbf{a}\cdot\mathbf{A}}$ means the product (in any order) of the operators $e^{ia_j A_j}$, and similarly for $e^{i\mathbf{b}\cdot\mathbf{B}}$.

Recall that A_j's and B_j's are defined as the infinitesimal generators of the groups U_j and V_j in Theorem 14.16, which in turn are defined in terms of the operators $T_\mathbf{a}$. If we use (14.31) to compute the right-hand side of (14.35), we obtain

$$\begin{aligned}
(e^{i\mathbf{b}\cdot\mathbf{B}}e^{i\mathbf{a}\cdot\mathbf{A}}V\psi)(0) &= (T_{\mathbf{b}/\sqrt{2}}T_{i\mathbf{a}/\sqrt{2}}V\psi)(0) \\
&= e^{i\hbar\mathbf{a}\cdot\mathbf{b}/2}(T_{(\mathbf{b}+i\mathbf{a})/\sqrt{2}}V\psi)(0) \\
&= e^{i\hbar\mathbf{a}\cdot\mathbf{b}/2}e^{-\hbar(|\mathbf{a}|^2+|\mathbf{b}|^2)/4}(V\psi)(\hbar(\mathbf{b}+i\mathbf{a})/\sqrt{2}).
\end{aligned}$$

Thus, if we apply (14.35) with $\mathbf{a} = -\sqrt{2}\mathbf{y}_0/\hbar$ and $\mathbf{b} = \sqrt{2}\mathbf{x}_0/\hbar$, we obtain

$$\begin{aligned}
\langle e^{-i\sqrt{2}\mathbf{y}_0\cdot\mathbf{X}/\hbar}e^{-i\sqrt{2}\mathbf{x}_0\cdot\mathbf{P}/\hbar}\phi_0,\psi\rangle \\
= e^{i\mathbf{x}_0\cdot\mathbf{y}_0/\hbar}e^{-(|\mathbf{x}_0|^2+|\mathbf{y}_0|^2)/(2\hbar)}(V\psi)(\mathbf{x}_0+i\mathbf{y}_0). \qquad (14.36)
\end{aligned}$$

Solving (14.36) for $(V\psi)(\mathbf{x}_0+i\mathbf{y}_0)$ gives

$$\begin{aligned}
(V\psi)(\mathbf{x}_0+i\mathbf{y}_0) &= (\pi\hbar)^{-n/4}e^{-i\mathbf{x}_0\cdot\mathbf{y}_0/\hbar}e^{(|\mathbf{x}_0|^2+|\mathbf{y}_0|^2)/(2\hbar)} \\
&\times \int_{\mathbb{R}^n} e^{i\sqrt{2}\mathbf{y}_0\cdot\mathbf{x}/\hbar}e^{-|\mathbf{x}-\sqrt{2}\mathbf{x}_0|^2/(2\hbar)}\psi(\mathbf{x})\, d\mathbf{x},
\end{aligned}$$

which simplifies to the claimed formula for $V\psi$. ∎

14.5 Exercises

1. Show that if operators A and B satisfy the exponentiated commutation relations of Sect. 14.2, they satisfy the "semi-exponentiated" commutation relations, that is, the hypotheses of Theorem 12.8.

Hint: For any $a, s \in \mathbb{R}$ and $\psi \in \text{Dom}(A)$, rearrange the expression

$$\frac{e^{isA}(e^{iaB}\psi) - (e^{iaB}\psi)}{s}$$

using the exponentiated commutation relations. Then let s tend to zero and apply Stone's theorem.

2. (a) Suppose $\alpha : \mathbb{R} \to \mathcal{B}(\mathbf{H})$ is a differentiable map, meaning that

$$\lim_{h \to 0} \frac{\alpha(t + h) - \alpha(t)}{h}$$

exists in the norm topology of $\mathcal{B}(\mathbf{H})$ for each t. Show that if $d\alpha/dt = 0$ for all t, then α is constant.

(b) Suppose $\alpha : \mathbb{R} \to \mathcal{B}(\mathbf{H})$ is a differentiable map such that

$$\frac{d\alpha}{dt} = \alpha(t)A$$

for some fixed $A \in \mathcal{B}(\mathbf{H})$. Show that $\alpha(t) = \alpha(0)e^{tA}$ for all t.

3. Show that the operators $A_j := P_j$ and $B_j := -X_j$ on $L^2(\mathbb{R}^n)$ satisfy the exponentiated commutation relations. Determine the unitary operator $U : L^2(\mathbb{R}^n) \to L^2(\mathbb{R}^n)$ (unique up to a constant) such that

$$U e^{itA_j} U^{-1} = e^{itX_j}$$
$$U e^{itB_j} U^{-1} = e^{itP_j}.$$

4. Verify that the operators $U_{\mathbf{a},\mathbf{b}}(t)$ in (14.9) form a strongly continuous one-parameter unitary group.

5. In this exercise, we develop a discrete version of (the $n = 1$ case of) the Stone–von Neumann theorem. Let p be a prime number, let $\mathbb{Z}/p$ denote the field of integers modulo p, and let h be a nonzero element of $\mathbb{Z}/p$. Consider the finite-dimensional Hilbert space $L^2(\mathbb{Z}/p)$, taken with respect to the counting measure on $\mathbb{Z}/p$. Let U denote the "modulation" operator

$$(Uf)(n) = e^{2\pi in/p} f(n)$$

and let V denote the "translation" operator on $L^2(\mathbb{Z}/p)$, given by

$$(Vf)(n) = f(n + h).$$

In the case of the modulation operator, note that the expression $e^{2\pi in/p}$ descends unambiguously from $n \in \mathbb{Z}$ to $n \in \mathbb{Z}/p$.

(a) Verify that $U^p = V^p = I$ and that, for all l and m in $\mathbb{Z}$,

$$U^l V^m = e^{-2\pi i l m/p} V^m U^l.$$

(b) Suppose now that A and B are unitary operators on a finite-dimensional Hilbert space $\mathbf{H}$ satisfying $A^p = B^p = I$ and

$$A^l B^m = e^{-2\pi i l m/p} B^m V^l.$$

Suppose also that the only subspaces of $\mathbf{H}$ invariant under both A and B are $\{0\}$ and $\mathbf{H}$. Show that there is a unitary map W from $\mathbf{H}$ to $L^2(\mathbb{Z}/p)$ such that

$$W A W^{-1} = U$$
$$W B W^{-1} = V.$$

Hint: Show that if $v \in \mathbf{H}$ is an eigenvector for A, then so is $B^l v$ for any l. Show that each eigenspace for A has dimension 1 and identify the associated eigenvectors with the "δ-functions" in $L^2(\mathbb{Z}/p)$.

6. Given a constant $u \in \mathbb{C}$ with $|u| = 1$ and a pair of vectors $\mathbf{a}, \mathbf{b} \in \mathbb{R}^n$, let $U_{u,\mathbf{a},\mathbf{b}}$ be the unitary operator on $L^2(\mathbb{R}^n)$ given by

$$(U_{u,\mathbf{a},\mathbf{b}} \psi)(\mathbf{x}) = u e^{i \mathbf{a} \cdot \mathbf{x}} \psi(\mathbf{x} + \hbar \mathbf{b}).$$

(a) Verify that the set of operators of this form a group under the operation of composition, and denote this group by H_n.

(b) Let $\tilde{H}_n$ denote the set of $(n + 2) \times (n + 2)$ matrices of the form

$$A = \begin{pmatrix} 1 & a_1 & \cdots & a_n & c \\ & 1 & & & b_1 \\ & & \ddots & & \vdots \\ & & & 1 & b_n \\ & & & & 1 \end{pmatrix},$$

with $a_1, \ldots, a_n$ and $b_1, \ldots, b_n$ in $\mathbb{R}$. (The only nonzero entries in A are on the main diagonal, in the first row, and in the last column.) Verify that $\tilde{H}_n$ forms a group under matrix multiplication. Show that there is a surjective group homomorphism $\Phi : \tilde{H}_n \to H_n$ with discrete kernel.

Hint: Compare the formulas for group multiplication in H_n and $\tilde{H}_n$.

Note: In the language of Chap. 16, $\tilde{H}_n$ is the universal covering group of H_n. The group $\tilde{H}_n$ is called the *Heisenberg group*.

7. Show by direct computation that the operators $T_{\mathbf{a}}$ in (14.30) satisfy the relations (14.31).

8. Using dominated convergence, show that for every holomorphic polynomial F on $\mathbb{C}^n$, we have

$$\lim_{\mathbf{a}\to\mathbf{b}} \|T_{\mathbf{a}}F - T_{\mathbf{b}}F\|^2_{L^2(\mathbb{C}^n,\mu_\hbar)} = 0,$$

where $T_{\mathbf{a}}$ is as in (14.30).

15

The WKB Approximation

15.1 Introduction

The WKB method, named for Gregor Wentzel, Hendrik Kramers, and Léon Brillouin, gives an approximation to the eigenfunctions and eigenvalues of the Hamiltonian operator $\hat{H}$ in one dimension. The approximation is best understood as applying to a fixed range of energies as $\hbar$ tends to zero. (It is also reasonable in many cases to think of the approximation as applying to a fixed value of h as the energy tends to infinity.)

The idea of the WKB approximation is that the potential function $V(x)$ can be thought of as being "slowly varying," with the result that solutions to the time-independent Schrödinger equation will look locally like the solutions in the case of a constant potential. In the classically allowed region, this line of thinking will yield an approximation consisting of a rapidly oscillating complex exponential multiplied by a slowly varying amplitude. We make the "local frequency" of the exponential equal to what it would be if V were constant. Having made this choice, there is a unique choice for the amplitude that yields an error that is of order $\hbar^2$. This amplitude, however, tends to infinity as we approach the "turning points," that is, the points where the classical particle changes directions. Similarly, in the classically forbidden region, we obtain approximate solutions that are rapidly growing or decaying exponentials, multiplied by a slowly varying factor. Again, there is a unique choice for the slowly varying factor that gives errors of order $\hbar^2$, and again, this factor blows up at the turning points.

B.C. Hall, *Quantum Theory for Mathematicians*, Graduate Texts
in Mathematics 267, DOI 10.1007/978-1-4614-7116-5_15,
© Springer Science+Business Media New York 2013

The difficulty near the turning points means that we cannot directly "match" the approximate solutions in different regimes the way we did in Chap. 5. Instead, we will use the Airy function to approximate the solution to the Schrödinger equation near the turning points. Asymptotics of the Airy function will then yield the appropriate matching condition, which turns out to be a corrected form of the Bohr–Sommerfeld rule that appears in the "old" quantum theory.

15.2   The Old Quantum Theory and the Bohr–Sommerfeld Condition

The old quantum theory, developed by Bohr, Sommerfeld, and de Broglie, among others, may be pictured as follows. Consider, for simplicity, a particle with one degree of freedom, and let C be a level set in phase space of the Hamiltonian,

$$C = \left\{ (x,p) \in \mathbb{R}^2 \,\middle|\, H(x,p) = E \right\}, \tag{15.1}$$

which we assume to be a closed curve. We now imagine drawing a "wave" on C, that is, some oscillatory function defined over C. Following the *de Broglie hypothesis* (Sect. 1.2.2), we postulate that the local frequency k of the wave as a function of x is $p/\hbar$. This means that the phase of our wave should be obtained by integrating the 1-form

$$\frac{1}{\hbar} p \, dx \tag{15.2}$$

along the curve. Thus, the wave itself can be pictured as a function on C of the form

$$\cos\left(\frac{1}{\hbar} \int_{x_0}^{x} p \, dx - \delta \right), \tag{15.3}$$

where x_0 is some arbitrary starting point on the curve C and where δ is an arbitrary phase. Note that the old quantum theory did not offer a physical interpretation of this wave; it was simply a crude attempt to introduce waves into the picture.

The *Bohr–Sommerfeld condition* is simply the requirement that the function in (15.3) should match up with itself when we go all the way around the curve. This will happen precisely if

$$\frac{1}{\hbar} \int_C p \, dx = 2\pi n, \tag{15.4}$$

for some integer n. The energy levels in the old quantum theory were taken to be those numbers E for which the corresponding level curve C satisfies the Bohr–Sommerfeld condition (15.4). Although Bohr–Sommerfeld quantization had some successes, notably explaining the energy levels of the hydrogen atom, it ultimately failed to correctly predict the energies of complex systems.

For systems with one degree of freedom, a vestige of the Bohr–Sommerfeld approach survives in modern quantum theory, with two modifications. First, the condition (15.4) has to be corrected by replacing the n by $n+1/2$ on the right-hand side of (15.4). (The replacement of n by $n+1/2$ is known as the *Maslov correction*.) Second, this condition does not (in most cases) give the exact energy levels, but only the leading-order semiclassical approximation to the energy levels. The preceding discussion leads to the following definition.

Condition 15.1 *A number E is said to satisfy the Maslov-corrected Bohr–Sommerfeld condition if*

$$\frac{1}{\hbar} \int_C p \, dx = 2\pi(n + 1/2) \tag{15.5}$$

for some integer n, where C is the classical energy curve in (15.1). In light of Green's theorem, this condition may be rewritten as

$$\frac{1}{2\pi\hbar}(Area \ enclosed \ by \ C) = n + \frac{1}{2}.$$

When the Maslov correction is included, the Bohr–Sommerfeld condition can be stated as saying that the wave with phase given by integrating the 1-form in (15.2) should be 180° out of phase with itself after one trip around the energy curve. Figure 15.1 shows an example, which should be contrasted with Fig. 1.3. (Note also that Fig. 1.3 is drawn in the configuration space, whereas Fig. 15.1 is in the phase space.)

In our analysis in the subsequent sections, we will see that the Maslov correction—that is, the extra $1/2$ in (15.5), as compared to (15.4)—actually consists of a contribution of $1/4$ from each of the two "turning points" of the classical particle. (The turning points are the points where the classical particle changes directions.) Specifically, in the WKB approximation, the phase of the wave function will be computed as the integral of $(p \, dx)/\hbar$ along one "branch" of the classical energy curve C. Using the Airy function to approximate the wave function near the turning points, we will obtain an "extra" $\pi/4$ of phase between each turning point and the last local maximum or minimum of the wave function. Because of the two branches of C, the extra $\pi/4$ of phase near each of the two turning points actually contributes an extra π to the integral on the left-hand side of (15.5).

The reader may wonder why there is no comparable correction term in our discussion of the Bohr–de Broglie model of the hydrogen atom in Sect. 1.2.2. One way to answer this question is as follows. As we will see in Sect. 18.1, the Schrödinger operator for the hydrogen atom can be reduced

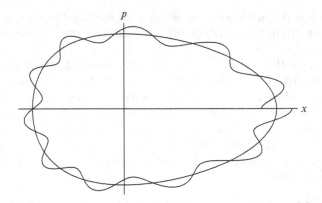

FIGURE 15.1. A trajectory satisfying the corrected Bohr–Sommerfeld condition with $n = 10$.

to a one-dimensional Schrödinger operator with an *effective potential* of the form

$$V_{\text{eff}}(r) = -\frac{Q^2}{r} + \frac{\hbar^2 l(l+1)}{2mr^2}.$$

Here l is a non-negative integer that labels the "total angular momentum" of the wave function. At least when $l > 0$, one can analyze this Schrödinger operator using a WKB-type analysis very similar to the one in the current chapter, with one important modification: The radial wave function [the quantity $h(r)$ in (18.5)] must be zero at $r = 0$ in order for the wave function to be in the domain of the Hamiltonian.

If one analyzes the situation carefully, it turns out that the zero boundary condition at $r = 0$ introduces *another* correction into the Bohr–Sommerfeld condition in the amount of $1/2$. There is still also a correction of $1/4$ for each of the two turning points, leading to the condition

$$\frac{1}{\hbar} \int_C p \, dx = 2\pi \left(n + \frac{1}{4} + \frac{1}{4} + \frac{1}{2} \right) = 2\pi(n+1).$$

Since $n + 1$ is again an integer, we are effectively back to the uncorrected Bohr–Sommerfeld condition. See Chap. 11 of [8] for a discussion of different approaches to the WKB approximation for radial potentials.

15.3   Classical and Semiclassical Approximations

We are interested in finding approximate solutions to the time-independent Schrödinger equation,

$$-\frac{\hbar^2}{2m} \frac{d^2\psi}{dx^2} + (V(x) - E)\psi(x) = 0 \tag{15.6}$$

for small values of $\hbar$. Ultimately, we will need to analyze the behavior of solutions in three different regions, the classically allowed region [points where $V(x) < E$], the classically forbidden region (points where $V(x) > E$), and the region near the "turning points," that is, the points where $V(x) = E$.

Let us consider at first the classically allowed region. Given a potential V and an energy level E, we can solve (up to a choice of sign) for the momentum of a classical particle as a function of position as

$$p(x) = \sqrt{2m(E - V(x))}.$$

We look for approximate solutions ψ to (15.6) of the form

$$\psi(x) = A(x)e^{\pm iS(x)/\hbar}, \tag{15.7}$$

where S satisfies $S'(x) = p(x)$. Note that we are taking the phase of our wave function to be

$$\text{phase} = \pm\frac{1}{\hbar} \int p(x) \, dx,$$

as in the old quantum theory in Sect. 15.2. The "amplitude function" $A(x)$ will be chosen to be independent of $\hbar$ and thus "slowly varying" (for small $\hbar$) compared to the exponent $S(x)/\hbar$.

Our first, elementary, result is that for *any* number E for which there is a classically allowed region and for *any* reasonable choice of the amplitude $A(x)$ in (15.7), we obtain an approximate eigenvector solution to the time-independent Schrödinger equation, with an error term of order $\hbar$.

Proposition 15.2 *For any two numbers E_1 and E_2 with $E_1 > \inf_{x \in \mathbb{R}} V(x)$, there exists a constant C and a nonzero function $A \in C_c^\infty(\mathbb{R})$ with the following property. For every $E \in [E_1, E_2]$, the support of A is contained in the classically allowed region at energy E and the function ψ given by*

$$\psi(x) = A(x) \exp\left\{\pm\frac{i}{\hbar} \int p(x) \, dx\right\}$$

satisfies

$$\|\hat{H}\psi - E\psi\| \leq C\hbar \|\psi\|. \tag{15.8}$$

Proof. For any $E \in [E_1, E_2]$, the classically allowed region for energy E contains the classically allowed region for energy E_1. We choose, then, A to be any nonzero element of $C_c^\infty(\mathbb{R})$ with support in the classically allowed region for energy E_1. If we evaluate $\hat{H}\psi - E\psi$ by direct calculation, there will a term in which two derivatives fall on the exponential factor, bringing down a factor involving $p(x)^2$. The definition of $p(x)$ is such that the term involving $p(x)^2$ will cancel the term involving $V(x) - E$, leaving us with

$$\hat{H}\psi - E\psi = -\frac{\hbar^2}{2m}\left(A''(x) \pm \frac{i}{\hbar}2A'(x)p(x) \pm \frac{i}{\hbar}p'(x)A(x)\right)$$

$$\times \exp\left\{\pm\frac{i}{\hbar}\int p(x)\,dx\right\}. \tag{15.9}$$

(Here, each occurrence of the symbol $\pm$ has the same value, either all pluses or all minuses.) Thus,

$$\|\hat{H}\psi - E\psi\| \leq \frac{\hbar^2}{2m}\|A''\| + \frac{\hbar}{2m}\|2A'p + Ap'\|. \tag{15.10}$$

Since $\|\psi\|$ is independent of $\hbar$, the right-hand side of (15.10) is of order $\hbar\|\psi\|$. It is easy to check that $\|2A'p + Ap'\|$ is bounded as a function of E for any E in the range $[E_1, E_2]$ and the result follows. ∎

Proposition 15.2, along with elementary spectral theory, tells us that for any E larger than the minimum of V, there is a point $\tilde{E}$ in the spectrum of $\hat{H}$ such that

$$|E - \tilde{E}| \leq c\hbar. \tag{15.11}$$

(See Exercise 4 in Chap. 10.) If we assume that $V(x)$ tends to $+\infty$ as $x \to \pm\infty$, then $\hat{H}$ will have discrete spectrum and we can say that $\tilde{E}$ is an eigenvalue for $\hat{H}$. The conclusion, for such potentials, is this: Given *any* number $E \in [E_1, E_2]$, there is an eigenvalue of $\hat{H}$ within $C\hbar$ of E. Thus, as $\hbar$ tends to zero, the eigenvalues of $\hat{H}$ "fill up" the entire range of values of the classical energy function.

Proposition 15.2 is one manifestation of the "classical limit" of quantum mechanics: the quantum energy spectrum is, in a certain sense, approximating the classical energy spectrum as $\hbar$ gets small. Notice, however, that this result tells us only that the eigenvalues are at most order $\hbar$ apart and nothing further about the location of the individual eigenvalues.

In this chapter, we will show that if E satisfies the corrected Bohr–Sommerfeld condition, then there exists an eigenvalue $\tilde{E}$ of $\hat{H}$ such that

$$|E - \tilde{E}| \leq C\hbar^{9/8}. \tag{15.12}$$

An estimate of the form (15.12) locates eigenvalues with an error bound that is small compared to the expected average spacing between the eigenvalues, which is of order $\hbar$. On the other hand, the approximate energy levels E are determined by Condition 15.1, which is a condition on the *classical* energy curve. Thus, (15.12) can be described as a *semiclassical* estimate: It is estimating quantum mechanical quantities (the individual energy levels) in classical terms (the level curves of the classical Hamiltonian).

15.4 The WKB Approximation Away from the Turning Points

We consider only the simplest interesting case of the WKB approximation, in which the following assumption holds. See the book of Miller [30] for much about this sort of asymptotic analysis.

Assumption 15.3 *Consider a smooth, real-valued potential $V(x)$, with $V(x) \to +\infty$ as $x \to \pm\infty$. Assume that the functions $V'(x)/V(x)$ and $V''(x)/V(x)$ are bounded for x near $\pm\infty$.*

Consider also a range of energies of the form $E_1 \leq E \leq E_2$. Assume that for each E in this range, there are exactly two points, $a(E)$ and $b(E)$, with $a(E) < b(E)$, for which $V(x) = E$. Further assume that the derivative of V is nonzero at $a(E)$ and $b(E)$, for all $E \in [E_1, E_2]$.

See Fig. 15.2 for a typical example. Since V is locally bounded and tends to $+\infty$ at infinity, $\hat{H}$ is essentially self-adjoint on $C_c^\infty(\mathbb{R})$ (Theorem 9.39) and has purely discrete spectrum (Theorem XIII.16 in Volume IV of [34]). The assumption that V'/V and V''/V be bounded near infinity is stronger than necessary, but still applies to most of the interesting cases.

We refer to $a(E)$ and $b(E)$ as the *turning points*, since these are the points where a classical particle with energy E changes direction. When the energy E is understood as being fixed, we will write the turning points simply as a and b.

15.4.1 The Classically Allowed Region

As in Sect. 15.3, we seek approximate solutions to the time independent Schrödinger equation having the following form in the classically allowed region:

$$\psi = A(x) \exp\left\{ \pm \frac{i}{\hbar} \int p(x)\, dx \right\}, \tag{15.13}$$

where $p(x) = \sqrt{2m(E - V(x))}$ is the momentum of a classical particle with energy E and position x. According to (15.9), this form for ψ gives

$$\hat{H}\psi - E\psi = -\frac{\hbar^2}{2m} \left(A''(x) \pm \frac{i}{\hbar} 2A'(x)p(x) \pm \frac{i}{\hbar} p'(x)A(x) \right)$$

$$\times \exp\left\{ \pm \frac{i}{\hbar} \int p(x)\, dx \right\}. \tag{15.14}$$

Since we want to obtain an approximate solution with an error smaller than $\hbar$, we require that the second and third terms in parentheses in (15.14) cancel. This cancellation will occur if A satisfies

$$2A'(x)p(x) = -p'(x)A(x)$$

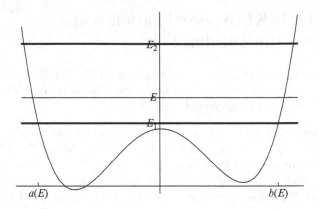

FIGURE 15.2. A potential satisfying Assumption 15.3.

or

$$\frac{A'(x)}{A(x)} = -\frac{1}{2}\frac{p'(x)}{p(x)}, \tag{15.15}$$

which we can easily solve (Exercise 3) as

$$A(x) = C(p(x))^{-1/2}. \tag{15.16}$$

If A is given by (15.16), we will have

$$\hat{H}\psi - E\psi = -\frac{\hbar^2}{2m}\frac{A''(x)}{A(x)}\psi(x), \tag{15.17}$$

indicating that our error is of order $\hbar^2$. This expression, however, is only local, in that it applies only in the classically allowed region. Furthermore, $p(x)$ tends to zero at the turning points, which means that $A(x)$ becomes unbounded at these points. This blow-up of the amplitude is a substantial complicating factor in the analysis.

We can get an approximate solution to the Schrödinger equation by taking a linear combination of the function in (15.13) with two different choices for the sign in the exponent, with constants c_1 and c_2. It is convenient to take the basepoint of our integration to be the left-hand turning point $a = a(E)$. Furthermore, since the Schrödinger operator $\hat{H}$ commutes with complex conjugation, the real and imaginary parts of any solution to the time-independent Schrödinger equation is again a solution. We will therefore consider only real-valued approximate solutions, i.e., those in which $c_2 = \overline{c_1}$. Using Exercise 1, we can then write our approximate solution as follows.

Summary 15.4 *Suppose ψ is a real-valued solution to the time-independent Schrödinger equation. Then in the classically allowed region but away from the turning points, we expect that ψ is well approximated by an expression of the form*

$$\frac{R}{\sqrt{p(x)}} \cos\left\{\frac{1}{\hbar} \int_a^x p(y)\, dy - \delta\right\}, \qquad (15.18)$$

where $p(x) = \sqrt{2m(E - V(x))}$ is the momentum of a classical particle with energy E and position x. Here R and δ are real constants, referred to as the amplitude and the phase of the approximate solution.

We refer to the function in (15.18) as the *oscillatory WKB function*. In integrating the square of the oscillatory WKB function over some interval, we may apply the identity $\cos^2\theta = (1 + \cos(2\theta))/2$ to the cosine factor. The rapidly oscillating $\cos(2\theta)$ term will be small for small $\hbar$ because of cancellation between positive and negative values. Thus, the integral of $\psi^2(x)$ over an interval will be, to leading order, just a constant times the integral of $1/p(x)$, or, equivalently, a constant times $1/v(x)$, where v is the velocity of the classical particle. But the integral of $1/v(x) = dt/dx$ with respect to x is just the time t that the classical particle spends in the interval. We obtain, then, the following result.

Conclusion 15.5 *If the amplitude R in (15.18) is chosen so that ψ has L^2 norm 1 over $[a, b]$, then the probability of finding the quantum particle in an interval $[c, d] \subset [a, b]$ is approximately the fraction of time the classical particle spends in $[c, d]$ over one period of classical motion.*

15.4.2   The Classically Forbidden Region

In the classically forbidden region, let us introduce the quantity

$$q(x) := \sqrt{2m(V(x) - E)}.$$

We look for approximate solutions to the Schrödinger equation (15.6) of the form

$$\psi(x) = A(x) \exp\left\{\pm\frac{1}{\hbar} \int_{x_0}^x q(y)\, dy\right\}.$$

If we analyze approximate solutions of this form precisely as in the classically allowed region, we again find that there is a unique choice for A (up to multiplication by a constant) that causes the order-$\hbar$ terms in $\hat{H}\psi - E\psi$ to cancel, namely $A(x) = C(q(x))^{-1/2}$. If we are hoping to approximate a *square-integrable* solution of the Schrödinger equation, we want to take a minus sign in the exponent on the interval (b, ∞), and it is convenient to the basepoint of our integration to be b. In the region $(-\infty, a)$, we want to take a plus sign in the exponent; it is then convenient to take the basepoint of our integration to be a and to reverse the direction of integration, which changes the sign in the exponent back to being negative.

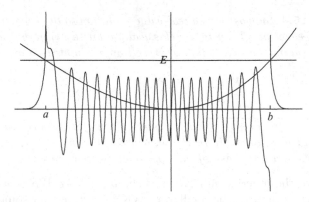

FIGURE 15.3. The WKB functions, extended all the way to the turning points.

Summary 15.6 *If $\psi_1(x)$ is a solution to the time-independent Schrödinger equation that tends to zero as x approaches $-\infty$, we expect that ψ_1 will be well approximated on $(-\infty, a)$, but away from the turning point, by the expression*

$$\frac{c_1}{\sqrt{q(x)}} \exp\left\{-\frac{1}{\hbar} \int_x^a q(y)\, dy\right\}, \tag{15.19}$$

where $q(x) = \sqrt{2m(V(x) - E)}$. Meanwhile, if $\psi_2(x)$ is a solution to the time-independent Schrödinger equation that tends to zero as x approaches $+\infty$, we expect that ψ will be well approximated on $(b, +\infty)$, but away from the turning point, by the expression

$$\frac{c_2}{\sqrt{q(x)}} \exp\left\{-\frac{1}{\hbar} \int_b^x q(y)\, dy\right\}. \tag{15.20}$$

We refer to the functions in (15.19) and (15.20) as the *exponential WKB functions*. The general theory of ordinary differential equations tells us that any solution to the time-independent Schrödinger equation for a smooth potential is smooth. Thus, the singularity at the turning points is an artifact of our approximation method. Nevertheless, for small values of $\hbar$, the true solution will "track" the WKB approximation until x gets very close to the turning point, with the result that the true solution will be large, but finite, near the turning points.

Figure 15.3 plots a potential function $V(x)$, an energy level E, and the WKB functions in both the classically allowed and classically forbidden regions. In the figure, the WKB functions have been (improperly) used all the way up to the turning points.

15.5   The Airy Function and the Connection Formulas

For any constant c_1 and any energy level E, we expect that there is a unique solution ψ_1 of the Schrödinger equation (15.6) that is well approximated for x tending to $-\infty$ by a function of the form (15.19). We expect that this solution will be well approximated in the classically allowed region (but not too close to the turning points) by a function of the form (15.18) for a unique pair of constants R and δ. In this section, we will see that the correct choices for R and δ are

$$R = 2c_1, \quad \delta = \frac{\pi}{4}. \tag{15.21}$$

The formula (15.21) for R and δ is called a *connection formula*; there is a similar formula connecting an approximate solution that tends to zero as x tends to $+\infty$ to an approximate solution in the classically allowed region. By comparing the two connection formulas, we will obtain conditions on the energy E under which the two approximate solutions (one that decays near $-\infty$ and one that decays near $+\infty$) agree up to a constant in the classically allowed region. The condition on E will turn out to be precisely Condition 15.1.

The discussion in the previous paragraph should be compared to the analysis in Chap. 5, where we determined the constants for the solution inside the well in terms of the energy level and the constant in front of the exponentially decaying solution outside the well. Here, of course, the analysis is more complicated because neither of the approximations (15.19) or (15.18) is valid near the turning point. The connection formula will be obtained, then, by using the Airy equation to approximate the Schrödinger equation near the turning points.

To get a reasonable approximation of our wave function near the turning points, we approximate V locally by a linear function. (By contrast, in the WKB functions, we are essentially thinking of V as being locally constant.) Thus, for example, near the turning point a, we write $V(x) \approx (a - x)F_0$, where $F_0 = -V'(a)$, yielding the approximate equation

$$-\frac{\hbar^2}{2m}\frac{d^2\psi}{dx^2} + (a - x)F_0\psi = 0.$$

By making the change of variable

$$u = \left(\frac{2mF_0}{\hbar^2}\right)^{1/3}(a - x) \tag{15.22}$$

we can reduce the equation to

$$\frac{d^2\psi}{du^2} - u\psi(u) = 0, \tag{15.23}$$

which is the Airy equation.

Equation (15.23) has two linearly independent solutions, denoted $\text{Ai}(u)$ and $\text{Bi}(u)$. We are interested in the solution $\text{Ai}(u)$, since this is the one that decays for $u > 0$, that is, for $x < a$. The function $\text{Ai}(u)$ is defined by the following convergent improper integral

$$\text{Ai}(u) = \frac{1}{\pi} \int_0^\infty \cos\left(\frac{t^3}{3} + ut\right) dt. \tag{15.24}$$

Intuitively, convergence is due to the very rapid oscillation of the integrand for large t, which produces a cancellation between the positive and negative values of the cosine function. Rigorously, convergence can be proved using integration by parts, as in Exercise 6. By differentiating under the integral sign (Exercise 7), one can show that Ai indeed satisfies the Airy equation (15.23).

As $|u|$ gets large, the integrand in (15.24) becomes more and more rapidly oscillating, producing more cancellation. The only exception to this behavior is when the derivative (with respect to t) of the function $t^3/3 + ut$ is zero. Near such a point, the argument of the cosine function is changing slowly and there is little oscillation. If u is negative, there is a unique critical point of $t^3/3 + ut$, at $t = \sqrt{-u}$, and we expect that the main contribution to the integral in (15.24) will come from $t \approx \sqrt{-u}$. If u is positive, $t^3/3 + ut$ has no critical points, and we expect that the integral in (15.24) will become quite small as u tends to $+\infty$. This sort of reasoning can be used to determine the precise asymptotics of the Airy function as u tends to $+\infty$ and as u tends to $-\infty$; see the discussion following (15.32) and (15.33).

We now state our main result, which will be derived in the remainder of this section. The result is not rigorous, because we have not estimated any of errors involved; such error estimates will be performed in Sect. 15.6.

Claim 15.7 *If ψ_1 is a solution of the Schrödinger equation (15.6) that tends to zero near $-\infty$, then ψ_1 can be normalized so that the following approximations hold*

$$\psi_1(x) \approx \frac{1}{2\sqrt{q(x)}} \exp\left\{-\frac{1}{\hbar} \int_x^a q(y)\, dy\right\} \quad \text{(near } -\infty\text{)} \tag{15.25}$$

$$\psi_1(x) \approx \frac{\sqrt{\pi}}{(2mF_0\hbar)^{1/6}} \text{Ai}\left(\left(\frac{2mF_0}{\hbar^2}\right)^{1/3}(a - x)\right) \quad \text{(near } x = a\text{)} \tag{15.26}$$

$$\psi_1(x) \approx \frac{1}{\sqrt{p(x)}} \cos\left\{\frac{1}{\hbar} \int_a^x p(y)\, dy - \frac{\pi}{4}\right\} \quad (a < x < b). \tag{15.27}$$

Here $F_0 = -V'(a)$ and in the case of (15.27), x should not be too close to a or to b.

Similarly, if ψ_2 is a solution of the Schrödinger equation (15.6) that tends to zero near $+\infty$, then ψ_2 can be normalized so that the following approximations hold

$$\psi_2(x) \approx \frac{1}{\sqrt{p(x)}} \cos\left\{-\frac{1}{\hbar}\int_x^b p(y)\,dy + \frac{\pi}{4}\right\} \quad (a < x < b) \tag{15.28}$$

$$\psi_2(x) \approx \frac{\sqrt{\pi}}{(2mF_1\hbar)^{1/6}} \mathrm{Ai}\left(\left(\frac{2F_1 m}{\hbar^2}\right)^{1/3}(x-b)\right) \quad (near\ x = b) \tag{15.29}$$

$$\psi_2(x) \approx \frac{1}{2\sqrt{q(x)}} \exp\left\{-\frac{1}{\hbar}\int_b^x q(y)\,dy\right\} \quad (near\ +\infty). \tag{15.30}$$

Here $F_1 = V'(b)$ and in the case of (15.28), x should not be too close to a or to b.

The approximate formulas for ψ_1 and ψ_2 will agree, up to multiplication by a constant, in the classically allowed region if and only if we have

$$\frac{1}{\hbar}\int_a^b p(x)\,dx = \left(n + \frac{1}{2}\right)\pi \tag{15.31}$$

for some non-negative integer n.

More specifically, (15.27) and (15.28) are equal when the integer n in (15.31) is even and they are negatives of each other when n is odd. Note that there is a factor of 2 in the denominator in (15.25) but not in (15.27); this factor accounts for the expression $R = 2c_1$ in (15.21).

Since the classical energy curve consists of two "branches," of the form $(x, p(x))$ and $(x, -p(x))$, the compatibility condition (15.31) is equivalent to Condition 15.1. Since the phase of the approximate wave function in the classically allowed region is given by $1/\hbar$ times the integral of $p\,dx$, the condition (15.31) says that the wave function goes through *a little more than n half-cycles* between the two turning points, where a half-cycle corresponds to a change in the phase in the amount of π, or the interval between two critical points of the wave function. In particular, the wave function has exactly $n+1$ critical points inside the classically allowed region. The first and last critical points occur slightly inside the turning points, leaving a change in phase of roughly $\pi/4$ between the extreme critical point and the turning point.

Figure 15.4 considers the same potential as in Fig. 15.3. The figure shows the WKB functions (15.25) and (15.27), together with the scaled Airy function (15.26), near the turning point $x = a$. Note that there is a good match between the WKB functions and the scaled Airy function when x is close to, but not *too* close to, the turning point. Meanwhile, Fig. 15.5 then shows the full approximate wave function with $\hbar$ chosen so that (15.31) holds with $n = 39$, obtained by using the WKB functions away from the turning points and the scaled Airy functions near the turning points. Finally,

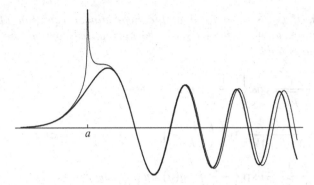

FIGURE 15.4. Plots of the scaled Airy function (*thick curve*) and the WKB functions, near the turning point $x = a$.

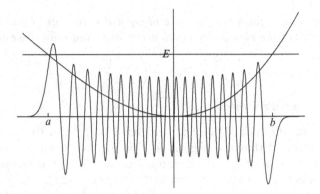

FIGURE 15.5. The approximate wave function with $n = 39$.

Fig. 15.6 shows the probability distribution associated to the approximate wave function, plotted together with the function $1/p(x)$. (Compare the discussion preceding Conclusion 15.5.)

We now derive the results in Claim 15.7. The Airy function $\mathrm{Ai}(u)$ is known to have the following asymptotic behavior:

$$\mathrm{Ai}(u) \approx \frac{1}{2\sqrt{\pi}u^{1/4}} \exp\left\{-\frac{2}{3}u^{3/2}\right\}, \quad u \to +\infty, \tag{15.32}$$

and

$$\mathrm{Ai}(u) \approx \frac{1}{\sqrt{\pi}(-u)^{1/4}} \cos\left(\frac{2}{3}(-u)^{3/2} - \frac{\pi}{4}\right), \quad u \to -\infty. \tag{15.33}$$

For u tending to $-\infty$, the asymptotics in (15.33) can be obtained by a straightforward application of the "method of stationary phase," as explained in Exercise 9. For u tending to $+\infty$, repeated integrations by parts (Exercise 8) show that $\mathrm{Ai}(u)$ decays faster than any power of u, which is all

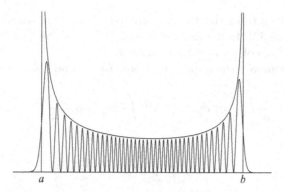

FIGURE 15.6. The probability distribution of the approximate wave function, plotted against the function $1/p(x)$.

that is strictly required for the main theorem of Sect. 15.6. To obtain the precise asymptotics in (15.32), one should deform the contour of integration to obtain a different integral representation of Ai(u), and then apply some variant of the method of stationary phase, such as Laplace's method or the method of steepest descent. See Sect. 4.7 of [30] for one approach to this analysis.

We will use the Airy function on an interval around the turning points with a length that goes to zero as $\hbar$ tends to zero (so that the linear approximation to the potential gets better and better) but with a length that is large compared to $\hbar^{2/3}$ (so that the value of u at the ends of the interval will be large, putting us into the asymptotic region of the Airy function). See Sect. 15.6 for more information.

We use the linear approximation $V(x) \approx (a - x)F_0$ to the potential near $x = a$, where $F_0 = -V'(a)$, which turns the Schrödinger equation (15.6) into the Airy equation, as previously noted. Now, the linear approximation to V yields

$$p \approx \sqrt{2mF_0}\sqrt{x - a} \qquad (15.34)$$

and

$$\frac{1}{\hbar}\int_a^x p(y)\,dy \approx \frac{\sqrt{2mF_0}}{\hbar}\frac{(x-a)^{3/2}}{3/2} = \frac{2}{3}(-u)^{3/2}. \qquad (15.35)$$

From here it is a simple matter to check, using (15.33), that

$$\frac{\sqrt{\pi}}{(2mF_0\hbar)^{1/6}}\mathrm{Ai}(u) \approx \frac{1}{\sqrt{p(x)}}\cos\left(\frac{1}{\hbar}\int_a^x p(y)\,dy - \frac{\pi}{4}\right)$$

for $x > a$, where the approximation holds in an intermediate region where x is close to a but not *too* close to a. Thus, if we scale our solution ψ_1 to the Schrödinger equation so that it is approximated by $\pi^{1/2}(2mF_0\hbar)^{-1/6}$ times Ai(u) near $x = a$, it should satisfy (15.27) in the classically allowed region (but away from the turning points). It is then straightforward to verify, using (15.32), that this multiple of $\text{Ai}(u)$ satisfies (15.25) for x near $-\infty$. The analysis of ψ_2 is entirely similar.

Finally, to compare the approximations (15.27) and (15.28), we note that

$$-\frac{1}{\hbar}\int_x^b p(y)\,dy + \frac{\pi}{4} = \left(\int_a^x p(y)\,dy - \frac{\pi}{4}\right) - \phi,$$

where

$$\phi = \frac{1}{\hbar}\int_a^b p(y)\,dy - \pi/2.$$

Now, if ϕ is an odd multiple of π, then $\cos(\theta - \phi) = -\cos\theta$ and if ϕ is an even multiple of π, then $\cos(\theta - \phi) = \cos\theta$. For all other values of ϕ (Exercise 4), $\cos(\theta - \phi)$ is not a constant multiple of $\cos\theta$. Thus, (15.31) is a necessary and sufficient condition for the two approximate solutions to agree up to a constant in the classically allowed region.

15.6   A Rigorous Error Estimate

The preceding sections give a treatment of the WKB approximation that is typical of many books in the literature. This treatment gives the idea that energies E satisfying the corrected Bohr–Sommerfeld Condition (Condition 15.1) should be approximate eigenvalues for the Hamiltonian operator $\hat{H}$, without specifying the sense in which this approximation holds. In this section, we prove a rigorous estimate, as follows.

Theorem 15.8 *For any potential V and range $[E_1, E_2]$ of energies satisfying Assumption 15.3, there is a constant C such that the following holds. For any energy $E \in [E_1, E_2]$ satisfying Condition 15.1, there exists a nonzero function ψ belonging to $\text{Dom}(\hat{H})$ such that*

$$\|\hat{H}\psi - E\psi\| < C\hbar^{9/8}\|\psi\|. \tag{15.36}$$

As noted already in Sect. 15.3, an estimate of the form $\|\hat{H}\psi - E\psi\| < \varepsilon\|\psi\|$ implies that there is a point $\tilde{E}$ in the spectrum of $\hat{H}$ with $|E - \tilde{E}| < \varepsilon$. (See Exercise 4 in Chap. 10.) Since, under our assumptions on V, the spectrum of $\hat{H}$ is purely discrete, we conclude that for each number $E \in [E_1, E_2]$ satisfying Condition 15.1, there is an actual eigenvalue $\tilde{E}$ for $\hat{H}$ with

$$|E - \tilde{E}| < C\hbar^{9/8}. \tag{15.37}$$

If E satisfies Condition 15.1, then the estimate (15.37) actually holds with $\hbar^{9/8}$ replaced by $\hbar^2$ on the right-hand side. It is not, however, possible to obtain such an optimal estimate by the methods we are using in this chapter. Specifically, the approximate eigenvector ψ constructed in the proof of Theorem 15.8 *does not* satisfy an estimate of the form $\|\hat{H}\psi - E\psi\| < C\hbar^2$. One can, however, construct an approximate eigenvector by different methods—for example, the method in [31]—that satisfies an order-$\hbar^2$ error estimate, for any E satisfying the corrected Condition 15.1. Nevertheless, the error bound in (15.37) is small compared to the typical spacing between the energy levels, which is of order $\hbar$.

Recall, as we noted at the beginning of Sect. 15.4, that a Schrödinger operator with potential V that is smooth and tends to $+\infty$ at $\pm\infty$ is essentially self-adjoint on $C_c^\infty(\mathbb{R})$. The operator $\hat{H}$ in Theorem 15.8 is, more precisely, the unique self-adjoint extension of the Schrödinger operator defined on $C_c^\infty(\mathbb{R})$.

15.6.1 Preliminaries

Our construction of the approximate eigenfunction ψ will be essentially by the WKB approximation as outlined in Claim 15.7. That is to say, we will define ψ using scaled Airy functions near the turning points and by the standard WKB functions in the classically allowed and classically forbidden regions. There is, however, a difficulty with this approach, which is that at the boundary between different regions, the scaled Airy function does not *exactly* match the WKB functions, but only approximately. What this means is that if we define ψ by the WKB formula in, say, an interval of the form $(-\infty, a - \varepsilon)$ and we define ψ by a scaled Airy function on $(a - \varepsilon, a + \varepsilon)$, then ψ may be discontinuous at $a - \varepsilon$. Even if we scale ψ by a constant on one of these intervals to eliminate the discontinuity in ψ itself, the derivative of ψ will still probably be discontinuous. But if the derivative of ψ is discontinuous, ψ is not actually in the domain of $\hat{H}$, and the left-hand side of (15.36) does not make sense. (Compare Sect. 5.2.)

The condition that ψ' be continuous is not just a technicality: If we did not worry about continuity of ψ', then we could *always* match the scaled Airy function to the WKB functions, just by multiplying the various functions by constants, regardless of whether or not the energy satisfies the corrected Bohr–Sommerfeld Condition. In that case, we would be claiming that *any* number $E \in [E_1, E_2]$ is within $C\hbar^{9/8}$ of an eigenvalue of $\hat{H}$, which is false already for the harmonic oscillator.

To work around the difficulty described in the previous paragraphs, we must put in a transition region over which we smoothly pass from one function to the other, using the "join" construction described in Sect. 15.6.4. Thus, we define the function ψ in Theorem 15.8 as follows. We use the formulas in Claim 15.7 in the indicated intervals, except that multiply the functions (15.28), (15.29), and (15.30) by -1 when n is odd. We use the scaled Airy functions (15.26) and (15.29) on intervals of the form $(a - \varepsilon, a + \varepsilon)$ and $(b - \varepsilon, b + \varepsilon)$, respectively, for some ε depending on $\hbar$ in a manner to be determined later. We then put in four transition regions, each

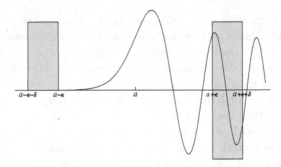

FIGURE 15.7. The approximate eigenfunction ψ, with the transition regions shaded.

having length δ, where δ also depends on $\hbar$ in a manner to be determined later. The first transition region, for example, is the interval $(a - \varepsilon - \delta, a - \varepsilon)$ between the first classically forbidden region and the first turning point. In each transition region, we change over smoothly from one function to another. See Fig. 15.7 for an illustration of the transition regions around the turning point $x = a$.

Suppose $\hat{H}_0$ denotes the Schrödinger operator with potential V, with domain equal to $C_c^\infty(\mathbb{R})$. Then, as we have noted, $\hat{H}_0$ is essentially self-adjoint, and we are letting $\hat{H}$, which coincides with the adjoint operator $\hat{H}_0^*$, denote the unique self-adjoint extension of $\hat{H}_0$. Now, the domain of $\hat{H}_0^*$ consists of all functions $\psi \in L^2(\mathbb{R})$ such that the Schrödinger operator, computed in the distributional sense, again belongs to $L^2(\mathbb{R})$. In particular, if ψ is smooth, then ψ belongs to the domain of $\hat{H} = \hat{H}_0^*$ if and only if ψ is in $L^2(\mathbb{R})$ and $-(\hbar^2/2m)\psi'' + V\psi$ is also in $L^2(\mathbb{R})$.

Because of the joins, our approximate eigenfunction is ψ actually infinitely differentiable on all of $\mathbb{R}$. And since $V(x)$ tends to $+\infty$ at $\pm\infty$, the exponential WKB functions (15.25) and (15.30) have rapid decay at infinity, which shows that ψ is in $L^2(\mathbb{R})$. Furthermore, for x near $\pm\infty$, the calculation (15.17) applies, with $A(x) = Cq(x)^{-1/2}$. We obtain, after a short calculation,

$$
-\frac{\hbar^2}{2m}\psi''(x) + V(x)\psi(x)
$$
$$
= -\frac{\hbar^2}{2m}\left(\frac{5}{16}\left(\frac{V'(x)}{V(x) - E}\right)^2 - \frac{1}{4}\frac{V''(x)}{V(x) - E}\right)\psi(x). \tag{15.38}
$$

Since V'/V and V''/V are assumed to be bounded near infinity and $\psi(x)$ tends to $+\infty$ at $\pm\infty$, we see that the Schrödinger operator applied to ψ is bounded by a constant times ψ near infinity and is thus square integrable. This shows that ψ is in the domain $\hat{H}$.

In Sect. 15.6.2, we will take the width 2ε of the region around the turning points to be of order $\hbar^{1/2}$. In that case, the L^2 norm of our approximate wave function is of order 1 (bounded and bounded away from zero) as $\hbar$ tends to zero, despite the blow-up of order $\hbar^{-1/6}$ very near the turning points. Although this result is not hard to verify (Exercise 10), if anything, the norm would be blowing up as $\hbar$ tends to zero, which would only help us in showing that $\|\hat{H}\psi - E\psi\|$ is small compared to $\|\psi\|$.

To prove Theorem 15.8, we must estimate the contributions to the quantity $\|\hat{H}\psi - E\psi\|$ from four different types of regions: the classically allowed region, the classically forbidden regions, the regions near the turning points, and the transition regions. These estimates will occupy the remainder of this section, with the analysis in the transition regions being the most involved. In particular, it is essential that the derivative of scaled Airy function *almost* match the derivative of the WKB function in the transition region, as in the second part of Lemma 15.9.

15.6.2   The Regions Near the Turning Points

We use a scaled Airy function in an interval around each turning point. [We use (15.26) near $x = a$ and either (15.29) or the negative thereof near $x = b$, depending on whether n is even or odd.] We now verify that taking these intervals to have length of order $\hbar^{1/2}$ will give satisfactory estimates. If ψ denotes one of the scaled Airy functions, then ψ satisfies a Schrödinger equation in which the potential V is replaced by a linear approximation $\tilde{V}$ near one of the turning points, which means that

$$\hat{H}\psi - E\psi = (V(x) - \tilde{V}(x))\psi. \tag{15.39}$$

The difference between $V(x)$ and its linear approximation $\tilde{V}(x)$ grows at most quadratically with the distance from the turning point. Meanwhile, the asymptotics of the Airy function tell us that it can be bounded as $|\mathrm{Ai}(u)| \leq Cu^{-1/4}$. (This is terrible estimate for small u, but still true.) Now u, as defined in (15.22), is of order $\hbar^{-2/3}$ times the distance to the turning point. Since, also, there is factor of $\hbar^{-1/6}$ in (15.26) and the distance from the turning point is at most of order $\hbar^{1/2}$, we find that

$$|\hat{H}\psi - E\psi| \leq C(\hbar^{1/2})^2 \hbar^{-1/6}(\hbar^{-2/3}\hbar^{1/2})^{-1/4} = C\hbar^{7/8}$$

over the interval around each turning point. Finally, if a function f satisfies $|f| \leq D$ on an interval of length L, then the L^2 norm of f over that interval will be at most $D\sqrt{L}$. Thus, over the interval around the turning points,

$$\|\hat{H}\psi - E\psi\| = O(\hbar^{7/8}\hbar^{1/4}) = O(\hbar^{9/8}).$$

15.6.3   The Classically Allowed and Classically Forbidden Regions

The expression (15.38) for $\hat{H}\psi - E\psi$, derived from (15.17), applies both in the classically allowed region and in the classically forbidden regions. Let us consider first the classically allowed region. Although (15.38) is nominally of order $\hbar^2$, we use this expression on an interval whose ends get closer and closer to the turning point as $\hbar$ tends to zero. Since, also, the expression in (15.38) is blowing up at the turning points, the contribution to $\|\hat{H}\psi - E\psi\|$ from this interval is of order larger than $\hbar^2$.

We have taken the interval around the turning point to have length 2ε that is of order $\hbar^{1/2}$, and we will also take (Sect. 15.6.4) the transition regions to have length δ that is of order $\hbar^{1/2}$. Thus, we use the oscillatory WKB function on an interval of the form $(a + \gamma, b - \gamma)$, where $\gamma = \varepsilon + \delta$ is of order $\hbar^{1/2}$. Now, the formula for ψ in the classically allowed regions has a factor of $1/\sqrt{p(x)}$ times a bounded quantity (the cosine factor). Since $V'(a)$ is assumed to be nonzero, $V(x) - E$ behaves like a constant times $(x - a)$ and so $1/\sqrt{p(x)}$ behaves like a constant time $(x - a)^{-1/4}$ for x approaching a, with similar behavior near the other turning point.

Meanwhile, the more problematic term in (15.38) is the term having $(V(x) - E)^2$ in the denominator. Keeping in mind the $1/\sqrt{p}$ blowup of ψ itself, this term behaves like $(x - a)^{-9/4}$ as x approaches a. Thus, we may estimate the norm of $\hat{H}\psi - E\psi$ over the left half of the classically allowed region as

$$\|\hat{H}\psi - E\psi\| \leq C\hbar^2 \left(\int_{(a+b)/2}^{a+\gamma} (x - a)^{-9/2}\, dx \right)^{1/2}$$
$$= C'\hbar^2 (\gamma^{-7/2} - ((a + b)/2)^{7/2})^{1/2}.$$

Since γ is of order $\hbar^{1/2}$, the contribution to $\|\hat{H}\psi - E\psi\|$ from the interval $(a + \gamma, (a + b)/2)$ will consist of a term of order $\hbar^2 \hbar^{-7/8} = \hbar^{9/8}$, plus lower-order terms. The estimate over the other half of the classically allowed region is similar.

Meanwhile, in the first classically forbidden region, we also apply (15.38). By Assumption 15.3, V'/V and V''/V are bounded near infinity. Thus, $V'/(V - E)$ and $V''/(V - E)$ will also be bounded near infinity, and thus also bounded on $(-\infty, a-1)$, since $V - E$ is strictly positive on this interval and tends to $+\infty$ as x tends to $-\infty$. We see, then, that the norm of $\hat{H}\psi - E\psi$ over $(-\infty, a - 1)$ is bounded by a constant times $\hbar^2 \|\psi\|$.

The norm of $\hat{H}\psi - E\psi$ over an interval of the form $(a - 1, a - \gamma)$ can be analyzed similarly to the classically allowed region. The estimates from this region are better, however, because of the exponentially decaying factor in the definition of the WKB function. Thus, the contribution to $\|\hat{H}\psi - E\psi\|$ from the classically forbidden region $(-\infty, a - \gamma)$ is certainly no larger than order $\hbar^{9/8}$, and similarly for the other classically forbidden region.

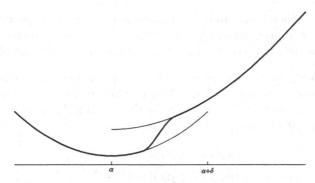

FIGURE 15.8. The join of two functions over the interval $[\alpha, \alpha + \delta]$ (*thick curve*).

15.6.4   *The Transition Regions*

Given two smooth functions ψ_1 and ψ_2 and some interval of the form $[\alpha, \alpha + \delta]$, we now define a "join" $\psi_1 \sqcup \psi_2$ of ψ_1 and ψ_2, where $\psi_1 \sqcup \psi_2(x)$ is equal to $\psi_1(x)$ for $x < \alpha$ and equal to $\psi_2(x)$ for $x > \alpha + \delta$, and where $\psi_1 \sqcup \psi_2$ is smooth everywhere. Let χ be a smooth function on $[0,1]$ that is identically equal to 0 in a neighborhood of 0 and identically equal to 1 in a neighborhood of 1. Then define $\psi_1 \sqcup \psi_2$ by

$$(\psi_1 \sqcup \psi_2)(x) = \psi_1(x) + (\psi_2(x) - \psi_1(x))\chi((x - \alpha)/\delta).$$

(See Fig. 15.8.) By direct calculation, we have

$$(\hat{H} - EI)(\psi_1 \sqcup \psi_2) = (\hat{H}\psi_1 - E\psi_1) \sqcup (\hat{H}\psi_2 - E\psi_2)$$
$$- \frac{1}{\delta} \frac{\hbar^2}{m}(\psi_2'(x) - \psi_1'(x))\chi'((x - a)/\delta)$$
$$- \frac{1}{\delta^2} \frac{\hbar^2}{2m}(\psi_2(x) - \psi_1(x))\chi''((x - a)/\delta). \qquad (15.40)$$

In our constructing our approximate eigenfunction, we use five different formulas in five different regions: the two classically forbidden regions, the classically allowed region, and the regions near the two turning points. Since none of these functions exactly matches the function in the next interval, we put in a total of four joins in order to produce a function that is in the domain of $\hat{H}$. We choose the width δ of the interval on which the join takes place to be of the same size as the intervals around the turning points, namely, order $\hbar^{1/2}$.

The most critical case is the transition from the region near the turning points to the classically allowed region. Consider, for example, the scaled Airy function ψ_1 in (15.26) and the oscillatory WKB function ψ_2 in (15.27). There are two contributions to the mismatch between these two functions. First, there is a discrepancy between the Airy function and its leading-order asymptotics. Second, there is an error in the approximations (15.34) and (15.35), which come from the discrepancy between the potential $V(x)$ and its linear approximation $\tilde{V}(x)$ near $x = a$. We need to consider both contributions to the mismatch in our estimation of $\psi_1 - \psi_2$ and of $\psi_1' - \psi_2'$.

Lemma 15.9 *Let ψ_1 denote the scaled Airy function in (15.26), let $\tilde{\psi}_1$ denote the same function with the Airy function replaced by the right-hand side of (15.33), and let ψ_2 denote the oscillatory WKB function in (15.27). If $x - a$ is positive and of order $\hbar^{1/2}$, we have*

$$|\psi_1(x) - \tilde{\psi}_1(x)| = O(\hbar^{1/8})$$
$$|\tilde{\psi}_1(x) - \psi_2(x)| = O(\hbar^{1/8})$$

and

$$|\psi_1'(x) - \tilde{\psi}_1'(x)| = O(\hbar^{-5/8})$$
$$|\tilde{\psi}_1'(x) - \psi_2(x)| = O(\hbar^{-5/8}).$$

Before giving the proof of this lemma, let us verify that these estimates are sufficient to control the contribution to $\|\hat{H}\psi - E\psi\|$ from the transition region $(a + \varepsilon, a + \varepsilon + \delta)$ between the first turning point and the classically allowed region, where both ε and δ are taken to be of order $\hbar^{1/2}$. We must consider each of the three lines in (15.40). The L^2 norm of the first line is of order at most $\hbar^{9/8}$, by precisely the same argument as in Sect. 15.6.3.

For the second and third lines, we recall that if a function f is bounded by C, then the L^2 norm of f over an interval of length L is at most $C\sqrt{L}$. Since we are taking the length δ of our transition interval to be of order $\hbar^{1/2}$, the L^2 norm of the second line of (15.40) is of order

$$\frac{1}{\hbar^{1/2}} \hbar^2 \hbar^{-5/8} \hbar^{1/4} = \hbar^{9/8}.$$

Meanwhile, the contribution from the third line of (15.40) is of order

$$\frac{1}{\hbar} \hbar^2 \hbar^{1/8} \hbar^{1/4} = \hbar^{11/8}.$$

Thus, the contribution to $\|\hat{H}\psi - E\psi\|$ from the transition region $(a + \varepsilon, a + \varepsilon + \delta)$ is of order at most $\hbar^{9/8}$.

The analysis of the transition between the classically allowed region and the region around $x = b$ is entirely similar. The analysis of the transitions between the regions near the turning points and the classically forbidden regions is also similar, but much less delicate, because all of the functions involved are very small in the transition region. When $(a - x)$ is positive and of order $\hbar^{1/2}$, for example, u, as defined in (15.22) will be of order $\hbar^{-1/6}$ and so $u^{3/2}$ is of order $\hbar^{-1/4}$. Thus, the exponential factor in leading-order asymptotics of the Airy function for $u > 0$ will behave like $\exp(-C\hbar^{-1/4})$, which is very small for small $\hbar$, certainly smaller than any power of $\hbar$. Since all the factors in front of the exponential will behave like $\hbar$ to a power, the overall contribution to $\|\hat{H}\psi - E\psi\|$ from the transition between the region near the turning points and the classically forbidden region is smaller than any power of $\hbar$. Thus, none of the transition regions contributes an error worse that $O(\hbar^{9/8})$.

Proof of Lemma 15.9. We consider only the estimates for the derivatives of the functions involved. The analysis of the functions themselves is similar (but easier) and is left as an exercise to the reader (Exercise 11).

We begin by considering $\psi_1' - \tilde{\psi}_1'$. With a little algebra, we compute that

$$\frac{d\psi_1}{dx} - \frac{d\tilde{\psi}_1'}{dx} = -\sqrt{\pi}(2mF_0)^{1/6}\hbar^{-5/6}(\mathrm{Ai}'(u) - \widetilde{\mathrm{Ai}}'(u)) \tag{15.41}$$

where u is as in (15.22) and where $\widetilde{\mathrm{Ai}}$ is the function on the right-hand side of (15.33).

Now, $\mathrm{Ai}(u)$ has an asymptotic expansion for $u \to -\infty$ given by

$$\mathrm{Ai}(u) = \widetilde{\mathrm{Ai}}(u)(1 + Cu^{-3/2} + \cdots),$$

and $\mathrm{Ai}'(u)$ has the asymptotic expansion obtained by formally differentiating this with respect to u. [See Eq. (7.64) in [30].] From this, we obtain

$$\mathrm{Ai}'(u) - \widetilde{\mathrm{Ai}}'(u) = \widetilde{\mathrm{Ai}}'(u)O((-u)^{-3/2}) + \widetilde{\mathrm{Ai}}(u)O((-u)^{-5/2}). \tag{15.42}$$

From the explicit formula for $\widetilde{\mathrm{Ai}}$, we see that $\widetilde{\mathrm{Ai}}(u)$ is of order $(-u)^{-1/4}$. Meanwhile, the formula $\widetilde{\mathrm{Ai}}'(u)$ will contain two terms, the larger of which will be of order $u^{1/4}$. Thus, the slower-decaying term on the right-hand side of (15.42) is the first one, which is of order $(-u)^{-5/4}$. Now, in the transition regions, u behaves like $\hbar^{-2/3}\hbar^{1/2} = \hbar^{-1/6}$. Thus, (15.42) goes like $\hbar^{5/24}$ and so (15.41) goes like $\hbar^{-5/6+5/24} = \hbar^{-5/8}$, as claimed.

We now consider $\tilde{\psi}_1' - \psi_2'$. By direct calculation, the derivatives of $\tilde{\psi}_1$ and ψ_2 each consist of two terms, a "dominant" obtained by differentiating the cosine factor and a "subdominant" term obtained by differentiating the coefficient of the cosine factor. In the case of $\tilde{\psi}_1'$, the dominant term in the derivative may be simplified to

$$-\frac{1}{\hbar}((2mF_0)(x-a))^{1/4}\sin\left(\frac{2}{3}(-u)^{3/2} - \frac{\pi}{4}\right). \tag{15.43}$$

According to Exercise 12, we have, when $x - a$ is of order $\hbar^{1/2}$, the estimates

$$((2mF_0)(a-x))^{1/4} = \sqrt{p} + \sqrt{p}O(\hbar^{1/2}) \tag{15.44}$$

and

$$\frac{2}{3}(-u)^{3/2} = \frac{1}{\hbar}\int_a^x p(y)\,dy + O(\hbar^{1/4}). \tag{15.45}$$

Since the derivative of $\sin\theta$ is bounded, a change of order $\hbar^{1/4}$ in the argument of a sine function produces a change of order $\hbar^{1/4}$ in the value of the sine. Thus, if we substitute (15.44) and (15.45) into (15.43), we find that the difference between the dominant term in $\tilde{\psi}_1'$ and the dominant term in ψ_1' is

$$\frac{1}{\hbar}\sqrt{p}\,O(\hbar^{1/4}) + \text{lower-order terms.}$$

Since $\sqrt{p}$ is of order $(x-a)^{1/4}$ or $\hbar^{1/8}$, we get an error of order $\hbar^{-5/8}$, as claimed.

Finally, the subdominant terms in the derivatives of $\tilde{\psi}_1$ and ψ_2 are easily seen to be separately of order $\hbar^{-5/8}$. Thus, even without taking into account the cancellation between these terms, they do not change the order of the estimate. ∎

15.6.5 Proof of the Main Theorem

We have estimated the contributions to $\|\hat{H}\psi - E\psi\|$ from each type of region: classically allowed and classically forbidden regions, the regions around the turning points, and the transition regions. In each case, we have found a contribution that is of order at most $\hbar^{9/8}\|\psi\|$. Thus, it remains only to verify that the constants in all estimates are bounded uniformly over the given range $E_1 \le E \le E_2$ of energies.

This verification is straightforward. Near the turning point $x = a$, for example, we need to estimate the difference between the potential $V(x)$ and its linear approximation $\tilde{V}(x)$ near $x = a$. As a consequence of the Taylor remainder formula, $|V(x) - \tilde{V}(x)|$ will be bounded by $C\,|x-a|^2/2$, where C is the maximum of $|V''(x)|$ over the interval from a to x. As E varies over $[E_1, E_2]$, the set of points where we have to evaluate $|V''(x)|$ will be bounded, meaning that C can be taken to be independent of E, for E in such a range.

Similarly, in the classically allowed region, the blow-up of $1/(V(x)-E)^2$ near $x = a(E)$ can be controlled by the minimum of $|V'(y)|$ for y between a and x. By assumption, $|V'(x)| > 0$ at all the turning points $a(E)$ and $b(E)$ with $E_1 \le E \le E_2$, and thus, by continuity, in some neighborhood of that set of turning points. Thus, blow-up of $1/(V(x)-E)^2$ will be controlled by the minimum of $|V'(x)|$ on an interval of the form $[a(E_2) + \alpha, a(E_1) + \alpha]$ for some small $\alpha > 0$. The remaining details of this verification are left to the reader.

15.7   Other Approaches

The main complicating factor in the WKB approximation is the singular behavior near the turning points. The turning points, meanwhile, are only problematic because we are working in the position representation. The turning points, after all, are the points on the classical trajectory where the *position* of the particle achieves a maximum or a minimum. If we were to work in the momentum representation, the points where the *momentum* achieves a maximum or a minimum would instead be the problematic points. A. Voros [42] has proposed working in the Segal–Bargmann representation (Sect. 14.4). In Voros's analysis, there are no turning points and, thus, the analysis is much simpler. The problem with Voros's approach is that he only gives an approximation to the wave function on the classical energy curve. Even in simple cases, Voros's expression does not admit a holomorphic extension to the whole plane, but has branching behavior inside the classical energy curve. Thus, Voros's formula does not define an element of the quantum Hilbert space (which is a space of *entire* holomorphic functions), let alone an element of the domain of the Hamiltonian.

Nevertheless, it is possible to build approximate eigenfunctions as superpositions of coherent states, using formulas similar to those in Voros. This approach avoids dealing with turning points but still yields a rigorous eigenvalue estimate, with the same corrected Bohr–Sommerfeld condition as in Condition 15.1. See [31, 23, 7], or (in greater generality) [26].

15.8   Exercises

1. Show that if c_1 is any complex number, then we have an identity of the form
$$c_1 e^{i\theta} + \overline{c_1} e^{-i\theta} = R\cos(\theta - \delta)$$
for some real numbers R and δ.

2. Let $H(x,p) = p^2/2m + m\omega^2 x^2/2$ be the Hamiltonian for a harmonic oscillator having mass m and classical frequency ω. Show that a positive number E satisfies the corrected Bohr Sommerfeld condition (Condition 15.1) if and only if E is of the form $(n + 1/2)\hbar\omega$, where n is a non-negative integer.

   Note: In light of the results of Chap. 11, this calculation means that, in this very special case, the corrected Bohr–Sommerfeld condition gives the *exact* eigenvalues of the quantum Hamiltonian $\hat{H}$.

3. Suppose A and p are two nonzero, smooth functions satisfying (15.15). Show that $A(x) = C(p(x))^{-1/2}$ for some constant C.

   Hint: Think in terms of the logarithms of the functions involved.

4. Show that $\cos(\theta - \delta)$, viewed as a function of θ, agrees, up to multiplication by a constant, with $\cos(\theta - \delta')$ if and only if $\delta - \delta'$ is an integer multiple of π.

5. If ψ is an eigenvector for $\hat{H}$ that is approximated by (15.25) near $-\infty$, one might hope to find an approximate expression for ψ in the classically allowed region by analytically continuing around the turning point in the complex plane. Even assuming V is analytic, however, it is fairly evident that analytic continuation in the upper half-plane does not give the same answer as in the lower half-planes. Nevertheless, one could use the average of the upper and lower half-plane results as a (totally nonrigorous) guess for the behavior of ψ in the classically allowed region.

   Show that the above approach gives the correct phase δ in the connection formula (15.21) but is off by a factor of 2 in the amplitude R.

6. Using integration by parts, show that the limit

$$\lim_{A \to +\infty} \int_0^A \cos\left(\frac{t^3}{3} + ut\right) dt$$

   exists.

   Hint: Multiply and divide by $t^2 + u$ (avoiding points where $t^2 + u = 0$ in the case $u < 0$).

7. In this exercise, we sketch an argument that the Airy function in (15.24) satisfies the differential equation $\psi''(u) - u\psi(u) = 0$. For the purposes of this exercise, let us say that $\int_0^\infty f(t)\, dt = C$ if $\int_0^A f(t)\, dt = C + g(A)$, where the function g is bounded and oscillates around an average value of zero.

   Assuming that it is legal to differentiate under the integral sign, verify that Ai(u) satisfies the stated equation.

   Hint: After differentiating under the integral, look for a term that can be integrated explicitly.

   Note: A more rigorous approach to this verification would be to integrate by parts as in Exercise 6 and *then* differentiate under the integral. This approach is, however, a bit messier.

8. By integrating by parts repeatedly in (15.24), show that Ai(u) decays faster than any power of u as u tends to $+\infty$.

   Hint: A key point is to show that the boundary terms in the integration by parts vanish at every stage. After performing the integrations by parts, estimate the resulting integral by using the inequality

$$\frac{1}{(t^2 + u)^n} < \frac{1}{(t^2 + 1)^k} \frac{1}{u^{n-k}}, \quad u > 1,$$

   for some appropriate choice of k.

9. (a) For $u < 0$, make the change-of-variable $\tau = t/\sqrt{-u}$ in the integral formula for the Airy function, to obtain the expression

$$\text{Ai}(u) = \frac{\sqrt{-u}}{\pi} \int_0^\infty \cos\left(\alpha\left(\frac{\tau^3}{3} - \tau\right)\right) d\tau, \qquad (15.46)$$

where $\alpha = (-u)^{3/2}$.

(b) Suppose f is a smooth function on $[a, b]$ having a unique critical point x_0. Assuming that x_0 is in the interior of $[a, b]$ and that $f''(x_0) \neq 0$, the method of stationary phase asserts that

$$\int_a^b g(x)e^{i\alpha f(x)}\, dx = g(x_0)e^{i\alpha f(x_0)}e^{\pm i\pi/4}\sqrt{\frac{2\pi}{\alpha\,|f''(x_0)|}} + O\left(\frac{1}{\alpha}\right)$$

for α tending to $+\infty$, where the plus sign in the exponent is taken when $f''(x_0) > 0$ and the minus sign is taken when $f''(x_0) < 0$. (See, e.g., Eq. (5.12) in [30].)

Using this result, obtain the asymptotic formula (15.33).

Hint: Divide the integral in (15.46) into an integral over $[0, 2]$ and an integral over $[2, \infty)$. Use stationary phase for the first interval and integration by parts (as in Exercise 6) for the second interval.

10. Let ψ be the approximate eigenfunction for $\hat{H}$ defined in the beginning of Sect. 15.6. Show that the norm of ψ is bounded and bounded away from zero as $\hbar$ tends to zero.

Hint: First show that the L^2 norm of ψ over the intervals around the turning points goes like $\hbar^{-1/6}\hbar^{1/4}$. Then check that the functions $p(x)^{-1/2}$ and $q(x)^{-1/2}$ are square integrable near the turning points.

11. By imitating the arguments in the proof of Lemma 15.9, prove the estimates for ψ_1 $\tilde{\psi}_1$ and $\tilde{\psi}_1 - \psi_2$ in the lemma.

12. By writing $V(x)$ as $F_0(a-x)$ plus an error term of order $(x-a)^2$, verify that the estimates (15.44) and (15.45) in the proof of Lemma 15.9 hold in the transition region. (Assume that $x - a$ is of order $\hbar^{1/2}$ in the transition region.)

Hint: The leading-order Taylor expansion of $(1+z)^a$ is $1+az+O(z^2)$, for any real number a.

16

Lie Groups, Lie Algebras, and Representations

An important concept in physics is that of *symmetry*, whether it be rotational symmetry for many physical systems or Lorentz symmetry in relativistic systems. In many cases, the group of symmetries of a system is a *continuous group*, that is, a group that is parameterized by one or more real parameters. More precisely, the symmetry group is often a *Lie group*, that is, a smooth manifold endowed with a group structure in such a way that operations of inversion and group multiplication are smooth. The tangent space at the identity in a Lie group has a natural "bracket" operation that makes the tangent space into a Lie algebra. The Lie algebra of a Lie group encodes many of the properties of the Lie group, and yet the Lie algebra is easier to work with because it is a linear space.

In quantum mechanics, the way symmetry is encoded is usually through a unitary action of the group on the relevant Hilbert space. That is, we assume we are given a *unitary representation* of the relevant symmetry group G, that is, a continuous homomorphism of G into $\mathsf{U}(\mathbf{H})$, the group of unitary operators on the quantum Hilbert space $\mathbf{H}$. Actually, since two unit vectors in $\mathbf{H}$ that differ only by a constant represent the same physical state, we should more properly consider *projective* unitary representations. A projective representation is a homomorphism of a group G into $\mathsf{U}(\mathbf{H})/\mathsf{U}(1)$, where $\mathsf{U}(1)$ is the group of complex numbers of magnitude 1, thought of multiples of I in $\mathsf{U}(\mathbf{H})$. An ordinary or projective representation of a Lie group gives rise to an ordinary or projective representation of its Lie algebra. The angular momentum operators, for example, form a representation of the Lie algebra of the rotation group.

B.C. Hall, *Quantum Theory for Mathematicians*, Graduate Texts
in Mathematics 267, DOI 10.1007/978-1-4614-7116-5_16,
© Springer Science+Business Media New York 2013

Saying that, for example, the Hamiltonian operator of a quantum system is invariant under rotations means that $\hat{H}$ commutes with the relevant representation of the rotation group and thus also with the associated Lie algebra operators. This commutativity, in turn, implies that the eigenspaces for $\hat{H}$ are invariant under rotations. We will use this commutativity in Chap. 18 to help us in determining the energy eigenvectors for the hydrogen atom.

In this chapter, we will make a brief survey of Lie groups, Lie algebras, and their representations. For our purposes, it suffices to consider *matrix Lie groups*, those that can be realized as closed subgroups of the group of $n \times n$ invertible matrices. Inevitably, I have had to present some of the deeper results without proof. Proofs of all results stated here can be found in [21]. The results of this chapter will be put to use in Chap. 17, in our study of angular momentum, and in Chap. 18, in our study of the hydrogen atom.

16.1   Summary

In this chapter, we will consider a matrix Lie group G, which is, by definition, a (topologically) closed subgroup of some $\mathsf{GL}(n; \mathbb{C})$, where $\mathsf{GL}(n; \mathbb{C})$ is the group of $n \times n$ invertible matrices with complex entries. To each such G, we will associate the Lie algebra $\mathfrak{g}$ of G, where $\mathfrak{g}$ is a real subspace of $M_n(\mathbb{C})$, the space of all $n \times n$ matrices. We will see that G is automatically an embedded real submanifold of $M_n(\mathbb{C})$ and that $\mathfrak{g}$ is the tangent space of G at the identity matrix.

Now, $\mathfrak{g}$ is not just a real vector space, but comes with a "bracket" operation mapping $\mathfrak{g} \times \mathfrak{g}$ into $\mathfrak{g}$. Specifically, we will show that for all X and Y in $\mathfrak{g}$, the matrix $XY - YX$ belongs again to $\mathfrak{g}$. Thus, we define our bracket by setting $[X, Y]$ equal to $XY - YX$. As it turns out, the Lie algebra $\mathfrak{g}$, as a vector space with the bracket operation, encodes a lot of information about the group G. On the other hand, computing at the level of the Lie algebra is generally easier than computing at the group level, simply because $\mathfrak{g}$ is a linear space.

We will be interested in unitary *representations* of our group G, that is, continuous homomorphisms of G into $\mathsf{U}(\mathbf{H})$, the group of unitary operators on a Hilbert space. If we restrict attention, at first, to the case in which $\mathbf{H}$ is finite dimensional, then each representation Π of G gives rise to a representation π of the Lie algebra $\mathfrak{g}$ of G. That is to say, π is a linear map of $\mathfrak{g}$ into the space of linear maps of V to V, satisfying $\pi([X, Y]) = [\pi(X), \pi(Y)]$. A deeper question is whether every representation π of $\mathfrak{g}$ comes from a representation Π of G. As it turns out, the answer in general is no, but the answer is yes if G is simply connected.

We may consider, for example, the case $G = SO(3)$. This group is not simply connected. On the other hand, the Lie algebra $so(3)$ of $SO(3)$ is isomorphic to the Lie algebra $su(2)$ of $SU(2)$, and $SU(2)$ is simply connected. [That is, $SU(2)$ is the "universal cover" of $SO(3)$.] Thus, given a representation π of $so(3)$, there may or may not be an associated representation Π of $SO(3)$. Even if there is not, however, there is always a representation Π' of the group $SU(2)$.

In quantum mechanics, the vector $e^{i\theta}\psi$ represents the same physical state as ψ. Thus, it is natural to consider "projective" unitary representations, that is, homomorphisms of G into the quotient group $U(\mathbf{H})/\{e^{i\theta}I\}$. In the finite-dimensional case, each projective representation can be "deprojectivized" at the level of the Lie algebra $\mathfrak{g}$ of G. We can then pass from the Lie algebra to the universal cover of G, that is, the simply connected group with Lie algebra $\mathfrak{g}$. In particular, in the finite-dimensional case, the irreducible *projective* unitary representations of $SO(3)$ are in one-to-one correspondence with irreducible *ordinary* unitary representations of the universal cover $SU(2)$ of $SO(3)$. Although the Hilbert spaces of physical systems are usually infinite dimensional, for compact groups such as $SO(3)$, general unitary representations can be decomposed as direct sums of finite-dimensional ones. (See, e.g., Proposition 17.19 and the discussion following it.)

16.2 Matrix Lie Groups

Let $M_n(\mathbb{C})$ denote the space of $n \times n$ matrices with complex entries. We identify $M_n(\mathbb{C})$ with $\mathbb{C}^{n^2}$, equipped with the usual topology. Thus, a sequence A_m in $M_n(\mathbb{C})$ *converges* to a matrix $A \in M_n(\mathbb{C})$ if $(A_m)_{jk}$ converges to A_{jk} as m tends to infinity, for all $1 \leq j, k \leq n$. Let $GL(n; \mathbb{C})$ denote the *general linear group*, consisting of all invertible $n \times n$ matrices with complex entries. Then $GL(n; \mathbb{C})$ forms a group under the operation of matrix multiplication. Furthermore, $GL(n; \mathbb{C})$—that is, the set of $A \in M_n(\mathbb{C})$ with $\det A \neq 0$—is an open subset of $M_n(\mathbb{C})$. Since $M_n(\mathbb{C})$ is a complex vector space of dimension n^2, it may be identified with $\mathbb{C}^{n^2} \cong \mathbb{R}^{2n^2}$. Since $GL(n; \mathbb{C})$ is an open subset of $M_n(\mathbb{C})$, it looks locally like $\mathbb{R}^{2n^2}$ and is therefore a real manifold of dimension $2n^2$.

Definition 16.1 *A subgroup G of $GL(n; \mathbb{C})$ is **closed** if for each sequence A_m in G that converges to a matrix A, either A is again in G or A is not invertible. A **matrix Lie group** is a closed subgroup of some $GL(n; \mathbb{C})$.*

A subgroup G of $GL(n; \mathbb{C})$ is closed if it is topologically closed *as a subset of* $GL(n; \mathbb{C})$—but not necessarily as a subset of $M_n(\mathbb{C})$. We will see that each matrix Lie group is a real embedded submanifold of $GL(n; \mathbb{C})$ and thus is a Lie group.

Definition 16.2 *If G_1 and G_2 are matrix Lie groups, then a **Lie group homomorphism** of G_1 to G_2 is a continuous group homomorphism of G_1 into G_2. A Lie group homomorphism is called a **Lie group isomorphism** if it is one-to-one and onto with continuous inverse. Two matrix Lie groups are called **isomorphic** if there exists a Lie group isomorphism between them.*

Example 16.3 *The real **general linear group**, denoted $\mathsf{GL}(n, \mathbb{R})$, is the group of invertible $n \times n$ matrices with real entries. The groups $\mathsf{SL}(n, \mathbb{C})$ and $\mathsf{SL}(n, \mathbb{R})$ are, respectively, the groups of complex and real matrices with determinant 1. They are called the **special linear groups**.*

Example 16.4 *An $n \times n$ matrix $U \in M_n(\mathbb{C})$ is said to be **unitary** if $U^*U = UU^* = I$. A matrix U is unitary if and only if*

$$\langle Uv, Uw \rangle = \langle v, w \rangle$$

*for all $v, w \in \mathbb{C}^n$. The group of unitary matrices is denoted $\mathsf{U}(n)$ and called the $(n \times n)$ **unitary group**. The **special unitary group**, denoted $\mathsf{SU}(n)$, is the subgroup of $\mathsf{U}(n)$ consisting of unitary matrices with determinant 1.*

The condition $(U^*U)_{jk} = \delta_{jk}$ is equivalent to the condition that the columns of U form an orthonormal set in $\mathbb{C}^n$, as can be seen by direct computation. Geometrically, the condition $U^*U = I$ is equivalent to the condition that $\langle Uv_1, Uv_2 \rangle = \langle v_1, v_2 \rangle$ for all $v_1, v_2 \in \mathbb{C}^n$, i.e., that U preserves the inner product on $\mathbb{C}^n$. By taking the determinant of the condition $U^*U = I$, we see that $|\det U| = 1$ for all $U \in \mathsf{U}(n)$.

In this, the finite-dimensional case, the condition $U^*U = I$ implies that U^* is the inverse of U and thus that $UU^* = I$. This result does not hold in the infinite-dimensional case.

Example 16.5 *An $n \times n$ real matrix $R \in M_n(\mathbb{R})$ is said to be orthogonal if $R^{tr}R = RR^{tr} = I$. A matrix R is orthogonal if and only if*

$$\langle Rv, Rw \rangle = \langle v, w \rangle$$

*for all $v, w \in \mathbb{R}^n$. The group of orthogonal matrices is denoted $\mathsf{O}(n)$ and is called the $(n \times n)$ **orthogonal group**. The **special orthogonal group**, denoted $\mathsf{SO}(n)$, is the subgroup of $\mathsf{O}(n)$ consisting of orthogonal matrices with determinant 1.*

As in the unitary case, the condition $R^{tr}R = I$ implies that $RR^{tr} = I$ and that the columns of R form an orthonormal set in $\mathbb{R}^n$. Geometrically, a real matrix R is in $\mathsf{O}(n)$ if and only if $\langle Rv_1, Rv_2 \rangle = \langle v_1, v_2 \rangle$ for all $v_1, v_2 \in \mathbb{R}^n$, i.e., if and only if R preserves the inner product on $\mathbb{R}^n$. By taking the determinant of the condition $R^{tr}R = I$ we see that $\det R = \pm 1$ for all $R \in \mathsf{O}(n)$.

It is easy to verify that all the groups in Examples 16.3, 16.4, and 16.5 are, indeed, subgroups of $\mathsf{GL}(n, \mathbb{C})$ and that they are closed.

Definition 16.6 *A matrix Lie group G is* **connected** *if for all $A, B \in G$ there is a continuous path $A : [0, 1] \to M_n(\mathbb{C})$ such that $A(0) = A$ and $A(1) = B$ and such that $A(t)$ lies in G for all t. A matrix Lie group G is* **simply connected** *if it is connected and every continuous loop in G can be shrunk continuously to a point in G. A matrix Lie group G is* **compact** *if it is* **compact** *as a subset of $M_n(\mathbb{C}) \cong \mathbb{R}^{2n^2}$.*

By the Heine–Borel theorem (e.g., Proposition 0.26 of [12]), a matrix Lie group G is compact if and only if it is a closed and bounded subset of $M_n(\mathbb{C})$. The condition we are calling "connected" is, more properly, the condition of being *path connected*. We will see, however, that each matrix Lie group is an embedded real submanifold of $M_n(\mathbb{C})$ and is, therefore, locally path connected. For matrix Lie groups, then, connectedness and path connectedness are equivalent.

To prove that a matrix Lie group G is connected, it suffices to prove that for all $A \in G$, there is a continuous path in G connecting A to I. After all, if both A and B can be connected to I, then they can be connected to each other.

Example 16.7 *The groups* $\mathsf{O}(n)$, $\mathsf{SO}(n)$, $\mathsf{U}(n)$, *and* $\mathsf{SU}(n)$ *are compact.*

Proof. The conditions defining these groups are obtained by setting certain continuous functions equal to a constant. The group $\mathsf{SU}(n)$, for example, is defined by setting $(U^*U)_{jk} = \delta_{jk}$ for each j and k and by setting $\det U = 1$. These groups are thus closed not just as subsets of $\mathsf{GL}(n; \mathbb{C})$ but also as subsets of $M_n(\mathbb{C})$. Furthermore, each of these groups has the property that each column of any matrix in the group is a unit vector. Thus, each group is a bounded subset of $M_n(\mathbb{C})$. ∎

Example 16.8 *The group* $\mathsf{U}(n)$ *is connected.*

Proof. If $U \in M_n(\mathbb{C})$ is unitary, then U has an orthonormal basis of eigenvectors with eigenvalues of absolute value 1. Thus, there is another unitary matrix V (the change of basis matrix) such that

$$
U = V \begin{pmatrix} e^{i\theta_1} & & & \\ & e^{i\theta_2} & & \\ & & \ddots & \\ & & & e^{i\theta_n} \end{pmatrix} V^{-1},
$$

for some real numbers $\theta_1, \theta_2, \ldots, \theta_n$. Thus, we can define a family $U(t)$ of unitary matrices by setting

$$
U(t) = V \begin{pmatrix} e^{it\theta_1} & & & \\ & e^{it\theta_2} & & \\ & & \ddots & \\ & & & e^{it\theta_n} \end{pmatrix} V^{-1}.
$$

Then $U(\cdot)$ is a continuous path lying in $\mathsf{U}(n)$ with $U(0) = I$ and $U(1) = U$.
∎

Example 16.9 *The group* $\mathsf{SU}(2)$ *is simply connected.*

Proof. We claim that

$$\mathsf{SU}(2) = \left\{ \begin{pmatrix} \alpha & -\bar{\beta} \\ \beta & \bar{\alpha} \end{pmatrix} \middle| \alpha, \beta \in \mathbb{C}, \ |\alpha|^2 + |\beta|^2 = 1 \right\}.$$

It is easy to see that each matrix of the indicated form is indeed unitary and has determinant 1. On the other hand, if U is any element of $\mathsf{SU}(2)$, then the first column of U is a unit vector $(\alpha, \beta) \in \mathbb{C}^2$. The second column of U must then be orthogonal to (α, β). Since $(-\bar{\beta}, \bar{\alpha})$ is orthogonal to (α, β) and $\mathbb{C}^2$ is 2-dimensional, the second column of U must be a multiple of $(-\bar{\beta}, \bar{\alpha})$. But the only multiple that produces a matrix with determinant 1 is 1.

We see, then, that $\mathsf{SU}(2)$ is, topologically, the unit sphere S^3 inside $\mathbb{C}^2 \cong \mathbb{R}^4$ and is, therefore, simply connected. ∎

16.3   Lie Algebras

We now introduce the general algebraic concept of a Lie algebra. Once this is done, we will show how to associate a real Lie algebra with an arbitrary matrix Lie group.

Definition 16.10 *A **Lie algebra** over a field* $\mathbb{F}$ *is a vector space* $\mathfrak{g}$ *over* $\mathbb{F}$, *together with a "bracket" map* $[\cdot, \cdot] : \mathfrak{g} \times \mathfrak{g} \to \mathfrak{g}$ *having the following properties:*

 1. $[\cdot, \cdot]$ *is bilinear*

 2. $[Y, X] = -[X, Y]$ *for all* $X, Y \in \mathfrak{g}$

 3. $[X, X] = 0$ *for all* $X \in \mathfrak{g}$

 4. *For all* $X, Y, Z \in \mathfrak{g}$ *we have the **Jacobi identity***

$$[X, [Y, Z]] + [Y, [Z, X]] + [Z, [X, Y]] = 0.$$

If the characteristic of $\mathbb{F}$ is not equal to 2, then Property 3 is a consequence of Property 2. If $\mathbb{F} = \mathbb{R}$, then we say that $\mathfrak{g}$ is a real Lie algebra. An example of a real Lie algebra is the vector space $\mathbb{R}^3$ with the bracket equal to the cross product. Properties 1, 2, and 3 are evident from the definition of the cross product, while the Jacobi identity is a known property of the cross product that can be verified by direct calculation.

A large class of Lie algebras may be obtained by the following procedure.

Example 16.11 *Let $\mathcal{A}$ be an associative algebra and let $\mathfrak{g}$ be a subspace of $\mathcal{A}$ with the property that for all x, y in $\mathfrak{g}$, $xy - yx$ is again in $\mathfrak{g}$. Then the bracket*

$$[x, y] := xy - yx$$

makes $\mathfrak{g}$ into a Lie algebra.

In Example 16.11, we may take, for example, $\mathfrak{g} = \mathcal{A}$. It is evident that this bracket satisfies Properties 1, 2, and 3 of a Lie algebra, and the Jacobi identity is easily verified by direct calculation. As it turns out, every Lie algebra is isomorphic to a Lie algebra of this type. (This claim is a consequence of the Poincaré–Birkhoff–Witt theorem, which is proved, for example, in Sect. 5.2 of [25]. The algebra $\mathcal{A}$ in the Poincaré–Birkhoff–Witt theorem is the so-called *universal enveloping algebra* of $\mathfrak{g}$.)

Definition 16.12 *If $\mathfrak{g}_1$ and $\mathfrak{g}_2$ are Lie algebras, a map $\phi : \mathfrak{g}_1 \to \mathfrak{g}_2$ is called a **Lie algebra homomorphism** if ϕ is linear and ϕ satisfies*

$$\phi([X, Y]) = [\phi(X), \phi(Y)]$$

*for all $X, Y \in \mathfrak{g}_1$. A Lie algebra homomorphism is called a **Lie algebra isomorphism** if it is one-to-one and onto.*

Definition 16.13 *If $\mathfrak{g}$ is a Lie algebra, a **subalgebra** of $\mathfrak{g}$ is a subspace $\mathfrak{h}$ of $\mathfrak{g}$ with the property that $[X, Y] \in \mathfrak{h}$ for all X and Y in $\mathfrak{h}$. An **ideal** in $\mathfrak{g}$ is a subalgebra $\mathfrak{h}$ of $\mathfrak{g}$ with the stronger property that $[X, Y] \in \mathfrak{h}$ for all X in $\mathfrak{g}$ and Y in $\mathfrak{h}$.*

The notion of a subalgebra of a Lie algebra is analogous to the notion of a subgroup of a group, while the notion of an ideal in a Lie algebra is analogous to the notion of a *normal* subgroup of a group. In particular, the kernel of any Lie algebra homomorphism is an ideal, just as the kernel of a group homomorphism is a normal subgroup.

Definition 16.14 *The **direct sum** of Lie algebras $\mathfrak{g}_1$ and $\mathfrak{g}_2$, denoted $\mathfrak{g}_1 \oplus \mathfrak{g}_2$, is the direct sum of $\mathfrak{g}_1$ and $\mathfrak{g}_2$ as a vector space, equipped with the bracket given by*

$$[(X_1, Y_1), (X_2, Y_2)] = ([X_1, X_2], [Y_1, Y_2])$$

for all $X_1, X_2 \in \mathfrak{g}_1$ and $Y_1, Y_2 \in \mathfrak{g}_2$.

16.4 The Matrix Exponential

In the next section, we will associate a Lie algebra with each matrix Lie group. To describe this association, we need the notion of the exponential of a matrix. Given a matrix $X \in M_n(\mathbb{C})$, we define the *matrix exponential* of X, denoted by e^X or $\exp(X)$, by the usual power series,

$$e^X = \sum_{m=0}^{\infty} \frac{X^m}{m!},$$

where $X^0 = I$ (the identity matrix). This series converges absolutely for all $X \in M_n(\mathbb{C})$, as can easily be seen using the inequality $\|X^m\| \le \|X\|^m$, where $\|X\|$ is the operator norm of X; see Definition A.35. (In this, the finite-dimensional case, we could just as well use the Hilbert–Schmidt norm, which amounts to using the usual Euclidean norm on $M_n(\mathbb{C}) \cong \mathbb{C}^{n^2}$. See Exercise 3.) The matrix exponential shares some but not all of the properties of the exponential of a number.

Theorem 16.15 *The matrix exponential has the following properties for all $X, Y \in M_n(\mathbb{C})$.*

1. $e^0 = I$

2. $e^{X^{tr}} = (e^X)^{tr}$ *and* $e^{X^*} = (e^X)^*$

3. *If A is an invertible $n \times n$ matrix, then*

$$e^{AXA^{-1}} = Ae^X A^{-1}.$$

4. $\det(e^X) = e^{\text{trace}(X)}$

5. *If $XY = YX$ then* $e^{X+Y} = e^X e^Y$

6. e^X *is invertible and* $(e^X)^{-1} = e^{-X}$

7. *Even if $XY \ne YX$, we have*

$$e^{X+Y} = \lim_{m \to \infty} \left(e^{X/m} e^{Y/m}\right)^m.$$

Here X^{tr} and X^* denote the transpose and adjoint (conjugate transpose) of X, respectively. Property 7 is known as the Lie Product Formula and is a special case of the Trotter Product formula (Theorem 20.1). Properties 1, 2, and 3 are easily verified using term-by-term computation. Property 6 follows from Property 5 by taking $Y = -X$ and applying Property 1. The proofs of Properties 4, 5, and 7 are outlined in Exercises 5, 6, and 7.

Suppose a matrix X is diagonalizable, meaning that

$$X = A \begin{pmatrix} \lambda_1 & & 0 \\ & \ddots & \\ 0 & & \lambda_n \end{pmatrix} A^{-1},$$

for some invertible matrix A and complex numbers $\lambda_1, \lambda_2, \ldots, \lambda_n$. Then using Property 3 of Theorem 16.15, it is easy to see that

$$
e^X = A \begin{pmatrix} e^{\lambda_1} & & 0 \\ & \ddots & \\ 0 & & e^{\lambda_n} \end{pmatrix} A^{-1}.
$$

If X is not diagonalizable, e^X can be computed in terms of the SN decomposition of X. See Sect. 2.2 of [21] for details.

Example 16.16 *If*

$$
X = \begin{pmatrix} 0 & a \\ -a & 0 \end{pmatrix}
$$

then

$$
e^X = \begin{pmatrix} \cos a & \sin a \\ -\sin a & \cos a \end{pmatrix}.
$$

Proof. The eigenvalues of X are $\pm ia$ and the corresponding eigenvectors are $(1, \pm i)$. Thus, we may calculate that

$$
e^X = \begin{pmatrix} 1 & 1 \\ i & -i \end{pmatrix} \begin{pmatrix} e^{ia} & 0 \\ 0 & e^{-ia} \end{pmatrix} \frac{1}{(-2i)} \begin{pmatrix} -i & -1 \\ -i & 1 \end{pmatrix}
$$

$$
= -\frac{1}{2i} \begin{pmatrix} -i(e^{ia} + e^{-ia}) & -e^{ia} + e^{-ia} \\ e^{ia} - e^{-ia} & -i(e^{ia} + e^{-ia}) \end{pmatrix},
$$

which simplifies to the desired result. ∎

The relation $e^{X+Y} = e^X e^Y$ certainly does not hold for general (noncommuting) matrices X and Y. Nevertheless, for any $X \in M_n(\mathbb{C})$ we have

$$
e^{(s+t)X} = e^{sX} e^{tX}
$$

for all s and t in $\mathbb{R}$, since sX commutes with tX. Thus, for each X, the set of matrices of the form e^{tX}, $t \in \mathbb{R}$, forms a subgroup of $GL(n; \mathbb{C})$. It is not hard to show (Exercise 4), using term-by-term differentiation, that

$$
\frac{d}{dt} e^{tX} \bigg|_{t=0} = X. \tag{16.1}
$$

Here, the derivative of a matrix-valued function is defined as being *entrywise*. [That is, if $f(t)$ is a matrix-valued function, df/dt is the matrix-valued function whose (j, k) entry is $d(f(t)_{jk})/dt$.]

Definition 16.17 *A **one-parameter subgroup** of* $\mathsf{GL}(n; \mathbb{C})$ *is a continuous homomorphism of* $\mathbb{R}$ *into* $\mathsf{GL}(n; \mathbb{C})$, *that is, a continuous map* $A : \mathbb{R} \to \mathsf{GL}(n; \mathbb{C})$ *such that* $A(0) = I$ *and* $A(s + t) = A(s)A(t)$ *for all* $s, t \in \mathbb{R}$.

Theorem 16.18 *If $A(\cdot)$ is a one-parameter subgroup of $\mathsf{GL}(n;\mathbb{C})$, there exists a unique $X \in M_n(\mathbb{C})$ such that*

$$A(t) = e^{tX}$$

for all $t \in \mathbb{R}$.

This is Theorem 2.13 in [21].

16.5   The Lie Algebra of a Matrix Lie Group

We now associate a Lie algebra $\mathfrak{g}$ to each matrix Lie group G.

Definition 16.19 *If $G \subset \mathsf{GL}(n;\mathbb{C})$ is a matrix Lie group, then the **Lie algebra** $\mathfrak{g}$ of G is defined as follows:*

$$\mathfrak{g} = \left\{ X \in M_n(\mathbb{C}) \,\big|\, e^{tX} \in G \text{ for all } t \in \mathbb{R} \right\}.$$

That is to say, X belongs to $\mathfrak{g}$ if and only if the one-parameter subgroup generated by X lies entirely in G. Note that to have X belong to $\mathfrak{g}$, we need only have e^{tX} belong to G for all *real* numbers t.

Proposition 16.20 *For any matrix Lie group G, the Lie algebra $\mathfrak{g}$ of G has the following properties.*

1. *The zero matrix 0 belongs to $\mathfrak{g}$.*

2. *For all X in $\mathfrak{g}$, tX belongs to $\mathfrak{g}$ for all real numbers t.*

3. *For all X and Y in $\mathfrak{g}$, $X + Y$ belongs to $\mathfrak{g}$.*

4. *For all $A \in G$ and $X \in \mathfrak{g}$ we have $AXA^{-1} \in \mathfrak{g}$.*

5. *For all X and Y in $\mathfrak{g}$, the commutator $[X, Y] := XY - YX$ belongs to $\mathfrak{g}$.*

The first three properties of $\mathfrak{g}$ say that $\mathfrak{g}$ is a real vector space. Since $M_n(\mathbb{C})$ is an associative algebra under the operation of matrix multiplication, the last property of $\mathfrak{g}$ shows that $\mathfrak{g}$ is a real Lie algebra (Example 16.11).

Proof. Points 1 and 2 are elementary, and Point 3 follows from the Lie product formula, using the assumption that G is closed. Point 4 follows from Property 3 in Theorem 16.15. To verify Point 5, we observe that the commutator $[X, Y]$ may be computed as

$$[X, Y] = \frac{d}{dt} e^{tX} Y e^{-tX} \bigg|_{t=0},$$

using (4) and an easily verified product rule for differentiation of matrix-valued functions. For $X, Y \in \mathfrak{g}$, $e^{tX}Ye^{-tX}$ belongs to $\mathfrak{g}$ for all $t \in \mathbb{R}$, by Point 4. Furthermore, we have already shown that $\mathfrak{g}$ is a real subspace of $M_n(\mathbb{C})$ and therefore a closed subset of $M_n(\mathbb{C})$. Thus,

$$[X, Y] = \lim_{h \to 0} \frac{e^{hX}Ye^{-hX} - Y}{h}$$

belongs to $\mathfrak{g}$. ∎

Example 16.21 *Let* $\mathsf{gl}(n;\mathbb{C})$, $\mathsf{gl}(n;\mathbb{R})$, $\mathsf{sl}(n;\mathbb{C})$, *and* $\mathsf{sl}(n;\mathbb{R})$ *denote the Lie algebras of* $\mathsf{GL}(n;\mathbb{C})$, $\mathsf{GL}(n;\mathbb{R})$, $\mathsf{SL}(n;\mathbb{C})$, *and* $\mathsf{SL}(n;\mathbb{R})$, *respectively. Then we have*

$$\mathsf{gl}(n;\mathbb{C}) = M_n(\mathbb{C})$$
$$\mathsf{gl}(n;\mathbb{R}) = M_n(\mathbb{R})$$
$$\mathsf{sl}(n;\mathbb{C}) = \{X \in M_n(\mathbb{C}) \,|\, \mathrm{trace}(X) = 0\}$$
$$\mathsf{sl}(n;\mathbb{R}) = \{X \in M_n(\mathbb{R}) \,|\, \mathrm{trace}(X) = 0\}.$$

Proof. Let us consider, for example, the case of $\mathsf{sl}(n;\mathbb{C})$. By Property 4 of Theorem 16.15, if $\mathrm{trace}(X) = 0$, then

$$\det(e^{tX}) = e^{t\,\mathrm{trace}(X)} = e^0 = 1,$$

so that $e^{tX} \in \mathsf{SL}(n;\mathbb{C})$. In the other direction, if $X \in \mathsf{sl}(n;\mathbb{C})$, then by the above calculation, we must have $e^{t\,\mathrm{trace}(X)} = 0$ for all $t \in \mathbb{R}$, which is possible only if $\mathrm{trace}(X) = 0$. The proofs of the other cases are similar and are omitted. ∎

Example 16.22 *The Lie algebras* $\mathsf{u}(n)$ *and* $\mathsf{su}(n)$ *of* $\mathsf{U}(n)$ *and* $\mathsf{SU}(n)$ *are given by*

$$\mathsf{u}(n) = \{X \in M_n(\mathbb{C}) \,|\, X^* = -X\}$$
$$\mathsf{su}(n) = \{X \in u(n) \,|\, \mathrm{trace}(X) = 0\}.$$

The Lie algebra $\mathsf{so}(n)$ *of* $\mathsf{SO}(n)$ *is given by*

$$\mathsf{so}(n) = \{X \in M_n(\mathbb{R}) \,|\, X^{tr} = -X\}.$$

Finally, the Lie algebra of $\mathsf{O}(n)$ *is equal to* $\mathsf{so}(n)$.

Proof. If $X^* = -X$, then by Property 2 of Theorem 16.15,

$$(e^{tX})^* = e^{tX^*} = e^{-tX} = (e^{tX})^{-1},$$

showing that e^{tX} is unitary. In the other direction, if e^{tX} is unitary for all $t \in \mathbb{R}$, then $(e^{tX})^* = (e^{tX})^{-1} = e^{-tX}$. Thus, $e^{tX^*} = e^{-tX}$. Differentiating this relation at $t = 0$, using (16.1), gives $X^* = -X$. Thus, the Lie algebra of

$U(n)$ consists exactly of the matrices with the property that $X^* = -X$. For the Lie algebra of $\mathsf{SU}(n)$, we add the trace-zero condition, as in the proof of Example 16.21. The calculations for $\mathsf{SO}(n)$ are similar and are omitted. Note that if $X \in M_n(\mathbb{R})$ satisfies $X^{tr} = -X$, then the diagonal entries of X are zero and, thus, trace(X) is automatically 0. This observation explains why the Lie algebras of $\mathsf{O}(n)$ and $\mathsf{SO}(n)$ are the same. ∎

Specializing Proposition 16.22 the case $n = 3$ gives

$$\mathsf{so}(3) = \left\{ \begin{pmatrix} 0 & a & b \\ -a & 0 & c \\ -b & -c & 0 \end{pmatrix} \middle| a, b, c \in \mathbb{R} \right\}.$$

We can use the following basis for $\mathsf{so}(3)$:

$$F_1 := \begin{pmatrix} 0 & 0 & 0 \\ 0 & 0 & -1 \\ 0 & 1 & 0 \end{pmatrix}; \; F_2 := \begin{pmatrix} 0 & 0 & 1 \\ 0 & 0 & 0 \\ -1 & 0 & 0 \end{pmatrix}; \; F_3 := \begin{pmatrix} 0 & -1 & 0 \\ 1 & 0 & 0 \\ 0 & 0 & 0 \end{pmatrix}.$$

$$(16.2)$$

Direct calculation establishes the following commutation relations for the F_j's:

$$[F_1, F_2] = F_3$$
$$[F_2, F_3] = F_1$$
$$[F_3, F_1] = F_2.$$

$$(16.3)$$

More concisely, we have $[F_1, F_2] = F_3$, together with relations obtained from this one by cyclic permutation of the indices. Note that all remaining commutation relations follow from (16.3) by means of the skew-symmetry of the bracket; we have, for example, $[F_2, F_1] = -F_3$ and $[F_1, F_1] = 0$.

16.6   Relationships Between Lie Groups and Lie Algebras

In this section, we explore the relationships between matrix Lie groups and their Lie algebras. In particular, we investigate the question of the extent to which a matrix Lie group is determined (up to isomorphism) by its Lie algebra. We begin by showing that every Lie group homomorphism gives rise to a Lie algebra homomorphism in a natural way.

Theorem 16.23 *Suppose G_1 and G_2 are matrix Lie groups with Lie algebras $\mathfrak{g}_1$ and $\mathfrak{g}_2$, respectively, and suppose $\Phi : G_1 \to G_2$ is a Lie group homomorphism. Then there exists a unique linear map $\phi : \mathfrak{g}_1 \to \mathfrak{g}_2$ such that*

$$\Phi(e^{tX}) = e^{t\phi(X)}$$

for all $t \in \mathbb{R}$ and $X \in \mathfrak{g}$. This linear map has the following additional properties:

1. $\phi([X,Y]) = [\phi(X), \phi(Y)]$ *for all $X, Y \in \mathfrak{g}$*

2. $\phi(AXA^{-1}) = \Phi(A)\phi(X)\Phi(A)^{-1}$ *for all $A \in G$ and $X \in \mathfrak{g}$*

3. $\phi(X)$ *may be computed as*

$$\phi(X) = \frac{d}{dt}\Phi\left(e^{tX}\right)\Big|_{t=0}.$$

Point 1 shows that ϕ is a Lie algebra homomorphism. Part of the assertion of Point 3 of the theorem is that $\Phi(e^{tX})$ is a smooth function of t for each X.

To construct ϕ, note that since Φ is a continuous homomorphism, the map $t \mapsto \Phi(e^{tX})$ is a one-parameter subgroup. By Theorem 16.18, there exists a unique Y such that $\Phi(e^{tX}) = e^{tY}$ for all $t \in \mathbb{R}$. We then set $\phi(X) = Y$. An argument similar to the proof of Proposition 16.20 then establishes the desired properties of ϕ. See the proof of Theorem 2.21 in [21] for the details.

Corollary 16.24 *Suppose that G_1 and G_2 are matrix Lie groups with Lie algebras $\mathfrak{g}_1$ and $\mathfrak{g}_2$, respectively. If G_1 is isomorphic to G_2, then $\mathfrak{g}_1$ is isomorphic to $\mathfrak{g}_2$.*

Proof. See Exercise 11. ∎

Our next task is to show that for any matrix Lie group G, the Lie algebra $\mathfrak{g}$ of G is large enough to capture what is happening in a neighborhood of the identity in G. This will show, for example, that for connected matrix Lie groups, a Lie group homomorphism is determined by the corresponding Lie algebra homomorphism.

Theorem 16.25 *Let G be a matrix Lie group with Lie algebra $\mathfrak{g}$. Then there exists a neighborhood U of 0 in $M_n(\mathbb{C})$ and a neighborhood V of I in $M_n(\mathbb{C})$ such that the matrix exponential maps U diffeomorphically onto V and such that for all $X \in U$, we have that X belongs to $\mathfrak{g}$ if and only if e^X belongs to G.*

See Theorem 2.27 in [21]. This result has a number of important consequences.

Corollary 16.26 *Every matrix Lie group $G \subset \mathsf{GL}(n; \mathbb{C})$ is a real embedded submanifold of $M_n(\mathbb{C})$ with the dimension of G equal to the dimension of $\mathfrak{g}$ as a real vector space.*

The claim means, more precisely, that for each $A \in G$, there exists a neighborhood U of A and a diffeomorphism Φ of U with a neighborhood V of 0 in $\mathbb{R}^{2n^2}$ such that $\Phi(U \cap G) = V \cap \mathbb{R}^d$, where $d = \dim \mathfrak{g}$. That is to say, after a change of coordinates, G "looks" locally like a little piece of $\mathbb{R}^d$ sitting inside $M_n(\mathbb{C}) \cong \mathbb{R}^{2n^2}$.

Proof. We use exponential coordinates in the neighborhood V of I in $M_n(\mathbb{C})$, meaning that we write each element A of V as $A = e^X$, with $X \in U$. Theorem 16.25 says that near the identity, in these coordinates, G "looks like" the real vector space $\mathfrak{g}$ inside $M_n(\mathbb{C})$. Given any other point $A \in G$, we can use left multiplication by A^{-1} to move the action to the identity (Exercise 17), with the result that G looks like $\mathfrak{g} \subset M_n(\mathbb{C})$ near A. Thus, G is a real embedded submanifold of dimension $d = \dim \mathfrak{g}$. ∎

Corollary 16.27 *The Lie algebra $\mathfrak{g}$ of a matrix Lie group G is the tangent space to G at I. That is to say, $\mathfrak{g}$ coincides with the set of those X in $M_n(\mathbb{C})$ for which there exists a smooth curve $\gamma : \mathbb{R} \to M_n(\mathbb{C})$ lying entirely in G and such that $\gamma(0) = I$ and $\gamma'(0) = X$.*

Proof. If $X \in \mathfrak{g}$, then X is the derivative of e^{tX} at $t = 0$, so $\mathfrak{g}$ is contained in the tangent space at I. In the other direction, if γ is any smooth curve in $M_n(\mathbb{C})$ that lies entirely in G and passes through I at $t = 0$, then by Theorem 16.25, we can express γ as $\gamma(t) = e^{\delta(t)}$ (at least for small t), where δ is a smooth curve in $\mathfrak{g}$ with $\delta(0) = 0$. It is then easy to see (Exercise 8) that $\gamma'(0) = \delta'(0)$. But if δ lies in $\mathfrak{g}$, then $\delta'(0)$, which equals $\gamma'(0)$, also lies in $\mathfrak{g}$, as in the proof of Proposition 16.20. Thus, the tangent space at I is contained in $\mathfrak{g}$. ∎

Corollary 16.28 *If a matrix Lie group G is connected, then for all $A \in G$ there exists a finite sequence $X_1, X_2, \ldots, X_N$ of elements of $\mathfrak{g}$ such that*

$$A = e^{X_1} e^{X_2} \cdots e^{X_N}.$$

Proof. If G is connected in the sense of Definition 16.6 (which really means that G is path connected), then G is certainly connected in the usual topological sense of having no nontrivial sets that are both open and closed. Let U denote the set of points in G that can be expressed as a product of exponentials of elements of $\mathfrak{g}$. This set is open in G because if $A \in U$ and $B \in G$ is close to A, then $A^{-1}B$ is close to I in G, and therefore $A^{-1}B = e^X$ for some $X \in \mathfrak{g}$. Thus, $B = Ae^X$, which means that B is also a product of exponentials. In the other direction, if $B \in G$ is in the closure of U, then there is some element A of U that is close to B. We then have, again, that $B = Ae^X$ for some $X \in \mathfrak{g}$, which, again, means that $B \in U$. Now, G is connected and U is both open and closed. Since U is nonempty ($I \in U$), we have $U = G$. ∎

Corollary 16.29 *Suppose that G_1 and G_2 are matrix Lie groups with Lie algebras $\mathfrak{g}_1$ and $\mathfrak{g}_2$, respectively. Suppose that $\Phi_1 : G_1 \to G_2$ and $\Phi_2 : G_1 \to G_2$ are Lie group homomorphisms, with associated Lie algebra homomorphisms ϕ_1 and ϕ_2, respectively. If G_1 is connected and $\phi_1 = \phi_2$, then $\Phi_1 = \Phi_2$.*

Proof. The result follows from Corollary 16.28 and the condition $\Phi_j(e^X) = e^{\phi_j(X)}$, $j = 1, 2$. ∎

We have seen that a homomorphism of matrix Lie groups gives rise to a homomorphism of the associated Lie algebra, and (Corollary 16.29) that if the domain group is connected, the Lie algebra homomorphism determines the Lie group homomorphism. A more difficult question is whether we can go in the opposite direction, from a Lie algebra homomorphism to a Lie group homomorphism. That is to say, given a Lie algebra homomorphism between the Lie algebras of two matrix Lie groups, does there exist a Lie group homomorphism related in the usual way to the Lie algebra homomorphism? The answer turns out to be yes, *provided* that the domain group G_1 is connected and *simply connected* (i.e., that every continuous loop in G_1 can be shrunk continuously in G_1 to a point).

Theorem 16.30 *Suppose that G_1 and G_2 are matrix Lie groups with Lie algebras $\mathfrak{g}_1$ and $\mathfrak{g}_2$, respectively, and suppose that $\phi : \mathfrak{g}_1 \to \mathfrak{g}_2$ is a Lie algebra homomorphism. If G_1 is connected and simply connected, then there exists a unique Lie group homomorphism $\Phi : G_1 \to G_2$ such that Φ and ϕ are related as in Theorem 16.23.*

One way to prove this deep result is to make use of the *Baker–Campbell–Hausdorff formula*. (See, e.g., Chap. 3 of [21].) This formula states that for all sufficiently small X and Y in $M_n(\mathbb{C})$ we have

$$e^X e^Y = e^{X+Y+\frac{1}{2}[X,Y]+\frac{1}{12}[X,[X,Y]]-\frac{1}{12}[Y,[X,Y]]+\cdots}.$$

Here $\cdots$ denotes terms that are expressible in terms of repeated commutators involving X and Y, with coefficients that are "universal," that is, independent of n (the size of the matrices) and of the choice of X and Y in $M_n(\mathbb{C})$. Given a Lie algebra homomorphism $\phi : \mathfrak{g}_1 \to \mathfrak{g}_2$, one can use the Baker–Campbell–Hausdorff formula to construct a "local homomorphism," mapping a neighborhood of the identity in G_1 into G_2. If G_1 is connected and simply connected, it is possible to extend this local representation to a global representation. See Sect. 3.6 of [21] for the details of this construction.

Corollary 16.31 *Suppose that G_1 and G_2 are matrix Lie groups with Lie algebras $\mathfrak{g}_1$ and $\mathfrak{g}_2$, respectively. If G_1 and G_2 are connected and simply connected and $\mathfrak{g}_1$ is isomorphic to $\mathfrak{g}_2$, then G_1 is isomorphic to G_2.*

Proof. Suppose $\phi : \mathfrak{g}_1 \to \mathfrak{g}_2$ is a Lie algebra isomorphism. Since G_1 is connected and simply connected, there exists a Lie group homomorphism $\Phi : G_1 \to G_2$ related in the usual way to ϕ. Since G_2 is connected and simply connected, there exists a Lie group homomorphism $\Psi : G_2 \to G_1$ related in the usual way to ϕ^{-1}. Consider now the homomorphism $\Psi \circ \Phi : G_1 \to G_1$.

By the composition property of Lie algebra homomorphisms (Exercise 10), the Lie algebra homomorphism associated with $\Psi \circ \Phi$ is $\phi^{-1} \circ \phi = I$. It then follows from Corollary 16.29 that $\Psi \circ \Phi = I$. A similar argument shows that $\Phi \circ \Psi = I$, which means that Φ is a Lie group isomorphism. ∎

Corollary 16.31 does not hold without the assumption that both groups are simply connected, as the following important example shows.

Example 16.32 *The Lie algebras* su(2) *and* so(3) *are isomorphic, but the groups* SU(2) *and* SO(3) *are not isomorphic.*

Since SU(2) is simply connected (Example 16.9), SO(3) must fail to be simply connected. Indeed, $\pi_1(\mathsf{SO}(3)) \cong \mathbb{Z}/2$, as can be seen from Example 16.34.

Proof. The Lie algebra su(2) of SU(2) is the space of 2×2 skew-self-adjoint matrices with trace zero. Explicitly,

$$\mathsf{su}(2) = \left\{ \begin{pmatrix} ia & b + ic \\ -b + ic & -ia \end{pmatrix} \middle| a, b, c \in \mathbb{R} \right\}.$$

We may consider the following basis for su(2):

$$E_1 = \frac{1}{2} \begin{pmatrix} i & 0 \\ 0 & -i \end{pmatrix}; \quad E_2 = \frac{1}{2} \begin{pmatrix} 0 & 1 \\ -1 & 0 \end{pmatrix}; \quad E_3 = \frac{1}{2} \begin{pmatrix} 0 & i \\ i & 0 \end{pmatrix}. \quad (16.4)$$

Direct calculation shows that $[E_1, E_2] = E_3$ and relations obtained from this by cyclic permutation of the indices. These are the same relations as those satisfied by the basis elements F_j, $j = 1, 2, 3$, for so(3) in (16.2) and (16.3). Thus, there is a Lie algebra isomorphism $\phi : \mathsf{su}(2) \to \mathsf{so}(3)$ such that $\phi(E_j) = F_j$, $j = 1, 2, 3$.

On the other hand, there can be no isomorphism between SU(2) and SO(3), since SU(2) has a nontrivial center (containing at least I and $-I$), whereas the center of SO(3) is trivial (Exercise 14). ∎

Definition 16.33 *Suppose G is a connected matrix Lie group with Lie algebra $\mathfrak{g}$. A **universal cover** of G is an ordered pair $(\tilde{G}, \Phi)$ consisting of a simply connected matrix Lie group $\tilde{G}$ and a Lie group homomorphism $\Phi : \tilde{G} \to G$ such that the associated Lie algebra homomorphism $\phi : \tilde{\mathfrak{g}} \to \mathfrak{g}$ is an isomorphism of the Lie algebra $\tilde{\mathfrak{g}}$ of $\tilde{G}$ with $\mathfrak{g}$. The map Φ is called the **covering map** for $\tilde{G}$.*

Although each Lie group has a universal cover that is again a Lie group, the universal cover of a *matrix* Lie group may not be isomorphic to any matrix Lie group. [The universal cover of SL(2; $\mathbb{R}$), e.g., is not a matrix Lie group.] It can be shown, however, that if a matrix Lie group G is compact, then the universal cover of G is again a matrix Lie group (not necessarily compact).

Suppose $\tilde{G}$ is any simply connected Lie group with a Lie algebra $\tilde{\mathfrak{g}}$ that is isomorphic to $\mathfrak{g}$. The choice of a particular isomorphism $\phi : \tilde{\mathfrak{g}} \to \mathfrak{g}$ gives rise, by Theorem 16.30, to a Lie group homomorphism $\Phi : \tilde{G} \to G$, so that $(\tilde{G}, \Phi)$ is a universal cover of G.

If $(\tilde{G}, \Phi)$ is a universal cover of G, it is often convenient to use the isomorphism ϕ to identify $\tilde{\mathfrak{g}}$ with $\mathfrak{g}$. If we follow this convention, we may say that a universal cover of G is a simply connected group $\tilde{G}$ having "the same" Lie algebra as G.

If $(\tilde{G}_1, \Phi_1)$ and $(\tilde{G}_2, \Phi_2)$ are two universal covers of a given matrix Lie group G, then there is a unique Lie group isomorphism $\Psi : \tilde{G}_1 \to \tilde{G}_2$ such that $\Phi_2(\Psi(A)) = \Phi_1(A)$ for all $A \in \tilde{G}_1$. (This result follows easily from Corollary 16.31.) In light of this uniqueness result, we will often speak of "the" universal cover of G.

Example 16.34 *Let $\Phi : \mathsf{SU}(2) \to \mathsf{SO}(3)$ be the unique Lie group homomorphism for which the associated Lie algebra homomorphism ϕ satisfies $\phi(E_j) = F_j$, $j = 1, 2, 3$. Then $\ker \Phi = \{I, -I\}$ and $(\mathsf{SU}(2), \Phi)$ is a universal cover of $\mathsf{SO}(3)$.*

Proof. Since E_1 is diagonal, it is easy to see that $e^{2\pi E_1} = -I$ in $SU(2)$. On the other hand, by a trivial extension of Example 16.16, we have

$$e^{aF_1} = \begin{pmatrix} 1 & 0 & 0 \\ 0 & \cos a & -\sin a \\ 0 & \sin a & \cos a \end{pmatrix}$$

for all $a \in \mathbb{R}$. In particular, $e^{2\pi F_1} = I$. Thus,

$$\Phi(-I) = \Phi(e^{2\pi E_1}) = e^{2\pi F_1} = I.$$

This shows that $-I$ belongs to the kernel of Φ.

Now, since ϕ is injective, Φ is injective in a neighborhood of I. After all, given distinct elements A and B of $SU(2)$ near I, Theorem 16.25 tells us that we can express A as e^X and B as e^Y, with X and Y being distinct small elements of $su(2)$. Then $\phi(X)$ and $\phi(Y)$ are distinct small elements of $so(3)$. Applying Theorem 16.25 again tells us that $\Phi(A) = e^{\phi(X)}$ and $\Phi(B) = e^{\phi(Y)}$ are distinct.

We see, then, that $\ker \Phi$ is a discrete normal subgroup of $SU(2)$. But a standard exercise (Exercise 1) shows that a discrete normal subgroup of a connected group is automatically central. On the other hand, it is easily verified (Exercise 2) that the center of $SU(2)$ is $\{I, -I\}$, so $\ker \Phi$ cannot be larger than $\{I, -I\}$.

To show that Φ maps *onto* $SO(3)$, we first verify (Exercise 13) that each element R of $SO(3)$ can be expressed as $R = e^X$, with $X \in so(3)$. Since ϕ is surjective and $\Phi(e^X) = e^{\phi(X)}$, Φ maps onto $SO(3)$. ∎

16.7   Finite-Dimensional Representations of Lie Groups and Lie Algebras

A *representation* of a group G is a homomorphism Π of G into $\mathsf{GL}(V)$, the group of invertible linear transformations on some vector space. If Π is injective then G is isomorphic to its image under Π; thus, Π serves to "represent" G concretely as a group of invertible linear transformations. (We continue to use the term "representation" even if Π is not injective.) Similarly, a representation of a Lie algebra $\mathfrak{g}$ is a Lie algebra homomorphism of $\mathfrak{g}$ into $\mathfrak{gl}(V)$, the space of all linear transformations of V, where we equip $\mathfrak{gl}(V)$ with the bracket $[X, Y] := XY - YX$.

Recall that an *action* of a group G on a set X is a map from $G \times X$ to X, denoted $(g, x) \mapsto g \cdot x$ satisfying $e \cdot x = x$ for all $x \in X$ and $g \cdot (h \cdot x) = (gh) \cdot x$ for all $g, h \in G$ and $x \in X$. A representation Π of G on some vector space V gives rise to a *linear action* of G on V, given by $g \cdot v = \Pi(g)v$. (A linear action is an action for which the map $v \mapsto g \cdot v$ is linear for each g.) Thus, we may use $g \cdot v$ as an alternative notation to $\Pi(g)v$, when convenient.

16.7.1   Finite-Dimensional Representations

If G is a matrix Lie group, then G is already represented as a group of matrices. Nevertheless, it is of interest [as we will see in Chap. 17 in the case $G = \mathsf{SO}(3)$] to explore other representations of G. Since a matrix Lie group has a topological structure (inherited from $M_n(\mathbb{C})$), it is natural to require representations to be continuous. It is also simpler to deal at first with *finite-dimensional* representations, that is, those where the vector space in question is finite dimensional, although eventually we will need to consider infinite-dimensional representations as well. This discussion leads to the following definition.

Definition 16.35 *Let $G \subset \mathsf{GL}(n; \mathbb{C})$ be a matrix Lie group. A finite-dimensional* **representation** *of G is a continuous homomorphism of G into $\mathsf{GL}(V)$, the group of invertible linear transformations of a finite-dimensional vector space V.*

We will assume that all of our vector spaces are over the field $\mathbb{C}$, even though it is occasionally of interest to consider also representations over $\mathbb{R}$. The topology on $\mathsf{GL}(V)$ is defined by picking a basis, and thereby identifying the space of linear maps of V to V with $M_n(\mathbb{C})$. We then use the subset topology on $\mathsf{GL}(V) \cong \mathsf{GL}(n; \mathbb{C}) \subset M_n(\mathbb{C})$. This topology is easily seen to be independent of the choice of basis.

An important example of representations in quantum theory arises from the time-independent Schrödinger equation in $\mathbb{R}^n$, namely the equation $\hat{H}\psi = E\psi$, for a fixed constant $E \in \mathbb{R}$. If $\hat{H}$ is invariant under rotations, then the *space* of solutions to this equation is invariant under rotations.

Note that an individual solution ψ to this equation may or may not be a rotationally invariant (i.e., radial) function. But if $\hat{H}$ is rotationally invariant, then rotating a solution to $\hat{H}\psi = E\psi$ will give another solution of this equation. Even if the quantum Hilbert space is infinite dimensional, the solution spaces to $\hat{H}\psi = E\psi$ are typically finite dimensional and constitute finite dimensional representations of the group $\mathsf{SO}(n)$ of rotations. If we can understand what all possible finite-dimensional representations of $\mathsf{SO}(n)$ look like, we will have made a lot of progress in understanding solutions to $\hat{H}\psi = E\psi$ in the rotationally invariant case. This line of reasoning will be explored in detail in Chap. 18.

We may consider as well finite-dimensional representations of Lie algebras. Assuming our Lie algebra $\mathfrak{g}$ is finite dimensional (which is the only case we will consider in this chapter), there is no need to impose a requirement of continuity, since a linear map of one finite-dimensional real or complex vector space to another is automatically continuous.

Definition 16.36 *A finite-dimensional **representation** of a Lie algebra* $\mathfrak{g}$ *is a Lie algebra homomorphism of* $\mathfrak{g}$ *into* $\mathsf{gl}(V)$*, the space of all linear transformations of* V*. Here* $\mathsf{gl}(V)$ *is considered as a Lie algebra with bracket given by* $[X,Y] = XY - YX$*.*

We typically consider Lie algebras defined over the field $\mathbb{R}$, since the Lie algebra of a matrix Lie group is in general only a real subspace of $M_n(\mathbb{C})$. Nevertheless, it is convenient to consider vector spaces over $\mathbb{C}$. If $\mathfrak{g}$ is a real Lie algebra and V, and therefore also $\mathsf{gl}(V)$, is a complex vector space, then we require only that $\pi : \mathfrak{g} \to \mathsf{gl}(V)$ be *real* linear, which is the only requirement that makes sense.

In the interest of simplifying the terminology, we will sometimes speak of "a representation V," without making explicit mention of the homomorphism Π or π.

Definition 16.37 *If* $\Pi : G \to \mathsf{GL}(V)$ *is a representation of a matrix Lie group* G*, then a subspace* W *of* V *is called an **invariant subspace** if* $\Pi(g)w \in W$ *for all* $g \in G$ *and* $w \in W$*. Similarly, if* $\pi : \mathfrak{g} \to \mathsf{gl}(V)$ *is a representation of a Lie algebra* $\mathfrak{g}$*, then a subspace* W *of* V *is called an invariant subspace if* $\pi(X)w \in W$ *for all* $X \in \mathfrak{g}$ *and* $w \in W$*. A representation of a group or Lie algebra is called **irreducible** if the only invariant subspaces are* $W = V$ *and* $W = \{0\}$*.*

Definition 16.38 *If* (Π, V_1) *and* (Σ, V_2) *are representations of a matrix Lie group* G*, a map* $\Phi : V_1 \to V_2$ *is called an **intertwining map** (or **morphism**) if* $\Phi(\Pi(g)v) = \Sigma(g)\Phi(v)$ *for all* $v \in V_1$*, with an analogous definition for intertwining maps of Lie algebra representations. If an intertwining map is an invertible linear map, it is called an **isomorphism**. Two representations are said to be **isomorphic** (or **equivalent**) if there exists an isomorphism between them.*

In the "action" notation, the requirement on an intertwining map Φ is that $\Phi(g \cdot v) = g \cdot \Phi(v)$, meaning that Φ commutes with the action of G. A typical goal of representation theory is to classify all finite-dimensional irreducible representations of G up to isomorphism.

Given a representation $\Pi : G \to \mathsf{GL}(V)$ of a matrix Lie group G, we can identify $\mathsf{GL}(V)$ with $\mathsf{GL}(N; \mathbb{C})$ and $\mathsf{gl}(V)$ with $\mathsf{gl}(n; \mathbb{C})$ by picking a basis for V. We may then apply Theorem 16.23 to obtain a representation $\pi : \mathfrak{g} \to \mathsf{gl}(V)$ such that

$$\Pi(e^X) = e^{\pi(X)}$$

for all $X \in \mathfrak{g}$.

Proposition 16.39 *Suppose G is a connected matrix Lie group with Lie algebra $\mathfrak{g}$. Suppose that $\Pi : G \to \mathsf{GL}(V)$ is a finite-dimensional representation of G and $\pi : \mathfrak{g} \to \mathsf{gl}(V)$ is the associated Lie algebra representation. Then a subspace W of V is invariant under the action of G if and only if it is invariant under the action of $\mathfrak{g}$. In particular, Π is irreducible if and only if π is irreducible. Furthermore, two representations of G are isomorphic if and only if the associated Lie algebra representations are isomorphic.*

In general, given an representation π of $\mathfrak{g}$, there may be no representation Π such that π and Π are related in the usual way. If, however, G is simply connected, Theorem 16.30 tells us that there is, in fact, a Π associated with every π.

Proof. Suppose $W \subset V$ is invariant under $\pi(X)$ for all $X \in \mathfrak{g}$. Then W is invariant under $\pi(X)^m$ for all m. Since V is finite dimensional, any subspace of it is automatically a closed subset and thus W is invariant under

$$\Pi(e^X) = e^{\pi(X)} = \sum_{m=0}^{\infty} \frac{\pi(X)^m}{m!}.$$

Since G is connected, every element of G is (Corollary 16.28) a product of exponentials of elements of $\mathfrak{g}$, and so W is invariant under $\Pi(A)$ for all $A \in G$.

In the other direction, if W is invariant under $\Pi(A)$ for all $A \in G$, then since W is closed, it is invariant under

$$\pi(X) = \lim_{h \to 0} \frac{e^{hX} - I}{h},$$

for all $X \in \mathfrak{g}$.

Now suppose Π_1 and Π_2 are two representations of G, acting on vector spaces V_1 and V_2, respectively. If $\Phi : V_1 \to V_2$ is an invertible linear map, then an argument similar to the above shows $\Phi \Pi_1(A) = \Pi_2(A) \Phi$ for all $A \in G$ if and only if $\Phi \pi_1(X) = \pi_2(X) \Phi$ for all $X \in \mathfrak{g}$. Thus, Φ is an isomorphism of group representations if and only if it is an isomorphism of Lie algebra representations. ∎

Theorem 16.40 (Schur's Lemma) *If V_1 and V_2 are two irreducible representations of a group or Lie algebra, then the following hold.*

1. *If $\Phi : V_1 \to V_2$ is an intertwining map, then either $\Phi = 0$ or Φ is an isomorphism.*

2. *If $\Phi : V_1 \to V_2$ and $\Psi : V_1 \to V_2$ are nonzero intertwining maps, then there exists a nonzero constant $c \in \mathbb{C}$ such that $\Phi = c\Psi$. In particular, if Φ is an intertwining map of V_1 to itself then $\Phi = cI$.*

Although the first part of Schur's lemma holds for representations over an arbitrary field, the second part holds only for representations over algebraically closed fields.

Proof. It is easy to see that $\ker \Phi$ is an invariant subspace of V_1. Since V_1 is irreducible, this means that either $\ker \Phi = V_1$, in which case $\Phi = 0$, or $\ker \Phi = \{0\}$, in which case Φ is injective. Similarly, the range of Φ is invariant, and thus equal to either $\{0\}$ or V_2. If Φ is not zero, then the range of Φ is not zero, hence all of V_2. Thus, if Φ is not zero, it is both injective and surjective, establishing Point 1.

For Point 2, since Φ and Ψ are nonzero, they are isomorphisms, by Point 1. It suffices to prove that $\Gamma := \Phi^{-1}\Psi$ is a multiple of the identity, where Γ is an intertwining map of V_1 to itself. Since we are working over $\mathbb{C}$, Γ must have at least one eigenvalue λ. If W denotes the λ-eigenspace of Γ, then W is invariant under the action of the group or Lie algebra. After all, if $\Gamma w = \lambda w$, then (in the notation of the group case) $\Gamma(\Pi(A)w) = \Pi(A)\Gamma w = \lambda \Pi(A)w$. Since λ is an eigenvector of Γ, the invariant subspace W is nonzero and thus $W = V_1$, which means precisely that $\Gamma = \lambda I$. ∎

16.7.2   Unitary Representations

In quantum mechanics, we are interested not only in vector spaces, but, more specifically, in Hilbert spaces, since expectation values are defined in terms of an inner product. We wish to consider, then, actions of a group that preserve the inner product as well as the linear structure. Although the Hilbert spaces in quantum mechanics are generally infinite dimensional, we restrict our attention in this section to the finite-dimensional case.

Definition 16.41 *Suppose V is a finite-dimensional Hilbert space over $\mathbb{C}$. Denote by $\mathsf{U}(V)$ the group of invertible linear transformations of V that preserve the inner product. A (finite-dimensional) **unitary representation** of a matrix Lie group G is a continuous homomorphism of $\Pi : G \to \mathsf{U}(V)$, for some finite-dimensional Hilbert space V.*

Proposition 16.42 *Let $\Pi : G \to \mathsf{GL}(V)$ be a finite-dimensional representation of a connected matrix Lie group G, and let π be the associated representation of the Lie algebra $\mathfrak{g}$ of G. Let $\langle \cdot, \cdot \rangle$ be an inner product on V.*

Then Π is unitary with respect to $\langle \cdot, \cdot \rangle$ if and only if $\pi(X)$ is skew-self-adjoint with respect to $\langle \cdot, \cdot \rangle$ for all $X \in \mathfrak{g}$, that is, if and only if

$$\pi(X)^* = -\pi(X)$$

for all $X \in \mathfrak{g}$.

In a slight abuse of notation, we will refer to a representation π of a Lie algebra $\mathfrak{g}$ on a finite-dimensional inner product space as *unitary* if $\pi(X)^* = -\pi(X)$ for all $X \in \mathfrak{g}$.

Proof. Suppose first that $\Pi(A)$ is unitary for all $A \in G$. Then for all $X \in \mathfrak{g}$ and $t \in \mathbb{R}$ we have

$$\Pi(e^{tX})^* = \Pi(e^{tX})^{-1} = \Pi(e^{-tX}) = e^{-t\pi(X)}.$$

On the other hand,

$$\Pi(e^{tX})^* = (e^{t\pi(X)})^* = e^{t\pi(X)^*}.$$

Thus,

$$e^{t\pi(X)^*} = e^{-t\pi(X)}$$

for all t. Differentiating at $t = 0$ yields $\pi(X)^* = -\pi(X)$.

In the other direction, if $\pi(X)^* = -\pi(X)$ for all $X \in \mathfrak{g}$, then

$$\Pi(e^X)^* = e^{\pi(X)^*} = e^{-\pi(X)} = \Pi(e^{-X}) = \Pi(e^X)^{-1},$$

meaning that $\Pi(e^X)$ is unitary. Since G is connected, Corollary 16.28 tells us that each element A of G is expressible as a product of exponentials, from which it follows that $\Pi(A)$ is unitary. ∎

16.7.3 Projective Unitary Representations

In quantum mechanics, two unit vectors in the quantum Hilbert space that differ by multiplication by a constant are considered to represent the same physical state. Thus, an operator of the form $e^{i\theta}I$, with $\theta \in \mathbb{R}$, will act as the identity at the level of the physical states. Suppose that V is a Hilbert space over $\mathbb{C}$, assumed for the moment to be finite dimensional. Then it is natural to consider homomorphisms not into $\mathsf{U}(V)$ but rather into the quotient group $\mathsf{U}(V)/\{e^{i\theta}I\}$. Of course, given a homomorphism Π of G into $\mathsf{U}(V)$, we can always turn Π into a homomorphism of G into the quotient group, just by composing Π with the quotient map. Not every homomorphism into the quotient group, however, arises from a homomorphism into $\mathsf{U}(V)$.

Definition 16.43 *Suppose V is a finite-dimensional Hilbert space over $\mathbb{C}$. Then the **projective unitary group** over V, denoted $\mathsf{PU}(V)$, is the quotient group*

$$\mathsf{PU}(V) = \mathsf{U}(V)/\{e^{i\theta}I\},$$

where $\{e^{i\theta}I\}$ denotes the group of matrices of the form $e^{i\theta}I$, $\theta \in \mathbb{R}$.

Note that $\{e^{i\theta}I\}$ is a closed normal subgroup of $\mathsf{U}(V)$. Now, $\mathsf{U}(V)$ is (isomorphic to) a matrix Lie group, since we can identify it with $\mathsf{U}(n)$ by picking an orthonormal basis for V. In general, the quotient of a matrix Lie group by a closed normal subgroup may not be a matrix Lie group. In this case, however, it is not hard to realize the quotient $\mathsf{U}(n)/\{e^{i\theta}I\}$ as a matrix Lie group.

Proposition 16.44 *If V is a finite-dimensional Hilbert space over $\mathbb{C}$, then* $\mathsf{PU}(V)$ *is isomorphic to a matrix Lie group.*

Let $Q : \mathsf{U}(V) \to \mathsf{PU}(V)$ be the quotient homomorphism and let $q : \mathsf{u}(V) \to \mathsf{pu}(V)$ be the associated Lie algebra homomorphism. Then q maps $\mathsf{u}(V)$ onto $\mathsf{pu}(V)$ and the kernel of q is the space of matrices of the form iaI with $a \in \mathbb{R}$. Thus, $\mathsf{pu}(V)$ is isomorphic to $\mathsf{u}(V)/\{iaI\}$.

The Lie algebra $\mathsf{u}(V)$ of $\mathsf{U}(V)$ is the space of skew-self-adjoint operators on V. In Proposition 16.44, the space $\{iaI\}$ is an ideal in $\mathsf{u}(V)$ and the quotient is in the sense of Lie algebras over $\mathbb{R}$; see Exercise 9. If $\dim V = N$, then it is not hard to see that the Lie algebra $\mathsf{pu}(V) \cong \mathsf{u}(V)/\{iaI\}$ is isomorphic to the Lie algebra $\mathsf{su}(N)$. The *group* $\mathsf{PU}(V)$ is not, however, isomorphic to the group $\mathsf{SU}(N)$. See Exercise 16.

Proof. If $\dim V = N$, then $\mathsf{gl}(V)$, the space of all linear maps of V to V, has dimension N^2. Given $U \in \mathsf{U}(V)$, we can define

$$C_U : \mathsf{gl}(V) \to \mathsf{gl}(V)$$

by

$$C_U(X) = UXU^{-1}.$$

(That is to say, C_U is conjugation by U.) Note that $(C_U)^{-1} = C_{U^{-1}}$ and $C_{UV} = C_U C_V$. Thus, C (i.e., the map $U \mapsto C_U$) is a homomorphism of $\mathsf{U}(V)$ into $\mathsf{GL}(\mathsf{gl}(V))$, and this homomorphism is clearly continuous. If U is a multiple of the identity, then C_U is the identity operator on $\mathsf{gl}(V)$. Conversely, if C_U is the identity, then $UX = XU$ for all $X \in \mathsf{gl}(V)$, which implies (Exercise 18) that U is a multiple of the identity. Thus, the kernel of C consists precisely of those scalar multiples of the identity that are in $\mathsf{U}(V)$; that is, $\ker C = \{e^{i\theta}I\}$.

We have constructed, then, a homomorphism of $\mathsf{U}(V)$ into $\mathsf{GL}(\mathsf{gl}(V)) \cong \mathsf{GL}(N^2; \mathbb{C})$ with a kernel that is precisely $\{e^{i\theta}I\}$. The image of $\mathsf{U}(V)$ under this homomorphism is, therefore, isomorphic to the quotient group $\mathsf{U}(V)/\{e^{i\theta}I\}$. Furthermore, since $\mathsf{U}(V)$ is compact, the image of $\mathsf{U}(V)$ under C is compact and thus closed. This image is, then, a matrix Lie group isomorphic to $\mathsf{PU}(V)$.

Let c be the associated Lie algebra homomorphism associated with the homomorphism C. Using Point 3 of Theorem 16.23, we may calculate that

$$
\begin{aligned}
c_X(Y) &= \frac{d}{dt} e^{tX} Y e^{-tX} \Big|_{t=0} \\
&= XY - YX \\
&= [X, Y].
\end{aligned}
$$

Using Exercise 18 again, we see that $c_X = 0$ if and only if X is a multiple of the identity. Thus, the kernel of c consists of all the scalar multiples of I in $\mathsf{u}(V)$, namely $\{iaI\}$.

Now, the image of $\mathsf{U}(V)$ under C is (isomorphic to) $\mathsf{PU}(V)$; in particular, C maps $\mathsf{U}(V)$ *onto* $\mathsf{PU}(V)$. It follows that c must map $\mathsf{u}(V)$ *onto* $\mathsf{pu}(V)$. (This claim follows from Theorem 3.15 in [21].) Thus, $\mathsf{pu}(V) \cong \mathsf{u}(V)/\{iaI\}$. ∎

Definition 16.45 *A finite-dimensional **projective unitary representation** of a matrix Lie group G is a continuous homomorphism Π of G into $\mathsf{PU}(V)$, where V is a finite-dimensional Hilbert space over $\mathbb{C}$. A subspace W of V is said to be **invariant** under Π if for each $A \in G$, W is invariant under U for every $U \in \mathsf{U}(V)$ such that $[U] = \Pi(A)$. A projective unitary representation (Π, V) is **irreducible** if the only invariant subspaces are $\{0\}$ and V.*

Given an ordinary unitary representation, $\Sigma : G \to \mathsf{U}(V)$, we can always form a projective representation, $\Pi : G \to \mathsf{PU}(V)$, simply by setting $\Pi = Q \circ \Sigma$. Not every projective representation, however, arises in this fashion. Thus, considering projective representations gives us more flexibility than considering ordinary unitary representations.

Proposition 16.46 *Let $\Pi : G \to \mathsf{PU}(V)$ be a finite-dimensional projective unitary representation of a matrix Lie group G, and let $\pi : \mathfrak{g} \to \mathsf{pu}(V)$ be the associated Lie algebra homomorphism. Then there exists a Lie algebra homomorphism $\sigma : \mathfrak{g} \to \mathsf{u}(V)$ such that $\pi(X) = q(\sigma(X))$ for all $X \in \mathfrak{g}$. It is possible to choose σ so that $\mathrm{trace}(\sigma(X)) = 0$ for all $X \in \mathfrak{g}$, and σ is unique if we require this condition.*

That is to say, every finite-dimensional projective representation can be "de-projectivized" at the Lie algebra level. In general, σ is not unique, because there may be σ's for which $\mathrm{trace}(\sigma(X))$ is nonzero for some X. On the other hand, if $\mathfrak{g}$ has the property that every $X \in \mathfrak{g}$ is a linear combination of commutators—which is true if $\mathfrak{g} = \mathsf{so}(3)$—then σ is unique. See Exercise 15.

Proof. Recall that $\mathsf{pu}(V) \cong \mathsf{u}(V)/\{iaI\}$. That is, for each $X \in \mathfrak{g}$, $\pi(X)$ denotes a whole family of operator that differ by adding iaI. If $Y \in \mathsf{u}(n)$ is any representative of $\pi(X)$, then since $Y^* = -Y$, the trace of Y will be pure imaginary. Thus, there is a unique pure-imaginary constant $c =$

$-\mathrm{trace}(Y)/\dim V$ such that the trace of $Y + cI$ is zero. Let us then set $\sigma(X) = Y + cI$. Since π is a Lie algebra homomorphism, $\sigma([X,Y])$ will equal $[\sigma(X), \sigma(Y)] + iaI$, for some $a \in \mathbb{R}$. Since $\mathrm{trace}(\sigma([X,Y])) = 0$ by construction and since the commutator of any two matrices has trace zero, we see that actually $a = 0$. Thus, a σ as in the proposition exists, and it is unique if we require that $\sigma(X)$ have trace zero. ∎

Theorem 16.47 *Suppose G is a matrix Lie group and $\tilde{G}$ is a universal cover of G, with covering map Φ. Then the following hold.*

1. *Let $\Pi : G \to \mathsf{PU}(V)$ be a finite-dimensional projective unitary representation of G. Then there is an ordinary unitary representation $\Sigma : \tilde{G} \to \mathsf{U}(V)$ of $\tilde{G}$ such that $\Pi \circ \Phi = Q \circ \Sigma$. Any such Σ is irreducible if and only if Π is irreducible. It is possible to choose Σ so that $\det(\Sigma(A)) = 1$ for all $A \in \tilde{G}$, and Σ is unique if we require this condition.*

2. *Let Σ be a finite-dimensional irreducible unitary representation of $\tilde{G}$. Then the kernel of the associated projective unitary representation $Q \circ \Sigma$ contains the kernel of the covering map Φ. Thus, $Q \circ \Sigma$ factors through G and gives rise to a projective unitary representation of G.*

In the finite-dimensional case, then, there is a one-to-one correspondence between irreducible projective unitary representations of G and irreducible, determinant-one ordinary unitary representations of $\tilde{G}$. Point 1 of the theorem means that any finite-dimensional projective unitary representation of the group G can be "de-projectivized" at the expense of passing to the universal cover $\tilde{G}$ of G.

Note that Theorem 16.47 applies only to *finite-dimensional* projective unitary representations. Example 16.56 will provide an infinite-dimensional example in which Point 1 of the theorem fails.

Proof. If $\mathfrak{g}$ is the Lie algebra of G, Proposition 16.46 tells us that we can find an ordinary representation $\sigma : \mathfrak{g} \to \mathsf{u}(V)$ such that $q \circ \sigma = \pi$. We then define a representation $\tilde{\sigma} : \tilde{\mathfrak{g}} \to \mathsf{u}(V)$ of the Lie algebra $\tilde{\mathfrak{g}}$ of $\tilde{G}$ by setting $\tilde{\sigma}(X) = \sigma(\phi(X))$, $X \in \tilde{\mathfrak{g}}$. Since $\tilde{G}$ is simply connected, we can then find a unique representation $\Sigma : \tilde{G} \to \mathsf{U}(V)$ such that $\Sigma(e^X) = e^{\tilde{\sigma}(X)}$ for all $X \in \tilde{\mathfrak{g}}$. Since

$$q \circ \tilde{\sigma} = q \circ \sigma \circ \phi = \pi \circ \phi,$$

it follows that $Q \circ \Sigma = \Pi \circ \Phi$. Furthermore, if Σ maps into $\mathsf{SU}(V)$, $\sigma = \tilde{\sigma} \circ \phi^{-1}$ maps into $\mathsf{su}(n)$. This condition uniquely determines σ and thus also $\tilde{\sigma}$ and Σ, establishing Point 1 of the theorem.

For Point 2, observe that $\ker \Phi$ is a discrete normal subgroup of $\tilde{G}$, which is therefore central (Exercises 1 and 12). Thus, for all $A \in \ker \Phi$, we have

$$\Sigma(A)\Sigma(B) = \Sigma(AB) = \Sigma(BA) = \Sigma(B)\Sigma(A)$$

for all $B \in \tilde{G}$. That is to say, $\Sigma(A)$ is an intertwining map of V to itself. Since V is also irreducible as a representation of $\tilde{G}$, Schur's lemma tells us that $\Sigma(A) = cI$, where $|c| = 1$ because $\Sigma(A) \in \mathsf{U}(V)$. Thus, A is in the kernel of the associated projective representation $Q \circ \Sigma$. ∎

16.8   New Representations from Old

In this section, we consider three basic mechanisms for combining representations to produce new representations: direct sums, tensor products, and duals. This section assumes familiarity with these notions at the level of vector spaces; a brief review is provided in Appendix A.1.

Definition 16.48 *Suppose* (Π_1, V_1) *and* (Π_2, V_2) *are representations of a matrix Lie group* G. *The* **direct sum** *of these two representations is the representation* $\Pi_1 \oplus \Pi_2 : G \rightarrow \mathsf{GL}(V_1 \oplus V_2)$ *given by*

$$(\Pi_1 \oplus \Pi_2)(A) = \Pi_1(A) \oplus \Pi_2(A).$$

The **tensor product** *of* Π_1 *and* Π_2 *is the representation* $\Pi_1 \otimes \Pi_2 : G \rightarrow \mathsf{GL}(V_1 \otimes V_2)$ *given by*

$$(\Pi_1 \otimes \Pi_2)(A) = \Pi_1(A) \otimes \Pi_2(A).$$

Finally, the **dual** *of* Π_1 *is the representation* $\Pi_1^{tr} : G \rightarrow \mathsf{GL}(V^*)$ *given by*

$$\Pi_1^{tr}(A) = \Pi_1(A^{-1})^{tr} = \left(\Pi_1(A)^{tr}\right)^{-1}.$$

Similarly, the direct sum, tensor product, and dual of Lie algebra representations can be defined by

$$(\pi_1 \oplus \pi_2)(X) = \pi_1(X) \oplus \pi_2(X)$$
$$(\pi_1 \otimes \pi_2)(X) = \pi_1(X) \otimes I + I \otimes \pi_2(X)$$
$$\pi_1^{tr}(X) = -\pi_1(X)^{tr}.$$

It is important to note the differences in formulas between the group and the Lie algebra in the case of tensor products and dual representations. It is easy to motivate the definitions for the Lie algebra: If G acts on $V_1 \otimes V_2$ by $\Pi_1(A) \otimes \Pi_2(A)$, then the associated Lie algebra action will be given by

$$\frac{d}{dt}\Pi_1(e^{tX}) \otimes \Pi_2(e^{tX})\Big|_{t=0} = \pi_1(X) \otimes I + I \otimes \pi_2(X).$$

Of course, we continue to use this last formula for tensor products of Lie algebra representations, even if there is no associated group representations.

Remark 16.49 *If* (Π_1, V_1) *and* (Π_2, V_2) *are representations of a group* G, *it is possible to view* $V_1 \otimes V_2$ *as a representation of the direct product group* $G \times G$, *by setting*

$$(\Pi_1 \otimes \Pi_2)(A, B) = \Pi_1(A) \otimes \Pi_2(B).$$

Similarly, if (π_1, V_1) *and* (π_2, V_2) *are representations of a Lie algebra* $\mathfrak{g}$, *it is possible to view* $V_1 \otimes V_2$ *as a representation of* $\mathfrak{g} \oplus \mathfrak{g}$ *by setting*

$$(\pi_1 \otimes \pi_2)(X, Y) = \pi_1(X) \otimes I + I \otimes \pi_2(Y).$$

Nevertheless, it is, in most cases, more natural to view $V_1 \otimes V_2$ as a representation of G itself, rather than of $G \times G$. Even if V_1 and V_2 are irreducible representations of G, the space $V_1 \otimes V_2$ will in most cases fail to be irreducible as a representation of G. If, for example, we take $V_1 = V_2 = V$, then the space of symmetric tensors inside $V \otimes V$ will form a nontrivial invariant subspace, unless $\dim V = 1$. An important problem in representation theory is to decompose $V_1 \otimes V_2$ as a direct sum of irreducible representations, where V_1 and V_2 are irreducible representations of a fixed group or Lie algebra. In the case of the Lie algebra $\mathrm{su}(2)$, this decomposition is discussed in Sect. 17.9.

Definition 16.50 *A finite-dimensional representation of a group or Lie algebra is said to be* **completely reducible** *if it is isomorphic to a direct sum of irreducible representations.*

Proposition 16.51 *Every finite-dimensional unitary representation of a group or Lie algebra is completely reducible.*

Proof. Suppose (Π, V) is a unitary representation of a matrix Lie group G. If W is a subspace of V invariant under each $\Pi(A)$, then $W^\perp$ is invariant under each $\Pi(A)^*$, as the reader may easily verify. But since Π is unitary,

$$\Pi(A)^* = \Pi(A)^{-1} = \Pi(A^{-1}).$$

Thus, $W^\perp$ is invariant under $\Pi(A^{-1})$ for all $A \in G$, hence under $\Pi(A)$ for all $A \in G$. We conclude that, in the unitary case, the orthogonal complement of an invariant subspace is always invariant.

If V is irreducible, there is nothing to prove. If not, we pick a nontrivial invariant subspace W and decompose V as $W \oplus W^\perp$. The restriction of Π to W or to $W^\perp$ is again a unitary representation, so we can repeat this procedure for each of these subspaces. Since V is finite dimensional, the process must eventually terminate, yielding an orthogonal decomposition of V as a direct sum of irreducible invariant subspaces.

If we consider a unitary representation π of a Lie algebra $\mathfrak{g}$, we have the same argument, but with the identity $\Pi(A)^* = \Pi(A^{-1})$ replaced by $\pi(X)^* = -\pi(X)$. ∎

Proposition 16.52 *Suppose K is a compact matrix Lie group. For any finite-dimensional representation (Π, V) of K, there exists an inner product on V such that $\Pi(A)$ is unitary for all $A \in G$. In particular, every finite-dimensional representation of K is completely reducible.*

See Proposition 4.36 in [21].

16.9   Infinite-Dimensional Unitary Representations

For the applications we have in mind, we need to consider representations that are infinite dimensional. The theory of such representations is inevitably more complicated than that of finite-dimensional representations. For our purposes, it suffices to consider the nicest sort of infinite-dimensional representations—unitary representations in a Hilbert space.

16.9.1   Ordinary Unitary Representations

We begin by considering ordinary representations and then turn to projective representations.

Definition 16.53 *Suppose G is a matrix Lie group. Then a **unitary representation** of G is a strongly continuous homomorphism $\Pi : G \to \mathsf{U}(\mathbf{H})$, where $\mathbf{H}$ is a separable Hilbert space and $\mathsf{U}(\mathbf{H})$ is the group of unitary operators on $\mathbf{H}$. Here, strong continuity of Π means that if a sequence A_m in G converges to $A \in G$, then*

$$\lim_{m \to \infty} \|\Pi(A_m)\psi - \Pi(A)\psi\| = 0$$

for all $\psi \in \mathbf{H}$.

We can attempt to associate to a unitary representation Π of G some sort of representation π of the Lie algebra $\mathfrak{g}$ of G, by imitating the construction in Theorem 16.23. For any $X \in \mathfrak{g}$, the map $t \mapsto \Pi(e^{tX})$ is a strongly continuous one-parameter unitary group. Thus, Stone's theorem (Theorem 10.15) tells us that there exists a unique self-adjoint operator A such that $\Pi(e^{tX}) = e^{itA}$ for all $t \in \mathbb{R}$. If we let $\pi(X)$ denote the skew-self-adjoint operator iA, we will have

$$\Pi(e^{tX}) = e^{t\pi(X)}. \tag{16.5}$$

The operators $\pi(X)$, $X \in \mathfrak{g}$, are in general unbounded and defined only on a dense subspace of $\mathbf{H}$. Nevertheless, it can be shown (see, e.g., [43]) that there exists a dense subspace V of $\mathbf{H}$ contained in the domain of each $\pi(X)$ and that is invariant under each $\pi(X)$, and on which we have $\pi([X, Y]) = [\pi(X), \pi(Y)]$. In the case of the particular representation that we will consider in the next chapter, we can avoid these difficulties by looking at finite-dimensional invariant subspaces.

Proposition 16.54 *Suppose G is a matrix Lie group and $\Pi : G \to \mathsf{U}(\mathbf{H})$ is a unitary representation of G. For each $X \in \mathfrak{g}$, let $\pi(X)$ denote the operator in (16.5). Suppose $V \subset \mathbf{H}$ is a finite-dimensional subspace of $\mathbf{H}$ such that $\Pi(A)$ maps V into V, for all $A \in G$. Then for all $X \in \mathfrak{g}$, $V \subset \mathrm{Dom}(\pi(X))$, $\pi(X)$ maps V into V, and we have*

$$\pi([X,Y])v = [\pi(X), \pi(Y)]v \tag{16.6}$$

for all $v \in V$.

In the other direction, suppose G is connected and suppose V is any finite-dimensional subspace of $\mathbf{H}$ such that for all $X \in \mathfrak{g}$, $V \subset \mathrm{Dom}(\pi(X))$ and $\pi(X)$ maps V into V. Then $\Pi(A)$ also maps V into V, for all $A \in G$.

Proof. Since V is invariant under both $\Pi(A)$ and $\Pi(A)^* = \Pi(A^{-1})$, the restriction to V of each $\Pi(A)$ is unitary. The operators $\Pi(A)|_V$ form a finite-dimensional unitary representation of G that is strongly continuous and thus continuous. (In the finite-dimensional case, all reasonable notions of continuity for representations coincide.) For each $X \in \mathfrak{g}$, Theorem 16.18 tells us that there is an operator $\tilde{X}$ on V such that

$$\Pi(e^{tX})\big|_V = e^{t\tilde{X}}.$$

Thus, for any $v \in V$, we have

$$\lim_{t \to 0} \frac{\Pi(e^{tX})v - v}{t} = \lim_{t \to 0} \frac{e^{t\tilde{X}}v - v}{t} = \tilde{X}v.$$

This calculation shows that v is in the domain of the infinitesimal generator $\pi(X)$ of the unitary group $\Pi(e^{tX})$, and that $\pi(X)v = \tilde{X}v$. Since the operators $\tilde{X}$, $X \in \mathfrak{g}$, form a representation of $\mathfrak{g}$, we have the relation (16.6).

In the other direction, if V is invariant under $\pi(X)$, the restriction of $\pi(X)$ to V is automatically bounded. Thus, there is a constant C such that

$$\|\pi(X)^m v\| \leq C^m \|v\| \tag{16.7}$$

for all $v \in V$. If we use the direct-integral form of the spectral theorem for the self-adjoint operator $A := -i\pi(X)$, it is easy to see that (16.7) can only hold if v, viewed as an element of the direct integral, is supported on a bounded interval inside the spectrum of A. Since the power series of the function $\lambda \mapsto e^{t\lambda}$ converges to $e^{t\lambda}$ uniformly on any finite interval, we will have

$$\Pi(e^{tX})v = e^{itA}v = \sum_{m=0}^{\infty} \frac{t^m \pi(X)^m}{m!} v.$$

Each term in the above power series belongs to V, which is finite dimensional and thus closed. We conclude that $\Pi(e^{tX})v$ belongs to V for all $X \in \mathfrak{g}$. Since G is connected, each element of G is a product of exponentials of Lie algebra elements, and we have the claim. ∎

16.9.2 Projective Unitary Representations

Given a Hilbert space $\mathbf{H}$, let $S^{\mathbf{H}}$ denote the unit sphere in $\mathbf{H}$, that is, the set of vectors with norm 1. Let $P\mathbf{H}$ be the quotient space $(S^{\mathbf{H}})/\sim$, where "$\sim$" denotes the equivalence relation in which $u \sim v$ if and only if $u = e^{i\theta}v$ for some $\theta \in \mathbb{R}$. The quotient map $q : S^{\mathbf{H}} \to P\mathbf{H}$ induces a topology on $P\mathbf{H}$ in which a set $U \subset P\mathbf{H}$ is open if and only if $q^{-1}(U)$ is open as a subset of the metric space $S^{\mathbf{H}} \subset \mathbf{H}$.

As in the finite-dimensional case, we can form the quotient group

$$\mathsf{PU}(\mathbf{H}) := \mathsf{U}(\mathbf{H})/\{e^{i\theta}I\}.$$

The action of $\mathsf{U}(\mathbf{H})$ on $S^{\mathbf{H}}$ descends to a well-defined action of $\mathsf{PU}(\mathbf{H})$ on $P\mathbf{H}$.

Definition 16.55 *A **projective unitary representation** of a matrix Lie group G is a homomorphism $\Pi : G \to \mathsf{PU}(\mathbf{H})$, for some Hilbert space $\mathbf{H}$, with the property that if a sequence A_m in G converges to A in G, then*

$$\Pi(A_m)x \to \Pi(A)x$$

for all $x \in P\mathbf{H}$.

Recall that in the finite-dimensional case, every projective unitary representation of G can be "de-projectivized" at the expense of possibly having to pass to the universal cover $\tilde{G}$ of G (Theorem 16.47). The de-projectivization proceeds by passing to the Lie algebra, choosing the trace-zero representative of each equivalence class, and then exponentiating back to the universal cover of the original group. This approach does not work in the infinite-dimensional case. After all, even assuming we can construct a Lie algebra homomorphism $\pi(X)$ for each $X \in \mathfrak{g}$, the representatives of $\pi(X)$ are typically *unbounded* operators on $\mathbf{H}$, for which the notion of trace does not make sense. This difficulty is not just a technicality; the corresponding result in the infinite-dimensional case is false, as we will now see.

Example 16.56 *For all $(a, b) \in \mathbb{R}^2$, define an operator $T_{(a,b)}$ on $L^2(\mathbb{R})$ by*

$$(T_{(a,b)}\psi)(x) = e^{iax}\psi(x - b).$$

Then $T_{(a,b)}$ is unitary for all $(a, b) \in \mathbb{R}^2$ and we have

$$\left(T_{(a,b)}T_{(a',b')}\psi\right)(x) = e^{iax}e^{ia'(x-b)}\psi(x - (b + b'))$$
$$= e^{-ia'b}\left(T_{(a+a',b+b')}\psi\right)(x). \tag{16.8}$$

The map $(a, b) \mapsto [T_{(a,b)}]$ is a homomorphism of $\mathbb{R}^2$ into $\mathsf{PU}(L^2(\mathbb{R}))$, and this homomorphism is continuous in the sense of Definition 16.55. There does not, however, exist any homomorphism $S : \mathbb{R}^2 \to \mathsf{U}(L^2(\mathbb{R}))$ such that $[S_{(a,b)}] = [T_{(a,b)}]$ for all $(a, b) \in \mathbb{R}^2$.

Thus, even though $\mathbb{R}^2$ is simply connected (and thus its own universal cover), there is no way to de-projectivize the projective unitary representation $(a, b) \mapsto [T_{(a,b)}]$ of $\mathbb{R}^2$.

Proof. The map $(a, b) \to T_{(a,b)}$ is easily seen to be strongly continuous, and thus the map $(a, b) \mapsto [T_{(a,b)}]$ is continuous in the sense of Definition 16.55. If a homomorphism S with the indicated properties existed, then there would be constants $\theta_{a,b}$ such that $S_{(a,b)} = e^{i\theta_{a,b}} T_{(a,b)}$. But then since S is a homomorphism from the *commutative* group $\mathbb{R}^2$ into $\mathsf{U}(L^2(\mathbb{R}))$, the operator $S_{(a,b)}$ would have to commute with $S_{(a',b')}$ for all (a, b) and (a', b'). But then the operators $T_{(a,b)}$ and $T_{(a',b')}$, being constant multiples of commuting operators, would need to commute as well. But this is not the case; for example, $T_{(a,0)}$ does not commute with $T_{(0,b')}$, as is easily verified using (16.8). ∎

Despite the negative result in Example 16.56, there is a positive result in this direction: If G is connected and "semi-simple," every projective unitary representation of G can be de-projectivized after passing to the universal cover. Here, a Lie algebra $\mathfrak{g}$ is said to be *simple* if $\mathfrak{g}$ has no nontrivial ideals *and* $\dim \mathfrak{g} \geq 2$. A Lie algebra is said to be *semi-simple* if it is a direct sum of simple algebras. Finally, a Lie group G is said to be *semi-simple* if the Lie algebra $\mathfrak{g}$ of G is semi-simple.

For any connected Lie group G, a projective unitary representation Π of G can be de-projectivized by passing to a *one-dimensional central extension*. A one-dimensional central extension of G is a Lie group G' together with a surjective homomorphism $\Phi : G' \to G$ such that the kernel of Φ is one-dimensional and contained in the center of G'. See the article [1] of V. Bargmann for more information about these issues.

16.10   Exercises

1. Suppose that G is a connected matrix Lie group and that N is a discrete normal subgroup of G, meaning that there is some neighborhood U of I in G such that $U \cap N = \{I\}$. Show that N is contained in the center of G.

   Hint: Consider the quantity gng^{-1} for $g \in G$ and $n \in N$.

2. (a) Suppose two elements U and V of $\mathsf{SU}(2)$ commute. Show that each eigenspace for U is invariant under V and vice versa.

   (b) Show that if U is in the center of $\mathsf{SU}(2)$, then $U = I$ or $U = -I$.

3. Define the *Hilbert–Schmidt norm* of a matrix $X \in M_n(\mathbb{C})$ by the formula
$$\|X\|_{\mathrm{HS}}^2 = \sum_{j,k=1}^{n} |X_{jk}|^2 .$$

Using the Cauchy–Schwarz inequality, show that

$$\|XY\|_{\mathrm{HS}} \le \|X\|_{\mathrm{HS}} \|Y\|_{\mathrm{HS}} \tag{16.9}$$

for all $X, Y \in M_n(\mathbb{C})$.

4. Using term-by-term differentiation of power series, show that for all $X \in M_n(\mathbb{C})$ and all $1 \le j, k \le n$, we have

$$\frac{d}{dt}\left[\left(e^{tX}\right)_{jk}\right]\bigg|_{t=0} = X_{jk}.$$

5. Verify Property 4 of Theorem 16.15. This should be easy in the case that X is diagonalizable. In the general case, either use the Jordan canonical form or appeal to the fact that diagonalizable matrices are dense in $M_n(\mathbb{C})$.

6. Suppose X and Y are *commuting* $n \times n$ matrices. Show that

$$e^X e^Y = e^{X+Y}.$$

This is Property 5 of Theorem 16.15.

Hint: Multiply together the power series for e^X and e^Y and then group terms where the total power of X and Y is n.

7. For $A \in M_n(\mathbb{C})$, define the *logarithm* of A by the power series

$$\log A = A - I - \frac{(A-I)^2}{2} + \frac{(A-I)^3}{3} - \cdots$$

whenever this series converges. Assume the following result: If A is sufficiently close to I, then $\log A$ is defined and $\exp(\log A) = A$. [This can be seen easily when A is diagonalizable, and the set of diagonalizable matrices is dense in $M_n(\mathbb{C})$.]

(a) Show that there exists a constant C such that for all A with $\|A - I\| < 1/2$ we have

$$\|\log A - (A - I)\| \le C \|A - I\|^2.$$

(b) Show that for all $X, Y \in M_n(\mathbb{C})$ we have

$$\log\left(e^{X/m} e^{Y/m}\right) = \frac{X}{m} + \frac{Y}{m} + O\left(\frac{1}{m^2}\right). \tag{16.10}$$

Note that $e^{X/m} e^{Y/m}$ tends to I as m tends to infinity, so that the left-hand side of (16.10) is defined for all sufficiently large m.

(c) Prove the Lie Product Formula.

8. (a) Show that for all $X, Y \in M_n(\mathbb{C})$,

$$\left\| \frac{d}{dt}(X + tY)^m \bigg|_{t=0} \right\| \leq m \, \|X\|^{m-1} \, \|Y\| \, .$$

   (b) Show that the map $X \mapsto e^{tX}$ is a continuously differentiable map of $M_n(\mathbb{C}) \cong \mathbb{R}^{2n^2}$ to itself.

   (c) Using Exercise 4, show that the differential of the map $X \mapsto e^X$ at $X = 0$ is the identity map of $M_n(\mathbb{C})$ to itself. (Recall that the differential of smooth map of $\mathbb{R}^j$ to $\mathbb{R}^k$, evaluated at a point in $\mathbb{R}^j$, is a linear map of $\mathbb{R}^j$ to $\mathbb{R}^k$.)

9. Suppose $\mathfrak{g}$ is a Lie algebra and $\mathfrak{h}$ is an ideal in $\mathfrak{g}$. Let $\mathfrak{g}/\mathfrak{h}$ denote the vector space quotient of $\mathfrak{g}$ by $\mathfrak{h}$. Show that the bracket on $\mathfrak{g}$ descends unambiguously to a bilinear map on $\mathfrak{g}/\mathfrak{h}$, and that $\mathfrak{g}/\mathfrak{h}$ forms a Lie algebra under this map.

10. Suppose that G_1, G_2, and G_3 are matrix Lie groups with Lie algebras $\mathfrak{g}_1$, $\mathfrak{g}_2$, and $\mathfrak{g}_3$, respectively. Suppose that $\Phi : G_1 \to G_2$ and $\Psi : G_2 \to G_3$ are Lie group homomorphisms with associated Lie algebra homomorphisms ϕ and ψ, respectively. Show that the Lie algebra homomorphism associated to $\Psi \circ \Phi : G_1 \to G_3$ is $\psi \circ \phi$.

11. Show that isomorphic matrix Lie groups have isomorphic Lie algebras.

12. Suppose G_1 and G_2 are matrix Lie groups with Lie algebras $\mathfrak{g}_1$ and $\mathfrak{g}_2$, respectively. Suppose $\Phi : G_1 \to G_2$ is a Lie group homomorphism with the property that the associated Lie algebra homomorphism $\phi : \mathfrak{g}_1 \to \mathfrak{g}_2$ is injective. Show that there exists a neighborhood U of the identity in G_1 such that $U \cap \ker \Phi = \{I\}$.

   Hint: Use Theorem 16.25.

13. (a) Show that every $R \in \mathrm{SO}(3)$ has an eigenvalue of 1.

   (b) Show that every $R \in \mathrm{SO}(3)$ is conjugate in $\mathrm{SO}(3)$ to matrix of the form

$$\begin{pmatrix} 1 & 0 & 0 \\ 0 & \cos\theta & -\sin\theta \\ 0 & \sin\theta & \cos\theta \end{pmatrix}$$

   for some $\theta \in \mathbb{R}$.

   (c) Show that the exponential map from so(3) to SO(3) is surjective.

   (d) Show that SO(3) is connected.

14. Show that the center of SO(3) is trivial.

   Hint: Use Part (a) of Exercise 13.

15. Given a Lie algebra $\mathfrak{g}$, let $[\mathfrak{g}, \mathfrak{g}]$ denote the space of linear combinations of commutators, that is, the space spanned by elements of the form $[X, Y]$ with $X, Y \in \mathfrak{g}$.

    (a) Show that $[\mathfrak{g}, \mathfrak{g}]$ is an ideal in $\mathfrak{g}$ and that the quotient $\mathfrak{g}/[\mathfrak{g}, \mathfrak{g}]$ is commutative. (The ideal $[\mathfrak{g}, \mathfrak{g}]$ is called the *commutator ideal* of $\mathfrak{g}$.)

    (b) If $\mathfrak{g} = \mathsf{so}(3)$, show that $[\mathfrak{g}, \mathfrak{g}] = \mathfrak{g}$.

    (c) If $\pi : \mathfrak{g} \to \mathsf{gl}(V)$ is any finite-dimensional representation of $\mathfrak{g}$, show that $\pi([\mathfrak{g}, \mathfrak{g}])$ is contained in $\mathsf{sl}(V)$, the space of endomorphisms of V with trace zero.

16. (a) Show that the Lie algebra $\mathsf{pu}(n) \cong \mathsf{u}(n)/\{ia\mathbb{R}\}$ is isomorphic to the Lie algebra $\mathsf{su}(n)$.

    (b) Let $\{e^{2\pi i k/n}I\}$ denote the group of matrices that are of the form of an nth root of unity times the identity. Show that the group $PU(n)$ is isomorphic to $SU(n)/\{e^{2\pi i k/n}I\}$.

17. Suppose that G is a matrix Lie group with Lie algebra $\mathfrak{g}$ and that A is an element of G. Show that the operation of left multiplication by A^{-1} is a diffeomorphism of $M_n(\mathbb{C})$. Now show that there exist neighborhoods U of 0 in $M_n(\mathbb{C})$ and V of A in $M_n(\mathbb{C})$ such that the map $X \mapsto Ae^X$ maps U diffeomorphically onto V and such that for $X \in U$, we have $X \in \mathfrak{g}$ if and only if $Ae^X \in V$. (Use Theorem 16.25.)

18. Suppose that $Z \in M_n(\mathbb{C})$ has the property that $ZX = XZ$ for all $X \in M_n(\mathbb{C})$. Show that $Z = cI$ for some $c \in \mathbb{C}$.

19. Suppose $(\Pi, \mathbf{H})$ is a unitary representation of a matrix Lie group G, and suppose V_1 and V_2 are finite-dimensional irreducible invariant subspaces of $\mathbf{H}$. Show that if V_1 and V_2 are not isomorphic as representations of G, then V_1 is orthogonal to V_2 inside $\mathbf{H}$.

    Hint: Show that the orthogonal projection of $\mathbf{H}$ onto V_1 or V_2 is an intertwining map, and use Schur's lemma.

17
Angular Momentum and Spin

17.1 The Role of Angular Momentum in Quantum Mechanics

Classically, angular momentum may be thought of as the Hamiltonian generator of rotations (Proposition 2.30). Angular momentum is a particularly useful concept when a system has rotational symmetry, since in that case the angular momentum is a conserved quantity (Proposition 2.18). Quantum mechanically, angular momentum is still the "generator" of rotations, meaning that it is the infinitesimal generator of a one-parameter group of unitary rotation operators, in the sense of Stone's theorem (Theorem 10.15). The quantum angular momentum is again conserved in systems with rotational symmetry. This means that if the Hamiltonian $\hat{H}$ is invariant under rotations, then $\hat{H}$ commutes with the angular momentum operators, in which case, the angular momentum operators are constants of motion in the quantum mechanical sense.

The various components of the classical angular momentum vector for a particle in $\mathbb{R}^3$ satisfy certain simple commutation relations under the Poisson bracket (Exercise 19 in Chap. 2). We will see that those relations are the commutation relations for the Lie algebra $\mathsf{so}(3)$ of the rotation group $\mathsf{SO}(3)$. If $\hat{H}$ commutes with each component of the angular momentum, each eigenspace for $\hat{H}$ (the solution space to $\hat{H}\psi = \lambda\psi$ for a given λ) is invariant under the angular momentum operators. Thus, the eigenspace constitutes a *representation* of the Lie algebra $\mathsf{so}(3)$. By classifying the irreducible (finite-dimensional) representations of $\mathsf{so}(3)$, we can obtain a lot

B.C. Hall, *Quantum Theory for Mathematicians*, Graduate Texts in Mathematics 267, DOI 10.1007/978-1-4614-7116-5_17, © Springer Science+Business Media New York 2013

of information about the structure of the solution spaces to the equation $\hat{H}\psi = \lambda\psi$, in the case that $\hat{H}$ is invariant under rotations. Specifically, the representation theory of so(3) allows us to determine completely the angular dependence of a solution $\psi(x)$, leaving only the radial dependence of ψ to be determined. This has the effect of reducing the number of independent variables from three to one (just the radius r in polar coordinates), thereby reducing the problem to solving an *ordinary* differential equation.

Understanding angular momentum from the point of view of representations of a Lie algebra also prepares us to understand the concept of *spin*. The Hilbert space for a particle in $\mathbb{R}^3$ with spin is the tensor product of $L^2(\mathbb{R}^3)$ with a finite-dimensional vector space V, where V carries an irreducible action of the rotation group SO(3). In this setting, the proper notion of "action" is a *projective* representation of SO(3), meaning a family of operators satisfying the relations of SO(3) up to phase factors (constants of absolute value one). These phase factors are permitted because, physically, two vectors that differ only by a constant represent the same physical state. By Proposition 16.46, every projective representation of SO(3) can be de-projectivized at the level of the Lie algebra so(3). Conversely, every irreducible ordinary representation of the Lie algebra so(3) gives rise to a representation of the universal cover SU(2) of SO(3), which in turn gives rise (Theorem 16.47) to a projective representation of SO(3). Thus, the possibilities for the space V are in one-to-one correspondence with the irreducible representations of the Lie algebra so(3). In the case of "half-integer spin," the space V does not carry an ordinary representation of the group SO(3).

17.2   The Angular Momentum Operators in $\mathbb{R}^3$

Recall from Sect. 2.4 that the classical angular momentum for a particle in $\mathbb{R}^3$ is given by $\mathbf{J} = \mathbf{x} \times \mathbf{p}$, so that, say, $J_3 = x_1p_2 - x_2p_1$. As in Sect. 3.10, we introduce the quantum mechanical counterpart, a "vector" $\hat{\mathbf{J}}$ with components that are operators,

$$\hat{\mathbf{J}} = \mathbf{X} \times \mathbf{P}.$$

Thus, for example, $\hat{J}_1 = X_2P_3 - X_3P_2$. Note that each component of the angular momentum involves products of *distinct* components of the position and momentum operators $\mathbf{X}$ and $\mathbf{P}$, which commute. Thus, in the expression for, say, $\hat{J}_3$, it does not matter whether we write X_2P_3 or P_3X_2.

The angular momentum operators are unbounded operators and are defined only on a dense subspace of $L^2(\mathbb{R}^3)$. For the moment, we will not specify the domain of these operators, leaving that until the next section. We will see, however, that the domain of each angular momentum operator contains the Schwartz space $\mathcal{S}(\mathbb{R}^3)$ (Definition A.15).

As in Exercise 10 in Chap. 3, we can use the canonical commutation relations to obtain $[\hat{J}_1, \hat{J}_2] = i\hbar \hat{J}_3$. We may similarly compute $[\hat{J}_2, \hat{J}_3]$ and $[\hat{J}_1, \hat{J}_2]$ to obtain the complete set of commutation relations among the $\hat{J}$'s:

$$\frac{1}{i\hbar}[\hat{J}_1, \hat{J}_2] = \hat{J}_3; \quad \frac{1}{i\hbar}[\hat{J}_2, \hat{J}_3] = \hat{J}_1; \quad \frac{1}{i\hbar}[\hat{J}_3, \hat{J}_1] = \hat{J}_2.$$

These relations compare well with the Poisson bracket relations among the various components of the classical angular momentum vector (Exercise 19 in Chap. 2).

Writing out $\hat{J}_3$ explicitly, we have

$$(\hat{J}_3\psi)(\mathbf{x}) = -i\hbar\left(x_1\frac{\partial}{\partial x_2} - x_2\frac{\partial}{\partial x_1}\right)\psi(\mathbf{x}) \tag{17.1}$$

$$-i\hbar\left.\frac{d}{d\theta}\psi(R_\theta\mathbf{x})\right|_{\theta=0}, \tag{17.2}$$

where R_θ denotes a counterclockwise rotation by angle θ in the (x_1, x_2) plane, with similar expression for $\hat{J}_1$ and $\hat{J}_2$. This description of the angular momentum operators demonstrates that they—like the components of the classical angular momentum—are closely connected to *rotations* (recall Propositions 2.18 and 2.30). The connection between angular momentum and rotations will be made more explicit in the following sections by recognizing that they make up the Lie algebra action associated with the natural action of the rotation group on $L^2(\mathbb{R}^3)$.

We may define a new version of the angular momentum operators $\tilde{J}_j$, given by

$$\tilde{J}_j = \frac{1}{\hbar}\hat{J}_j. \tag{17.3}$$

Since Planck's constant and angular momentum have the same units, the $\tilde{J}_j$'s do not depend on the choice of units; we refer to them as the *dimensionless* versions of the angular momentum operators.

17.3 Angular Momentum from the Lie Algebra Point of View

We begin this section by looking at the natural action of the rotation group SO(3) on $L^2(\mathbb{R}^3)$.

Definition 17.1 *For each $R \in$ SO(3), define $\Pi(R) : L^2(\mathbb{R}^3) \to L^2(\mathbb{R}^3)$ by*

$$(\Pi(R)\psi)(x) = \psi(R^{-1}x). \tag{17.4}$$

Proposition 17.2 *For each $R \in$ SO(3), the map $\Pi(R) : L^2(\mathbb{R}^3) \to L^2(\mathbb{R}^3)$ is unitary. Furthermore, the map $\Pi :$ SO(3) $\to$ U$(L^2(\mathbb{R}^3))$ is a strongly continuous homomorphism.*

Proof. Since the Lebesgue measure on $\mathbb{R}^3$ is invariant under rotations, $\Pi(R)$ is unitary for all $R \in \mathsf{SO}(3)$. It is easily checked that $\Pi(R_1 R_2) = \Pi(R_1)\Pi(R_2)$; for this to be true, we need to have $\psi(R^{-1}x)$ rather than $\psi(Rx)$ in the definition of $\Pi(R)$. Arguing as in the proof of Example 10.12, we can easily verify that Π is strongly continuous. ∎

Recall the computation of the Lie algebra $\mathsf{so}(3)$ of $\mathsf{SO}(3)$ in Sect. 16.5, and the basis $\{F_1, F_2, F_3\}$ for $\mathsf{so}(3)$ in (16.2) in that section.

Proposition 17.3 *For each* $X \in \mathsf{so}(3)$, *let* $\pi(X)$ *denote the skew-self-adjoint operator such that*

$$\Pi(e^{tX}) = e^{t\pi(X)}. \tag{17.5}$$

Then the domain of each $\pi(F_j)$ *contains the Schwartz space* $\mathcal{S}(\mathbb{R}^3)$ *and on* $\mathcal{S}(\mathbb{R}^3)$ *we have the relation*

$$\hat{J}_j = i\hbar\pi(F_j).$$

In the notation of Stone's theorem (Theorem 10.15), the operator $\pi(X)$ in (17.5) is i times the infinitesimal generator of the one-parameter unitary group $t \mapsto \Pi(e^{tX})$.

Proof. In the case of $\hat{J}_3$, we compute as in Example 16.16 that e^{tF_3} is a counterclockwise rotation in the (x_1, x_2)-plane. If ψ belongs to $\mathcal{S}(\mathbb{R}^3)$ then the limit defining the derivative in (17.2) is easily seen to hold in the L^2 sense. Thus, recalling the inverse on the right-hand side of (17.4), we see that $\hat{J}_3$ coincides with $i\hbar\pi(F_3)$, as claimed. Similar calculations apply to $\hat{J}_1$ and $\hat{J}_2$. ∎

Although it is not easy to determine the precise domain of each angular momentum operator, we can see from Proposition 16.54 that if ψ belongs to a finite-dimensional subspace of $L^2(\mathbb{R}^3)$ that is invariant under rotations, then ψ belongs to the domain of each $\hat{J}_j$.

17.4   The Irreducible Representations of $\mathsf{so}(3)$

In this section, we classify the irreducible finite-dimensional representations of the Lie algebra $\mathsf{so}(3)$, up to isomorphism. (See Sect. 16.7 for the definitions and elementary properties of representations.) All representations are taken over the field of complex numbers and assumed to have dimension at least one. We continue to use the basis $\{F_1, F_2, F_3\}$ for $\mathsf{so}(3)$ in (16.2).

Theorem 17.4 *Let* $\pi : \mathsf{so}(3) \to \mathsf{gl}(V)$ *be a finite-dimensional irreducible representation of* $\mathsf{so}(3)$. *Define operators* L^+, L^-, *and* L_3 *on* V *by*

$$L^+ = i\pi(F_1) - \pi(F_2)$$
$$L^- = i\pi(F_1) + \pi(F_2)$$
$$L_3 = i\pi(F_3).$$

Let $l = \frac{1}{2}(\dim V - 1)$, *so that* $\dim V = 2l + 1$. *Then there exists a basis* $v_0, v_1, \ldots, v_{2l}$ *of V such that*

$$L_3 v_j = (l - j)v_j$$

$$L^- v_j = \begin{cases} v_{j+1} & \text{if } j < 2l \\ 0 & \text{if } j = 2l \end{cases} \tag{17.6}$$

$$L^+ v_j = \begin{cases} j(2l + 1 - j)v_{j-1} & \text{if } j > 0 \\ 0 & \text{if } j = 0 \end{cases}.$$

Thus, the quantity l completely determines the structure of an irreducible representation of so(3). Since $\dim V$ is a positive integer, l has to have one of the following values:

$$l = 0, \frac{1}{2}, 1, \frac{3}{2}, \ldots. \tag{17.7}$$

The proof of Theorem 17.4 is given later in this section.

Definition 17.5 *If* (π, V) *is an irreducible finite-dimensional representation of* so(3), *then the* **spin** *of* (π, V) *is the largest eigenvalue of the operator* $L_3 := i\pi(F_3)$. *Equivalently,* l *is the unique number such that* $\dim V = 2l + 1$.

Our next result says that *all* the values of l in (17.7) actually arise as spins of irreducible representations of so(3).

Theorem 17.6 *For any* $l = 0, \frac{1}{2}, 1, \frac{3}{2}, \ldots$ *there exists an irreducible representation of* so(3) *of dimension* $2l + 1$, *and any two irreducible representations of* so(3) *of dimension* $2l + 1$ *are isomorphic.*

Note that the theorem is only asserting the existence, for each l, of a representation of the *Lie algebra* so(3). As we will see in the next section, an irreducible representation π of so(3) comes from a representation Π of SO(3) if and only if l is an integer. Nevertheless, the representations of so(3) with *half-integer* values of l—the ones where l is half of an integer but not an integer—still play an important role in quantum physics, as discussed in Sect. 17.8. (Although it would be clearer to refer to the case $l = 1/2, 3/2, 5/2, \ldots$ as "integer plus a half," the terminology "half-integer" is firmly established.)

By comparison to Proposition 17.3, we may think of L_3 as the analog of the third component of the dimensionless angular momentum operator on the space V. Indeed, we will eventually be interested in applying Theorem 17.4 to the case in which V is a subspace of $L^2(\mathbb{R}^3)$ that is invariant under the action of SO(3). In that case, L_3 will be precisely (the restriction to V of) the dimensionless angular momentum operator $\tilde{J}_3$.

Observe that Theorem 17.4 bears a strong similarity to our analysis of the quantum harmonic oscillator. In both cases, we have a "chain" of eigenvectors for a certain operator, along with raising and lowering operators that raise and lower the eigenvalue of that operator. In the case of the harmonic oscillator, we have a chain that begins with a ground state and then extends infinitely in one direction. In the case of so(3) representations, we have a chain that is finite in both directions. The chain begins with an eigenvector v_0 for L_3 with maximal eigenvalue, so that v_0 is annihilated by the raising operator L^+. A key step in the proof of Theorem 17.4 is to determine how the chain can terminate (in the direction of lower eigenvalues for L_3) without violating the commutation relations among L_3, L^+, and L^-.

Proof of Theorem 17.4. Since π is a Lie algebra homomorphism, the $\pi(F_j)$'s satisfy the same commutation relations as the F_j's themselves. From this we can easily verify the following relations among the operators L^+, L^-, and L_3:

$$[L_3, L^+] = L^+ \tag{17.8}$$

$$[L_3, L^-] = -L^- \tag{17.9}$$

$$[L^+, L^-] = 2L_3. \tag{17.10}$$

Now, since we are working over the algebraically closed field $\mathbb{C}$, the operator L_3 has at least one eigenvector v with eigenvalue λ. Consider, then, $L^+ v$. Using (17.8), we compute that

$$L_3 L^+ v = (L^+ L_3 + L^+)v = L^+(\lambda v) + L^+ v = (\lambda + 1)L^+ v. \tag{17.11}$$

Thus, either $L^+ v = 0$ or $L^+ v$ is an eigenvector for L_3 with eigenvalue $\lambda + 1$. We call L^+ the "raising operator," since it has the effect of raising the eigenvalue of L_3 by 1.

If we apply L^+ repeatedly to v, we obtain eigenvectors for L_3 with eigenvalues increasing by 1 at each step, as long as we do not get the zero vector. Eventually, though, we must get 0, since the operator L_3 has only finitely many eigenvalues. Thus, there exists $k \geq 0$ such that $(L^+)^k v \neq 0$ but $(L^+)^{k+1} v = 0$. By applying (17.11) repeatedly, we see that $(L^+)^k v$ is an eigenvector for L_3 with eigenvalue $\lambda + k$.

Let us now introduce the notation $v_0 := (L^+)^k v$ and $\mu = \lambda + k$. Then v_0 is a nonzero vector with $L^+ v_0 = 0$ and $L_3 v_0 = \mu v_0$. We now forget about the original vector v and eigenvalue λ and consider only v_0 and μ. Define vectors v_j by

$$v_j = (L^-)^j v_0, \quad j = 0, 1, 2, \ldots.$$

Arguing as in (17.11), but using (17.9) in place of (17.8), we see that L^- has the effect of either lowering the eigenvalue of L_3 by 1 or of giving the zero vector. Thus, $L_3 v_j = (\mu - j)v_j$.

Next, we claim that for $j \geq 1$ we have

$$L^+ v_j = j(2\mu + 1 - j)v_j, \quad j = 1, 2, 3, \ldots, \tag{17.12}$$

which is easily proved by induction on j, using (17.10) (Exercise 2). Since, again, L_3 has only finitely many eigenvectors, v_j must eventually be zero. Thus, there exists some $N \geq 0$ such that $v_N \neq 0$ but $v_{N+1} = 0$. Since $v_{N+1} = 0$, applying (17.12) with $j = N$ gives

$$0 = L^+ v_{N+1} = (N+1)(2\mu - N)v_N.$$

Since $v_N \neq 0$ and $N + 1 > 0$, we must have $(2\mu - N) = 0$. This means that μ must equal $N/2$.

Letting $l = N/2$ and putting $\mu = N/2 = l$, we have the formulas recorded in (17.6). Meanwhile, since the v_j's are eigenvectors for L_3 with distinct eigenvalues, the v_j's are automatically linearly independent. Furthermore, the span of the v_j's is invariant under L^+, L^-, and L_3, hence under all of so(3). Since V is assumed to be irreducible, the span of the v_j's must be all of V. Thus, the v_j's form a basis for V. The dimension of V is therefore equal to the number of v_j's, which is $N + 1 = 2l + 1$. ∎

Proof of Theorem 17.6. We construct V simply by *defining* a space V with basis $v_0, v_1, \ldots, v_{2l}$ and defining the action of so(3) by (17.6). It is a simple matter (Exercise 4) to check that L^+, L^-, and L_3, defined in this way, have the correct commutation relations, so that V is indeed a representation of so(3).

It remains to show that V is irreducible. Suppose that W is an invariant subspace of V and that $W \neq \{0\}$. We need to show that $W = V$. To this end, suppose that w is some nonzero element of W, which we can decompose as $w = \sum_{j=0}^{2l} a_j v_j$. Let j_0 be the largest index for which a_j is nonzero. According to the formula for L^+ in (17.6), applying L^+ to any of the vectors $v_1, \ldots, v_{2l}$ gives a *nonzero* multiple of the previous element in our chain. Thus, $(L^+)^{j_0} w$ will be a nonzero multiple of v_0. Since W is invariant, this means that v_0 belongs to W. But then by applying L^- repeatedly, we see that v_j belongs to W for each j, so that $W = V$.

Theorem 17.4 tells us that any irreducible representation of so(3) of dimension $2l + 1$ has a basis as in (17.6). We can then construct an isomorphism between any two irreducible representations by mapping this basis in one space to the corresponding basis in the other space. ∎

In the rest of this section, we look at some additional properties of representations of so(3).

Proposition 17.7 *Let* π : so(3) $\to$ gl(V) *be an irreducible representation of* so(3). *Then there exists an inner product on* V, *unique up to multiplication by a constant, such that* $\pi(X)$ *is skew-self-adjoint for all* $X \in$ so(3).

Proof. Recalling how the operators L_3, L^+, and L^- are defined, we can see that the assertion that each $\pi(X)$, $X \in$ so(3), is skew-self-adjoint is equivalent to the assertion that L_3 is self-adjoint and that L^+ and L^- are adjoints of each other. Since the v_j's are eigenvectors for L_3 with distinct eigenvalues, if L_3 is to be self-adjoint, the v_j's must be orthogonal.

Conversely, if we have any inner product for which the v_j's are orthogonal, then L_3 will be self-adjoint, as is easily verified.

It remains to investigate the consequences of the condition $(L^+)^* = L^-$. Assuming this condition, we compute that

$$\langle v_j, v_j \rangle = \langle L^- v_{j-1}, L^- v_{j-1} \rangle = \langle v_{j-1}, L^+ L^- v_{j-1} \rangle.$$

But $L^+ L^- = L^- L^+ + 2L_3$. Furthermore, $L_3 v_{j-1} = (l - j + 1)v_{j-1}$ and $L^+ v_{j-1} = (j - 1)(2l - j + 2)v_{j-1}$ and, thus,

$$\langle v_j, v_j \rangle = \langle v_{j-1}, L^+ L^- v_{j-1} \rangle$$
$$= (j - 1)(2l - j + 2)\langle v_{j-1}, L^- v_{j-2} \rangle + 2(l - j + 1)\langle v_{j-1}, v_{j-1} \rangle.$$

Recalling that $L^- v_{j-2} = v_{j-1}$ and simplifying gives

$$\langle v_j, v_j \rangle = j(2l - j + 1)\langle v_{j-1}, v_{j-1} \rangle. \tag{17.13}$$

It is easy to see that if the v_j's are orthogonal, then L^+ and L^- are adjoints of each other if and only if the normalization condition (17.13) holds for $j = 1, 2, \ldots, 2l$. Since $j(2l - j + 1)$ is positive for each such j, there is no obstruction to normalizing the v_j's so that this condition holds, and so an inner product with the desired property exists. Since the only freedom of choice in defining the inner product is the normalization of v_0, the inner product is unique up to multiplication by a constant. ∎

Proposition 17.8 *Suppose (π, V) is an irreducible representation of so(3) of dimension $2l + 1$. Define the Casimir operator $C_\pi \in \mathrm{End}(V)$ by the formula*

$$C_\pi = \pi(F_1)^2 + \pi(F_2)^2 + \pi(F_3)^2.$$

Then for all $v \in V$, we have

$$C_\pi v = -l(l + 1)v.$$

Proof. See Exercise 3. ∎

If we look at the proof of Theorem 17.4, we see that the only place in which irreducibility was used is in showing that the span of $v_0, v_1, \ldots, v_{2l}$ is equal to V. We can therefore obtain the following result, which will be used in Sect. 17.9.

Proposition 17.9 *Let (π, V) be any finite-dimensional representation of so(3), not necessarily irreducible. Suppose v_0 is a nonzero element of V such that $L^+ v_0 = 0$ and $L_3 v_0 = \lambda v_0$ for some $\lambda \in \mathbb{C}$. Then λ is equal to a nonnegative integer or half-integer l. Furthermore, the vectors $v_0, v_1, \ldots, v_{2l}$ defined by*

$$v_j = (L^-)^j v_0, \quad j = 0, 1, \ldots, 2l,$$

span an irreducible invariant subspace of V of dimension $2l + 1$, and L^+, L^-, and L_3 act on these vectors according to the formulas in Theorem 17.4.

In general, given a finite-dimensional representation (π, V) of a Lie algebra and a nonzero vector $v_0 \in V$, we say that v_0 is a *cyclic vector* for V if the smallest invariant subspace of V containing v_0 is all of V. In Proposition 17.9, the vector v_0 is certainly a cyclic vector for $W := \text{span}(v_0, \ldots, v_{2l})$. It should be noted, however, that a representation's having a cyclic vector *does not*, in general, mean that the representation is irreducible (Exercise 5). Thus, the irreducibility of W is not the result of some general result about cyclic vectors, but holds only because of the assumed special properties of the vector v_0.

17.5   The Irreducible Representations of SO(3)

Having classified the irreducible representations of the *Lie algebra* so(3), we now turn to the classification of the representations of the *group* SO(3). Since SO(3) is connected (Exercise 13 in Chap. 16), Proposition 16.39 tells us that a representation of SO(3) is irreducible if and only if the associated Lie algebra representation is irreducible, and that two representations of SO(3) are isomorphic if and only if the associated Lie algebra representations are isomorphic. Thus, to classify the irreducible representations of SO(3) up to isomorphism, we merely have to determine which irreducible representations of the Lie algebra so(3) come from a representation of the group SO(3).

Proposition 17.10 *Let $\pi_l :$ so(3) $\to$ gl(V) be an irreducible representation of* so(3), *with spin* $l := \frac{1}{2}(\dim V - 1)$. *If l is an integer (i.e., if the dimension of V is odd), then there exists a representation $\Pi_l :$ SO(3) $\to$ GL(V) such that Π_l and π_l are related as in Theorem 16.23. If l is a half-integer (i.e., if the dimension of V is even) then no such representation Π_l exists.*

It follows from this result and Proposition 16.39 that the irreducible representations of the *group* SO(3) are precisely the Π_l's for which l is an integer.

Proof. If l is a half-integer, then L_3 is diagonal in the basis $\{v_j\}$, with eigenvalues being half-integers. Thus,

$$e^{2\pi \pi_l(F_3)} = e^{2\pi i L_3} = -I.$$

(Here the "π" in front of π_l is the number $\pi = 3.14\ldots$.) On the other hand, by a simple modification of Example 16.16, we can see that the matrix $F_3 \in$ so(3) satisfies $e^{2\pi F_3} = I$. Thus, if a corresponding representation Π_l of SO(3) existed, we would have

$$\Pi_l(I) = \Pi_l\left(e^{2\pi F_3}\right) = e^{2\pi \pi_l(F_3)} = -I,$$

which is a contradiction.

If l is an integer, we make use of the isomorphism ϕ between $\mathsf{su}(2)$ and $\mathsf{so}(3)$ described in the proof of Example 16.32, which maps the basis $\{E_1, E_2, E_3\}$ of $\mathsf{su}(2)$ to the basis $\{F_1, F_2, F_3\}$ of $\mathsf{so}(3)$. We obtain a representation π'_l of $\mathsf{su}(2)$ by setting $\pi'_l(X) = \pi_l(\phi(X))$. Since $\mathsf{SU}(2)$ is simply connected, Theorem 16.30 tell us that there is a representation Π'_l of $\mathsf{SU}(2)$ related to π'_l in the usual way. We then compute that

$$\Pi'_l(-I) = \Pi'_l\left(e^{2\pi E_1}\right) = e^{2\pi \pi'_l(E_1)} = e^{2\pi \pi_l(F_1)} = e^{2\pi i L_3} = I,$$

since the eigenvalues of L_3 are integers.

Now, by Example 16.34, there is a surjective homomorphism Φ from $\mathsf{SU}(2)$ onto $\mathsf{SO}(3)$ for which the associated Lie algebra homomorphism is ϕ, and $\ker \Phi = \{I, -I\}$. Since the kernel of Π'_l contains $\{I, -I\}$, the map Π'_l factors through $\mathsf{SO}(3)$, giving a representation Π_l of $\mathsf{SO}(3)$ such that $\Pi'_l = \Pi_l \circ \Phi$. By Exercise 10 in Chap. 16, the associated Lie algebra representation σ_l of $\mathsf{so}(3)$ satisfies $\pi'_l = \sigma_l \circ \phi$, so that $\sigma_l = \pi'_l \circ \phi^{-1} = \pi_l$. Thus, Π_l is the desired representation of $\mathsf{SO}(3)$. ∎

17.6   Realizing the Representations Inside $L^2(S^2)$

In this section, we deviate from the traditional treatment in the physics literature by thinking of the "spherical harmonics" as restrictions to the unit sphere of certain polynomials on $\mathbb{R}^3$, rather than describing the spherical harmonics in angular coordinates on the sphere. Our approach avoids some messy computations in polar coordinates and it also generalizes readily to higher dimensions.

Recall from Sect. 17.3 that there is a natural unitary representation Π : $\mathsf{SO}(3) \to L^2(\mathbb{R}^3)$ given by $\Pi(R)\psi(x) = \psi(R^{-1}x)$. In solving rotationally invariant problems such as the quantum hydrogen atom, it will be useful to understand the structure of finite-dimensional subspaces V of $L^2(\mathbb{R}^3)$ such that V is invariant under Π and such that the restriction of Π to V is irreducible.

If we write functions on $\mathbb{R}^3$ in polar coordinates, then $\mathsf{SO}(3)$ acts only on the angle variables. Thus, it is useful to consider also the action of $\mathsf{SO}(3)$ on $L^2(S^2)$, given by the same formula as for $L^2(\mathbb{R}^3)$, namely

$$(\Pi(R)\psi)(\mathbf{x}) = \psi(R^{-1}\mathbf{x}), \quad \mathbf{x} \in S^2.$$

In computing the norm for $L^2(S^2)$, we use the surface area measure on S^2, which is invariant under the action of $\mathsf{SO}(3)$. Once we have found invariant subspaces inside $L^2(S^2)$, it is a simple matter to produce invariant subspaces inside $L^2(\mathbb{R}^3)$ as well, as we will see in the next section.

We will be interested in this section in harmonic polynomials on $\mathbb{R}^3$, that is, polynomials p satisfying $\Delta p = 0$, where Δ is the Laplacian. Since we always consider representations over $\mathbb{C}$, we allow these polynomials to have complex coefficients.

Definition 17.11 *Let l be a non-negative integer. Define a subspace V_l of $L^2(S^2)$ by setting V_l equal to the space of restrictions to S^2 of harmonic polynomials on $\mathbb{R}^3$ that are homogeneous of degree l. Then V_l is called the space of **spherical harmonics** of degree l.*

Note that if p is a *homogeneous* polynomial on $\mathbb{R}^3$ of some degree l, then the restriction of p to S^2 is identically zero only if p itself is identically zero. After all, if p is homogeneous of degree l and zero on S^2, then

$$p(\mathbf{x}) = |\mathbf{x}|^l \, p\left(\frac{\mathbf{x}}{|\mathbf{x}|}\right) = 0$$

for all $\mathbf{x} \neq 0$, and hence, by continuity, for all $\mathbf{x} \in \mathbb{R}^3$. (By contrast, the nonzero, nonhomogeneous polynomial $p(\mathbf{x}) := x_1^2 + x_2^2 + x_3^2 - 1$ is identically zero on S^2.) We are therefore free to shift back and forth between thinking of the elements of V_l as functions on S^2 or as functions on $\mathbb{R}^3$.

It is well known that the Laplacian Δ commutes with rotations. It follows that each V_l is invariant under the action of the rotation group. We will eventually see that V_l is irreducible under this action.

Every homogeneous polynomial of degree 0 or 1 is harmonic. Thus, V_0 consists of the constant functions on S^2 and V_1 is spanned by the restrictions to S^2 of the functions x_1, x_2, and x_3. Meanwhile, the space of homogeneous polynomials of degree 2 is 6-dimensional, and the space of harmonic polynomials that are homogeneous of degree 2 is spanned by the following five polynomials: $x_1 x_2$, $x_2 x_3$, $x_3 x_1$, $x_1^2 - x_2^2$, and $x_2^2 - x_3^2$. (The polynomial $x_1^2 - x_3^2$ is also harmonic, but it is just the sum $x_1^2 - x_2^2$, and $x_2^2 - x_3^2$.)

Theorem 17.12 *The spaces V_l have the following properties.*

1. *Each V_l has dimension $2l + 1$.*

2. *Each V_l is invariant under the action of the rotation group and irreducible under this action.*

3. *For $l \neq m$, the spaces V_l and V_m are orthogonal in $L^2(S^2)$.*

4. *The Hilbert space $L^2(S^2)$ decomposes as the orthogonal direct sum of the V_l's, as l ranges over the non-negative integers.*

The remainder of this section will be devoted to the proof of Theorem 17.12. We proceed in a series of lemmas, along with some corollaries of those lemmas.

Lemma 17.13 *Let $\mathcal{P}$ denote the space of polynomials on $\mathbb{R}^3$ with complex coefficients. There exists an inner product $\langle \cdot, \cdot \rangle$ on $\mathcal{P}$ with the property that*

$$\langle p, \Delta q \rangle_{\mathcal{P}} = \langle x^2 p, q \rangle_{\mathcal{P}},$$

where

$$x^2 = x_1^2 + x_2^2 + x_3^2.$$

Proof. Although it is possible to give a combinatorial construction of the desired inner product, we can also give an analytic construction. Every polynomial p on $\mathbb{R}^3$ certainly has a holomorphic extension to $\mathbb{C}^3$, denoted $p_\mathbb{C}$. We may define, then,

$$\langle p, q \rangle_{\mathcal{P}} = \int_{\mathbb{C}^3} \overline{p_\mathbb{C}(\mathbf{z})} q_\mathbb{C}(\mathbf{z}) \frac{e^{-|z|^2/2}}{\pi^{3/2}} \, d^6 z,$$

which is nothing but the inner product of $p_\mathbb{C}$ and $q_\mathbb{C}$ as elements of the Segal–Bargmann space $\mathcal{H}L^2(\mathbb{C}^3, \mu_1)$. According to Lemma 14.12, we have

$$\int_{\mathbb{C}^3} \overline{p_\mathbb{C}(\mathbf{z})} \frac{\partial q_\mathbb{C}}{\partial z_j}(\mathbf{z}) \frac{e^{-|z|^2/2}}{\pi^{3/2}} \, d^6 z = \int_{\mathbb{C}^3} \overline{z_j p_\mathbb{C}(\mathbf{z})} q_\mathbb{C}(\mathbf{z}) \frac{e^{-|z|^2/2}}{\pi^{3/2}} \, d^6 z$$

for all $p, q \in \mathcal{P}$ and all $j = 1, 2, 3$. This relation means that

$$\left\langle p, \frac{\partial q}{\partial x_j} \right\rangle_{\mathcal{P}} = \langle x_j p, q \rangle_{\mathcal{P}},$$

from which we readily obtain the desired property of our inner product. ∎

A standard bit of elementary combinatorics shows that the number of ordered triples (l_1, l_2, l_3) with $l_1 + l_2 + l_3 = l$ is equal to $(l + 2)(l + 1)/2$. Since the monomials $x_1^{l_1} x_2^{l_2} x_3^{l_3}$ with $l_1 + l_2 + l_3 = l$ form a basis for $\mathcal{P}_l$, we have $\dim \mathcal{P}_l = (l + 2)(l + 1)/2$.

Corollary 17.14 *If $\mathcal{P}_l$ denotes the space of polynomials on $\mathbb{R}^3$ that are homogeneous of degree l, then the Laplacian Δ maps $\mathcal{P}_l$ onto $\mathcal{P}_{l-2}$ for all $l \geq 2$. Thus, for all $l \geq 2$, we have*

$$\begin{aligned} \dim V_l &= \dim \mathcal{P}_l - \dim \mathcal{P}_{l-2} \\ &= \frac{(l+2)(l+1)}{2} - \frac{l(l-1)}{2} \\ &= 2l + 1. \end{aligned}$$

Proof. Let us equip the finite-dimensional spaces $\mathcal{P}_l$ and $\mathcal{P}_{l-2}$ with the inner product from Lemma 17.13. It is easy to see that the statement, "The orthogonal complement of the image is the kernel of the adjoint," applies to linear maps of one finite-dimensional inner product space to another. Applying this to $\Delta : \mathcal{P}_l \to \mathcal{P}_{l-2}$, we note that the adjoint of Δ is multiplication by x^2, which is clearly injective, since $x_1^2 + x_2^2 + x_3^2$ is zero only at the origin. Thus, the orthogonal complement of the image of Δ is $\{0\}$. Since the spaces are finite-dimensional, this means that Δ maps $\mathcal{P}_l$ onto $\mathcal{P}_{l-2}$. ■

Corollary 17.15 *Let l be a non-negative integer and let $k = l/2$ if l is even and let $k = (l-1)/2$ if l is odd. Then each $p \in \mathcal{P}_l$ can be decomposed in the form*

$$p(\mathbf{x}) = p_0(\mathbf{x}) + |\mathbf{x}|^2 p_1(\mathbf{x}) + |\mathbf{x}|^4 p_2(\mathbf{x}) + \cdots + |\mathbf{x}|^{2k} p_k(\mathbf{x}),$$

where each $p_j(\mathbf{x})$ is a harmonic polynomial that is homogeneous of degree $l - 2j$. In particular, the restriction of p to S^2 satisfies

$$p|_{S^2} = (p_0 + p_1 + \cdots + p_k)|_{S^2},$$

where $p_0 + p_1 + \cdots + p_k$ is a (nonhomogeneous) harmonic polynomial.

Given any polynomial p, not necessarily homogeneous, we can apply Corollary 17.15 to each homogeneous piece of p. We see, then, that given any polynomial p, there exists a *harmonic* polynomial $\tilde{p}$ such that p and $\tilde{p}$ have the same restriction to S^2.

Proof. We proceed by induction on l. If $l = 0$ or $l = 1$, then all $p \in \mathcal{P}_l$ are harmonic and the desired decomposition is simply $p = p_0$. Consider, then, some $l \geq 2$ and assume the result holds for all degrees less than l. Lemma 17.13 tells us that $\mathcal{P}_l$ decomposes as an orthogonal direct sum of the kernel of Δ and the image of $\mathcal{P}_{l-2}$ under multiplication by $|\mathbf{x}|^2$. Thus, any $p \in \mathcal{P}_l$ can be decomposed as $p = p_0 + |\mathbf{x}|^2 q_0$, where p_0 is harmonic and q_0 belongs to $\mathcal{P}_{l-2}$. By induction, q_0 has a decomposition of the desired form; substituting this in for q_0 in the decomposition $p = p_0 + |\mathbf{x}|^2 q_0$ gives the desired decomposition of p. ■

To show that V_l is irreducible under the action Π of $SO(3)$, we pass to the Lie algebra. Since, as we have remarked, restriction to the sphere is injective on homogeneous polynomials, we may think of the elements of V_j as polynomials on $\mathbb{R}^3$, in which case, the Lie algebra action π associated with Π is given in terms of the usual angular momentum operators.

Lemma 17.16 *As in Theorem 17.4, let $L_3 = i\pi(F_3) = \tilde{J}_3$ and let $L^+ = i\pi(F_1) - \pi(F_2) = \tilde{J}_1 + i\tilde{J}_2$. For any non-negative integer l, the polynomial $p(x_1, x_2, x_3) := (x_1 + ix_2)^l$ belongs to V_l and satisfies*

$$L_3 p = lp$$

and

$$L^+ p = 0.$$

Proof. Since it is independent of x_3 and holomorphic as a function of $z := x_1 + ix_2$, the polynomial p is automatically harmonic, which can also be verified by direct calculation. Meanwhile, applying L_3 to p gives

$$
-i\left(x_1\frac{\partial}{\partial x_2} - x_2\frac{\partial}{\partial x_1}\right)(x_1 + ix_2)^l
$$
$$
= -i\left[x_1 l(x_1 + ix_2)^{l-1}(i) - x_2 l(x_1 + ix_2)^{l-1}\right]
$$
$$
= l(x_1 + ix_2)^l.
$$

Finally, applying $L^+ := i\pi(F_1) - \pi(F_2)$ to p gives

$$
-i\left(x_2\frac{\partial}{\partial x_3} - x_3\frac{\partial}{\partial x_2}\right)p + \left(x_3\frac{\partial}{\partial x_1} - x_1\frac{\partial}{\partial x_3}\right)p
$$
$$
= -i(-x_3 l(x_1 + ix_2)^{l-1}(i)) + x_3 l(x_1 + ix_2)^{l-1}(1)
$$
$$
= 0,
$$

as claimed. ∎

Corollary 17.17 *The space V_l is irreducible under the action of* SO(3).

Proof. By Proposition 17.9, if we apply L^- repeatedly to the polynomial p, we obtain a "chain" of eigenvectors of length $2l + 1$. These eigenvectors span an irreducible invariant subspace of dimension $2l + 1$. Since we have already established that dim $V_l = 2l + 1$, the elements of the chain must span V_l, which implies that V_l is irreducible. ∎

We have now assembled all the pieces necessary for a proof of the main result of this section.

Proof of Theorem 17.12. We have already proved Points 1 and 2 of the theorem in Corollaries 17.14 and 17.17, respectively. Now, each V_l is an irreducible representation of SO(3), and no two of the V_l's can be isomorphic, because they all have different dimensions. Thus, by Exercise 19 in Chap. 16, V_l and V_m must be orthogonal inside $L^2(S^2)$ for $l \neq m$, which is Point 3.

Finally, by the Stone–Weierstrass theorem and the density results of Theorem A.10, the restrictions to S^2 of polynomials on $\mathbb{R}^3$ form a dense subspace of $L^2(S^2)$. But Corollary 17.15 shows that the space of restrictions to S^2 of polynomials coincides with the space of restrictions to S^2 of *harmonic* polynomials. Thus, the span of the V_j's is dense in $L^2(S^2)$, establishing Point 4. ∎

17.7   Realizing the Representations Inside $L^2(\mathbb{R}^3)$

Recall that for homogeneous polynomials on $\mathbb{R}^3$, the restriction map from $\mathbb{R}^3$ to S^2 is injective. Thus, we may think of the space V_l equally well as a space of functions on S^2 (as in the previous section) or as a space of functions on $\mathbb{R}^3$. In this section, then, we will let V_l denote the space of harmonic polynomials on $\mathbb{R}^3$ that are homogeneous of degree l.

Definition 17.18 *Suppose l is a non-negative integer and f is a measurable function on $(0, \infty)$ such that*

$$\int_0^\infty |f(r)|^2 \, r^{2l+2} \, dr < \infty. \tag{17.14}$$

Let $V_{l,f} \subset L^2(\mathbb{R}^3)$ denote the space of functions ψ of the form

$$\psi(\mathbf{x}) = p(\mathbf{x}) f(|\mathbf{x}|), \tag{17.15}$$

where $p \in V_l$.

The condition on $f(r)$ is precisely what one needs to make $\psi(\mathbf{x})$ a square-integrable function on $\mathbb{R}^3$ (compute the L^2 norm in spherical coordinates).

Definition 17.18 is not the one that physicists typically use. In the physics literature, one sees a functions of the form

$$\psi(\mathbf{x}) = Y_{lm}(\theta, \phi) g(r), \tag{17.16}$$

where r, θ, and ϕ are the usual spherical coordinates. Here Y_{lm} is the *restriction to the sphere* of a particular harmonic polynomial that is homogeneous of degree l, written in spherical coordinates. (Up to a normalization factor, the Y_{lm}'s are obtained by using the basis for V_l in Theorem 17.4.) Thus, if we move along a ray from the origin in $\mathbb{R}^3$, only the value of $g(r)$ changes. By contrast, in (17.15), as we move along a ray, the $p(\mathbf{x})$ factor contributes a factor of r^l. We can write the physics expression in rectangular coordinates as

$$\psi(\mathbf{x}) = Y_{lm}\left(\frac{\mathbf{x}}{|\mathbf{x}|}\right) g(|\mathbf{x}|)$$
$$= Y_{lm}(\mathbf{x}) \frac{g(|\mathbf{x}|)}{|\mathbf{x}|^l}. \tag{17.17}$$

For computational purposes, the expression (17.15) is more convenient than (17.17); in fact, in the analysis of the hydrogen atom, physicists multiply by r^l at some later point in the calculation, just so that the relevant differential equation will take on a simpler form.

Proposition 17.19 *Every space of the form $V_{l,f} \subset L^2(\mathbb{R}^3)$ is invariant and irreducible under the action of* SO(3). *Conversely, every finite-dimensional, irreducible,* SO(3)-*invariant subspace of $L^2(\mathbb{R}^3)$ is of the form $V_{l,f}$ for some non-negative integer l and some f satisfying (17.14).*

Proof. Since the factor $f(|\mathbf{x}|)$ is invariant under rotations, the action of SO(3) only affects the function p. Thus, $V_{l,f}$ is isomorphic, as a representation of SO(3), to the space V_l, which is irreducible by Theorem 17.12.

For the other direction, the Lebesgue measure on $\mathbb{R}^3$ decomposes as a product of the surface area measure on S^2 with the measure $4\pi r^2 \, dr$ on $(0, \infty)$. Thus, by a standard measure-theoretic result (Proposition 19.12), $L^2(\mathbb{R}^3)$ decomposes canonically as the Hilbert tensor product of $L^2(S^2)$ and $L^2((0, \infty))$, where a vector of the form $f \otimes g$ in the tensor product corresponds to the function $f(\theta, \phi)g(r)$ in $L^2(\mathbb{R}^3)$, as in (17.16). Since $L^2(S^2)$ decomposes (Theorem 17.12) as the sum of the spaces V_l, $l = 0, 1, 2, \ldots$, we can decompose $L^2(\mathbb{R}^3)$ as sum of spaces of the form

$$V_{l,k} := V_l \otimes g_k,$$

where the g_k's form an orthonormal basis for $L^2((0, \infty))$.

Now, let V be any finite-dimensional, irreducible, SO(3)-invariant subspace of $L^2(\mathbb{R}^3)$. Let $\pi_{l,k} : L^2(\mathbb{R}^3) \to V_{l,k}$ be the orthogonal projection operator, and let $\rho_{l,k}$ be the restriction of $\pi_{l,k}$ to V. This map is easily seen to be an intertwining map for the action of SO(3). Thus, since both V and $V_{l,k}$ are irreducible, Schur's lemma tells us that each $\rho_{l,k}$ is either zero or an isomorphism. Furthermore, since the spaces $V_{l,k}$ are nonisomorphic for different values of l, we cannot have both $\rho_{k,l}$ and $\rho_{k',l'}$ being nonzero for $l \neq l'$. On the other hand, $\rho_{k,l}$ cannot be zero for all k and l, since the $V_{k,l}$'s span $L^2(\mathbb{R}^3)$. Thus, there must be some value l_0 of l such that ρ_{l_0,k_0} is nonzero for some k_0 but such that $\rho_{l,k} = 0$ for all $l \neq l_0$.

Applying Schur's lemma again, we see that $\rho_{l_0,k}(\rho_{l_0,k_0})^{-1}$ must be of the form $c_k I$ for each k. Given any $\psi \in V$, let v be the unique element of V such that $\rho_{l_0,k_0}(\psi) = v \otimes g_{k_0}$. Then we have

$$\rho_{l_0,k}(\psi) = c_k(v \otimes g_k)$$

for every k. Since also $\rho_{l,k}(\psi) = 0$ for $l \neq l_0$, we conclude that ψ must be of the form $v \otimes g$, where

$$g = \sum_k c_k g_k.$$

Since this holds for each $\psi \in V$ (with the same set of constants c_k), we see that $V = V_{l_0} \otimes g$, which is nothing but the form in (17.16). Then V is of the form claimed in the proposition, where $f(r) = g(r)/r^{l_0}$. ∎

It can further be shown that each closed, SO(3)-invariant subspace of $L^2(\mathbb{R}^3)$ decomposes as an orthogonal direct sum of finite-dimensional, irreducible, SO(3)-invariant subspaces. This result is just a special case of a general result for strongly continuous unitary representations of compact topological groups. (See, e.g., Chap. 5 of [10].) Since we already know that $L^2(\mathbb{R}^3)$ is a direct sum of finite-dimensional, irreducible invariant subspaces, it is probably possible to give an elementary proof of this result, but we will not pursue that approach here.

17.8 Spin

We classified irreducible finite-dimensional representations of the Lie algebra so(3) by their "spin" l, where l is the largest eigenvalue for the operator $L_3 = i\pi(F_3)$. The possible values for l are non-negative integers $(0, 1, 2, \ldots)$ and the positive half-integers $(1/2, 3/2, \ldots)$. Inside $L^2(S^2)$ and $L^2(\mathbb{R}^3)$, however, we found only irreducible representations of so(3) with *integer* spin. It is easy to understand why the half-integer spin representations do not occur: They do not correspond to any representation of the *group* SO(3). Since $L^2(S^2)$ and $L^2(\mathbb{R}^3)$ both carry a natural unitary action Π of the group SO(3), any finite-dimensional subspace that is invariant under the associated Lie algebra representation π will also be invariant under Π and thus constitute a representation of SO(3).

Although the half-integer representations π_l of the Lie algebra so(3) cannot be exponentiated to representations of SO(3), they can be exponentiated to representations of the universal cover SU(2) of SO(3), as in the proof of Proposition 17.10. For a half-integer l, the associated representation Π'_l of SU(2) satisfies $\Pi'_l(-I) = -I$, which means that Π'_l does not factor through SO(3) $\cong$ SU(2)/$\{I, -I\}$. If, however, we think about *projective* representations, we see that $[-I]$ *is* the identity element in PU(V). Thus, even when l is a half-integer, we get a well-defined projective representation Π_l of SO(3) that satisfies

$$\Pi_l(e^{tX}) = [e^{t\pi_l(X)}]$$

for all $X \in$ so(3), where $[U]$ denotes the image of $U \in$ U(V) in PU(V).

It is generally believed that the physics of the universe is invariant under the rotation group SO(3). This does not mean that one never considers models without rotational symmetry, because the local environment of, say, a hydrogen atom in a magnetic field breaks the rotational symmetry of the hydrogen atom. Nevertheless, if we were to rotation *both* the hydrogen atom *and* the magnetic field, the physics of the problem would not change. In quantum mechanics, rotational symmetry means that there should be a projective unitary representation of SO(3) on the Hilbert space of the universe that commutes with the Hamiltonian operator. Now, the Hilbert space of the universe (if there is such a thing) is built up out of Hilbert spaces for each type of particle. Thus, we expect that the Hilbert space for a single particle will also carry a projective unitary representation of SO(3).

The simplest possibility for the Hilbert space of a single particle is the Hilbert space $L^2(\mathbb{R}^3)$, which certainly carries an (ordinary) unitary action of SO(3), as we have been discussing in this chapter. Based on various experimental observations, however, physicists have proposed a modification to the Hilbert space for an individual particle that incorporates "internal degrees of freedom." The proposal is that for each type of particle, the quantum Hilbert space should be of the form $L^2(\mathbb{R}^3)\hat{\otimes}V$, where V

is a finite-dimensional Hilbert space that carries an irreducible projective unitary representation of $SO(3)$. Here $\hat{\otimes}$ is the Hilbert tensor product (Appendix A.4.5). The (projective) action of $SO(3)$ on V describes the action of the rotation group on the internal degrees of freedom of the particle.

Now, according to Proposition 16.46, the space V carries a (trace-zero) ordinary representation π of the Lie algebra $so(3)$. In customary physics terminology, the largest eigenvalue l of the operator $L_3 := i\pi(F_3)$ in V is then called the *spin* of the particle. We then denote the space V by V_l to indicate the value of the spin. Electrons, for example, are "spin 1/2" particles, meaning that the Hilbert space for a single electron is $L^2(\mathbb{R}^3)\hat{\otimes}V_{1/2}$, where $V_{1/2}$ is a two-dimensional projective representation of $SO(3)$.

It is easy to see that the tensor product of two projective unitary representations of a given group is again a projective unitary representation of that group. (By contrast, the direct sum of two projective unitary representations is in general not again a projective unitary representation.) In the case at hand, we can think of $L^2(\mathbb{R}^3)$ as carrying a unitary representation Π of $SU(2)$ that factors through $SO(3)$, that is, for which $\Pi(-I) = I$. Meanwhile, we can think of V_l as a carrying a unitary representation Π_l of $SU(2)$ in which $\Pi_l(-I) = \pm I$, with the plus sign if l is an integer and the minus sign if l is a half-integer. Thus, $L^2(\mathbb{R}^3)\hat{\otimes}V_l$ carries a unitary representation $\Pi \otimes \Pi_l$ of $SU(2)$ in which $(\Pi \otimes \Pi_l)(-I) = \pm I$. Thus, in the projective sense, $\Pi \otimes \Pi_l$ factors through $SO(3)$.

Summary 17.20 (Spin) *Each type of particle has a "spin" l, which is a non-negative integer or half-integer. The Hilbert space for such a particle is $L^2(\mathbb{R}^3)\hat{\otimes}V_l$, where V_l is an irreducible projective representation of $SO(3)$ of dimension $2l + 1$.*

Since V_l is finite dimensional, the Hilbert tensor product $L^2(\mathbb{R}^3)\hat{\otimes}V_l$ coincides with the algebraic tensor product of $L^2(\mathbb{R}^3)$ with V_l.

Definition 17.21 *A particle for which the spin is an integer is called a **boson**, and a particle for which the spin is a half-integer is called a **fermion**.*

To see the significance of the distinction between integer and half-integer spin, one needs to look at the structure of the Hilbert space describing *multiple* particles of a given type, such as the Hilbert space for five electrons. This topic is discussed in Chap. 19.

17.9 Tensor Products of Representations: "Addition of Angular Momentum"

Let V_l and V_m be irreducible representations of $so(3)$ with dimensions $2l+1$ and $2m + 1$, respectively. As discussed in Sect. 16.8, the tensor product space $V_l \otimes V_m$ can be viewed as another representation of $so(3)$. Unless one of l and m is zero, $V_l \otimes V_m$ is *not* irreducible. It is of interest, then, to decompose $V_l \otimes V_m$ as a direct sum of irreducible invariant subspaces. This decomposition—in the case that V_l is an irreducible SO(3)-invariant subspace of $L^2(\mathbb{R}^3)$ and V_m is the space of internal degrees of freedom of a particle—will help us in decomposing the Hilbert space for a particle with spin into irreducible, SO(3)-invariant subspaces.

Proposition 17.22 *Let $V_{1/2}$ be an irreducible representation of* so(3) *of dimension 2, and let V_l be an irreducible representation of* so(3) *of dimension $2l + 1$, where l is a non-negative integer or half-integer. If $l = 0$, $V_l \otimes V_{1/2}$ is irreducible. If $l > 0$, then we have*

$$V_l \otimes V_{1/2} \cong V_{l+1/2} \oplus V_{l-1/2},$$

where "$\cong$" denotes an isomorphism of representations.

Proof. If $l = 0$, then it is easy to see that $V_l \otimes V_{1/2}$ is isomorphic to $V_{1/2}$, which is irreducible. Assume, then, that $l > 0$.

Let L^+, L^-, and L_3 be the operators in Theorem 17.4, constructed using the representation π_l, and let σ^+, σ^-, and σ_3 be the analogous operators constructed using the representation $\pi_{1/2}$. As in Sect. 16.8, we define operators J^+, J^-, and J_3 on $V_l \otimes V_{1/2}$ by

$$
\begin{aligned}
J^+ &= L^+ \otimes I + I \otimes \sigma^+ \\
J^- &= L^- \otimes I + I \otimes \sigma^- \\
J_3 &= L_3 \otimes I + I \otimes \sigma_3.
\end{aligned}
\tag{17.18}
$$

Let $\{v_0, \ldots, v_{2l}\}$ be a basis for V_l as in Theorem 17.4, and let $\{e_0, e_1\}$ be a similar basis for $V_{1/2}$. Then the vectors of the form $v_j \otimes e_k$ form a basis for $V_l \otimes V_{1/2}$. The eigenvalues of J_3 are the numbers of the form

$$(l - j) + \left(\frac{1}{2} - k\right),$$

$j = 0, 1, \ldots, 2l$, $k = 0, 1$. Thus, the eigenvalues of J_3 range from $l + 1/2$ to $-(l + 1/2)$. The numbers $l + 1/2$ and $-(l + 1/2)$ occur as eigenvalues only once. All other eigenvalues λ occur twice, once as $(\lambda - 1/2) + 1/2$ and once as $(\lambda + 1/2) - 1/2$.

The vector $v_0 \otimes e_0$ is an eigenvector for J_3 with the largest possible eigenvalue $l + 1/2$, so that $J^+(v_0 \otimes e_0) = 0$. According to Proposition 17.9, if we apply J^- repeatedly, we will obtain a "chain" of eigenvectors of length $2l + 2$, and the span of these vectors forms an irreducible invariant subspace W_0 isomorphic to $V_{l+1/2}$.

Now, by Proposition 17.7, there exist inner products on V_l and $V_{1/2}$ that make π_l and $\pi_{1/2}$ "unitary," meaning that $\pi(X)^* = -\pi(X)$ for all $X \in$ so(3). If we use on $V_l \otimes V_{1/2}$ the natural inner product, obtained from the inner products on V_l and $V_{1/2}$ as in Appendix A.4.5, then $\pi_l \otimes \pi_{1/2}$ is also unitary. Thus, the orthogonal complement of the invariant subspace W_0 is also invariant. Since all eigenvalues for J_3 except the largest and smallest have multiplicity 2, we see that the largest eigenvalue for J_3 in $W_0^\perp$ is $l - 1/2$. Let $w_0 \in W_0^\perp$ be an eigenvector for J_3 with eigenvalue $l - 1/2$. If we repeatedly apply the lowering operator $J^- : L^- \otimes I + I \otimes \sigma^-$ to w_0, we will obtain a chain of eigenvectors of length $2l$. These eigenvectors span an irreducible invariant subspace W_1 of $V_l \otimes V_{1/2}$ of dimension $2l$. Since

$$\dim W_0 + \dim W_1 = 4l + 2 = \dim(V_l \otimes V_{1/2}),$$

we must have $W_1 = W_0^\perp$, completing the proof. ∎

Since an electron is a "spin 1/2" particle, the Hilbert space for a single electron is, according to Sect. 17.8, $L^2(\mathbb{R}^3) \hat{\otimes} V_{1/2}$, where $V_{1/2}$ is an irreducible projective unitary representation of $\mathsf{SO}(3)$ of dimension 2. Meanwhile, in Sect. 17.7, we saw how to find irreducible, $\mathsf{SO}(3)$-invariant subspaces $V_{l,f}$ of $L^2(\mathbb{R}^3)$ of dimension $2l + 1$, for $l = 0, 1, 2, \ldots$, where f is an arbitrary radial function. By applying Proposition 17.22 to the case $V_l = V_{l,f}$, we obtain irreducible $\mathsf{SO}(3)$-invariant subspaces of the Hilbert space $L^2(\mathbb{R}^3) \hat{\otimes} V_{1/2}$. Finding such subspaces is essential in, for example, analyzing the fine structure of the hydrogen atom.

In the case that V_l is an $\mathsf{SO}(3)$-invariant subspace of $L^2(\mathbb{R}^3)$, the formula for, say, the operator J_3 in (17.18) 17.22 is written in the physics literature as

$$J_3 = L_3 + \sigma_3, \tag{17.19}$$

where it is understood that L_3 acts on the first factor in the tensor product and σ_3 acts on the second factor. (That is to say, the tensor product with the identity operator is understood and thus not written.) Here L_3 is the ordinary angular momentum operator and σ_3 describes the action of the basis element $F_3 \in \mathsf{so}(3)$ on the space $V_{1/2}$. Formulas such as (17.19) account for the physics terminology "addition of angular momentum" to describe the analysis of tensor products of representations of $\mathsf{so}(3)$. In this context, the operator L_3 $(= L_3 \otimes I)$ is called an *orbital angular momentum* operator, and the operator σ_3 $(= I \otimes \sigma_3)$ is called a *spin angular momentum* operator, and similarly for $L^\pm$ and $\sigma^\pm$.

We now record the general result for tensor products of irreducible representations of $\mathsf{so}(3)$.

Proposition 17.23 *For any $j = 0, 1/2, 1, \ldots$, let V_j denote the unique irreducible representation of $\mathsf{so}(3)$ of dimension $2j + 1$. Then for any l and m with $l \geq m$, we have*

$$V_l \otimes V_m \cong V_{l+m} \oplus V_{l+m-1} \oplus \cdots \oplus V_{l-m+1} \oplus V_{l-m}. \tag{17.20}$$

The proof of this result is similar to that of Proposition 17.22, and is omitted; see Theorem D.1 in Appendix D of [21]. An important property of this decomposition is that each irreducible representation that occurs on the right-hand side of (17.20) occurs only once. This property of the representations of so(3) is the key idea in the proof of the Wigner–Eckart theorem. See Appendix D of [21] for details.

17.10   Vectors and Vector Operators

Definition 17.24 *A function* $\mathbf{c} : \mathbb{R}^3 \times \mathbb{R}^3 \to \mathbb{R}^3$ *is said to* **transform like a vector** *if*

$$\mathbf{c}(R\mathbf{x}, R\mathbf{p}) = R(\mathbf{c}(\mathbf{x}, \mathbf{p})) \tag{17.21}$$

for all $R \in \mathsf{SO}(3)$.

In the physics literature, the expression "is a vector" is sometimes used in place of "transforms like a vector."

Note that in Definition 17.24, we only consider the transformation property of $\mathbf{c}$ under elements of $\mathsf{SO}(3)$ rather than under a general element of $\mathsf{O}(3)$. If $\mathbf{c}$ transforms like a vector, one says that $\mathbf{c}$ is an "true vector" if $\mathbf{c}$ satisfies (17.21) for all R in $\mathsf{O}(3)$ [not just in $\mathsf{SO}(3)$] and one says that $\mathbf{c}$ is a "pseudovector" if $\mathbf{c}$ satisfies $\mathbf{c}(R\mathbf{x}, R\mathbf{p}) = -R(\mathbf{c}(\mathbf{x}, \mathbf{p}))$ for $R \in \mathsf{O}(3)\backslash\mathsf{SO}(3)$. For our purposes, it is not necessary to distinguish between true vectors and pseudovectors.

The position function $\mathbf{c}_1(\mathbf{x}, \mathbf{p}) := \mathbf{x}$, the momentum function $\mathbf{c}_2(\mathbf{x}, \mathbf{p}) := \mathbf{p}$, and the angular momentum function $\mathbf{c}_3(\mathbf{x}, \mathbf{p}) := \mathbf{x} \times \mathbf{p}$ are simple examples of functions that transform like vectors. (Transformation under rotations is one of the standard properties of the cross product.) A typical example of a function transforming like a vector is $\mathbf{c}(\mathbf{x}, \mathbf{p}) = (\mathbf{x} \cdot \mathbf{p}) |\mathbf{x}| (\mathbf{x} \times \mathbf{p})$.

Proposition 17.25 *Let* $\mathbf{j}(\mathbf{x}, \mathbf{p}) = \mathbf{x} \times \mathbf{p}$ *denote the angular momentum function on* $\mathbb{R}^3 \times \mathbb{R}^3$. *Suppose a smooth function* $\mathbf{c} : \mathbb{R}^3 \times \mathbb{R}^3 \to \mathbb{R}^3$ *transforms like a vector. Then we have*

$$\{c_k, j_k\} = 0 \tag{17.22}$$

for $k = 1, 2, 3$. *Furthermore, we have*

$$\{c_1, j_2\} = \{j_1, c_2\} = c_3 \tag{17.23}$$

and other relations obtained from (17.23) by cyclically permuting the indices.

Proof. Let $R(\theta)$ denote a counterclockwise rotation by angle θ in the (x_1, x_2)-plane. Applying (17.21) with $R = R(\theta)$ and looking only at the first component of the vectors, we have

$$c_1(R(\theta)\mathbf{x}, R(\theta)\mathbf{p}) = c_1(\mathbf{x}, \mathbf{p}) \cos\theta - c_2(\mathbf{x}, \mathbf{p}) \sin\theta. \tag{17.24}$$

Now, as in the proof of Proposition 2.30, the Poisson bracket $\{c_1, j_3\}$ is precisely the derivative of the left-hand side of (17.24) with respect to θ, evaluated at $\theta = 0$. Thus,

$$\{c_1, j_3\} = -c_2$$

and so $\{j_3, c_1\} = c_2$, which is one of the relations obtained from (17.23) by cyclically permuting the indices.

Meanwhile, if we again apply (17.21) with $R = R(\theta)$ but look now at the third component of the vectors, we have that

$$c_3(R(\theta)\mathbf{x}, R(\theta)\mathbf{p}) = c_3(\mathbf{x}, \mathbf{p}).$$

Differentiating this relation with respect to θ at $\theta = 0$ gives $\{c_3, j_3\} = 0$. All other brackets are computed similarly. ∎

We now turn to the quantum counterpart of a function that transforms like a vector.

Definition 17.26 *For any ordered triple* $\mathbf{C} := (C_1, C_2, C_3)$ *of operators on* $L^2(\mathbb{R}^3)$ *and any vector* $\mathbf{v} \in \mathbb{R}^3$, *let* $\mathbf{v} \cdot \mathbf{C}$ *be the operator*

$$\mathbf{v} \cdot \mathbf{C} = \sum_{j=1}^{3} v_j C_j. \tag{17.25}$$

Then an ordered triple $\mathbf{C}$ *of operators on* $L^2(\mathbb{R}^3)$ *is called a **vector operator** if*

$$(R\mathbf{v}) \cdot \mathbf{C} = \Pi(R)(\mathbf{v} \cdot \mathbf{C})\Pi(R)^{-1} \tag{17.26}$$

for all $R \in \mathsf{SO}(3)$.

Here $\Pi(\cdot)$ is the natural unitary action of $\mathsf{SO}(3)$ on $L^2(\mathbb{R}^3)$ in Definition 17.1. Let us try to understand what this definition is saying in the case of, say, the angular momentum, which is (as we shall see) a vector operator. The operators $\hat{J}_1$, $\hat{J}_2$, and $\hat{J}_3$ represent the components of $\hat{\mathbf{J}}$ in the directions of $\mathbf{e}_1$, $\mathbf{e}_2$, and $\mathbf{e}_3$, respectively. More generally, we can consider the component of $\hat{\mathbf{J}}$ in the direction of any unit vector $\mathbf{v}$, which will be nothing but $\mathbf{v} \cdot \hat{\mathbf{J}}$, as defined in (17.25). Since there is no preferred direction in space, we expect that for any two unit vectors $\mathbf{v}_1$ and $\mathbf{v}_2$, the operators $\mathbf{v}_1 \cdot \hat{\mathbf{J}}$ and $\mathbf{v}_2 \cdot \hat{\mathbf{J}}$ should be "the same operator, up to rotation." Specifically, if R is some rotation with $R\mathbf{v}_1 = \mathbf{v}_2$, then $\mathbf{v}_1 \cdot \hat{\mathbf{J}}$ and $\mathbf{v}_2 \cdot \hat{\mathbf{J}}$ should differ only by the action of R on the Hilbert space $L^2(\mathbb{R}^3)$. But this is precisely what (17.26) says, with $\mathbf{v} = \mathbf{v}_1$ and $\mathbf{C} = \hat{\mathbf{J}}$:

$$\mathbf{v}_2 \cdot \hat{\mathbf{J}} = \Pi(R)(\mathbf{v}_1 \cdot \hat{\mathbf{J}})\Pi(R)^{-1}$$

We will not concern ourselves with the question of whether (17.26) continues to hold for $R \in \mathsf{O}(3)\backslash\mathsf{SO}(3)$. The position and momentum operators $\mathbf{X}$ and $\mathbf{P}$ are easily seen to be vector operators. As in the classical case, the cross product of two vector operators is again a vector operator. (See Exercise 7 in Chap. 18.) In particular, the angular momentum, $\hat{\mathbf{J}} = \mathbf{X} \times \mathbf{P}$ is a vector operator.

If the operators C_1, C_2, and C_3 are unbounded, we should say something in Definition 17.26 about the domains of the operators in question. The simplest approach is to find some dense subspace V of $L^2(\mathbb{R}^3)$ that is contained in the domain of each C_j and such that V is invariant under rotations. In that case, the equality in (17.26) is understood to hold when applied to a vector in V. In many cases, we can take V to be the Schwartz space $\mathcal{S}(\mathbb{R}^3)$. In the following proposition, the space V should satisfy certain technical domain conditions that permit differentiation of (17.29) when applied to a vector ψ in V. We will not pursue the details of such conditions here.

Proposition 17.27 *If* $\mathbf{C}$ *is a vector operator, then the components of* $\mathbf{C}$ *satisfy*

$$\frac{1}{i\hbar}[C_j, \hat{J}_j] = 0 \tag{17.27}$$

for $j = 1, 2, 3$. *Furthermore, we have*

$$\frac{1}{i\hbar}[C_1, \hat{J}_2] = \frac{1}{i\hbar}[\hat{J}_1, C_2] = C_3, \tag{17.28}$$

and other relations obtained from (17.28) by cyclically permuting the indices.

Proof. As in the proof of Proposition 17.25, $R(\theta)$ denote a rotation in the (x_1, x_2)-plane, and let $\mathbf{e}_1 = (1, 0, 0)$. Applying (17.26) with $R = R(\theta)$ and $\mathbf{v} = \mathbf{e}_1$, we have

$$\Pi(R(\theta))C_1\Pi(R(\theta))^{-1} - C_1 \cos\theta + C_2 \sin\theta. \tag{17.29}$$

But $R(\theta) = e^{\theta F_3}$, where $\{F_j\}$ is the basis for $\mathsf{so}(3)$ described in Sect. 16.5. Thus, differentiating (17.29) with respect to θ at $\theta = 0$ gives

$$\pi(F_3)C_1 - C_1\pi(F_3) = C_2.$$

Since $\hat{J}_3 = i\hbar\pi(F_3)$ (Proposition 17.3), we obtain $(1/(i\hbar))[\hat{J}_3, C_1] = C_2$, which is one of the relations obtained from (17.28) by cyclically permuting the variables.

Meanwhile, applying (17.26) with $R = R(\theta)$ and $\mathbf{v} = \mathbf{e}_3$ gives

$$\Pi(R(\theta))C_3\Pi(R(\theta))^{-1} = C_3.$$

Differentiating this relation with respect to θ at $\theta = 0$ gives $[\pi(F_3), C_3] = 0$. All other relations are obtained similarly. ∎

For more information about vector operators, including the Wigner–Eckart theorem, see Appendix D of [21]. See also Exercise 7.

17.11   Exercises

1. Verify the expression (17.2) for the vector field $x_1 \partial/\partial x_2 - x_2 \partial/\partial x_1$.

2. Verify the relation (17.12) in the proof of Theorem 17.4, using induction on j and the commutation relation (17.10).

3. This exercise provides a proof of Proposition 17.8. Let (π, V_l) denote an irreducible representation of $\mathsf{so}(3)$ of dimension $2l + 1$ and let C_π denote the Casimir operator as defined in the proposition.

   (a) Show that $[\pi(F_j), C_\pi] = 0$ for all $j = 1, 2, 3$.

   (b) Using Schur's lemma, show that there is some $\lambda \in \mathbb{C}$ such that $C_\pi v = \lambda v$ for all $v \in V$.

   (c) Show that
   $$C_\pi = -\left(L_3^2 + L^- L^+ + L_3\right),$$
   where L^+, L^-, and L_3 are as in Theorem 17.4.

   (d) By computing C_π on some suitably chosen vector in V, show that the constant λ in Part (b) has the value $-l(l+1)$.

4. Let l be any non-negative integer or half-integer. Construct a vector space V by decreeing that vectors $\{v_0, v_1, \ldots, v_{2l}\}$ form a basis for V. Define operators L^+, L^-, and L_3 on V by the expressions in (17.6). Show that these operators satisfy the commutation relations (17.8), (17.9), and (17.10).

   Hint: In the case of L^-, treat the vector v_{2l} separately from the other basis vectors. In the case of the L^+, treat the vector v_0 separately from the other basis vectors.

5. Let (π, V) be an irreducible representation of $\mathsf{so}(3)$ of dimension 2, with basis $\{v_0, v_1\}$ as in (17.6). Consider $V \oplus V$ as a representation of $\mathsf{so}(3)$ as in Sect. 16.8. Let $v = (v_0, v_1)$. Show that the smallest invariant subspace of $V \oplus V$ containing v is $V \oplus V$.

   Note: This shows that $V \oplus V$ has a cyclic vector, even though $V \oplus V$ is not irreducible.

6. Compute explicit bases for the two irreducible invariant subspaces $W_0 \cong V_{3/2}$ and $W_0^\perp \cong V_{1/2}$ of $V_1 \otimes V_{1/2}$. Each basis element for W_0 or $W_0^\perp$ should be expressed as a linear combination of the elements $v_j \otimes e_k$ in the proof of Proposition 17.22.

7. Let V_l, V_m, and V_n be irreducible representation of $\mathsf{so}(3)$ of dimension $2l + 1$, $2m + 1$, and $2n + 1$, respectively. Suppose that Φ and Ψ are nonzero intertwining maps of V_l into $V_m \otimes V_n$. Show that $\Phi = c\Psi$ for some $c \in \mathbb{C}$.

Hint: Use Proposition 17.23 and Schur's lemma.

Note: This result is closely related to the Wigner–Eckart theorem for "irreducible tensor operators."

18

Radial Potentials and the Hydrogen Atom

18.1 Radial Potentials

If V is any radial function on $\mathbb{R}^3$, let $\hat{H} = -(\hbar^2/(2m))\Delta + V$ be the corresponding Hamiltonian operator, acting on $L^2(\mathbb{R}^3)$. We will look for solutions to the time-independent Schrödinger equation $\hat{H}\psi = E\psi$ of the form $\psi(\mathbf{x}) = p(\mathbf{x})f(|\mathbf{x}|)$, where f is a smooth function on $(0, \infty)$ and p is a harmonic polynomial on $\mathbb{R}^3$ that is homogeneous of degree l.

Proposition 18.1 *Let p be a harmonic polynomial on $\mathbb{R}^3$ that is homogeneous of degree l and let f be a smooth function on $(0, \infty)$. Let ψ be the function on $\mathbb{R}^3 \backslash \{0\}$ given by*

$$\psi(\mathbf{x}) = p(\mathbf{x})f(|\mathbf{x}|). \tag{18.1}$$

Then on $\mathbb{R}^3 \backslash \{0\}$ we have

$$\Delta\psi(\mathbf{x}) = p(\mathbf{x})\left[\frac{d^2f}{dr^2} + \frac{2(l+1)}{r}\frac{df}{dr}\right].$$

Proof. We begin with the case $l = 0$, so that p is a constant—which we take to be 1—and ψ is just the radial function $f(|\mathbf{x}|)$. Then

$$\frac{\partial}{\partial x_j}f(|\mathbf{x}|) = \frac{df}{dr}\frac{d}{dx_j}\sqrt{x_1^2 + x_2^2 + x_3^2}$$

$$= \frac{df}{dr}\frac{x_j}{|\mathbf{x}|}$$

B.C. Hall, *Quantum Theory for Mathematicians*, Graduate Texts in Mathematics 267, DOI 10.1007/978-1-4614-7116-5_18, © Springer Science+Business Media New York 2013

and so

$$\sum_{j=1}^{3} \frac{\partial^2}{\partial x_j^2} f(|\mathbf{x}|) = \sum_{j=1}^{3} \left[\frac{d^2 f}{dr^2} \frac{x_j^2}{|\mathbf{x}|^2} + \frac{df}{dr} \left(\frac{1}{|\mathbf{x}|} - \frac{x_j^2}{|\mathbf{x}|^3} \right) \right]$$
$$= \frac{d^2 f}{dr^2} + \frac{2}{r} \frac{df}{dr}.$$

For the general case, the product rule for the Laplacian gives

$$\Delta \psi = (\Delta p) f(|\mathbf{x}|) + 2\nabla p \cdot \nabla f(|\mathbf{x}|) + p \Delta f(|\mathbf{x}|).$$

Now, $\Delta p = 0$ by assumption. Furthermore, since $f(|\mathbf{x}|)$ is radial, its gradient points in the radial direction. Thus, only the radial component of ∇p is relevant. Moreover, on each ray through the origin, p behaves like a constant times r^l. Thus, the r-derivative of p is $(l/r)p$, giving

$$\Delta \psi = \frac{2l}{r} p \frac{df}{dr} + p \frac{d^2 f}{dr^2} + \frac{2}{r} p \frac{df}{dr},$$

which simplifies to the desired expression. $\blacksquare$

Although the decomposition of functions in Definition 17.18 is for many purposes the most convenient one, it is not quite the customary way of turning spherical harmonics into functions on $\mathbb{R}^3$. Conventionally, one works in polar coordinates and considers functions of the form

$$\psi(r, \theta, \phi) = p(\theta, \phi) g(r),$$

where p is the restriction to S^2 of an element of V_l. We can express this decomposition in rectangular coordinates as

$$\psi(\mathbf{x}) = p\left(\frac{\mathbf{x}}{|\mathbf{x}|} \right) g(|\mathbf{x}|) = \frac{p(\mathbf{x})}{|\mathbf{x}|^l} g(|\mathbf{x}|).$$

We can then obtain a more customary form of Proposition 18.1 as follows.

Proposition 18.2 *Suppose $p \in V_l$ and f is a smooth function on $(0, \infty)$, and let ψ by the function on $\mathbb{R}^3 \backslash \{0\}$ given by*

$$\psi(\mathbf{x}) = p\left(\frac{\mathbf{x}}{|\mathbf{x}|} \right) g(|\mathbf{x}|).$$

Then

$$(\Delta \psi)(r\mathbf{x}) = p(\mathbf{x}) \left[\frac{d^2 g}{dr^2} + \frac{2}{r} \frac{dg}{dr} - \frac{l(l+1)}{r^2} g(r) \right] \qquad (18.2)$$

for all $\mathbf{x} \in S^2$ and $r \in (0, \infty)$.

Proof. Since p is homogeneous of degree l,

$$p\left(\frac{\mathbf{x}}{|\mathbf{x}|}\right) = \frac{p(\mathbf{x})}{|\mathbf{x}|^l}.$$

Thus,

$$\psi(\mathbf{x}) = p(\mathbf{x})\left(\frac{f(|\mathbf{x}|)}{|\mathbf{x}|^l}\right).$$

Applying Proposition 18.1 gives

$$\Delta\psi(\mathbf{x}) = p(\mathbf{x})\left[\frac{d^2}{dr^2} + \frac{2(l+1)}{r}\frac{d}{dr}\right]\left(\frac{f(r)}{r^l}\right).$$

From here it is straightforward but unilluminating calculation to verify the formula in the proposition. ∎

Still another way to write functions on $\mathbb{R}^3$ is in the form

$$\psi(\mathbf{x}) = \frac{1}{|\mathbf{x}|}p\left(\frac{\mathbf{x}}{|\mathbf{x}|}\right)h(|\mathbf{x}|), \qquad (18.3)$$

so that $h(r) = rg(r)$. If we replace $g(r)$ by $h(r)/r$ in (18.2), we obtain, after a short calculation,

$$(\Delta\psi)(r\mathbf{x}) = \frac{1}{|\mathbf{x}|}p(\mathbf{x})\left[\frac{d^2h}{dr^2} - \frac{l(l+1)}{r^2}h(r)\right], \qquad \mathbf{x} \in S^2. \qquad (18.4)$$

Writing wave functions in the form (18.3) is convenient because we then have, for any radial potential,

$$-\frac{\hbar^2}{2m}\Delta\psi + V(|\mathbf{x}|)\psi = \frac{1}{|\mathbf{x}|}p(\mathbf{x})\left[-\frac{\hbar^2}{2m}\frac{d^2h}{dr^2} + V_{\text{eff}}(r)h(r)\right], \qquad (18.5)$$

where V_{eff} is the *effective potential* given by

$$V_{\text{eff}}(r) = V(r) + \frac{\hbar^2 l(l+1)}{2mr^2}. \qquad (18.6)$$

Note that the quantity in square brackets in (18.5) is just an ordinary one-dimensional Schrödinger operator, since the first derivative term in (18.2) has been eliminated. Despite the naturalness of the form (18.3), it is the form (18.1) that is ultimately most convenient for finding the bound states of the hydrogen atom Hamiltonian.

Now, as the discussion following Proposition 9.34 illustrates, even if ψ is square-integrable over $\mathbb{R}^3\backslash\{0\}$ and $\Delta\psi$ is square-integrable over $\mathbb{R}^3\backslash\{0\}$, ψ may not be in the domain of the Laplacian, since the distributional Laplacian of ψ may contain a term that is supported at the origin. In the case of hydrogen atom, however, we will consider functions ψ of the form (18.1) where f and df/dr are bounded near the origin and have exponential decay near infinity. Proposition 9.35 then tells us that ψ is in the domain of Δ.

18.2   The Hydrogen Atom: Preliminaries

A hydrogen atom is formed out of a single electron that is "bound" to a proton by means of the electromagnetic attraction between the oppositely charged particles. The study of the hydrogen atom is a very important test case in quantum mechanics, and the ability of the Schrödinger equation to explain the observed energy levels of hydrogen was a crucial early success of the theory.

A proton is approximately 1,800 times as massive as an electron. Thus, to first approximation, we may think of the location of the proton as being fixed, with the electron "orbiting" around this location. A more careful analysis considers both the proton and the electron as orbiting around their center of mass. The Hamiltonian for the relative position of the two particles is precisely that of a particle orbiting around a fixed center, *except* that the mass of the electron is replaced by the reduced mass μ of the electron–proton system. (See Exercise 1.) Here, as in Proposition 2.16 in the classical case,

$$\mu = \frac{m_e m_p}{m_e + m_p},$$

where m_e and m_p are the masses of the proton and electron, respectively. Since $m_p \gg m_e$, the reduced mass is nearly the same as the mass of the electron.

After separating out the motion of the center of mass, we are left with the following Hamiltonian for the relative position of the electron:

$$\hat{H} = -\frac{\hbar^2}{2\mu}\Delta - \frac{Q^2}{|\mathbf{x}|}, \tag{18.7}$$

where Q is the charge of the electron. (We use a system of units, such as "electrostatic" or "Gaussian" units, in which the Coulomb constant is equal to 1.) It follows from Theorem 9.38 that $\hat{H}$ is self-adjoint on $\mathrm{Dom}(\Delta)$ and that $\hat{H}$ is bounded below.

Note that the classical Hamiltonian $H(\mathbf{x}, \mathbf{p})$ for a hydrogen atom is *not* bounded below. After all, we can simply take $\mathbf{p} = 0$ and take $\mathbf{x}$ very close to the origin. This unboundedness would cause strange behavior for a hypothetical classical hydrogen atom. After all, modeling a hydrogen atom using the $1/r$ potential is only an approximation. We are using an electrostatic formula for the force, the correct one when the positions of the particles are held fixed, in a dynamical situation. A more realistic model of hydrogen takes into account radiation, that is, the interaction of the charged electron with the electromagnetic fields. Classically, a negatively charge particle orbiting a positively charged nucleus would radiate, thus giving up energy to the electromagnetic fields. The classical particle would spiral rapidly toward the origin, with the particle's energy going to $-\infty$ and the energy of the electromagnetic field going to $+\infty$. Thus, if hydrogen were made up of *classical* charged particles, the electron would go into a "death spiral" and emit a giant burst of electromagnetic radiation.

Fortunately for us, this is not how real particles behave! In actuality, the electron is a quantum particle. A quantum electron "orbiting" a proton can still give up energy to the electromagnetic field. The Hamiltonian for the *quantum* hydrogen atom, however, is bounded below, as a consequence of Theorem 9.38. Thus, the electron can only drop to its ground state (the state of lowest energy), at which point it becomes stable.

18.3   The Bound States of the Hydrogen Atom

Our goal in this section is to find the eigenvectors for the Hamiltonian $\hat{H}$ in (18.7) with negative eigenvalues. Such eigenvectors constitute "bound states," that is, states in which the electron is bound to the proton. For each negative number E, we look at the eigenspace V_E for $\hat{H}$ with eigenvalue E, that is, the space of all $\psi \in \mathrm{Dom}(\hat{H})$ satisfying $\hat{H}\psi = E\psi$. Since $\hat{H}$ is self-adjoint and, therefore, closed, this eigenspace will be a closed subspace of $L^2(\mathbb{R}^3)$. Since, also, $\hat{H}$ commutes with rotations, V_E will be invariant under the usual action (Definition 17.1) of $\mathrm{SO}(3)$ on $L^2(\mathbb{R}^3)$. Thus, by the discussion at the end of Sect. 17.7, V_E decomposes as a direct sum of *finite-dimensional*, irreducible $\mathrm{SO}(3)$-invariant subspaces.

We now look for such subspaces of V_E. In the following theorem, we assume that the radial part of the wave function (the function f in the notation $V_{l,f}$ in Definition 17.18) has a certain very special form. After analyzing this case, we argue that we have found in this way all of the eigenvectors for $\hat{H}$ with negative eigenvalues.

Theorem 18.3 *For each positive integer n, let*

$$E_n = -\frac{\mu Q^4}{2\hbar^2} \frac{1}{n^2} \tag{18.8}$$

where Q is the charge of the electron and μ is the reduced mass of the electron–proton system, and let

$$\rho_n(\mathbf{x}) = \frac{\sqrt{8\mu |E_n|}}{\hbar} |\mathbf{x}|.$$

Then for each $l = 0, 1, \ldots, n-1$, there exists a polynomial $L_{n,l}$ such that for each homogeneous harmonic polynomial q of degree l, the function

$$\psi(\mathbf{x}) = q(\mathbf{x})e^{-\rho_n(\mathbf{x})/2}L_{n,l}(\rho_n(\mathbf{x})) \tag{18.9}$$

satisfies

$$\hat{H}\psi = E_n\psi.$$

It follows from Proposition 9.35 that the functions ψ in (18.9) belong to $\text{Dom}(\Delta)$ and thus, by Theorem 9.38, to $\text{Dom}(\hat{H})$. The polynomials $L_{n,l}$ are the Laguerre polynomials. The coefficient of $-1/n^2$ in the formula (18.8) for E_n is the Rydberg constant (compare Sect. 1.2.1).

Let us see how to connect Theorem 18.3 to the usual expression for the hydrogen atom eigenvectors in the physics literature. In the first place, physicists choose a certain basis $q_{l,m}$ for the space of harmonic polynomials, which is—up to normalization constants—the basis in Theorem 17.4. In the second place, physicists write the solutions in spherical coordinates. When changing to spherical coordinates, we should keep in mind that $q_{l,m}$ is homogeneous of degree l and that $\rho_n(\mathbf{x})$ is just a constant multiple of the distance from the origin. We obtain, then, the following expression:

$$\psi_{n,l,m}(r,\theta,\phi) = Y_{l,m}(\theta,\phi)\rho_n^l e^{-\rho_n/2} L_{n,l}(\rho_n), \qquad (18.10)$$

where $Y_{l,m}(\theta,\phi)$ is the restriction to the unit sphere of $p_{l,m}$.

Proof. If E is a negative real number, we look for solutions to $\hat{H}\psi = E\psi$ of the form $q(\mathbf{x})f(|\mathbf{x}|)$, where $q \in V_l$. Provided that $f(r)$ and $f'(r)$ are bounded near the origin, Proposition 9.35 allows us to compute $\Delta\psi$ on $\mathbb{R}^3\backslash\{0\}$ without worrying about whether ψ is differentiable at the origin. Using Proposition 18.1, the equation for f is

$$-\frac{\hbar^2}{2\mu}\left[\frac{d^2f}{dr^2} + \frac{2(l+1)}{r}\frac{df}{dr}\right] - \frac{Q^2}{r}f(r) = Ef(r). \qquad (18.11)$$

For large r, where the two terms that involve a factor of $1/r$ become negligible, and so

$$-\frac{\hbar^2}{2\mu}\frac{d^2f}{dr^2} \approx Ef. \qquad (18.12)$$

Recalling that E is negative, (18.12) tells us that near infinity, f should behave like a combination of a growing and a decaying exponential. Since we want square-integrable solutions, we require that only the exponentially decaying term be present.

We therefore postulate a solution of the form

$$f(r) = \exp\left\{-\frac{\sqrt{2\mu|E|}}{\hbar}r\right\}g(r), \qquad (18.13)$$

for some function g. If we plug (18.13) into (18.11) for f, there are canceling terms equal to $Eg(r)$ on each side, leaving

$$-\frac{\hbar^2}{2\mu}\left[\frac{d^2g}{dr^2} - 2\frac{\sqrt{2\mu|E|}}{\hbar}\frac{dg}{dr} + \frac{2(l+1)}{r}\frac{dg}{dr} - \frac{2(l+1)}{r}\frac{\sqrt{2\mu|E|}}{\hbar}g(r)\right]$$

$$= \frac{Q^2}{r}g(r).$$

We now introduce the new variable $\rho = (\sqrt{8\mu |E|}/\hbar)r$. After making this change of variable, we find that each term in square brackets obtains a factor of $8\mu |E| / \hbar^2$, so that our equation becomes

$$-\frac{\hbar^2}{2\mu}\frac{8\mu |E|}{\hbar^2}\left[\frac{d^2g}{d\rho^2} - \frac{dg}{d\rho} + \frac{2(l+1)}{\rho}\frac{dg}{d\rho} - \frac{(l+1)}{\rho}g(\rho)\right] = \frac{2\sqrt{2\mu |E|}}{\hbar}\frac{Q^2}{\rho}g(\rho).$$

Multiplying through by ρ and simplifying yields the equation.

$$\rho\frac{d^2g}{d\rho^2} - \rho\frac{dg}{d\rho} + 2(l+1)\frac{dg}{d\rho} + \left[\frac{Q^2\sqrt{\mu}}{\hbar\sqrt{2|E|}} - (l+1)\right]g(\rho) = 0. \qquad (18.14)$$

If we postulate for g a power series $\sum_{k=0}^{\infty} a_k\rho^k$, we obtain the following recurrence relations for the coefficients:

$$a_{k+1} = a_k\frac{[k + l + 1 - \lambda]}{k[(k+1) + 2(l+1)]} \qquad (18.15)$$

where

$$\lambda = \frac{Q^2\sqrt{\mu}}{\hbar\sqrt{2|E|}}.$$

The series for g will terminate, yielding a polynomial solution to (18.14), provided that λ is an integer n with $n \geq l + 1$. We can then solve for the energy in terms of n as follows:

$$|E| = \frac{\mu Q^4}{2n^2\hbar^2}.$$

Recalling that E is negative, we have obtained the desired form for the energy levels. Furthermore, the condition $n \geq l+1$ is the same as $l \leq n-1$. Finally, if we plug in the formula for ρ in terms of r and the formula for f in terms of g, we obtain the form of the solution stated in the theorem. ∎

It is important to emphasize that the functions in Theorem 18.3 *do not* span the entire Hilbert space $L^2(\mathbb{R}^3)$. After all, these functions are all eigenvectors for $\hat{H}$ with *negative* eigenvalues. If these vectors spanned $L^2(\mathbb{R}^3)$, then the expectation value of the energy would always be negative. But it is easy to produce functions ψ in the domain of $\hat{H}$ for which $\langle\psi, \hat{H}\psi\rangle > 0$. Simply take ψ to be a Gaussian wave packet with mean position far from the origin and with very large mean momentum. Then $\langle\psi, V\psi\rangle$ will be close to zero but $\langle\psi, P^2\psi\rangle$ will be large and positive. Nevertheless, it can be shown that the functions in Theorem 18.3 span the negative energy subspace of $L^2(\mathbb{R}^3)$. It is possible to analyze also the positive part of the spectrum of $\hat{H}$, but the spectrum above zero is purely continuous and represents a hydrogen atom that has ionized, that is, in which the electron has escaped from the proton.

Theorem 18.4 *As n varies over all positive integers, l varies from 0 to $n - 1$, and g varies over all homogeneous harmonic polynomials of degree l, the eigenvectors in Theorem 18.3 span the negative-energy subspace of $L^2(\mathbb{R}^3)$, that is, the range of the projection $\mu^{\hat{H}}((-\infty, 0))$, where $\mu^{\hat{H}}$ is the projection-valued measure associated to $\hat{H}$ by the spectral theorem.*

Proof. The proof requires results from spectral theory that go beyond the machinery that we have developed in Chaps. 9 and 10, and which we cannot reproduce in full here. Specifically, we make use of Theorem V.5.7 of [27], which tells us that the negative-energy portion of the spectrum of $\hat{H}$ is discrete, consisting of eigenvalues of finite multiplicity accumulating only at zero.

We indicate briefly why the above result holds. If A and B are unbounded self-adjoint operators, let us say that B is a *relatively compact* perturbation of A if $A(B - \lambda I)^{-1}$ is a compact operator for every λ in the resolvent set of B. According to Lemma V.5.8 of [27], the potential energy operator for the hydrogen atom is a relatively compact perturbation of the kinetic energy operator. This is a strengthening of what we showed in the proof of Theorem 9.38, namely that the potential energy operator is relatively bounded with respect to the kinetic energy operator, with relative bound less than 1. The proof of relative compactness relies on the fact that the potential for the hydrogen atom goes to zero at infinity.

Meanwhile, let us say that λ belongs to the *essential spectrum* of an unbounded self-adjoint operator A if either λ is a nonisolated point in $\sigma(A)$ or λ is an eigenvalue for A with infinite multiplicity. According to Theorem IV.5.35 of [27], a relatively compact perturbation of a self-adjoint operator does not change the essential spectrum. Thus, the essential spectrum of $\hat{H}$ is equal to the essential spectrum of the kinetic energy operator, which is certainly contained in $[0, \infty)$, since the kinetic energy operator is non-negative. It follows that any point in the negative-energy part of the spectrum of $\hat{H}$ must be an isolated point in $\sigma(\hat{H})$ and an eigenvalue of finite multiplicity.

In light of the preceding result, there is no continuous spectrum for $\hat{H}$ below zero, and we need only look for *square-integrable* eigenvectors. Since, also, each eigenspace for $\hat{H}$ with eigenvalue $E < 0$ is finite dimensional, it will decompose as a direct sum of irreducible, SO(3)-invariant subspaces. Such subspaces, according to Proposition 17.19, are always of the form $V_{l,f}$ for some l and f, where $V_{l,f}$ is as in Definition 17.18. Thus, we look for functions ψ of the form $\psi(\mathbf{x}) = p(\mathbf{x})f(|\mathbf{x}|)$ such that $\hat{H}\psi = E\psi$ for some $E < 0$.

Now, if a function of the form $p(\mathbf{x})f(|\mathbf{x}|)$ is to be an eigenfunction of the Hamiltonian, f must satisfy the differential equation (18.11). By elementary results from the theory of linear ordinary differential equations, this equation has precisely two linearly independent solutions, for any value of E. Both solutions can be constructed by postulating a solution of the form (18.13), introducing the new variable ρ, and then using a power series expansion for $g(\rho)$ (Exercise 9). One of the solutions for $g(\rho)$ will have a power series starting with $\rho^{-(2l+1)}$, in which case $\psi(\mathbf{x})$ will blow up like $1/|\mathbf{x}|^{(l+1)}$ near the origin; such a function is not in the domain of the Hamiltonian (Exercise 14 in Chap. 9). The other solution for $g(\rho)$ will start with ρ^0 and may be obtained by using the form (18.13), changing from the variable r to the variable ρ, and then using the recurrence relation (18.15) to define the coefficients of a power series. If the resulting series does not terminate, it is not hard to see that the terms will behave for large k like the series for e^ρ. Since the function f is equal to $e^{-\rho/2}g(\rho)$, this function will grow like $e^{\rho/2}$ near infinity, which means that ψ will not be in $L^2(\mathbb{R}^3)$. Thus, to get a square-integrable solution, the series for $g(\rho)$ must terminate, in which case ψ is one of the functions in Theorem 18.3. ∎

Corollary 18.5 *Each eigenvalue E_n, as given in Theorem 18.3, has multiplicity n^2.*

Proof. According to Theorem 18.4, the eigenvectors in Theorem 18.3 constitute all of the eigenvectors for $\hat{H}$ with eigenvalue E_n. The number of independent eigenvectors with eigenvalue E_n is thus the sum of the dimensions of the spaces V_l of spherical harmonics, with $l = 0, 1, \ldots, n-1$. This number is, by Theorem 17.12,

$$\sum_{l=0}^{n-1}(2l+1) = n^2,$$

as claimed. ∎

18.4 The Runge–Lenz Vector in the Quantum Kepler Problem

In Sect. 2.6, we showed that the classical Kepler problem can be solved almost completely by making use of the Runge–Lenz vector, which is a conserved quantity. The quantum version of the Runge–Lenz vector commutes with the Hamiltonian and can elucidate a number of special properties of the quantum Kepler problem, which we typically think of as describing a hydrogen atom. In particular, the Runge–Lenz vector will help to explain (1) the simple form $-R/n^2$ of the negative energies of the hydrogen atom and (2) the apparent coincidence by which energy of the states in (18.9) is independent of l for a given n. Note that the rotational symmetry of the problem explains why the energy of the states in (18.9) is independent of the choice of the harmonic polynomial q. Nevertheless, rotational symmetry cannot explain why states for different values of l—and thus different radial dependence in the wave function—have the same energy. This apparent coincidence will be explained by an additional symmetry of the problem, that is expressible in terms of the Runge–Lenz vector. See also Sect. 7 of [17] for a somewhat different (but related) explanation for the structure of the eigenvalues of the hydrogen atom and their multiplicities.

There are several computations involving the Runge–Lenz vector that, while elementary, are laborious. Those computations are deferred to Sect. 18.6.

18.4.1 Some Notation

To keep the notation as simple as possible, we will adopt in this section Einstein's *summation convention*, which states that repeated indices are always summed on, even if there is no summation sign written. In this section, the sum will always range from 1 to 3. Using this convention, we write, say, the dot product of two vectors $\mathbf{u}, \mathbf{v}$ in $\mathbb{R}^3$ as $\mathbf{u} \cdot \mathbf{v} = u_j v_j$, where the summation convention frees us from having to write out explicitly the sum over j.

We will make frequent use of the *totally antisymmetric symbol* ε_{jkl}, where j, k, and l range from 1 to 3, defined as follows,

Definition 18.6 *For $j, k, l \in \{1, 2, 3\}$, define ε_{jkl} by the formula*

$$\varepsilon_{jkl} = \begin{cases} 1 & \text{if } (j, k, l) \text{ is an even permutation of } (1, 2, 3) \\ -1 & \text{if } (j, k, l) \text{ is an odd permutation of } (1, 2, 3) \\ 0 & \text{if any two of } j, k, l \text{ are equal} \end{cases} \quad .$$

Thus, for example, $\varepsilon_{321} = -1$ and $\varepsilon_{212} = 0$. The commutation relations for the basis $\{F_1, F_2, F_3\}$ for $\mathsf{so}(3)$ may be written (using the summation convention!) as

$$[F_j, F_k] = \varepsilon_{jkl} F_l. \tag{18.16}$$

For instance, if we take $j = 1$ and $k = 2$ in (18.16), then the sum on l gives a nonzero value only when $l = 3$, and we recover the relation $[F_1, F_2] = F_3$.

18.4.2 The Classical Runge–Lenz Vector, Revisited

We have already introduced, in Sect. 2.6, the Runge–Lenz vector $\mathbf{A}$ in the classical mechanics of a particle moving in a $1/r$ potential. We require a few more properties of $\mathbf{A}$ before turning to the quantum version. We consider a classical particle in $\mathbb{R}^3$ with Hamiltonian given by

$$H(\mathbf{x}, \mathbf{p}) = \frac{|\mathbf{p}|^2}{2\mu} - \frac{Q^2}{|\mathbf{x}|}. \tag{18.17}$$

This is just the Hamiltonian for the classical Kepler problem, except that we replace the mass m of the planet by the reduced mass μ of the electron–proton system, and we replace the constant $k := mMG$ by Q^2.

For the Hamiltonian in (18.17), the Runge–Lenz vector is given by the formula

$$\mathbf{A}(\mathbf{x},\mathbf{p}) = \frac{1}{\mu Q^2}\mathbf{p}\times\mathbf{J} - \frac{\mathbf{x}}{|\mathbf{x}|},$$

where $\mathbf{J} := \mathbf{x}\times\mathbf{p}$ is the angular momentum. By Proposition 2.34, the Runge–Lenz vector is a conserved quantity for the classical Kepler problem, in addition to H and $\mathbf{J}$, which are conserved quantities for any radial potential. By results of Sect. 2.6, we have the following relations among these conserved quantities:

$$\mathbf{A}\cdot\mathbf{J} = 0$$

$$|\mathbf{A}|^2 = 1 + \frac{2H}{\mu Q^4}|\mathbf{J}|^2.$$

Lemma 18.7 *The Runge–Lenz vector $\mathbf{A}$ and the Hamiltonian H in (18.17) satisfy the following Poisson bracket relations:*

$$\{A_j, H\} = 0$$

$$\{A_j, A_m\} = -\frac{2}{\mu Q^4}\varepsilon_{jml}J_l H. \qquad (18.18)$$

We have already shown that the Runge–Lenz vector is a conserved quantity (Proposition 2.34), which is equivalent (Proposition 2.25) to saying that the Poisson bracket of A_j with H is zero, as claimed. The proof of (18.18) is deferred to Sect. 18.6. We now introduce certain combinations of the Runge–Lenz vector, the angular momentum, and the Hamiltonian that form a Lie algebra under the Poisson bracket. In the construction of these functions, we need to take a square root of the Hamiltonian, which necessitates separating the positive-energy and negative-energy parts of the phase space. Our interest is primarily in the negative-energy case.

Definition 18.8 *Let U^- denote the negative-energy part of the classical phase space,*

$$U^- = \left\{(\mathbf{x},\mathbf{p})\in\mathbb{R}^6\,\middle|\, H(\mathbf{x},\mathbf{p}) < 0\right\}.$$

Consider on U^- the normalized Runge–Lenz vector $\mathbf{B}$ given by

$$\mathbf{B} = \sqrt{\frac{\mu Q^4}{2\,|H|}}\,\mathbf{A}.$$

Define also vector-valued functions $\mathbf{I}$ and $\mathbf{K}$ on U^- by

$$\mathbf{I} = \frac{\mathbf{J}+\mathbf{B}}{2};\quad \mathbf{K} = \frac{\mathbf{J}-\mathbf{B}}{2}.$$

Theorem 18.9 *The functions* **I** *and* **K** *Poisson-commute with the Hamiltonian and satisfy the following Poisson-bracket relations on the negative-energy set* U^-:

$$\{I_j, I_k\} = \varepsilon_{jkl} I_l$$
$$\{K_j, K_k\} = \varepsilon_{jkl} K_l$$
$$\{I_j, K_k\} = 0.$$

The functions **I** *and* **K** *also satisfy the following algebraic relations:*

$$|\mathbf{I}|^2 = |\mathbf{K}|^2 = \frac{\mu Q^4}{8|H|}.$$

In Theorem 18.9, we use the summation convention introduced in the previous subsection. The proof of this theorem is elementary but rather laborious, and is deferred to Sect. 18.6.

The span of the functions I_1, I_2, I_3 and K_1, K_2, K_3 on U^-, which is the same as the span of the functions B_1, B_2, B_3 and J_1, J_2, J_3, forms a 6-dimensional Lie algebra under the Poisson bracket. Comparing the Poisson-bracket relations among the I's and among the K's to the relations among the basis elements F_1, F_2, F_3 for $\mathsf{so}(3)$, we see that the span of the I's and the span of the K's are both isomorphic to $\mathsf{so}(3)$ [or, if you prefer, to $\mathsf{su}(2)$]. Since also each I_j commutes with each K_k, the 6-dimensional Lie algebra spanned by the I's and the K's is isomorphic to $\mathsf{so}(3) \oplus \mathsf{so}(3)$. Meanwhile, as demonstrated in Exercise 4, $\mathsf{so}(3) \oplus \mathsf{so}(3)$ is isomorphic to the Lie algebra $\mathsf{so}(4)$. Since all the I's and K's Poisson-commute with the Hamiltonian, we say that the Kepler problem has $\mathsf{so}(4)$ symmetry. This is in contrast to the dynamics of a particle moving in $\mathbb{R}^3$ in the force generated by a typical radial potential, which has only $\mathsf{so}(3)$ symmetry.

To be more precise, "$\mathsf{so}(4)$ symmetry" prevails only on the negative-energy subset U^- of the classical phase space. On the positive-energy subset U^+, the span of the functions B_1, B_2, B_3 and J_1, J_2, J_3 again forms a 6-dimensional Lie algebra. This Lie algebra, however, is *not* isomorphic to $\mathsf{so}(4)$, but rather to $\mathsf{so}(3,1)$, where $\mathsf{so}(3,1)$ is the Lie algebra of the group of 4×4 matrices that preserve the quadratic form $x_1^2 + x_2^2 + x_3^2 - x_4^2$. The reason the formulas on U^+ are different from those on U^- is that calculations of the relevant Poisson brackets involves the function $H/|H|$, which has the value 1 on U^+ and the value -1 on U^-. (The factor of H comes from Lemma 18.7 and the factor of $|H|$ from the factor of $\sqrt{|H|}$ in the definition of **B**.)

18.4.3   The Quantum Runge–Lenz Vector

We now introduce the quantum counterpart $\hat{\mathbf{A}}$ of the classical Runge–Lenz vector **A**. The quantum Runge–Lenz satisfies most of the same properties as the classical version, with a few small but crucial "quantum corrections."

Definition 18.10 *Define the* **quantum Runge–Lenz vector** *by*

$$\hat{\mathbf{A}} = \frac{1}{\mu Q^2} \frac{1}{2} \left(\mathbf{P} \times \hat{\mathbf{J}} - \hat{\mathbf{J}} \times \mathbf{P} \right) - \frac{\mathbf{X}}{|\mathbf{X}|}.$$

Note that in the quantum case, $-\hat{\mathbf{J}} \times \mathbf{P}$ is not the same as $\mathbf{P} \times \hat{\mathbf{J}}$, because of the noncommutativity of the factors. The particular combination of $\mathbf{P} \times \hat{\mathbf{J}}$ and $\hat{\mathbf{J}} \times \mathbf{P}$ in Definition 18.10 is used because it is yields a self-adjoint operator. The Runge–Lenz vector can also be computed as

$$\hat{\mathbf{A}} = \frac{1}{\mu Q^2} \left(\mathbf{P} \times \hat{\mathbf{J}} - i\hbar \mathbf{P} \right) - \frac{\mathbf{X}}{|\mathbf{X}|}, \tag{18.19}$$

as will be verified in Sect. 18.6.

In the interests of keeping the exposition manageable, we will not concern ourselves in what follows with determining the precise domains on which various identities hold.

Proposition 18.11 *The quantum Runge–Lenz vector* $\hat{\mathbf{A}}$ *satisfies the following relations:*

$$\hat{\mathbf{A}} \cdot \hat{\mathbf{J}} = \hat{\mathbf{J}} \cdot \hat{\mathbf{A}} = 0$$

$$\hat{\mathbf{A}} \cdot \hat{\mathbf{A}} = 1 + \frac{2\hat{H}}{\mu Q^4} \left(\hat{\mathbf{J}} \cdot \hat{\mathbf{J}} + \hbar^2 \right). \tag{18.20}$$

Note that there is a "quantum correction" in (18.20); the factor of $\mathbf{J} \cdot \mathbf{J}$ in the classical expression for $\mathbf{A} \cdot \mathbf{A}$ is replaced by $\hat{\mathbf{J}} \cdot \hat{\mathbf{J}} + \hbar^2$. This correction gives rise to a quantum correction in (18.22), which in turn is essential to getting the correct value for the energy eigenvalues in Corollary 18.17. The proof of this result and the other results of this section are deferred to Sect. 18.6.

Lemma 18.12 *The quantum Runge–Lenz vector* $\hat{\mathbf{A}}$ *and the Hamiltonian* $\hat{H}$ *satisfy the following commutation relations:*

$$\frac{1}{i\hbar}[\hat{A}_j, \hat{H}] = 0$$

$$\frac{1}{i\hbar}[\hat{A}_j, \hat{A}_m] = -\frac{2}{\mu Q^4} \varepsilon_{jml} \hat{J}_l \hat{H}. \tag{18.21}$$

Note that since $\hat{H}$ commutes with rotations, it commutes with the angular momentum operators $\hat{J}_l$. Thus, in (18.21), we could just as well write $\hat{H}\hat{J}_l$ in place of $\hat{J}_l\hat{H}$. As in the classical case, if we normalize the components of the Runge–Lenz vector by dividing by the square root of the Hamiltonian, then these operators together with the angular momentum operators form a 6-dimensional Lie algebra.

Definition 18.13 *Let V^- denote the negative-energy subspace of $L^2(\mathbb{R}^3)$, that is, the range of the spectral projection $\mu^{\hat{H}}((-\infty, 0))$. Let $|\hat{H}|$ denote the restriction to V^- of the operator $-\hat{H}$. On V^-, define operators $\hat{\mathbf{B}}$ by*

$$\hat{\mathbf{B}} = \frac{\mu Q^2}{\sqrt{2\mu|\hat{H}|}} \hat{\mathbf{A}}.$$

Define also operators $\hat{\mathbf{I}}$ and $\hat{\mathbf{K}}$, as in the classical case, by

$$\hat{\mathbf{I}} = \frac{\hat{\mathbf{J}} + \hat{\mathbf{B}}}{2}; \quad \hat{\mathbf{K}} = \frac{\hat{\mathbf{J}} - \hat{\mathbf{B}}}{2}.$$

It is possible to define the absolute value of any self-adjoint operator by means of the functional calculus. However, since the restriction of $\hat{H}$ to V^- is, by definition, negative definite, the restriction of $|\hat{H}|$ to V^- coincides with the restriction to V^- of $-\hat{H}$. The operator $1/\sqrt{|\hat{H}|}$ is the operator with a restriction to the energy eigenspace with eigenvalue E_n that is $1/\sqrt{|E_n|}I$. The components of $\hat{\mathbf{B}}$ are unbounded operators, defined on suitable dense subspaces of the Hilbert space V^-.

Theorem 18.14 *The operators $\hat{\mathbf{I}}$ and $\hat{\mathbf{K}}$ commute with the Hamiltonian $\hat{H}$ and satisfy the following commutation relations:*

$$\frac{1}{i\hbar}[\hat{I}_j, \hat{I}_k] = \varepsilon_{jkl}\hat{I}_l$$

$$\frac{1}{i\hbar}[\hat{K}_j, \hat{K}_k] = \varepsilon_{jkl}\hat{K}_l$$

$$\frac{1}{i\hbar}[\hat{I}_j, \hat{K}_k] = 0.$$

These operators also satisfy the following algebraic relations:

$$\hat{\mathbf{I}} \cdot \hat{\mathbf{I}} = \hat{\mathbf{K}} \cdot \hat{\mathbf{K}} = \frac{\mu Q^4}{8|\hat{H}|} - \frac{\hbar^2}{4}. \tag{18.22}$$

18.4.4 Representations of so(4)

In light of the commutation relations in Theorem 18.14, we can define a representation π of the Lie algebra $so(4) \cong so(3) \oplus so(3)$ on the negative-energy subspace V^- as follows:

$$\pi(F_j, 0) = \frac{1}{i\hbar}\hat{I}_j; \quad \pi(0, F_j) = \frac{1}{i\hbar}\hat{K}_j. \tag{18.23}$$

It is therefore desirable to classify the irreducible finite-dimensional representations of $so(3) \oplus so(3)$, which we do in the following proposition.

Proposition 18.15 *Suppose V_k and V_l are irreducible representations of* so(3) *of dimensions $2k+1$ and $2l+1$, respectively. Then $V_k \otimes V_l$ is irreducible when viewed as a representation of* so(3) $\oplus$ so(3) *as in Remark 16.49. Furthermore, every irreducible finite-dimensional representation of* so(3)$\oplus$so(3) *is isomorphic to $V_k \otimes V_l$ for a unique ordered pair (k,l).*

For any representation $V_k \otimes V_l$ of so(3) $\oplus$ so(3)*, define Casimir operators C_1 and C_2 by the formula*

$$C_1 = \sum_{j=1}^{3} \pi_k(F_j)^2 \otimes I; \quad C_2 = \sum_{j=1}^{3} I \otimes \pi_l(F_j)^2.$$

Then we have

$$C_1 = -k(k+1)I; \quad C_2 = -l(l+1)I.$$

Proof. To classify the irreducible representations of so(3) $\oplus$ so(3), we could appeal to the general theory of representations of direct sums of Lie algebras. It is not hard, however, to give a direct proof using the same sort of reasoning we used in the classifications of irreducible representations of so(3). We will omit the details of this computation. The result on the Casimir operators follows easily from Proposition 17.8. ∎

In any finite-dimensional subspace of V^- that is invariant and irreducible under the action of so(3) $\oplus$ so(3) in (18.23), the Casimir operators are given by $C_1 = -\hat{\mathbf{I}} \cdot \hat{\mathbf{I}}/\hbar^2$ and $C_2 = -\hat{\mathbf{K}} \cdot \hat{\mathbf{K}}/\hbar^2$. Since, by Theorem 18.14, $\hat{\mathbf{I}} \cdot \hat{\mathbf{I}} = \hat{\mathbf{K}} \cdot \hat{\mathbf{K}}$ on V^-, all of the irreducible representations of so(3)$\oplus$so(3) that arise inside V^- will be of the form $V_k \otimes V_k$.

Theorem 18.16 *Let $W^{(n)}$ denote the eigenspace for the Hamiltonian with eigenvalue E_n. Then $W^{(n)}$ is invariant and irreducible under the action of* so(3) $\oplus$ so(3) *in (18.23). More specifically, we have the isomorphism*

$$W^{(n)} \cong V_k \otimes V_k,$$

as representations of so(3) $\oplus$ so(3)*, where $k = (n-1)/2$ and where V_k is the irreducible representation of* so(3) *of dimension $2k + 1 = n$.*

Corollary 18.17 *If n, k, and $W^{(n)}$ are as in Theorem 18.16, then for all $\psi \in W^{(n)}$, we have*

$$\hat{\mathbf{I}} \cdot \hat{\mathbf{I}}\psi = \hat{\mathbf{J}} \cdot \hat{\mathbf{J}}\psi = \hbar^2 k(k+1).$$

Using (18.22), the eigenvalue E_n of $\hat{H}$ on $W^{(n)}$ can be solved for as

$$E_n = -\frac{\mu Q^4}{8\hbar^2 \left(k + \frac{1}{2}\right)^2} = -\frac{\mu Q^2}{2\hbar^2 n^2}.$$

The expression for E_n in Corollary 18.17 is the same as in Theorem 18.3. The remarkable thing about the proof of Theorem 18.17 is that it is purely algebraic, relying only on the commutation relations among the operators $\hat{I}_k$ and $\hat{K}_l$, along with the relationship (18.22) between the Hamiltonian operator $\hat{H}$ and the $\hat{I}_k$'s and $\hat{K}_l$'s.

Proof of Corollary 18.17. It is easily seen that the operators $\hat{\mathbf{I}} \cdot \hat{\mathbf{I}}$ and $\hat{\mathbf{K}} \cdot \hat{\mathbf{K}}$, when restricted to an irreducible subspace for the action of $\mathsf{so}(3) \oplus \mathsf{so}(3)$, are equal to $-\hbar^2 C_1$ and $-\hbar^2 C_2$, where C_1 and C_2 are the Casimir operators appearing in Proposition 18.15. Thus, if $W^{(n)}$ is isomorphic to $V_k \otimes V_k$, with $k = (n-1)/2$, then $\hat{\mathbf{I}} \cdot \hat{\mathbf{I}}$ and $\hat{\mathbf{K}} \cdot \hat{\mathbf{K}}$ will be equal to $\hbar^2 k(k+1)I$, as claimed. On the other hand, $\hat{\mathbf{I}} \cdot \hat{\mathbf{I}}$ and $\hat{\mathbf{K}} \cdot \hat{\mathbf{K}}$ are related to the Hamiltonian $\hat{H}$ by (18.22), from which we can solve for E_n. ∎

Proof of Theorem 18.16. Since each component of $\mathbf{A}$ and $\hat{\mathbf{J}}$ commutes with $\hat{H}$, each component of $\hat{\mathbf{I}}$ and $\hat{\mathbf{K}}$ will also commute with $\hat{H}$. Each eigenspace of $\hat{H}$ is therefore invariant under the action of $\hat{\mathbf{I}}$ and $\hat{\mathbf{K}}$. Since the $\hat{I}$'s and $\hat{K}$'s are self-adjoint and $W^{(n)}$ is finite dimensional, $W^{(n)}$ will decompose as a direct sum of irreducible invariant subspaces. By Proposition 18.15, these irreducible subspaces will be of the form $V_k \otimes V_l$, where V_k and V_l are irreducible representations of $\mathsf{so}(3)$ of dimension $2k + 1$ and $2l + 1$, respectively. But now, the operators $\hat{\mathbf{I}} \cdot \hat{\mathbf{I}}$ and $\hat{\mathbf{K}} \cdot \hat{\mathbf{K}}$, when restricted to one of the irreducible subspaces of $W^{(n)}$, are equal to $-\hbar^2 C_1$ and $-\hbar^2 C_2$, where C_1 and C_2 are the Casimir operators appearing in Proposition 18.15. Since $\hat{\mathbf{I}} \cdot \hat{\mathbf{I}} = \hat{\mathbf{K}} \cdot \hat{\mathbf{K}}$ on all of V^-, the eigenvalues of C_1 and C_2 must be equal on each irreducible subspace of $W^{(n)}$. Thus, we must have $k = l$, meaning that only irreducible subspaces of the form $V_k \otimes V_k$ arise.

Now, under the isomorphism of some irreducible subspace of $W^{(n)}$ with $V_k \otimes V_k$, the operators $\hat{I}_k$ and $\hat{K}_k$ act as $i\hbar F_k \otimes I$ and $i\hbar I \otimes F_k$, respectively, where the F_k's are the usual basis for $\mathsf{so}(3)$. Since $\hat{\mathbf{J}} = \hat{\mathbf{I}} + \hat{\mathbf{K}}$, each $\hat{J}_k$ acts as $i\hbar(F_k \otimes I + I \otimes F_k)$. This means that $V_k \otimes V_k$, under the action of the $\hat{J}_k$'s, can be thought of as a tensor product of two representations of $\mathsf{so}(3)$, viewed as another representation of $\mathsf{so}(3)$ as in Definition 16.48. Viewed this way, $V_k \otimes V_k$ decomposes as in Proposition 17.23 as

$$V_k \otimes V_k \cong V_0 \oplus V_1 \oplus \cdots \oplus V_{2k}. \tag{18.24}$$

On the other hand, we know from Theorem 18.3 that $W^{(n)}$ decomposes under the action of $\mathsf{so}(3)$ as

$$V_0 \oplus V_1 \oplus \cdots \oplus V_{n-1}. \tag{18.25}$$

Thus, the space of the form $V_k \otimes V_k$ must be all of $W^{(n)}$; if there were another term then the trivial representation V_0 would occur more than once in $W^{(n)}$. This being the case, matching the decompositions (18.24) and (18.25) requires that $2k = n - 1$, as claimed in the theorem. ∎

The proof of Theorem 18.16 relies to some extent on the results of Sect. 18.3. Using only algebraic manipulations involving the Runge–Lenz vector, however, we could still argue that the eigenvalues of $\hat{H}$ must be of the form given in Corollary 18.17. We would not, however, know that for every positive integer n, the number E_n is actually an eigenvalue for $\hat{H}$. We would also not know that each eigenspace $W^{(n)}$ is irreducible under the action of so(4); conceivably, based only on the algebra, $W^{(n)}$ could have, say, dimension $2n^2$ instead of n^2.

18.5   The Role of Spin

The spin of the electron is $1/2$. As discussed in Sect. 17.8, this means that the Hilbert space for an electron is $L^2(\mathbb{R}^3)\hat{\otimes}V_{1/2}$, where $V_{1/2}$ is a 2-dimensional vector space that carries an irreducible projective unitary representation of SO(3). Up to now, we have neglected the spin in our calculations. The reason for this omission is simple: to first approximation, the spin plays no role in the calculation. Specifically, in the simplest model of a hydrogen atom with spin, the Hamiltonian is simply $\hat{H} \otimes I$, where $\hat{H}$ is the operator in (18.7), acting on $L^2(\mathbb{R}^3)$. For any $n > 0$, we can obtain a basis of eigenvectors for $\hat{H} \otimes I$ with eigenvalue E_n by taking vectors of the form $\psi_{n,l,m} \otimes e_j$, where the $\psi_{n,l,m}$'s are as in (18.10) and where $\{e_1, e_2\}$ forms a basis for $V_{1/2}$.

Now, from the point of view of rotational symmetry, the basis $\psi_{n,l,m} \otimes e_j$ is not the most natural one. Rather, we should decompose the eigenspaces into irreducible invariant subspaces for the (projective) action of SO(3), where SO(3) acts on *both* $L^2(\mathbb{R}^3)$ and $V_{1/2}$. We have already decomposed the eigenspaces inside $L^2(\mathbb{R}^3)$ into irreducible invariant subspaces, namely the span of $\psi_{n,l,m}$ where n and l are fixed and m varies. Thus, to obtain the irreducible invariant subspaces inside $L^2(\mathbb{R}^3)\hat{\otimes}V_{1/2}$, we use the method of "addition of angular momentum" from Sect. 17.9. According to Proposition 17.22, $V_l \otimes V_{1/2}$ is irreducible if $l = 0$ and isomorphic to $V_{l+1/2} \oplus V_{l-1/2}$ if $l > 0$. Consider, for example, the case $n = 3$, $l = 1$, the so-called "$3p$ states" in traditional chemistry terminology. Since $V_1 \otimes V_{1/2}$ decomposes as $V_{3/2} \oplus V_{1/2}$, when we take spin into account, we obtain a 4-dimensional space and a 2-dimensional space. We can obtain bases for these spaces by tracing through the proof of Proposition 17.22.

The decomposition described in the previous paragraph is essential when considering the "fine structure" of hydrogen. Our model of hydrogen using the Hamiltonian (18.7) is only a first approximation. More realistic models take into account various corrections, including radiative corrections, a finite size for the nucleus, and "spin–orbit coupling," among other things. The notion of spin–orbit coupling adds a term into the Hamiltonian involving the operator $\hat{\mathbf{J}} \cdot \sigma$, where σ_1, σ_2, and σ_3 are the operators describing the action of so(3) on $V_{1/2}$. When this term is included, the Hamiltonian is no longer of the form $A \otimes I$ for some operator A on $L^2(\mathbb{R}^3)$. Thus, we can no longer simply append the spin to the end of the computation, but must take it into account from the beginning.

The various corrections to the Hamiltonian for the hydrogen atom have the effect of reducing the multiplicities of the eigenvalues. Almost any correction we make, for example, will destroy the independence of the eigenvalue on l for a given n, simply because the correction terms in the Hamiltonian will not commute with the quantum Runge–Lenz vector. Nevertheless, all of the corrections that make up the fine structure of hydrogen preserve the rotational symmetry of the problem. Thus, the same irreducible representations of $SO(3)$ that we had in the simple model will appear after the corrections are made. For $n = 2$, $l = 1$, for example, we will still have a 4-dimensional space and 2-dimensional space, but these two spaces will no longer have the same energy.

18.6   Runge–Lenz Calculations

In this section, we fill in many of the computations that we passed over without proof in Sect. 18.4. Although all the calculations are, in principle, elementary, there are a number of nonobvious tricks that help simplify the algebra. We will make frequent use of the concepts of functions that transform like vectors (on the classical side) and of vector operators (on the quantum side), including Propositions 17.25 and 17.27 (Sect. 17.10). In particular, we note that the position $\mathbf{x}$, the momentum $\mathbf{p}$, the angular momentum $\mathbf{j}$, and the Runge–Lenz vector $\mathbf{A}$ all transform like vectors, and that the corresponding quantum quantities are all vector operators. (Compare Exercise 7.) In the "ε" notation of Sect. 18.4.1, Proposition 17.27 takes the form

$$\frac{1}{i\hbar}[C_j, \hat{J}_k] = \frac{1}{i\hbar}[\hat{J}_j, C_k] = \varepsilon_{jkl}C_l. \tag{18.26}$$

In the quantum mechanical calculations, there are a number of "quantum corrections," in which dot products and cross products of vector operators do not behave as they do in the classical case.

Lemma 18.18 *The ε-function in Definition 18.6 satisfies the relations*

$$\varepsilon_{jkl}\varepsilon_{jmn} = \delta_{km}\delta_{ln} - \delta_{kn}\delta_{lm}$$
$$\varepsilon_{jkl}\varepsilon_{jkm} = 2\delta_{lm}.$$

The proof of these results is not difficult and is left to the reader (Exercise 6). The following identities involving the cross product of vector operators will be useful to us.

Lemma 18.19 *If* $\mathbf{C}$, $\mathbf{D}$, *and* $\mathbf{E}$ *are arbitrary vector operators, we have*

$$\mathbf{C} \cdot (\mathbf{D} \times \mathbf{E}) = (\mathbf{C} \times \mathbf{D}) \cdot \mathbf{E} \tag{18.27}$$
$$\mathbf{C} \times \mathbf{D} + \mathbf{D} \times \mathbf{C} = \varepsilon_{jkl}[C_k, D_l] \tag{18.28}$$
$$\mathbf{C} \times \mathbf{C} = \frac{1}{2}\varepsilon_{jkl}[C_k, C_l]. \tag{18.29}$$

In particular, if the different components of **C** *commute, then* **C** × **C** = 0. *Finally,*

$$(\mathbf{C} \times (\mathbf{D} \times \mathbf{E}))_j = C_k D_j E_k - C_k D_k E_j. \tag{18.30}$$

As special cases of these results, we have

$$\hat{\mathbf{J}} \times \mathbf{P} + \mathbf{P} \times \hat{\mathbf{J}} = 2i\hbar\mathbf{P} \tag{18.31}$$

$$\hat{\mathbf{J}} \times \hat{\mathbf{J}} = i\hbar\hat{\mathbf{J}} \tag{18.32}$$

Note that if the entries of **D** and **E** commute, then the right-hand side of (18.30) reduces to the classical expression, $(\mathbf{C} \cdot \mathbf{E})\mathbf{D} - (\mathbf{C} \cdot \mathbf{D})\mathbf{E}$. Using (18.31), we can easily verify the alternative expression (18.19) for the Runge–Lenz vector.

Proof. The right-hand side of (18.27) is computed as $\varepsilon_{jkl} C_k D_l E_j$. If we note that $\varepsilon_{jkl} = \varepsilon_{klj}$ and then relabel the indices, we obtain $\varepsilon_{jkl} C_j D_k E_l$, which is equal to the left-hand side of (18.27). For (18.28), we compute that

$$
\begin{aligned}
(\mathbf{C} \times \mathbf{D} + \mathbf{D} \times \mathbf{C})_j &= \varepsilon_{jkl} C_k D_l + \varepsilon_{jkl} D_k C_l \\
&= \varepsilon_{jkl} C_k D_l + \varepsilon_{jkl} C_l D_k - \varepsilon_{jkl}[C_l, D_k].
\end{aligned} \tag{18.33}
$$

If we note that $\varepsilon_{jkl} = -\varepsilon_{jlk}$ and then relabel the indices k and l, we see that $\varepsilon_{jkl} C_l D_k = -\varepsilon_{jkl} C_k D_l$, so that the first two terms in the second line of (18.33) cancel. The remaining term can be put into the claimed form by relabeling the indices k and l. The identity (18.29) is just the $\mathbf{D} = \mathbf{C}$ case of (18.28). Finally, (18.30) follows easily from Lemma 18.18.

To obtain (18.31) and (18.32), we apply (18.28) and (18.29), respectively. Since both $\hat{\mathbf{J}}$ and $\mathbf{P}$ are vector operators, the desired result follows easily from Lemma 18.18. ∎

We now turn to the proofs of the results of Sect. 18.4. We prove only the quantum versions of the results, since the classical results are extremely similar, except that certain quantum corrections can be ignored.

Proof of Lemma 18.12, First Part. We begin by showing that $\hat{A}_j$ commutes with $\hat{H}$ for each j. Since $\hat{H}$ commutes with $\hat{\mathbf{J}}$, we have

$$[\hat{A}_j, \hat{H}] = \frac{1}{\mu Q^2} \frac{1}{2} \left(\varepsilon_{jkl}[P_k, \hat{H}]\hat{J}_l - \hat{J}_k[P_l, \hat{H}] \right) - \left[\frac{X_j}{|\mathbf{X}|}, \hat{H} \right].$$

Meanwhile, since the P's commute among themselves, we have

$$[P_k, \hat{H}] = -Q^2 \left[P_k, \frac{1}{|\mathbf{X}|} \right] = -i\hbar Q^2 \frac{X_k}{|\mathbf{X}|^3}.$$

Thus,

$$
\begin{aligned}
\varepsilon_{jkl}[P_k, \hat{H}]\hat{J}_l &= -i\hbar Q^2 \varepsilon_{jkl}\varepsilon_{lmn}\frac{X_k}{|\mathbf{X}|^3}X_m P_n \\
&= -i\hbar Q^2 (\delta_{jm}\delta_{kn} - \delta_{jn}\delta_{km})\frac{X_k}{|\mathbf{X}|^3}X_m P_n \\
&= -i\hbar Q^2 \frac{1}{|\mathbf{X}|^3}(X_n X_j P_n - X_m X_m P_j) \\
&= -i\hbar Q^2 \frac{1}{|\mathbf{X}|^3}(X_j(\mathbf{X} \cdot \mathbf{P}) - (\mathbf{X} \cdot \mathbf{X})P_j).
\end{aligned}
\tag{18.34}
$$

We compute $\varepsilon_{jkl}\hat{J}_k[P_l, \hat{H}]$ in a similar way. Note that $\hat{J}_k = \varepsilon_{kmn}X_m P_n = \varepsilon_{kmn}P_n X_m$, since X_m and P_n commute except when $m = n$, in which case $\varepsilon_{kmn} = 0$. The result is

$$
\varepsilon_{jkl}\hat{J}_k[P_l, \hat{H}] = -i\hbar(P_j(\mathbf{X} \cdot \mathbf{X}) - (\mathbf{P} \cdot \mathbf{X})X_j)\frac{1}{|\mathbf{X}|^3}.
$$

Meanwhile, since the X's commute among themselves, we have

$$
\begin{aligned}
&\left[\frac{X_j}{|\mathbf{X}|}, \hat{H}\right] \\
&= \left[\frac{X_j}{|\mathbf{X}|}, \frac{P^2}{2\mu}\right] \\
&= \frac{1}{2\mu}\left[\frac{X_j}{|\mathbf{X}|}, P_k\right]P_k + \frac{1}{2\mu}P_k\left[\frac{X_j}{|\mathbf{X}|}, P_k\right] \\
&= \frac{i\hbar}{2\mu}\left(\frac{1}{|\mathbf{X}|}\delta_{jk} - \frac{X_j X_k}{|\mathbf{X}|^3}\right)P_k + \frac{i\hbar}{2\mu}P_k\left(\frac{1}{|\mathbf{X}|}\delta_{jk} - \frac{X_j X_k}{|\mathbf{X}|^3}\right) \\
&= \frac{i\hbar}{2\mu}\left(\frac{1}{|\mathbf{X}|}P_j - \frac{X_j}{|\mathbf{X}|^3}(\mathbf{X} \cdot \mathbf{P})\right) + \frac{i\hbar}{2\mu}\left(P_j\frac{1}{|\mathbf{X}|} - (\mathbf{P} \cdot \mathbf{X})\frac{X_j}{|\mathbf{X}|^3}\right).
\end{aligned}
\tag{18.35}
$$

It is now a simple matter to compute $[\hat{A}_j, \hat{H}]$ by combining (18.34) and (18.35) and verify that everything cancels. We have, for example, a term involving $(X_j/|\mathbf{X}|^3)(\mathbf{X} \cdot \mathbf{P})$ in (18.34) and a canceling term in (18.35). ∎

Before proceeding with the remaining results concerning the Runge–Lenz vector, we verify some results that will be needed later. There are some quantum corrections compared to the corresponding classical results.

Lemma 18.20 *As in the classical case, the following "orthogonality" relations among vector operators hold:*

$$
\hat{\mathbf{J}} \cdot \mathbf{P} = \mathbf{P} \cdot \hat{\mathbf{J}} = 0
\tag{18.36}
$$

$$
\hat{\mathbf{J}} \cdot \mathbf{X} = \mathbf{X} \cdot \hat{\mathbf{J}} = 0
\tag{18.37}
$$

$$
(\mathbf{P} \times \hat{\mathbf{J}}) \cdot \hat{\mathbf{J}} = \hat{\mathbf{J}} \cdot (\mathbf{P} \times \hat{\mathbf{J}}) = 0.
\tag{18.38}
$$

Meanwhile, there is a quantum correction in the dot product between **P** *and* **P** × **Ĵ**, *as follows:*

$$\mathbf{P} \cdot (\mathbf{P} \times \hat{\mathbf{J}}) = 0 \tag{18.39}$$

$$(\mathbf{P} \times \hat{\mathbf{J}}) \cdot \mathbf{P} = 2i\hbar(\mathbf{P} \cdot \mathbf{P}). \tag{18.40}$$

Finally, we have

$$(\mathbf{P} \times \hat{\mathbf{J}}) \cdot (\mathbf{P} \times \hat{\mathbf{J}}) = (\mathbf{P} \cdot \mathbf{P})(\hat{\mathbf{J}} \cdot \hat{\mathbf{J}}) \tag{18.41}$$

$$\mathbf{X} \cdot (\mathbf{P} \times \hat{\mathbf{J}}) = \hat{\mathbf{J}} \cdot \hat{\mathbf{J}} \tag{18.42}$$

$$(\mathbf{P} \times \hat{\mathbf{J}}) \cdot \mathbf{X} = \hat{\mathbf{J}} \cdot \hat{\mathbf{J}} + 2i\hbar\mathbf{P} \cdot \mathbf{X}. \tag{18.43}$$

Proof. By (18.27) and (18.29), we have

$$\hat{\mathbf{J}} \cdot \mathbf{P} = (\mathbf{X} \times \mathbf{P}) \cdot \mathbf{P} = \mathbf{X} \cdot (\mathbf{P} \times \mathbf{P}) = 0,$$

since the different components of **P** commute. The same reasoning shows that $\mathbf{P} \cdot \hat{\mathbf{J}}$, $\hat{\mathbf{J}} \cdot \mathbf{X}$, and $\mathbf{X} \cdot \hat{\mathbf{J}}$ are all zero. To compute $(\mathbf{P} \times \hat{\mathbf{J}}) \cdot \hat{\mathbf{J}}$, we first use (18.27), then use (18.32), and then use that $\mathbf{P} \cdot \hat{\mathbf{J}} = 0$. For $\hat{\mathbf{J}} \cdot (\mathbf{P} \times \hat{\mathbf{J}})$, we rewrite $\mathbf{P} \times \hat{\mathbf{J}}$ in terms of $\hat{\mathbf{J}} \times \mathbf{P}$, using (18.31). The correction term involves **P**, which has a dot product of zero with $\hat{\mathbf{J}}$, and so the answer is again zero.

We use (18.27) and (18.29) again to establish (18.39). To get (18.40), we first rewrite $\mathbf{P} \times \hat{\mathbf{J}}$ in terms of $\hat{\mathbf{J}} \times \mathbf{P}$ using (18.31) and then apply (18.39). To establish (18.41), we apply (18.27) and then (18.30), giving

$$(\mathbf{P} \times \hat{\mathbf{J}}) \cdot (\mathbf{P} \times \hat{\mathbf{J}}) = P_j \hat{J}_k P_j \hat{J}_k - P_j \hat{J}_k P_k \hat{J}_j. \tag{18.44}$$

The second term on the right-hand side of (18.44) is zero because $\hat{\mathbf{J}} \cdot \mathbf{P} = 0$. For the first term, we move $\hat{J}_k$ to the right past P_j. This generates the term we want plus a correction term equal to $i\hbar\varepsilon_{kjl}P_j P_l \hat{J}_k$. The correction term is zero because P_j and P_l commute and ε_{kjl} is changes sign under interchange of j and l. The identity (18.42) follows immediately from (18.27) and the definition of $\hat{\mathbf{J}}$. The identity (18.43) follows from (18.27) and (18.28). ∎

Lemma 18.21 *For all j and m, we have*

$$[(\mathbf{P} \times \hat{\mathbf{J}})_j, (\mathbf{P} \times \hat{\mathbf{J}})_m] = -i\hbar(\mathbf{P} \cdot \mathbf{P})\varepsilon_{jml} \hat{J}_l.$$

Proof. In computing $[P_k \hat{J}_l, P_n \hat{J}_o]$, we use repeatedly the product rule for commutators (Point 3 of Proposition 3.15). We obtain four terms, one of which is zero (the term involving $[P_k, P_n]$). We use Proposition 17.27 (in the form (18.26)) to evaluate all remaining terms, giving

$$\frac{1}{i\hbar}[\varepsilon_{jkl}P_k \hat{J}_l, \varepsilon_{mno}P_n \hat{J}_o]$$

$$= \varepsilon_{jkl}\varepsilon_{mno}\left(P_k[\hat{J}_l, P_n]\hat{J}_o + P_n P_k[\hat{J}_l, \hat{J}_o] + P_n[P_k, \hat{J}_o]\hat{J}_l\right). \tag{18.45}$$

Let us compute the first of the three terms on the right-hand side of (18.45). Using Lemma 18.18 and the fact that $\mathbf{P}$ is a vector operator, we get

$$\varepsilon_{jkl}\varepsilon_{mno}P_k[\hat{J}_l, P_n]\hat{J}_o = \varepsilon_{jkl}(\delta_{op}\delta_{ml} - \delta_{ol}\delta_{mp})P_kP_p\hat{J}_o$$
$$= \varepsilon_{jkm}P_kP_p\hat{J}_p - \varepsilon_{jko}P_kP_m\hat{J}_o$$
$$= \varepsilon_{jkm}P_k(\mathbf{P}\cdot\hat{\mathbf{J}}) - P_m(\mathbf{P}\times\hat{\mathbf{J}})_j.$$

If we compute the second and third terms similarly, we obtain

$$\frac{1}{i\hbar}[\varepsilon_{jkl}P_k\hat{J}_l, \varepsilon_{mno}P_n\hat{J}_o] = \varepsilon_{jkm}P_k(\mathbf{P}\cdot\hat{\mathbf{J}}) - P_m(\mathbf{P}\times\hat{\mathbf{J}})_j$$
$$+ (\mathbf{P}\times\mathbf{P})_j\hat{J}_m - \varepsilon_{jkm}P_k(\mathbf{P}\cdot\hat{\mathbf{J}}) + P_m(\mathbf{P}\times\hat{\mathbf{J}})_j - (\mathbf{P}\cdot\mathbf{P})\varepsilon_{jml}\hat{J}_l.$$

Three of the above terms are zero (those involving $\mathbf{P}\cdot\hat{\mathbf{J}}$ or $\mathbf{P}\times\mathbf{P}$) and two other terms cancel, leaving us with

$$\frac{1}{i\hbar}[\varepsilon_{jkl}P_k\hat{J}_l, \varepsilon_{mno}P_n\hat{J}_o] = -(\mathbf{P}\cdot\mathbf{P})\varepsilon_{jml}\hat{J}_l,$$

as claimed. ∎

We now continue with the proof of the properties of the Runge–Lenz vector.

Proof Proposition 18.11. From the first set of orthogonality relations in Lemma 18.20, we can see easily that $\hat{\mathbf{J}}\cdot\hat{\mathbf{A}} = \hat{\mathbf{A}}\cdot\hat{\mathbf{J}} = 0$. Meanwhile, using the expression (18.19) for $\hat{\mathbf{A}}$ and expanding out $\hat{\mathbf{A}}\cdot\hat{\mathbf{A}}$ yields, after a little simplification,

$$\hat{\mathbf{A}}\cdot\hat{\mathbf{A}} = 1 + \frac{1}{\mu^2Q^4}(\mathbf{P}\cdot\mathbf{P})\left(\hat{\mathbf{J}}\cdot\hat{\mathbf{J}} + \hbar^2\right)$$
$$- \frac{1}{\mu Q^2}\left(2\hat{\mathbf{J}}\cdot\hat{\mathbf{J}}\frac{1}{|\mathbf{X}|} + i\hbar\left(\mathbf{P}\cdot\frac{\mathbf{X}}{|\mathbf{X}|} - \frac{\mathbf{X}}{|\mathbf{X}|}\cdot\mathbf{P}\right)\right).$$

Now,

$$\frac{\mathbf{X}}{|\mathbf{X}|}\cdot\mathbf{P} - \mathbf{P}\cdot\frac{\mathbf{X}}{|\mathbf{X}|} = i\hbar\left(\frac{\delta_{kk}}{|\mathbf{X}|} - \frac{X_k}{|\mathbf{X}|^2}\frac{X_k}{|\mathbf{X}|}\right) = 2i\hbar\frac{1}{|\mathbf{X}|}.$$

Thus,

$$\hat{\mathbf{A}}\cdot\hat{\mathbf{A}} = 1 + \left((\hat{\mathbf{J}}\cdot\hat{\mathbf{J}}) + \hbar^2\right)\frac{2}{\mu Q^4}\left(\frac{(\mathbf{P}\cdot\mathbf{P})}{2\mu} - Q^2\frac{1}{|\mathbf{X}|}\right),$$

as claimed. ∎

Proof of Lemma 18.12, Second Part. We write $\hat{\mathbf{A}}$ in the form given in (18.19). In computing the commutator of $\hat{A}_j$ with $\hat{A}_m$, we get several different types of terms, which we compute one at a time. Of course, the commutator of $X_j/|\mathbf{X}|$ with $X_m/|\mathbf{X}|$ is zero. The commutator of the $\mathbf{P}\times\hat{\mathbf{J}}$ terms has been computed in Lemma 18.21.

Meanwhile, to compute the commutator of $P_k \hat{J}_l$ with $X_m(1/|\mathbf{X}|)$, we again get four terms and, again, one of these is zero, namely the one involving $\{\hat{J}_l, 1/|\mathbf{X}|\}$, since $1/|\mathbf{X}|$ is invariant under rotations. We have, then,

$$\frac{1}{i\hbar}\left[\varepsilon_{jkl}P_k\hat{J}_l, X_m\frac{1}{|\mathbf{X}|}\right]$$

$$= \varepsilon_{jkl}[P_k, X_m]\hat{J}_l\frac{1}{|\mathbf{X}|} + \varepsilon_{jkl}P_k[\hat{J}_l, X_m]\frac{1}{|\mathbf{X}|} + \varepsilon_{jkl}X_m\left[P_k, \frac{1}{|\mathbf{X}|}\right]\hat{J}_l$$

$$= -\varepsilon_{jkl}\delta_{km}\hat{J}_l\frac{1}{|\mathbf{X}|} + \varepsilon_{jkl}\varepsilon_{lmn}P_kX_n\frac{1}{|\mathbf{X}|} + \varepsilon_{jkl}X_m\frac{X_k}{|\mathbf{X}|^3}\varepsilon_{lno}X_nP_o.$$

If we apply Lemma 18.18 and carry out some computations similar to ones we have already performed, we obtain

$$\frac{1}{i\hbar}\left[\varepsilon_{jkl}P_k\hat{J}_l, X_m\frac{1}{|\mathbf{X}|}\right] = -\varepsilon_{jml}\hat{J}_l\frac{1}{|\mathbf{X}|} + \delta_{jm}(\mathbf{P}\cdot\mathbf{X})\frac{1}{|\mathbf{X}|}$$

$$+ X_mX_j\frac{1}{|\mathbf{X}|^3}(\mathbf{X}\cdot\mathbf{P}) - \left(P_m\frac{X_j}{|\mathbf{X}|} + \frac{X_m}{|\mathbf{X}|}P_j\right). \tag{18.46}$$

In a commutator of the form $[\alpha_j + \beta_j, \alpha_m + \beta_m]$, the terms involving the commutator of an α with a β will be $[\alpha_j, \beta_m] + [\beta_j, \alpha_m]$, which is equal to $[\alpha_j, \beta_m] - [\alpha_m, \beta_j]$. This quantity is skew-symmetric j with m, meaning that it changes sign when we interchange j with m. Thus, terms in (18.46) that are symmetric in j and m will disappear when we compute the full commutator of $\hat{A}_j$ with $\hat{A}_m$. Thus, the second and third terms in (18.46) can be ignored. In the last term, we can commute P_m past X_j to obtain

$$P_m\frac{X_j}{|\mathbf{X}|} + \frac{X_m}{|\mathbf{X}|}P_j = \frac{X_j}{|\mathbf{X}|}P_m + \frac{X_m}{|\mathbf{X}|}P_j - i\hbar\left(\frac{\delta_{jm}}{|\mathbf{X}|} - \frac{X_jX_m}{|\mathbf{X}|^3}\right), \tag{18.47}$$

which is also symmetric. Thus, only the first term in (18.46) contributes to the computation of $[\hat{A}_j, \hat{A}_m]$. This term is skew-symmetric in j and m and will be doubled when we compute $[\hat{A}_j, \hat{A}_m]$.

Now, it is straightforward to compute $[\varepsilon_{jkl}P_k\hat{J}_l, P_m]$ and $[P_j, X_m/|\mathbf{X}|]$ and to verify that these commutators are symmetric in j and m (Exercise 8) and therefore do not contribute to the computation of $[\hat{A}_j, \hat{A}_m]$. We are left, then, with the following

$$\frac{1}{i\hbar}[\hat{A}_j, \hat{A}_m] = -\frac{1}{\mu^2Q^4}\varepsilon_{jml}(\mathbf{P}\cdot\mathbf{P})\hat{J}_l + \frac{1}{\mu Q^2}2\varepsilon_{jml}\hat{J}_l\frac{1}{|\mathbf{X}|}$$

$$= -\frac{2}{\mu Q^4}\varepsilon_{jml}\hat{J}_l\left(\frac{\mathbf{P}\cdot\mathbf{P}}{2\mu} - \frac{Q^2}{|\mathbf{X}|}\right),$$

which is what is claimed in the lemma. ∎

Proof of Theorem 18.14. Since the Hamiltonian $\hat{H}$ is invariant under rotations, $\hat{H}$ commutes with each component of the angular momentum. We have also established that $\hat{H}$ commutes with each component of the Runge–Lenz vector. From this it follows easily that $\hat{\mathbf{I}}$ and $\hat{\mathbf{K}}$ commute with the Hamiltonian.

Since A_k commutes with $\hat{H}$, it also commutes with any function of $\hat{H}$. It then follows from Lemma 18.12 that

$$\frac{1}{i\hbar}[\hat{B}_k, \hat{B}_l] = \frac{\mu Q^4}{2|\hat{H}|}[\hat{A}_k, \hat{A}_l] = -\frac{\mu Q^4}{2|\hat{H}|}\frac{2}{\mu Q^4}\varepsilon_{jml}\hat{J}_l\hat{H}.$$

Since $\hat{H}/|\hat{H}| = -I$ on the negative-energy subspace V^-, the above expression reduces to $\varepsilon_{jml}\hat{J}_l$. (The result on the positive-energy subspace will differ by a crucial minus sign from what we have on V^-.)

Meanwhile, since both $\hat{\mathbf{B}}$ and $\hat{\mathbf{J}}$ are vector operators, we have, by Proposition 17.27, $(1/(i\hbar))[\hat{B}_j, \hat{J}_k] = \varepsilon_{jkl}\hat{B}_l$ and $(1/(i\hbar))[\hat{J}_j, \hat{J}_k] = \varepsilon_{jkl}\hat{J}_l$. From the commutation relations among the $\hat{B}_j$'s and $\hat{J}_j$'s, it is an easy calculation to verify the claimed commutation relations among the components of $\hat{\mathbf{I}}$ and $\hat{\mathbf{K}}$. ∎

18.7   Exercises

1. Consider the quantum Hamiltonian for two particles in $\mathbb{R}^3$ interacting by means of a $1/r$ potential:

$$\hat{H} = -\frac{\hbar^2}{2m_1}\Delta_1 - \frac{\hbar^2}{2m_2}\Delta_2 - \frac{Q^2}{|\mathbf{x}^1 - \mathbf{x}^2|}.$$

Here, as in Sect. 3.11, Δ_1 is the Laplacian with respect to the variable $\mathbf{x}^1$ and Δ_2 is the Laplacian with respect to the variable $\mathbf{x}^2$. As in Sect. 2.3.3, introduce new variables consisting of the center of mass, $\mathbf{c} = (m_1\mathbf{x}^1 + m_2\mathbf{x}^2)/(m_1 + m_2)$, and the relative position, $\mathbf{y} = \mathbf{x}^1 - \mathbf{x}^2$. Show that $\hat{H}_2$ can be expressed in these variables as

$$-\frac{\hbar^2}{2(m_1 + m_2)}\Delta_\mathbf{c} - \frac{\hbar^2}{2\mu}\Delta_\mathbf{y} - \frac{Q^2}{|\mathbf{y}|},$$

where μ is the reduced mass, given by $\mu = m_1 m_2/(m_1 + m_2)$.

Note: In the new variables, $\hat{H}$ is the sum of two terms, one of which involves only the variable $\mathbf{c}$ and one of which involves only the variable $\mathbf{y}$. The term involving only $\mathbf{c}$ is the Hamiltonian for a free particle with mass $m_1 + m_2$, whereas the term involving only $\mathbf{y}$ is the Hamiltonian for a particle of mass μ moving in a $1/r$ potential.

2. Let $H(\mathbf{x}, \mathbf{p}) = |\mathbf{p}|^2 / (2\mu) - Q^2 / |\mathbf{x}|$ denote the Hamiltonian for the classical Kepler problem in $\mathbb{R}^3$. Show that for every $\varepsilon > 0$, the region in $\mathbb{R}^6$ given by $\{(\mathbf{x}, \mathbf{p}) \,|\, H(\mathbf{x}, \mathbf{p}) < -\varepsilon\}$ has finite volume.

3. Let $\mathbb{H}$ denote the real span of the following four elements of $M_2(\mathbb{C})$:

$$\mathbf{1} := \begin{pmatrix} 1 & 0 \\ 0 & 1 \end{pmatrix}; \quad \mathbf{i} := \begin{pmatrix} i & 0 \\ 0 & -i \end{pmatrix};$$

$$\mathbf{j} := \begin{pmatrix} 0 & 1 \\ -1 & 0 \end{pmatrix}; \quad \mathbf{j} := \begin{pmatrix} 0 & i \\ i & 0 \end{pmatrix}.$$

(a) Show that $\mathbb{H}$ forms an associative algebra over $\mathbb{R}$, under the operation of matrix multiplication, and that the following relations are satisfied:

$$\mathbf{i}^2 = \mathbf{j}^2 = \mathbf{k}^2 = -1$$
$$\mathbf{ij} = -\mathbf{ji} = \mathbf{k}$$
$$\mathbf{jk} = -\mathbf{kj} = \mathbf{i}$$
$$\mathbf{ki} = -\mathbf{ik} = \mathbf{j}.$$

The algebra $\mathbb{H}$ is (one particular realization of) the *quaternion algebra*.

(b) Show that each nonzero element of $\mathbb{H}$ has a multiplicative inverse.

Hint: Imitate the argument that each nonzero complex number has a multiplicative inverse.

4. Let $\mathbb{H}$ denote the quaternion algebra defined in Exercise 3. This exercise establishes explicitly an isomorphism between the Lie algebras $\mathsf{so}(4)$ and $\mathsf{so}(3) \oplus \mathsf{so}(3)$ (compare Definition 16.14).

(a) Let V be the subspace of $\mathbb{H}$ spanned by $\mathbf{i}$, $\mathbf{j}$, and $\mathbf{k}$. Show that V forms a Lie algebra under the bracket $[\alpha, \beta] = \alpha\beta - \beta\alpha$ and that V is isomorphic as a Lie algebra to $\mathsf{so}(3)$.

(b) Let $\mathrm{End}(\mathbb{H})$ denote the algebra of real-linear maps of $\mathbb{H}$ to itself. Given $\alpha \in V$, let $L_\alpha \in \mathrm{End}(\mathbb{H})$ be the "left multiplication by α" map, $L_\alpha(\beta) = \alpha\beta$, and let $R_\alpha \in \mathrm{End}(\mathbb{H})$ be the "right multiplication by α" map, $R_\alpha(\beta) = \beta\alpha$. Show that the maps $\alpha \mapsto L_\alpha$ and $\alpha \mapsto -R_\alpha$ are Lie algebra homomorphisms of V into $\mathrm{End}(\mathbb{H})$.

(c) Consider the inner product on $\mathbb{H}$ in which $\{1, \mathbf{i}, \mathbf{j}, \mathbf{k}\}$ forms an orthonormal basis. Given $\alpha \in V$, show that

$$\langle L_\alpha \beta, \gamma \rangle = -\langle \beta, L_\alpha \gamma \rangle$$
$$\langle R_\alpha \beta, \gamma \rangle = -\langle \beta, R_\alpha \gamma \rangle.$$

That is to say, L_α and R_α belong to $so(4)$, which we identify with the space of elements of $\text{End}(\mathbb{H})$ that are skew-symmetric with respect to the inner product in Part (c).

(d) Show that the map $(\alpha, \beta) \mapsto L_\alpha - R_\beta$ is a Lie algebra isomorphism of $so(3) \oplus so(3)$ to $so(4)$.

(e) Let D denote the diagonal subalgebra of $so(3) \oplus so(3)$, that is, the set of elements of the form (X, X). Show that the image of D under the isomorphism in Part (d) is the set of elements Y of $so(4) \subset \text{End}(\mathbb{H})$ having the following form with respect to the basis in Part (c):

$$Y = \begin{pmatrix} 0 & 0 \\ 0 & Z \end{pmatrix},$$

where $Z \in so(3)$.

5. Describe explicitly the two subalgebras of $so(4)$ corresponding to the two copies of $so(3)$ in the isomorphism

$$so(4) \cong so(3) \oplus so(3)$$

in Exercise 4.

6. Verify Lemma 18.18.

   Hint: First show that $\varepsilon_{jkl}\varepsilon_{jmn} = 0$ unless $(k, l) = (m, n)$ or $(k, l) = (n, m)$.

7. In this exercise, we use the summation convention of Sect. 18.4.1.

   (a) Show that for any 3×3 matrix M and any indices $j, k, l \in \{1, 2, 3\}$, we have

   $$\varepsilon_{mno}M_{jm}M_{kn}M_{lo} = \varepsilon_{jkl}(\det M).$$

   (b) Show that if $\mathbf{C}$ is a vector operator, then for all $R \in SO(3)$, we have

   $$\Pi(R)C_k\Pi(R)^{-1} = R_{lk}C_l.$$

   (c) Show that the cross product of two vector operators is a vector operator.

   Hint: Write the definition of a vector operator in the equivalent form

   $$\mathbf{v} \cdot \mathbf{C} = \Pi(R)((R^{-1}\mathbf{v}) \cdot \mathbf{C})\Pi(R)^{-1}.$$

8. Compute $[\varepsilon_{jkl}P_k\hat{J}_l, P_m]$ and $[P_j, X_m/|\mathbf{X}|]$ and show that both of these quantities are symmetric in j and m, meaning that the value is unchanged if we interchange j and m.

9. Show that the Eq. (18.14) has two power series solutions for $g(\rho)$, one starting with $\rho^{-(2l+1)}$ and one starting with ρ^0.

19

Systems and Subsystems, Multiple Particles

19.1 Introduction

Up to this point, we have considered the state of a quantum system to be described by a unit vector in the corresponding Hilbert space, or more properly, an equivalence class of unit vectors under the equivalence relation $\psi \sim e^{i\theta}\psi$. We will see in this section that this notion of the state of a quantum system is too limited. We will introduce a more general notion of the state of a system, described by a *density matrix*. The special case in which the system can be described by a unit vector will be called a *pure state*.

One way to see the inadequacy of the notion of state as a unit vector is to consider systems and subsystems. We will examine this topic in greater detail in Sect. 19.5, but for now let us consider the example of a system of two spinless "distinguishable" particles moving in $\mathbb{R}^3$. (For now, the reader need not worry about the notion of distinguishable particles; just think of them as being two different types of particles, with, say, different masses or charges.) Let us assume the combined state of the two particles can be described by a unit vector in the corresponding Hilbert space, which is (according to Sect. 3.11) $L^2(\mathbb{R}^6)$. We have, then, a wave function $\psi(\mathbf{x}, \mathbf{y})$, where $\mathbf{x}$ is the position of the first particle and $\mathbf{y}$ is the position of the second particle.

Given a wave function $\psi(\mathbf{x}, \mathbf{y})$ for the combined system, what is the wave function describing the state of the first particle only? *If* the wave function of the combined system happens to be a product, say, $\psi(\mathbf{x}, \mathbf{y}) =$

B.C. Hall, *Quantum Theory for Mathematicians*, Graduate Texts in Mathematics 267, DOI 10.1007/978-1-4614-7116-5_19,
© Springer Science+Business Media New York 2013

$\psi_1(\mathbf{x})\psi_2(\mathbf{y})$, then, naturally, we would say that the state of the first particle is simply ψ_1. Of course, one might object that we could rewrite ψ as $\psi(\mathbf{x}, \mathbf{y}) = [c\psi_1(\mathbf{x})][\psi_2(\mathbf{y})/c]$ for any constant c, but this only affects the wave function for the first particle by a constant, which does not affect the physical state.

In general, however, the wave function of the combined system need not be a product. Already when ψ is a linear combination of two products, $\psi(\mathbf{x}, \mathbf{y}) = \psi_1(\mathbf{x})\psi_2(\mathbf{y}) + \phi_1(\mathbf{x})\phi_2(\mathbf{y})$, it is unclear what the correct wave function is for the first particle. At first glance, it might seem natural to try $\psi_1(\mathbf{x}) + \phi_1(\mathbf{x})$, but upon closer examination, this is not an unambiguous proposal. After all, we can just as well write $\psi(\mathbf{x}, \mathbf{y}) = [c_1\psi_1(\mathbf{x})][\psi_2(\mathbf{y})/c_1] + [c_2\phi_1(\mathbf{x})][\phi_2(\mathbf{y})/c_2]$, but then the resulting wave functions for the first particle, $\psi_1(\mathbf{x}) + \psi_2(\mathbf{x})$ and $c_1\psi_1(\mathbf{x}) + c_2\psi_2(\mathbf{x})$, are not scalar multiples of one another. For a general unit vector ψ in $L^2(\mathbb{R}^6)$, the situation is even worse. The conclusion is this: There does not seem to be any way to associate to ψ a general unit vector ψ' in $L^2(\mathbb{R}^3)$ such that ψ' could sensibly be described as "the state of the first particle."

Although we cannot associate with ψ a wave function ψ' for the first particle, there is no difficulty in taking expectation values of observables related to the first particle. We can make perfect sense of, say, the expected position of the first particle, as

$$\left\langle \psi, X_j^{(1)}\psi \right\rangle = \int_{\mathbb{R}^6} x_j \, |\psi(\mathbf{x}, \mathbf{y})|^2 \; d\mathbf{x} \, d\mathbf{y}.$$

Here $X_j^{(1)}$ indicates the operator of multiplication by the jth component of the *first* vector in the function $\psi(\cdot, \cdot) : \mathbb{R}^3 \times \mathbb{R}^3 \to \mathbb{C}$. That is to say, the operator X_j acting on $L^2(\mathbb{R}^3)$ can be "promoted" to an operator on $L^2(\mathbb{R}^6)$ by having it act in the first variable only. Similarly, the momentum operator P_j on $L^2(\mathbb{R}^3)$ can be promoted to an operator $P_j^{(1)}$ on $L^2(\mathbb{R}^6)$, by letting it act on the first variable, meaning that $P_j^{(1)}\psi$ is $-i\hbar$ times the partial derivative with respect to the jth component of the first vector in $\psi(\cdot, \cdot)$. In fact, as we will see in Sect. 19.5, given *any* self-adjoint operator on $L^2(\mathbb{R}^3)$, there is a natural way to promote it into an operator on $L^2(\mathbb{R}^6)$, where its expectation value may then be defined.

Thus, although there is no natural way to associate with a unit vector ψ in $L^2(\mathbb{R}^6)$ a unit vector in $L^2(\mathbb{R}^3)$, there *is* a natural way to associate with ψ expectation values of observables on $L^2(\mathbb{R}^3)$. This suggests that we should introduce a more general notion of the "state" of a quantum system, a notion in which with each "reasonable" family of expectation values for the quantum observables there is associated a quantum state. This notion turns out to be that of *density matrices* (positive, self-adjoint operators with trace 1).

In Sect. 19.3, we introduce the notion of a density matrix. Theorem 19.9 in that section will tell us that, given any reasonable assignment ϕ of expectation values to observables, there is a unique density matrix ρ such that $\phi(A) = \text{trace}(\rho A)$ for all observables A. In the special case in which the state of the system is given by a unit vector ψ in the Hilbert space, then ρ will be just the projection onto ψ and $\text{trace}(\rho A)$ will be equal to the familiar expression $\langle \psi, A\psi \rangle$. In Sect. 19.5, we will consider composite quantum systems and introduce a method (the partial trace) of defining a density matrix for a subsystem from a density matrix for the whole system. Finally, in Sect. 19.6, we will consider the important special case of composite systems made up of multiple identical particles.

19.2   Trace-Class and Hilbert–Schmidt Operators

In this section, we explore notions related to the trace of an operator on a Hilbert space. The results of this section are presented without proof; see Chap. VI in Volume I of [34] for proofs and additional information.

Proposition 19.1 *Suppose $A \in \mathcal{B}(\mathbf{H})$ is non-negative and self-adjoint. Then for any two orthonormal bases $\{e_j\}$ and $\{f_j\}$ for $\mathbf{H}$, we have*

$$\sum_j \langle e_j, Ae_j \rangle = \sum_j \langle f_j, Af_j \rangle .$$

Note that since A is non-negative, $\langle e_j, Ae_j \rangle$ and $\langle f_j, Af_j \rangle$ are non-negative real numbers. Thus, the sums are always well defined, but may have the value of $+\infty$.

Definition 19.2 *If $A \in \mathcal{B}(\mathbf{H})$ is non-negative and self-adjoint, the value of $\sum_j \langle e_j, Ae_j \rangle$, for any arbitrarily chosen orthonormal basis, is called the* **trace** *of A. If $\text{trace}(A) < +\infty$, then we say that A is* **trace class***.*

For a general $A \in \mathcal{B}(\mathbf{H})$, we say that A is **trace class** *if the non-negative self-adjoint operator $\sqrt{A^*A}$ is a trace class.*

Note that for any $A \in \mathcal{B}(\mathbf{H})$, A^*A is self-adjoint and non negative. Thus, the square root of A^*A may be defined by the functional calculus (Definition 7.13 or Proposition 8.4).

Proposition 19.3

1. *If $A \in \mathcal{B}(\mathbf{H})$ is trace class, then for any orthonormal basis $\{e_j\}$, the sum $\sum_j \langle e_j, Ae_j \rangle$ is absolutely convergent. Furthermore, the value of this sum, which we denote as $\text{trace}(A)$, is independent of the choice of orthonormal basis.*

2. *If $A \in \mathcal{B}(\mathbf{H})$ is trace class, then A^* is also trace class and*

$$\text{trace}(A^*) = \overline{\text{trace}(A)}.$$

3. *If $A \in \mathcal{B}(\mathbf{H})$ is trace class, then for all $B \in \mathcal{B}(\mathbf{H})$, the operators AB and BA are also trace class, and*

$$\text{trace}(AB) = \text{trace}(BA).$$

Recall that $A \in \mathcal{B}(\mathbf{H})$ is said to be compact if A maps every bounded set in $\mathbf{H}$ to a set with compact closure. If a self-adjoint operator A is trace class, it is necessarily compact and thus has an orthonormal basis $\{e_j\}$ of eigenvectors, for which the associated eigenvalues λ_j are real and tend to zero as j tends to infinity. (See Theorem VI.16 in Volume I of [34]. One can deduce the result from, say, the direct integral form of the spectral theorem for bounded self-adjoint operators by verifying that unless A has point spectrum with eigenvalues tending to zero, the operator of multiplication by λ in the direct integral will not be compact.) Point 1 of Proposition 19.3 then tells us that $\sum_j |\lambda_j| < \infty$ and that $\text{trace}(A) = \sum_j \lambda_j$. Conversely, if A is a self-adjoint operator having an orthonormal basis of eigenvectors for which the associated eigenvalues satisfy $\sum_j |\lambda_j| < \infty$, then A is trace class.

Definition 19.4 *An operator $A \in \mathcal{B}(\mathbf{H})$ is said to be **Hilbert–Schmidt** if $\text{trace}(A^*A) < \infty$.*

Since A^*A is self-adjoint and non-negative, $\text{trace}(A^*A)$ is defined (but possibly infinite) for any $A \in \mathcal{B}(\mathbf{H})$. If A is trace class, then (by definition) the trace of $\sqrt{A^*A}$ is finite, in which case, the trace of $\sqrt{A^*A}\sqrt{A^*A}$ is also finite, by Point 3 of Proposition 19.3. Thus, every trace-class operator is Hilbert–Schmidt (but not vice versa).

Proposition 19.5 *If $A \in \mathcal{B}(\mathbf{H})$ is Hilbert–Schmidt, so is A^*. If $A, B \in \mathcal{B}(\mathbf{H})$ are Hilbert–Schmidt, then AB and BA are trace class and $\text{trace}(AB)$ equals $\text{trace}(BA)$.*

If A and B are Hilbert–Schmidt operators, the *Hilbert–Schmidt inner product* of A and B is $\langle A, B \rangle_{HS} := \text{trace}(A^*B)$ and the *Hilbert–Schmidt norm* of A satisfies $\|A\|_{HS}^2 = \langle A, A \rangle_{HS}$. The space of Hilbert–Schmidt operators is a Hilbert space with respect to $\langle \cdot, \cdot \rangle_{HS}$.

19.3 Density Matrices: The General Notion of the State of a Quantum System

Typically, we think of the quantum observables—the ones with expectations values that we wish to take—as being unbounded self-adjoint operators. But of course we can also take expectation values of bounded self-adjoint operators, and indeed expectations for bounded operators determine those for unbounded operators. After all, suppose A is an unbounded self-adjoint operator and suppose we know the expectation value for $1_E(A)$ for every Borel set $E \subset \mathbb{R}$, where 1_E is the indicator function of E and

$1_E(A)$ is defined by the functional calculus (Definition 7.13). The expectation value for $1_E(A)$ is the probability of obtaining a value in E for a measurement of the observable A. If we know this probability for each E, then we know the full probability distribution of the measurements, and thus we can compute the expectation value of A. Furthermore, we can always introduce expectation values for (bounded) non-self-adjoint operators. Each such operator A is of the form $A = A_1 + iA_2$ with A_1 and A_2 self-adjoint, and so we may reasonably define the expectation value of A to be the expectation value of A_1 plus i times the expectation value of A_2.

We then postulate that the general notion of the "state" of a quantum system should be simply a "list" of expectation values for all bounded operators, satisfying some reasonable hypotheses.

Definition 19.6 *A linear map $\Phi : \mathcal{B}(\mathbf{H}) \to \mathbb{C}$ is a **family of expectation values** if the following conditions hold.*

1. *$\Phi(I) = 1$.*

2. *$\Phi(A)$ is real whenever A is self-adjoint.*

3. *$\Phi(A) \geq 0$ whenever A is self-adjoint and non-negative.*

4. *For any sequence A_n in $\mathcal{B}(\mathbf{H})$, if $\|A_n\psi - A\psi\| \to 0$ for all $\psi \in \mathbf{H}$, then $\Phi(A_n) \to \Phi(A)$.*

Point 4 in the definition says that Φ is continuous with respect to the strong (sequential) convergence in $\mathcal{B}(\mathbf{H})$. By Exercise 3, any linear map on $\mathcal{B}(\mathbf{H})$ satisfying Points 1, 2, and 3 is automatically continuous with respect to the operator norm topology, meaning that if $\|A_n - A\| \to 0$ then $\Phi(A_n) \to \Phi(A)$. However, to establish our characterization of families of expectation values in terms of density matrices, we need continuity of Φ under a more general sort of convergence, where we only assume that $\|A_n\psi - A\psi\| \to 0$ for each ψ. This stronger continuity property does not follow from Properties 1–3. Exercise 5 gives an example of a linear functional on $\mathcal{B}(\mathbf{H})$ that satisfies Points 1–3 of Definition 19.6, but not Point 4.

Definition 19.7 *An operator $\rho \in \mathcal{B}(\mathbf{H})$ is a **density matrix** if ρ is self-adjoint and non-negative and $\operatorname{trace}(\rho) = 1$.*

Of course, since the trace of a density matrix is assumed to be finite, every density matrix is trace class. The next two results give a precise characterization of families of expectation values in terms of density matrices.

Proposition 19.8 *Suppose ρ is a density matrix on $\mathbf{H}$. Then the map $\Phi_\rho : \mathcal{B}(\mathbf{H}) \to \mathbb{C}$ given by*

$$\Phi_\rho(A) = \operatorname{trace}(\rho A) = \operatorname{trace}(A\rho)$$

is a family of expectation values.

Proof. If we define $\Phi_\rho(A) = \text{trace}(\rho A)$, then $\Phi_\rho(I) = \text{trace}(\rho) = 1$. For any $A \in \mathcal{B}(\mathbf{H})$, we have,

$$\text{trace}(\rho A^*) = \text{trace}(A^* \rho) = \text{trace}((\rho A)^*) = \overline{\text{trace}(\rho A)}.$$

It follows that $\text{trace}(\rho A)$ is real when A is self-adjoint. Let $\rho^{1/2}$ be the non-negative self-adjoint square root of ρ. Then $\rho^{1/2}$ and $A\rho^{1/2}$ are Hilbert–Schmidt (in the latter case, by Point 3 of Proposition 19.3). It follows that $\text{trace}(A\rho^{1/2}\rho^{1/2}) = \text{trace}(\rho^{1/2}A\rho^{1/2})$, by Proposition 19.5. Thus, if A is self-adjoint and non-negative,

$$\text{trace}(\rho A) = \text{trace}(\rho^{1/2}\rho^{1/2}A) = \text{trace}(\rho^{1/2}A\rho^{1/2}) \geq 0, \qquad (19.1)$$

because $\rho^{1/2}A\rho^{1/2}$ is self-adjoint and non-negative. We have established that Φ_ρ satisfies Points 1, 2, and 3 of Definition 19.6.

Meanwhile, suppose $A_n\psi$ converges in norm to $A\psi$, for each ψ in $\mathbf{H}$. Then $\|A_n\psi\|$ is bounded as a function of n for each fixed ψ. Thus, by the principle of uniform boundedness (Theorem A.40), there is a constant C such that $\|A_n\| \leq C$. Now, if $\{e_j\}$ is an orthonormal basis for $\mathbf{H}$, we have

$$\left| \left\langle e_j, \rho^{1/2}A_n\rho^{1/2}e_j \right\rangle \right| = \left| \left\langle \rho^{1/2}e_j, A_n\rho^{1/2}e_j \right\rangle \right| \leq C \left\| \rho^{1/2}e_j \right\|^2,$$

and,

$$\sum_j \left\| \rho^{1/2}e_j \right\|^2 = \sum_j \left\langle \rho^{1/2}e_j, \rho^{1/2}e_j \right\rangle = \sum_j \left\langle e_j, \rho e_j \right\rangle = \text{trace}(\rho) < \infty.$$

Furthermore, since $A_n(\rho^{1/2}e_j)$ converges to $A(\rho^{1/2}e_j)$ for each j, dominated convergence tells us that

$$\text{trace}(\rho^{1/2}A\rho^{1/2}) = \sum_j \left\langle e_j, \rho^{1/2}A\rho^{1/2}e_j \right\rangle$$

$$= \lim_{n\to\infty} \sum_j \left\langle e_j, \rho^{1/2}A_n\rho^{1/2}e_j \right\rangle$$

$$= \lim_{n\to\infty} \text{trace}(\rho^{1/2}A_n\rho^{1/2}).$$

As in (19.1), we can shift the second factor of $\rho^{1/2}$ to the front of the trace to obtain Point 4 in Definition 19.6. ∎

Theorem 19.9 *For any family of expectation values* $\Phi : \mathcal{B}(\mathbf{H}) \to \mathbb{C}$, *there is a unique density matrix* ρ *such that* $\Phi(A) = \text{trace}(\rho A)$ *for all* $A \in \mathcal{B}(\mathbf{H})$.

Proof. Recall from Sect. 3.12 the Dirac notation, in which the expression $|\phi\rangle\langle\psi|$ denotes the linear operator taking any vector $\chi \in \mathbf{H}$ to the vector $|\phi\rangle\langle\psi|\chi\rangle$ (in physics notation), that is, the vector $\langle\psi, \chi\rangle \phi$ (in math notation). If ρ is trace class, then by Exercise 2,

$$\text{trace}(\rho |\phi\rangle\langle\psi|) = \langle\psi, \rho\phi\rangle.$$

Thus, if an operator ρ with the desired properties is to exist, we must have

$$\langle \psi, \rho\phi \rangle = \Phi(|\phi\rangle\langle\psi|).$$

Now, by Exercise 3, Φ satisfies $\|\Phi(A)\| \leq \|A\|$. From this, we can see that the map

$$L_\Phi(\phi, \psi) := \Phi(|\phi\rangle\langle\psi|)$$

is a bounded sesquilinear form, so that (by Proposition A.63), there is a unique bounded operator ρ such that $\Phi(|\phi\rangle\langle\psi|) = \langle \psi, \rho\phi \rangle$ for all ϕ and ψ. Since $|\phi\rangle\langle\phi|$ is self-adjoint and non-negative, $L_\Phi(\phi, \phi)$ is real and non-negative, which means that ρ is self-adjoint (by Proposition A.63) and non-negative.

Meanwhile, if $\{e_j\}$ is an orthonormal basis for $\mathbf{H}$, then by Definition 19.2,

$$\begin{aligned}
\mathrm{trace}(\rho) &= \lim_{N\to\infty} \sum_{j=1}^{N} \langle e_j, \rho e_j \rangle \\
&= \lim_{N\to\infty} \Phi\left(|e_1\rangle\langle e_1| + \cdots + |e_N\rangle\langle e_N| \right) \\
&= \Phi(I) = 1.
\end{aligned}$$

In passing from the second line to the third, we have used Point 4 of Definition 19.6. Thus, ρ is a density matrix.

We have now found a density matrix ρ such that $\Phi(|\phi\rangle\langle\psi|)$ agrees with $\mathrm{trace}(\rho\,|\phi\rangle\langle\psi|)$ for all $\phi, \psi \in \mathbf{H}$. By linearity, $\Phi(A) = \mathrm{trace}(\rho A)$ for all finite-rank operators A (see Exercise 4). Now, if $\{e_j\}$ is an orthonormal basis for $\mathbf{H}$, let P_N be the orthogonal projection onto the span of $e_1, \ldots, e_N$. Then for any $A \in \mathcal{B}(\mathbf{H})$, the operator $P_N A$ has finite rank and $P_N A\psi \to A\psi$ for all $\psi \in \mathbf{H}$. Thus, for all $A \in \mathcal{B}(\mathbf{H})$,

$$\Phi(A) = \lim_{N\to\infty} \Phi(P_N A) = \lim_{N\to\infty} \mathrm{trace}(\rho P_N A) = \mathrm{trace}(\rho A),$$

by Proposition 19.8 ∎

Our next result shows that our new notion of the state of a system includes our old notion.

Proposition 19.10 *For any unit vector $\psi \in \mathbf{H}$, let $|\psi\rangle\langle\psi|$, in accordance with Notation 3.29, denote the orthogonal projection onto the span of ψ. Then $|\psi\rangle\langle\psi|$ is a density matrix and for all $A \in \mathcal{B}(\mathbf{H})$, we have*

$$\mathrm{trace}(|\psi\rangle\langle\psi|\, A) = \langle \psi, A\psi \rangle.$$

Note that if $\psi_2 = e^{i\theta}\psi_1$, then $|\psi_1\rangle\langle\psi_1| = |\psi_2\rangle\langle\psi_2|$. Thus, from our new point of view, we may say that the reason ψ_1 and ψ_2 represent the same "physical state" is that they determine the same density matrix.

Proof. Since it is an orthogonal projection, $|\psi\rangle\langle\psi|$ is bounded, self-adjoint, and non-negative. To compute its trace, we choose an orthonormal basis

$\{e_j\}$ for $\mathbf{H}$ with $e_1 = \psi$, which gives $\mathrm{trace}(|\psi\rangle\langle\psi|) = 1$. Using the same orthonormal basis, we compute that, for any $A \in \mathcal{B}(\mathbf{H})$,

$$\mathrm{trace}(|\psi\rangle\langle\psi|\, A) = \sum_j \langle e_j, \psi\rangle\langle\psi, Ae_j\rangle = \langle\psi, A\psi\rangle,$$

as desired. ∎

Definition 19.11 *A density matrix $\rho \in \mathcal{B}(\mathbf{H})$ is a* **pure state** *if there exists a unit vector $\psi \in \mathbf{H}$ such that ρ is equal to the orthogonal projection onto the span of ψ. The density matrix ρ is called a* **mixed state** *if no such unit vector ψ exists.*

An isolated system that is in a pure state initially will remain in a pure state for all later times, since the initial state ψ_0 evolves to the pure state $e^{-i\hat{H}t/\hbar}\psi_0$, where $\hat{H}$ is the Hamiltonian for the system. But if a system is interacting with its environment, then as discussed in Sect. 19.5, the system may move into a mixed state at a later time.

There are several different ways of characterizing the pure states as a subset of the density matrices. First, it is not hard to see (Exercise 6) that a density matrix ρ is a pure state if and only if $\mathrm{trace}(\rho^2) = 1$. Second, the set of density matrices is a convex set, since if ρ_1 and ρ_2 are non-negative and have trace 1, then so is $\lambda\rho_1 + (1 - \lambda)\rho_2$, for $0 < \lambda < 1$. According to Exercise 7, the pure states are precisely the extreme points of this set. That is, a density matrix ρ is a pure state if and only if it *cannot* be expressed as $\rho = \lambda\rho_1 + (1 - \lambda)\rho_2$ where ρ_1 and ρ_2 are distinct density matrices and λ belongs to $(0, 1)$. Third, we may define the *von Neumann entropy* $S(\rho)$ of a density matrix ρ by

$$S(\rho) = \mathrm{trace}(-\rho\log\rho),$$

where $\rho\log\rho$ is defined by the functional calculus. (Since $\lim_{\lambda\to 0+}\lambda\log\lambda = 0$, we interpret $0\log 0$ as being 0.) Since the eigenvalues of ρ are all between 0 and 1, we see that $-\rho\log\rho$ is a non-negative self-adjoint operator, which has a well-defined trace, which may have the value $+\infty$. According to Exercise 8, a density matrix ρ is a pure state if and only if $S(\rho) = 0$.

Suppose that we have two pure states, coming from unit vectors ψ_1 and ψ_2. Then there are two different senses in which we can take a *superposition*, that is, linear combination, of the corresponding quantum states. If we use our old point of view, in which the states are vectors in $\mathbf{H}$, then we may take the linear combination $c_1\psi_1 + c_2\psi_2$, and then normalize this vector to be a unit vector. If we use our new point of view, in which the states are density matrices, then we may take the linear combination $c_1\,|\psi_1\rangle\langle\psi_1| + c_2\,|\psi_2\rangle\langle\psi_2|$, where in this case c_1 and c_2 should be non-negative and should add to 1. These two notions of superposition are different, since

$$C\,|c_1\psi_1 + c_2\psi_2\rangle\langle c_1\psi_1 + c_2\psi_2| \neq c_1\,|\psi_1\rangle\langle\psi_1| + c_2\,|\psi_2\rangle\langle\psi_2|, \tag{19.2}$$

no matter how the constant C is chosen. After all, the state on the left-hand side of (19.2) is a pure state, whereas (unless ψ_2 is a multiple of ψ_1), the state on the right-hand side of (19.2) is a mixed state, since the range of this operator is 2-dimensional rather than 1-dimensional.

Physicists call the first sort of superposition (in which we take a linear combination of vectors in **H**) *coherent superposition* or *quantum superposition*, and they call the second sort of superposition (in which we take a linear combination of the associated density matrices) *incoherent superposition*. The reason for the term "coherent" is that coherent superposition depends on the *phases* of the coefficients. That is, if ψ_1 and ψ_2 are linearly independent, the vector $c_1 e^{i\theta}\psi_1 + c_2 e^{i\phi}\psi_2$ does not represent the same quantum state as $c_1\psi_1 + c_2\psi_2$, unless $e^{i\theta} = e^{i\phi}$. By contrast, the density matrix associated with $e^{i\theta}\psi$ is the same as the density matrix associated with ψ, and so the phases have no effect when taking linear combinations of the density matrices associated to vectors in **H**. When taking a coherent superposition, there is no simple relationship between the expectation value of an observable in the states ψ_1 and ψ_2 and the expectation value of the same observable in the state $c_1\psi_1 + c_2\psi_2$. On the other hand, when taking an incoherent superposition, expectation values in the new state are just linear combinations of the original expectation values:

$$\mathrm{trace}\left((c_1\,|\psi_1\rangle\langle\psi_1| + c_2\,|\psi_2\rangle\langle\psi_2|)A\right) = c_1\,\langle\psi_1, A\psi_1\rangle + c_2\,\langle\psi_2, A\psi_2\rangle.$$

19.4 Modified Axioms for Quantum Mechanics

We may now modify the axioms of quantum mechanics introduced in Sect. 3.6 to incorporate density matrices, beginning with our revised notion of a state.

Axiom 6 *The state of a quantum system is described by a density matrix ρ on an appropriate Hilbert space* **H**. *If A is any bounded operator on* **H**, *the expectation value of A in the state ρ is given by the quantity* $\mathrm{trace}(\rho A) - \mathrm{trace}(A\rho)$.

In Axiom 6, we assume that A is bounded, so that $\mathrm{trace}(\rho A)$ and $\mathrm{trace}(A\rho)$ are defined and equal by Proposition 19.3. If A is unbounded and self-adjoint, we can construct a probability measure μ_ρ^A describing the probabilities for measurements of A in the state ρ, by the formula

$$\mu_\rho^A(E) = \mathrm{trace}(\rho 1_E(A)),$$

where $1_E(A)$ is defined by the functional calculus.

We then define the expectation value of A in the state ρ as $\int_\mathbb{R} \lambda\,d\mu_\rho^A(\lambda)$, provided the integral is absolutely convergent. If the integral is absolutely convergent, it is reasonable to hope that both ρA and $A\rho$ will be densely defined and bounded, that (the bounded extension to **H** of) these operators will be trace class, and that both trace(ρA) and trace($A\rho$) will coincide with $\int_{\mathbb{R}} \lambda \, d\mu_\rho^A(\lambda)$. We will not, however, enter into an investigation of this issue.

Next, we propose a variant of Axiom 4, describing the "collapse of the wave function."

Axiom 7 *Suppose a quantum system is initially in a state ρ and a measurement of a self-adjoint operator A with point spectrum is performed. If the measurement results in the value λ for A, then immediately after the measurement, the system will be in the state ρ', where*

$$\rho' = \frac{1}{Z} P_\lambda \rho P_\lambda.$$

Here P_λ is the orthogonal projection onto the λ-eigenspace of A and $Z = $ trace($P_\lambda \rho P_\lambda$).

Note that if ρ is non-negative, self-adjoint, and trace class, then $P_\lambda \rho P_\lambda$ is also non-negative, self-adjoint, and trace class. Implicit in Axiom 7 is the assumption that the measurement can only result in values λ for which $P_\lambda \rho P_\lambda$ is nonzero. In particular, λ must be an eigenvalue for A.

Finally, we introduce the notion of time-evolution for our new notion of "state."

Axiom 8 *The time evolution of the state of the system is described by the following equation for a time-dependent density matrix $\rho(t)$:*

$$\frac{d\rho}{dt} = -\frac{1}{i\hbar}[\rho, \hat{H}]. \tag{19.3}$$

This equation may be solved, formally, by setting

$$\rho(t) = e^{-it\hat{H}/\hbar} \rho_0 e^{it\hat{H}/\hbar}, \tag{19.4}$$

where ρ_0 is the state of the system at time $t = 0$.

There are some domain issues involved in the interpretation of the equation (19.3). Rather than entering into an examination of those issues here, we will simply take (19.4) as the definition of the time-evolution of a density matrix. Presumably, if ρ_0 is nice enough, then the map $t \mapsto \rho(t)$ will be differentiable as a curve in the Banach space $\mathcal{B}(\mathbf{H})$ and its derivative will be (an extension of) the operator on the right-hand side of (19.3). By comparison, it follows from Stone's theorem and Lemma 10.17 that the family of pure states $\psi(t) := e^{-it\hat{H}/\hbar}\psi_0$ satisfies the Schrödinger equation in the natural Hilbert space sense if and only if ψ_0 belongs to the domain of $\hat{H}$. To see that the time-evolution in (19.4) is consistent with the previously defined time-evolution of pure states, observe that

$$e^{-it\hat{H}/\hbar} |\psi_0\rangle\langle\psi_0| e^{it\hat{H}/\hbar} = |e^{-it\hat{H}/\hbar}\psi_0\rangle\langle e^{-it\hat{H}/\hbar}\psi_0| = |\psi(t)\rangle\langle\psi(t)|,$$

since $(e^{it\hat{H}/\hbar})^* = e^{-it\hat{H}/\hbar}$.

It should be noted that (19.3) differs by a minus sign from the time-evolution in the Heisenberg picture of quantum mechanics (Definition 3.20). Although this difference may seem strange, keep in mind that in Axiom 8, we are *not* adopting the Heisenberg point of view, in which the states are independent of time and the observables evolve in time. Rather, we are adopting a modified version of the Schrödinger picture, in which it is the *states* that evolve in time, but where the states are now certain sorts of operators. Even though both the states and the observables are now operators, the observables (in the Heisenberg picture) and the states (in the Schrödinger picture) must evolve in *opposite directions* in time, in order for the expectation values of the observables to be the same in the two pictures.

19.5 Composite Systems and the Tensor Product

As discussed in Sect. 3.11, the Hilbert space for two (nonidentical, spinless) particles moving in $\mathbb{R}^3$ is $L^2(\mathbb{R}^6)$. Given a unit vector (i.e., a pure state) ψ in $L^2(\mathbb{R}^6)$, the quantity $\left|\psi(\mathbf{x}^1, \mathbf{x}^2)\right|^2$ represents the joint probability distribution for the position $\mathbf{x}^1$ of the first particle and the position $\mathbf{x}^2$ of the second particle. The following result shows that $L^2(\mathbb{R}^6)$ is naturally isomorphic to the Hilbert tensor product of two copies of the Hilbert space for the individual particles, namely $L^2(\mathbb{R}^3)$.

Proposition 19.12 *Suppose that (X_1, μ_1) and (X_2, μ_2) are σ-finite measure spaces. Then there is a unique unitary map*

$$p : L^2(X_1, \mu_1) \hat{\otimes} L^2(X_2, \mu_2) \to L^2(X_1 \times X_2, \mu_1 \times \mu_2)$$

such that

$$p(\phi \otimes \psi)(x, y) = \phi(x)\psi(y)$$

for all $\phi \in L^2(X_1, \mu_1)$ and $\psi \in L^2(X_2, \mu_2)$.

Here $\hat{\otimes}$ denotes the Hilbert tensor product defined in Appendix A.4.5.
Proof. For simplicity of notation, we suppress the dependence of L^2 spaces on the measure, writing, say, $L^2(X_1)$ rather than $L^2(X_1, \mu_1)$. Consider first the algebraic (i.e., uncompleted) tensor product $L^2(X_1) \otimes L^2(X_2)$. Using the universal property of tensor products, we can construct a linear map p of $L^2(X_1) \otimes L^2(X_2) \to L^2(X_1 \times X_2)$ determined uniquely by the requirement that

$$p(\phi \otimes \psi)(x, y) = \phi(x)\psi(y).$$

Now, every element of the algebraic tensor product $L^2(X_1) \otimes L^2(X_2)$ can be expressed as a linear combination of elements of the form $\phi_j \otimes \psi_j$, with

$\phi_j \in L^2(X_1)$ and ψ_j in $L^2(X_2)$. By computing on such linear combinations, we can easily verify that p is isometric. Thus, by the bounded linear transformation (BLT) theorem (Theorem A.36), p has a unique isometric extension to a map of the completed tensor product $L^2(X_1)\hat{\otimes}L^2(X_2)$ into $L^2(X_1 \times X_2)$.

It remains only to show that p is surjective. Since both measures are σ-finite, it is a simple exercise to reduce the problem to the case where μ_1 and μ_2 are finite, which we henceforth assume. Suppose $\psi \in L^2(X_1 \times X_2)$ is orthogonal to the image of p. Then ψ is orthogonal to the indicator function of every measurable rectangle, and hence to the indicator function of any finite disjoint union of measurable rectangles. The collection $\mathcal{A}$ of such disjoint unions is an algebra of sets. Let $\mathcal{M}$ denote the collection of measurable subsets E of $X_1 \times X_2$ such that the integral of ψ over E is zero. Then $\mathcal{M}$ is a monotone class containing $\mathcal{A}$. By the monotone class lemma (Theorem A.8), $\mathcal{M}$ contains the σ-algebra generated by $\mathcal{A}$, which is the σ-algebra on which $\mu_1 \times \mu_2$ is defined. Thus, the integral of ψ over every measurable set is zero, which implies that ψ is zero almost everywhere. ∎

The preceding example suggests the following general principle.

Axiom 9 *The Hilbert space for a composite system made up of two subsystems is the Hilbert tensor product $\mathbf{H}_1\hat{\otimes}\mathbf{H}_2$ of the Hilbert spaces $\mathbf{H}_1$ and $\mathbf{H}_2$ describing the subsystems.*

If A and B are bounded operators on $\mathbf{H}_1$ and $\mathbf{H}_2$, respectively, then there is a unique bounded operator $A \otimes B$ on $\mathbf{H}_1\hat{\otimes}\mathbf{H}_2$ such that

$$(A \otimes B)(\phi \otimes \psi) = (A\phi) \otimes (B\psi)$$

for all $\phi \in \mathbf{H}_1$ and $\psi \in \mathbf{H}_2$. (See Appendix A.4.5.)

Theorem 19.13 *Suppose that ρ is a density matrix on $\mathbf{H}_1\hat{\otimes}\mathbf{H}_2$. Then there exists a unique density matrix $\rho^{(1)}$ on $\mathbf{H}_1$ with the property that*

$$\operatorname{trace}(\rho^{(1)}A) = \operatorname{trace}(\rho(A \otimes I)) \tag{19.5}$$

for all $A \in \mathcal{B}(\mathbf{H}_1)$. We call $\rho^{(1)}$ the partial trace of ρ with respect to $\mathbf{H}_2$. If $\{f_k\}$ is an orthonormal basis for $\mathbf{H}_2$, then the operator $\rho^{(1)}$ satisfies

$$\langle \phi, \rho^{(1)}\psi \rangle = \sum_k \langle \phi \otimes f_k, \rho(\psi \otimes f_k) \rangle \tag{19.6}$$

for all $\phi, \psi \in \mathbf{H}_1$. Similarly, there is a unique density matrix $\rho^{(2)}$ on $\mathbf{H}_2$ satisfying $\operatorname{trace}(\rho^{(2)}B) = \operatorname{trace}(\rho(I \otimes B))$ for all $B \in \mathcal{B}(\mathbf{H}_2)$. If $\{e_j\}$ is an orthonormal basis for $\mathbf{H}_1$, then $\rho^{(2)}$ satisfies

$$\langle \phi, \rho^{(2)}\psi \rangle = \sum_j \langle e_j \otimes \phi, \rho(e_j \otimes \psi) \rangle \tag{19.7}$$

for all $\phi, \psi \in \mathbf{H}_2$.

The motivation for the terminology "partial trace" is provided by (19.6) and (19.7), which are similar to the formula for the trace of an operator, except that the sums range only over a basis for one of the two Hilbert spaces. One special case of Theorem 19.13 is the one in which the density matrix ρ is of the form $\rho = \rho_1 \otimes \rho_2$, where ρ_1 and ρ_2 are density matrices on $\mathbf{H}_1$ and $\mathbf{H}_2$, respectively. (Any operator ρ of this form is a density matrix on $\mathbf{H}_1 \times \mathbf{H}_2$.) In that case, it is not hard to see that $\rho^{(1)} = \rho_1$ and $\rho^{(2)} = \rho_2$. We may describe this case by saying that the state of the first system is "independent" of the state of the second system.

Lemma 19.14 *For any sequence $A_n \in \mathcal{B}(\mathbf{H}_1)$, if $\|A_n \psi - A\psi\| \to 0$ for some $A \in \mathcal{B}(\mathbf{H})$ and all $\psi \in \mathbf{H}_1$, then*

$$\|(A_n \otimes I)\phi - (A \otimes I)\phi\| \to 0$$

for all $\phi \in \mathbf{H}_1 \otimes \mathbf{H}_2$. A similar result holds for operators of the form $I \otimes B_n$.

Proof. See Exercise 9. ∎

Proof of Theorem 19.13. The existence and uniqueness of $\rho^{(1)}$ and $\rho^{(2)}$ follow from Lemma 19.14 and Theorem 19.9. Meanwhile, if $\{e_j\}$ is an orthonormal basis for $\mathbf{H}_1$ and $\{f_k\}$ is an orthonormal basis for $\mathbf{H}_2$, we have

$$
\begin{aligned}
\langle \phi, \rho^{(1)} \psi \rangle &= \mathrm{trace}(\rho^{(1)} |\psi\rangle\langle\phi|) \\
&= \sum_{j,k} \langle e_j \otimes f_k, \rho(|\psi\rangle\langle\phi| \otimes I)(e_j \otimes f_k)\rangle \\
&= \sum_{j,k} \langle e_j \otimes f_k, \rho(\psi \langle \phi, e_j\rangle \otimes f_k)\rangle \\
&= \sum_{k} \left\langle \left(\sum_{j} \langle e_j, \phi\rangle\, e_j\right) \otimes f_k, \rho(\psi \otimes f_k) \right\rangle \\
&= \sum_{k} \langle \phi \otimes f_k, \rho(\psi \otimes f_k)\rangle .
\end{aligned}
$$

This is the desired formula for $\langle \phi, \rho^{(1)} \psi \rangle$. Note that because ρ is trace class and $|\psi\rangle\langle\phi| \otimes I$ is bounded, $\rho(|\psi\rangle\langle\phi| \otimes I)$ is trace class, in which case the sum in the second line is absolutely convergent, by Proposition 19.3. Thus, we are allowed to rearrange the sum freely. ∎

Suppose we have two quantum systems with Hilbert spaces $\mathbf{H}_1$ and $\mathbf{H}_2$ and Hamiltonians $\hat{H}_1$ and $\hat{H}_2$. If the two systems do not interact with each other and the composite system is initially in a (pure) state of the form $\phi_0 \otimes \psi_0$, then we expect that at some later time, the composite system will be in the state $\phi(t) \otimes \psi(t)$, where $\phi(t) = e^{-it\hat{H}_1/\hbar}\psi_0$ and $\psi(t) = e^{-it\hat{H}_2/\hbar}$. Ignoring domain considerations, we may compute that

$$i\hbar\frac{d}{dt}\left[\phi(t) \otimes \psi(t)\right] = (\hat{H}_1\phi(t)) \otimes \psi(t) + \phi(t) \otimes (\hat{H}_2\psi(t))$$

$$= (\hat{H}_1 \otimes I + I \otimes \hat{H}_2)(\phi(t) \otimes \psi(t)).$$

This calculation suggests that the correct Hamiltonian for a noninteracting composite system is the operator $\hat{H}_1 \otimes I + I \otimes \hat{H}_2$.

It is not, however, obvious how to select a domain for $\hat{H}_1 \otimes I + I \otimes \hat{H}_2$ in such a way that this operator will be self-adjoint. (The reader is invited to try to choose such a domain "by hand.") The easiest way to deal with this issue is to use Stone's theorem, as in the following definition.

Definition 19.15 *If A and B are self-adjoint operators on $\mathbf{H}_1$ and $\mathbf{H}_2$, define the operator $A \otimes I + I \otimes B$ to be the infinitesimal generator of the strongly continuous one-parameter unitary group $e^{itA} \otimes e^{itB}$. Thus, by Stone's theorem, $A \otimes I + I \otimes B$ is self-adjoint.*

It is not hard to check that $e^{itA} \otimes e^{itB}$ is indeed strongly continuous. In the case $B = 0$, the operator $A \otimes I$ is defined as the infinitesimal generator of $e^{itA} \otimes I$. If A and B happen to be bounded, then $A \otimes I + I \otimes B$ defined by Definition 19.15 coincides with $A \otimes I + I \otimes B$ defined as the sum of tensor products of bounded operators, as in Sect. A.4.5.

Axiom 10 *Suppose $\mathbf{H}_1$ and $\mathbf{H}_2$ are the Hilbert spaces for two quantum systems, with Hamiltonians $\hat{H}_1$ and $\hat{H}_2$, respectively. Then the Hamiltonian for the noninteracting composite system is $\hat{H}_1 \otimes I + I \otimes \hat{H}_2$, where the domain of $\hat{H}_1 \otimes I + I \otimes \hat{H}_2$ is as in Definition 19.15.*

A physicist would write $\hat{H}_1 \otimes I + I \otimes \hat{H}_2$ simply as $\hat{H}_1 + \hat{H}_2$, with the understanding that $\hat{H}_1$ acts only on the first factor in the tensor product and $\hat{H}_2$ acts only on the second factor.

In general, the two components of a composite system will interact, in which case the Hamiltonian for the composite system is typically of the form

$$\hat{H} = \hat{H}_1 \otimes I + I \otimes \hat{H}_2 + \hat{H}_{\text{int}},$$

where $\hat{H}_{\text{int}}$ is an "interaction term." Often, the interaction term may be considered "small" compared with the other terms in the Hamiltonian. Consider, for example, a system consisting of particles in a box, with a barrier dividing the box in half. Suppose the particles interact by means of a two-particle potential of the form $\sum_{j<k} V(\mathbf{x}^j - \mathbf{x}^k)$ (Sect. 2.3.2) and that $V(\mathbf{x}^j - \mathbf{x}^k)$ is very small unless the two particles are close together. There will typically be far more pairs of nearby particles in which the two particles are on the same side of the box than nearby pairs on opposite sides. Thus, even though the interaction between the two systems may substantially affect the behavior of the composite system over long periods of time, it is

still reasonable to think of $\hat{H}_1 \otimes I$ as "the energy of the first subsystem" and $I \otimes \hat{H}_2$ as "the energy of the second subsystem."

Suppose we start out in a state ρ of the composite system for which the state $\rho^{(1)}$ of the first subsystem is a pure state. If the system is an interacting one, the first subsystem will probably not remain in a pure state at later times. Indeed, suppose that the second subsystem is very large system having temperature T. Then, according to the postulates of quantum statistical mechanics, we are supposed to believe that once the two systems have reached thermal equilibrium, the state of the first subsystem will be given by the following highly mixed state:

$$\rho^{(1)} = \frac{1}{Z(\beta)} e^{-\beta \hat{H}_1}. \tag{19.8}$$

Here $\beta = 1/(k_B T)$, where k_B is Boltzmann's constant, and $Z(\beta)$ is a normalization constant, known as the *partition function* of the theory, given by $Z(\beta) = \text{trace}(e^{-\beta \hat{H}_1})$.

Of course, for this idea to make sense, $e^{-\beta \hat{H}_1}$ must be trace class. This will be the case provided that $\hat{H}_1$ has discrete spectrum with eigenvalues tending to $+\infty$ at some reasonable rate. Thus, in quantum statistical mechanics, the expectation value of some observable A for the first subsystem will be (once equilibrium is reached)

$$\langle A \rangle = \frac{1}{Z} \text{trace}(e^{-\beta \hat{H}_1} A). \tag{19.9}$$

In particular, when $A = \hat{H}_1$, (19.9) provides a natural generalization of Planck's model of blackbody radiation; compare Exercise 2 in Chap. 1.

19.6 Multiple Particles: Bosons and Fermions

As discussed in Sect. 17.8, each type of particle (electron, proton, neutron, etc.) has a *spin* l, where the possible value for l are

$$l = 0, \frac{1}{2}, 1, \frac{3}{2}, \ldots.$$

The Hilbert space for a particle moving in $\mathbb{R}^3$ and having spin l is $L^2(\mathbb{R}^3) \hat{\otimes} V_l$, where V_l is a finite-dimensional Hilbert space that carries an irreducible *projective* unitary representation of $SO(3)$ of dimension $2l + 1$. There is a natural unitary identification of $L^2(\mathbb{R}^3) \hat{\otimes} V_l$ with $L^2(\mathbb{R}^3; V_l)$, the space of square-integrable functions on $\mathbb{R}^3$ with values in V_l, in which the element $\psi \otimes v$ of $L^2(\mathbb{R}^3) \hat{\otimes} V_l$ is identified with the function

$$\mathbf{x} \mapsto \psi(\mathbf{x}) v$$

in $L^2(\mathbb{R}^3; V_l)$.

Now, we have already mentioned, in Sect. 3.11, the idea that in quantum mechanics, *identical particles are indistinguishable.* Let us think about this in the case of two identical particles with spin l. Our first guess as to the Hilbert space for such a system is the tensor product of two copies of $L^2(\mathbb{R}^3; V_l)$, which may be identified with

$$L^2(\mathbb{R}^6; V_l \otimes V_l).$$

If ψ is a unit vector in this space, thought of as a pure state, then saying that the two particles are "indistinguishable" means that $\psi(\mathbf{x}^2, \mathbf{x}^1)$ should represent the same physical state as $\psi(\mathbf{x}^1, \mathbf{x}^2)$, that is, $\psi(\mathbf{x}^2, \mathbf{x}^1) = c\psi(\mathbf{x}^1, \mathbf{x}^2)$ for some nonzero constant c. Applying this rule twice shows that c must be either 1 or -1.

A variety of theoretical and experimental considerations suggest the following principle: For particles with integer spin ($l = 0, 1, \ldots$), the constant c in the preceding paragraph is 1, whereas for particles with half-integer spin ($l = 1/2, 3/2, \ldots$) the constant c is -1. Particles with integer spin are called *bosons* and particles with half-integer spin are called *fermions.* We encode the discussion in the two preceding paragraphs in the following axiom.

Axiom 11 *Consider a collection of N identical particles moving in $\mathbb{R}^3$ and having integer spin l. Then the Hilbert space for such a collection is the subspace of $L^2(\mathbb{R}^{3N}; (V_l)^{\otimes N})$ consisting of those square-integrable functions ψ for which*

$$\psi(\mathbf{x}^{\sigma(1)}, \mathbf{x}^{\sigma(2)}, \ldots, \mathbf{x}^{\sigma(N)}) = \psi(\mathbf{x}^1, \mathbf{x}^2, \ldots, \mathbf{x}^N)$$

for every permutation σ. Consider also a collection of N identical particles moving in $\mathbb{R}^3$ and having half-integer spin l. Then the Hilbert space for such a collection is the subspace of $L^2(\mathbb{R}^{3N}; (V_l)^{\otimes N})$ consisting of those square-integrable functions ψ for which

$$\psi(\mathbf{x}^{\sigma(1)}, \mathbf{x}^{\sigma(2)}, \ldots, \mathbf{x}^{\sigma(N)}) = \text{sign}(\sigma)\psi(\mathbf{x}^1, \mathbf{x}^2, \ldots, \mathbf{x}^N)$$

for every permutation σ.

One may well ask *why* Axiom 11 holds. More specifically, one may first ask why it is that identical particles are indistinguishable, and then separately ask why integer-spin particles are bosons and half-integer-spin particles are fermions. Both questions are best answered from the point of view of quantum field theory, to which ordinary nonrelativistic quantum mechanics is an approximation.

In field theory, one starts with a "classical" field theory, meaning a differential equation for functions $\phi(\mathbf{x}, t)$ on $\mathbb{R}^4$ with values in some finite-dimensional vector space. Electromagnetic fields, for example, are—at any one fixed time—functions on $\mathbb{R}^3$ with values in $\mathbb{R}^6$, where $\mathbb{R}^6$ describes the three components of the electric field and the three components of the magnetic field. These functions on $\mathbb{R}^3$ then evolve in time according to Maxwell's equation. In *quantum* field theory, one regards, say, Maxwell's equations as a sort of infinite-dimensional dynamical system, which we may quantize in something like the way we quantize Newton's equation to get ordinary nonrelativistic quantum mechanics. In the quantum version of Maxwell's equations, the energy in each mode of the fields is "quantized," meaning that one can only add energy to each mode in multiples of a certain unit (or "quantum") of energy. This is analogous to the quantum harmonic oscillator, in which the allowed energies differ by integer multiples of the $\hbar\omega$. In quantum field theory, then, a *particle* is one quantum of excitation of a certain field.

For simplicity, let us think of a field theory in which the classical fields take values in $\mathbb{R}$. Then even at the classical level, it is possible to think that we have something like particles, namely localized bumps in the field $\phi(\mathbf{x})$ located at several different points in space. These bumps might, for example, be in the shape of a Gaussian wave-packet, that is, a Gaussian envelope multiplied by a sinusoidally oscillating function. From this point of view, we can gain some understanding of why identical particles are indistinguishable. Suppose we have a Gaussian wave packet near a point $\mathbf{a}$ in $\mathbb{R}^3$ and then an identically shaped Gaussian wave packet near another point $\mathbf{b}$. The state $\phi(\mathbf{x})$ of the field is precisely the same as if we have a packet near $\mathbf{b}$ and then also a packet near $\mathbf{a}$. That is to say, there is no distinct state of the system that corresponds to interchanging the two particles; whichever bump we think of as the "first" particle, we have the same field $\phi(\mathbf{x})$. Even in the quantum version of such a system, there no meaning to asking which is the first particle and which is the second. Thus, even in nonrelativistic quantum mechanics, which is a low-energy approximation to quantum field theory, we expect identical particles to be indistinguishable.

Although the preceding discussion does not explain the distinction between bosons and fermions, that distinction also emerges from quantum field theory, through something called the *spin–statistics theorem* (see, e.g., [38]).

19.7 "Statistics" and the Pauli Exclusion Principle

The spin of an electron is equal to $1/2$ and electrons are, therefore, fermions. The famous *Pauli exclusion principle* is a consequence of the fermionic nature of electrons. Pauli's principle states that two electrons cannot be in the same state at the same time. This means that if ψ is a square-integrable, $\mathbb{C}^2$-valued function on $\mathbb{R}^3$ (which could describe the state of a single electron), then the function $\Psi : \mathbb{R}^6 \to \mathbb{C}^2 \otimes \mathbb{C}^2$ given by

$$\Psi(\mathbf{x}^1, \mathbf{x}^2) = \psi(\mathbf{x}^1) \otimes \psi(\mathbf{x}^2)$$

is *not* a possible state of a two-electron system, since Ψ does not satisfy Axiom 11. On the other hand, if ψ_1 and ψ_2 are two linearly independent elements of $L^2(\mathbb{R}^3; \mathbb{C}^2)$, then the function $\Phi : \mathbb{R}^6 \to \mathbb{C}^2 \otimes \mathbb{C}^2$ given by

$$\Phi(\mathbf{x}^1, \mathbf{x}^2) = \psi_1(\mathbf{x}^1)\psi_2(\mathbf{x}^2) - \psi_2(\mathbf{x}^1)\psi_1(\mathbf{x}^2) \qquad (19.10)$$

is a possible state of a two-electron system. [If ψ_1 and ψ_2 are independent, then Φ is a nonzero element of $L^2(\mathbb{R}^6; \mathbb{C}^2 \otimes \mathbb{C}^2)$, which can then be normalized to be a unit vector. See Exercise 10.]

Let us try to understand the implications of the Pauli exclusion principle for multielectron atoms. Let us model an N-electron atom as having a nucleus with positive charge Nq, where the charge of a single electron is $-q$. Since the nucleus is much more massive than the electrons, we can treat the nucleus as being fixed and the electrons as moving in potential of the form $-Nq/|\mathbf{x}|$. As a *very* rough approximation to the structure of such an atom, we can ignore the electron–electron interaction and take a Hamiltonian of the form

$$\hat{H} = \sum_{j=1}^{N} \left(-\frac{\hbar^2}{2m}\Delta^j - \frac{Nq^2}{|\mathbf{x}^j|} \right),$$

where Δ^j is the Laplacian acting on the jth variable. That is, we are taking our Hamiltonian to be simply

$$(\hat{H} \otimes I \otimes I \otimes \cdots \otimes I) + (I \otimes \hat{H} \otimes I \otimes \cdots \otimes I) + (I \otimes I \otimes \hat{H} \otimes \cdots \otimes I) + \cdots,$$

where $\hat{H}$ is the Hamiltonian for a single electron.

If, say, N is even, the lowest-energy state for this Hamiltonian in the antisymmetric subspace of $L^2(\mathbb{R}^{3N}; (\mathbb{C}^2)^{\otimes N})$ will be

$$\Psi_0(\mathbf{x}^1, \mathbf{x}^2, \ldots, \mathbf{x}^N)$$
$$= \mathrm{AS}\left(\psi_0^+(\mathbf{x}^1) \otimes \psi_0^-(\mathbf{x}^2) \otimes \psi_1^+(\mathbf{x}^3) \otimes \cdots \otimes \psi_{N/2}^+(\mathbf{x}^{N-1}) \otimes \psi_{N/2}^-(\mathbf{x}^N) \right).$$
$$(19.11)$$

If N is odd, the product ends with $\psi_{(N+1)/2}^+(\mathbf{x}^N)$. The notation in (19.11) is as follows. First, AS is the antisymmetrization operator, given by

$$\mathrm{AS}(f)(\mathbf{x}^1, \ldots, \mathbf{x}^N) = \sum_{\sigma \in S^N} \mathrm{sign}(\sigma) f(\mathbf{x}^{\sigma(1)}, \mathbf{x}^{\sigma(2)}, \cdots, \mathbf{x}^{\sigma(N)}).$$

Second, the functions $\psi_0, \psi_1, \psi_2, \ldots$ are the eigenvectors in $L^2(\mathbb{R}^3)$ for the Hamiltonian of a single particle in $\mathbb{R}^3$ moving in a potential of the form $-Nq^2/|\mathbf{x}|$, arranged so that the eigenvalues of ψ_j are weakly increasing with j. The ψ_j's are just the states computed in Chap. 18, but with q replaced by $\sqrt{N}q$. Third, $\psi_j^+(\mathbf{x})$ denotes $\psi_j(\mathbf{x}) \otimes e_1$ and $\psi_j^-(\mathbf{x})$ denotes $\psi_j(\mathbf{x}) \otimes e_2$, where $\{e_1, e_2\}$ is the standard basis for $\mathbb{C}^2$.

What the expression for Ψ_0 means is that, if we ignore (at first) the interaction between the electrons, but retain the Pauli exclusion principle, then we put the first electron into the ground state of the single-electron system, with "spin up" (i.e., tensored with e_1). Then we put the second electron into the ground state with "spin down" (tensored with e_2). Then the third electron goes into the first excited state of the single-electron system with spin up, and so on. Of course, this model of a multielectron atom is very rough, since the interaction between the electrons actually plays a significant role. Nevertheless, this model highlights the critical role played by the exclusion principle, which forces successive electrons to go into higher and higher energy states. In particular, this crude approximation suggests (correctly!) that even for more realistic models of a multielectron atom, the lowest energy level in the antisymmetric subspace of $L^2(\mathbb{R}^{3N}; (\mathbb{C}^2)^{\otimes N})$ is much higher than the lowest energy level of the same Hamiltonian in all of $L^2(\mathbb{R}^{3N}; (\mathbb{C}^2)^{\otimes N})$.

Meanwhile, in quantum statistical mechanics, one considers a large collection of identical particles confined to some finite region of space. If the system is isolated (rather than in thermal equilibrium with its environment), the goal of statistical mechanics is to "count" the number $N(E)$ of quantum states with energy less than E, as a function of E. [That is, $N(E)$ is number of eigenvalues for the Hamiltonian less than E, counted with their multiplicity.] As the preceding discussion of the Pauli exclusion principle suggests, we will get very different answers for $N(E)$ if the particles are fermions than if they are bosons. Bosons are said to follow *Bose–Einstein statistics*, whereas fermions are said to follow *Fermi–Dirac statistics*. The term "statistics" here refers to the different behavior of the two types of particles in quantum statistical mechanics. The spin–statistics theorem in quantum field theory tells us that particles with integer spin have to be bosons (obeying Bose–Einstein statistics) and particles with half-integer spin have to be fermions (obeying Fermi–Dirac statistics).

One fascinating example of quantum statistical mechanics occurs when the particles are bosons and the interaction between particles is negligible. In that case, the lowest energy state will simply be

$$\Psi_0(\mathbf{x}^1, \mathbf{x}^2, \ldots, \mathbf{x}^N) = \psi_0(\mathbf{x}^1) \otimes \psi_0(\mathbf{x}^2) \otimes \cdots \otimes \psi_0(\mathbf{x}^N),$$

where ψ_0 is the ground state of the single-particle system. Now, quantum statistical mechanics tells us that at a given temperature, the state of the system will be an (incoherent) superposition of the ground state and the various excited states. If the temperature is low enough, then the coefficient of the ground state will be close to 1, and thus, "all the particles are in the ground state." A system in such a state is called a *Bose–Einstein condensate*, a state that was predicted on theoretical grounds by Satyendra Nath Bose and Einstein in the 1920s. Bose–Einstein condensates were first observed experimentally in laser-cooled gases in June 1995 by Eric Cornell and Carl Wieman, in work for which they, along with Wolfgang Ketterle, were awarded the 2001 Nobel Prize in physics.

19.8   Exercises

1. Suppose that X is a Hilbert–Schmidt operator on $\mathbf{H}$ and that $\{e_j\}$ is an orthonormal basis for $\mathbf{H}$. Show that

$$\|X\|_{HS}^2 = \sum_{j,k} |\langle e_j, X e_k \rangle|^2 .$$

2. Given $\phi, \psi \in \mathbf{H}$, let $|\phi\rangle\langle\psi|$ denote the operator defined in Notation 3.28. Show that if $A \in \mathcal{B}(\mathbf{H})$ is trace class, then

$$\mathrm{trace}(A\,|\phi\rangle\langle\psi|) = \langle \psi, A\phi \rangle .$$

   Hint: If $\{e_j\}$ is an orthonormal basis for $\mathbf{H}$, then for any $\chi \in \mathbf{H}$, we have $\chi = \sum_j \langle e_j, \chi \rangle \, e_j$.

3. Suppose $\Phi : \mathcal{B}(\mathbf{H}) \to \mathbb{C}$ is a linear functional with the properties (1) that $\Phi(A)$ is real whenever A is self-adjoint and (2) that $\Phi(A)$ is real and non-negative whenever A is self-adjoint and non-negative. Show that if A is self-adjoint, then

$$- \|A\|\, \Phi(I) \le \Phi(A) \le \|A\|\, \Phi(I).$$

   Conclude that Φ is bounded relative to the operator norm on $\mathcal{B}(\mathbf{H})$.

   Hint: Show that if A is self-adjoint, then $\|A\|\,I + A$ and $\|A\|\,I - A$ are non-negative.

4. An operator $A \in \mathcal{B}(\mathbf{H})$ is said to have *finite rank* if range(A) is finite dimensional.

   (a) Show that if $A \in \mathcal{B}(\mathbf{H})$ has finite rank, then so does A^*.

   (b) Given $A \in \mathcal{B}(\mathbf{H})$, show that A has finite rank if and only if there exist vectors $\phi_1, \ldots, \phi_N$ and $\psi_1, \ldots, \psi_N$ such that

   $$A = |\phi_1\rangle\langle\psi_1| + \cdots + |\phi_N\rangle\langle\psi_N| .$$

   (c) Let A be any element of $\mathcal{B}(\mathbf{H})$, let $\{e_j\}$ be an orthonormal basis for $\mathbf{H}$, and let P_N be the orthogonal projection onto the span of $e_1, \ldots, e_N$. Show that $P_N A$ has finite rank and that for all $\psi \in \mathbf{H}$, we have

   $$\lim_{N \to \infty} \|P_N A\psi - A\psi\| = 0.$$

Note: This result shows that each bounded operator can be expressed as a strong limit of finite-rank operators. By contrast, if $\dim \mathbf{H} = \infty$, then Part (a) of Exercise 5 shows that not every bounded operator can be expressed as an operator-norm limit of finite-rank operators.

5. In this exercise, assume that $\dim \mathbf{H} = \infty$.

   (a) Show that if A has finite rank, then $\|A + cI\| \geq |c|$ for any $c \in \mathbb{C}$. (With $c = -1$, this shows that I is not an operator-norm limit of finite-rank operators.)

   (b) Let $\mathcal{K}(\mathbf{H})$ denote the closure of the finite-rank operators with respect to the operator norm on $\mathcal{B}(\mathbf{H})$. Let V denote the space of operators of the form $B + cI$, with $B \in \mathcal{K}(\mathbf{H})$. Define a linear functional $\Phi : V \to \mathbb{C}$ by $\Phi(B + cI) = c$ for all $B \in \mathcal{K}(\mathbf{H})$. Show that $|\Phi(A)| \leq \|A\|$ for all $A \in V$.

   Note: It can be shown that $\mathcal{K}(\mathbf{H})$ is precisely the space of compact operators on $\mathbf{H}$.

   (c) Let $\Psi_1 : \mathcal{B}(\mathbf{H}) \to \mathbb{C}$ be any linear functional such that $\Psi_1 = \Phi$ on V and such that $|\Psi_1(A)| \leq \|A\|$ for all $A \in \mathcal{B}(\mathbf{H})$. (Such a functional exists by the Hahn–Banach theorem.) Let $\Psi_2 : \mathcal{B}(\mathbf{H}) \to \mathbb{C}$ be defined by

   $$\Psi_2(A) = \frac{1}{2}(\Psi_1(A) + \overline{\Psi_1(A^*)}).$$

   Show that Ψ_2 satisfies Properties 1, 2, and 3 of Definition 19.6, but that there does not exist any density matrix ρ such that $\Psi_2(A) = \mathrm{trace}(\rho A)$ for all $A \in \mathcal{B}(\mathbf{H})$. (Thus, in light of Theorem 19.9, Ψ_2 must not satisfy Property 4 of Definition 19.6.)

6. In Exercises 6, 7, and 8, assume that each density matrix ρ is compact, so that ρ has an orthonormal basis $\{e_j\}$ of eigenvectors, for which the associated eigenvalues $\{\lambda_j\}$ are real and tend to zero as j tends to infinity. (Compare Theorem VI.16 in [34].)

   Show that a density matrix ρ is a pure state if and only if $\mathrm{trace}(\rho^2) = 1$.

7. (a) Show that each mixed state ρ is a nontrivial convex combination of other density matrices.

   (b) Show that a pure state cannot be expressed as a nontrivial convex combination of other density matrices.

   Hint: Show that the function $f(\lambda) := \mathrm{trace}\left((\lambda\rho_1 + (1 - \lambda)\rho_2)^2\right)$ is a convex function of λ.

8. For any density matrix ρ, show that the von Neumann entropy $S(\rho) :=$ trace$(-\rho \log \rho)$ is zero if and only if ρ is a pure state.

9. Prove Lemma 19.14.

   Hint: First use the principle of uniform boundedness (Theorem A.40) to show that there exists a constant C with $\|A_n\| \leq C$ for all n. Then, if $\{f_j\}$ is an orthonormal basis for $\mathbf{H}_2$, decompose $\mathbf{H}_1 \hat{\otimes} \mathbf{H}_2$ as the Hilbert space direct sum of the subspaces $\mathbf{H}_1 \otimes f_j$, where each of these subspaces is isometrically identified with $\mathbf{H}_1$ in the obvious way.

10. Suppose that ψ_1 and ψ_2 are two linearly independent elements of $L^2(\mathbb{R}^3; \mathbb{C}^2)$. Show that the function Φ in (19.10) is a nonzero element of $L^2(\mathbb{R}^6; \mathbb{C}^2 \otimes \mathbb{C}^2)$.

20

The Path Integral Formulation of Quantum Mechanics

We turn now to a topic that is important already for ordinary quantum mechanics and essential in quantum field theory: the so-called path integral. In the setting of ordinary quantum mechanics (of the sort we have been considering in this book), the integrals in question are over spaces of "paths," that is, maps of some interval $[a, b]$ into $\mathbb{R}^n$. In the setting of quantum field theory, the integrals are integrals over spaces of "fields," that is, maps of some region inside $\mathbb{R}^d$ into $\mathbb{R}^n$. Formal integrals of this sort abound in the physics literature, and it is typically difficult to make rigorous mathematical sense of them—although much effort has been expended in the attempt! In this chapter, we will develop a rigorous integral over spaces of paths by using the *Wiener measure*, resulting in the Feynman–Kac formula.

We begin with the Trotter product formula, which will be our main tool in deriving the path integral formulas. From there we turn to the (heuristic) path integral formula of Feynman, and then to the rigorous version of Feynman's result obtained by M. Kac, the so-called Feynman–Kac formula. Although it is not feasible to give complete proofs of all results presented here, we give enough proofs to get a flavor of the mathematics involved. We will prove a version of the Trotter product formula and, assuming the existence of the Wiener measure, a version of the Feynman–Kac formula.

B.C. Hall, *Quantum Theory for Mathematicians*, Graduate Texts
in Mathematics 267, DOI 10.1007/978-1-4614-7116-5_20,
© Springer Science+Business Media New York 2013

20.1   Trotter Product Formula

The Lie product formula (Point 7 of Theorem 16.15) says that for all X and Y in $M_n(\mathbb{C})$, we have

$$e^{X+Y} = \lim_{m \to \infty} (e^{X/m} e^{Y/m})^m.$$

The Trotter product formula asserts that a similar result holds for certain classes of *unbounded* operators on Hilbert spaces.

Theorem 20.1 (Trotter Product Formula) *Suppose that A and B are self-adjoint operators on* $\mathbf{H}$ *and that $A+B$ is densely defined and essentially self-adjoint on* $\mathrm{Dom}(A) \cap \mathrm{Dom}(B)$. *Then the following results hold.*

   1. For all $\psi \in \mathbf{H}$, we have

$$\lim_{N \to \infty} \left\| e^{it(A+B)}\psi - (e^{itA/N} e^{itB/N})^N \psi \right\|. \qquad (20.1)$$

   2. If A and B are bounded below, then for all $\psi \in \mathbf{H}$, we have

$$\lim_{N \to \infty} \left\| e^{-t(A+B)}\psi - (e^{-tA/N} e^{-tB/N})^N \psi \right\|. \qquad (20.2)$$

In both results, the expression $A + B$ refers to the unique self-adjoint extension of the operator defined on $\mathrm{Dom}(A) \cap \mathrm{Dom}(B)$.

In the usual terminology of functional analysis, (20.1) asserts that the operators $(e^{itA/N} e^{itB/N})^N$ converge to $e^{it(A+B)}$ in the "strong operator topology," and similarly with (20.2).

We will give a proof of this result in the special case in which $A + B$ is densely defined and self-adjoint on $\mathrm{Dom}(A) \cap \mathrm{Dom}(B)$. This condition holds, for example, whenever the Kato–Rellich theorem (Theorem 9.37) applies. See Sect. A.5 of [14] for a proof of the version stated above.

Proof. Since all the operators in Point 1 of the theorem are unitary, it is easy to see that if the result holds on some dense subspace W of $\mathbf{H}$, it holds on all of $\mathbf{H}$. In Point 2 of the theorem, we first make a simple reduction to the case where A and B are non-negative, and then have the same conclusion, since all operators involved will then be contractions.

We will prove Point 1 of the theorem, with the proof of Point 2 being similar. Let us introduce the notation $S_s := e^{is(A+B)}$ and $T_s := e^{isA} e^{isB}$. What we want to prove is that for each $\psi \in \mathbf{H}$, the quantity $\left\|(S_t - (T_{t/N})^N)\psi\right\|$ tends to zero as N tends to infinity. Now, a simple calculation shows that

$$\left\|(S_t - (T_{t/N})^N)\psi\right\| = \left\| \sum_{j=0}^{N-1} (T_{t/N})^j (S_{t/N} - T_{t/N})(S_{t/N})^{N-j-1}\psi \right\|. \qquad (20.3)$$

Since S. is a one-parameter unitary group, $(S_{t/N})^{N-j-1}\psi = S_s\psi$, where $s = (N - j - 1)t/N$. Thus, if we let $\psi_s = S_s\psi$, we have

$$\left\|(S_t - (T_{t/N})^N)\psi\right\| \leq N \sup_{0\leq s\leq t} \left\|(S_{t/N} - T_{t/N})\psi_s\right\|. \qquad (20.4)$$

Now, for any ψ in $\mathrm{Dom}(A + B)$, we have

$$\lim_{N\to\infty} N(S_{t/N}\psi - \psi) = it(A + B)\psi,$$

by Stone's theorem. Meanwhile, according to Exercise 2, we have

$$\lim_{s\to 0} \frac{1}{s}(T_s - I)\psi = iA\psi + iB\psi, \qquad (20.5)$$

for all $\psi \in \mathrm{Dom}(A) \cap \mathrm{Dom}(B)$. (This result is clear at the heuristic level.) Thus,

$$\lim_{N\to\infty} N(S_{t/N} - T_{t/N})\psi = \lim_{N\to\infty} N(S_{t/N} - I)\psi - \lim_{N\to\infty} N(T_{t/N} - I)\psi$$
$$= it(A + B)\psi - it(A + B)\psi = 0 \qquad (20.6)$$

for every $\psi \in \mathrm{Dom}(A) \cap \mathrm{Dom}(B)$.

Let $W = \mathrm{Dom}(A) \cap \mathrm{Dom}(B)$, which is, by assumption, dense in $\mathbf{H}$, equipped with the norm $\|\cdot\|_1$ given by

$$\|\psi\|_1 = \|\psi\| + \|(A + B)\psi\|.$$

Since we are assuming $A + B$ is self-adjoint, and thus also closed, on W, we see that W is a Banach space with respect $\|\cdot\|_1$ (Exercise 6 in Chap. 9). Now, the operators $N(S_{t/N} - T_{t/N})$ are certainly bounded from W to $\mathbf{H}$, for each N. Furthermore, (20.6) shows that for each $\psi \in W$, we have

$$\sup_N \left\|N(S_{t/N} - T_{t/N})\psi\right\| < \infty.$$

Thus, by the principle of uniform boundedness (Theorem A.40), there is a constant C such that

$$\left\|N(S_{t/N} - T_{t/N})\psi\right\| \leq C \|\psi\|_1$$

for all $\psi \in W$. It then follows (Exercise 3) that $\left\|N(S_{t/N} - T_{t/N})\psi\right\|$ tends to zero *uniformly* on every compact subset of W.

Suppose, now, that for each $\psi \in W$, the $s \mapsto \psi_s$ is continuous in W. If so, the image of the compact interval $[0, t]$ under $s \mapsto \psi_s$ will be compact in W, and so $\left\|N(S_{t/N} - T_{t/N})\psi_s\right\|$ will tend to zero uniformly in s. Thus, by (20.4), we will have Point 1 of the theorem. To establish the desired continuity, we first note that by Lemma 10.17, the operators $S_s = e^{is(A+B)}$

preserve $\mathrm{Dom}(A+B)$, which is equal to W, by assumption. Then for any $s, r \in [0, t]$ and $\psi \in W$, we have

$$\left\| e^{is(A+B)}\psi - e^{ir(A+B)}\psi \right\|_1$$
$$= \left\| e^{is(A+B)}\psi - e^{ir(A+B)}\psi \right\| + \left\| (A+B)(e^{is(A+B)}\psi - e^{ir(A+B)}\psi) \right\|$$
$$= \left\| (e^{is(A+B)} - e^{ir(A+B)})\psi \right\| + \left\| (e^{is(A+B)} - e^{ir(A+B)})(A+B)\psi \right\|, \quad (20.7)$$

where we have used Lemma 10.17 again in the second equality. The strong continuity of $e^{is(A+B)}$ (Proposition 10.14) then ensures that the right-hand side of (20.7) tends to zero as s approaches r. ∎

20.2 Formal Derivation of the Feynman Path Integral

In this section, we apply Point 1 of the Trotter product formula to the operator

$$-\frac{1}{\hbar}\hat{H} = \frac{\hbar}{2m}\Delta - \frac{1}{\hbar}V(X). \quad (20.8)$$

Let us call the operators on the right-hand side of (20.8) A and B, respectively, and let us assume V is sufficiently nice that $\hat{H}$ is essentially self-adjoint on $\mathrm{Dom}(A) \cap \mathrm{Dom}(B)$. Any bounded potential certainly has this property, as do many unbounded potentials. (See, e.g., Theorem 9.38.)

Point 1 of Theorem 20.1 then tells us that

$$e^{-it\hat{H}/\hbar}\psi = \lim_{N \to \infty} \left(\exp\left\{ \frac{it\hbar\Delta}{2mN} \right\} \exp\left\{ -\frac{itV(X)}{N\hbar} \right\} \right)^N \psi.$$

Under mild assumptions on ψ, Theorem 4.5 (extended to n dimensions) tells us that $\exp(it\hbar\Delta/(2mN))$ may be computed as

$$e^{it\hbar\Delta/(2mN)}\psi(\mathbf{x}_0) = \left(\frac{mN}{it\hbar} \right)^{n/2} \int_{\mathbb{R}^n} \exp\left\{ i\frac{mN}{2t\hbar} |\mathbf{x}_1 - \mathbf{x}_0|^2 \right\} \psi(\mathbf{x}_1)\, d\mathbf{x}_1.$$

Meanwhile, $\exp(-itV(X)/(N\hbar))$ is simply a multiplication operator.

Thus, assuming that Theorem 4.5 applies at each stage, we have

$$
\left[\left(\exp\left\{\frac{it\hbar\Delta}{2mN}\right\}\exp\left\{-\frac{itV(X)}{N\hbar}\right\}\right)^N\psi\right](\mathbf{x}_0)
$$

$$
= C\int_{\mathbb{R}^n}\exp\left\{i\frac{mN}{2t\hbar}|\mathbf{x}_1-\mathbf{x}_0|^2\right\}\exp\left\{-\frac{itV(\mathbf{x}_1)}{N\hbar}\right\}
$$

$$
\times\int_{\mathbb{R}^n}\exp\left\{i\frac{mN}{2t\hbar}|\mathbf{x}_{N-1}-\mathbf{x}_{N-2}|^2\right\}\exp\left\{-\frac{itV(\mathbf{x}_{N-1})}{N\hbar}\right\}
$$

$$
\times\cdots\times\int_{\mathbb{R}^n}\exp\left\{i\frac{mN}{2t\hbar}|\mathbf{x}_N-\mathbf{x}_{N-1}|^2\right\}\exp\left\{-\frac{itV(\mathbf{x}_N)}{N\hbar}\right\}
$$

$$
\times\psi(\mathbf{x}_N)\,d\mathbf{x}_N\,d\mathbf{x}_{N-1}\cdots d\mathbf{x}_1,
$$

where $C = (mN/(it\hbar))^{nN/2}$. Letting $\varepsilon = t/N$ and assuming we can freely rearrange the order of integration, we obtain

$$
(e^{-it\hat{H}/\hbar}\psi)(\mathbf{x}_0)
$$

$$
= \lim_{N\to\infty} C\int_{(\mathbb{R}^n)^N}\exp\left\{\frac{i}{\hbar}\sum_{j=1}^{N}\varepsilon\left[\frac{m}{2}\left|\frac{\mathbf{x}_j-\mathbf{x}_{j-1}}{\varepsilon}\right|^2 - V(\mathbf{x}_{j-1})\right]\right\}
$$

$$
\times\psi(\mathbf{x}_N)\,d\mathbf{x}_1\,d\mathbf{x}_2\cdots d\mathbf{x}_N. \tag{20.9}
$$

So far, the argument is mostly rigorous, coming from the Trotter product formula and Theorem 4.5. The nonrigorous part comes in attempting to evaluate the limit on the right-hand side of (20.9). Let us think of the values $\mathbf{x}_j$, $j = 0,\ldots,N$ as constituting the values of a path $\mathbf{x}(s)$ at the points $s_j := j\varepsilon = jt/N$:

$$
\mathbf{x}_j = \mathbf{x}(jt/N).
$$

Since the distance between s_{j-1} and s_j is ε, the quantity $|\mathbf{x}_j-\mathbf{x}_{j-1}|/\varepsilon$ is an approximation to the derivative of $x(s)$ with respect to s. Meanwhile, the sum over j in the right-hand side of (20.9) is an approximation to an integral. Thus, if we then take the limit of the right-hand of (20.9) in a totally nonrigorous fashion, we obtain

$$
(e^{-it\hat{H}/\hbar}\psi)(\mathbf{x}_0)
$$

$$
= C\int_{\substack{\text{paths with}\\ \mathbf{x}(0)=\mathbf{x}_0}}\exp\left\{\frac{i}{\hbar}\int_0^t\left[\frac{m}{2}\left|\frac{d\mathbf{x}}{ds}\right|^2 - V(\mathbf{x}(s))\right]\,ds\right\}\psi(\mathbf{x}(t))\,\mathcal{D}\mathbf{x}. \tag{20.10}
$$

Here, C is a normalization constant and $\mathcal{D}\mathbf{x}$ is something like "Lebesgue measure" on the space of all paths $\mathbf{x}(\cdot)$ mapping $[0,t]$ into $\mathbb{R}^n$. (The quantity $\mathbf{x}$ in the expression $\mathcal{D}\mathbf{x}$ is a *path*, not a point in $\mathbb{R}^n$.)

The reader who is familiar with the Lagrangian approach to mechanics will recognize the expression in square brackets in the exponent on the right-hand side of (20.10) as the *Lagrangian* of the particle, $L = T - V$, where $T = (1/2)m |v|^2$ is the kinetic energy and V is the potential energy. The integral of the Lagrangian over some time interval is called the *action functional*, denoted by the letter $\mathcal{S}$. That is to say, given a path $\mathbf{x}(\cdot)$, we define the action functional of $\mathbf{x}(\cdot)$ over a time-interval $[a, b]$ as follows:

$$\mathcal{S}(\mathbf{x}(\cdot), a, b) := \int_a^b \left[\frac{m}{2} \left| \frac{d\mathbf{x}}{ds} \right|^2 - V(\mathbf{x}(s)) \right] ds. \qquad (20.11)$$

In Lagrangian mechanics, one shows that the solutions to Newton's law are precisely the *stationary points* of the action functional. Using the notation in (20.11), we may rewrite (20.10) as

$$(e^{-it\hat{H}/\hbar}\psi)(\mathbf{x}_0) = C \int_{\substack{\text{paths with} \\ \mathbf{x}(0)=\mathbf{x}_0}} \exp \left\{ \frac{i}{\hbar} \mathcal{S}(\mathbf{x}(\cdot), 0, t) \right\} \psi(\mathbf{x}(t)) \, \mathcal{D}\mathbf{x}. \qquad (20.12)$$

This formula is the *Feynman path integral formula*.

Now, knowledge of Lagrangian mechanics is not directly relevant to the derivation of the Feynman path integral formula. Nevertheless, it is intriguing that the an important quantity from *classical* mechanics should appear in the Feynman path integral formula in *quantum* mechanics. Indeed, this appearance raises the possibility that one can use the path integral formula to make connections between quantum mechanics and classical mechanics. Indeed, the "method of stationary phase" (when applied, formally, in an infinite-dimensional setting) asserts that for small values of $\hbar$, the main contribution to the right-hand side of (20.12) comes from regions near the stationary points of the action functional, namely the classical trajectories. Using this method, Gutzwiller was able to derive his famous trace formula, which provides predictions of typical eigenvalue spacings for Schrödinger operators based on the behavior of the underlying classical system. More information about this fascinating subject can be found in books on "quantum chaos," including [19] by Gutzwiller himself.

It is notoriously difficult to attach a rigorous meaning to the right-hand side of the Feynman path integral formula. Note that the formal expression "$\mathcal{D}\mathbf{x}$" is the limit as N tends to infinity of the integral over $(\mathbb{R}^n)^N$ in (20.9) with respect to the Lebesgue measure (i.e., the measure given by $d\mathbf{x}_1 \, d\mathbf{x}_2 \cdots d\mathbf{x}_N$). Thus, "$\mathcal{D}\mathbf{x}$" should be something like Lebesgue measure on the space of all paths (maps from $[0, t]$ into $\mathbb{R}^n$). However, it is known that an infinite-dimensional vector space (say, a Banach space) does not have any "reasonable" (say, σ-finite) translation-invariant measure that could play the role of Lebesgue measure. Furthermore, the absolute value of the constant C is easily seen to be infinite. Thus, we certainly cannot take the right-hand side of (20.12) literally.

A better approach is to avoid looking at the component parts of the Feynman path integral and instead to look at the whole expression against which the function $\psi(\mathbf{x}(t))$ is being integrated. If we could attach a rigorous meaning to the expression

$$C \exp\left\{\frac{i}{\hbar}\mathcal{S}(\mathbf{x}(\cdot), 0, t)\right\} \mathcal{D}\mathbf{x}, \tag{20.13}$$

as, say, a complex-valued measure on the space of continuous paths, then this could serve to give a meaning to the path integral. It is known, however, that there is no complex measure on the space of paths that makes the Feynman path integral formula true. The oscillatory behavior produced by the i in the exponent in (20.13) makes it difficult to give a rigorous meaning to the Feynman path integral in its original form.

20.3 The Imaginary-Time Calculation

In trying to give a rigorous meaning to the path integral formula of Feynman, Kac proceeded by considering the "imaginary time" time-evolution operator $\exp(-t\hat{H}/\hbar)$, which is just the usual time-evolution operator $\exp(-it\hat{H}/\hbar)$ evaluated with t replaced by $-it$. The idea is that if one can use path integrals to understand the operators $\exp(-t\hat{H}/\hbar)$, one can go back to the "real time" operator $\exp(-it\hat{H}/\hbar)$ by analytic continuation with respect to t.

The counterpart of Theorem 4.5 for $\exp(-t\hbar\Delta/(2m))$ (proved in the same way) is

$$(e^{-t\hbar\Delta/(2m)}\psi)(\mathbf{x}_0) = \left(\frac{m}{2\pi t\hbar}\right)^{n/2} \int_{\mathbb{R}^n} \exp\left\{-\frac{m}{2t\hbar}|\mathbf{x}_1 - \mathbf{x}_0|^2\right\} \psi(\mathbf{x}_1)\, d\mathbf{x}_1.$$

Unlike Theorem 4.5, however, the above expression holds for all $\psi \in L^2(\mathbb{R}^n)$, with absolute convergence of the integral for every $\mathbf{x}_0 \in \mathbb{R}^n$. Applying the Trotter product formula and rearranging the integral as before gives

$$(e^{-t\hat{H}/\hbar}\psi)(\mathbf{x}_0)$$
$$= \lim_{N\to\infty} C \int_{(\mathbb{R}^n)^N} \exp\left\{-\frac{1}{\hbar}\sum_{j=1}^{N}\varepsilon\left[\frac{m}{2}\left|\frac{\mathbf{x}_j - \mathbf{x}_{j-1}}{\varepsilon}\right|^2 + V(\mathbf{x}_{j-1})\right]\right\}$$
$$\times \psi(\mathbf{x}_N)\, d\mathbf{x}_1\, d\mathbf{x}_2 \cdots d\mathbf{x}_N. \tag{20.14}$$

If V is, say, bounded below, then there is no difficulty in changing the order of integration, because of the rapid decay of the integrand. Note that there is a relative sign change between the two terms in square brackets, compared to (20.9). Taking a formal limit as before gives

$$(e^{-t\hat{H}/\hbar}\psi)(\mathbf{x})$$

$$= C \int_{\substack{\text{paths with} \\ \mathbf{x}(0)=\mathbf{x}_0}} \exp\left\{ -\frac{1}{\hbar} \int_0^t \left[\frac{m}{2}\left|\frac{d\mathbf{x}}{ds}\right|^2 + V(\mathbf{x}(s)) \right] ds \right\} \psi(\mathbf{x}(t)) \, \mathcal{D}\mathbf{x}.$$

$$(20.15)$$

Note that the integral in the exponent on the right-hand side is *not* the classical action in (20.11), because the potential term has the wrong sign.

Kac's idea was to separate out the quadratic part of the exponent on the right-hand side of (20.15) and attempt to interpret the expression

$$C \exp\left\{ -\frac{1}{\hbar} \int_0^t \frac{m}{2}\left|\frac{d\mathbf{x}}{ds}\right|^2 ds \right\} \mathcal{D}\mathbf{x} \qquad (20.16)$$

as a measure on the space of paths. Specifically, this is a *Gaussian* measure, one with a (formal) density with respect to the Lebesgue measure that is the exponential of a quadratic expression. There is a well-developed theory of Gaussian measures on infinite-dimensional spaces. Although there is no Lebesgue measure in the infinite-dimensional case, one can construct Gaussian measures as limits of Gaussian measures on spaces of large finite dimension.

20.4   The Wiener Measure

Kac identified the formal expression in (20.16) as the *Wiener measure*. To be precise, for each fixed $\mathbf{x}_0 \in \mathbb{R}$, there is a Wiener measure $\mu_{\mathbf{x}_0}$, where $\mu_{\mathbf{x}_0}$ is supported on the set of paths $\mathbf{x} : [0, t] \to \mathbb{R}$ with $\mathbf{x}(0) = \mathbf{x}_0$. The Wiener measure was developed by Norbert Wiener as a rigorous embodiment of Albert Einstein's mathematical model of Brownian motion. Einstein, in one of his 1905 papers, had proposed that the random motion of a very small particle in water was due to collisions between the particle and the water molecules. Einstein postulated that the *increments* of a Brownian path $\mathbf{x}$ [quantities of the form $\mathbf{x}(t) - \mathbf{x}(s)$] should be independent for disjoint time intervals and should be normal random variables with mean zero and variance proportional to $t - s$. The following theorem shows that there is a unique measure on the space of continuous paths satisfying Einstein's criteria. Let $\mathcal{C}_{\mathbf{x}_0}([0, t]; \mathbb{R}^n)$ denote the space of continuous maps $\mathbf{x}(\cdot)$ of $[0, t]$ into $\mathbb{R}^n$ satisfying $\mathbf{x}(0) = \mathbf{x}_0$, equipped with the supremum norm.

Theorem 20.2 (Wiener) *For each vector $\mathbf{x}_0 \in \mathbb{R}^n$ and each pair of positive numbers σ and t, there exists a unique measure $\mu_{\mathbf{x}_0}^\sigma$ on the Borel σ-algebra in $\mathcal{C}_{\mathbf{x}_0}([0, t]; \mathbb{R}^n)$ such that the following condition holds. For each* *sequence $0 = t_0 < t_1 < \cdots < t_N \leq t$ of real numbers and each non-negative measurable function f on $(\mathbb{R}^n)^N$, we have*

$$\int_{C_{\mathbf{x}_0}([0,t];\mathbb{R}^n)} f(\mathbf{x}(t_1), \mathbf{x}(t_2), \ldots, \mathbf{x}(t_N))\, d\mu^{\sigma}_{\mathbf{x}_0}(\mathbf{x})$$

$$= C \int_{\mathbb{R}^N} \exp\left\{ -\frac{1}{2\sigma} \sum_{j=1}^{N} \frac{|\mathbf{x}_j - \mathbf{x}_{j-1}|^2}{t_j - t_{j-1}} \right\} f(\mathbf{x}_1, \mathbf{x}_2, \ldots, \mathbf{x}_N)\, d\mathbf{x}_1 \cdots d\mathbf{x}_N,$$

$$(20.17)$$

where

$$C = \prod_{j=1}^{N} \frac{1}{\sqrt{2\pi\sigma(t_j - t_{j-1})}}.$$

Note that the right-hand side of (20.17) is extremely similar to the right-hand side of (20.14), except that there are no terms involving the potential V in the exponent in (20.17). Thus, it is reasonable to think that the Wiener measure is a rigorous version of the formal expression in (20.16). It should be noted, however, that the heuristic expression (20.16) is misleading in one important respect. That expression suggests that the measure is supported on paths $\mathbf{x}(\cdot)$ for which $d\mathbf{x}/dt$ belongs to $L^2([0,t];\mathbb{R}^n)$, since the exponential factor would seemingly "damp out" any paths for which this is not the case. This conclusion is, however, incorrect. [One should, in general, be extremely cautious in drawing conclusions based on purely formal expressions such as the one in (20.16).] Actually, the "typical" path with respect to the Wiener measure is nowhere differentiable; that is, the set of paths $\mathbf{x}(t)$ that are differentiable for even one value of t form a set of measure zero.

This discrepancy is actually a general feature of Gaussian measures on infinite-dimensional spaces: They are always supported on a larger space than the formal expression would suggest. In the case of the Wiener measure, the space on which the measure actually lives (the space of continuous functions) is nice enough that no difficulties arise in the formulation of our main result, the Feynman–Kac formula. In the setting of quantum field theory, however, issues concerning the support of a Gaussian measure become serious difficulties. See Sect. 20.6 for more information.

20.5 The Feynman–Kac Formula

The Wiener measure gives a rigorous interpretation to the expression in (20.16). Thus, the Wiener measure encapsulates everything in (20.15) except for the term involving V in the exponent and the factor of $\psi(\mathbf{x}(t))$. This reasoning accounts for the form of the following result.

Theorem 20.3 (Feynman–Kac Formula) *Suppose $V : \mathbb{R}^3 \to \mathbb{R}$ can be expressed as the sum of a function in $L^2(\mathbb{R}^3)$ and a bounded function. Then for all $\mathbf{x}_0 \in \mathbb{R}^3$, we have*

$$(e^{-t\hat{H}/\hbar}\psi)(\mathbf{x}_0)$$

$$= \int_{\mathcal{C}_{\mathbf{x}_0}([0,t];\mathbb{R}^3)} \exp\left\{-\frac{1}{\hbar}\int_0^t V(\mathbf{x}(s))\,ds\right\} \psi\,(\mathbf{x}(t))\,\,d\mu_{\mathbf{x}_0}^\sigma(\mathbf{x}),$$

where $\mu_{\mathbf{x}_0}^\sigma$ is the Wiener measure on $\mathcal{C}_{\mathbf{x}_0}([0,t];\mathbb{R}^3)$ and where $\sigma = \hbar/m$.

Of course, similar results hold in other dimensions, under suitable assumptions on the potential. We refer the interested reader to [37] or [14] for details on different versions of the Feynman–Kac formula. Theorem 20.3 cannot be obtained directly from the Trotter product formula, because the limit in (20.14) is an L^2 limit rather than a pointwise limit. We will content ourselves with proving an "integrated" version of the Feynman–Kac formula for nice potentials; Theorem 20.3 is Theorem 6.5 of [37].

Definition 20.4 *Let $\mathcal{C}([0,t];\mathbb{R}^n)$ denote the space of all continuous paths on $[0,t]$ with values in $\mathbb{R}^n$. For all $\sigma > 0$, let μ^σ be the measure on $\mathcal{C}([0,t];\mathbb{R}^n)$ given by*

$$\mu(E) = \int_{\mathbb{R}^n} \mu_{\mathbf{x}_0}^\sigma(E)\,dx_0.$$

Proposition 20.5 *Suppose $V : \mathbb{R}^n \to \mathbb{R}$ is bounded and continuous. Then for all $\phi, \psi \in L^2(\mathbb{R}^n)$, we have*

$$\langle \phi, e^{-t\hat{H}/\hbar}\psi \rangle$$

$$= \int_{\mathcal{C}([0,t];\mathbb{R}^n)} \overline{\phi(\mathbf{x}(0))} \exp\left\{-\frac{1}{\hbar}\int_0^t V(\mathbf{x}(s))\,ds\right\} \psi\,(\mathbf{x}(t))\,\,d\mu^\sigma(\mathbf{x}),$$

where μ^σ is as in Definition 20.4 and where $\sigma = \hbar/m$.

Proof. We begin with (20.14) and apply Theorem 20.2 with parameters chosen as follows. We take $\sigma = \hbar/m$, we take the sequence $\langle t_j \rangle$ to be given by $t_j = jt/N$, and we take f to be the function given by

$$f(\mathbf{x}_1, \mathbf{x}_2, \ldots, \mathbf{x}_N) = \psi(\mathbf{x}_N).$$

Theorem 20.2 then allows us to express the right-hand side of (20.14) as an integral against the Wiener measure, giving

$$(e^{-t\hat{H}/\hbar}\psi)(\mathbf{x}_0)$$

$$= \lim_{N\to\infty} \int_{\mathcal{C}_{\mathbf{x}_0}([0,t];\mathbb{R}^n)} \exp\left\{-\frac{1}{\hbar}\sum_{j=1}^N \frac{t}{N}V\left(\mathbf{x}\left(\frac{jt}{N}\right)\right)\right\} \psi(\mathbf{x}(t))\,\,d\mu_{\mathbf{x}_0}^\sigma(\mathbf{x}).$$

Since the limit in the above equation is an L^2 limit, we may move the inner product with ϕ inside the limit on the right-hand side. The integral with respect to $\mu^\sigma_{\mathbf{x}_0}$ and the integral with respect to $d\mathbf{x}_0$ may then be combined into a single integral with respect to μ^σ, giving

$$\langle \phi, e^{-t\hat{H}/\hbar} \psi \rangle = \lim_{N \to \infty} \int_{\mathcal{C}([0,t];\mathbb{R}^n)} \overline{\phi(\mathbf{x}(0))}$$

$$\times \exp\left\{ -\frac{1}{\hbar} \sum_{j=1}^{N} \frac{t}{N} V\left(\mathbf{x}\left(\frac{jt}{N}\right) \right) \right\} \psi(\mathbf{x}(t)) \; d\mu^\sigma(\mathbf{x}). \qquad (20.18)$$

Now, since V is continuous,

$$\lim_{N \to \infty} \sum_{j=1}^{N} \frac{t}{N} V\left(\mathbf{x}\left(\frac{jt}{N}\right) \right) = \int_0^t V(\mathbf{x}(s)) \; ds,$$

for every continuous path $\mathbf{x}$. Furthermore, it is easily seen that the "distribution" of the quantity $\mathbf{x}(s)$ with respect to the measure μ^σ is the Lebesgue measure on $\mathbb{R}^n$, for any $s \in [0,t]$. Thus, the function $\mathbf{x} \mapsto \phi(\mathbf{x}(0))$ is square-integrable with respect to μ^σ, with L^2 norm equal to the L^2 norm of ϕ over $\mathbb{R}^n$, and similarly for $\mathbf{x} \mapsto \psi(\mathbf{x}(t))$. It follows that the quantity $\overline{\phi(\mathbf{x}(0))}\psi(\mathbf{x}(t))$ is an L^1 function on $\mathcal{C}([0,t];\mathbb{R}^n)$. Since V is bounded, we may apply dominated convergence to move the limit inside the integral, at which point we obtain the desired result. ∎

20.6 Path Integrals in Quantum Field Theory

In this section, we briefly discuss the path integral approach to quantum field theory. We consider quantum field theory in a space–time of dimension d, so that space has dimension $d-1$. The configuration space for the classical version of the theory is the collection of "spatial" fields, that is, maps $\phi(\mathbf{x})$ of $\mathbb{R}^{d-1}$ into some finite-dimensional vector space V. A *path* in the space of fields is then a map $\phi(\mathbf{x},t)$ of $\mathbb{R}^{d-1} \times \mathbb{R} \cong \mathbb{R}^d$ into V. In the path integral approach to quantum field theory (which is the most commonly used approach to the subject), one considers integrals over the space of such paths.

Let us consider, as a simple example, what is called ϕ^4 theory. In this theory, the fields ϕ map into $\mathbb{R}$ and we consider a path integral of the form

$$C \int_{\mathcal{F}_d} \exp\left\{ -\frac{1}{\hbar} \int_{\mathbb{R}^d} \left[c_1 \|\nabla\phi(\mathbf{x})\|^2 + c_2\phi(\mathbf{x})^2 + c_3\phi(\mathbf{x})^4 \right] d\mathbf{x} \right\}$$

$$\times F(\phi) \, \mathcal{D}\phi, \qquad (20.19)$$

for some functional $F(\phi)$ on the space of fields. [The expression in (20.19) is, more precisely, a "Euclidean" or "imaginary time" path integral. Such an integral is the counterpart in quantum field theory of the integral occurring in the Feynman–Kac formula in quantum mechanics.] In (20.19), $\mathcal{F}_d$ represents the space of all "fields" (i.e., functions) mapping our space–time $\mathbb{R}^d$ into $\mathbb{R}$. In an attempt to make sense of this heuristic expression, we may follow the strategy we used in deriving the Feynman–Kac formula by separating out the quadratic part of the exponent. We look, then, for a measure μ on $\mathcal{F}_d$ given by the heuristic expression

$$d\mu(\phi) \text{ "="} C \exp\left\{-\frac{1}{\hbar}\int_{\mathbb{R}^d}\left[c_1\|\nabla\phi(\mathbf{x})\|^2 + c_2\phi(\mathbf{x})^2\right]\,d\mathbf{x}\right\}\mathcal{D}\phi. \quad (20.20)$$

Using the theory of Gaussian measures, one can construct a rigorously defined measure corresponding to the heuristic expression in (20.20). There is, however, a serious difficulty with this approach: The measure μ is supported on very "rough" fields, much rougher than the heuristic expression suggests. In fact, we have the following result.

Proposition 20.6 *For all $d \geq 1$, there exists a Gaussian measure on the space $\mathcal{F}_d$ of fields on $\mathbb{R}^d$ corresponding to the heuristic expression (20.20). For $d \geq 2$, however, this measure is not supported on any space of ordinary functions, but rather on a space of distributions.*

We will not prove this result here; see Sect. 8.5 of [14] for more information. Here, then, is the problem with the path integral approach to quantum field theory on space–times of dimension $d \geq 2$: The functional $\int_{\mathbb{R}^d}\phi(x)^4\,dx$ does not make sense for a "typical" field with respect to the measure μ in (20.20). As a result, we cannot make sense of (20.19) simply by absorbing all the Gaussian part into the definition of the measure μ, since what is left over is not a μ-almost everywhere defined functional of ϕ. Indeed, even a local integral, of the form $\int_U \phi(x)^4\,dx$ for some bounded region U in $\mathbb{R}^d$, fails to be almost-everywhere defined with respect to μ. After all, if $\int_U \phi(x)^4\,dx$ made sense, then ϕ would be a locally L^4 function, rather than a distribution.

It should be emphasized that the difficulty described in the previous paragraph is not just a technicality that can be swept away by some simple trick. Furthermore, this difficulty is not specific to ϕ^4 theory, but is present in all "nontrivial" field theories. In all interesting field theories, the fields defined by the Gaussian part of the path integral are fundamentally "too rough" to allow us to make sense of the non-Gaussian part of the integral. This phenomenon is *the* fundamental mathematical difficulty in the path integral approach to quantum field theory.

To have a chance to make rigorous sense of path integrals in quantum field theory, one has to employ a complicated regularization process known as renormalization. This process has, so far, been carried out in a rigorous fashion only for a very small number of field theories. One of the Clay Millennium Prize problems is to make rigorous sense out of the Yang–Mills field theory in four space–time dimensions. See [14] for a detailed survey of the mathematical issues connected with the path integral approach to quantum field theory. See also [13] for a treatment of quantum field theory and renormalization with a greater eye toward the physical content.

Since the roughness of the fields is a major problem in trying to give a rigorous meaning to path integrals, let us think for moment why it arises. Suppose we wish to construct a Gaussian measure from a certain heuristic expression of the form $\mu = Ce^{-Q(x)}\mathcal{D}x$, where Q is a positive-definite quadratic functional of x. A reasonable approach is to consider the (real) Hilbert space $\mathbf{H}$ for which $\|x\|_H^2 = Q(x)$. [In the case of (20.20), $\mathbf{H}$ would be the "Sobolev space" of fields having one derivative in L^2.] The heuristic expression for the Gaussian measure then takes the form

$$d\mu(x) = Ce^{-\|x\|_\mathbf{H}^2}\,\mathcal{D}x. \tag{20.21}$$

One might now try to approximate μ by Gaussian measures μ_N on Hilbert spaces $\mathbf{H}_N$ of dimension $N < \infty$. If $\dim \mathbf{H} < \infty$, then the expression (20.21) is perfectly rigorous, where the constant C may be taken to normalize μ to be a probability measure. A simple calculation (Exercise 4), however, shows that for any R, we have

$$\lim_{N\to\infty} \mu_N(B_{R,N}) = 0,$$

where $B_{R,N}$ denotes the ball of radius R in $\mathbf{H}_N$. This means that in the $N \to \infty$ limit, all of the "mass" of the measure is outside the ball of radius R, for *every* R. Thus, in the limit, the measure is supported entirely on points x where $\|x\|_H = \infty$, that is, on points that are not actually in $\mathbf{H}$. The measures μ_N *do* converge to a measure μ as N tends to infinity, but μ *does not* live on $\mathbf{H}$, but on some larger space $B \supset \mathbf{H}$. The original space $\mathbf{H}$ is a set of μ-measure zero inside B. See [16] for more information. In the case of the measure μ corresponding to the heuristic expression in (20.20), μ does not—as the expression suggests—live on the Sobolev space of fields with one derivative in L^2, but on a larger space, which turns out to be a space of distributions.

20.7   Exercises

1. Verify the identity (20.3) in the proof of the Trotter product formula.

2. Verify (20.5) in the proof of the Trotter product formula, using Stone's theorem and the following identity:

$$\frac{1}{s}(e^{isA}e^{isB} - I)\psi = e^{isA}(iB\psi) + e^{isA}\left(\frac{1}{s}(e^{isB} - I)\psi - iB\psi\right)$$

$$+ \frac{1}{s}(e^{isA} - I)\psi.$$

3. Suppose $\{A_N\}$ is a family of bounded operators mapping a Banach space W_1 to a Banach space W_2. Suppose that for some constant C, we have $\|A_N\| \leq C$ for all N. Finally, suppose that $\|A_N\psi\| \to 0$ as $N \to \infty$, for every $\psi \in W$.

   (a) Show that for each $\psi \in W$ and each $\varepsilon > 0$, there exists a neighborhood U of ψ and an integer M such that

$$\|A_N\phi\| < \varepsilon$$

   for all $\phi \in U$ and $N \geq M$.

   (b) If K is a compact subset of W, show that $\|A_N\psi\|$ tends to zero *uniformly* for $\psi \in K$.

4. (a) Let $\mathbf{H}_N$ be an N-dimensional Hilbert space. Show that the measure

$$d\mu_N(x) := \pi^{-N/2} e^{-\|x\|^2} dx$$

   is a probability measure. Here dx is the Lebesgue measure on $\mathbf{H}_N$, normalized to that the unit cube has volume 1.

   Hint: Use Proposition A.22.

   (b) Let $B_{R,N}$ denote the ball of radius R in $\mathbf{H}_N$. Show that for each $R < \infty$, there exists number $a_R < 1$ such that

$$\mu_N(B_{R,N}) < (a_R)^N.$$

   Thus, $\lim_{N\to\infty} \mu_N(B_{R,N}) = 0$.

   Hint: The ball $B_{R,N}$ is contained in a cube centered at the origin with side length $2R$.

21
Hamiltonian Mechanics on Manifolds

In this chapter, we generalize the Hamiltonian approach to mechanics (introduced already in the Euclidean case in Sect. 2.5) to general manifolds. The chapter assumes familiarity with the basic notions of smooth manifolds, including tangent and cotangent spaces, vector fields, and differential forms. These notions are reviewed very briefly in Sect. 21.1, mainly in the interest of fixing the notation. See, for example, Chap. 2 of [40] for a concise treatment of manifolds and [29] for a detailed account. Throughout the chapter, we will use the *summation convention*, that repeated indices are always summed on.

21.1 Calculus on Manifolds

Throughout this section, M will denote a smooth, n-dimensional manifold.

21.1.1 Tangent Spaces, Vector Fields, and Flows

For each $x \in M$, we have the *tangent space* to M at x, denoted $T_x M$. Given a smooth coordinate system $x_1, \ldots, x_n$ on M, the vectors

$$\frac{\partial}{\partial x_1}, \ldots, \frac{\partial}{\partial x_n} \tag{21.1}$$

form a basis for the tangent space at each point. A *vector field* X on M is map assigning to each point $x \in M$ an element X_x of $T_x M$. A vector

B.C. Hall, *Quantum Theory for Mathematicians*, Graduate Texts
in Mathematics 267, DOI 10.1007/978-1-4614-7116-5_21,
© Springer Science+Business Media New York 2013

field X is *smooth* if the coefficients of X in a basis of the form (21.1) are smooth functions, for every smooth coordinate system. As in Exercise 14 in Chap. 2, we think of a vector field as a first-order differential operator satisfying the Leibniz rule:

$$X(fg) = X(f)g + fX(g).$$

Given a smooth vector field X on M and a point $x \in M$, there exists a curve $\gamma_x : (a, b) \to M$ such that $\gamma_x(0) = x$ and

$$\frac{d\gamma_x}{dt} = X_{\gamma_x(t)}.$$

Any two such curves agree on the intersection of their intervals of definition. There is a largest interval $(a_x^{\max}, b_x^{\max})$ on which such a curve can be defined. If, for each $x \in M$, we have $a_x^{\max} = -\infty$ and $b_x^{\max} = +\infty$, we say that the vector field X is *complete*. If M is compact, then each smooth vector field on M is complete. We may assemble the curves γ_x into the *flow* Φ generated by X, defined as

$$\Phi_t(x) = \gamma_x(t),$$

whenever $a_x^{\max} < t < b_x^{\max}$. If t does not belong to $(a_x^{\max}, b_x^{\max})$, then $\Phi_t(x)$ is not defined. The flow Φ satisfies

$$\Phi_0(x) = x. \tag{21.2}$$

Furthermore, if x is in the domain of Φ_t and $\Phi_t(x)$ is in the domain of Φ_s, then x is in the domain of Φ_{s+t} and

$$\Phi_s(\Phi_t(x)) = \Phi_{s+t}(x). \tag{21.3}$$

In the other direction, given a family of maps Φ satisfying (21.2) and (21.3) and appropriate domain properties, there is a unique vector field X such that Φ is the flow generated by X. In particular, if $\Phi_t(x)$ is defined for all x and t, is smooth as a map of $M \times \mathbb{R}$ into M, and satisfies (21.2) and (21.3), there is a unique complete vector field X such that Φ is the flow generated by X.

21.1.2 Differential Forms

For each x, the tangent space $T_x M$ is an n-dimensional real vector space. The dual vector space to $T_x M$ is the *cotangent space* to M at x, denoted $T_x^* M$. Given a smooth function f on M and a point $x \in M$, the *differential* of f at x is the element of $T_x^* M$ given by

$$df(X) = X(f)$$

for each $X \in T_x f$. In particular, in any local coordinate system $x_1, \ldots, x_n$, the elements $dx_1, \ldots, dx_n$ satisfy

$$dx_j \left(\frac{\partial}{\partial x_k} \right) = \delta_{jk}.$$

Thus, the elements $dx_1, \ldots, dx_n$ form a basis for $T_x^* M$ at each point. For any smooth function f, we have

$$df = \frac{\partial f}{\partial x_j} dx_j. \tag{21.4}$$

A *k-form* α on M is a mapping assigning to each point $x \in M$ a k-linear, alternating functional α_x on $T_x M$. A k-form is *smooth* if $\alpha(X_1, \ldots, X_k)$ is a smooth function on M for each k-tuple of smooth vector fields $X_1, \ldots, X_k$ on M. In particular, if f is a smooth function, then df is a smooth 1-form. If α is a smooth k-form and X a smooth vector field, we may define the *contraction* of α with X, which is the $(k-1)$-form $i_X \alpha$ given by

$$(i_X \alpha)(X_1, \ldots, X_{k-1}) = \alpha(X, X_1, \ldots, X_{k-1}).$$

Given a k-linear form ϕ on a vector space V, define the *antisymmetrization* AS(ϕ) of ϕ by

$$\mathrm{AS}(\phi)(v_1, \ldots, v_k) = \sum_{\sigma \in S_k} \mathrm{sign}(\sigma) \phi(v_{\sigma(1)}, v_{\sigma(2)}, \ldots, v_{\sigma(k)}),$$

where S_k denotes the permutation group on k elements. Given a k-form α and an l-form β on M, let $\alpha \otimes \beta$ be the $(k+l)$-linear form on each $T_x M$ given by

$$(\alpha \otimes \beta)(X_1, \ldots, X_{k+l}) = \alpha(X_1, \ldots, X_k) \beta(X_{k+1}, \ldots, X_{k+l}).$$

Then let $\alpha \wedge \beta$ denote the $(k+l)$-form given by

$$\alpha \wedge \beta = \mathrm{AS}(\alpha \otimes \beta).$$

In particular, if α and β are 1-forms, then $\alpha \wedge \beta$ is the 2-form given by

$$(\alpha \wedge \beta)(X, Y) = \alpha(X)\beta(Y) - \alpha(Y)\beta(X).$$

In a smooth coordinate system $x_1, \ldots, x_n$, a smooth k-form α can be expressed uniquely as

$$\alpha = a_{j_1, \ldots, j_k}(x) \, dx_{j_1} \wedge \cdots \wedge dx_{j_k}.$$

A 2-form ω on M is said to be *nondegenerate* if ω defines a nondegenerate bilinear form on each $T_x M$. More explicitly, this means that for each $x \in M$ and each nonzero $X \in T_x M$, there exists a $Y \in T_x M$ such that

$$\omega(X, Y) \neq 0.$$

Suppose α is a smooth k-form on M and S is a compact, oriented, k-dimensional submanifold-with-boundary of M. Then one can define the *integral* of α over M. There is a map d, called the *exterior derivative*, mapping smooth k-forms to smooth $(k+1)$-forms and having the property that

$$\int_S d\beta = \int_{\partial S} \beta \tag{21.5}$$

for every compact, oriented, k-dimensional submanifold-with-boundary S of M and every $(k-1)$-form β on M. Here ∂S is the boundary of S, with the natural orientation induced by the orientation on M. The relation (21.5) is known as *Stoke's theorem*. A k-form α is said to be *closed* if $d\alpha = 0$.

The exterior derivative may be computed in coordinates by the formula

$$d(f\, dx_{j_1} \wedge \cdots \wedge dx_{j_k}) = \frac{\partial f}{\partial x_l} dx_l \wedge dx_{j_1} \wedge \cdots \wedge dx_{j_k}.$$

A coordinate-invariant formula for the exterior derivative of a k-form α is:

$$d\alpha(X_1, \ldots, X_{k+1}) = \sum_{j=1}^{k+1} (-1)^{j+1} \alpha(X_1, \ldots, \widehat{X_j}, \ldots, X_{k+1})$$
$$+ \sum_{j<l} (-1)^{j+l} \alpha([X_j, X_l], X_1, \ldots, \widehat{X_j}, \ldots, X_{k+1}),$$

where $\widehat{X_j}$ indicates that the X_j term is omitted and where $[X_j, X_l]$ is the commutator of X_j and X_l as first-order differential operators. In particular, if α is a 1-form, we have

$$(d\alpha)(X, Y) = X(\alpha(Y)) - Y(\alpha(X)) - \alpha([X, Y]). \tag{21.6}$$

A key identity satisfied by the exterior derivative is

$$d(d\alpha) = 0$$

for all k-forms α. Conversely, if β is a closed $(k+1)$-form (i.e., $d\beta = 0$), then β can be expressed *locally* in the form $\beta = d\alpha$ for some k-form α. More precisely, if β is closed, then for any $x \in M$ there exists a neighborhood U of x and a k-form α defined on U such that $\beta = d\alpha$ on U. If M satisfies certain topological conditions, then each closed k-form α on M can be expressed *globally* in the form $\alpha = d\beta$. In particular, if M is simply connected, then each closed 1-form β can be expressed globally in the form $\beta = df$ for some smooth function (i.e., 0-form) f.

If X is a vector field and α is a k-form, we may define the *Lie derivative* of α in the direction of X, denoted $\mathcal{L}_X \alpha$, as follows:

$$\mathcal{L}_X \alpha = \frac{d}{dt}(\Phi_t^*)(\alpha)\Big|_{t=0},$$

where Φ_t is the flow generated by X and $(\Phi_t^*)(\alpha)$ is the *pullback* of α by Φ_t. The Lie derivative may be computed using the formula

$$\mathcal{L}_X = i_X \circ d + d \circ i_X. \tag{21.7}$$

21.2 Mechanics on Symplectic Manifolds

The reader is warned that sign conventions in the subject of Hamiltonian mechanics are not consistent from author to author.

21.2.1 Symplectic Manifolds

A symplectic manifold is, roughly, a manifold with enough additional structure to allow one to define the Poisson bracket of two functions.

Definition 21.1 *A **symplectic manifold** is a smooth manifold N together with a closed, nondegenerate 2-form ω on N. If (N_1, ω_1) and (N_2, ω_2) are symplectic manifolds, a map $\Phi : N_1 \to N_2$ is a **symplectomorphism** if Φ is a diffeomorphism and in addition*

$$\Phi^*(\omega_2) = \omega_1.$$

It is not hard to see that every symplectic manifold must be even dimensional, for the simple reason that an odd-dimensional vector space does not admit a nondegenerate, skew-symmetric bilinear form.

Throughout this chapter, N will always denote a symplectic manifold of dimension $2n$ with symplectic form ω.

We now show that the cotangent bundle of any manifold has the structure of a symplectic manifold in a canonical way. Suppose $x_1, \ldots, x_n$ is a coordinate system defined on an open set $U \subset M$. Then at each point $x \in U$, an element ϕ of $T_x^* M$ can be expressed uniquely in the form

$$\phi = p_j \, dx_j$$

for a sequence $p_1, \ldots, p_n$ of real numbers. The quantities $x_1, \ldots, x_n$ and $p_1, \ldots, p_n$ constitute a coordinate system on $\pi^{-1}(U)$. We refer to a coordinate system of this sort as a *standard coordinate system* on $T^* M$.

Example 21.2 *For any smooth manifold M, define a 1-form θ on the cotangent bundle $T^* M$ by*

$$\theta(X)_{(x,\phi)} = \phi(\pi_*(X))$$

for each tangent vector $X \in T_{(x,\phi)}(T^* M)$, where $\pi : T^* M \to M$ is the canonical projection. Then the 2-form $\omega := d\theta$ is closed and nondegenerate. We refer to θ and ω as the canonical 1-form and the canonical 2-form on $T^* M$, respectively.

Proof. Using a coordinate system $\{x_j\}$ on X and the associated standard coordinate system $\{x_j, p_j\}$ on T^*M, the projection π is given by $\pi(x, p) = x$. Meanwhile, a tangent vector X to T^*M is expressible as a linear combination the $\partial/\partial x_j$'s and $\partial/\partial p_j$'s. Thus,

$$\theta\left(a_k\frac{\partial}{\partial x_k} + b_k\frac{\partial}{\partial p_k}\right) = (p_j\,dx_j)\left(a_k\frac{\partial}{\partial x_k}\right).$$

What this means is that

$$\theta = p_j\,dx_j,$$

where the x_j's are now viewed as functions on T^*M rather than on M. We have, then,

$$\omega = d\theta = dp_j \wedge dx_j.$$

It is now easy to see that ω is nondegenerate (Exercise 1). ∎

21.2.2   Poisson Brackets and Hamiltonian Vector Fields

If ω is nondegenerate, then it gives a canonical identification of $T_z N$ with $T_z^* N$ at each point, by identifying a vector X in $T_z N$ with the linear functional $\omega(X, \cdot)$ in $T_z^* N$. We can then transfer the bilinear form ω from $T_z N$ to $T_z^* N$ by means of this identification. We denote the resulting bilinear form on $T_z^* N$ by ω^{-1}.

Definition 21.3 *If f and g are smooth functions on N, define the **Poisson bracket** $\{f, g\}$ of f and g by*

$$\{f, g\} = -\omega^{-1}(df, dg).$$

In particular, if $\mathbf{1}$ denotes the constant function on N, then $\{\mathbf{1}, f\} = \{f, \mathbf{1}\} = 0$ for all smooth functions f.

Example 21.4 *If ω is the canonical 2-form on T^*M, then the associated Poisson bracket may be computed in standard coordinates as*

$$\{f, g\} = \frac{\partial f}{\partial x_j}\frac{\partial g}{\partial p_j} - \frac{\partial f}{\partial p_j}\frac{\partial g}{\partial x_j}$$

for all smooth functions f and g on T^*M.

Proof. The linear functional

$$\omega\left(\frac{\partial}{\partial x_j}, \cdot\right)$$

has a value of -1 on the vector $\partial/\partial p_j$ and a value of 0 on all the other basic partial derivatives. This means that $\omega(\partial/\partial x_j, \cdot) = -dp_j$. Similarly,

$\omega(\partial/\partial p_j, \cdot) = dx_j$. We may thus compute, for example, that

$$-1 = \omega\left(\frac{\partial}{\partial x_j}, \frac{\partial}{\partial p_j}\right)$$
$$= \omega^{-1}(-dp_j, dx_j)$$
$$= \omega^{-1}(dx_j, dp_j).$$

Meanwhile, $\omega^{-1}(dx_j, dx_k) = \omega^{-1}(dp_j, dp_k) = 0$ and $\omega^{-1}(dp_j, dx_k) = 0$ when $j \neq k$. Thus, we compute that

$$\{f, g\} = -\omega^{-1}\left(\frac{\partial f}{\partial x_j}dx_j + \frac{\partial f}{\partial p_j}dp_j, \frac{\partial g}{\partial x_k}dx_k + \frac{\partial g}{\partial p_k}dp_k\right)$$
$$= \frac{\partial f}{\partial x_j}\frac{\partial g}{\partial p_k}\delta_{jk} - \frac{\partial f}{\partial p_j}\frac{\partial g}{\partial x_k}\delta_{jk},$$

which reduces to the claimed expression. ∎

Proposition 21.5 *For any smooth functions f, g, h on N, we have*

$$\{g, f\} = -\{f, g\}$$

and

$$\{f, gh\} = \{f, g\}h + g\{f, h\}.$$

Proof. Since ω is skew-symmetric on the tangent space to N at each point and ω^{-1} is obtained from ω by means of an isomorphism of tangent and cotangent space, ω^{-1} is a skew-symmetric form on the cotangent space. The skew-symmetry of the Poisson bracket follows. The second relation follows from the Leibniz product rule for $d(gh)$ together with the bilinearity of ω^{-1}. ∎

Definition 21.6 *If f is a smooth function on N, let X_f be the unique vector field on N such that*

$$df = \omega(X_f, \cdot). \tag{21.8}$$

*We call X_f the **Hamiltonian vector field** associated to f.*

That is to say, X_f corresponds to df under the isomorphism between tangent and cotangent spaces established by ω.

Proposition 21.7 *For all f and g,*

$$X_f(g) = \{f, g\} = -X_g(f).$$

Furthermore,

$$\omega(X_f, X_g) = -\{f, g\}.$$

Proof. For each $z \in N$, we are using ω to identify $T_z N$ with $T_z^* N$. Equation (21.8) says that under this identification, X_f is identified with df. Thus,

$$-\omega^{-1}(df, dg) = -\omega(X_f, X_g) = -df(X_g) = -X_g(f).$$

Thus, $\{f, g\} = -X_g(f)$, as claimed. A similar argument with the roles of f and g reversed gives the claimed relationship between $X_f(g)$ and $\{g, f\}$. Finally,

$$\omega(X_f, X_g) = df(X_g) = X_g(f) = -\{f, g\},$$

as claimed. ∎

Definition 21.8 *For any smooth function f on N, the **Hamiltonian flow** generated by f, denoted Φ^f, is the flow generated by the vector field $-X_f$.*

In the case $N = T^* \mathbb{R}^n \cong \mathbb{R}^{2n}$, this definition agrees with our notation in Sect. 2.5.

Proposition 21.9 *For any smooth function f on N, the Hamiltonian flow Φ^f preserves ω.*

Proof. In general, a flow Φ preserves a differential form α if and only if the Lie derivative $L_X \alpha = 0$, where X is the vector field generating Φ. In our case, since ω is closed, we have, by (21.7),

$$\mathcal{L}_{X_f} \omega = d[i_{X_f} \omega] = d^2 f = 0,$$

since $i_{X_f} \omega$ is, by the definition of X_f, equal to df. ∎

Proposition 21.10 *For any smooth functions f, g, h on N, the Jacobi identity holds:*

$$\{f, \{g, h\}\} + \{g, \{h, f\}\} + \{h, \{f, g\}\} = 0.$$

This result shows that the space of smooth function on N forms a Lie algebra under the Poisson bracket. The proof of Proposition 21.10 relies on Proposition 21.9, which in turn relies on the fact that ω is closed.

Proof. Since the Hamiltonian flow Φ^f preserves ω, it also preserves ω^{-1} and thus

$$\omega^{-1}(d(g \circ \Phi_t^f), d(h \circ \Phi_t^f)) = \omega^{-1}(dg, dh) \circ \Phi_t^f,$$

or, equivalently,

$$\{g \circ \Phi_t^f, h \circ \Phi_t^f\} = \{g, h\} \circ \Phi_t^f.$$

Differentiating this relation with respect to t at $t = 0$ gives

$$\{-X_f(g), h\} + \{g, -X_f(h)\} = -X_f(\{g, h\}),$$

or, equivalently,

$$-\{\{f,g\},h\} + \{g,\{f,h\}\} = -\{f,\{g,h\}\}.$$

After moving $-\{f,\{g,h\}\}$ to the left-hand side of the equation and using the skew-symmetry of the Poisson bracket, we obtain the Jacobi identity. ∎

Proposition 21.11 *For any smooth functions f and g on N, the Hamiltonian vector fields X_f and X_g satisfy*

$$[X_f, X_g] = X_{\{f,g\}}.$$

Proof. See Exercise 3. ∎

21.2.3 Hamiltonian Flows and Conserved Quantities

We have seen (Proposition 21.9) that if f is a smooth function, then the flow generated by X_f preserves ω. We have the following partial converse to this result.

Proposition 21.12 *Suppose Φ is the flow generated by a vector field $-X$ on N. If Φ preserves ω then X can be represented locally in the form $X = X_f$ for some smooth function f on N. If N is simply connected, the function f exists globally on N.*

Proof. The statement that Φ preserves ω can be expressed infinitesimally as

$$\mathcal{L}_X \omega = 0.$$

Since also ω is closed, (21.7) tells us that

$$d(i_X \omega) = 0.$$

Since $i_X \omega$ is closed, this 1-form can be expressed locally as $i_X \omega = df$ for some smooth function f, which says precisely that $X = X_f$. If N is simply connected, then every closed 1-form can be expressed globally as df, for some smooth function f. ∎

A flow of the sort in Proposition 21.12 is said to be *locally Hamiltonian*. Such a flow is said to be (globally) *Hamiltonian* if the function f in the proposition can be defined on all of N. (Compare Definition 21.8.) If Φ is a Hamiltonian flow, the function f such that $\Phi = \Phi^f$ is called a *Hamiltonian generator* of Φ. If N is connected, then any two Hamiltonian generators of Φ must differ by a constant.

To see that, in general, f is only defined locally, consider the symplectic manifold $S^1 \times \mathbb{R}$, with symplectic form $\omega = d\phi \wedge dx$, where ϕ is the angular coordinate on S^1 and x is the linear coordinate on $\mathbb{R}$. Note that the 1-form

$d\phi$ is independent of the choice of a local angle variable on S^1, since any two such angle functions differ by a constant (an integer multiple of 2π). Thus, $d\phi$ is a globally defined, smooth 1-form, even though there is no globally defined, smooth angle function ϕ. Define a flow Φ by

$$\Phi_t(\phi, x) = (\phi, x + t).$$

This flow certainly preserves ω, since dx is invariant under translations.

The flow Φ is generated by the vector field $-X = \partial/\partial x$, and

$$\omega(-\partial/\partial x, \cdot) = d\phi.$$

As we have already noted, however, there is no globally defined function ϕ whose differential is $d\phi$.

Although any smooth function on a symplectic manifold N generates a Hamiltonian flow, in physical examples there is usually one distinguished function with a Hamiltonian flow that is thought of as "the" time-evolution of the system.

Definition 21.13 *A **Hamiltonian system** is a symplectic manifold N together with a distinguished Hamiltonian flow Φ^H, generated by smooth function H on N, called the **Hamiltonian** of the system. A function f is called a **conserved quantity** for a Hamiltonian system (N, Φ^H) if $f(\Phi_t^H(x))$ is independent of t for each fixed $x \in N$.*

As in the $\mathbb{R}^{2n}$ case, conserved quantities are useful in understanding the nature of the dynamics. See the discussion following Corollary 2.26.

Proposition 21.14 *For any Hamiltonian system (N, Φ^H), we have*

$$\frac{d}{dt} f(\Phi_t^H(z)) = \{f, H\}(\Phi_t^H(z)),$$

for all $z \in N$, or, more concisely,

$$\frac{df}{dt} = \{f, H\}.$$

In particular, a smooth function f on N is a conserved quantity for a Hamiltonian system Φ^H if and only if $\{f, H\} = 0$.

Proof. For the flow generated by any vector field X, we have

$$\frac{d}{dt} f(\Phi_t(z)) = X_{\Phi_t(z)} f.$$

If $X = -X_f$, then by Proposition 21.7, we have the claimed result. ∎

Proposition 21.15 *A smooth function f is a conserved quantity for a Hamiltonian system (N, Φ^H) if and only if H is invariant under the Hamiltonian flow Φ^f generated by f.*

Proof. By the previous proposition, H is invariant under the flow generated by f if and only if $\{H, f\} = 0$, which holds if and only if $\{f, H\} = 0$, which holds if and only if f is a conserved quantity. ∎

21.2.4   The Liouville Form

A symplectic manifold N has a natural volume form, which allows us to formulate an analog on N of Liouville's theorem (Theorem 2.27).

Definition 21.16 *If N is a $2n$-dimensional symplectic manifold, the* **Liouville form** *on N is the $2n$-form λ given by*

$$\lambda = \frac{1}{n!}\omega^n,$$

where $\omega^n = \omega \wedge \cdots \wedge \omega$.

Since ω is, by assumption, a nondegenerate form on each tangent space $T_z N$, it is not hard to check that λ is a nonvanishing $(2n)$-linear form on each $T_z N$. Thus, λ determines an orientation on N. Given a compactly supported continuous function f on N, we can define the integral of f over N, computed with respect to the orientation determined by λ itself. Using the version of the Riesz representation theorem for *locally* compact topological spaces, one can show that there is a unique *measure*, called the Liouville volume measure, for which the integral of every continuous compactly supported function f is given by $\int_N f \, \lambda$.

We are now ready to state the general form of Liouville's theorem.

Theorem 21.17 (Liouville's Theorem) *For any smooth function f on N, the Hamiltonian flow Φ^f preserves λ.*

Proof. The flow Φ^f will preserve λ if and only if the vector field X_f satisfies $\mathcal{L}_{X_f}\lambda = 0$. But

$$\mathcal{L}_{X_f}\lambda = \frac{1}{n!}[(\mathcal{L}_{X_f}\omega) \wedge \omega \wedge \cdots \wedge \omega$$
$$+ \omega \wedge (\mathcal{L}_{X_f}\omega) \wedge \omega \wedge \cdots \wedge \omega$$
$$+ \cdots + \omega \wedge \cdots \wedge \omega \wedge (\mathcal{L}_{X_f}\omega)].$$

Since we have already shown (Proposition 21.9) that $\mathcal{L}_{X_f}\omega = 0$, we see that $\mathcal{L}_{X_f}\lambda = 0$. ∎

21.3   Exercises

1. Show that the canonical 2-form ω on T^*M is nondegenerate.

   Hint: Work in standard coordinates $\{x_j, p_j\}$.

2. Show that if $\Phi : M \to M$ is a diffeomorphism, then the induced map $\Phi^* : T^*M \to T^*M$ is a symplectomorphism.

3. Using Proposition 21.7 and the Jacobi identity for the Poisson bracket, verify that

$$[X_f, X_g] = X_{\{f,g\}}$$

for all smooth functions f and g on N.

4. If N is compact, show that

$$\int_N \{f, g\} \, \lambda = 0$$

for all smooth function f and g on N.

Hint: Apply Liouville's theorem to the flow Φ_t^f.

22

Geometric Quantization on Euclidean Space

22.1 Introduction

In this chapter, we consider the geometric quantization program in the setting of the symplectic manifold $\mathbb{R}^{2n}$, with the canonical 2-form $\omega = dp_j \wedge dx_j$. We begin with the "prequantum" Hilbert space $L^2(\mathbb{R}^{2n})$ and define "prequantum" operators $Q_{\text{pre}}(f)$. These operators satisfy

$$Q_{\text{pre}}(\{f, g\}) = \frac{1}{i\hbar}[Q_{\text{pre}}(f), Q_{\text{pre}}(g)]$$

for *all* f and g. Nevertheless, there are several undesirable aspects to the prequantization map that make it physically unreasonable to interpret it as "quantization." To obtain the quantum Hilbert space, we reduce the number of variables from $2n$ to n. Depending on how we do this reduction, we will obtain either the position Hilbert space, the momentum Hilbert space, or the Segal–Bargmann space. Each of these subspaces is preserved by the prequantized position and momentum operators, and by certain other operators of the form $Q_{\text{pre}}(f)$.

Although the material in this chapter is a special case of what we do in Chap. 23, doing this case first allows us to get a feeling for the methods and results of geometric quantization quickly, without needing to develop the full machinery of line bundles, connections, and polarizations over general symplectic manifolds. In any case, we would need to carry out most of the calculations in this chapter eventually, as standard examples of the general theory.

B.C. Hall, *Quantum Theory for Mathematicians*, Graduate Texts in Mathematics 267, DOI 10.1007/978-1-4614-7116-5_22, © Springer Science+Business Media New York 2013

Although this chapter does not require the full machinery of symplectic manifolds, we will make use of the notions of 1-forms and 2-forms on $\mathbb{R}^{2n}$, along with the notion of the differential of a 1-form. In particular, the expression (21.6) for the differential of a 1-form will be used.

The reader should be warned that sign conventions in geometric quantization are not consistent from author to author. The sign conventions used here are chosen to maintain consistency with the physics literature. In particular, we could eliminate an annoying minus sign in the definition of the holomorphic subspace if we were willing to allow the function p_j to quantize to $i\hbar \, \partial/\partial x_j$. Since, however, the convention $P_j = -i\hbar \, \partial/\partial x_j$ is universal in the physics literature, we have chosen to be consistent with that convention and to accept some slightly inconvenient sign choices elsewhere. We continue to follow the *summation convention*, in which repeated indices are always summed on.

22.2   Prequantization

Ideally, a quantization procedure Q, mapping functions on a symplectic manifold N to operators on some Hilbert space $\mathbf{H}$, should satisfy the following properties. First, $Q(f)$ should be self-adjoint whenever f is real valued. Second, we should have $Q(\mathbf{1}) = I$, where $\mathbf{1}$ is the constant function. Third, $Q(\{f,g\})$ should be equal to $[Q(f), Q(g)]/(i\hbar)$. Fourth, there should be some sort of "smallness" assumption. In the case $N = \mathbb{R}^{2n}$, for example, we may require that $\mathbf{H}$ should be irreducible under the action of the (exponentiated) position and momentum operators. (See Definition 14.6.) Although Groenewold's theorem (Theorem 13.13) suggests that it is unrealistic to expect to find a quantization procedure that satisfies all of these properties exactly, we try to come as close as possible.

Throughout this chapter, we follow the convention of thinking of a "vector field" on $\mathbb{R}^N$ as a first-order differential operator, as in Exercise 14 in Chap. 2. Given, for example, the vector-valued function

$$X = (2x_1 + x_2, x_1 x_2)$$

on $\mathbb{R}^2$, we identify X with the operator of "differentiation in the direction of X," that is, with the following first-order differential operator:

$$X = (2x_1 + x_2)\frac{\partial}{\partial x_1} + x_1 x_2 \frac{\partial}{\partial x_2}.$$

In particular, given a smooth function f on $\mathbb{R}^{2n}$, the Hamiltonian vector field X_f associated to f is thought of as a differential operator:

$$X_f = \{f, \cdot\} = \frac{\partial f}{\partial x_j}\frac{\partial}{\partial p_j} - \frac{\partial f}{\partial p_j}\frac{\partial}{\partial x_j}, \tag{22.1}$$

acting on $C^\infty(\mathbb{R}^{2n})$. (Compare Proposition 21.7.) By Proposition 21.11, the commutator (as differential operators) of two Hamiltonian vector fields X_f and X_g is $X_{\{f,g\}}$. Thus, the operators $i\hbar X_f$ satisfy the desired commutation relations:

$$[i\hbar X_f, i\hbar X_g] = (i\hbar)^2 X_{\{f,g\}} = (i\hbar)(i\hbar X_{\{f,g\}}).$$

It is tempting, then, to define a (pre)quantization map simply by taking $Q(f) = i\hbar X_f$, viewed as a self-adjoint operator on the Hilbert space $L^2(\mathbb{R}^{2n})$. This map, however, does not satisfy $Q(1) = I$. If we to correct our definition to $Q(f) = i\hbar X_f + f$, where f means the operator of multiplication by f, then $Q(1) = I$ but the desired commutation property is destroyed.

It is possible to achieve both $Q(1) = I$ and the desired commutation relations by adding one more term as follows. If $\omega = dp_j \wedge dx_j$ is the canonical 2-form on $\mathbb{R}^{2n}$, let θ be any *symplectic potential* for ω, that is, any one-form with

$$d\theta = \omega. \tag{22.2}$$

(We may, e.g., take $\theta = p_j dx_j$.) For a smooth function f on $\mathbb{R}^{2n}$, define an operator $Q_{\mathrm{pre}}(f)$, acting on $C^\infty(\mathbb{R}^{2n})$, by

$$Q_{\mathrm{pre}}(f) = i\hbar\left(X_f - \frac{i}{\hbar}\theta(X_f)\right) + f. \tag{22.3}$$

The expression f on the right-hand side of (22.3) means, more precisely, the operator of multiplication by f, and similarly for the function $\theta(X_f)$. Note that since θ is a 1-form and X_f is a vector field, $\theta(X_f)$ is a function on $\mathbb{R}^{2n}$. The operator $Q_{\mathrm{pre}}(f)$ is the *prequantization* of f and is to be viewed as an unbounded operator on $L^2(\mathbb{R}^{2n})$, where we refer to $L^2(\mathbb{R}^{2n})$ as the *prequantum Hilbert space*.

According to Exercise 1, any divergence free vector field on $\mathbb{R}^N$ is a skew-symmetric operator on $C_c^\infty(\mathbb{R}^N) \subset L^2(\mathbb{R}^N)$. Meanwhile, each Hamiltonian vector field is divergence free, as we have already remarked in the proof of Liouville's theorem (Theorem 2.27). Thus, for any smooth, real-valued function f on $\mathbb{R}^{2n}$, the operator $Q_{\mathrm{pre}}(f)$ is at least symmetric. It can be shown that if X_f is complete, meaning that the associated Hamiltonian flow is defined for all times, then $Q_{\mathrm{pre}}(f)$ is actually self-adjoint on a natural domain. (See the discussion following the proof of Proposition 23.13.)

As it turns out, the $\theta(X_f)$ term in (22.3) is precisely what is needed to restore the desired commutation relations, while still allowing $Q_{\mathrm{pre}}(1)$ to equal the identity.

Proposition 22.1 *For all $f, g \in C^\infty(\mathbb{R}^{2n})$, we have*

$$\frac{1}{i\hbar}[Q_{\mathrm{pre}}(f), Q_{\mathrm{pre}}(g)] = Q_{\mathrm{pre}}(\{f, g\}),$$

where the identity is to be understood as an equality of operators on C^∞ $(\mathbb{R}^{2n})$.

Before proving this result, it is useful to understand the behavior of the expression $X_f - (i/\hbar)\theta(X_f)$ occurring in the definition of $Q_{\mathrm{pre}}(f)$.

Definition 22.2 *For any symplectic potential θ and vector field X on $\mathbb{R}^{2n}$, let ∇_X denote the* **covariant derivative** *operator, acting on $C^\infty(\mathbb{R}^{2n})$, given by*

$$\nabla_X = X - \frac{i}{\hbar}\theta(X). \qquad (22.4)$$

Note that our prequantized operators can be written as

$$Q_{\mathrm{pre}}(f) = i\hbar\nabla_{X_f} + f.$$

Proposition 22.3 *For any symplectic potential θ, let ∇_X denote the associated covariant derivative in (22.4). Then for all smooth vector fields X and Y on $\mathbb{R}^{2n}$, we have*

$$[\nabla_X, \nabla_Y] = \nabla_{[X,Y]} - \frac{i}{\hbar}\omega(X,Y). \qquad (22.5)$$

In particular, if $X = X_f$ and $Y = X_g$, we have

$$[\nabla_{X_f}, \nabla_{X_g}] = \nabla_{X_{\{f,g\}}} + \frac{i}{\hbar}\{f,g\}.$$

According to standard differential geometric definitions, the 2-form $\omega/\hbar$ on the right-hand side of (22.5) is the *curvature* of the covariant derivative ∇. For our purposes, the fact that $[\nabla_{X_f}, \nabla_{X_g}]$ in not simply $\nabla_{X_{\{f,g\}}}$ is an advantage. The extra term in the formula for the commutator is just what we need to compensate for the failure of the operators $i\hbar X_f + f$ to have the desired commutation relations.

Proof. Using the easily verified identity $[\nabla_X, f] = X(f)$, we obtain

$$[\nabla_X, \nabla_Y] - \nabla_{[X,Y]} = -\frac{i}{\hbar}[X(\theta(Y)) - Y(\theta(X)) - \theta([X,Y])].$$

In light of (21.6), the right-hand side becomes $-(i/\hbar)(d\theta)(X,Y)$, where $d\theta = \omega$. ∎

We may now easily prove Proposition 22.1.

Proof of Proposition 22.1. Using Proposition 22.3, we obtain

$$\frac{1}{i\hbar}\left[i\hbar\nabla_{X_f} + f, i\hbar\nabla_{X_g} + g\right]$$

$$= (i\hbar)\left(\nabla_{X_{\{f,g\}}} + \frac{i}{\hbar}\{f,g\}\right) + X_f(g) - X_g(f)$$

$$= i\hbar\nabla_{X_{\{f,g\}}} - \{f,g\} + \{f,g\} + \{f,g\},$$

which reduces to what we want. ∎

Example 22.4 *If* $\theta = p_j dx_j$, *the prequantized position and momentum operators are given by*

$$Q_{\mathrm{pre}}(x_j) = x_j + i\hbar \frac{\partial}{\partial p_j}$$

$$Q_{\mathrm{pre}}(p_j) = -i\hbar \frac{\partial}{\partial x_j}.$$

These operators are essentially self-adjoint on $C_c^\infty(\mathbb{R}^{2n})$ and their self-adjoint extensions satisfy the exponentiated commutation relations of Definition 14.2.

Proof. We compute that $X_{x_j} = \partial/\partial p_j$ and that $\theta(X_{x_j}) = 0$, giving the indicated expression for $Q_{\mathrm{pre}}(x_j)$. Meanwhile, $X_{p_j} = -\partial/\partial x_j$ and $\theta(X_{p_j}) = -p_j$. There is a cancellation of the $\theta(X_{p_j})$ term in the definition of $Q_{\mathrm{pre}}(p_j)$ with the p_j term, leaving $Q_{\mathrm{pre}}(p_j) = i\hbar X_{p_j}$.

The essential self-adjointness of the operators follows from Proposition 9.40. To verify the exponentiated commutation relations, we calculate the associated one-parameter unitary groups as

$$(e^{itQ_{\mathrm{pre}}(x_j)}\psi)(\mathbf{x}, \mathbf{p}) = e^{itx_j}\psi(\mathbf{x}, \mathbf{p} - t\hbar\mathbf{e}_j)$$

$$(e^{itQ_{\mathrm{pre}}(p_j)}\psi)(\mathbf{x}, \mathbf{p}) = \psi(\mathbf{x} + t\hbar\mathbf{e}_j, \mathbf{p}), \tag{22.6}$$

where we now let $Q_{\mathrm{pre}}(x_j)$ and $Q_{\mathrm{pre}}(p_j)$ denote the unique self-adjoint extensions of the given operators on $C_c^\infty(\mathbb{R}^{2n})$. (Compare Proposition 13.5.) The exponentiated commutation relations can now be easily verified by direct calculation. ∎

As we have presented things so far, the concept of covariant derivative, and thus also of prequantization, depends on the choice of symplectic potential θ. This dependence is, however, illusory; we will now show that the prequantum maps obtained with two different symplectic potentials are unitarily equivalent.

Proposition 22.5 *Suppose that θ_1 and θ_2 are two different symplectic potentials for the canonical 2-form ω, so that $d(\theta^1 - \theta^2) = 0$. Let the associated covariant derivatives be denoted by ∇^1 and ∇^2. Choose a real-valued function γ so that $d\gamma = \theta^1 - \theta^2$ and let U_γ be the unitary map of $L^2(\mathbb{R}^{2n})$ to itself given by*

$$U_\gamma\psi = e^{-i\gamma/\hbar}\psi.$$

Then for every vector field X, we have

$$U_\gamma \nabla_X^1 U_\gamma^{-1} = \nabla_X^2. \tag{22.7}$$

If $Q_{\mathrm{pre}}^j(f)$, $j = 1, 2$, are the associated prequantization maps, it follows that

$$U_\gamma Q_{\mathrm{pre}}^1(f)U_\gamma^{-1} = Q_{\mathrm{pre}}^2(f). \tag{22.8}$$

*The map U_γ is called a **gauge transformation**.*

Proof. The operation of multiplication by $\theta^1(X)$ commutes with multiplication by $e^{-i\gamma/\hbar}$, whereas

$$X(e^{i\gamma/\hbar}\psi) = e^{i\gamma/\hbar}X\psi + \frac{i}{\hbar}e^{i\gamma/\hbar}X(\gamma)\psi.$$

Since $X(\gamma) = (d\gamma)(X) = \theta^1(X) - \theta^2(X)$, we obtain

$$\nabla_X^1(e^{i\gamma/\hbar}\psi) = e^{i\gamma/\hbar}\left(X + \frac{i}{\hbar}X(\gamma) - \frac{i}{\hbar}\theta^1(X_f)\right)\psi$$

$$= e^{i\gamma/\hbar}\left(X - \frac{i}{\hbar}\theta^2(X_f)\right)\psi$$

$$= e^{i\gamma/\hbar}\nabla_X^2\psi.$$

Multiplying both sides of this equality by $e^{-i\gamma/\hbar}$ gives (22.7). Equation (22.8) follows by observing that multiplication by f commutes with multiplication by $e^{-i\gamma/\hbar}$. $\blacksquare$

22.3 Problems with Prequantization

Given the naturalness of the prequantization construction, it is tempting to think that prequantization could actually be considered as quantization. Why not take our Hilbert space to be $L^2(\mathbb{R}^{2n})$ and the quantized operators to be $Q_{\mathrm{pre}}(f)$? To answer this question, we now examine some undesirable properties of prequantization.

In the first place, the Hilbert space $L^2(\mathbb{R}^{2n})$ is very far from irreducible under the action of the quantized position and momentum operators, in contrast to the ordinary Schrödinger Hilbert space $L^2(\mathbb{R}^n)$, which is irreducible, by Proposition 14.7. Indeed, in Sect. 22.4, we will construct a large family of invariant subspaces. (See Proposition 22.13.)

In the second place, the prequantization map is very far from being multiplicative. Of course, since quantum operators do not commute, we cannot expect any quantization scheme Q to satisfy $Q(fg) = Q(f)Q(g)$ for *all* f and g. Nevertheless, the standard quantization schemes we have considered in Chap. 13 do satisfy this relation for certain classes of observables f and g. In the Weyl quantization, for example, we have multiplicativity if f and g are both functions of $\mathbf{x}$ only, independent of $\mathbf{p}$ (or functions of $\mathbf{p}$, independent of $\mathbf{x}$). For the prequantization map, however, we almost never have multiplicativity, for the simple reason that $Q_{\mathrm{pre}}(fg)$ is a first-order differential operator, whereas $Q_{\mathrm{pre}}(f)Q_{\mathrm{pre}}(g)$ is second-order, provided there is at least one point where X_f and X_g are both nonzero.

In the third place, the prequantization map badly fails to map positive functions to positive operators. Although most of the quantization schemes in Chap. 13 do not *always* map positive functions to positive operators, they somehow come close to doing so. Indeed, Q_{Weyl}, Q_{Wick}, and $Q_{\text{anti}-\text{Wick}}$ all map the harmonic oscillator Hamiltonian to a non-negative operator, since $a^*a + (1/2)I$, a^*a, and aa^* are all non-negative. (See Exercise 4 in Chap. 13.) By contrast, the prequantized harmonic oscillator Hamiltonian has spectrum that is unbounded below, as we now demonstrate.

Proposition 22.6 *Consider a harmonic oscillator Hamiltonian of the form*

$$H(x,p) = \frac{1}{2m} \left(p^2 + (m\omega x)^2 \right).$$

Then for each integer n, the number $n\hbar\omega$ is an eigenvalue for $Q_{\text{pre}}(H)$.

Note that n in the proposition is allowed to be negative, so that the spectrum of $Q_{\text{pre}}(H)$ is not even bounded below. On the other hand, in Sect. 22.5, we will consider a certain closed subspace $\mathbf{H}_\alpha$ of the prequantum Hilbert space, which is one candidate for the *quantum* Hilbert space. For appropriate choice of α, the space $\mathbf{H}_\alpha$ is invariant under $Q_{\text{pre}}(H)$ and the restriction of $Q_{\text{pre}}(H)$ is self-adjoint with spectrum $n\hbar\omega$, where n ranges over the *non-negative* integers. See Proposition 22.14. And finally, when we introduce half-forms in Sect. 23.7, we will finally restore the spectrum $(n+1/2)\hbar\omega$, where n ranges over the non-negative integers, that we found in Chap. 11.

Proof. We can write H as

$$H(x,p) = \frac{1}{2m}(p^2 + y^2),$$

where $y = m\omega x$. The flow associated to this Hamiltonian consists of rotations in the (y,p)-plane. If we choose our symplectic potential to be

$$\theta = \frac{1}{2}(p\,dx - x\,dp) = \frac{1}{2m\omega}(p\,dy - y\,dp),$$

then the $\theta(X_H)$ term in $Q_{\text{pre}}(H)$ cancels with the H term, leaving

$$Q_{\text{pre}}(H) = i\hbar X_H$$
$$= i\hbar \left(m\omega^2 x \frac{\partial}{\partial p} - \frac{p}{m}\frac{\partial}{\partial x} \right)$$
$$= i\hbar\omega \left(y \frac{\partial}{\partial p} - p\frac{\partial}{\partial y} \right).$$

Now, if ϕ denotes the angular variable for polar coordinates in the (y,p)-plane, then $y\,\partial/\partial p - p\,\partial/\partial y$ is just $\partial/\partial\phi$. Thus, we can find eigenvectors for $Q_{\text{pre}}(H)$ of the form

$$\psi_n(r,\phi) = f(r)e^{-in\phi}.$$

where n is an integer and f is an arbitrary function with $\int_0^\infty |f(r)|^2 \, r \, dr < \infty$.

∎

The conclusion of the matter is that it is not physically reasonable to use prequantization as our quantization scheme. Instead, we will pass to a "smaller" Hilbert space on which the position and momentum operators act irreducibly.

22.4   Quantization

To obtain a Hilbert space that can be thought of as giving us a "quantization" (as opposed to a prequantization) of $\mathbb{R}^{2n}$, we restrict ourselves to a subspace of the prequantum Hilbert space. The idea is that we should be using only half of the variables on $\mathbb{R}^{2n}$. We might, for example, restrict ourselves to functions that depend only on the position variables and are independent of the momentum variables. Now, the space of functions ψ that are, say, independent of $\mathbf{p}$ in the ordinary sense (i.e., $\psi(\mathbf{x}, \mathbf{p}) = \psi(\mathbf{x}, \mathbf{p}')$) is not invariant under gauge transformations (the maps U_γ in Proposition 22.5). The gauge-invariant notion of being independent of $\mathbf{p}$ is that the *covariant* derivatives of ψ should be zero in the $\mathbf{p}$-directions. Similarly, we may consider spaces of functions with covariant derivatives that are are zero in some other set of n directions.

Definition 22.7 *Fix a symplectic potential θ. Define the **position subspace** as the subspace of $C^\infty(\mathbb{R}^{2n})$ consisting of functions ψ for which*

$$\nabla_{\partial/\partial p_j}\psi = 0$$

*for all j. Similarly, define the **momentum subspace** as the subspace of $C^\infty(\mathbb{R}^{2n})$ consisting of functions ψ for which*

$$\nabla_{\partial/\partial x_j} = 0$$

*for all j. Finally, define the **holomorphic subspace** with parameter α to be the subspace of $C^\infty(\mathbb{R}^{2n})$ consisting of functions ψ for which*

$$\nabla_{\partial/\partial \bar{z}_j}\psi = 0$$

for all j, where $z_j = x_j - i\alpha p_j$ and where $\partial/\partial z_j$ and $\partial/\partial \bar{z}_j$ are defined by

$$\frac{\partial}{\partial z_j} = \frac{1}{2}\left(\frac{\partial}{\partial x_j} + \frac{i}{\alpha}\frac{\partial}{\partial p_j}\right); \quad \frac{\partial}{\partial \bar{z}_j} = \frac{1}{2}\left(\frac{\partial}{\partial x_j} - \frac{i}{\alpha}\frac{\partial}{\partial p_j}\right), \qquad (22.9)$$

The operators $\partial/\partial z_j$ and $\partial/\partial \bar{z}_j$ are nothing but the usual complex derivative operators on $\mathbb{C}^n$ written in terms of the variables $\mathbf{x}$ and $\mathbf{p}$, where we identify $\mathbb{R}^{2n}$ with $\mathbb{C}^n$ by the map $(\mathbf{x}, \mathbf{p}) \mapsto \mathbf{x} - i\alpha\mathbf{p}$.

Of course, the exact form of the various subspaces in Definition 22.7 depends on the choice of symplectic potential. It is convenient to use the symplectic potential $\theta = p_j \, dx_j$.

Proposition 22.8 *Take the symplectic potential* $\theta = p_j\, dx_j$. *Then the position, momentum, and holomorphic subspaces may be computed as follows. The position subspace consists of smooth functions* ψ *on* $\mathbb{R}^{2n}$ *of the form*

$$\psi(\mathbf{x}, \mathbf{p}) = \phi(\mathbf{x}),$$

where ϕ *is an arbitrary smooth function on* $\mathbb{R}^n$. *The momentum subspace consists of smooth functions* ψ *of the form*

$$\psi(\mathbf{x}, \mathbf{p}) = e^{i\mathbf{x}\cdot\mathbf{p}/\hbar}\phi(\mathbf{p}), \tag{22.10}$$

where ϕ *is an arbitrary smooth function on* $\mathbb{R}^n$. *Finally, the holomorphic subspace consists of functions of the form*

$$\psi(\mathbf{x}, \mathbf{p}) = F(z_1, \ldots, z_n)e^{-\alpha|\mathbf{p}|^2/(2\hbar)}, \tag{22.11}$$

where F *is an arbitrary holomorphic function on* $\mathbb{C}^n$ *and where* $z_j = x_j - i\alpha p_j$.

Proof. Since $\theta(\partial/\partial p_j) = 0$, we have $\nabla_{\partial/\partial p_j} = \partial/\partial p_j$, so that functions that are covariantly constant in the $\mathbf{p}$-directions are actually constant in the $\mathbf{p}$-directions. Meanwhile, $\theta(\partial/\partial x_j) = p_j$ and so

$$\nabla_{\partial/\partial x_j} = \frac{\partial}{\partial x_j} - \frac{i}{\hbar}p_j.$$

Now, any function ψ on $\mathbb{R}^{2n}$ can be written in the form $e^{i\mathbf{x}\cdot\mathbf{p}/\hbar}\phi(\mathbf{x}, \mathbf{p})$ for some other function ϕ. If we use this form to compute $\nabla_{\partial/\partial p_j}\psi$, there is a convenient cancellation, giving

$$(\nabla_{\partial/\partial x_j}\psi)(\mathbf{x}, \mathbf{p}) = e^{i\mathbf{x}\cdot\mathbf{p}/\hbar}\frac{\partial\phi}{\partial x_j}.$$

Thus, $\nabla_{\partial/\partial x_j}\psi = 0$ for all j if and only if ϕ is independent of $\mathbf{x}$.

Finally, we note that $\theta(\partial/\partial z_j) = p_j/2$, so that

$$\nabla_{\partial/\partial z_j} = \frac{\partial}{\partial \bar{z}_j} - \frac{i}{2\hbar}p_j.$$

Any function ψ on $\mathbb{R}^{2n}$ can be written in the form $\psi(\mathbf{x}, \mathbf{p}) = e^{-\alpha|\mathbf{p}|^2/(2\hbar)}F$ for some other function F, where we note that

$$e^{-\alpha|\mathbf{p}|^2/(2\hbar)} = \exp\left(\sum_j (\bar{z}_j - z_j)^2/(8\alpha\hbar)\right).$$

Thus,

$$\frac{\partial}{\partial \bar{z}_j}e^{-\alpha|\mathbf{p}|^2/(2\hbar)} = \frac{\bar{z}_j - z_j}{4\alpha\hbar}e^{-\alpha|\mathbf{p}|^2/(2\hbar)} = \frac{i}{2\hbar}p_je^{-\alpha|\mathbf{p}|^2/(2\hbar)}.$$

When we compute $\nabla_{\partial/\partial \bar{z}_j}\psi$ using the indicated form, there is another convenient cancellation, giving

$$(\nabla_{\partial/\partial \bar{z}_j}\psi)(\mathbf{x},\mathbf{p}) = e^{-\alpha|\mathbf{p}|^2/(2\hbar)}\frac{\partial F}{\partial \bar{z}_j}.$$

Thus, $\nabla_{\partial/\partial \bar{z}_j}\psi = 0$ for all j if and only if F is holomorphic as a function of the variables $z_j = x_j - i\alpha p_j$. ∎

From the physical standpoint, we do not merely want a vector space of functions, but a Hilbert space. It is natural, then, to look at functions of the forms computed in Proposition 22.8 that belong to $L^2(\mathbb{R}^{2n})$. In the case of the position and momentum subspaces, we encounter a serious problem: There are no nonzero functions of the indicated form that are square integrable over $\mathbb{R}^{2n}$. After all, if ψ is in the position subspace, then $\psi(\mathbf{x},\mathbf{p})$ is independent of $\mathbf{p}$ and the integral of $|\psi|^2$ over the $\mathbf{p}$-variables will be infinite, unless ψ is zero almost everywhere. If ψ is in the momentum subspace, $|\psi|^2$ is independent of $\mathbf{x}$ and we have a similar problem.

The solution to this problem is to integrate not over $\mathbb{R}^{2n}$ but over $\mathbb{R}^n$. Although the "proper" way to make this change of integration is to introduce the notion of "half-forms," as in Chap. 23, we will content ourselves in this chapter with the following simplistic rule: integrate only over the variables on which $|\psi|^2$ depends. If we want to get a Hilbert space (not just an inner product space), we must also allow functions of the specified form that are square integrable but not necessarily smooth. We may therefore identify the *position Hilbert space* and *momentum Hilbert space* as follows.

Conclusion 22.9 *The position Hilbert space is the space of functions on* $\mathbb{R}^{2n}$ *of the form*

$$\psi(\mathbf{x},\mathbf{p}) = \phi(\mathbf{x}),$$

where $\phi \in L^2(\mathbb{R}^n)$. *The norm of such a function is computed as*

$$\|\psi\|^2 = \int_{\mathbb{R}^n} |\phi(\mathbf{x})|^2 \, d\mathbf{x}.$$

The momentum Hilbert space is the space of functions on $\mathbb{R}^{2n}$ *of the form*

$$\psi(\mathbf{x},\mathbf{p}) = e^{i\mathbf{x}\cdot\mathbf{p}/\hbar}\phi(\mathbf{p}),$$

where $\phi \in L^2(\mathbb{R}^n)$. *The norm of such a function is computed as*

$$\|\psi\|^2 = \int_{\mathbb{R}^n} |\phi(\mathbf{p})|^2 \, d\mathbf{p}.$$

If we consider the holomorphic subspace, we find that it behaves better than the position and momentum subspaces, in that there exist nonzero functions of the form (22.11) that are square integrable over $\mathbb{R}^{2n}$, as we will see shortly. Furthermore, the space of functions of the form (22.11) that are square integrable over $\mathbb{R}^{2n}$ form a closed subspace of $L^2(\mathbb{R}^{2n})$, by the same argument as in the proof of Proposition 14.15.

Conclusion 22.10 *The* **holomorphic Hilbert space** *consists of those functions ψ of the form (22.11) that are square integrable over $\mathbb{R}^{2n}$. If ψ is identified with the holomorphic function F in (22.11), then this Hilbert space may be identified with $\mathcal{H}L^2(\mathbb{C}^n, \nu)$, where*

$$\nu(\mathbf{z}) = e^{-|\mathrm{Im}\, \mathbf{z}|^2/(\alpha\hbar)}.$$

The space $\mathcal{H}L^2(\mathbb{C}^n, \nu)$ is nothing but an invariant form of the Segal–Bargmann space (Definition 14.14), where here "invariant" means that the density ν is invariant under translations in the real directions. This space can be identified unitarily with the ordinary Segal–Bargmann space $\mathcal{H}L^2(\mathbb{C}^n, \mu_{2\alpha\hbar})$ as follows. Define a map $\Psi : \mathcal{H}L^2(\mathbb{C}^n, \mu_{2\alpha\hbar}) \rightarrow \mathcal{H}L^2(\mathbb{C}^n, \nu)$ by

$$\Psi(F)(\mathbf{z}) = (2\pi\alpha\hbar)^{-n/2} e^{-\mathbf{z}^2/(4\alpha\hbar)} F(\mathbf{z}), \tag{22.12}$$

where $\mathbf{z}^2 = z_1^2 + \cdots + z_n^2$. Then a simple calculation shows that

$$\|\Psi(F)\|_{L^2(\mathbb{C}^n, \nu)}^2 = \int_{\mathbb{C}^n} |F(\mathbf{z})|^2\, \mu_{2\alpha\hbar}(\mathbf{z})\, d\mathbf{z}.$$

Since also $e^{-\mathbf{z}^2/(4\alpha\hbar)}$ is holomorphic as a function of $\mathbf{z}$, we see that Ψ maps $\mathcal{H}L^2(\mathbb{C}^n, \mu_{2\alpha\hbar})$ isometrically into $\mathcal{H}L^2(\mathbb{C}^n, \nu)$. The map Ψ has an inverse given by multiplication by $(2\pi\alpha\hbar)^{n/2} e^{\mathbf{z}^2/(4\alpha\hbar)}$, showing that Ψ is actually unitary. In particular, there exist many nonzero holomorphic functions on $\mathbb{C}^n$ that belong to $\mathcal{H}L^2(\mathbb{C}^n, \nu)$.

We will regard any of the Hilbert spaces in Conclusions 22.9 and 22.10 as our *quantum Hilbert space*. These spaces are to be compared to the prequantum Hilbert space $L^2(\mathbb{R}^{2n})$, which is in some sense "bigger," consisting of functions of twice as many variables. Note there are multiple possibilities for the quantum Hilbert space. To reduce from the prequantum Hilbert space to the quantum Hilbert space, we have to *choose* a set of n variables, and then we look at functions that depend only on those n variables. Indeed, there are many other possibilities for the quantum Hilbert space; we have considered only the most common choices. We defer a discussion of the general theory until Chap. 23.

The reader may wonder why we are using the definition $z_j = x_j - i\alpha p_j$ ($\alpha > 0$) rather than $z_j = x_j + i\alpha p_j$. If we repeated the preceding calculations with $z_j = x_j + i\alpha p_j$, with a corresponding sign change in the definition of $\partial/\partial\bar{z}_j$, we would find that ψ satisfies $\nabla_{\partial/\partial\bar{z}_j}\psi$ for all j if and only if ψ is of the form

$$\psi(\mathbf{x}, \mathbf{p}) = F(z_1, \ldots, z_n) e^{\alpha|\mathbf{p}|^2/(2\hbar)}, \tag{22.13}$$

where F is holomorphic on $\mathbb{C}^n$. The change in sign in the exponent between (22.11) and (22.13) has a drastic effect: There are no nonzero holomorphic functions F for which the function ψ in (22.13) is square integrable over $\mathbb{R}^{2n}$. (See Exercise 3.) Unlike the situation with the position and momentum

Hilbert spaces, there is no natural way to alter the domain of integration to make a function of the form (22.13) have finite norm.

We see, then, that there is a big difference between the definitions $z_j = x_j - i\alpha p_j$ and $z_j = x_j + i\alpha p_j$. In the general framework of geometric quantization, we will have a similar distinction, where complex structures satisfying a certain positivity condition behave well, whereas the "opposite" complex structures behave badly. (See Definition 23.19 in Sect. 23.4.)

22.5   Quantization of Observables

Now that we have constructed our quantum (as opposed to prequantum) Hilbert spaces, we need to construct operators on these spaces. According to the standard geometric quantization program, the quantum operator associated with a function f is supposed to be simply the restriction to the quantum Hilbert space of the prequantum operator $Q_{\mathrm{pre}}(f)$, *provided* that $Q_{\mathrm{pre}}(f)$ leaves the quantum Hilbert space invariant.

Proposition 22.11 *The position, momentum, and holomorphic subspaces in Definition 22.7 are all invariant under the prequantum operators $Q_{\mathrm{pre}}(x_j)$ and $Q_{\mathrm{pre}}(p_j)$. Specifically, in the position subspace, we have*

$$Q_{\mathrm{pre}}(x_j)\phi(\mathbf{x}) = x_j\phi(\mathbf{x})$$
$$Q_{\mathrm{pre}}(p_j)\phi(\mathbf{x}) = -i\hbar\frac{\partial\phi}{\partial x_j},$$

in the momentum subspace, we have

$$Q_{\mathrm{pre}}(x_j)(e^{i\mathbf{x}\cdot\mathbf{p}/\hbar}\phi(\mathbf{p})) = e^{i\mathbf{x}\cdot\mathbf{p}/\hbar}\left(i\hbar\frac{\partial\phi}{\partial p_j}(\mathbf{p})\right)$$
$$Q_{\mathrm{pre}}(p_j)(e^{i\mathbf{x}\cdot\mathbf{p}/\hbar}\phi(\mathbf{p})) = e^{i\mathbf{x}\cdot\mathbf{p}/\hbar}(p_j\phi(\mathbf{p})),$$

and in the holomorphic subspace, we have

$$Q_{\mathrm{pre}}(x_j)(F(\mathbf{z})e^{-\alpha|\mathbf{p}|^2/(2\hbar)}) = \left(\alpha\hbar\frac{\partial F}{\partial z_j} + z_j F(\mathbf{z})\right)e^{-\alpha|\mathbf{p}|^2/(2\hbar)}$$
$$Q_{\mathrm{pre}}(p_j)(F(\mathbf{z})e^{-\alpha|\mathbf{p}|^2/(2\hbar)}) = \left(-i\hbar\frac{\partial F}{\partial z_j}\right)e^{-\alpha|\mathbf{p}|^2/(2\hbar)}.$$

Proof. See Exercise 4. ∎

The invariance of the three subspaces under the prequantized position and momentum operators follows from a general result in geometric quantization, that for a real-valued function f, the prequantum operator $Q_{\mathrm{pre}}(f)$ preserves a given quantum space if and only if the Hamiltonian flow generated by f preserves the polarization defining the quantum space. The term "polarization" refers to the set of directions in which the elements of the quantum space are covariantly constant. In the case of the position, momentum, and holomorphic spaces, the set of such directions is the same at every point, which means that the polarization is invariant under translations. But the Hamiltonian flows generated by x_j and p_j are nothing but translations in the $-p_j$-directions and the x_j-directions, respectively. Of course, in this simple example, we can verify the invariance by direct computation, which also gives the indicated form of the operators on each subspace.

Note also that in each case, the "preferred" functions act simply as multiplication operators. In the position subspace, for example, the position operator $Q_{\mathrm{pre}}(x_j)$ acts simply as multiplication by x_j, whereas in the momentum subspace, the operator $Q_{\mathrm{pre}}(p_j)$ acts as multiplication by p_j. Finally, in the holomorphic subspace, the operator

$$Q_{\mathrm{pre}}(z_j) \left(F(\mathbf{z}) e^{-\alpha|\mathbf{p}|^2/(2\hbar)} \right) = (z_j F(\mathbf{z})) e^{-\alpha|\mathbf{p}|^2/(2\hbar)},$$

where $z_j = x_j - i\alpha p_j$, since the terms involving $\partial F/\partial z_j$ cancel.

We now focus on the position Hilbert space and look for operators of the form $Q_{\mathrm{pre}}(f)$ that leave the position subspace invariant.

Proposition 22.12 *The position subspace is invariant under $Q_{\mathrm{pre}}(f)$ whenever f is of the form*

$$f(\mathbf{x}, \mathbf{p}) = a(\mathbf{x}) + b_j(\mathbf{x})p_j \qquad (22.14)$$

for some smooth functions a and $b_1, \ldots, b_n$ on $\mathbb{R}^n$. On the other hand, the position subspace in not invariant under the operator $Q_{\mathrm{pre}}(p_j^2)$.

Proof. If f is of the form (22.14), calculation shows that $\theta(X_f) + f = a(\mathbf{x})$. If we drop any terms in X_f involving $\partial/\partial p_j$, since these are zero on the position subspace, we end up with

$$Q_{\mathrm{pre}}(f)(\phi(\mathbf{x})) = -i\hbar b_j(\mathbf{x})\frac{\partial\phi}{\partial x_j} + a(\mathbf{x})\phi(\mathbf{x}), \qquad (22.15)$$

which is again in the position subspace. [There is no **p**-dependence in the coefficient of $\partial/\partial x_j$ in (22.15) because $\partial f/\partial p_j$ is independent of **p**.] On the other hand, direct calculation shows that the restriction to the position subspace of $Q_{\mathrm{pre}}(f)$ is

$$-2i\hbar p_j\frac{\partial}{\partial x_j} - p_j^2,$$

which does not preserve the space of functions on $\mathbb{R}^{2n}$ that are independent of **p**. ∎

It should be noted that the expression on the right-hand side of (22.15) is not a self-adjoint, or even symmetric, operator on $L^2(\mathbb{R}^n)$, unless the vector field $\mathbf{b}(\mathbf{x})$ happens to be divergence free. (Even though the vector field X_f is divergence free on $\mathbb{R}^{2n}$, the way X_f acts on functions that are independent of $\mathbf{p}$ is not necessarily a divergence free vector field on $\mathbb{R}^n$.) This undesirable feature of our quantization scheme is the result of our simplistic method of passing from $L^2(\mathbb{R}^{2n})$ to $L^2(\mathbb{R}^n)$ in our derivation of Conclusion 22.9. When we do this reduction properly, using half-forms, we will obtain a self-adjoint operator. See Sect. 23.6.

We now consider the behavior of the holomorphic subspace under the prequantized position and momentum operators.

Proposition 22.13 *For any* $\alpha > 0$, *let* $\mathbf{H}_\alpha$ *be the subspace of* $L^2(\mathbb{R}^{2n})$ *consisting of smooth functions* ψ *that satisfy* $\nabla_{\partial/\partial \bar{z}_j} \psi = 0$, *where* $\partial/\partial \bar{z}_j$ *is as in (22.9). Then* $\mathbf{H}_\alpha$ *is a closed subspace of* $L^2(\mathbb{R}^{2n})$ *and* $\mathbf{H}_\alpha$ *is invariant under the one-parameter unitary groups generated by* $Q_{\mathrm{pre}}(x_j)$ *and* $Q_{\mathrm{pre}}(p_j)$. *Furthermore,* $Q_{\mathrm{pre}}(x_j)$ *and* $Q_{\mathrm{pre}}(p_j)$ *act irreducibly on* $\mathbf{H}_\alpha$ *in the sense of Definition 14.6.*

For each $\alpha > 0$, the holomorphic Hilbert space is a subspace of the prequantum Hilbert space invariant under the exponentiated position and momentum operators. Thus, the prequantum Hilbert space is far from being irreducible under the action of those operators.

Proof. The invariance of $\mathbf{H}_\alpha$ is a simple calculation (Exercise 5). Irreducibility can be established by reducing to the previously established irreducibility of the Segal–Bargmann space under the operators $T_\mathbf{a}$ in Theorem 14.16. To this end, we should check that the unitary map Ψ in (22.12) intertwines products of exponentials of $Q_{\mathrm{pre}}(x_j)$ and $Q_{\mathrm{pre}}(p_j)$ with operators of the form $T_\mathbf{a}$ (with $\hbar$ replaced by $2\alpha\hbar$). This is a straightforward but tedious calculation, and we omit the details. ■

We conclude this section with an example of a quantum subspace that is invariant under the (pre)quantized Hamiltonian of a harmonic oscillator.

Proposition 22.14 *Consider a harmonic oscillator with Hamiltonian*

$$H = \frac{1}{2m}\left(p^2 + (m\omega x)^2\right).$$

Consider also the subspace $\mathbf{H}_\alpha$ *in Proposition 22.13, with* $\alpha = 1/(m\omega)$. *Then the operator* $Q_{\mathrm{pre}}(H)$ *leaves* $\mathbf{H}_\alpha$ *invariant. Furthermore, the restriction of* $Q_{\mathrm{pre}}(H)$ *to* $\mathbf{H}_\alpha$ *has non-negative spectrum consisting of eigenvalues of the form* $n\hbar\omega$, *where* n *ranges over the* non-negative *integers.*

Proposition 22.14 is a much more physically reasonable result for the spectrum of the quantization of the non-negative function H than on the full prequantum Hilbert space, where (Proposition 22.6) the spectrum of $Q_{\mathrm{pre}}(H)$ is not even bounded below. When we introduce the "half-form correction" in Sect. 23.7, we will finally be able to obtain the "correct" spectrum for the quantum harmonic oscillator, consisting of numbers of the form $(n + 1/2)\hbar\omega$, $n = 0, 1, 2, \ldots$ See Example 23.53.

Proof. As in the proof of Proposition 22.6, we introduce the variable $y = m\omega x$. With $\alpha = 1/(m\omega)$, this gives $z = (y - ip)/(m\omega)$. We use the symplectic potential

$$\theta = \frac{1}{2}(p \, dx - x \, dp) = \frac{1}{2m\omega}(p \, dy - y \, dp).$$

Then

$$\theta\left(\frac{\partial}{\partial \bar{z}}\right) = \frac{1}{2}\left(p + \frac{i}{\alpha}x\right) = \frac{i}{2\alpha}z$$

and so $\nabla_{\partial/\partial z} = \partial/\partial \bar{z} + z/(2\alpha\hbar)$. From this, we can easily check that the holomorphic subspace consists of functions of the form

$$F(\mathbf{z})e^{-|z|^2/(2\alpha\hbar)} = F(\mathbf{z})\exp\left\{-\frac{(y^2 + p^2)}{2m\omega\hbar}\right\}, \tag{22.16}$$

where F is holomorphic.

Meanwhile, as in the proof of Proposition 22.6, we have

$$Q_{\mathrm{pre}}(H) = i\hbar\omega\left(y\frac{\partial}{\partial p} - p\frac{\partial}{\partial y}\right),$$

which is just an angular derivative in the (y, p)-plane. Since the exponential factor in (22.16) is rotationally invariant, $Q_{\mathrm{pre}}(H)$ only hits F. Meanwhile,

$$\left(y\frac{\partial}{\partial p} - p\frac{\partial}{\partial y}\right)F\left(\frac{y - ip}{m\omega}\right) = y\frac{dF}{dz}\left(-\frac{i}{m\omega}\right) - p\frac{dF}{dz}\frac{1}{m\omega}$$

$$= -\frac{i}{m\omega}(y - ip)\frac{dF}{dz}$$

$$= -iz\frac{dF}{dz}.$$

Thus,

$$Q_{\mathrm{pre}}(H)(F(\mathbf{z})e^{-|z|^2/(2\alpha\hbar)}) = \left(\hbar\omega z\frac{dF}{dz}\right)e^{-|z|^2/(2\alpha\hbar)},$$

which is again in the holomorphic subspace.

Finally, as in Proposition 14.15, the functions z^n, $n = 0, 1, 2, \ldots$, form an orthogonal basis for the Hilbert space $\mathbf{H}_\alpha$. Each monomial z^n is an eigenvector for the operator $z \, d/dz$ with eigenvalue n. This establishes the claim about the spectrum of the restriction to $\mathbf{H}_\alpha$ of $Q_{\mathrm{pre}}(H)$. ∎

The operator $F \mapsto \hbar\omega z \, dF/dz$ is self-adjoint on the holomorphic Hilbert space, in contrast to the operators in (22.15) in the case of the position Hilbert space. Indeed, self-adjointness is "automatic" in this case, because the holomorphic Hilbert space is actually a subspace of the prequantum Hilbert space, and the restriction of a self-adjoint operator to an invariant subspace is self-adjoint.

22.6 Exercises

1. Consider the vector field

$$X := a_j(\mathbf{x})\frac{\partial}{\partial x_j}$$

   on $\mathbb{R}^{2n}$, where the a_j's are smooth, real-valued functions. Show that X is skew-self-adjoint on $C_c^\infty(\mathbb{R}^N)$ if and only if the divergence of X (i.e., the quantity $\partial a_j/\partial x_j$) is identically zero.

2. Using the symplectic potential $\theta = p\,dx$, compute $Q_{\mathrm{pre}}(xp^2)$. Show that $Q_{\mathrm{pre}}(xp^2)$ is not in the algebra of operators generated by $Q_{\mathrm{pre}}(x)$ and $Q_{\mathrm{pre}}(p)$.

   Hint: Consider how $Q_{\mathrm{pre}}(xp^2)$ acts on functions that are independent of p.

3. (a) Suppose F is a holomorphic function on $\mathbb{C}$ such that

$$\int_{\mathbb{C}} |F(z)|^2 \; dz < \infty,$$

   where here dz denotes the 2-dimensional Lebesgue measure on $\mathbb{C} \cong \mathbb{R}^2$. Show that F is identically zero.

   Hint: If F is not identically zero, use a power series argument to show that the L^2 norm of F over a disk of radius R tends to infinity as R tends to infinity.

   (b) Show that if a function of the form (22.13), with F holomorphic on $\mathbb{C}^n$, is square integrable, then F must be identically zero.

4. Prove Proposition 22.11, using the explicit form of $Q_{\mathrm{pre}}(x_j)$ and $Q_{\mathrm{pre}}(p_j)$ in Example 22.4.

   Hint: In the case of the holomorphic subspace, express the operators $\partial/\partial x_j$ and $\partial/\partial p_j$ in terms of the operators $\partial/\partial z_j$ and $\partial/\partial \bar{z}_j$ in (22.9).

5. Show that the space of functions of the form in (22.11), where F is holomorphic on $\mathbb{C}^n$, is invariant under the operators $e^{itQ_{\mathrm{pre}}(x_j)}$ and $e^{itQ_{\mathrm{pre}}(p_j)}$ computed in (22.6), for all $t \in \mathbb{R}$ and $j = 1, 2, \ldots, n$.

23

Geometric Quantization on Manifolds

23.1 Introduction

Geometric quantization is a type of *quantization*, which is a general term for a procedure that associates a quantum system with a given classical system. In practical terms, if one is trying to deduce what sort of quantum system should model a given physical phenomenon, one often begins by observing the classical limit of the system. Electromagnetic radiation, for example, is describable on a macroscopic scale by Maxwell's equations. On a finer scale, quantum effects (photons) become important. How should one determine the correct *quantum* theory of electromagnetism? It seems that the only reasonable way to proceed is to "quantize" Maxwell's equations— and then to compare the resulting quantum system to experiment.

Meanwhile, not every physically interesting system has $\mathbb{R}^{2n}$ as its phase space. Geometric quantization, then, is an attempt to construct a quantum Hilbert space, together with appropriate operators, starting from a physical system having an arbitrary $2n$-dimensional symplectic manifold N as its phase space. To perform geometric quantization on N, one must first choose a polarization, that is, roughly, a choice of n directions on N in which the wave functions will be constant. If $N = T^*M$, then one may use the "vertical polarization," in which the wave functions are constant along the fibers of T^*M. For cotangent bundles with the vertical polarization, geometric quantization reproduces the "half-density quantization" of Blattner [4]. (See Examples 23.45 and 23.48.) Even for cotangent bundles, however, it is of interest to use polarizations other than the vertical polarization, as

B.C. Hall, *Quantum Theory for Mathematicians*, Graduate Texts in Mathematics 267, DOI 10.1007/978-1-4614-7116-5_23,
© Springer Science+Business Media New York 2013

we have seen already in the $\mathbb{R}^n$ case. In the case of the cotangent bundle of a compact Lie group, for example, the paper [20] shows how quantization with a complex polarization gives rise to a generalized Segal–Bargmann transform.

Some phase spaces, meanwhile, may not even be in the form of a cotangent bundle. In the orbit method in representation theory, for example, the relevant symplectic manifolds are "coadjoint orbits," which typically are not cotangent bundles. [In the SU(2) case, for instance, these orbits are 2-spheres with the natural rotationally invariant symplectic form.] In quantum field theory, meanwhile, one encounters Lagrangians that are linear, rather than quadratic, in the "velocity" variables. In such cases, the initial velocity is determined by the initial position, and one cannot think of the space of initial conditions as a (co)tangent bundle. Systems of this form *can* still be symplectic, but they are not cotangent bundles. Furthermore, it is common to think of compact symplectic manifolds (such as S^2 with a rotationally invariant symplectic form) as classical models of internal degrees of freedom, such as spin.

To quantize these more general symplectic manifolds, one needs a more general approach to quantization. Given a symplectic manifold (N, ω) satisfying a certain integrality condition, one can construct a line bundle L over N along with a connection ∇ on L which has a curvature of $\omega/\hbar$. One can then define "prequantum" operators, acting on sections of L, in much the same way we did in the Euclidean case in Chap. 22, and these operators will have the desired relationship between Poisson brackets and commutators. One then chooses a polarization on N and defines the quantum Hilbert space to be the space of sections that are covariantly constant in the directions of that polarization. If the Hamiltonian flow generated by a function f preserves the relevant polarization, then $Q_{\mathrm{pre}}(f)$ will preserve the quantum Hilbert space. In the case of real polarizations, there may fail to be any nonzero *square-integrable* sections that are covariantly constant in the directions of the polarization, a possibility that forces us to introduce the machinery of "half-forms."

Let us end this introduction with a brief critique of the framework of geometric quantization. In the first place, geometric quantization has too many definitions (bundles, connections, curvature, polarizations, half-forms) and too few theorems. In the second place, the class of functions that geometric quantization allows us to quantize—those functions for which the associated Hamiltonian flow preserves the polarization—is often dishearteningly small. In the case $N = T^*M$, for example, with the natural "vertical" polarization, geometric quantization does not allow us to quantize the kinetic energy function, at least not by the "standard procedure" of geometric quantization. Nevertheless, geometric quantization is the only game in town if one wants to quantize general symplectic manifolds in a way that produces an actual Hilbert space and operators thereon.

This chapter lays out in an orderly fashion all the ingredients needed to "do" geometric quantization. Furthermore, although this approach increases length, the chapter fills in the details of several arguments that are only sketched in the standard reference on the subject, the book [45] of Woodhouse. The presentation assumes basic results about symplectic manifolds from Chap. 21. Besides the basic results about manifolds reviewed in Sect. 21.1, we will make use of the Frobenius theorem (see, e.g., Chap. 19 of [29]).

As we have noted already in the introduction to Chap. 22, sign conventions in the subject of geometric quantization are not consistent from author to author.

23.2 Line Bundles and Connections

In this section, we develop the necessary machinery to extend the prequantization construction of Sect. 22.2 to arbitrary symplectic manifolds. We introduce the notion of a line bundle over a manifold and sections thereof, which look locally like complex-valued functions. We then introduce the notion of covariant derivatives of sections of a line bundle, where locally these covariant derivatives take the form $\nabla_X = X - i\theta(X)$ for a certain 1-form θ. We then introduce the curvature 2-form, which is a globally defined, closed 2-form that can be computed locally as $d\theta$. We continue to observe the summation convention, in which repeated indices are always summed on.

Definition 23.1 *If X is a smooth manifold, a **complex line bundle** over X is a smooth manifold L together with the following additional structures. First, we have a smooth, surjective map $\pi : L \to X$. Second, for each $x \in X$, the set $\pi^{-1}(\{x\})$ is equipped with the structure of a complex vector space of dimension 1. For each $x \in N$, the vector space $\pi^{-1}(\{x\})$ is called the **fiber** of L over x.*

*These structures are assumed to satisfy the **local triviality property**, namely that each $x \in X$ has a neighborhood U such that there exists a diffeomorphism $\chi : \pi^{-1}(U) \to U \times \mathbb{C}$ with the following properties. First,*

$$\pi(p) = \pi_1(\chi(p)),$$

where $\pi_1 : U \times \mathbb{C} \to U$ is projection onto the first factor. Second, for each $x \in U$, the map $p \mapsto \pi_2(\chi(p))$ is a vector space isomorphism of $\pi^{-1}(\{x\})$ with $\mathbb{C}$.

*A **section** of a line bundle L over X is a map $s : X \to L$ such that $\pi(s(p)) = p$ for all $p \in X$.*

For any manifold X, we can form the trivial line bundle $X \times \mathbb{C}$, where $\pi(x, z) = z$ and where the vector space structure on $\{x\} \times \mathbb{C}$ is just the usual vector space structure on $\mathbb{C}$. The local triviality property for a general line bundle L means that L "looks" locally like the trivial line bundle.

Definition 23.2 *A **connection** ∇ on a line bundle L over N is a map associating to each vector field X on N and section s of L another section $\nabla_X(s)$ of L satisfying the following properties. First, for each smooth function f on N, we have*

$$\nabla_{fX}(s) = f\nabla_X(s) \tag{23.1}$$

*for all vector fields X and sections s. Second, for each smooth function f on N, we have the **product rule***

$$\nabla_X(fs) = (X(f))s + f\nabla_X(s) \tag{23.2}$$

for all vector fields X and sections s.

Note that for any section s of L and any function f on N, the quantity fs is a section of s. Given a connection ∇ and a vector field X, the operator ∇_X is called the *covariant derivative* in the direction of X.

Definition 23.3 *A **Hermitian structure** on a line bundle L over N is a choice of an inner product $(\cdot, \cdot)$ on each fiber $\pi^{-1}(\{x\})$ of L such that for each smooth section s of L, (s, s) is a smooth function on N. A line bundle L together with a choice of a Hermitian structure on L will be called a **Hermitian line bundle**. A connection ∇ on a Hermitian line bundle L is called **Hermitian** if for every vector field on X, we have*

$$(\nabla_X(s_1), s_2) + (s_1, \nabla_X(s_2)) = X(s_1, s_2) \tag{23.3}$$

for all smooth sections s_1 and s_2 of L.

We will let the expression "Hermitian line bundle with connection" refer to a Hermitian line bundle L together with a Hermitian connection on L; that is, in this expression, "Hermitian" applies both to the bundle and to the connection.

Given a Hermitian line bundle L with connection, it is always possible to choose a locally defined smooth section s_0 near any point such that $(s_0, s_0) \equiv 1$. We call s_0 a *local isometric trivialization* of L. Any section s of L can be written locally as $s = fs_0$ for a unique complex-valued function f. Given a vector field X, let $\theta(X)$ be the unique function such that

$$\nabla_X(s_0) = -i\theta(X)s_0.$$

Using the assumption $\nabla_{fX} = f\nabla_X$, it can be shown (Exercise 1) that the value of $\theta(X)$ at a point p depends only on the value of X at p. Thus, θ defines a 1-form on N. Using the assumption that ∇ is Hermitian, it can be shown (Exercise 2) that $\theta(X)$ is always real valued.

Now, using the product rule (23.2) for covariant derivatives, we have

$$\nabla_X(fs_0) = X(f)s_0 + f\nabla_X(s_0)$$
$$= (X(f) - i\theta(X)f)s_0.$$

Thus, if we identify sections of L locally with the coefficient function f, we have

$$\nabla_X(f) = X(f) - i\theta(X)f, \tag{23.4}$$

as in Sect. 22.2. We call θ the *connection 1-form* associated to the particular local isometric trivialization.

Definition 23.4 *For any Hermitian line bundle* (L, ∇) *with connection, define the* **curvature 2-form** ω *of* ∇ *by requiring that*

$$\omega(X, Y)s = i\left(\nabla_X\nabla_Y - \nabla_Y\nabla_X - \nabla_{[X,Y]}\right)(s)$$

for all sections s *and vector fields* X *and* Y.

Of course, one should check that the given expression for ω is really a 2-form, meaning that the value of $\omega(X, Y)$ at a point z depends only on the values of X and Y at z, and that it does not depend on the choice of section s, provided only that $s(z) \neq 0$. One way to do this is to compute ω in a local isometric trivialization, as in the following result. (See Exercise 3 for a different approach.)

Proposition 23.5 *Let* s_0 *be a local isometric trivialization of* L *and let* θ *be the associated connection 1-form. Then the curvature 2-form* ω *of* ∇ *is expressed locally as*

$$\omega = d\theta.$$

In particular, ω *is a* closed *2-form.*

Proof. The computation is precisely the same as in the proof of Proposition 22.3 in the Euclidean case. ∎

A locally defined 1-form θ satisfying $d\theta = \omega$ is called a (local) *symplectic potential* for ω. Our next result says that *every* symplectic potential is the connection 1-form for some local isometric trivialization of L.

Proposition 23.6 *Let* (L, ∇) *be a Hermitian line bundle with connection over* N *with curvature 2-form* ω. *For each point* $z_0 \in N$ *and 1-form* θ *defined in a neighborhood* U *of* z_0 *satisfying* $d\theta = \omega$, *there is a subneighborhood* $V \subset U$ *of* z_0 *and a local isometric trivialization of* L *over* V *such that the connection 1-form of the trivialization is* θ.

Proof. Let s_0 be any isometric trivializing section defined in a neighborhood of z_0 and let η be the associated connection 1-form. Since $d(\eta - \theta) = 0$, there is a subneighborhood $V \subset U$ of z_0 on which $\eta - \theta = df$, for some smooth function f. If $s_1 = e^{if} s_0$, then

$$\begin{aligned} \nabla_X(s_1) &= iX(f)e^{if} s_0 + e^{if} \nabla_X(s_0) \\ &= iX(f)e^{if} s_0 - i\eta(X)e^{if} s_0 \\ &= -i(\eta(X) - df(X))s_1. \end{aligned}$$

Thus, the connection 1-form associated with the local isometric trivialization s_1 is $\eta - df = \theta$. ∎

Proposition 23.7 *If (L_1, ∇^1) and (L_2, ∇^2) are Hermitian line bundles with connection over N, let $L_1 \otimes L_2$ denote the line bundle over N for which the fiber over x is $L_{1,x} \otimes L_{2,x}$, with the natural inner product induced by the inner products on $L_{1,x}$ and $L_{2,x}$. Then there is a unique Hermitian connection ∇ on $L_1 \otimes L_2$ with the property that*

$$\nabla_X(s_1 \otimes s_2) = (\nabla^1_X s_1) \otimes s_2 + s_1 \otimes (\nabla^2_X s_2),$$

for all vector fields X on N and all smooth sections s_1 of L_1 and s_2 of L_2. The curvature 2-form ω for $(L_1 \otimes L_2, \nabla)$ is given by

$$\omega = \omega_1 + \omega_2,$$

where ω_1 and ω_2 are the curvature 2-forms for (L_1, ∇^1) and (L_2, ∇^2), respectively.

The proof of this proposition is a straightforward exercise in "definition chasing" and is left as an exercise to the reader.

Suppose that L is a Hermitian line bundle over N with connection ∇ and curvature 2-form ω. Given a loop $\gamma : [a, b] \to N$, we can construct a section s of L that is defined over γ such that the covariant derivative of s in the directions along γ is zero. Indeed, in a local isometric trivialization, such a section can be constructed as

$$s(\gamma(T)) = \exp\left\{ i \int_{\gamma(a)}^{\gamma(T)} \theta(\gamma(t))\, dt \right\}. \tag{23.5}$$

The value of s at the endpoint of the loop will in general not agree with the value at the starting point, but will differ by multiplication by a constant of absolute value 1.

Definition 23.8 *The **holonomy** of a loop $\gamma : [a, b] \to N$ is the unique constant α (of absolute value 1) such that $s(\gamma(b)) = \alpha s(\gamma(a))$, where s is a nonzero section defined over γ that is covariantly constant in the directions of γ.*

The value of the holonomy of γ is easily seen to be independent of the value of s at the starting point, provided this starting value is nonzero.

Suppose that S is a compact, oriented surface with boundary in N whose boundary ∂S is a loop. It is not hard to show that the holonomy around ∂S can be computed as

$$\text{holonomy}(\partial S) = \exp\left\{i\int_S \omega\right\}. \tag{23.6}$$

Indeed, if S is contained in the domain of a local isometric trivialization, then this result follows from (23.5) by means of Stoke's theorem (Sect. 21.1.2).

Now, if S is a closed (i.e., boundaryless) surface, its boundary is the trivial loop, which has a holonomy that is trivial, that is, equal to 1. (Think of approximating S by a surface for which the boundary is a very small loop.) Thus, for any closed surface S, (23.6) gives

$$\exp\left\{i\int_S \omega\right\} = 1, \quad \partial S = \varnothing. \tag{23.7}$$

Equivalently, we have

$$\frac{1}{2\pi}\int_S \omega \in \mathbb{Z}. \tag{23.8}$$

The condition (23.8) says that $\omega/(2\pi)$ is an *integral* 2-form. Clearly, not every closed 2-form satisfies this property.

The closedness of ω (Proposition 23.5) and the condition (23.8) represent necessary conditions that the curvature of a Hermitian connection must satisfy. It turns out that these two conditions are also sufficient.

Theorem 23.9 *Suppose ω is a closed 2-form on a manifold N for which $\omega/(2\pi)$ is integral in the sense of (23.8). Then there exists a Hermitian line bundle L over N with Hermitian connection ∇ such that the curvature of ∇ is equal to ω. If, in addition, N is simply connected, then (L, ∇) is unique up to equivalence.*

See Sect. 8.3 of [45] for a proof of this result. An equivalence of two Hermitian line bundles L_1 and L_2 with Hermitian connection over N is a diffeomorphism $\Phi : L_1 \to L_2$ such that for each $x \in N$, the restriction of Φ to $\pi_1^{-1}(\{x\})$ is an isometric linear map onto $\pi_2^{-1}(\{x\})$ and such that for each section s of L_1, we have

$$\Phi(\nabla_X(s)) = \nabla_X(\Phi(s)).$$

We now have the necessary tools to proceed with the program of geometric quantization on symplectic manifolds.

23.3 Prequantization

The first step in the program of geometric quantization for a symplectic manifold (N, ω) is to construct a Hermitian line bundle L over N with Hermitian connection for which the curvature 2-form is equal to $\omega/\hbar$. Theorem 23.9 gives the condition for the existence of such a bundle.

Definition 23.10 *A symplectic manifold (N, ω) is **quantizable** (for a particular value of $\hbar$) if*

$$\frac{1}{2\pi\hbar} \int_S \omega \in \mathbb{Z}$$

for every closed surface S in N.

Note that if (N, ω) is quantizable for a given value $\hbar_0$ of Planck's constant, then (N, ω) is also quantizable for $\hbar = \hbar_0/k$ for every positive integer k. Indeed, according to Proposition 23.7, if L is a Hermitian line bundle with connection having curvature $\omega/\hbar_0$, then $L^{\otimes k}$ (the tensor product of L with itself k times) is a Hermitian line bundle with connection having curvature $\omega/(\hbar_0/k)$.

For the remainder of this chapter, we will assume that N is a quantizable symplectic manifold with symplectic form ω and that (L, ∇) is a fixed Hermitian line bundle with connection of N with curvature $\omega/\hbar$.

If L is a Hermitian line bundle over a symplectic manifold N, we say that a measurable section s of L is *square integrable* if

$$\|s\| := \left(\int_N (s_1(x), s_1(x)) \, \lambda(x) \right)^{1/2}$$

is finite, where λ is the Liouville volume form on N. Given two square-integrable sections s_1 and s_2 of L, we define the *inner product* of s_1 and s_2 by

$$\langle s_1, s_2 \rangle = \int_N (s_1(x), s_2(x)) \, \lambda(x). \tag{23.9}$$

We use parentheses to denote the *pointwise* inner product $(s_1(x), s_2(x))$ of two sections s_1 and s_2, which is a function on N, and we use angled brackets to denote the *global* inner product $\langle s_1, s_2 \rangle$ of the sections, which is a number.

Definition 23.11 *The **prequantum Hilbert space** for N is the space of equivalence classes of square-integrable sections of L, where two sections are equivalent if they are equal almost everywhere with respect to the Liouville volume measure.*

Definition 23.12 *If f is a smooth complex-valued function on N, the **prequantum operator** $Q_{\mathrm{pre}}(f)$ is the unbounded operator on the prequantum*

Hilbert space given by

$$Q_{\text{pre}}(f) = i\hbar \nabla_{X_f} + f,$$

where f represents the operation of multiplying a section by f.

Proposition 23.13 *If f is real-valued, then $Q_{\text{pre}}(f)$ is symmetric on the space of smooth compactly supported sections of L.*

Proof. Let s_1 and s_2 be smooth, compactly supported sections of L and let Φ^f denote the Hamiltonian flow generated by f. For all sufficiently small t, every point in the supports of s_1 and s_2 will contained in the domain of Φ_t^f. Furthermore, by Liouville's theorem, the value of

$$\int_N [(s_1, s_2) \circ \Phi_t] \, \lambda$$

is independent of t. If we differentiate this relation with respect to t and evaluate at $t = 0$, we obtain, by (23.3),

$$0 = \int_N [(\nabla_{X_f}(s_1), s_2) + (s_1, \nabla_{X_f}(s_2))] \, \lambda.$$

Thus, ∇_{X_f} is a skew-symmetric operator on the space of smooth, compactly supported sections, from which it follows that $Q_{\text{pre}}(f)$ is symmetric. ∎

By the product rule for covariant derivatives and the identity $X_f(f) = \{f, f\} = 0$, we see that the two terms in the definition of $Q_{\text{pre}}(f)$ commute. We would then expect the exponential $e^{itQ_{\text{pre}}(f)}$ to decompose as a product of two exponentials. One of these exponentials is just e^{itf} and the other may be constructed as "parallel transport along the flow generated by X_f." Thus, if the flow generated by X_f is complete, it is possible to use Stone's theorem to construct $Q_{\text{pre}}(f)$ as a self-adjoint operator on a domain that includes the space of smooth compactly supported sections.

Proposition 23.14 *For any $f, g \in C^\infty(X)$, we have*

$$\frac{1}{i\hbar}[Q_{\text{pre}}(f), Q_{\text{pre}}(g)] = Q_{\text{pre}}(\{f, g\}),$$

where the equality holds as operators on the space of smooth sections of L.

Proof. The argument is precisely the same as in Proposition 22.1 in the $\mathbb{R}^{2n}$ case. ∎

As we have seen already in Sect. 22.3 in the $\mathbb{R}^{2n}$ case, the prequantum Hilbert space is "too large" to be considered the quantization of N.

23.4 Polarizations

In the $\mathbb{R}^n$ case, we have the position, momentum, and holomorphic subspaces (Definition 22.7), consisting of functions that depend only on $\mathbf{x}$, $\mathbf{p}$, or $\mathbf{z}$, in the sense that the covariant derivatives of functions in the directions of $\mathbf{p}$, $\mathbf{x}$, and $\bar{\mathbf{z}}$ are zero. In each case, the "basic observables" of the particular representation (the x_j's, the p_j's, and the z_j's, respectively) act simply as multiplication operators.

To generalize this to a symplectic manifold N of dimension $2n$, we may think of choosing n functions $\alpha_1, \ldots, \alpha_n$ on N that are "independent," in the sense that $d\alpha_1, \ldots, d\alpha_n$ are linearly independent at each point. We assume that the functions α_j Poisson commute ($\{\alpha_j, \alpha_k\} = 0$), which makes it reasonable to hope that the quantizations of the α_j's could act as (commuting) multiplication operators. For each $z \in N$, we let P_z be the n-dimensional space of directions in which the α_j's are constant, that is, the intersection of the kernels of $d\alpha_1, \ldots, d\alpha_n$. Since we wish to allow the functions α_j to be complex valued, P_z should be thought of as a subspace of the *complexified* tangent space $T_z^{\mathbb{C}}(N)$. The idea is that our quantum Hilbert space should consist of sections of a prequantum line bundle that are covariantly constant in the directions of P.

Now, at each point z, the Hamiltonian vector field X_{α_j} will belong to P_z, because

$$d\alpha_j(X_{\alpha_k}) = X_{\alpha_k}(\alpha_j) = \{\alpha_k, \alpha_j\} = 0.$$

Furthermore, since the $d\alpha_j$'s are linearly independent, the X_{α_j}'s are also independent, since X_{α_j} is obtained from $d\alpha_j$ by an isomorphism of tangent and cotangent spaces. Thus, the X_{α_j}'s must actually span P_z at each point, by a dimension count. Since also $\omega(X_{\alpha_j}, X_{\alpha_k}) = -\{\alpha_j, \alpha_k\} = 0$, we conclude that ω is identically zero on P_z. Furthermore, if X and Y are vector fields lying in P at each point, we can express them as

$$X = a_j(z)X_{\alpha_j}, \quad Y = b_j(z)X_{\alpha_j},$$

for some smooth functions a_j and b_j. Then

$$[X, Y] = a_j(z)X_{\alpha_j}(b_k)X_{\alpha_k} - b_k(z)X_{\alpha_k}(a_j)X_{\alpha_j},$$

because $[X_{\alpha_j}, X_{\alpha_k}] = X_{\{\alpha_j, \alpha_k\}} = 0$. Thus, the commutator of two vector fields lying in P will again lie in P.

Definition 23.15 *For any $z \in N$, a subspace P of $T_z N$ is said to be* **Lagrangian** *if* $\dim P = n$ *and* $\omega(X, Y) = 0$ *for all* $X, Y \in P$.

Definition 23.16 *A* **polarization** *of a symplectic manifold N is a choice at each point $z \in N$ of a Lagrangian subspace $P_z \subset T_z^{\mathbb{C}}(X)$, satisfying the following two conditions.*

1. *If two complex vector fields X and Y lie in P_z at each point z, then so does $[X, Y]$.*

2. *The dimension of $P_z \cap \overline{P_z}$ is constant.*

The first condition is called *integrability*, and we have motivated this condition in the discussion preceding the definition. The second condition is a technical one that prevents problems with certain constructions, such as the pairing map. (Although, in practice, one sometimes needs to work with "polarizations" in which the second condition is violated, extra care is needed in such cases.)

There is one small inaccuracy in our discussion of polarizations: For purely conventional reasons, the quantum Hilbert space is defined as the space of sections that are covariantly constant in the direction of $\bar{P}$, rather than P. Thus, P should really be the *complex conjugate* of the space of directions in which the sections are constant. This convention, however, makes no difference to the definition of a polarization, since if P satisfies the conditions of Definition 23.16, so does $\bar{P}$.

Example 23.17 *If M is any smooth manifold, let $N = T^*M$ be the cotangent bundle of M, equipped with the canonical 2-form ω (Example 21.2). For each $z \in T^*M$, let P_z be the complexification of the tangent space to the fiber T_z^*M. Then P is a polarization on T^*M, called the **vertical polarization**.*

Proof. If $\{x_j\}$ is any local coordinate system on M, let $\{x_j, p_j\}$ be the associated local coordinate system on T^*M. The canonical 2-form is given by $\omega = dp_j \wedge dx_j$. At each point $z \in T^*M$, the vertical subspace P_z is spanned by the vectors $\partial/\partial p_j$. Since $\omega(\partial/\partial p_j, \partial/\partial p_k) = 0$, we see that P_z is Lagrangian. Furthermore, $P_z = \bar{P}_z$ at every point, and so $\dim P_z \cap \overline{P_z}$ has the constant value $n = \dim M$. Finally, the integrability of P follows by computing the commutator of two vector fields of the form $f_j(x, p)\, \partial/\partial p_j$, which will again be a linear combination of the $\partial/\partial p_j$'s. Integrability also follows from the easy direction of the Frobenius theorem, since the fibers of T^*M are integral submanifolds for P. ∎

We may identify two special classes of polarizations, those that are *purely real* (i.e., $\overline{P_z} = P_z$ for all $z \in N$) and those that are *purely complex* (i.e., $P_z \cap \overline{P_z} = \{0\}$ for all $z \in N$). The vertical polarization, for example, is purely real.

If P is purely real, the integrability of P implies, by the Frobenius theorem, that every point in N is contained in a unique submanifold R that is maximal in the class of connected integral submanifolds for P. [An integral submanifold R for P is submanifold for which $T_z^{\mathbb{C}}(R) = P_z$ for all $z \in R$.] We will refer to the maximal connected, integral submanifolds of a purely real polarization as the *leaves* of the polarization.

In general, the leaves may not be embedded submanifolds of N. Suppose, for example, that $N = S^1 \times S^1$, with $\omega = d\theta \wedge d\phi$, where θ and ϕ are angular coordinates on the two copies of S^1. Then the tangent space to N at any point may be identified with $\mathbb{R}^2$ by means of the basis $\{\partial/\partial\theta, \partial/\partial\phi\}$. We may define a polarization P on N by defining P_z to be the span of the vector

$$\frac{\partial}{\partial\theta} + a\frac{\partial}{\partial\phi},$$

for some fixed irrational number a. Each leaf of P is then a set of the form

$$\left\{(e^{i\theta_0}e^{it}, e^{iat}) \in S^1 \times S^1 \,\middle|\, t \in \mathbb{R}\right\},$$

for some θ_0, which is an "irrational line" in $S^1 \times S^1$. Each leaf is then dense in $S^1 \times S^1$ and, thus, not embedded. We will need to avoid such pathological examples if we hope to successfully carry out the program of geometric quantization with respect to a real polarization. Much more information about the structure of real polarizations may be found in Sects. 4.5–4.7 of [45].

We now consider some elementary results concerning purely complex polarizations.

Proposition 23.18 *Suppose P is a purely complex polarization on N. For each $z \in N$, let $J_z : T_z^{\mathbb{C}}N \to T_z^{\mathbb{C}}N$ be the unique linear map such that $J_z = iI$ on P_z and $J_z = -iI$ on $\overline{P_z}$. Then J_z is real (i.e., it maps the real tangent space to itself) and ω is J_z-invariant [i.e., $\omega(J_zX_1, J_zX_2) = \omega(X_1, X_2)$ for all $X_1, X_2 \in T_z^{\mathbb{C}}N$].*

Proof. Since the restriction of J_z to $\overline{P_z}$ is the complex-conjugate of its restriction to P_z, the map J_z commutes with complex conjugation and thus maps real vectors (those satisfying $\bar{X} = X$) to real vectors. Meanwhile, since P_z is Lagrangian and ω is real, $\overline{P_z}$ is also Lagrangian. Given two vectors $X_1 = Y_1 + Z_1$ and $X_2 = Y_2 + Z_2$, with $Y_j \in P_z$ and $Z_j \in \overline{P_z}$, we compute that

$$\begin{aligned}
&\omega(J_zX_1, J_zX_2) \\
&= \omega(iY_1, iY_2) + \omega(iY_1, -iZ_2) + \omega(-iZ_1, iY_2) + \omega(-iZ_1, -iZ_2) \\
&= \omega(Y_1, Z_2) + \omega(Z_1, Y_2).
\end{aligned}$$

A similar calculation gives the same value for $\omega(X_1, X_2)$, showing that ω is J_z-invariant. ∎

A *complex structure* on a $2n$-dimensional manifold N is a collection of "holomorphic" coordinate systems that cover N and such that the transition maps between coordinate systems are holomorphic as maps between open sets in $\mathbb{R}^{2n} \cong \mathbb{C}^n$. At each point $z \in N$, there is a linear map $J_z : T_zN \to T_zN$ defined by the expression

$$J_z\left(\frac{\partial}{\partial x_j}\right) = \frac{\partial}{\partial y_j}; \quad J_z\left(\frac{\partial}{\partial y_j}\right) = -\frac{\partial}{\partial x_j},$$

where the x_j's and y_j's are the real and imaginary parts of holomorphic coordinates. This map is independent of the choice of holomorphic coordinates and satisfies $J_z^2 = -I$. At each point $z \in N$, the complexified tangent space $T_z^{\mathbb{C}} N$ can be decomposed into eigenspaces for J_z with eigenvalues i and $-i$; these are called the $(1,0)$- and $(0,1)$-tangent spaces, respectively.

Meanwhile, if N is any $2n$-dimensional manifold and J is a smoothly varying family of linear maps on each tangent space satisfying $J_z^2 = -I$ for all z, then J is called an *almost-complex structure*. Given an almost complex structure, we can divide the complexified tangent space into $\pm i$ eigenspaces for J. The Newlander–Nirenberg theorem asserts that if the family of $+i$ eigenspaces is integrable (in the sense of Point 1 of Definition 23.16), then there exists a unique complex structure on N for which these are the $(1,0)$-tangent spaces.

A purely complex polarization P gives rise to a complex structure on N, as follows. By Proposition 23.18 and the Newlander–Nirenberg theorem, there is a unique complex structure on N for which P_z is the $(1,0)$-tangent space, for all $z \in N$.

Now, we have already seen in the $\mathbb{R}^{2n}$ case that some purely complex polarizations behave better than others. [Compare (22.11) to (22.13)]. The geometric condition that characterizes the "good" polarizations is the following.

Definition 23.19 *For any purely complex polarization P, let J be the unique almost-complex structure on N such that $J_z = iI$ on P_z and $J_z = -iI$ on $\overline{P_z}$. We say that P is a **Kähler polarization** if the bilinear form*

$$g(X, Y) := \omega(X, J_z Y) \tag{23.10}$$

is positive definite for each $z \in N$.

For any purely complex polarization, the bilinear form g in (23.10) is symmetric, as the reader may easily verify using the J_z-invariance of ω.

Suppose, for example, that we identify $\mathbb{R}^2$ with $\mathbb{C}$ by the map $z = x - i\alpha p$, for some fixed $\alpha > 0$. If we define a purely complex polarization on $\mathbb{R}^2$ by taking P_z to be the span of the vector $\partial/\partial z$ in (22.9), then (Exercise 4), P is a Kähler polarization.

23.5 Quantization Without Half-Forms

To construct a prequantum Hilbert space, we must choose a line bundle (L, ∇) over (N, ω) having curvature $\omega/\hbar$. Such a bundle exists if $\omega/\hbar$ is an integral 2-form and is unique (up to equivalence) if N is simply connected. To pass to the *quantum* Hilbert space, we must make a substantial additional choice, that of a polarization P on N. In our first attempt at defining the quantum Hilbert space associated with P, we consider the space of sections of (L, ∇) that are covariantly constant in the directions of $\overline{P}$. Although this approach works reasonably well for a purely complex polarization, in the case of a purely real polarization, there typically are no square-integrable sections satisfying this condition. (Indeed, we have seen this problem already in the $\mathbb{R}^{2n}$ case, in Sect. 22.4.) In the next section, we will introduce half-forms to address this problem.

In the remainder of the chapter, we will let P denote a fixed polarization on N.

23.5.1   The General Case

As we have remarked, it is customary to consider sections that are covariantly constant in the directions of $\overline{P}$ rather than in the directions of P.

Definition 23.20 *A smooth section s of L is **polarized** (with respect to P) if*

$$\nabla_X s = 0 \tag{23.11}$$

*for every vector field X lying in $\overline{P}$. The **quantum Hilbert space** associated with P is the closure in the prequantum Hilbert space of the space of smooth, square-integrable, polarized sections of L.*

As in the Euclidean case, we will simply restrict the prequantum operators to the quantum Hilbert space, in those cases where $Q_{\mathrm{pre}}(f)$ preserves the space of polarized sections.

Definition 23.21 *A smooth, complex-valued function f on N is **quantizable** with respect to P if $Q_{\mathrm{pre}}(f)$ preserves the space of smooth sections that are polarized with respect to P.*

The following definition will provide a natural geometric condition guaranteeing quantizability of a function.

Definition 23.22 *A possibly complex vector field X **preserves** a polarization P if for every vector field Y lying in P, the vector field $[X, Y]$ also lies in P.*

Note that if X lies in P, then X preserves P, by the integrability assumption on P. There will typically be, however, many vector fields that do not lie in P but nevertheless preserve P.

If X is a real vector field, then $[X, Y]$ is the same as the Lie derivative $\mathcal{L}_X(Y)$. It is then not hard to show that X preserves P if and only if the flow generated by X preserves P, that is, if and only if $(\Phi_t)_*(P_z) = P_{\Phi_t(z)}$ for all z and t, where Φ is the flow of X. Furthermore, if X is real, then X preserves P if and only if X preserves $\overline{P}$.

Example 23.23 *If $N = T^*M$ for some manifold M and P is the vertical polarization on N, then a Hamiltonian vector field X_f preserves P if and only if $f = f_1 + f_2$, where f_1 is constant on each fiber and f_2 is linear on each fiber.*

Proof. In local coordinates $\{x_j, p_j\}$, a vector field X lying in P has the form $X = g_j \, \partial/\partial p_j$. Thus,

$$[X_f, X] = \left[\frac{\partial f}{\partial p_j}\frac{\partial}{\partial x_j}, g_k\frac{\partial}{\partial p_k}\right] - \left[\frac{\partial f}{\partial x_j}\frac{\partial}{\partial p_j}, g_k\frac{\partial}{\partial p_k}\right].$$

This commutator will consist of three "good" terms, which involve only p-derivatives, along with the following "bad" term:

$$-g_k\frac{\partial^2 f}{\partial p_k \partial p_j}\frac{\partial}{\partial x_j}.$$

If $\partial^2 f/\partial p_k \partial p_j$ is 0 for all j and k, then the bad term vanishes and $[X_f, X]$ again lies in P. Conversely, if we want the bad term to vanish for each choice of the coefficient functions g_j, we must have $\partial^2 f/\partial p_k \partial p_j = 0$ for all j and k. Thus, for each fixed value of x, f must contain only terms that are independent of p and terms that are linear in p. ∎

We now identify the condition for quantizability of functions.

Theorem 23.24 *For any smooth, complex-valued function f on N, if the Hamiltonian vector field X_f preserves $\bar{P}$, then f is quantizable.*

Since we do not assume that f is real-valued, the condition that X_f preserve $\bar{P}$ is not equivalent to the condition that X_f preserve P.

Proof. Given a polarized section s, we apply $Q_{\mathrm{pre}}(f)$ to s and then test whether $Q_{\mathrm{pre}}(f)s$ is still polarized, by applying ∇_X for some vector field X lying in $\bar{P}$. To this end, it is useful to compute the commutator of ∇_X and $Q_{\mathrm{pre}}(f)$, as follows:

$$[\nabla_X, Q_{\mathrm{pre}}(f)] = i\hbar\left[\nabla_X, \nabla_{X_f}\right] + [\nabla_X, f]$$

$$= i\hbar\left(\nabla_{[X,X_f]} - \frac{i}{\hbar}\omega(X, X_f)\right) + X(f)$$

$$= i\hbar\nabla_{[X,X_f]}, \tag{23.12}$$

where we have used that

$$\omega(X, X_f) = -\omega(X_f, X) = -df(X) = -X(f),$$

by Definition 21.6. Since X_f preserves $\bar{P}$, the vector field $[X, X_f]$ again lies in $\bar{P}$ and, thus,

$$\nabla_X(Q_{\mathrm{pre}}(f)s) = Q_{\mathrm{pre}}(f)\nabla_X s + i\hbar\nabla_{[X,X_f]}s = 0,$$

for every polarized section s, showing that $Q_{\mathrm{pre}}(f)s$ is again polarized. ∎

The converse of Theorem 23.24 is false in general. After all, as we will see in the following subsections, for a given polarization, there may not be any nonzero globally defined polarized sections, in which case, *any* function is quantizable. On the other hand, it can be shown that if $Q_{\mathrm{pre}}(f)$ preserves the space of *locally defined* polarized sections, then the Hamiltonian flow generated by f must preserve $\bar{P}$. This result follows by the same reasoning as in the proof of Theorem 23.24, once we know that there are sufficiently many locally defined polarized sections. We will establish such an existence result for purely real and purely complex polarizations in the following subsections; for the general case, see the discussion following Definition 9.1.1 in [45].

A special case of Theorem 23.24 is provided by "polarized functions," that is, functions f for which $X(f) = 0$ for all vector fields X lying in $\bar{P}$. For such an f, the action of $Q_{\mathrm{pre}}(f)$ on the quantum space is simply multiplication by f, as we anticipated in the introductory discussion in Sect. 23.4.

Proposition 23.25 *If f is a smooth, complex-valued function on N and the derivatives of f in the $\bar{P}$ directions are zero, then $Q_{\mathrm{pre}}(f)$ preserves the space P-polarized sections, and the restriction of $Q_{\mathrm{pre}}(f)$ to this space is simply multiplication by f.*

We have already seen special cases of this result in the $\mathbb{R}^{2n}$ case; see the discussion following Proposition 22.11.

Proof. If the derivatives of f in the direction of $\bar{P}$ are zero, then for $X \in \bar{P}$, we have

$$0 = X(f) = df(X) = \omega(X_f, X),$$

meaning that X_f is in the ω-orthogonal complement of $\bar{P}$. But since $\bar{P}$ is Lagrangian, this complement is just $\bar{P}$. Thus, X_f belongs to $\bar{P}$ and, in particular, X_f preserves $\bar{P}$, so that f is quantizable, by Theorem 23.24. Furthermore, $\nabla_{X_f} s = 0$ for any P-polarized section s, leaving only the fs term in the formula for $Q_{\mathrm{pre}}(f)s$. ∎

23.5.2   The Real Case

In the $\mathbb{R}^{2n}$ case, we have already computed the space of polarized sections for the vertical polarization in Proposition 22.8. As we observed there, there are no nonzero polarized sections that are square integrable over $\mathbb{R}^{2n}$. The same difficulty is easily seen to arise for the vertical polarization on any cotangent bundle $N = T^*M$. In Sect. 23.6, we will introduce half-forms to deal with this failure of square integrability.

We now examine properties of general real polarizations. We will see that polarized sections always exist locally, but not always globally.

Proposition 23.26 *If P is a purely real polarization on N, then for any $z_0 \in N$, there exist a neighborhood U of z_0 and a P-polarized section s of L defined over U such that $s(z_0) \neq 0$.*

Proof. According to the local form of the Frobenius theorem, we can find a neighborhood U of z_0 and a diffeomorphism Φ of U with a neighborhood V of the origin in $\mathbb{R}^n \times \mathbb{R}^n$ such that under Φ, the polarization P looks like the vertical polarization. That is to say, for each $z \in U$, the image of P_z under $\Phi_*(z)$ is just the span of the vectors $\partial/\partial y_1, \ldots, \partial/\partial y_n$, where the y's are the coordinates on the second copy of $\mathbb{R}^n$. By shrinking U if necessary, we can assume that L can be trivialized over U and that the open set V is the product of a ball B_1 centered at the origin in the first copy of $\mathbb{R}^n$ with a ball B_2 centered at the origin in the second copy of $\mathbb{R}^n$.

Let θ be the connection 1-form for an isometric trivialization of L over U and let $\tilde{\theta} = (\Phi^{-1})^*(\theta)$. Since the subspaces P_z are Lagrangian, the restriction of $\tilde{\theta}$ to the each set of the form $\{\mathbf{x}\} \times B_2$ is closed. Since B_2 is simply connected, there exists, for each $\mathbf{x} \in B_1$, a function $f_\mathbf{x}$ on B_2 such that the restriction of $\tilde{\theta}$ to $\{\mathbf{x}\} \times B_2$ equals $df_\mathbf{x}$. If we assume that $f_\mathbf{x}(0) = 0$, then $f_\mathbf{x}(\mathbf{y})$ will be smooth as a function of $(\mathbf{x}, \mathbf{y})$, since it is obtained simply by integrating $\tilde{\theta}$ from 0 to $\mathbf{y}$ in the vertical directions.

Now, let ϕ be any smooth function on B_1 with $\phi(0) \neq 0$ and define a function ψ on $B_1 \times B_2$ by

$$\psi(\mathbf{x}, \mathbf{y}) = \phi(\mathbf{x}) e^{if_\mathbf{x}(\mathbf{y})/\hbar}.$$

For any "vertical" vector field X (i.e., one where X is a linear combination of $\partial/\partial y_1, \ldots, \partial/\partial y_n$ with smooth coefficients), we compute that

$$X\psi = \frac{i}{\hbar}(Xf_\mathbf{x})\psi = \frac{i}{\hbar}df_\mathbf{x}(X)\psi = \frac{i}{\hbar}\tilde{\theta}(X)\psi.$$

Thus,

$$\left(X - \frac{i}{\hbar}\tilde{\theta}(X)\right)\psi = 0,$$

from which it follows that the function $\hat{\psi} := \psi \circ \Phi$ represents a polarized section on U in the given local trivialization of L. ∎

The existence of nonzero *global* polarized sections for a purely real polarization P is a more delicate question. If the leaves of P are not embedded, there is little chance of finding global polarized sections. Even if the leaves are embedded, there are obstructions. Since the tangent spaces to the leaves of P are Lagrangian subspaces, the restriction of L to R has zero curvature. There may, nevertheless, be loops in R for which the holonomy (Definition 23.8) is nontrivial. After all, if a loop γ in R is not the boundary of a surface S in R, then we cannot apply (23.6) to conclude that the holonomy of γ is trivial. The collection of holonomies for a leaf R of P can be understood as a homomorphism of $\pi_1(R)$ into S^1. If there is any loop in R with nontrivial holonomy, any polarized section of L must vanish on R.

Definition 23.27 *A submanifold R of N is said to be **Lagrangian** if* dim $R = n$ *and $T_z R$ is a Lagrangian subspace of $T_z N$ for each $z \in R$. A Lagrangian submanifold R of N is said to be **Bohr–Sommerfeld** (with respect to L) if the holonomy in L of every loop in R is trivial.*

We may summarize the preceding discussion as follows.

Conclusion 23.28 *For a purely real polarization P with embedded leaves, a polarized section vanishes on every leaf of P that is not Bohr–Sommerfeld.*

Our next example suggests that when the leaves are compact, the Bohr–Sommerfeld leaves typically form a discrete set within the set of all leaves.

Example 23.29 *Let $N = S^1 \times \mathbb{R}$, equipped with the symplectic form $\omega = dx \wedge d\phi$, where x is the linear coordinate on $\mathbb{R}$ and ϕ is the angular coordinate on S^1. Let L be the trivial line bundle on N, with sections that are identified with smooth functions. Let $\theta = x \, d\phi$ and define a connection ∇ on L by $\nabla_X = X - (i/\hbar)\theta(X)$, and let P be the purely real polarization of N for which the leaves are the sets of the form $S^1 \times \{x\}$, for $x \in \mathbb{R}$. Then a leaf $S^1 \times \{x\}$ is Bohr–Sommerfeld if and only if $x/\hbar$ is an integer.*

In particular, there are no nonzero, smooth polarized sections of L.

Proof. If we define a section locally on a given leaf $S^1 \times \{x\}$ as

$$s(\phi) = ce^{ix\phi/\hbar}$$

for some nonzero constant c, then it is easily verified that $\nabla_{\partial/\partial\phi} s = 0$. After one trip around the circle, the value of this section will be the starting value times $e^{2\pi i x/\hbar}$. Thus, the holonomy around $S^1 \times \{x\}$ is trivial if and only if $x/\hbar$ is an integer. A polarized section, then, would have to vanish on all the leaves where $x/\hbar$ is not an integer. Since such leaves form a dense subset of N, any smooth polarized section must be identically zero. ∎

Even in cases, such as Example 23.29, where there are no smooth polarized sections, one may still consider "distributional" polarized sections that are supported on the Bohr–Sommerfeld leaves, as on pp. 251–252 of [45].

23.5.3   The Complex Case

In Proposition 22.8, we computed the space of polarized sections for a certain positive, translation-invariant polarization on $\mathbb{R}^{2n}$, namely the one for which P_z is spanned by the vectors $\partial/\partial z_j$ in (22.9). The situation here is better than that for the vertical polarization, in that there are nonzero polarized sections that are square integrable over $\mathbb{R}^{2n}$. Recall, however, that if we take our polarization to be spanned by the vectors $\partial/\partial\bar{z}_j$, then [see (22.13)], then there are no nonzero square-integrable polarized sections. This example indicates the importance of the positivity condition in Definition 23.19.

For our next example, we consider the example of the unit disk D, equipped with the unique (up to a constant) symplectic form that is invariant under the group of fractional linear transformations that map D onto D. In this case, the quantum Hilbert space can be identified with a *weighted Bergman space*, that is, an L^2 space of holomorphic functions on D with respect to a measure of the form $(1 - |z|^2)^\nu dx\, dy$.

Example 23.30 *Let N be the unit disk $D \subset \mathbb{R}^2$ equipped with the following symplectic form:*

$$\omega = 4(1 - |z|^2)^{-2}\, dx \wedge dy = (1 - r^2)^{-2} r\, dr \wedge d\phi,$$

where (r, ϕ) are the usual polar coordinates. Let L be the trivial line bundle over D with connection $\nabla_X = X - (i/\hbar)\theta$, where θ is the symplectic potential for ω given by

$$\theta = 2\frac{r^2}{1 - r^2}\, d\phi.$$

Define a complex polarization on D by letting $P_z = \mathrm{Span}(\partial/\partial z)$, where $z = x - iy$. In that case, holomorphic sections s have the form

$$s(z) = F(z)(1 - |z|^2)^{1/\hbar},$$

where F is holomorphic. The norm of such a section is computed as

$$\|s\|^2 = 4 \int_D |F(z)|^2 (1 - |z|^2)^{2/\hbar - 2}\, dx\, dy.$$

As in the case of the plane, the seemingly unnatural definition $z = x - iy$ is necessary to obtain a Kähler polarization. If we used $z = x + iy$ instead, the holomorphic sections would have the form $F(z)(1 - |z|^2)^{-1/\hbar}$, in which case there would be no nonzero, square-integrable holomorphic sections. **Proof.** See Exercise 8. ∎

We now consider general purely complex polarizations. Recall that, by Proposition 23.18 and the Newlander–Nirenberg theorem, N has a unique complex structure for which P_z is the $(1, 0)$-subspace of $T_z^{\mathbb{C}} N$, for all $z \in N$. As in the purely real case, there always exist local polarized sections.

Theorem 23.31 *Suppose P is a purely complex polarization on N. Then for each $z_0 \in N$, there exists a P-polarized section s of L, defined in a neighborhood of z_0, such that $s(z_0) \neq 0$.*

We defer the proof of Theorem 23.31 until the end of this subsection.

Suppose s is as in the theorem and s' is any other locally defined P-polarized section. Then $s' = fs$ for some unique complex-valued function f, and by the product rule for covariant derivatives, $X(f) = 0$ for all $X \in \bar{P}_z$. This means that f is holomorphic with respect to the complex structure on N for which P is the $(1, 0)$-tangent space. Thus, we have a preferred family of local trivializations of L (the ones given by nonvanishing local polarized sections) such that the "ratio" of any two such trivializations is a holomorphic function. This means that we have given L the structure of a "holomorphic line bundle" over the complex manifold N in such a way that the holomorphic sections of L are precisely the polarized sections with respect to P.

Arguing as in the proof of Proposition 14.15, it is not hard to show that for a purely complex polarization, the space of square-integrable polarized sections of L forms a closed subspace of the prequantum Hilbert space. For any $z \in N$, if we choose a linear identification of the fiber of L over z with $\mathbb{C}$, then the map $s \mapsto s(z)$ is a linear functional on the quantum Hilbert space. It is not hard to show, as in the proof of Proposition 14.15, that this linear functional is continuous, and can therefore be represented as an inner product with a unique element of the quantum Hilbert space.

Definition 23.32 *Let P be a purely complex polarization on N. For each $z \in N$, choose a linear identification of the fiber of L over z with $\mathbb{C}$. Then the **coherent state** χ_z is the unique element of the quantum Hilbert space with respect to P such that*

$$s(z) = \langle \chi_z, s \rangle$$

for all s.

Suppose $N = \mathbb{R}^2$ with a polarization given by $P_z = \text{Span}(\partial/\partial z)$, where $z = x - i\alpha p$. If we use the symplectic potential $\theta = (p\,dx - x\,dp)/2$, then, as in the proof of Proposition 22.14, the quantum Hilbert space is naturally identifiable with the Segal–Bargmann space. In this case, the coherent states can be read off from Proposition 14.17.

It could happen that $\chi_z = 0$ for some $z \in N$, or even for *all* $z \in N$, depending on the choice of P. Even if χ_z is nonzero, χ_z is only well defined up to multiplication by a constant, because we must choose an identification of $L^{-1}(\{z\})$ with $\mathbb{C}$. But if $\chi_z \neq 0$, the one-dimensional subspace spanned by χ_z is independent of this choice. That is to say, whenever $\chi_z \neq 0$, the span of χ_z is a well-defined element of the projective space $\mathcal{P}(\mathbf{H})$, where $\mathbf{H}$ is the quantum Hilbert space.

Recall, meanwhile, that if (L, ∇) is a Hermitian line bundle with connection having curvature $\omega/\hbar$, then for any positive integer n, there is a natural Hermitian connection on $L^{\otimes k}$ having curvature $k\omega/\hbar$. This means that if L is a prequantum line bundle with one value $\hbar_0$ of Planck's constant, then $L^{\otimes k}$ is a prequantum line bundle with Planck's constant equal to $\hbar_0/k$. The following result shows that in the case of compact symplectic manifolds with Kähler polarizations, things behave nicely when k tends to infinity.

Theorem 23.33 *Assume N is compact and let P be a Kähler polarization on N. For each positive integer k, let $\mathbf{H}_k$ denote the space of polarized* sections of $L^{\otimes k}$. Then for all k, $\mathbf{H}_k$ is finite dimensional. Furthermore, for all sufficiently large k, we have the following results. First, the coherent state $\chi_z \in \mathbf{H}_k$ is nonzero for each $z \in N$. Second, the map

$$z \mapsto \mathrm{Span}(\chi_z)$$

is an antiholomorphic embedding of N into $\mathcal{P}(\mathbf{H}_k)$.

The finite dimensionality of $\mathbf{H}_k$ is a standard result in the theory of compact, complex manifolds. The embedding of N into $\mathcal{P}(\mathbf{H}_k)$ is the *Kodaira embedding theorem*, which we will not prove here. The Kodaira embedding theorem implies, in particular, that there exist nonzero, globally defined polarized sections of $L^{\otimes k}$, at least for large k. Since the value of Planck's constant for $L^{\otimes k}$ is $\hbar_0/k$, Planck's constant tends to zero as k tends to infinity. Thus, the study of holomorphic sections of $L^{\otimes k}$ for large k can be understood as being part of semiclassical analysis.

We now turn to the proof of Theorem 23.31, in which we will make use of basic properties of complex-valued differential forms on complex manifolds. ("Complex-valued" means that we allow the value of a k-form on a collection of k tangent vectors to be a complex number.) In a holomorphic local coordinate system $z_1, \ldots, z_n$, each form can be written as a wedge product of the dz_j's and $d\bar{z}_j$'s. A form is called a (p,q)-form if it is a linear combination of wedge products of p factors involving the dz_j's and q factors involving the $d\bar{z}_j$'s. Each form can be decomposed uniquely as a linear combination of (p,q)-forms for various values of p and q, and this decomposition does not depend on the choice of holomorphic coordinate system. If α is a (p,q)-form, then $d\alpha$ will be a linear combination of a $(p+1,q)$-form and a $(p,q+1)$-form. We define operators ∂ and $\bar{\partial}$ in such a way that ∂ maps (p,q)-forms to $(p+1,q)$-forms, $\bar{\partial}$ maps (p,q)-forms to $(p,q+1)$ forms, and $d = \partial + \bar{\partial}$. In particular,

$$\partial(f \, dz_{j_1} \wedge \cdots \wedge dz_{j_p} \wedge d\bar{z}_{k_1} \wedge \cdots \wedge dz_{k_q})$$
$$= \sum_l \frac{\partial f}{\partial z_l} dz_l \wedge dz_{j_1} \wedge \cdots \wedge dz_{j_p} \wedge d\bar{z}_{k_1} \wedge \cdots \wedge dz_{k_q}$$

and similarly for $\bar{\partial}$ with $(\partial f/\partial z_l) \, dz_l$ replaced by $(\partial f/\partial \bar{z}_l) \, d\bar{z}_l$.

The maps ∂ and $\bar{\partial}$ satisfy the identities:

$$\partial\partial = \bar{\partial}\bar{\partial} = 0$$
$$\partial\bar{\partial} = -\bar{\partial}\partial.$$

The Dolbeault lemma states that if a (p,q)-form α satisfies $\partial\alpha = 0$, then α can be expressed locally as $\partial\beta$ for some $(p-1,q)$-form, and if $\bar{\partial}\alpha = 0$, then α can be expressed locally as $\bar{\partial}\beta$ for some $(p,q-1)$-form. A $(p,0)$-form α is said to be *holomorphic* if it can be expressed in holomorphic coordinates as a sum of terms of the form

$$f(z) \, dz_{j_1} \wedge \cdots \wedge dz_{j_p},$$

where the coefficient functions f is holomorphic. A $(p, 0)$-form α is holomorphic if and only if $\bar{\partial}\alpha = 0$. If a holomorphic $(p, 0)$-form α satisfies $d\alpha = 0$ (or, equivalently, $\partial\alpha = 0$), then α can be written locally as $\alpha = d\beta$, for some holomorphic $(p - 1, 0)$-form.

Let P be a purely complex polarization on N and let J be the almost-complex structure for which P_z is the $(1, 0)$-tangent space at z. Since (Proposition 23.18), ω is J-invariant, it follows (Exercise 6) that ω is a $(1, 1)$-form.

Lemma 23.34 *Let N be a complex manifold with almost-complex structure J and let ω be a closed, J-invariant, real-valued $(1,1)$-form on N. Then for every point $z_0 \in N$, there exists a smooth, real-valued function κ defined in a neighborhood of z_0 such that $i\partial\bar{\partial}\kappa = \omega$.*

In the case that N is Kähler [i.e., the case where $\omega(X, JX) \geq 0$], a function κ as in the lemma is called a (local) *Kähler potential* for N.

Proof. By assumption, $d\omega = (\partial + \bar{\partial})\omega = 0$, from which it follows that $\partial\omega = \bar{\partial}\omega = 0$, because $\partial\omega$ is a $(2, 1)$-form and $\bar{\partial}\omega$ is a $(1, 2)$ form. Thus, by the Dolbeault lemma, there exists a $(1, 0)$-form α, defined in a neighborhood of z_0, such that $\bar{\partial}\alpha = \omega$. Then $\partial\alpha$ is a $(2, 0)$-form that satisfies

$$\bar{\partial}\partial\alpha = -\partial\bar{\partial}\alpha = -\partial\omega = 0.$$

This shows that $\partial\alpha$ is actually a *holomorphic* $(2, 0)$-form.

Since also $\partial\partial\alpha = 0$, we see that $\partial\alpha$ is closed, which means that there exists a holomorphic 1-form η, defined in a possibly smaller neighborhood of z_0, such that $d\eta = \partial\eta = \partial\alpha$. Thus, $\partial(\alpha - \eta) = 0$, and so by the Dolbeault lemma, there exists a function g, defined in a neighborhood of z_0, such that $\partial g = \alpha - \eta$. Thus, $\alpha = \eta + \partial g$ and so

$$\omega = \bar{\partial}\alpha = \bar{\partial}\partial g = -\partial\bar{\partial}g$$

since $\bar{\partial}\eta = 0$. The function $\kappa := ig$ then satisfies $i\partial\bar{\partial}\kappa = \omega$.

Now, a calculation in coordinates (Exercise 7) shows that the map $\kappa \mapsto i\partial\bar{\partial}f$ is real, that is, it maps real-valued functions to real-valued 2-forms. Since ω is real, the operator $i\partial\bar{\partial}$ must map the imaginary part of κ to zero. Thus, $i\partial\bar{\partial}\kappa$ is unchanged if κ is replaced by its real part. ∎

Proof of Theorem 23.31. Let κ be as in Lemma 23.34 and let θ be the real-valued 1-form given by

$$\theta = \text{Im}(\partial\kappa) = \frac{1}{2i}\left(\partial\kappa - \bar{\partial}\kappa\right). \tag{23.13}$$

Then because $\partial^2 = \bar{\partial}^2 = 0$, we have

$$d\theta = (\partial + \bar{\partial})\theta = \frac{1}{2i}(\bar{\partial}\partial\kappa - \partial\bar{\partial}\kappa) = \omega.$$

That is to say, θ is a symplectic potential for ω. Thus, by Proposition 23.6, we can find a local isometric trivialization s_0 of L for which the connection 1-form is $\theta/\hbar$.

For any vector X, we have

$$\nabla_X \left(e^{-\kappa/(2\hbar)} s_0 \right) = \left(-\frac{1}{2\hbar} X(\kappa) - \frac{i}{\hbar} \theta(X) \right) e^{-\kappa/\hbar} s_0, \qquad (23.14)$$

where $X(\kappa) = d\kappa(X) = \partial\kappa(X) + \bar{\partial}\kappa(X)$. Now, if X is of type $(0,1)$, then $\partial\kappa(X) = 0$, in which case, if we use (23.13), we find that the two terms on the right-hand side of (23.14) cancel. Thus, $e^{-\kappa/(2\hbar)} s_0$ is the desired local polarized section. ∎

23.6 Quantization with Half-Forms: The Real Case

In this section, we introduce a concept known as *half-forms*, which are designed to work around the problem that, in the case of real polarizations, there often do not exist any nonzero square-integrable polarized sections.

A polarized section s for a real polarization P tends to have infinite norm, because we may get infinity from integrating $|s|^2$ along the leaves of the polarization. To illustrate how half-forms work around this problem, consider the case of the vertical polarization on $\mathbb{R}^2 \cong T^*\mathbb{R}$. Elements of the half-form Hilbert space will be representable in the form $s \otimes \sqrt{dx}$, where s is a polarized section of L and where $\sqrt{dx}$ will be interpreted as a "section of the square root of the canonical bundle." To compute the norm of such an object, we first square it at each point to obtain the quantity $|s|^2 \, dx$. Since s is polarized, $|s|^2$ is a function of x only, independent of p. Thus, $|s|^2 \, dx$ may be thought of as a 1-form on $\mathbb{R}$, rather than on $\mathbb{R}^2$, which we may then integrate to obtain

$$\|s\|^2 := \int_{\mathbb{R}} |s|^2 (x) \, dx.$$

This procedure has two advantages over the one we used in Sect. 22.4, where we simply integrated $|s|^2$ itself over $\mathbb{R}$. First, a version of this procedure works for real polarizations on general symplectic manifolds. Second, the half-form approach will allow quantized observables to be self-adjoint, which was not the case in Sect. 22.5 when we simply restricted prequantized observables to the polarized subspace. (See the discussion following Proposition 22.12.)

Throughout this section, we assume that N is a quantizable symplectic manifold, that L is a fixed prequantum line bundle over N, and that P is a fixed purely real polarization on N.

23.6.1   The Space of Leaves

Recall that a *leaf* of P is a maximal connected, integral submanifold of P. We may then form the *leaf space* Ξ (the set of all leaves of P) and a quotient map $q : N \to \Xi$ sending each point $z \in N$ to the unique leaf containing z. We may topologize Ξ by defining a set U in Ξ to be open if $q^{-1}(U)$ is open in N.

In order to be able to carry out the program of geometric quantization with respect to P, we must assume that Ξ can be given the structure of a smooth, n-dimensional manifold in such a way that $q : N \to \Xi$ is smooth and such that the kernel of $q_{*,z}$ is equal to $P_z^{\mathbb{R}}$, the intersection of P_z with the real tangent space of P_z. We abbreviate this assumption on Ξ by saying that Ξ *is a smooth manifold*. In the case $N = T^*M$ with the vertical polarization (Example 23.17), the leaf space Ξ is a smooth manifold diffeomorphic to M.

It should be emphasized that even if Ξ is a smooth manifold, there is no canonical "volume measure" on Ξ. Thus, our half-form Hilbert space will be defined in such a way that the pointwise "square" of an element will be an n-form, rather than a function, on the leaf space, which can then be integrated over the n-manifold Ξ.

23.6.2   The Canonical Bundle

We now introduce the canonical bundle of a purely real polarization P, with sections that are a special sort of n-form on N, along with a notion of polarized section of the canonical bundle. If the leaf space Ξ is a smooth manifold, the space of polarized sections of the canonical bundle can be identified with the space of all n-forms on the n-manifold Ξ.

Definition 23.35 *The **canonical bundle** $\mathcal{K}_P$ of P is the* real *line bundle with sections that are n-forms α having the property that*

$$X \lrcorner \alpha = 0 \tag{23.15}$$

*for every vector field X lying in P. A section α of $\mathcal{K}_P$ is **polarized** if*

$$X \lrcorner (d\alpha) = 0 \tag{23.16}$$

for every vector field X lying in P.

If an n-form α satisfies (23.15), then $\alpha(X_1, \ldots, X_n) = 0$ if any of the X_j's belongs to P. Thus, the value of α at any point z can be viewed as an n-linear, alternating functional on the quotient vector space $T_z N / P_z^{\mathbb{R}}$, where $P_z^{\mathbb{R}}$ is the intersection of P_z with the real tangent space. Since this quotient space is n-dimensional, we see that at each point, the space of possible values for α is one dimensional.

Meanwhile, if α satisfies (23.16), then at each point, $d\alpha$ is an $(n+1)$-linear, alternating functional on $T_z N / P_z^{\mathbb{R}}$, which must be zero. Thus, for sections of $\mathcal{K}_P$, (23.16) is equivalent to the condition

$$d\alpha = 0. \tag{23.17}$$

We can also introduce the *complexified canonical bundle* $\mathcal{K}_P^{\mathbb{C}}$, the sections of which are complex-valued n-forms satisfying (23.15). We define a section of $\mathcal{K}_P^{\mathbb{C}}$ to be polarized if it satisfies (23.16).

Example 23.36 *Let $N = T^*\mathbb{R}^n \cong \mathbb{R}^{2n}$ and let P be the vertical polarization on N. Then an n-form α on $\mathbb{R}^{2n}$ is a section of $\mathcal{K}_P$ if and only if α is of the form*

$$\alpha = f(\mathbf{x}, \mathbf{p}) \, dx_1 \wedge \cdots \wedge dx_n, \tag{23.18}$$

and α is a polarized section of $\mathcal{K}_P$ if and only if α is of the form

$$\alpha = g(\mathbf{x}) \, dx_1 \wedge \cdots \wedge dx_n, \tag{23.19}$$

for smooth functions f on $\mathbb{R}^{2n}$ and g on $\mathbb{R}^n$.

Proof. If α contained any term involving dp_j, the contraction of α with $\partial/\partial p_j$ would not be zero, leaving (23.18) as the only possible form for a section of $\mathcal{K}_P$. Assuming α is of the form (23.18), if f is not independent of $\mathbf{p}$, then $d\alpha$ will contain a nonzero term of the form $dp_j \wedge dx_1 \wedge \cdots \wedge dx_n$, leaving (23.19) as the only possible form for a polarized section of $\mathcal{K}_P$. ∎

In Example 23.36, the polarized sections of $\mathcal{K}_P$ are effectively just n-forms on the configuration space $\mathbb{R}^n$. This conclusion is a special case of the following result.

Proposition 23.37 *If the leaf space Ξ of P is a smooth manifold and α is a polarized section of $\mathcal{K}_P$, then there exists a unique n-form $\tilde{\alpha}$ on Ξ such that*

$$\alpha = q^*(\tilde{\alpha}),$$

where $q : N \to \Xi$ is the quotient map. Conversely, if β is any n-form on Ξ, then $\alpha := q^*(\beta)$ is a polarized section of $\mathcal{K}_P$.

Proof. Suppose, first, that $\alpha = q^*(\beta)$, for an n-form β on Ξ. Then $X \lrcorner \alpha = 0$ whenever X lies in P, since P is the kernel of q_*. Furthermore, $d\alpha = q^*(d\beta) = 0$, since β is an n-form on an n-manifold, showing that α is a polarized section of $\mathcal{K}_P$.

In the other direction, we have already noted in the proof of Proposition 23.26 that N can be identified locally with a neighborhood $U \times V$ of the origin $\mathbb{R}^n \times \mathbb{R}^n$ in such a way that leaves of P correspond to the sets of the form $\{\mathbf{x}\} \times V$. We can use q to identify $U \cong U \times \{0\}$ with an open set $\tilde{U}$ in Ξ. Thus, P looks locally just like the vertical polarization on $\mathbb{R}^{2n}$, and so, by Example 23.36, any polarized section α of $\mathcal{K}_P$ will be of the form

(23.19). Thus, α determines an n-form $\hat{\alpha}$ on U and α is the pullback of $\hat{\alpha}$ by the projection map of $U \times V$ onto U. It follows that α is locally the pullback by q of an n-form $\tilde{\alpha}$ on $\tilde{U}$. We leave it to the reader to check that overlapping neighborhoods in N give the same form $\tilde{\alpha}$ on Ξ and that the desired result holds globally. ∎

Recall from Theorem 23.24 that $Q_{\mathrm{pre}}(f)$ preserves the space of polarized sections with respect to P, provided that the flow of X_f preserves $\bar{P}$ (which equals P, in this case). We now establish that for any such f, the Lie derivative $\mathcal{L}_{X_f}$ preserves the space of polarized sections of $\mathcal{K}_P$. This result will eventually allow us to define a quantum operator $Q(f)$ on the half-form Hilbert space associated to P.

Proposition 23.38 *Suppose X is a vector field on N that preserves P, in the sense of Definition 23.22, and suppose α is a smooth section of $\mathcal{K}_P$. Then the Lie derivative $\mathcal{L}_X \alpha$ is another section of $\mathcal{K}_P$ and if α is polarized, $\mathcal{L}_X \alpha$ is also polarized.*

Proof. Suppose $X_1, \ldots, X_n$ are smooth vector fields, with X_1 lying in $\bar{P} = P$. Then, by a standard formula for the Lie derivative,

$$
\begin{aligned}
(\mathcal{L}_X \alpha)&(X_1, \ldots, X_n) \\
&= X(\alpha(X_1, \ldots, X_n)) - \alpha([X, X_1], X_2, \ldots, X_n) \\
&\quad - \sum_{j=2}^{n} \alpha(X_1, \ldots, X_{j-1}, [X, X_j], X_{j+1}, \ldots, X_n).
\end{aligned}
\tag{23.20}
$$

Now, because α is a section of $\mathcal{K}_P$, the first and third terms on the right-hand side of (23.20) vanish. Because X preserves P, $[X, X_1]$ will again lie in P, and so the second term vanishes as well. Thus, $X_1 \lrcorner (\mathcal{L}_X \alpha) = 0$, which means that $\mathcal{L}_X \alpha$ is again a section of $\mathcal{K}_P$.

Since $\mathcal{L}_X \alpha = X \lrcorner d\alpha + d(X \lrcorner \alpha)$, if α satisfies (23.17), we have

$$
d(\mathcal{L}_X \alpha) = d^2 (X \lrcorner \alpha) = 0,
$$

showing that α is again polarized. ∎

Proposition 23.39 *Suppose the leaf space Ξ of P is a smooth manifold and that a vector field X on N preserves P. Then there exists a unique vector field Y on Ξ such that*

$$
q_{*,z}(X) = Y
\tag{23.21}
$$

for all $z \in N$. Furthermore, if $\alpha = q^*(\beta)$ is a polarized section of $\mathcal{K}_P$, as in Proposition 23.37, then

$$
\mathcal{L}_X(q^*(\beta)) = q^*(\mathcal{L}_Y(\beta)).
\tag{23.22}
$$

That is to say, under the identification in Proposition 23.37 of polarized sections of $\mathcal{K}_P$ with n-forms on Ξ, the operator $\mathcal{L}_X$ corresponds to the Lie derivative on Ξ in the direction of Y.

Proof. By Definition 23.22, $[X, Z]$ lies in P whenever the vector field Z lies in P. Thus, if a function ϕ is constant along P (i.e., annihilated by every vector field Z lying in P), the same will be true of $X\phi$. Thus, if ϕ is of the form $\phi = \psi \circ q$ for some function ψ on Ξ, then $X\phi$ is of the form $\hat{\psi} \circ q$ for some other function $\hat{\psi}$ on Ξ. The map $\psi \mapsto \hat{\psi}$ is easily seen to be a vector field, that is, a derivation of $C^\infty(\Xi)$. We conclude, then, that there is a unique vector field Y on Ξ such that

$$X(\psi \circ q) = (Y\psi) \circ q \qquad (23.23)$$

for every smooth function ψ on Ξ. It then follows from the definition of the differential that (23.21) holds for all $z \in N$. From (23.21), it follows easily that for any n-form β on Ξ, we have

$$X \lrcorner (q^*(\beta)) = q^*(Y \lrcorner \beta). \qquad (23.24)$$

Since β, being a top-degree form, is closed, $q^*(\beta)$ is also closed. Thus, one of the terms in the formula (21.7) for the Lie derivative of β and $q^*(\beta)$ is zero. Applying d to both sides of (23.24) then gives (23.22). ∎

Given a vector field Y and a nowhere-vanishing n-form β on Ξ, let $\mathrm{div}_\beta Y$ be the unique function on Ξ such that

$$\mathcal{L}_Y(\beta) = (\mathrm{div}_\beta Y)\beta.$$

Then by (23.22), we have

$$\mathcal{L}_X(q^*(\beta)) = ((\mathrm{div}_\beta Y) \circ q)q^*(\beta). \qquad (23.25)$$

The expression (23.25) will be helpful in analyzing the quantization of observables in Sect. 23.6.5.

23.6.3 Square Roots of the Canonical Bundle

We now assume that the leaf space Ξ of P is an *orientable* manifold, and we choose on particular orientation of Ξ.

Definition 23.40 *Choose a nowhere-vanishing, oriented n-form β on Ξ, so that $\alpha := q^*(\beta)$ is (Proposition 23.37) a nowhere-vanishing section of $\mathcal{K}_P$. A section of $\mathcal{K}_P$ is **non-negative** if it is, at each point, a non-negative multiple of α. This notion does not depend on the choice of oriented n-form β.*

Since Ξ is orientable, the canonical bundle $\mathcal{K}_P$ is trivializable, since the section α in Definition 23.40 is a globally trivializing section. Thus, we can find a square root of $\mathcal{K}_P$, that is, a line bundle δ_P such that $\delta_P \otimes \delta_P$ is isomorphic to $\mathcal{K}_P$. (We may, for example, take δ_P to be the trivial bundle.) When we speak of a *square root* of $\mathcal{K}_P$, we will mean, more precisely, a bundle δ_P together with a *particular* isomorphism of $\delta_P \otimes \delta_P$ with $\mathcal{K}_P$. Thus, if s_1 and s_2 are sections of δ_P, we think of $s_1 \otimes s_2$ as being a section of $\mathcal{K}_P$. We assume, further, that the isomorphism of $\delta_P \otimes \delta_P$ with $\mathcal{K}_P$ is chosen so that for any section s of δ_P, the section $s \otimes s$ of $\mathcal{K}_P$ is non-negative. (If the initial isomorphism of $\delta_P \otimes \delta_P$ with $\mathcal{K}_P$ does not have this property, compose it with $-I$ in the fibers of $\mathcal{K}_P$.)

We may consider the *complexification* of δ_P, that is, the line bundle $\delta_P^{\mathbb{C}}$ whose fiber at each point is the complexification of the fiber of δ_P. There is then a notion of *complex conjugation* for sections of $\delta_P^{\mathbb{C}}$, which fixes the fiber of δ_P inside the fiber of $\delta_P^{\mathbb{C}}$ at each point. If s_1 and s_2 are sections of $\delta_P^{\mathbb{C}}$, we think of $s_1 \otimes s_2$ as a section of the complexified canonical bundle $\mathcal{K}_P^{\mathbb{C}}$.

If α is a section of $\mathcal{K}_P$ and X is a vector field lying in P, let us define an n-form $\nabla_X \alpha$ by

$$\nabla_X \alpha = X \lrcorner (d\alpha). \tag{23.26}$$

Since α is a section of $\mathcal{K}_P$, we have $X \lrcorner \alpha = 0$, which means that $\nabla_X \alpha$ actually coincides with $\mathcal{L}_X \alpha$, by (21.7). Since it lies in P, the vector field X preserves P, and thus $\nabla_X \alpha = \mathcal{L}_X \alpha$ is again a section of $\mathcal{K}_P$, by Proposition 23.38. The operator ∇ in (23.26) has all the properties of a connection on $\mathcal{K}_P$ except that it is only defined in the directions of P. [Note that $\mathcal{L}_X$ does not, in general, satisfy the condition $\mathcal{L}_{fX} = f\mathcal{L}_X$, as required by Definition 23.2. Since, however, $\mathcal{L}_X \alpha$ can also be computed as in (23.26), for any section α of $\mathcal{K}_P$, the map ∇ does satisfy $\nabla_{fX} = f\nabla_X$.]

We call ∇ the natural *partial connection* on $\mathcal{K}_P$. According to Definition 23.35, a section α of $\mathcal{K}_P$ is polarized if and only if $\nabla_X \alpha = 0$ for each vector field X lying in P. We now show that both the partial connection and the Lie derivative "descend" to sections of δ_P in a natural way. This result will, in particular, allow us to define a notion of polarized sections of δ_P.

Proposition 23.41 *Let δ_P be a fixed square root of $\mathcal{K}_P$. For any vector field X lying in P, there is a unique linear operator ∇_X mapping sections of δ_P to sections of δ_P, such that*

$$\nabla_X(fs_1) = X(f)s_1 + f\nabla_X s_1 \tag{23.27}$$

$$\nabla_X(s_1 \otimes s_2) = (\nabla_X s_1) \otimes s_2 + s_1 \otimes (\nabla_X s_2) \tag{23.28}$$

for all smooth functions f and all sections s_1 and s_2 of δ_P. On the left-hand side of (23.28), ∇_X is the partial connection on $\mathcal{K}_P$ given by (23.26).

If X is a vector field on N that preserves P, then there is a unique linear operator $\mathcal{L}_X$, mapping sections of δ_P to sections of δ_P such that

$$\mathcal{L}_X(fs_1) = X(f)s_1 + f\mathcal{L}_X s_1$$
$$\mathcal{L}_X(s_1 \otimes s_2) = (\mathcal{L}_X s_1) \otimes s_2 + s_1 \otimes (\mathcal{L}_X s_2)$$

for all smooth functions f and all sections s_1 and s_2 of δ_P.

Both of these constructions extend naturally from sections of δ_P to sections of $\delta_P^{\mathbb{C}}$.

We may then say that a section s of $\delta_P^{\mathbb{C}}$ is *polarized* if $\nabla_X s = 0$ for every smooth vector field X lying in P.

Proof. If V is a one-dimensional vector space, then the map $\otimes : V \times V \to V \otimes V$ is commutative: $u \otimes v = v \otimes u$ for all $u, v \in V$. Furthermore, if u_0 is a nonzero element of V, then the map $u \mapsto u \otimes u_0$ is an invertible linear map of V to $V \otimes V$. Suppose s_0 is a local nonvanishing section of δ_P. Applying (23.28) with $s_1 = s_2 = s_0$, we want

$$2(\nabla_X s_0) \otimes s_0 = \nabla_X(s_0 \otimes s_0). \tag{23.29}$$

Since the operation of tensoring with s_0 is invertible, there is a unique section "$\nabla_X s_0$" of δ_P for which (23.29) holds.

Locally, any section s of δ_P can be written as $s = gs_0$ for a unique function g. We then define $\nabla_X s$ by

$$\nabla_X s = X(g)s_0 + g\nabla_X s_0, \tag{23.30}$$

in which case, (23.27) is easily seen to hold. If $s_1 = g_1 s_0$ and $s_2 = g_2 s_0$, then using (23.29) and the symmetry of the tensor product, it is easy to verify that (23.28) holds, with both sides of the equation equal to

$$X(g_1 g_2)\nabla_X(s_0 \otimes s_0).$$

Uniqueness of ∇_X holds because both (23.29) and (23.30) are required by the definition of ∇_X. The action of ∇_X extends to sections of $\delta_P^{\mathbb{C}}$, by writing such sections as complex-valued functions times s_0. The analysis of the Lie derivative is similar and is omitted. ∎

23.6.4 The Half-Form Hilbert Space

We continue to assume that the leaf space Ξ of P is an orientable manifold, and that we have chosen an orientation on Ξ. We assume that we have chosen a square root δ_P of $\mathcal{K}_P$, as in Sect. 23.6.3. If L is a prequantum line bundle over N, we now form the tensor product bundle $L \otimes \delta_P^{\mathbb{C}}$. Given two sections s_1 and s_2 of $L \otimes \delta_P^{\mathbb{C}}$, we decompose them locally as $s_j = \mu_j \otimes \nu_j$, where μ_j is a section of L and ν_j is a section of $\delta_P^{\mathbb{C}}$, and where, say, the

μ_j's are taken to be nonvanishing. Then we can combine these sections to form the quantity

$$(s_1, s_2) := (\mu_1, \mu_2)\overline{\nu_1} \otimes \nu_2, \tag{23.31}$$

where (μ_1, μ_2) is the pointwise inner product given by the Hermitian structure on L. Since (μ_1, μ_2) is a scalar-valued function and $\overline{\nu_1} \otimes \nu_2$ is a section of $\mathcal{K}_P^{\mathbb{C}}$, the quantity (s_1, s_2) is a section of $\mathcal{K}_P^{\mathbb{C}}$. Any other decomposition of s_j as the tensor product of a nonvanishing section of a L and a section of δ_P is of the form $(f\mu_j) \otimes (\nu_j/f)$ for some nonvanishing function f, and the value of (s_1, s_2) is the same as for the original decomposition. Since it is independent of the choice of local decomposition, (s_1, s_2) is actually defined globally.

Given the connection on L and the partial connection (23.41) on $\delta_P^{\mathbb{C}}$, we can form a partial connection on $L \otimes \delta_P^{\mathbb{C}}$ with the following property. For any vector field X lying in P, and any section s of $L \otimes \delta_P^{\mathbb{C}}$, if we decompose s locally as $s = \mu \otimes \nu$, where μ is a nonvanishing section of L and ν is a section of δ_P, then

$$\nabla_X(s) = (\nabla_X \mu) \otimes \nu + \mu \otimes (\nabla_X \nu). \tag{23.32}$$

The reader may verify that if $\mu \otimes \nu$ is replaced by $(f\mu) \otimes (\nu/f)$ for some nonvanishing function f, the value of $\nabla_X(s)$ is unchanged. Thus, as with the quantity (s_1, s_2) in (23.31), $\nabla_X(s)$ is defined globally. We then define a section s of $L \otimes \delta_P^{\mathbb{C}}$ to be *polarized* if $\nabla_X s = 0$ for each vector field X lying in P. If s_1 and s_2 are polarized sections of $L \otimes \delta_P^{\mathbb{C}}$, then the section (s_1, s_2) in (23.31) is easily seen to be a polarized section of $\mathcal{K}_P$.

As in the case without half-forms there is an obstruction to the existence of globally defined polarized sections of $L \otimes \delta_P^{\mathbb{C}}$. We say that a leaf R is *Bohr–Sommerfeld* (in the half-form sense, with respect to a particular choice of δ_P) if there exists a nonzero section s of $L \otimes \delta_P^{\mathbb{C}}$ defined over R such that $\nabla_X s = 0$ for each tangent vector to R. As in the case without half-forms, if the leaves are topologically nontrivial, the Bohr–Sommerfeld leaves will in general be a discrete set in the space of all leaves.

The Bohr–Sommerfeld leaves in the half-form sense need not be the same as the Bohr–Sommerfeld leaves in the sense of Definition 23.27. In the setting of Example 23.29, for instance, the canonical bundle $\mathcal{K}_P$ is trivial, but the square-root bundle δ_P may be chosen to be nontrivial, by putting in a twist by 180 degrees over each copy of S^1. (That is to say, we think of S^1 as the interval $[0, 2\pi]$ with the ends identified, and we attach a copy of $\mathbb{R}$ to each point. But when identifying the fiber at 2π with the fiber at 0, we use the negative of the identity map.) As Exercise 9 shows, in this example, the Bohr–Sommerfeld leaves are the sets of the form $\{x\} \times S^1$, where $x/\hbar = n + 1/2$ for some integer n.

Definition 23.42 *For any purely real polarization P and any square root δ_P of $\mathcal{K}_P$, the **half-form space** is the space of smooth, polarized sections* *of $L \otimes \delta_P^{\mathbb{C}}$. For a polarized section s of $L \otimes \delta_P^{\mathbb{C}}$, define the norm of s by*

$$\|s\|^2 = \int_\Xi \widetilde{(s,s)}, \tag{23.33}$$

where (s,s) is as in (23.31) and where $\widetilde{(s,s)}$ is the n-form on Ξ given by Proposition 23.37. If s_1 and s_2 are elements of the half-form space with $\|s_1\| < \infty$ and $\|s_2\| < \infty$, define the inner product of s_1 and s_2 by

$$\langle s_1, s_2 \rangle = \int_\Xi \widetilde{(s_1, s_2)}.$$

*The **half-form Hilbert space** is the completion with respect to the norm (23.33) of the space of polarized sections s for which $\|s\|^2 < \infty$.*

The integral of n-forms on Ξ is taken with respect to the chosen orientation on Ξ. We can always decompose s locally as $s = \mu \otimes \nu$ with ν being a section of δ_P (as opposed to $\delta_P^{\mathbb{C}}$) and μ being a section of L. Then

$$(s,s) = (\mu, \mu)\nu \otimes \nu,$$

from which we see that (s,s) is a non-negative section of $\mathcal{K}_P$ (Definition 23.40). (Recall that we have chosen the identification of $\delta_P \otimes \delta_P$ with $\mathcal{K}_P$ in a particular way, so that $\nu \otimes \nu$ is always the pullback by q of an oriented form on Ξ.) Thus, the integral on the right-hand side of (23.33) is non-negative, but possibly infinite.

Example 23.43 *Let $N = T^*\mathbb{R} \cong \mathbb{R}^2$ and let L be the trivial bundle on N, with connection $\nabla_X = X - (i/\hbar)\theta(X)$, where $\theta = p\,dx$. Let P be the vertical polarization on N and orient $\mathbb{R}$ so that oriented 1-forms are positive multiples of dx. Let δ_P to be the trivial bundle and with a trivializing section "$\sqrt{dx}$" of δ_P such that $\sqrt{dx} \otimes \sqrt{dx} = dx$. Then every polarized section s of $L \otimes \delta_P^{\mathbb{C}}$ has the form*

$$s = \psi(x) \otimes \sqrt{dx} \tag{23.34}$$

for some function ψ on $\mathbb{R}$. The norm of such a section is computed as

$$\|s\|^2 = \int_\mathbb{R} |\psi(x)|^2 \, dx.$$

Proof. The sections of $\mathcal{K}_P$ are 1-forms that are zero on $\partial/\partial p$, that is, 1-forms of the form $\alpha = f(x,p)\,dx$. Such a 1-form satisfies $d\alpha = 0$ if and only if f is independent of p. Thus, dx is a globally defined polarized section of $\mathcal{K}_P$. If we choose δ_P to be trivial and let $\sqrt{dx}$ be such that $\sqrt{dx} \otimes \sqrt{dx} = dx$, then $\sqrt{dx}$ will be a polarized section of δ_P. Every section s of $L \otimes \delta_P^{\mathbb{C}}$ can be written uniquely as $s = \psi(x,p) \otimes \sqrt{dx}$ for some function ψ. Since $\sqrt{dx}$ is polarized and $\theta(\partial/\partial p) = 0$, we see that s is polarized if and only if ψ is independent of p. For a section of the form (23.34), we have $(s,s) = |\psi(x)|^2 \, dx$, in which case, $\widetilde{(s,s)}$ is given by the same formula as (s,s), but now interpreted as a 1-form on $\Xi \cong \mathbb{R}$ rather than $\mathbb{R}^2$. ∎

23.6.5   Quantization of Observables

Suppose f is a function on N for which X_f preserves P in the sense of Definition 23.22. We will now associate with f a self-adjoint (or, at least, symmetric) operator $Q(f)$ on the half-form Hilbert space of P. Operators of this sort will satisfy exactly the desired commutation relations.

Definition 23.44 *For any function f on N for which X_f preserves P, let $Q(f)$ be the operator on the half-form space of P given by*

$$Q(f)s = (Q_{\text{pre}}(f)\mu) \otimes \nu + i\hbar\, \mu \otimes \mathcal{L}_{X_f}\nu,$$

where s is decomposed locally as $s = \mu \otimes \nu$, with μ being a section of L and ν a section of $\delta_P^{\mathbb{C}}$.

The operator $Q(f)$ is well defined (i.e., independent of the choice of local trivialization) as may easily be verified. This independence holds, however, only because the coefficient $i\hbar$ of ∇_{X_f} in the first term exactly matches the coefficient $i\hbar$ of $\mathcal{L}_{X_f}$ in the second term.

Before describing the general properties of the operators $Q(f)$, we consider a simple example that illustrates the essential role of the Lie derivative term in Definition 23.44.

Example 23.45 *Let the notation be as in Example 23.43, and let $f : \mathbb{R}^2 \to \mathbb{R}$ be of the form*

$$f(x,p) = a(x) + b(x)p,$$

for some smooth functions a and b on $\mathbb{R}$. Then X_f preserves P and

$$Q(f)(\psi(x) \otimes \sqrt{dx}) = \tilde{\psi}(x) \otimes \sqrt{dx},$$

where

$$\tilde{\psi}(x) = -i\hbar\left(b(x)\psi'(x) + \frac{1}{2}b'(x)\psi(x)\right) + a(x)\psi(x).$$

In particular, if $f(x,p) = x$, then $\tilde{\psi}(x) = x\psi(x)$ and if $f(x,p) = p$, then $\tilde{\psi}(x) = -i\hbar\, \partial\psi/\partial x$. More generally, if a and b are polynomials, then the action of $Q(f)$ on ψ coincides with the Weyl quantization of f (Exercise 8 in Chap. 13).

The term involving $b'(x)$ comes from the presence of half-forms and is absent in the formula (22.15) for $Q_{\text{pre}}(f)$. The b' term, with the exact coefficient of $1/2$, is necessary for $Q(f)$ to be self-adjoint (or, at least, symmetric); see Exercise 10. Example 23.45 is actually quite representative of the general case. [Compare (23.38) in the proof of Theorem 23.47 and Example 23.48.]

Proof. We have computed $Q_{\text{pre}}(f)$ in (22.15) in the proof of Proposition 22.12. We compute that X_f is equal to $-b(x)\, \partial/\partial x$ plus a term involving $\partial/\partial p$. Since the 1-form dx is closed, we obtain, by (21.7),

$$\mathcal{L}_{X_f}(dx) = d(X_f \lrcorner dx) = -db(x) = -b'(x)\, dx.$$

Using Proposition 23.41, we then obtain

$$\mathcal{L}_{X_f}\left(\sqrt{dx}\right) \otimes \sqrt{dx} = -\frac{1}{2}b'(x)\ dx = -\frac{1}{2}b'(x)\sqrt{dx} \otimes \sqrt{dx}, \qquad (23.35)$$

which gives

$$\mathcal{L}_{X_f}\left(\sqrt{dx}\right) = -\frac{1}{2}b'(x)\sqrt{dx}.$$

Adding the $\mathcal{L}_{X_f}$ term to the previously computed expression for $Q_{\mathrm{pre}}(f)$ gives the desired result. ∎

Returning now to the setting of general real polarizations, we establish two key results for the quantized observables $Q(f)$, that they satisfy the desired commutation relations and that they are self-adjoint (or, at least, symmetric) whenever f is real valued. It can also be shown that when f is a polarized function (i.e., constant along each leaf of P), then $Q(f)$ acts on the quantum Hilbert space simply as multiplication by f. See Exercise 11.

Theorem 23.46 *Suppose f and g are functions on N for which X_f and X_g preserve P. Then the operators $Q(f)$ and $Q(g)$ satisfy*

$$\frac{1}{i\hbar}\left[Q(f), Q(g)\right] = Q(\{f, g\})$$

on the space of smooth, polarized sections of $L \otimes \delta_P^{\mathbb{C}}$.

Proof. Since $Q(h)$ is a local operator for any function h, it suffices to prove the result locally. Let us choose, then, a local nonvanishing section ν_0 of $\delta_P^{\mathbb{C}}$, so that, locally, each section s of $L \otimes \delta_P^{\mathbb{C}}$ can be decomposed uniquely as $s = \mu \otimes \nu_0$. For any vector field preserving P, we let $\gamma(X)$ be the function such that

$$\mathcal{L}_X(\nu_0) = \gamma(X)\nu_0.$$

We then have $Q(f)(\mu \otimes \nu_0) = \tilde{\mu} \otimes \nu_0$, where

$$\tilde{\mu} = [Q_{\mathrm{pre}}(f) + i\hbar\gamma(X_f)]\mu.$$

We now compute that

$$[Q_{\mathrm{pre}}(f) + i\hbar\gamma(X_f), Q_{\mathrm{pre}}(g) + i\hbar\gamma(X_g)]$$
$$= [Q_{\mathrm{pre}}(f), Q_{\mathrm{pre}}(g)] + i\hbar[Q_{\mathrm{pre}}(f), \gamma(X_g)] + i\hbar[\gamma(X_g), Q_{\mathrm{pre}}(f)]$$
$$= i\hbar Q_{\mathrm{pre}}(\{f, g\}) + (i\hbar)^2\left(X_f(\gamma(X_g)) - X_g(\gamma(X_f))\right).$$

The desired result will follow if we can verify that

$$X_f(\gamma(X_g)) - X_g(\gamma(X_f)) = \gamma(X_{\{f,g\}}). \qquad (23.36)$$

To verify (23.36), we use a standard identity for the Lie derivative on forms: $\mathcal{L}_{[X,Y]} = [\mathcal{L}_X, \mathcal{L}_Y]$. Using Proposition 23.41, we can easily show that this identity holds also on sections of $\delta_P^{\mathbb{C}}$, for vector fields that preserve P. It is then a simple calculation (Exercise 12) to verify (23.36). ∎

Theorem 23.47 *If $f \in C^\infty(N)$ is real valued and X_f preserves P, then the operator $Q(f)$ is symmetric on the space of smooth sections s in the half-form space for which $\widetilde{(s,s)}$ has compact support on Ξ.*

Proof. Suppose $\alpha = q^*(\beta)$ is polarized section of $\mathcal{K}_P$, so that there is, at least locally, a corresponding polarized section $\sqrt{q^*(\beta)}$ of δ_P. If X_f preserves P, then by Proposition 23.39, there is a unique vector field Y_f on Ξ such that $q_{*,z}(X_f) = Y_f$ for all $z \in N$. Using (23.25) and Proposition 23.41, we get

$$\mathcal{L}_{X_f}\left(\sqrt{q^*(\beta)}\right) = \frac{1}{2}((\operatorname{div}_\beta Y_f) \circ q)\sqrt{q^*(\beta)}.$$

Meanwhile, it is not hard to show (Exercise 13) that it is possible to choose a local symplectic potential θ that is zero in the directions of P. Thus, we can trivialize L locally in such a way that sections that are covariantly constant along P are simply functions that are constant along P in the ordinary sense. Thus, elements s of the half-form space have, locally, the form

$$s = (\psi \circ q) \otimes \sqrt{q^*(\beta)} \tag{23.37}$$

for some function ψ and n-form β on Ξ. Thus, if X_f preserves P, and a section s is decomposed locally as in (23.37), we have

$$Q(f)(s) = (\tilde{\psi} \circ q) \otimes \sqrt{q^*(\beta)},$$

where

$$\tilde{\psi} = i\hbar\left(Y_f\psi + \frac{1}{2}(\operatorname{div}_\beta Y_f)\psi\right) + (-\theta(X_f) - f)\psi. \tag{23.38}$$

It can be verified (Exercise 14) that the function $-\theta(X_f) - f$ is constant along P and thus may be thought of as a function on Ξ.

By multiplying elements of the half-form space by functions of the form $\chi \circ q$, with χ having compact support in Ξ, we can "localize" the calculations on Ξ. Suppose s_1 and s_2 are two elements of the half-form space decomposed as in (23.37) near a point $z \in N$, with the same β and two different functions ψ_1 and ψ_2 on Ξ. Then $\widetilde{(s_1, s_2)}$ has the form $\overline{\psi_1}\psi_2\beta$ in a neighborhood U of $q(z)$. By localization, we may assume that $\widetilde{(s_1, s_2)}$ has compact support in U, and we then have

$$\langle s_1, Q(f)s_2\rangle = -i\hbar \int_\Xi \overline{\psi_1}\tilde{\psi}_2\,\beta,$$

where $\tilde{\psi}_2$ is as in (23.38). "Integration by parts" (Exercise 15) with respect to β then shows that this quantity coincides with $\langle Q(f)s_1, s_2\rangle$. $\blacksquare$

Example 23.48 (Cotangent Bundles) *Let $N = T^*M$ for an oriented manifold M, let θ be the canonical 1-form on N, and let L be the trivial* *line bundle on N, with connection $\nabla_X = X - (i/\hbar)\theta(X)$. Let P be the vertical polarization on N, so that $\mathcal{K}_P$ is trivial, and let δ_P be chosen to be trivial. Let β be an arbitrary nowhere-vanishing, oriented n-form on M, so that $\alpha := \pi^*(\beta)$ is a nowhere-vanishing section of $\mathcal{K}_P$, and choose a trivializing section $\sqrt{\alpha}$ of δ_P with $\sqrt{\alpha} \otimes \sqrt{\alpha} = \alpha$. In that case, elements s of the half-form Hilbert space have the form $s = (\psi \circ \pi) \otimes \sqrt{\alpha}$, where ψ is a function on M, and*

$$\|s\|^2 = \int_M |\psi|^2 \, \beta.$$

The half-form Hilbert space may, thus, be identified with $L^2(M, \beta)$.

Suppose now that f is a function on T^*M of the form $f = f_1 + f_2$, where f_1 is constant on each fiber of T^*M and f_2 is linear on each fiber. Then f_2 may be thought of as a section of $T^{**}M \cong TM$, that is, as a vector field Y_f on M. In that case, X_f preserves P and $Q(f)$ acts on elements of the half-forms space as

$$Q(f)\left((\psi \circ \pi) \otimes \sqrt{\alpha}\right) = (\tilde{\psi} \circ \pi) \otimes \sqrt{\alpha},$$

where

$$\tilde{\psi} = i\hbar\left(Y_f\psi + \frac{1}{2}(\text{div}_\beta \, Y_f)\psi\right) + f_1\psi.$$

Here $\text{div}_\beta \, Y_f$ is the unique function such that $\mathcal{L}_{Y_f}\beta = (\text{div}_\beta \, Y_f)\beta$.

A simple calculation in coordinates shows that the vector field Y_f in the example satisfies $X_f(\psi \circ \pi) = (Y_f\psi) \circ \pi$, so that our notation is consistent with that in Proposition 23.39 [see (23.23)].

Proof. The calculation is precisely the same as in the proof of Theorem 23.47, except that the decomposition in (23.37) is now global. The claimed form of $Q(f)$ is nothing but the expression (23.38), where the reader may easily compute, using local coordinates, that $-\theta(X_f) - f = f_1$. ∎

It is an unfortunate feature of geometric quantization that in the case of the vertical polarization on cotangent bundles, it only permits us to quantize functions that are at most linear in the momentum variables. In a typical physical system having T^*M as its phase space, there will be a "kinetic energy" term in the classical Hamiltonian that is quadratic in p. To quantize such a system, one has to find a way to quantize the kinetic energy term, "by hook or by crook."

One approach to this problem is to allow the exponentiated quantized Hamiltonian to change the polarization, and then to use pairing maps (Sect. 23.8) to "project" back to the Hilbert space for the original polarization. As explained in Sect. 9.7 of [45], this approach succeeds in the case that the kinetic energy term is $g(p, p)/(2m)$, where g is the Riemannian structure on T^*M induced by a Riemannian structure on TM. The quantized kinetic energy operator turns out to be given by the map

$$\psi \mapsto -\frac{\hbar^2}{2m}\left((\Delta\psi)(x) - \frac{1}{6}R(x)\psi(x)\right), \tag{23.39}$$

where Δ is the Laplacian for M (taken to be a negative operator) and where $R(x)$ is the scalar curvature of the Riemannian structure on TM. The calculation in [45] glosses over one technical issue, which is that the time-evolved polarizations may not be everywhere transverse to the original polarization. Nevertheless, the calculation provides a reasonable geometric motivation for the formula (23.39).

It should be emphasized that, because of the projections involved in the computation of the quantized kinetic energy operator, it does *not* satisfy the desired commutation relations with the quantizations of functions whose flow preserves the vertical polarization. Nevertheless, this approach to quantizing the kinetic energy may simply be the best one can do.

23.7 Quantization with Half-Forms: The Complex Case

In the case of a purely complex polarization, half-forms are not "necessary," in that we typically have a nonzero Hilbert space even without them. Nevertheless, their inclusion gives advantages. In the first place, using half-forms makes the complex case more parallel to the real case. In the second place, complex quantization with half-forms simply gives better results than without half-forms. In the case of the harmonic oscillator, for example, the inclusion of half-forms allows (Example 23.53) geometric quantization to reproduce precisely the spectrum $(n+1/2)\hbar\omega, n = 0, 1, 2, \ldots$, that we found in the traditional treatment. This result should be compared to Proposition 22.14 without half-forms, where the spectrum is found to be $n\hbar\omega$.

Throughout this section, we assume that (N, ω) is a $2n$-dimensional quantizable symplectic manifold, that (L, ∇) is prequantum line bundle over N, and that P is a Kähler polarization on N (Definition 23.19). Since the definitions in the complex case are very similar to those in the real case (with a few important differences), we will run through them quickly. Since $\bar{P}$ is no longer equal to P, we need to replace P by $\bar{P}$ in may of the formulas from Sect. 23.6.

The *canonical bundle* $\mathcal{K}_P$ of P is the complex line bundle for which the sections are n-forms α satisfying

$$X \lrcorner \alpha$$

for each vector field X lying in $\bar{P}$. Sections of $\mathcal{K}_P$ are precisely the $(n, 0)$-forms on N. A section of $\mathcal{K}_P$ is said to be *polarized* if

$$X \lrcorner (d\alpha) = 0 \tag{23.40}$$

for every vector field lying in $\bar{P}$, or, equivalently, if $d\alpha = 0$. Polarized sections of $\mathcal{K}_P$ are precisely the holomorphic $(n, 0)$-forms on N. By a *square root* of $\mathcal{K}_P$ we will mean a complex line bundle δ_P over N such that $\delta_P \otimes \delta_P$ is isomorphic with $\mathcal{K}_P$, together with a particular isomorphism of $\delta_P \otimes \delta_P$ with $\mathcal{K}_P$. Thus, if s_1 and s_2 are sections of δ_P, we think of $s_1 \otimes s_2$ as being a section of $\mathcal{K}_P$. We assume that such a square root exists and we fix for the remainder of this section one particular square root δ_P.

If X is a vector field that preserves $\bar{P}$, in the sense of Definition 23.22, then $\mathcal{L}_X$ preserves the space of sections of $\mathcal{K}_P$ and also the space of polarized sections of $\mathcal{K}_P$. The condition (23.40) defining polarized sections of $\mathcal{K}_P$ can be understood as the vanishing of a *partial connection* ∇., defined for vector fields lying in $\bar{P}$, and given by $\nabla_X \alpha = X \lrcorner (d\alpha)$. Both the partial connection (for vector fields lying in $\bar{P}$) and the Lie derivative (for vector fields preserving $\bar{P}$) descend from $\mathcal{K}_P$ to δ_P, as in Proposition 23.41 in the real case. The connection on L and the partial connection on δ_P combine to give a partial connection on $L \otimes \delta_P$. A section s of $L \otimes \delta_P$ is said to be *polarized* if $\nabla_X s = 0$ for all vector fields X lying in $\bar{P}$.

Notation 23.49 *If β is any $2n$-form on N, let the expression*

$$\frac{\beta}{\lambda}$$

denote the unique function on N such that $\beta = (\beta/\lambda)\lambda$, where λ is the Liouville form in Definition 21.16.

Unlike the canonical bundle in the real case, the canonical bundle in the purely complex case carries a natural Hermitian structure.

Proposition 23.50 *If α is an $(n,0)$-form on N, then at each point the $2n$-form*

$$(-1)^{n(n-1)/2}(-i)^n \, \bar{\alpha} \wedge \alpha$$

is a non-negative multiple of the Liouville form λ. There is then a unique Hermitian structure on δ_P with the property that for each section s of δ_P we have

$$|s|^2 = \left(\frac{(-1)^{n(n-1)/2}(-i)^n}{2^n} \frac{\overline{(s \otimes s)} \wedge (s \otimes s)}{\lambda} \right)^{1/2} . \tag{23.41}$$

The factor of 2^n in the denominator in (23.41) is inserted for convenience, to make certain formulas come out more nicely.

Proof. See Exercise 17. ∎

Since, by assumption, there is Hermitian structure on L, the above Hermitian structure on δ_P gives rise in a natural way to a Hermitian structure on $L \otimes \delta_P$.

Definition 23.51 *The **half-form Hilbert space** for a Kähler polarization P on N is the space of square-integrable polarized sections of $L \otimes \delta_P$.*

In the $\mathbb{C}^n$ case, using the canonical 1-form as our symplectic potential, elements of the half-form Hilbert space take the form

$$e^{-|\operatorname{Im} z|^2/(2\alpha\hbar)} F(\mathbf{z}) \otimes \sqrt{dz_1 \wedge \cdots \wedge dz_n}.$$

In this special case, the norm of the half-form factor $\sqrt{dz_1 \wedge \cdots \wedge dz_n}$ is constant and the half-form Hilbert space is still identifiable with the space in Conclusion 22.10. In the case of the unit disk, on the other hand, the presence of half-forms alters the inner product; see Exercise 16.

We now define quantized observables on the half-form Hilbert space, using the same formula as in the real case.

Definition 23.52 *If f is a function on N for which X_f preserves $\bar{P}$, let $Q(f)$ be the operator on the half-form Hilbert space of P given by*

$$Q(f)s = (Q_{\mathrm{pre}}(f)\mu) \otimes \nu - i\hbar\, \mu \otimes \mathcal{L}_{X_f}\nu,$$

where s is decomposed locally as $s = \mu \otimes \nu$; with μ being a section of L and ν a section of δ_P.

These operators satisfy $[Q(f), Q(g)]/(i\hbar) = Q(\{f, g\})$ on the space of smooth polarized sections of $L \otimes \delta_P$, with the proof of this result being identical to the proof of Theorem 23.46 in the real case. If f is real-valued and X_f preserves $\bar{P}$, then $Q(f)$ will be at least symmetric, assuming we can find a dense subspace of the half-form Hilbert space consisting of "nice" functions. (Finding dense subspaces is more difficult in the holomorphic case than in the real case.) A proof of this claim is sketched in Exercise 18.

Example 23.53 *Consider $\mathbb{R}^2 \cong T^*\mathbb{R}$ with the Kähler polarization P given by the global complex coordinate $z = (x - ip/(m\omega))$, for some positive number ω. Take δ_P to be trivial with trivializing section $\sqrt{dz}$. Consider also the harmonic oscillator Hamiltonian $H := (p^2 + (m\omega x)^2)/(2m)$. Then X_H preserves the P and the operator $Q(H)$ on the half-form Hilbert space has spectrum consisting of numbers of the form $(n + 1/2)\hbar\omega$, where $n = 0, 1, 2, \ldots$.*

In this example, ω is the frequency of the oscillator and not the canonical 2-form.

Proof. The calculation is the same as in the proof of Proposition 22.14, except for the addition of the Lie derivative term. A simple calculation shows that $\mathcal{L}_{X_H}(dz) = i\omega\, dz$, from which it follows that $\mathcal{L}_{X_H}\sqrt{dz} = (i\omega/2)\sqrt{dz}$. It is then easy to see that the set of elements of the form $e^{-m\omega|\operatorname{Im} z|^2/(2\hbar)} z^n \otimes \sqrt{dz}$ form an orthonormal basis of eigenvectors for $Q(H)$, with eigenvalues $(n + 1/2)\hbar\omega$. ∎

23.8   Pairing Maps

Pairing maps are designed to allow us to compare the results of quantizing with respect to two different polarizations. We consider mainly the case of two "transverse" real polarizations; the case of two complex polarizations or one real and one complex polarization can be treated with minor modifications.

Suppose that P and P' are two purely real polarizations and that the associated leaf spaces Ξ_1 and Ξ_2 are oriented manifolds. Suppose also that P and P' are *transverse* at each point $z \in N$, meaning that $P_z \cap P'_z = \{0\}$. If α and β are polarized sections of $\mathcal{K}_P$ and $\mathcal{K}_{P'}$, respectively, the transversality assumption is easily shown to imply that $\alpha \wedge \beta$ is a nowhere-vanishing $2n$-form on N. Thus, for any point $z \in N$, we can define a bilinear "pairing" from $\delta_{P,z} \times \delta_{P',z} \to \mathbb{R}$ by

$$(\nu_1, \nu_2) = \left(\frac{(\nu_1 \otimes \nu_1) \wedge (\nu_2 \otimes \nu_2)}{\lambda} \right)^{1/2}. \tag{23.42}$$

(Recall Notation 23.49.) We can extend this pairing to a pairing $\delta^{\mathbb{C}}_{P,z} \times \delta^{\mathbb{C}}_{P',z} \to \mathbb{C}$ that is conjugate linear in the first factor and linear in the second factor. Finally, we extend to a pairing of $(L_z \otimes \delta^{\mathbb{C}}_{P,z}) \times (L_z \otimes \delta^{\mathbb{C}}_{P',z}) \to \mathbb{C}$ by setting $(\mu_1 \otimes \nu_1, \mu_2 \otimes \nu_2)$ equal to $(\mu_1, \mu_2)(\nu_1, \nu_2)$, where (μ_1, μ_2) is computed with respect to the Hermitian structure on L.

Let $\mathbf{H}_1$ and $\mathbf{H}_2$ denote the half-form Hilbert spaces for P and P', respectively. Given $s_1 \in \mathbf{H}_1$ and $s_2 \in \mathbf{H}_2$, we define the *pairing* of s_1 and s_2 by

$$\langle s_1, s_2 \rangle_{P,P'} = c \int_N (s_1, s_2)\, \lambda,$$

provided that the integral is absolutely convergent. Here (s_1, s_2) is the *pointwise* pairing of s_1 and s_2 defined in the previous paragraph and c is a certain "universal" constant, depending only on $\hbar$ and the dimension of n, that can be chosen to make certain examples work out nicely. We now look for a *pairing map* $\Lambda_{P,P'} : \mathbf{H}_1 \to \mathbf{H}_2$ with the property that

$$\langle s_1, s_2 \rangle_{P,P'} = \langle \Lambda_{P,P'} s_1, s_2 \rangle_{\mathbf{H}_2}. \tag{23.43}$$

If the pairing is bounded (i.e., it satisfies $|\langle s_1, s_2 \rangle_{P,P'}| \leq C \|s_1\| \|s_2\|$ for some constant C), there is a unique bounded operator $\Lambda_{P,P'}$ satisfying (23.43). Even if the pairing is unbounded, we may be able to define $\Lambda_{P,P'}$ as an unbounded operator.

If we were optimistic, we might hope that the pairing map for any two transverse polarizations would be unitary, or at least a constant multiple of a unitary map. *If* this were the case, it would suggest that quantization is independent of the choice of polarization, in the sense that there would be a natural unitary map between the Hilbert spaces for two different polarizations. As it turns out, however, the typical pairing map is *not* a constant multiple of a unitary map. Nevertheless, there are certain special cases where the pairing map is unitary (up to a constant), including the case of translation-invariant polarizations on $\mathbb{R}^{2n}$. See also [20] for an example of a pairing map between a real and a complex polarization that is a constant multiple of a unitary map.

We compute just one very special case of the pairing map between two real polarizations.

Example 23.54 *Consider* $N = \mathbb{R}^2 \cong T^*\mathbb{R}$ *and take* L *to be trivial with connection 1-form* $\theta = p\,dx$. *Let* P *be the vertical polarization, spanned at each point by* $\partial/\partial p$, *and let* P' *be the horizontal polarization, spanned at each point by* $\partial/\partial x$. *Then elements* s_1 *of the half-form space for* P *have the form*

$$s_1(x, p) = \phi(x) \otimes \sqrt{dx} \tag{23.44}$$

and elements s_2 *of the half-form space for* P' *have the form*

$$s_2(x, p) = \psi(p)e^{ixp/\hbar} \otimes \sqrt{dp}, \tag{23.45}$$

where ϕ *and* ψ *are functions on* $\mathbb{R}$. *If* $c = 1$, *the pairing is computed as*

$$\langle s_1, s_2 \rangle_{P,P'} = -\int_{\mathbb{R}^2} \overline{\phi(x)}\psi(p)e^{ixp/\hbar}\,dx\,dp. \tag{23.46}$$

If s_1 *has the form (23.44), then* $\Lambda_{P,P'}(s_1)$ *has the form (23.45), where*

$$\psi(p) = -\int_{\mathbb{R}} \phi(x)e^{-ixp/\hbar}\,dx.$$

Thus, $\Lambda_{P,P'}$ *is a scaled version of the Fourier transform and is, in particular, a constant multiple of a unitary map.*

The pairing should be defined initially on some dense subspace of the Hilbert spaces, such as the subspaces where ϕ and ψ are Schwartz functions. The pairing map can also be defined initially on the Schwartz space, recognized as being unitary (up to a constant), and then extended by continuity to all of $\mathbf{H}_1$. Once the pairing map is extended to $\mathbf{H}_1$, the pairing itself can be defined for all $s_1 \in \mathbf{H}_1$ and $s_2 \in \mathbf{H}_2$ by taking (23.43) as the definition of $\langle s_1, s_2 \rangle_{P,P'}$. Even though it is possible, as just described, to extend the pairing to all of $\mathbf{H}_1 \times \mathbf{H}_2$, the integral in (23.46) is not always absolutely convergent.

Proof. The forms (23.44) and (23.45) are obtained by a simple modification of the argument in the proof of Proposition 22.8. We can compute that the pointwise pairing of $\sqrt{dx}$ and $\sqrt{dp}$ is -1, which gives the indicated form of the pairing in (23.46). The pairing may be rewritten as

$$\int_{\mathbb{R}} \overline{\int_{\mathbb{R}} \phi(x)e^{-ixp/\hbar}\,dx}\,\psi(p)\,dp,$$

which gives the indicated form of the pairing map. ∎

23.9   Exercises

1. Let L be a line bundle with connection ∇ over N. Let s be a section of L and let X_1 and X_2 be two vector fields on N such that $X_1(z) = X_2(z)$ for some fixed point $z \in N$. Show that

$$\nabla_{X_1}(s)(z) = \nabla_{X_2}(s)(z).$$

   Hint: Use the assumption that $\nabla_{fX} = f\nabla_X$.

2. Let L be a Hermitian line bundle with Hermitian connection ∇ and let s_0 be a locally defined section of L such that $(s_0, s_0) = 1$. Given a vector field X, let $\theta(X)$ be the unique function such that

$$\nabla_X s_0 = -i\theta(X)s_0.$$

   Show that $\theta(X)$ is real valued.

   Hint: Use the Hermitian property of the connection.

3. Consider the definition of the curvature 2-form $\omega(X, Y)$ in Definition 23.4.

   (a) Show that the expression for ω is C^∞-linear in each of the variables X, Y, and s. That is to say, show that for all smooth functions f, we have $\omega(fX, Y)s = f\omega(X, Y)s$, and similarly for the variables Y and s.

   (b) Show that the value of $\omega(X, Y)s$ at a point z depends only on the values of X, Y, and s at the point z.

   (c) Show that the value of $\omega(X, Y)$ at a point z does not depend on the value of s at z, provided that $s(z) \neq 0$.

4. Consider the symplectic form $\omega = dp \wedge dx$ on $\mathbb{R}^2$. Define a purely complex polarization on $\mathbb{R}^2$ by taking P_z to be the span of the vector $\partial/\partial z$ in (22.9), for some fixed $\alpha > 0$. Show that P is a Kähler polarization.

5. Let P be the polarization on $\mathbb{R}^2$ in Exercise 4. Show that the function $\kappa(x, p) := \alpha p^2$ is a Kähler potential for P.

6. Suppose that ω is a J-invariant 2-form on a complex manifold N. Show that ω is a $(1, 1)$-form. (Recall the definitions preceding Lemma 23.34.)

   Hint: Write $\omega = \omega^1 + \omega^2$, where ω^1 is a $(1, 1)$-form and ω^2 is a sum of a $(2, 0)$-form and a $(0, 2)$-form. Show that

$$\omega^2(JX, JY) = -\omega^2(X, Y)$$

   for all tangent vectors X and Y.

7. Suppose that κ is a smooth, real-valued function on a complex manifold N. Show that the 2-form $i\partial\bar\partial\kappa$ is a real-valued 2-form.

8. In Example 23.30, verify that θ is a symplectic potential for ω, and compute $\theta(\partial/\partial\bar{z})$, where, with $z = x - iy$, we have $\partial/\partial\bar{z} = (\partial/\partial x - i\partial/\partial y)/2$. Then verify that $s_0(z) := (1 - |z|^2)^{1/\hbar}$ satisfies $\nabla_{\partial/\partial\bar{z}}s_0 = 0$ and thus constitutes a global trivializing holomorphic section.

9. Consider the situation in Example 23.29. Show that the canonical bundle for P is trivial, with trivializing section dx. Let δ_P be the (nontrivial) bundle described in the paragraph preceding Definition 23.42. Since the tensor product of any real line bundle with itself is trivial, $\delta_P \otimes \delta_P$ is isomorphic to $\mathcal{K}_P$. Let $\sqrt{dx}$ denote a *discontinuous* section defined over the set $0 < \phi < 2\pi$ such that $\sqrt{dx} \otimes \sqrt{dx} = dx$. Show that $\nabla_X(dx) = 0$ and $\nabla_X\sqrt{dx} = 0$ for every vector field lying in P. Now show that the Bohr–Sommerfeld leaves (in the half-form sense, for this choice of δ_P) are the sets of the form $\{x\} \times S^1$, where $x/\hbar = n + 1/2$ for some integer n.

10. Let b be a smooth, real-valued function on $\mathbb{R}$ and let c be a real constant. Show that an operator of the form

$$\psi \mapsto -i\hbar\left(b(x)\psi'(x) + cb'(x)\psi(x)\right)$$

is symmetric on $C_c^\infty(\mathbb{R}) \subset L^2(\mathbb{R})$ if and only if $c = 1/2$.

11. Let P be a real polarization and let f be a smooth polarized function on N, that is, one for which derivatives in the direction of P are zero. Show that $Q(f)$ acts on the half-form Hilbert space simply as multiplication by f. (Compare Proposition 23.25 in the case without half-forms.)

    Hint: Show that $\mathcal{L}_{X_f}\alpha = 0$ whenever α is a polarized section of $\mathcal{K}_P$.

12. Using the identities $\mathcal{L}_{[X,Y]} = [\mathcal{L}_X, \mathcal{L}_Y]$ and $X_{\{f,g\}} = [X_f, X_g]$, verify the identity (23.36).

13. Prove that if P is a real polarization on N, it is possible to choose a symplectic potential θ locally in such a way that θ is zero on P.

    Hint: Use functions $f_{\mathbf{x}}$ as in the proof of Proposition 23.26.

14. Suppose that P is a purely real polarization on N and θ is a local symplectic potential that vanishes on P. Suppose also that f is a real-valued function for which X_f preserves P. Show that the function $-\theta(X_f) - f$ is constant along the leaves of P.

    Hint: If X is a vector field lying in P, use (21.6) to show that $X(\theta(X_f)) = d\theta(X, X_f)$.

15. Suppose that β is a nowhere vanishing n-form on an oriented manifold Ξ, that X is a real vector field on Ξ, and that ϕ and ψ are smooth, compactly supported functions on Ξ. Verify the following formula for "integration by parts":

$$\int_\Xi (X\phi)\psi\,\beta = -\int_\Xi \phi(X\psi)\,\beta - \int_\Xi \phi\psi(\mathrm{div}_\beta X)\,\beta,$$

where $\mathrm{div}_\beta X$ is the function such that $\mathcal{L}_X\beta = (\mathrm{div}_\beta X)\beta$.

Hint: If Φ_t is the flow generated by X, then for all sufficiently small t, $\Phi_t(x)$ is defined for all x in the support of $\phi\psi$ and the integral of $(\Phi_t)^*(\phi\psi\beta)$ over Ξ is independent of t.

16. Let the notation be as in Exercise 8. Then the canonical bundle for P is trivial, with trivializing section dz. Take δ_P to be trivial, with trivializing section $\sqrt{dz}$. Show that every polarized section s of $L\otimes\delta_P$ is of the form

$$s = F(z)s_0(z) \otimes \sqrt{dz},$$

where F is holomorphic. Show that the norm of such a section is, up to a constant, the L^2 norm of F with respect to a measure of the form $(1 - |z|^2)^\nu$, but that the value of ν is not the same as when half-forms are not included.

17. Let P be a Kähler polarization on N, let $z_1,\ldots,z_n$ be holomorphic local coordinates on N, and let A be the matrix given by

$$A_{jk} = \omega\left(\frac{\partial}{\partial \bar{z}_j}, \frac{\partial}{\partial z_k}\right).$$

   (a) Show that the matrix iA is positive definite.
   (b) Show that $\omega = A_{jk}\,d\bar{z}_j \wedge dz_k$.
   (c) Show that the quantity $\omega^{\otimes n}/n!$ may be computed as

   $$\det(iA)(-1)^{n(n-1)/2}(-i)^n d\bar{z}_1 \wedge \cdots \wedge d\bar{z}_n \wedge dz_1 \wedge \cdots \wedge dz_n.$$

   (d) Verify Proposition 23.50.

18. Let P be a Kähler polarization on N, let δ_P be a fixed square root of $\mathcal{K}_P$, and let f be a smooth, real-valued function such that X_f preserves $\bar{P}$. Throughout this problem, if s_1 and s_2 are local sections of a line bundle, with s_2 nonvanishing, s_1/s_2 will denote the unique function such that $s_1 = (s_1/s_2)s_2$.

   (a) Show that for any continuous compactly supported function ψ on N, we have

   $$\int_N X_f(\psi)\,\lambda = 0.$$

Hint: Use Liouville's theorem.

Note: The same result holds if ψ is not compactly supported but is "sufficiently nice."

(b) If ν is a local nonvanishing section of δ_P, show that

$$\frac{\mathcal{L}_{X_f}\nu}{\nu} = \frac{1}{2}\frac{\mathcal{L}_{X_f}(\nu \otimes \nu)}{\nu \otimes \nu}.$$

(c) If α is any $2n$-form on N, show that

$$\frac{\mathcal{L}_{X_f}\alpha}{\lambda} = X_f\left(\frac{\alpha}{\lambda}\right).$$

(d) Suppose s_1 and s_2 are polarized sections of $L \otimes \delta_P$, decomposed locally as $s_j = \mu_j \otimes \nu_j$, $j = 1, 2$. Show that

$$iX_f(s_1, s_2) = (i(\nabla_{X_f}\mu_1) \otimes \nu_1, s_2) + (i\mu_1 \otimes (\mathcal{L}_{X_f}\nu_1) \otimes s_2) \\ + (s_1, i(\nabla_{X_f}\mu_2) \otimes \nu_2) + (s_1, i\mu_2 \otimes (\mathcal{L}_{X_f}\nu_2)),$$

where $(\cdot, \cdot)$ is computed with respect to the Hermitian structure on $L \otimes \delta_P$ described in Sect. 23.7.

Hint: Use the identity $\mathcal{L}_{X_f}(\alpha \wedge \beta) = (\mathcal{L}_{X_f}\alpha) \wedge \beta + \alpha \wedge (\mathcal{L}_{X_f}\beta)$.

(e) Suppose s_1 and s_2 are polarized sections of $L \otimes \delta_P$ belonging to the domain of $Q(f)$ and such that (s_1, s_2) is "sufficiently nice." Show that

$$\langle s_1, Q(f)s_2 \rangle = \langle Q(f)s_1, s_2 \rangle.$$

Appendix A
Review of Basic Material

A.1 Tensor Products of Vector Spaces

Given two vector spaces V_1 and V_2 over $\mathbb{C}$, the tensor product is a new vector space $V_1 \otimes V_2$, together with a bilinear "product" map $\otimes : V_1 \times V_2 \to V_1 \otimes V_2$. If V_1 and V_2 are finite dimensional with bases $\{u_j\}$ and $\{v_k\}$, then $V_1 \otimes V_2$ is finite dimensional with $\{u_j \otimes v_k\}$ forming a basis for $V_1 \otimes V_2$. In the finite-dimensional case, we could simply *define* the tensor product by this basis property, but then we would have to worry about whether the construction is basis independent. Instead, we define $V_1 \otimes V_2$ by a "universal property."

Definition A.1 *Suppose V_1 and V_2 are vector spaces over a field $\mathbb{F}$. Then a **tensor product** of V_1 and V_2 is a vector space W over $\mathbb{F}$ together with a bilinear map $T : V_1 \times V_2 \to W$ having the following "universal property": If U is any vector space over $\mathbb{F}$ and $\Phi : V_1 \times V_2 \to U$ is a bilinear map, then there exists a unique linear map $\tilde{\Phi} : W \to U$ such that the following diagram commutes:*

$$
\begin{array}{ccc}
V_1 \times V_2 & \xrightarrow{T} & W \\
\Phi \downarrow & \swarrow \tilde{\Phi} & \\
U & &
\end{array} .
$$

Proposition A.2 *For any two vector spaces V_1 and V_2, a tensor product of V_1 and V_2 exists and is unique up to "canonical isomorphism." That is, for two tensor products (W_1, T_1) and (W_2, T_2), there is a unique invertible linear map $\Psi : W_1 \to W_2$ such that $T_2 = \Psi \circ T_1$.*

B.C. Hall, *Quantum Theory for Mathematicians*, Graduate Texts in Mathematics 267, DOI 10.1007/978-1-4614-7116-5, © Springer Science+Business Media New York 2013

In light of the uniqueness result, we may speak of "the" tensor product of V_1 and V_2. We choose any one tensor product and we denote it by $V_1 \otimes V_2$. We also denote the linear map $T : V_1 \times V_2 \to V_1 \otimes V_2$ as $(u, v) \mapsto u \otimes v$. In this notation, the universal property reads as follows: Given any *bilinear* map Φ of $V_1 \times V_2$ into a vector space U, there exists a unique *linear* map $\tilde{\Phi} : V_1 \otimes V_2 \to U$ such that

$$\tilde{\Phi}(u \otimes v) = \Phi(u, v).$$

Proposition A.3 *If V_1 and V_2 are finite-dimensional vector spaces with bases $\{u_j\}_{j=1}^{n_1}$ and $\{v_k\}_{k=1}^{n_2}$, then $V_1 \otimes V_2$ is finite dimensional and the set of elements of the form $u_j \otimes v_k$, $1 \le j \le n_1$, $1 \le k \le n_2$, forms a basis for $V_1 \otimes V_2$. In particular,*

$$\dim(V_1 \otimes V_2) = (\dim V_1)(\dim V_2).$$

It should be emphasized that, in general, not every element of $V_1 \otimes V_2$ is of the form $u \otimes v$ with $u \in V_1$ and $v \in V_2$. All we can say is that each element of $V_1 \otimes V_2$ can be decomposed as a *linear combination* of elements of the form $u \otimes v$. This decomposition, furthermore, is far from canonical; even in the finite-dimensional case, it depends on a choice of bases for V_1 and V_2. Nevertheless, the universal property of the tensor product tells us that we can define linear maps from $V_1 \otimes V_2$ to any vector space U, simply by defining them on elements of the form $u \otimes v$. Provided that $\Phi(u, v)$ is bilinear in u and v, the universal property tells us that there is a unique linear map $\tilde{\Phi}$ on $V_1 \otimes V_2$ such that on element of the form $u \otimes v$, $\tilde{\Phi}$ is equal to $\Phi(u, v)$. A representative application of the universal property is in the following result.

Proposition A.4 *If $A \in \mathrm{End}(V_1)$ and $B \in \mathrm{End}(V_2)$, there exists a unique linear map $A \otimes B : V_1 \otimes V_2 \to V_1 \otimes V_2$ such that*

$$(A \otimes B)(u \otimes v) = (Au) \otimes (Bv).$$

For $A_1, A_2 \in \mathrm{End}(V_1)$ and $B_1, B_2 \in \mathrm{End}(V_2)$, we have

$$(A_1 \otimes B_1)(A_2 \otimes B_2) = (A_1 A_2) \otimes (B_1 B_2).$$

To construct $A \otimes B$, we apply the universal property with $U = V_1 \otimes V_2$ and $\Phi(u, v) = (Au) \otimes (Bv)$. Since A and B are linear and $\otimes$ is bilinear, Φ is bilinear. The linear map $\tilde{\Phi} : V_1 \otimes V_2 \to V_1 \otimes V_2$ is then the map that we denote $A \otimes B$.

The tensor product, as we have defined it in this section, applies to all vector spaces, whether finite dimensional or infinite dimensional. The construction, however, is purely algebraic; if there is a topology on V_1 and V_2, the tensor product takes no account of that topology. In the Hilbert space setting, then, we will have to refine the notion of the tensor product so that the tensor product of two Hilbert spaces will again be a Hilbert space. See Sect. A.4.5.

A.2   Measure Theory

It is assumed that the reader is familiar with the basic notions of measure theory, including the concepts of σ-algebras, measures, measurable functions, and the Lebesgue integral. A triple (X, Ω, μ), consisting of a set X, a σ-algebra Ω of subsets of X, and a (non-negative) measure μ on Ω is called a *measure space*. A measurable function $\psi : X \to \mathbb{C}$ is said to be *integrable* if $\int_X |\psi| \, d\mu < \infty$. The σ-algebra *generated by* any collection of subsets of a set X is the smallest σ-algebra of subsets of X containing that collection.

We assume those parts of measure theory that are entirely standard: the monotone convergence and dominated convergence theorems, L^p spaces, and Fubini's theorem. We briefly review a few other topics that might not be as familiar.

A measure μ on a measurable space (X, μ) is said to be σ-*finite* if X can be written as a countable union of measurable sets of finite measure.

Definition A.5 *Suppose μ and ν are two σ-finite measures on a measure space (X, Ω). Then we say that μ is **absolutely continuous** with respect to ν if for all $E \in \Omega$, if $\nu(E) = 0$ then $\mu(E) = 0$. We say that μ and ν are **equivalent** if each measure is absolutely continuous with respect to the other.*

Theorem A.6 (Radon–Nikodym) *Suppose μ and ν are two σ-finite measures on a measure space (X, Ω) and that μ is absolutely continuous with respect to ν. Then there exists a non-negative, measurable function ρ on X such that*

$$\mu(E) = \int_E \rho \, d\nu,$$

*for all $E \in \Omega$. The function ρ is called the **density** of μ with respect to ν.*

Definition A.7 *A collection $\mathcal{M}$ of subsets of a set X is called a **monotone class** if $\mathcal{M}$ is closed under countable increasing unions and countable decreasing intersections.*

A countable increasing union means the union of a sequence E_j of sets where E_j is contained in E_{j+1} for each j, with a similar definition for countable decreasing intersections.

Theorem A.8 (Monotone Class Lemma) *Suppose $\mathcal{M}$ is a monotone class of subsets of a set X and suppose $\mathcal{M}$ contains an algebra $\mathcal{A}$ of subsets of X. Then $\mathcal{M}$ contains the σ-algebra generated by $\mathcal{A}$.*

Corollary A.9 *Suppose μ and ν are two finite measures on a measure space (X, Ω). Suppose μ and ν agree on an algebra $\mathcal{A} \subset \Omega$. Then μ and ν agree on the σ-algebra generated by $\mathcal{A}$.*

Note that in general, the collection of sets on which two measures agree is *not* a σ-algebra, nor even an algebra.

Theorem A.10 *Suppose μ is a measure on the Borel σ-algebra in a locally compact, separable metric space X. Suppose also that $\mu(K) < \infty$ for each compact subset K of X. Then the space of continuous functions of compact support on X is dense in $L^p(X, \mu)$, for all p with $1 \leq p < \infty$.*

A word of clarification is in order here. If ψ is a continuous function on X with compact support, then $\int_X |\psi|^p \, d\mu$ is finite, since ψ is bounded and μ is finite on compact sets. Thus, we can define a map from $C_c(X)$ into $L^p(X, \mu)$ by mapping a continuous function ψ of compact support to the equivalence class $[\psi]$. The theorem is asserting, more precisely, that the *image* of $C_c(X)$ under this map is dense in $L^p(X, \mu)$. It should be noted, however, that the map $\psi \mapsto [\psi]$ need not be injective. After all, if there is a nonempty open set U inside X with $\mu(U) = 0$, then for any ψ with support contained in U, the equivalence class $[\psi]$ will be the zero element of $L^p(X, \mu)$. Nevertheless, we will allow ourselves a small abuse of terminology and say that $C_c(X)$ is dense in $L^p(X, \mu)$.

A.3   Elementary Functional Analysis

In this section, we briefly review some of the results from elementary functional analysis that we make use of the text. Most of these results can be found in the book of Rudin [32].

A.3.1   The Stone–Weierstrass Theorem

The Weierstrass theorem states that every continuous, real-valued function on an interval can be uniformly approximated by polynomials. A substantial generalization of this was obtained by Stone. If X is a compact metric space, let $\mathcal{C}(X; \mathbb{R})$ and $\mathcal{C}(X; \mathbb{C})$ denote the space of continuous real- and complex-valued continuous functions, respectively. A subset $\mathcal{A}$ of $\mathcal{C}(X; \mathbb{F})$ is called an *algebra* if it is closed under pointwise addition, pointwise multiplication, and multiplication by elements of $\mathbb{F}$, where $\mathbb{F} = \mathbb{R}$ or $\mathbb{C}$. An algebra $\mathcal{A}$ is said to *separate points* if for any two distinct points x and y in X, there exists $f \in \mathcal{A}$ such that $f(x) \neq f(y)$. We use on $\mathcal{C}(X; \mathbb{F})$ the *supremum norm*, given by

$$\|f\|_{\mathrm{sup}} := \sup_{x \in X} |f(x)|,$$

and $\mathcal{C}(X, \mathbb{F})$ is complete with respect to the associated distance function, $d(f, g) = \|f - g\|_{\mathrm{sup}}$.

Theorem A.11 (Stone–Weierstrass, Real Version) *Let X be a compact metric space and let $\mathcal{A}$ be an algebra in $\mathcal{C}(X; \mathbb{R})$. If $\mathcal{A}$ contains the constant functions and separates points, then $\mathcal{A}$ is dense in $\mathcal{C}(X; \mathbb{R})$ with respect to the supremum norm.*

Theorem A.12 (Stone–Weierstrass, Complex Version) *Let X be a compact metric space and let $\mathcal{A}$ be an algebra in $C(X; \mathbb{C})$. If $\mathcal{A}$ contains the constant functions, separates points, and is closed under complex conjugation, then $\mathcal{A}$ is dense in $C(X; \mathbb{C})$ with respect to the supremum norm.*

A consequence of the complex version of the Stone–Weierstrass theorem is the following: If K is a compact subset of $\mathbb{C}$, then every continuous, complex-valued function on K can be uniformly approximated by polynomials in z and $\bar{z}$.

A.3.2   The Fourier Transform

We now describe the Fourier transform on $\mathbb{R}^n$, in various forms.

Definition A.13 *For any $\psi \in L^1(\mathbb{R}^n)$, define the **Fourier transform** of ψ to be the function $\hat{\psi}$ on $\mathbb{R}^n$ given by*

$$\hat{\psi}(\mathbf{k}) = (2\pi)^{-n/2} \int_{-\infty}^{\infty} e^{-i\mathbf{k}\cdot\mathbf{x}} \psi(\mathbf{x}) \, d\mathbf{x}.$$

Proposition A.14 *For any $\psi \in L^1(\mathbb{R}^n)$, the Fourier transform $\hat{\psi}$ of ψ has the following properties: (1) $\left|\hat{\psi}(\mathbf{k})\right| \leq (2\pi)^{-n/2} \|\psi\|_{L^1}$, (2) $\hat{\psi}$ is continuous, and (3) $\hat{\psi}(\mathbf{k})$ tends to zero as $|\mathbf{k}|$ tends to ∞.*

The bound on $\hat{\psi}$ is obvious and the continuity of $\hat{\psi}$ follows from dominated convergence. To show that $\hat{\psi}$ tends to zero at infinity, we first establish this on a dense subspace of $L^1(\mathbb{R}^n)$ (e.g., the Schwartz space; see below) and then take uniform limits.

Definition A.15 *The **Schwartz space** $\mathcal{S}(\mathbb{R}^n)$ is the space of all C^∞ functions ψ on $\mathbb{R}^n$ such that*

$$\lim_{x \to \pm\infty} \left|\mathbf{x}^{\mathbf{j}} \partial^{\mathbf{k}} \psi(\mathbf{x})\right| = 0$$

for all n-tuples of non-negative integers $\mathbf{j}$ and $\mathbf{k}$. Here if $\mathbf{j} = (j_1, \ldots, j_n)$ then $\mathbf{x}^{\mathbf{j}} = x_1^{j_1} \cdots x_n^{j_n}$ and

$$\partial^{\mathbf{j}} = \left(\frac{\partial}{\partial x_1}\right)^{j_1} \cdots \left(\frac{\partial}{\partial x_n}\right)^{j_n}.$$

*An element of the Schwartz space is called a **Schwartz function**.*

Proposition A.16 *If ψ belongs to $\mathcal{S}(\mathbb{R}^n)$, then $\hat{\psi}$ also belongs to $\mathcal{S}(\mathbb{R}^n)$.*

The proof of this result hinges on the behavior of the Fourier transform under differentiation and under multiplication by x, results which are of interest in their on right.

Proposition A.17 *If ψ is a Schwartz function, the following properties hold*

1. *We have*

$$\widehat{\frac{\partial \psi}{\partial x_j}}(\mathbf{k}) = ik_j \hat{\psi}(\mathbf{k}).\tag{A.1}$$

2. *The function $\hat{\psi}$ is differentiable at every point and the Fourier transform of the function $x_j \psi(x)$ is given by*

$$\widehat{x_j \psi}(\mathbf{k}) = i\frac{\partial}{\partial k_j}\hat{\psi}(\mathbf{k}).\tag{A.2}$$

The first point is proved by integration by parts and the second by differentiation under the integral in the definition of $\hat{\psi}$.

Theorem A.18 (Fourier Inversion and Plancherel Formula, I) *The Fourier transform on $\mathcal{S}(\mathbb{R}^n)$ has the following properties.*

1. *The Fourier transform maps the Schwartz space onto the Schwartz space.*

2. *For all $\psi \in \mathcal{S}(\mathbb{R}^n)$, the function ψ can be recovered from its Fourier transform by the **Fourier inversion formula**:*

$$\psi(\mathbf{x}) = (2\pi)^{-n/2} \int_{-\infty}^{\infty} e^{i\mathbf{k}\cdot\mathbf{x}} \hat{\psi}(k) \, d\mathbf{k}.$$

3. *For all $\psi \in \mathcal{S}(\mathbb{R}^n)$, we have the **Plancherel theorem**:*

$$\int_{\mathbb{R}^n} |\psi(\mathbf{x})|^2 \, d\mathbf{x} = \int_{\mathbb{R}^n} |\hat{\psi}(\mathbf{k})|^2 \, d\mathbf{k}.$$

Since the Schwartz space is dense in $L^2(\mathbb{R}^n)$, the BLT theorem and Theorem A.18 imply that the Fourier transform extends uniquely to an isometric map of $L^2(\mathbb{R}^n)$ onto $L^2(\mathbb{R}^n)$.

Theorem A.19 (Fourier Inversion and Plancherel Theorem, II) *The Fourier transform extends to an isometric map $\mathcal{F}$ of $L^2(\mathbb{R}^n)$ onto $L^2(\mathbb{R}^n)$. This map may be computed as*

$$\mathcal{F}(\psi)(\mathbf{k}) = (2\pi)^{-n/2} \lim_{A\to\infty} \int_{|\mathbf{x}|\le A} e^{-i\mathbf{k}\cdot\mathbf{x}} \psi(\mathbf{x}) \, d\mathbf{x},\tag{A.3}$$

where the limit is in the norm topology of $L^2(\mathbb{R}^n)$. The inverse map $\mathcal{F}^{-1}$ may be computed as

$$\left(\mathcal{F}^{-1}f\right)(\mathbf{x}) = (2\pi)^{-n/2} \lim_{A\to\infty} \int_{|\mathbf{x}|\le A} e^{i\mathbf{k}\cdot\mathbf{x}} f(\mathbf{k}) \, d\mathbf{k}$$

If ψ belongs to $L^1(\mathbb{R}^n) \cap L^2(\mathbb{R}^n)$, then by dominated convergence, the limit in coincides with the L^1 Fourier transform in Definition A.13.

Definition A.20 *For two measurable functions ϕ and ψ, define the **convolution** $\phi * \psi$ of ϕ and ψ by the formula*

$$(\phi * \psi)(\mathbf{x}) = \int_{\mathbb{R}^n} \phi(\mathbf{x} - \mathbf{y})\psi(\mathbf{y}) \, d\mathbf{y},$$

provided that the integral is absolutely convergent for all $\mathbf{x}$.

Proposition A.21 *Suppose that ϕ and ψ belong to $L^1(\mathbb{R}^n) \cap L^2(\mathbb{R}^n)$. Then $\phi * \psi$ is defined and belongs to $L^1(\mathbb{R}^n) \cap L^2(\mathbb{R}^n)$ and we have*

$$(2\pi)^{-n/2}\mathcal{F}(\phi * \psi) = \mathcal{F}(\phi)\mathcal{F}(\psi).$$

This result is proved by plugging $\phi * \psi$ into the definition of the Fourier transform, writing $e^{-i\mathbf{k}\cdot\mathbf{x}}$ as $e^{-i\mathbf{k}\cdot\mathbf{y}}e^{-i\mathbf{k}\cdot(\mathbf{x}-\mathbf{y})}$, and using Fubini's theorem. We will have occasion to use the following Gaussian integral.

Proposition A.22 *For all $a > 0$ and $b \in \mathbb{C}$, we have*

$$\frac{1}{\sqrt{2\pi}} \int_{-\infty}^{\infty} e^{-x^2/(2a)} e^{bx} \, dx = \sqrt{a}e^{ab^2/2}.$$

Taking $b = ik$ in the last part of the proposition gives us the Fourier transform of the Gaussian function $e^{-x^2/(2a)}$. Taking $b = 0$ allows us to determine the proper normalization of the Gaussian probability density.

A.3.3 Distributions

In this section we give a brief account of the theory of distributions—what physicists call "generalized functions"—including the notion of "derivative in the distribution sense."

The idea is that we study functions by studying their integral against some class of very nice "test functions." Consider, for example, a locally integrable function f and consider integrals of the form

$$\int_{\mathbb{R}^n} \chi(\mathbf{x})f(\mathbf{x}) \, d\mathbf{x}, \tag{A.4}$$

where χ belongs to $C_c^\infty(\mathbb{R}^n)$, the space of smooth, compactly supported functions. We might think, for example, that χ is positive, has integral equal to 1, and is supported near some point $\mathbf{a} \in \mathbb{R}^n$. In that case, the integral (A.4) is an approximation to the value of f at $\mathbf{a}$, what physicists describe as a "smeared out" version of $f(\mathbf{a})$.

Proposition A.23 *Suppose f_1 and f_2 are locally integrable functions on $\mathbb{R}^n$. If*

$$\int_{\mathbb{R}^n} \chi(\mathbf{x}) f_1(\mathbf{x}) \; d\mathbf{x} = \int_{\mathbb{R}^n} \chi(\mathbf{x}) f_2(\mathbf{x}) \; d\mathbf{x}$$

for all $\chi \in C_c^\infty(\mathbb{R}^n)$, then $f_1(\mathbf{x}) = f_2(\mathbf{x})$ for almost every $\mathbf{x}$.

The idea now is that we allow objects that do not have values at points, but for which something like (A.4) makes sense. Mathematically, we think of (A.4) as a linear functional on $C_c^\infty(\mathbb{R}^n)$.

Definition A.24 *A sequence $\chi_m \in C_c^\infty(\mathbb{R}^n)$ is said to **converge** to $\chi \in C_c^\infty(\mathbb{R}^n)$ if (1) there exists a single compact set K containing the support of all the χ_n's, (2) χ_m converges uniformly to χ, and (3) each derivative of χ_m converges uniformly to the corresponding derivative of χ.*

Definition A.25 *A **distribution** on $\mathbb{R}^n$ is a linear map $T : C_c^\infty(\mathbb{R}^n) \to \mathbb{C}$ having the following continuity property: If χ_m converges to χ in the sense of Definition A.24, $T(\chi_m)$ converges to $T(\chi)$.*

The continuity condition on T should be regarded as a technicality, in that any functional that is well defined and linear on all of $C_c^\infty(\mathbb{R}^n)$ and is obtained in a reasonably constructive fashion will satisfy this property.

Example A.26 *The Dirac δ-"function" is the distribution δ defined by*

$$\delta(\chi) = \chi(0).$$

Definition A.27 *If T is a distribution and f is a locally integrable function, the expression "T is equal to f" or "T is given by f" means that*

$$T(\chi) = \int_{\mathbb{R}^n} \chi(\mathbf{x}) f(\mathbf{x}) \; d\mathbf{x}$$

for all $\chi \in C_c^\infty(\mathbb{R}^n)$.

Definition A.28 *If T is a distribution, define the distribution $\partial T / \partial x_j$ by the formula*

$$\frac{\partial T}{\partial x_j}(\chi) = -T\left(\frac{\partial \chi}{\partial x_j}\right).$$

It is easy to verify that if T has the continuity property in Definition A.25, then so does $\partial T / \partial x_j$. Furthermore, if T is given by a continuously differentiable function, then the derivative of T is in the distribution sense coincides with the derivative of T in the classical sense, as can easily be shown using integration by parts. If T is a distribution, we may define ΔT by repeated applications of Definition A.28, with the result that

$$(\Delta T)(\chi) = T(\Delta \chi).$$

Proposition A.29 *If ϕ and ψ are L^2 functions, the equation $\partial\psi/\partial x_j = \phi$ holds in the distribution sense if and only if*

$$-\left\langle \frac{\partial\chi}{\partial x_j}, \psi \right\rangle = \langle \chi, \phi \rangle$$

for all $\chi \in C_c^\infty(\mathbb{R}^n)$. Similarly, the equation $\Delta\psi = \phi$ holds in the distribution sense if and only if

$$\langle \Delta\chi, \psi \rangle = \langle \chi, \phi \rangle$$

for all $\chi \in C_c^\infty(\mathbb{R}^n)$.

Proposition A.30 *If T is a distribution on $\mathbb{R}$ and dT/dx is the zero distribution, then T is a constant, meaning that there is some constant c such that*

$$T(\chi) = \int_{-\infty}^{\infty} \chi(x)c \, dx. \tag{A.5}$$

Suppose, in particular, that if T is given by a locally integrable function f, and the derivative of T is zero. Then Proposition A.30 tells us that for some constant c, we have $\int_{-\infty}^{\infty} \chi(x)(f(x) - c) \, dx = 0$ for all $\chi \in C_c^\infty(\mathbb{R})$. Then Proposition A.23 tells us that $f(x) = c$ almost everywhere. This means that if the derivative of f is zero, even in the weak (or distributional) sense, then f must be constant.

A.3.4   Banach Spaces

In this section, we define Banach spaces and describe some of their elementary properties.

Definition A.31 *A **norm** on a vector space V over $\mathbb{F}$ ($\mathbb{F} = \mathbb{R}$ or $\mathbb{C}$) is a map from V into $\mathbb{R}$, denoted $\psi \mapsto \|\psi\|$, with the following properties.*

1. *For all $\psi \subset V$, $\|\psi\| \geq 0$, with equality if and only if $\psi = 0$.*

2. *For all $\psi \in V$ and $c \in \mathbb{F}$, we have $\|c\psi\| = |c| \, \|\psi\|$.*

3. *For all $\phi, \psi \in V$, we have $\|\phi + \psi\| \leq \|\phi\| + \|\psi\|$.*

If $\|\cdot\|$ is a norm on V, then we can define a distance function d on V by setting $d(\phi, \psi) = \|\psi - \phi\|$.

Definition A.32 *A normed vector space is said to be a **Banach space** if it is complete with respect to the associated distance function. A Banach space is said to be **separable** if contains a countable dense subset.*

One important class of examples of Banach spaces are the L^p spaces.

Definition A.33 *An infinite series, $\sum_{n=1}^{\infty} \psi_n$, with values in normed space V, is said to **converge** if there exists some $L \in V$ such that*

$$\lim_{N \to \infty} \|S_n - L\| = 0,$$

where $S_N = \sum_{n=1}^{N} \psi_n$.

Proposition A.34 *If V is a Banach space, then absolute convergence implies convergence in V. That is, if*

$$\sum_{n=1}^{\infty} \|\psi_n\| < \infty,$$

then $\sum_{n=1}^{\infty} \psi_n$ converges in V.

Definition A.35 *If V_1 and V_2 are normed spaces, a linear map $T : V_1 \to V_2$ is **bounded** if*

$$\sup_{\psi \in V_1 \setminus \{0\}} \frac{\|T\psi\|}{\|\psi\|} < \infty. \tag{A.6}$$

*If T is bounded, then the supremum in (A.6) is called the **operator norm** of T, denoted $\|T\|$.*

Theorem A.36 (Bounded Linear Transformation Theorem) *Let V_1 be a normed space and V_2 a Banach space. Suppose W is a dense subspace of V_1 and $T : W \to V_2$ is a bounded linear map. Then there exists a unique bounded linear map $\tilde{T} : V_1 \to V_2$ such that $\tilde{T}|_W = T$. Furthermore, the norm of $\tilde{T}$ equals the norm of T.*

Definition A.37 *If V is a normed space over $\mathbb{F}$ ($\mathbb{F} = \mathbb{R}$ or $\mathbb{C}$), then a **bounded linear functional** on V is a bounded linear map of V into $\mathbb{F}$, where on $\mathbb{F}$ we use the norm given by the absolute value. The collection of all bounded linear functionals, with the norm given by (A.6), is called the **dual space** to V, denoted V^*.*

Theorem A.38 *If V is a normed vector space, then the following results hold.*

1. *The dual space V^* is a Banach space.*

2. *For all $\psi \in V$, there exists a nonzero $\xi \in V^*$ such that*

$$|\xi(\psi)| = \|\xi\| \, \|\psi\|.$$

In particular, if $\xi(\psi) = 0$ for all $\xi \in V^*$, then $\psi = 0$.

Theorem A.39 (Closed Graph Theorem) *Suppose that V_1 is a Banach space and V_2 a normed vector space. For any linear map $T : V_1 \to V_2$, let $\text{Graph}(T)$ denote the set of pairs $(\psi, T\psi)$ in $V_1 \times V_2$ such that $\psi \in V_1$. If the graph of T is a closed subset of $V_1 \times V_2$, then T is bounded.*

Here is a simple example of how the closed graph theorem can be applied. Suppose V_1 and V_2 are Banach spaces and $T : V_1 \to V_2$ is a linear map that is one-to-one, onto, and bounded. Then the inverse map $T^{-1} : V_2 \to V_1$ is automatically bounded. To verify this, we first check that if T is bounded, then the graph of T is closed (easy). Then we observe that the graph of T^{-1} is also closed, since it is obtained from the graph of T by the map $(\phi, \psi) \mapsto (\psi, \phi)$. Thus, the theorem tells us that T^{-1} is bounded.

Theorem A.40 (Principle of Uniform Boundedness) *Suppose $\{T_\alpha\}$ is any family of bounded linear maps from a Banach space V_1 to a normed space V_2. Suppose that for each $\psi \in V_1$, there is a constant C_ψ such that $\|T_\alpha \psi\| \leq C_\psi$ for all α. Then there exists a constant C such that $\|T_\alpha\| \leq C$ for all α.*

That is, in contrapositive form, if the family $\{T_\alpha\}$ is unbounded, $\{T_\alpha \psi\}$ must be unbounded on ψ for some $\psi \in V_1$.

Corollary A.41 *Suppose V is a Banach space and E is a nonempty subset of V. Suppose that for all $\xi \in V^*$ there exists a constant C_ξ such that $|\xi(\psi)| \leq C_\xi$ for all $\psi \in E$. Then E is a bounded set.*

The corollary is obtained by identifying each $\psi \in V$ with the linear map $e_\psi : V^* \to \mathbb{C}$ given by evaluation on ψ; that is, $e_\psi(\xi) = \xi(\psi)$. Note that by Point 2 of Theorem A.38, the norm of e_ψ as an element of V^{**} is equal to the norm of ψ as an element of V.

A.4   Hilbert Spaces and Operators on Them

A.4.1   Inner Product Spaces and Hilbert Spaces

We now introduce a generalization to arbitrary vector spaces over $\mathbb{R}$ or $\mathbb{C}$ of the usual inner product (or dot product) on $\mathbb{R}^n$.

Definition A.42 *An **inner product** on a vector space over $\mathbb{F}$ ($\mathbb{F} = \mathbb{R}$ or $\mathbb{C}$) is a map $\langle \cdot, \cdot \rangle : V \times V \to \mathbb{F}$ with the following properties.*

1. *For all $\phi, \psi \in V$, we have $\langle \psi, \phi \rangle = \overline{\langle \phi, \psi \rangle}$.*

2. *For all $\phi \in V$, $\langle \phi, \phi \rangle$ is real and non-negative, and $\langle \phi, \phi \rangle = 0$ only if $\phi = 0$.*

3. *For all $\phi, \psi \in V$ and $c \in \mathbb{F}$, we have $\langle c\phi, \psi \rangle = \bar{c} \langle \phi, \psi \rangle$ and $\langle \phi, c\psi \rangle = c \langle \phi, \psi \rangle$.*

4. *For all $\phi, \psi, \chi \in V$, we have $\langle \phi + \psi, \chi \rangle = \langle \phi, \chi \rangle + \langle \psi, \chi \rangle$ and*

$$\langle \phi, \psi + \chi \rangle = \langle \phi, \psi \rangle + \langle \phi, \chi \rangle.$$

Note that we are following the physics convention of taking the complex conjugate in Point 3 of the definition on the *first* factor in the inner product.

Proposition A.43 *If V is an inner product space, then for all $\phi, \psi \in V$, we have the **Cauchy–Schwarz inequality**:*

$$|\langle \phi, \psi \rangle|^2 \leq \langle \phi, \phi \rangle \langle \psi, \psi \rangle .$$

Furthermore, if $\|\cdot\| : V \to \mathbb{R}$ is defined by

$$\|\psi\| = \sqrt{\langle \psi, \psi \rangle}, \tag{A.7}$$

then $\|\cdot\|$ is a norm on V.

Definition A.44 *A **Hilbert space** is a vector space $\mathbf{H}$ over $\mathbb{R}$ or $\mathbb{C}$, equipped with an inner product $\langle \cdot, \cdot \rangle$, such that $\mathbf{H}$ is complete in the norm given by (A.7).*

That is to say, a Hilbert space is a Banach space in which the norm comes from an inner product. In Appendix A.4 only, we allow $\mathbf{H}$ to denote an arbitrary Hilbert space over $\mathbb{R}$ or $\mathbb{C}$. (In the main body of the text, $\mathbf{H}$ denotes a separable complex Hilbert space.)

Definition A.45 *Suppose $\mathbf{H}_j$ is a sequence of separable Hilbert spaces. Then the **Hilbert space direct sum**, denoted*

$$\mathbf{H} := \bigoplus_{j=1}^{\infty} \mathbf{H}_j,$$

is the space of sequences $\psi = (\psi_1, \psi_2, \psi_3, \dots)$ such that $\psi_n \in \mathbf{H}_n$ and such that

$$\|\psi\|^2 := \sum_{j=1}^{\infty} \|\psi_j\|_j^2 < \infty. \tag{A.8}$$

*The **finite direct sum** of the $\mathbf{H}_j$'s is the set of $\psi = (\psi_1, \psi_2, \psi_3, \dots)$ such that $\psi_j = 0$ for all but finitely many values of j.*

We define an inner product on the direct sum by setting

$$\langle \phi, \psi \rangle = \sum_{j=1}^{\infty} \langle \phi_j, \psi_j \rangle \tag{A.9}$$

for all $\phi, \psi \in \mathbf{H}$. This inner product is well defined and $\mathbf{H}$ is complete with respect to this inner product, and hence a Hilbert space.

One important example of a Hilbert space is $L^2(X, \mu)$, where (X, μ) is a measure space.

Definition A.46 *If (X, μ) is a measure space, define an inner product on $L^2(X, \mu)$ by the formula*

$$\langle \phi, \psi \rangle = \int_X \overline{\phi(x)} \psi(x) \, d\mu(x). \qquad (A.10)$$

A standard result in measure theory states that the integral on the right-hand side of (A.10) is absolutely convergent for all ϕ and ψ in $L^2(X, \mu)$. It is then easy to verify that $\langle \cdot, \cdot \rangle$ is indeed an inner product on $L^2(X, \mu)$. Another standard result states that $L^2(X, \mu)$ is complete with respect to the norm associated with the inner product in (A.10); thus, $L^2(X, \mu)$ is a Hilbert space.

A.4.2 Orthogonality

One reason that Hilbert spaces are nicer to work with than general Banach spaces is that we have the concept of orthogonality.

Definition A.47 *Two elements ϕ and ψ of an inner product space are **orthogonal** if $\langle \phi, \psi \rangle = 0$.*

Definition A.48 *If V is any subspace of $\mathbf{H}$, define a subspace $V^\perp$ of $\mathbf{H}$ by*

$$V^\perp = \{ \phi \in \mathbf{H} | \langle \phi, \psi \rangle = 0 \text{ for all } \psi \in V \}.$$

*Then $V^\perp$ is called the **orthogonal space** of V.*

Proposition A.49

1. *If V is a closed subspace of $\mathbf{H}$, every $\psi \in \mathbf{H}$ can be decomposed uniquely as $\psi = \psi_1 + \psi_2$, with $\psi_1 \in V$ and $\psi_2 \in V^\perp$.*

2. *If V is any subspace of $\mathbf{H}$, then $(V^\perp)^\perp = \overline{V}$, where $\overline{V}$ is the closure of V. In particular, if V is closed, then $(V^\perp)^\perp = V$.*

If V is closed, we call $V^\perp$ the *orthogonal complement* of V.

Definition A.50 *A set $\{e_j\}$ of elements of $\mathbf{H}$, where j ranges over an arbitrary index set, is said to be **orthonormal** if*

$$\langle e_j, e_k \rangle = \begin{cases} 0 & j \neq k \\ 1 & j = k \end{cases}.$$

*An orthonormal set $\{e_j\}$ is an **orthonormal basis** for $\mathbf{H}$ if the space of finite linear combinations of the e_j's is dense in $\mathbf{H}$.*

If $\mathbf{H} = L^2([-L, L])$, for some positive number L, then the functions,

$$\psi_n = \frac{1}{\sqrt{2L}} e^{2\pi i n x / L}, \quad n \in \mathbb{Z}, \qquad (A.11)$$

form an orthonormal basis for $\mathbf{H}$.

Proposition A.51 *Suppose $\{e_j\}$ is an orthonormal basis for* **H**. *Then every ψ can be expressed uniquely as a convergent sum*

$$\psi = \sum_j a_j e_j, \tag{A.12}$$

where the coefficients are given by $a_j = \langle e_j, \psi \rangle$. If ψ is as in (A.12), then

$$\|\psi\|^2 = \sum_j |a_j|^2.$$

*Finally, if $\langle a_j \rangle$ is any sequence such that $\sum_j |a_j|^2 < \infty$, there exists a unique $\psi \in$ **H** such that $\langle e_j, \psi \rangle = a_j$ for all j.*

In the case that the orthonormal basis is the one in (A.11), the resulting series (A.12) is called the *Fourier series* of ψ.

A.4.3   The Riesz Theorem and Adjoints

We let $\mathcal{B}(\mathbf{H})$ denote the space of bounded linear maps of **H** to **H**. It is not hard to show that $\mathcal{B}(\mathbf{H})$ forms a Banach space under the operator norm.

Theorem A.52 (Riesz Theorem) *If $\xi : \mathbf{H} \to \mathbb{C}$ is a bounded linear functional, then there exists a unique $\chi \in \mathbf{H}$ such that*

$$\xi(\psi) = \langle \chi, \psi \rangle$$

*for all $\psi \in$ **H**. Furthermore, the operator norm of ξ as a linear functional is equal to the norm of χ as an element of* **H**.

We now turn to the concept of the *adjoint* of a bounded operator, along with the related concept of *quadratic forms* on **H**.

Proposition A.53 *For any $A \in \mathcal{B}(\mathbf{H})$, there exists a unique linear operator $A^* : \mathbf{H} \to \mathbf{H}$, called the **adjoint** of A, such that*

$$\langle \phi, A\psi \rangle = \langle A^*\phi, \psi \rangle$$

*for all $\phi, \psi \in$ **H**. For all $A, B \in \mathcal{B}(\mathbf{H})$ and $\alpha, \beta \in \mathbb{C}$ we have*

$$(A^*)^* = A$$
$$(AB)^* = B^*A^*$$
$$(\alpha A + \beta B)^* = \bar{\alpha}A^* + \bar{\beta}B^*$$
$$I^* = I.$$

The operator A^* is bounded and $\|A^*\| = \|A\|$.

Since A is a bounded operator, the map $\psi \mapsto \langle \phi, A\psi \rangle$ is a bounded linear functional for each fixed $\phi \in \mathbf{H}$. The Riesz theorem then tells us that there is a unique $\chi \in \mathbf{H}$ such that $\langle \phi, A\psi \rangle = \langle \chi, \psi \rangle$. The operator A^* is defined by setting $A^*\phi = \chi$. It is not hard to check that this definition makes A^* into a bounded linear operator.

Definition A.54 *An operator* $A \in \mathcal{B}(\mathbf{H})$ *is said to be* **self-adjoint** *if* $A^* = A$ *and* **skew-self-adjoint** *if* $A^* = -A$.

Definition A.55 *An operator* U *on* $\mathbf{H}$ *is* **unitary** *if* U *is surjective and preserves inner products, that is,* $\langle U\phi, U\psi \rangle = \langle \phi, \psi \rangle$ *for all* $\phi, \psi \in \mathbf{H}$.

If U is unitary, then U preserves norms ($\|U\psi\| = \|\psi\|$ for all $\psi \in \mathbf{H}$); therefore, U is bounded with $\|U\| = 1$. By the polarization identity (Proposition A.59), if U preserves norms, then it also preserves inner products.

Proposition A.56 *A bounded operator* U *is unitary if and only if* $U^* = U^{-1}$, *that is, if and only if* $UU^* = U^*U = I$.

Proposition A.57 *For any closed subspace* $V \subset \mathbf{H}$, *there is a unique bounded operator* P *such that* $P = I$ *on* V *and* $P = 0$ *on the orthogonal complement* $V^\perp$. *This operator is called the* **orthogonal projection** *onto* V *and it satisfies* $P^2 = P$ *and* $P^* = P$.

Conversely, if P *is any bounded operator on* $\mathbf{H}$ *satisfying* $P^2 = P$ *and* $P^* = P$, *then* P *is the orthogonal projection onto a closed subspace* V, *where* $V = \mathrm{range}(P)$.

A.4.4 Quadratic Forms

In this section, we develop the theory of quadratic forms on Hilbert spaces. Since this is customarily done only for the inner product itself, we include the proofs of the results.

Definition A.58 *A* **sesquilinear form** *on* $\mathbf{H}$ *is a map* $L : \mathbf{H} \times \mathbf{H} \to \mathbb{C}$ *that is conjugate linear in the first factor and linear in the second factor. A sesquilinear form is* **bounded** *if there exists a constant* C *such that*

$$|L(\phi, \psi)| \leq C \|\phi\| \|\psi\|$$

for all $\phi, \psi \in \mathbf{H}$.

Proposition A.59 *If* L *is a sesquilinear form on* $\mathbf{H}$, L *can be recovered from its values on the diagonal (i.e., the value of* $L(\psi, \psi)$ *for various* ψ*'s) as follows:*

$$L(\phi, \psi) = \frac{1}{2} \left[L(\phi + \psi, \phi + \psi) - L(\phi, \phi) - L(\psi, \psi) \right]$$
$$- \frac{i}{2} \left[L(\phi + i\psi, \phi + i\psi) - L(\phi, \phi) - L(i\psi, i\psi) \right]. \tag{A.13}$$

*This formula is known as the **polarization identity**.*

Note that we do not assume any relationship between $L(\phi, \psi)$ and $L(\psi, \phi)$.
Proof. Direct calculation. ∎

Definition A.60 *A **quadratic form** on a Hilbert space* **H** *is a map* $Q :$ **H** $\to \mathbb{C}$ *with the following properties: (1)* $Q(\lambda\psi) = |\lambda|^2 Q(\psi)$ *for all* $\psi \in$ **H** *and* $\lambda \in \mathbb{C}$, *and (2) the map* $L :$ **H** $\times$ **H** $\to \mathbb{C}$ *defined by*

$$L(\phi, \psi) = \frac{1}{2} [Q(\phi + \psi) - Q(\phi) - Q(\psi)]$$
$$- \frac{i}{2} [Q(\phi + i\psi) - Q(\phi) - Q(i\psi)]$$

is a sesquilinear form. A quadratic form Q *is **bounded** if there exists a constant* C *such that*

$$|Q(\phi)| \leq C \|\phi\|^2$$

for all $\phi \in$ **H**. *The smallest such constant* C *is the **norm** of* Q.

Proposition A.61 *If* Q *is a quadratic form on* **H** *and* L *is the associated sesquilinear form, we have the following results.*

1. *For all* $\psi \in$ **H**, *we have* $Q(\psi) = L(\psi, \psi)$.

2. *If* Q *is a bounded, then* L *is bounded.*

3. *If* $Q(\psi)$ *belongs to* $\mathbb{R}$ *for all* $\psi \in$ **H**, *then* L *is conjugate symmetric, that is,*

$$L(\phi, \psi) = \overline{L(\psi, \phi)}$$

for all $\phi, \psi \in$ **H**.

Proof. Point 1 of the proposition is verified by taking $\phi = \psi$ in the expression for $L(\phi, \psi)$ and then using the relation $Q(\lambda\psi) = |\lambda|^2 Q(\psi)$. For Point 2, suppose $|Q(\psi)| \leq C \|\psi\|^2$ for all $\psi \in$ **H**. If $\|\phi\| = \|\psi\| = 1$, then $\phi + \psi$ and $\phi + i\psi$ have norm at most 2, and so

$$|L(\phi, \psi)| \leq \frac{1}{2} C (4 + 1 + 1 + 4 + 1 + 1) = 6C.$$

Now, for any ϕ and ψ in **H**, we can find unit vectors $\tilde{\phi}$ and $\tilde{\psi}$ such that $\phi = \|\phi\| \tilde{\phi}$ and $\psi = \|\psi\| \tilde{\psi}$. Then since L is assumed to be sesquilinear, we have

$$|L(\phi, \psi)| = \|\phi\| \|\psi\| L\left(\tilde{\phi}, \tilde{\psi}\right) \leq 6C \|\phi\| \|\psi\|,$$

showing that L is bounded.

For Point 3, assume that $Q(\psi)$ is real for all $\psi \in$ **H** and define a map $M :$ **H** $\times$ **H** $\to \mathbb{R}$ by

$$M(\phi, \psi) = \frac{1}{2} [Q(\phi + \psi) - Q(\phi) - Q(\psi)] = \text{Re} [L(\phi, \psi)].$$

Then M is real-bilinear (because it is the real part of L) and symmetric (because of the expression for M in terms of Q). Furthermore, $M(i\phi, i\psi) = M(\phi, \psi)$. These properties of M show that $M(\phi, i\psi) = -M(\psi, i\phi)$, and so

$$L(\phi, \psi) = M(\phi, \psi) - iM(\phi, i\psi)$$
$$= M(\psi, \phi) + iM(\psi, i\phi)$$
$$= \overline{L(\psi, \phi)},$$

which is what we wanted to prove. ∎

Example A.62 *If A is a bounded operator on $\mathbf{H}$, one can construct a bounded quadratic form Q_A on $\mathbf{H}$ by setting*

$$Q_A(\psi) = \langle \psi, A\psi \rangle, \quad \psi \in \mathbf{H}.$$

The associated sesquilinear form L_A is then given by

$$L_A(\phi, \psi) = \langle \phi, A\psi \rangle, \quad \phi, \psi \in \mathbf{H}.$$

Proposition A.63 *If Q is a bounded quadratic form on $\mathbf{H}$, there is a unique $A \in \mathcal{B}(\mathbf{H})$ such that $Q(\psi) = \langle \psi, A\psi \rangle$ for all $\psi \in \mathbf{H}$. If $Q(\psi)$ belongs to $\mathbb{R}$ for all $\psi \in \mathbf{H}$, then the operator A is self-adjoint.*

Proof. Since Q is bounded, L is also bounded, meaning that there exists a constant C such that $|L(\phi, \psi)| \leq C \|\phi\| \|\psi\|$ for all $\phi, \psi \in \mathbf{H}$. Thus, for any $\phi \in \mathbf{H}$, the linear functional $\psi \mapsto L(\phi, \psi)$ is bounded, with norm at most $C \|\phi\|$. By the Riesz theorem, then, there exists a unique $\chi \in \mathbf{H}$, with $\|\chi\| \leq C \|\phi\|$, such that $L(\phi, \psi) = \langle \chi, \psi \rangle$. We now define a map $B : \mathbf{H} \to \mathbf{H}$ by defining $B\phi = \chi$. Direct calculation shows that B is linear, and the inequality $\|\chi\| \leq C \|\phi\|$ shows that B is bounded. Setting $A = B^*$ establishes the existence of the desired operator. Uniqueness of A follows from the observation that if $\langle \phi, A\psi \rangle = 0$ for all $\phi, \psi \in \mathbf{H}$, then A is the zero operator.

If $Q(\psi)$ is real for all $\psi \in \mathbf{H}$, then by Point 3 of Proposition A.61, L is conjugate symmetric. Thus,

$$\langle \phi, A\psi \rangle = L(\phi, \psi) = \overline{L(\psi, \phi)} = \overline{\langle \psi, A\phi \rangle} = \langle A\phi, \psi \rangle$$

for all $\phi, \psi \in \mathbf{H}$, showing that A is self-adjoint. ∎

A.4.5 Tensor Products of Hilbert Spaces

Recall from Appendix A.1 the concept of the tensor product of two vector spaces.

Proposition A.64 *Suppose V_1 and V_2 are inner product spaces, with inner products $\langle \cdot, \cdot \rangle_1$ and $\langle \cdot, \cdot \rangle_2$. Then there exists a unique inner product $\langle \cdot, \cdot \rangle$ on $V_1 \otimes V_2$ such that*

$$\langle u_1 \otimes v_1, u_2 \otimes v_2 \rangle = \langle u_1, u_2 \rangle_1 \langle v_1 \otimes v_2 \rangle_2$$

for all $u_1, u_2 \in V_1$ and $v_1, v_2 \in V_2$.

If $\mathbf{H}_1$ and $\mathbf{H}_2$ are Hilbert spaces, then we can equip the tensor product $\mathbf{H}_1 \otimes \mathbf{H}_2$ with the inner product in Proposition A.64. If $\mathbf{H}_1$ and $\mathbf{H}_2$ are both infinite dimensional, however, $\mathbf{H}_1 \otimes \mathbf{H}_2$ will not be complete with respect to this inner product. Nevertheless, we can complete $\mathbf{H}_1 \otimes \mathbf{H}_2$ with respect to this inner product, thus obtaining a new Hilbert space.

Definition A.65 *If $\mathbf{H}_1$ and $\mathbf{H}_2$ are Hilbert spaces, then the **Hilbert tensor product** of $\mathbf{H}_1$ and $\mathbf{H}_2$, denoted $\mathbf{H}_1 \hat{\otimes} \mathbf{H}_2$, is the Hilbert space obtained by completing $\mathbf{H}_1 \otimes \mathbf{H}_2$ with respect to the inner product in Proposition A.64.*

Proposition A.66 *If $\mathbf{H}_1$ and $\mathbf{H}_2$ are Hilbert spaces with orthonormal bases $\{e_j\}$ and $\{f_k\}$, respectively, then $\{e_j \otimes f_k\}$ is an orthonormal basis for the Hilbert space $\mathbf{H}_1 \hat{\otimes} \mathbf{H}_2$.*

Proposition A.67 *If A is a bounded operator on $\mathbf{H}_1$ and B is a bounded operator on $\mathbf{H}_2$, then there exists a unique bounded operator on $\mathbf{H}_1 \hat{\otimes} \mathbf{H}_2$, denoted $A \otimes B$, such that*

$$(A \otimes B)(\phi \otimes \psi) = (A\phi) \otimes (B\psi)$$

for all $\phi \in \mathbf{H}_1$ and $\psi \in \mathbf{H}_2$.

To see that $A \otimes B$ is bounded, first write $A \otimes B$ as $(A \otimes I)(I \otimes B)$. Then, given any orthonormal basis $\{f_j\}$ for $\mathbf{H}_2$, we can decompose $\mathbf{H}_1 \hat{\otimes} \mathbf{H}_2$ as the Hilbert space direct sum of subspaces of the form $\mathbf{H}_1 \otimes f_j$. The operator $A \otimes I$ acts on this decomposition as a block-diagonal operator with A in each diagonal block. From this, it is easy to verify that $\|A \otimes I\| = \|A\|$. A similar argument shows that $\|I \otimes B\| = \|B\|$, and so

$$\|A \otimes B\| \leq \|A \otimes I\| \|I \otimes B\| = \|A\| \|B\|.$$

Meanwhile, by taking a sequence of unit vector $\phi_n \in \mathbf{H}_1$ and $\psi_n \in \mathbf{H}_2$ with $\|A\phi_n\| \to \|A\|$ and $\|B\psi_n\| \to \|B\|$, we see that the reverse inequality holds, and thus that $\|A \otimes B\| = \|A\| \|B\|$.

References

[1] V. Bargmann, On unitary ray representations of continuous groups. Ann. Math. **59**(2), 1–46 (1954)

[2] V. Bargmann, On a Hilbert space of analytic functions and an associated integral transform Part I. Comm. Pure Appl. Math. **14**, 187–214 (1961)

[3] S.J. Bernau, The spectral theorem for unbounded normal operators. Pacific J. Math. **19**, 391–406 (1966)

[4] R.J. Blattner, Quantization and representation theory. In *Harmonic Analysis on Homogeneous Spaces* (Proceedings of Symposia in Pure Mathematics, vol. XXVI, Williams College, Williamstown, MA, 1972). (American Mathematical Society, Providence, RI, 1973), pp. 147 165

[5] P.A.M. Dirac, A new notation for quantum mechanics. Math. Proc. Cambridge Philosoph. Soc. **35**, 416–418 (1939)

[6] P.A.M. Dirac, *The Principles of Quantum Mechanics*, 4th edn. (Oxford University Press, Oxford, 1982)

[7] S. De Bièvre, J.-C. Houard, M. Irac-Astaud, Wave packets localized on closed classical trajectories. In *Differential Equations with Applications to Mathematical Physics*. Mathematics in Science and Engineering, vol. 192 (Academic, Boston, 1993), pp. 25–32

[8] S. Dong, *Wave Equations in Higher Dimensions* (Springer, New York, 2011)

B.C. Hall, *Quantum Theory for Mathematicians*, Graduate Texts in Mathematics 267, DOI 10.1007/978-1-4614-7116-5, © Springer Science+Business Media New York 2013

546   References

[9] V. Fock, Verallgemeinerung und Lösung der Diracschen statistischen Gleichung. Zeit. Phys. **49**, 339–350 (1928)

[10] G.B. Folland, *A Course in Abstract Harmonic Analysis* (CRC Press, Boca Raton, FL, 1995)

[11] G.B. Folland, *Harmonic Analysis in Phase Space*. Annals of Mathematics Studies, vol. 122 (Princeton University Press, Princeton, 1989)

[12] G.B. Folland, *Real Analysis: Modern Techniques and Their Applications*, 2nd edn. (Wiley, New York, 1999)

[13] G.B. Folland, *Quantum Field Theory: A Tourist Guide for Mathematicians*. Mathematical Surveys and Monographs, vol. 149 (American Mathematical Society, Providence, RI, 2008)

[14] J. Glimm, A. Jaffe, *Quantum Physics: A Functional Integral Point of View*, 2nd edn. (Springer, New York, 1987)

[15] M.J. Gotay, On the Groenewold-Van Hove problem for $\mathbb{R}^{2n}$. J. Math. Phys. **40**, 2107–2116 (1999)

[16] L. Gross, Abstract Wiener Spaces. In Proceedings of Fifth Berkeley Symposium on Mathematical Statistics and Probability (Berkeley, CA, 1965/1966), vol. II: Contributions to Probability Theory, Part 1 (University of California Press, Berkeley, CA, 1967), pp. 31–42

[17] V. Guillemin, S. Sternberg, *Variations on a Theme by Kepler*. Colloquium Publications, vol. 42 (American Mathematical Society, Providence, RI, 1990)

[18] A. Gut, *Probability: A Graduate Course* (Springer, New York, 2005)

[19] M. Gutzwiller, *Chaos in Classical and Quantum Mechanics* (Springer, Berlin, 1990)

[20] B.C. Hall, Geometric quantization and the generalized Segal–Bargmann transform for Lie groups of compact type. Comm. Math. Phys. **226**, 233–268 (2002)

[21] B.C. Hall, *Lie Groups, Lie Algebras, and Representations: An Elementary Introduction*. Graduate Texts in Mathematics, vol. 222 (Springer, New York, 2003)

[22] K. Hannabuss, *An Introduction to Quantum Theory*. Oxford Graduate Texts in Mathematics (Oxford University Press, Oxford, 1997)

[23] G. Hagedorn, S. Robinson, Bohr–Sommerfeld quantization rules in the semiclassical limit. J. Phys. A **31**, 10113–10130 (1998)

[24] K. Hoffman, R. Kunze, *Linear Algebra*, 2nd edn. (Prentice-Hall, Englewood Cliffs, NJ, 1971)

[25] N. Jacobson, *Lie Algebras* (Dover Publications, New York, 1979)

[26] M.V. Karasëv, Connections on Lagrangian submanifolds and some problems in quasiclassical approximation. I. (Russian); translation in J. Soviet Math. **59**, 1053–1062 (1992)

[27] T. Kato, *Perturbation Theory for Linear Operators* (Reprint of the 1980 edition). (Springer, Berlin, 1995)

[28] W.G. Kelley, A.C. Petersen, *The Theory of Differential Equations: Classical and Qualitative (Universitext)*, 2nd edn. (Springer, New York, 2010)

[29] J. Lee, *Introduction to Smooth Manifolds*, 2nd edn. (Springer, London, 2006)

[30] P. Miller, *Applied Asymptotic Analysis* (American Mathematical Society, Providence, RI, 2006)

[31] T. Paul, A. Uribe, A construction of quasi-modes using coherent states. Ann. Inst. H. Poincaré Phys. Théor **59**, 357–381 (1993)

[32] W. Rudin, *Real and Complex Analysis*, 3rd edn. (McGraw-Hill, New York, 1987)

[33] W. Rudin, *Functional Analysis*, 2nd edn. International Series in Pure and Applied Mathematics (McGraw-Hill, New York, 1991)

[34] M. Reed, B. Simon, *Methods of Modern Mathematical Physics*. Volume I: Functional Analysis, 2nd edn. (Academic, San Diego, 1980). Volume II: Fourier Analysis, Self-Adjointness (Academic, New York, 1975). Volume III: Scattering Theory (Academic, New York, 1979). Volume IV: Analysis of Operators (Academic, New York, 1978)

[35] K. Schmüdgen, *Unbounded Self-Adjoint Operators on Hilbert Space*. Graduate Texts in Mathematics, vol. 265 (Springer, Dordrecht, 2012)

[36] I.E. Segal, Mathematical problems of relativistic physics. In *Proceedings of the Summer Seminar, Boulder, Colorado, 1960*, ed. by M. Kac (American Mathematical Society, Providence, RI, 1963)

[37] B. Simon, *Functional Integration and Quantum Physics*, 2nd edn. (American Mathematical Society, Providence, RI, 2005)

[38] R.F. Streater, A.S. Wightman, *PCT, Spin and Statistics, and All That* (Corrected third printing of the 1978 edition). Princeton Landmarks in Physics (Princeton University Press, Princeton, NJ, 2000)

[39] L.A. Takhtajan, *Quantum Mechanics for Mathematicians*. Graduate Studies in Mathematics, vol. 95 (American Mathematical Society, Providence, RI, 2008)

[40] W. Thirring, *A Course in Mathematical Physics I: Classical Dynamical Systems* (Translated by Evans M. Harrell). (Springer, New York, 1978)

[41] J. von Neumann, Die Eindeutigkeit der Schrödingerschen operatoren. Math. Ann. **105**, 570–578 (1931)

[42] A. Voros, Wentzel–Kramers–Brillouin method in the Bargmann representation. Phys. Rev. A **40**(3), 6814–6825 (1989)

[43] N.R. Wallach, *Real Reductive Groups I* (Academic, San Diego, 1988)

[44] R.E. Williamson, R.H. Crowell, H.F. Trotter, *Calculus of Vector Functions*, 3rd edn. (Prentice-Hall, Englewood Cliffs, NJ, 1968)

[45] N. Woodhouse, *Geometric Quantization*, 2nd edn. (Oxford University Press, Oxford, 1992)

[46] K. Yosida, *Functional Analysis*, 4th edn. (Springer, New York, 1980)

Index

Action functional, 446
Adjoint
  of a bounded operator, 55, 540
  of an unbounded operator, 56, 170
Airy function, 315
Almost complex structure, 495
Angular momentum
  addition of, 384
  function, 31, 39
  operator, 83, 367
  vector, 32, 368, 369
Axioms of quantum mechanics, 64, 427

Baker–Campbell–Hausdorff
  formula, 262, 281, 347
Banach space, 535
Bargmann space, see
  Segal–Bargmann space
Bergman space, 501
Blackbody radiation, 4, 433
BLT theorem, 536
Bohr, Niels, 9

Bohr–de Broglie model of
  hydrogen, 9
Bohr–Sommerfeld condition,
  306, 307, 317, 500, 512
Born, Max, 13, 14
Bose–Einstein
  condensate, 437
  statistics, 437
Boson, 85, 384, 434, 437
Bounded operator, 55, 131
Bounded-below operator, 178
Bra-ket notation, 85
Brownian motion, 6, 448

Canonical
  1-form, 459
  2-form, 459
  bundle, 506, 518
  commutation relations, 63,
    83, 228, 229, 279
Canonical transformation, see
  Symplectomorphism
Casimir operator, 374, 407
Cauchy–Schwarz inequality, 538
Cayley transform, 220, 222

B.C. Hall, *Quantum Theory for Mathematicians*, Graduate Texts
in Mathematics 267, DOI 10.1007/978-1-4614-7116-5,
© Springer Science+Business Media New York 2013

Printed in the United States
By Bookmasters